SAN FRANCISCO BAY: THE ECOSYSTEM

Further Investigations into the Natural History of San Francisco Bay and Delta
With Reference to the Influence of Man

SAN FRANCISCO BAY AND ENVIRONS

This is a portion of a NASA Landsat Thematic Mapper digital image recorded on June 20, 1990 from an altitude of approximately 675 km. (Color infrared composite of bands 4, 3, and 1; contrast sketched; scene identifier 52302-18061, path/row 44-34.) Each pixel has a resolution of approximately 30 m. Image provided courtesy of William Acevedo of NASA and Richard Smith of the U.S. Geological Survey.

SEVENTY-FIFTH ANNUAL MEETING
of the
PACIFIC DIVISION/AMERICAN ASSOCIATION FOR THE ADVANCEMENT OF SCIENCE
held at
SAN FRANCISCO STATE UNIVERSITY, SAN FRANCISCO, CALIFORNIA
June 19-24, 1994

SAN FRANCISCO BAY: THE ECOSYSTEM

Further Investigations into the Natural History of San Francisco Bay and Delta
With Reference to the Influence of Man

Editor
James T. Hollibaugh
Romberg Tiburon Center
San Francisco State University

San Francisco, California
1996

ISBN 0-934394-11-3
Library of Congress Card Number 96-070569

This volume has been typeset in Adobe Systems 10pt Times Roman in Ventura Publisher® Windows-version 4.1 and camera-ready copy produced on a Lexmark™ Optra R^+ laser printer by the Pacific Division AAAS. Printing and binding by Friesen Printers, Altona, Manitoba, Canada/Richard Thunes, San Francisco, California

c/o California Academy of Sciences
Golden Gate Park
San Francisco, California 94118

Manufactured in Canada by Friesen Printers, Altona, Manitoba

TABLE OF CONTENTS

ECOSYSTEM MANAGEMENT

SAN FRANCISCO BAY
THE ECOSYSTEM

FOREWORD

T. JOHN CONOMOS
U.S. Geological Survey
345 Middlefield Road, Menlo Park, California 94025

The San Francisco Bay estuarine system, located at the confluence of the Sacramento and San Joaquin Rivers, is notable for it's scenic beauty, recreation and international commerce. With the progress of urbanization, however, has come major changes in the ecosystem. In response to environmental concerns during the last three decades, elected officials, resource managers and regulators, and the public, have agreed that the system should be protected against any further unwise exploitation. Further, they recognized the need for credible, unbiased, scientific information regarding the significance of particular human activities, which are responsible for unwanted ecological changes.

Previous to the late 1960s, the ecosystem received little scientific study. The environmental awareness movement beginning at that time, however, initiated interdisciplinary investigations by state, federal and university scientists that have increased both in quantity and quality. These activities were highlighted by scientific meetings and symposiums, most notably the fifty-eighth (1977) and two decades later, the seventy-fifth (1994) annual meetings of the Pacific Division of the American Association for the Advancement of Science.

This volume includes most of the papers presented at the 1994 symposium and some additional chapters written since then. It serves as a valuable sequel to the volume with the title *San Francisco Bay: The Urbanized Estuary*, produced from the earlier symposium, and presents samplings of the great advances of knowledge that has been since gained on how the ecosystem functions physically, chemically and biologically.

I wish to express my appreciation to James T. Hollibaugh for organizing the symposium and for creating this excellent volume. Without his vision and dedication, this contribution would not have been developed. Strong and warm acknowledgments are also due to Alan E. Leviton, Executive Director of the Pacific Division, American Association for the Advancement of Science, for his insights and labors in support of the effort as well as for efforts past.

T. John Conomos
U.S. Geological Survey
June 1996

INTRODUCTION

JAMES T. HOLLIBAUGH
Tiburon Center, San Francisco State University
P. O. Box 855, Tiburon, CA 94920

Alien invasions, droughts, floods, toxic pollution and water use conflicts - these were the headlines for San Francisco Bay in the 1980s and 90s. While they are eye-catching, they provided the reader with only a very narrow perspective of the news of the Bay during this period. The deeper story concerns the dramatic advances in our understanding of the Bay as an ecosystem during this period – an understanding both driven by and underscored by these headlines.

Almost 20 years ago, Dr. T. J. Conomos organized a symposium on San Francisco Bay for the 58th annual meeting of the Pacific Division of the American Association for the Advancement of Science. The product of that symposium, *San Francisco Bay: The Urbanized Estuary* (edited by Dr. Conomos and published in 1979 by the Pacific Division AAAS), was the first comprehensive collection of papers on San Francisco Bay. This immensely popular volume – I was unable to find one at any price when I moved back to the Bay Area in 1983 and have been forced to rely on borrowing from a colleague a copy that has since disappeared (not to my personal library) – rapidly became the "Bible" of research scientists, consultants, environmental lawyers, managers and others concerned with San Francisco Bay. Although the collection was updated in a 1985 issue of *Hydrobiologia* that focused on a detailed study of the Bay conducted in 1980 (Cloern, J. E. & F. H. Nichols, eds., *Temporal Dynamics of an Estuary: San Francisco Bay*), *The Urbanized Estuary* remains a primary reference on San Francisco Bay.

On both political and scientific fronts, the nearly two decades between publication of *The Urbanized Estuary* and the present have hosted a number of significant events. A series of hearings to determine water allocations to protect Bay aquatic resources focused the attentions of managers and the general public alike on the Bay during the later half of the 1980's. Their attention was sharpened by what at the time appeared to be an unending series of drought years, possibly attributable to broad-scale climate change brought on by anthropogenic carbon dioxide emissions. These fears were erased, at least temporarily, by torrential storms and record floods during the winters of 1993 and 1995.

At the height of the drought an exotic species, the Asiatic clam *Potamocorbula amurensis*, was introduced to northern San Francisco Bay. Unlike many of its predecessors, the establishment of this organism had devastating consequences for the Bay. In slightly over 1 year following its introduction in 1986, it became established throughout the Bay. Its populations became so large in some areas, notably Suisun Bay, that the standing crops of phytoplankton were decimated and primary production plummeted to approximately 20% of its previous value. In addition to competing directly with zooplankton for what appeared to be limited phytoplankton food resources, the clam was found to impact zooplankton populations directly by capturing juvenile stages of some species. Because these zooplankton were key food items for the larvae of some species of fish, this posed the prospect of a cascading series of indirect negative impacts on the already beleaguered fish populations of the Bay.

This series of events spurred new research and analysis of historical data focused on elucidating the relationship between freshwater inflows and ecological processes in San Francisco Bay and on understanding the Bay as an ecosystem. This scrutiny raised many questions about our conceptual models of how the Bay functions as an ecosystem. What are the sources of organic matter fueling Bay productivity? What is the relationship between estuarine physics and biology or chemistry? What are the links between the Bay and the Sacramento-San Joaquin Rivers and Delta? How important is freshwater flow versus fresh water quality? How does the Bay function physically and what controls exchange with the ocean or circulation in the northern reach? It also focused more process-oriented research on the North Bay, where environmental research previously had been dominated by mandated monitoring programs.

Given the determined efforts to understand the Bay in the nearly two decades since the symposium on which *The Urbanized Estuary* was based, it seemed appropriate to consider organizing a similar symposium with the intent of producing a companion volume which updated *The Urbanized Estuary* and presents the emerging new paradigms. The 75th Annual Meeting of the Pacific Division, AAAS, held in San Francisco in 1994, provided the opportunity and Drs. T. J. Conomos and R. I. Bowman provided the encouragement and moral support needed to overcome my trepidation about the magnitude of the project. The result is this volume which complements *The Urbanized Estuary*.

ACKNOWLEDGMENTS

It would not have been possible to produce *San Francisco Bay: The Ecosystem* without the participants of the 1994 San Francisco Bay Symposium, many of whom are represented by papers included in this volume. I am also indebted to the individuals that gave their time to provide anonymous peer review. Their efforts contributed to the overall quality of these papers. A generous donation from Mr. Howard Allen helped to defray costs and allowed us to publish this as a hard-bound book. The Romberg Tiburon Center of San Francisco State University helped by providing clerical assistance, office supplies, telecommunications and my support while editing this volume.

I would also like to acknowledge the contribution, at least in spirit, of Mr. John Krautkramer. John was an environmental lawyer working on San Francisco Bay and related California water policy issues who died tragically early in 1995. He loved the Bay. His tireless defense of its rights and his ability to build consensus between groups with completely disparate interests is in no small measure responsible for many of the environmental protections now in place. John contributed to this volume by helping to focusing the public's awareness and, as a result politicians' interest, on the Bay, resulting in allocation of the resources needed to conduct some of the studies reported here.

Lastly, I want to express my deepest appreciation to the Pacific Division of the American Association for the Advancement of Science and notably to Drs. Michele L. Aldrich, chair of the Division's Publications Committee, and Alan E. Leviton, Division Executive Director, for their encouragement, and for their editorial and other support that made this publication a reality. With this volume, the Pacific Division AAAS is responsible for publishing the three major contemporary works that present an up-to-date synthesis of studies of the San Francisco Bay and Estuary, thus setting the standard for future presentations on critical estuarine environments.

James T. Hollibaugh
Tiburon, California
April 17, 1996

PHYSICAL PROCESSES

SAN FRANCISCO BAY SALINITY: OBSERVATIONS, NUMERICAL SIMULATION, AND STATISTICAL MODELS

D. H. PETERSON, D. R. CAYAN, M. D. DETTINGER,
M. A. NOBLE, L. G. RIDDLE, L. E. SCHEMEL,
R. E. SMITH, R. J. UNCLES AND R. A. WALTERS
Water Resources Division, Western Region, U.S. Geological Survey
345 Middlefield Road, MS 496, Menlo Park, CA 94025

Understanding the variability of salinity in San Francisco Bay is a key to defining the bay's physics, chemistry, and biology. This paper is in part a literature review and in part uses decades of observations supported by results from statistical-dynamical and numerical models to describe salinity variability in San Francisco Bay. Findings include the following: (1) Fresh-water delta flow (DF) is the master control on mean-monthly salinity in the Bay, (2) Salinity fluctuations are reasonably well modeled on mean monthly and daily time scales, (3) Hysteresis is observed in both data and models, (4) Coastal ocean processes affect salinity in the Bay and vice versa, (5) statistical dynamical models driven by variations in hourly sea surface height near the mouth can estimate about 80% of the hourly salinity variance near mid- estuary at tidal time scales during low relatively uniform delta flow, and (6) Climate (small effect) as well as freshwater diversions (large effect) control long-term (decadal) salinity variations.

Knowledge of the salinity field is fundamental to many estuarine studies. To a large degree, how clearly we understand the details of salinity variability in the San Francisco Bay measures how well we understand the entire estuarine system. Beyond this are the aspects of societal interests. Approximately 20 million people depend on the fresh water supply of the bay's delta, a supply vulnerable to sea-salt contamination especially in times of drought. For these reasons salinity is a key element in estuarine research and is often proposed as a water quality standard for estuarine resource management.

This paper describes the nature and causes of salinity variability in San Francisco Bay focusing on time scales from days to years. We know that tidal forces dominate the short time scales of salinity variability (hours to days) when the fresh water flow is low, and winds contribute small amounts of variance. At intermediate time scales, days to fortnights, salinity variability must include both tidal and fresh-water flow effects (and wind effects, which are considered beyond the scope of this paper). We also know that over the longer periods, the months, seasons, and years, the primary cause of salinity variability is related to changes in the discharge of the major source of fresh water, the Sacramento-San Joaquin River (Peterson *et al.* 1989). Hereafter this flow is referred to as delta flow (DF; see Fig. 1).

In reviewing previous work, one soon learns that California has a rich history of interest in the salinity of San Francisco Bay's waters. However, much of that literature, perhaps even hundreds of documents, are in the limited distribution category, including in-house reports of various agencies and are difficult to access. For a descriptive overview of the hydrography and salinity of the Bay, see Conomos

(1979) and Cloern & Nichols (1985). Only major issues on salinity variability in the bay are summarized here.

One of the earliest salinity studies was concerned with the effects of drought, upland fresh water consumption, and river and bay channelization on increasing salinity. The study was initiated because increased salinity led to invasion of boring organisms that devastated the bay's pier facilities (Hill & Kofoid 1927). Similarly, a comprehensive compilation on salinity was motivated in the 1920s and early 1930s by droughts which caused low DF and, in response high salinity threatened to contaminate the fresh water in the delta (California, 1931).

These early studies showed that salinity problems in the bay are clearly linked to delta flow. Hence, there was increased interest in determining a more quantitative estimate of delta flow. In the 1950s a lively discussion was focused on the difficult problem of how you estimate fresh water flow under tidal conditions when a large fresh water reservoir existed in the system (Todd & Lau 1956, 1957; Pritchard 1957, 1958). Estimating the actual DF was even more difficult because local consumption was very poorly known. Thus the inverse problem of using the salt field to better estimate delta flows (rather than the delta flow to estimate the salinity) was considered. This information was then used to refine the estimates of local delta consumption in the water budget to the Bay (Glover 1955).

Probably one of the most controversial papers inferred that delta flow had a major influence even on the South Bay salt field (McCulloch & others 1970). This apparently radical view was followed up by bay-wide surveys that included a more holistic analysis of the bay as a system, including the two books noted above. These more recent surveys suggest that evaporation and local stream flow can be important to the salinity cycle in South Bay and must be considered when interpreting the annual variability in nutrient patterns there (Hager & Schemel, this volume).

Much of the work in the 1980's involved field observations and numerical analysis of mechanisms of salinity variability including tidal and spring-neap variations, gravitational circulation and floods (*e.g.*, Ford, Wang & Cheng 1990; Smith & Cheng 1987; Smith & Cheng 1991; Walters & Gartner 1985). Uncles & Peterson (in press and 1995) provide a review of some of the topics in these papers. In essence, San Francisco Bay, like other estuaries, responds to variations in the above forcing. Vertical mixing strengthens during spring tidal regimes. Salinity stratification, and apparently gravitational circulation, strengthen during neap tides; the latter strengthening is particularly noticeable in deeper channels. Salinity stratification is also enhanced during river floods.

Several biological consequences of changes in salinity stratification and the salinity field are especially worth noting. First, salinity stratification changes the normal growth patterns for phytoplankton in the South Bay. These changes are most noticeable in spring during periods when the smallest neap tides occur (Cloern 1991, 1984; Koseff *et al.* 1993) . A second biological affect is that benthic filter feeders invade northern San Francisco Bay in times of drought and persistent high salinity farther inland (*e.g.*, 1976-1977, 1981, 1985, 1987, Nichols 1985; Carlton *et al.* 1990; Nichols, Thompson, & Schemel 1990). At the other extreme, floods on the rivers can alter benthic community patterns. Partly because salinity plays such an important role in community structure, the distance (X) of the 2 psu near-bottom isohaline from the Golden Gate, defined as X2, is proposed as a quantitative parameter for purposes of managing the bay's salinity field (Kimmerer & Monismith 1993; Denton, R.A. 1993).

In the 1990s the issue of the changing salinity field in San Francisco Bay has broadened to include: (1) the effects of climate control on fresh water flow into the bay (Cayan & Peterson 1993; Dettinger & Cayan 1995); (2) the exchange between the coastal ocean and the bay (Largier, this volume); and (3) paleo-salinity (Ingram & Sloan 1991, and Ingram & Ingle, this volume). But much of the core research focuses on more standard, long-standing issues as salt penetration with discharge (Harder 1977; Winkler 1985; Denton, R.A. 1993). This research includes investigations of how changes in magnitude and phase of estuarine elevation, velocity, salinity, river flow and geometry are linked to processes occurring in intratidal and tidal deltas (*e.g.*, Burau, Monismith & Stacey, this volume; Simpson, Sharples & Rippeth 1991; Friedrichs & Aubrey 1994; Uncles & Jordan 1980). High resolution numerical models are also used to explore these processes (Cheng, Casulli & Gartner 1993). In the following, we will review the aspects of the above processes that involve changes in the salinity field at subtidal and tidal frequencies, discuss some new results from numerical models, and touch briefly on the relationship between large-scale atmospheric patterns and salinity patterns in the bay.

DATA AND MODELS

The salinity observations used in this paper were obtained from a variety of sources (Table 1, Fig. 1). The original data were sampled at several different rates, spanning periods from 15 minute to daily. Tidal frequencies were removed from the data sampled at hourly rates or less with a low-pass Godin filter (supplied by Lawrence Smith 1992). Gaps in daily salinity observations were filled using spline methods (de Boor 1990).

The relations between mean-monthly delta flow (input) and salinity (response) has previously been studied using linear statistical dynamical methods (Peterson *et al.* 1989). These methods were also used to study the relations between tidal forcing (estimated sea surface elevation) and salinity at tidal time scales. Sea surface elevation near the Golden Gate was estimated using the method given by Cheng & Gartner (1984). In both instances, the model structure and parameter values were identified and estimated using an instrumental variable method (Ljung 1987, 1988). A limitation with this method is that this statistical dynamical method only estimates deviations from a mean salinity in response to deviations from mean delta flow; the mean is not estimated. However, a "steady-state" approximation of the relation between salinity and delta flow including the mean is represented by an exponential equation.

In this paper, we also use a numerical model to simulate salinity in response to delta flow. The numerical model estimates a mean as well as the fluctuating response. In brief, the numerical model is a two-layer box model with the boxes defined in Fig. 1. The model includes external forcing from delta flow and tidal currents averaged over one day. Physically-based vertical and horizontal eddy diffusivities are increased or decreased by a parameter greater or less than one. In most instances the physically based estimates underestimated the horizontal and overestimated the vertical eddy diffusivities used in this paper as well as in the earlier efforts cited. A more detailed description of the model is described by (Uncles & Peterson, in press and 1995).

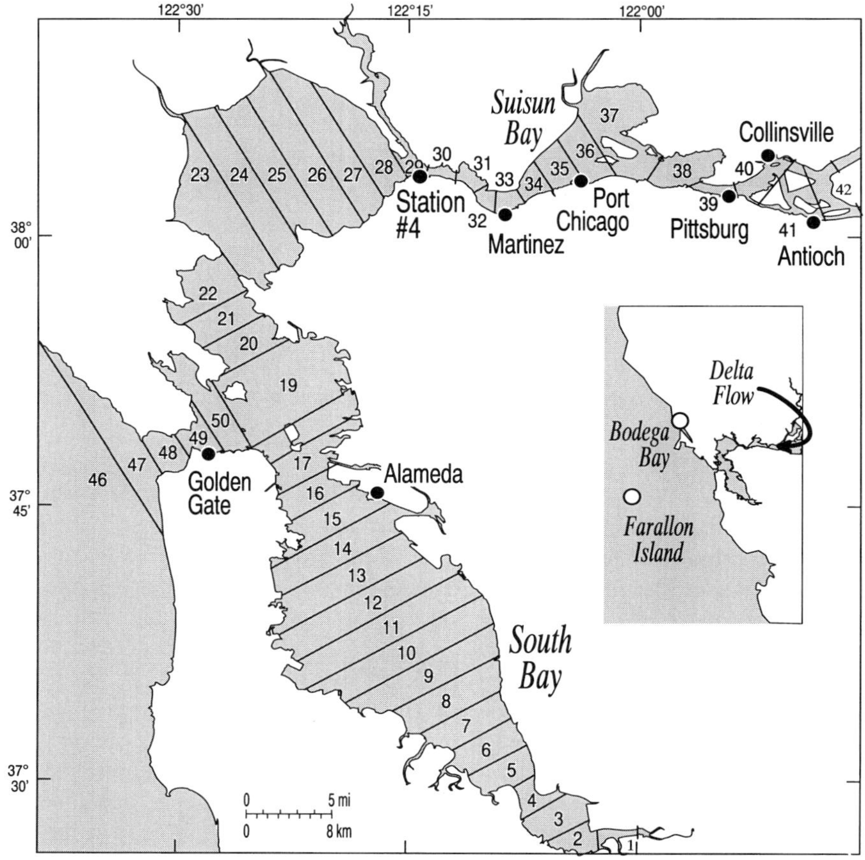

FIGURE 1. San Francisco Bay region and locations of observations in this study and the distribution of segments used in a numerical model of San Francisco Bay salinity (modified from Uncles & Peterson 1995).

EFFECTS OF SUBTIDAL FLUCTUATIONS IN DELTA FLOW

In the past a key issue in San Francisco Bay estuarine research, and we expect for years to come, is to more fully understand the effects of variations in delta flow on the bay. In this section, we will concentrate on how changes in mean-monthly DF modify salinity patterns in the bay. We will also discuss the link between

TABLE 1. Description of data, location and source parameter.

Parameter	*Period of Record (Water Year)*	*Location*	*Remarks/Source*
Delta Flow	1955-1992	Confluence of Sacramento-San Joaquin Rivers	Daily estimates, Sheila Greene California Dept. of Water Resources, unpublished
Salinity	1958-1992	Farallon Island	Daily grab samples, Patricia Walker Scripps Institution of Oceanography, unpublished
Salinity	1976-1992	Bodega Bay	Daily observations, Patricia Walker Scripps Institution of Oceanography, unpublished
Salinity	1922-1992	Fort Point	Daily grab samples, National Ocean Service, unpublished
Salinity	1941-1942	Alameda	Same as above
Salinity	1928-1956	Point Orient	Mean-monthly values from daily grab samples at maximum tide height (supplied by Phyllis Fox)
Salinity	1968-1988	Martinez	Mean-daily, Sheryl Baughman and Muriel Ferris, US Bureau of Reclamation, unpublished
Salinity	1968-1986	Port Chicago	Same as above
Salinity	1968-1988	Pittsburg	Same as above
Salinity	1968-1988	Collinsville	Same as above
Salinity	1968-1988	Antioch	Same as above
Salinity	1992	Station #4	15 minute near-surface and near-bottom, Larry Smith, U.S. Geological Survey, unpublished

salinity fluctuations in the coastal ocean and in the bay. Mean-monthly DF is used here because mean-daily delta flow is not measured directly. It is calculated via a series of source – sink terms, some involving highly managed water resources (California 1986). Mean-monthly estimates of DF are considered more reliable than daily estimates, presumably because some of the errors in daily estimates cancel.

San Francisco Bay Salinity

This first section on the Bay's salinity presents a simple relation between mean-monthly near-surface salinity as a decreasing exponential function of delta flow. This is followed by a section that studies bay-wide mean-daily (and mean-monthly) salinity using a numerical box model (NBM) largely forced by delta flow and tidal variations (Uncles & Peterson, in press and 1995).

An exponential model.

As a first-order approximation, the mean-monthly salinity in northern San Francisco Bay is defined as an exponential function of delta flow (Peterson *et al.* 1989):

$$S(x) = S_o + S_1 e^{-\beta(x)Q} \qquad (1)$$

$S(x)$ is the mean-monthly near surface salinity at a site, x, where x is the distance inland from the Golden Gate. It is a function of the mean-monthly delta flow Q (cubic meters per second) and the exponential coefficient, $\beta(x)$. S_o, an offset independent of the first-order model, and S_1, the apparent salinity at $Q = 0$, are coefficients of the empirical relation (note estuaries are not at steady state).

The exponential model predicts near-surface salinity reasonably well at locations in northern San Francisco Bay. The parameter $\beta(x)$ increases inland along the longitudinal axis of the estuary (Table 2) reflecting that the salinity response to delta flow increases from the mouth to the head of the estuary (*e.g.*, at some distance offshore there is no response). The parameters S_o and S_1 are reasonably consistent except for S_o at two stations, Point Orient and Martinez; we suspect this is partly from a problem with the sampling scheme at those stations as well as the simplistic nature of the model.

TABLE 2. Parameters of the exponential equation (1) for near-surface salinity in northern San Francisco Bay as a function of Delta flow.

Location	S_o		S_1	β
		(PSU)		*(seconds per cubic meter)*
Fort Point	0.15		33.5	0.000133
Point Orient	9[1]		23	0.00057
Martinez	0.5[2]		21	0.0018
Port Chicago	0.15		18	0.0033
Pittsburg	0.15		13	0.0082
Collinsville	0.15		13	0.012
Antioch	0.15		10	0.0119

[1] The mean-monthly values are based only on high-tide observations and may account, in part, for this disparate value.
[2] A possible artifact of the sampling location near the highest longitudinal salinity gradients.

One difficulty in using this model to represent the bay's salinity field is that we do not know quantitatively how the bay-wide mean-monthly salinity near the bottom compares with surface values throughout the bay. The difference is commonly estimated to be small at low salinities. For example, Kimmerer & Monismith (1993) assumed a surface salinity of 1.7 psu was equivalent to a near-bottom salinity of 2 psu when they calculated the location of the 2 psu at a height of one meter above the bottom (called X_2) as a function of varying levels in delta flow. It is reassuring that the location of the 2 psu value at the surface, calculated using equation (1), is close to the independently derived positions of the 2 psu value measured near the bed, at least in the interior estuary (Table 3).

Numerical Box Model

Although the above exponential relation between salinity and flow is useful, much of the detail remains obscure. The two-layer box model provides added insight into relationships between delta flow and salinity. The NBM results are

TABLE 3. Delta flow associated with salinity of 2 PSU at 1 meter above the bottom (X_2), northern San Francisco Bay Estuary.

X_2 Distance from Golden Gate (km) or, Location	*Delta Flow (cubic meters per second)*	*Reference*
60	1,200	K&M[1] 1993
Port Chicago	750	Table 2[2]
65	790	K&M[1] 1993
70	430	K&M[1] 1993
75	340	K&M[1] 1993
Pittsburg	260	Table 2[2]
80	210	K&M[1] 1993
Collinsville	210	Glover 1955[3]
Collinsville	190	Table 2[2]
85	140	K&M[1] 1993
Antioch	160	Table 2[2]
90	88	K&M[1] 1993

[1] Kimmerer and Monismith, 1993.
[2] Table 2 in this paper, assuming a surface salinity of 1.7 PSU is equivalent to a near-bottom salinity of 2 PSU (Kimmerer & Monismith 1993).
[3] Difficult to estimate because of variable flow, Fig. 3, p. 647, Glover, 1955.

tuned by time-series observations of salinity in the bay and the model skill is evaluated by subsequently comparing the simulated salinity to the observations. Model results serve several purposes: to develop a better understanding of salinity/DF interactions, to fill in large data gaps at specific measurement sites, and to predict salinities at sites where few observations have been taken.

The mean-monthly time series provide a start for comparing observed and simulated salinity near the two estuarine boundaries, the head, Collinsville, and mouth, Fort Point. One problem with the observational data, as stated above, is that we do not have a long record of daily-mean salinity observations at the mouth of the bay. In addition, the records of once-daily observations at the mouth are aliased with tidal noise. At the head, better data are available and we selected near-surface salinity sites located on both sides of the confluence of the Sacramento-San Joaquin River, one site slightly upstream from the other (Pittsburg and Collinsville, Fig. 1).

For the upper-estuary location, the simulated salinities are lower than the observed salinities for periods when salinity values are near zero (Fig. 2, Collinsville not shown). This disparity is partly due to the low values for river salinity specified in the model (Uncles & Peterson, in press and 1995). It is also possible that there is some uncertainty in the measured salinities because they are derived from the same linear transformation over the entire range of electrical conductivities and the signal to noise ratio probably decreases at very low conductivities. These small differences are of minor significance for this study; the basic salinity fluctuations are simulated reasonably well. The standard deviation in the residual salinity at Pittsburg was ± 0.45 psu and at Collinsville ± 0.44 psu. These residuals are low for this part of the bay partly because the mean salinity is low. Notice, however, that even though the model results are reasonable, the model skill is less

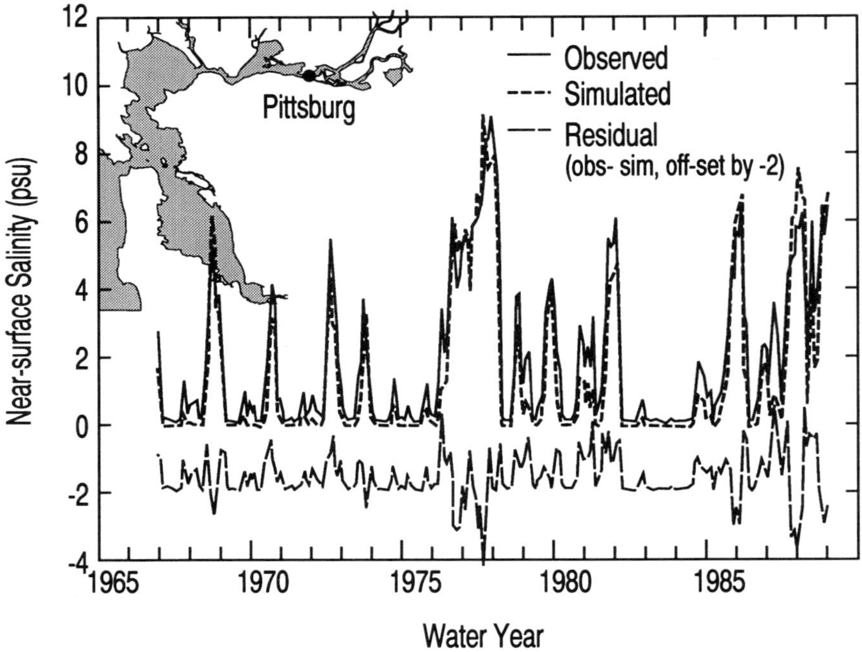

FIGURE 2. Observed and simulated mean-monthly near-surface salinity, Pittsburg; residual is the observed minus simulated value. Simulated values are from the numerical box model.

than the apparent accuracy in the observations (as is indicated by $R = 0.82$ in their residual cross-correlation). In other words, these data may be useful in guiding future model refinements.

Near the mouth of the estuary, at Fort Point and Alameda, the simulated salinities match observations fairly well (Figs. 3 and 4). The standard deviation of the residual at Fort Point is ±1.39 psu and at Alameda ±1.64 psu. This is much less than the standard deviation of the measured salinities, 3.5 and 4.7 respectively. The residual salinities between these two stations are not correlated ($R = 0.06$). Nor are they correlated with residual salinities near the head of the bay ($R = -0.15$ at zero lag). Hence, it may be difficult to use these data to further refine model predictions.

One application of the numerical model is to study time variations in residuals (observed minus simulated salinity). For example, visual inspection of the residual values for Fort Point (Fig. 3) suggests that the 1970s were less accurately simulated than the 1980s. One possible explanation is that the observations were of nonuniform quality; in fact the original field notes from the 1920s to present do show significant variations in completeness and in uniformity of sampling. In general the observations appear to be more complete in the early decades. The NBM results support the suggestion that data quality problems increased in the 1970s. But further, and perhaps even more convincing, in a separate unpublished study (J. Slack, pers. commun. 1988) simulated salinities from a regression model based on

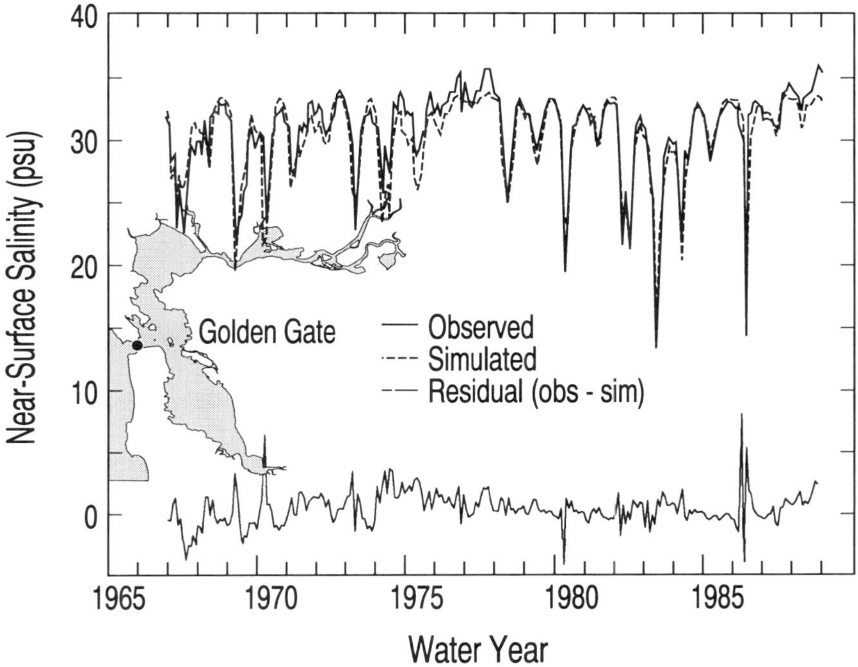

FIGURE 3. Observed and simulated mean-monthly near-surface salinity, Fort Point; residual is the observed minus simulated value. Simulated values are from the numerical box model.

observed salinity data from 1920 through 1940 represented the observed data in the 1970s and 1980s better than a similar model based on the 1970s and 1980s data alone (suggesting a lower signal-to-noise ratio in the 1970s and 1980s data compared to earlier data). The above suggests that during periods of sporadic observations, NBM may provide a better estimate of the actual values than the observations near the estuary mouth.

Another use of NBM is to provide a bay-wide frame-of-reference of the simulated salt field to compare with sparsely located observations. The mean residual salinity at Alameda is -2 psu (Fig. 4). This offset is significant and could be due to the east-side location of the observations. The model predicts salinity from an average across an entire segment. In this particular example, higher salinities may be expected on the relatively deep west side near the Golden Gate relative to the more distant and shallow east side location from coastal high salinities (see salinities for Hunters Point in Fig. 2, p. A4, McCulloch *et al.* 1970 compared to Alameda). Hence, the segment-averaged salinity could be slightly higher than the salinity observed on the east (but not necessarily 2 psu higher). Again, the model points out the complex spatial patterns for salinity in the estuary and a direction for possible future research.

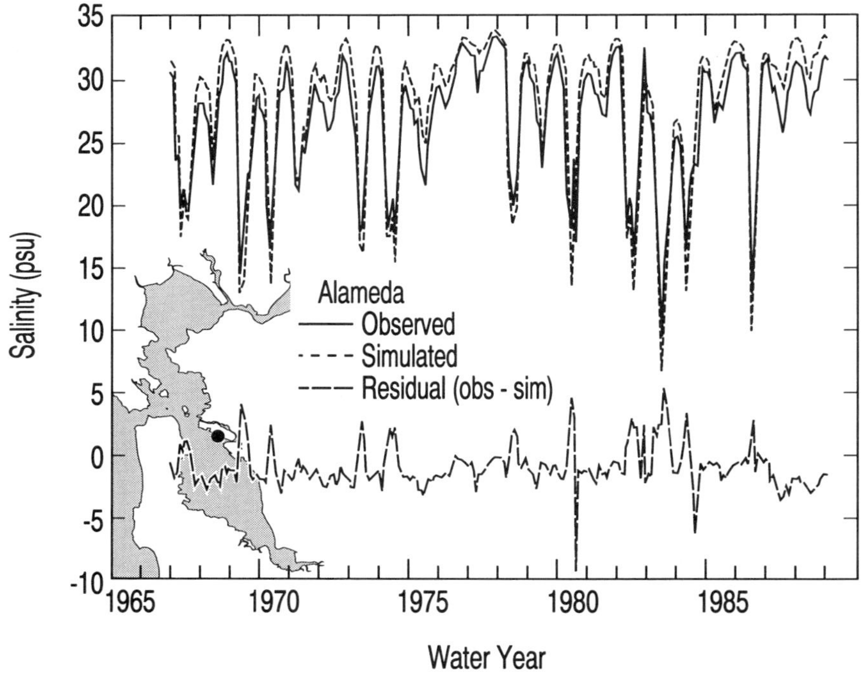

FIGURE 4. Observed and simulated mean-monthly near-surface salinity, Alameda; residual is the observed minus simulated value. Simulated values are from the numerical box model.

Hysteresis in the Salinity Response

In this subsection on the Bay's salinity we use linear statistical-dynamical models to study both time series observations and NBM results. As background, the annual cycle of mean-monthly salinity as a function of delta flow exhibits hysteresis (Fig. 5, and Peterson *et al.* 1989). Salinity is higher per unit discharge when delta flow is rising, lower per unit discharge when delta flow is declining. This is because the net discharge over the preceding months had changed the average salinity field in the bay. The bay is freshest after the highest discharge period.

There are also spatial differences in the effects of antecedent discharge. Salinity near the mouth of the estuary tends to be influenced most by past winter conditions because winter is the period of highest salinity variance. Salinity near the head of the estuary, where summer is the period of highest salinity variance, tends to be influenced most by past summer conditions.

These spatial and temporal patterns provide another test of the effectiveness of bay models. They also help identify where particular types of models work best. The response of salinity to discharge is near-linear at the mouth (over the observed range in discharge) and, therefore, linear models have a straight-forward applica-

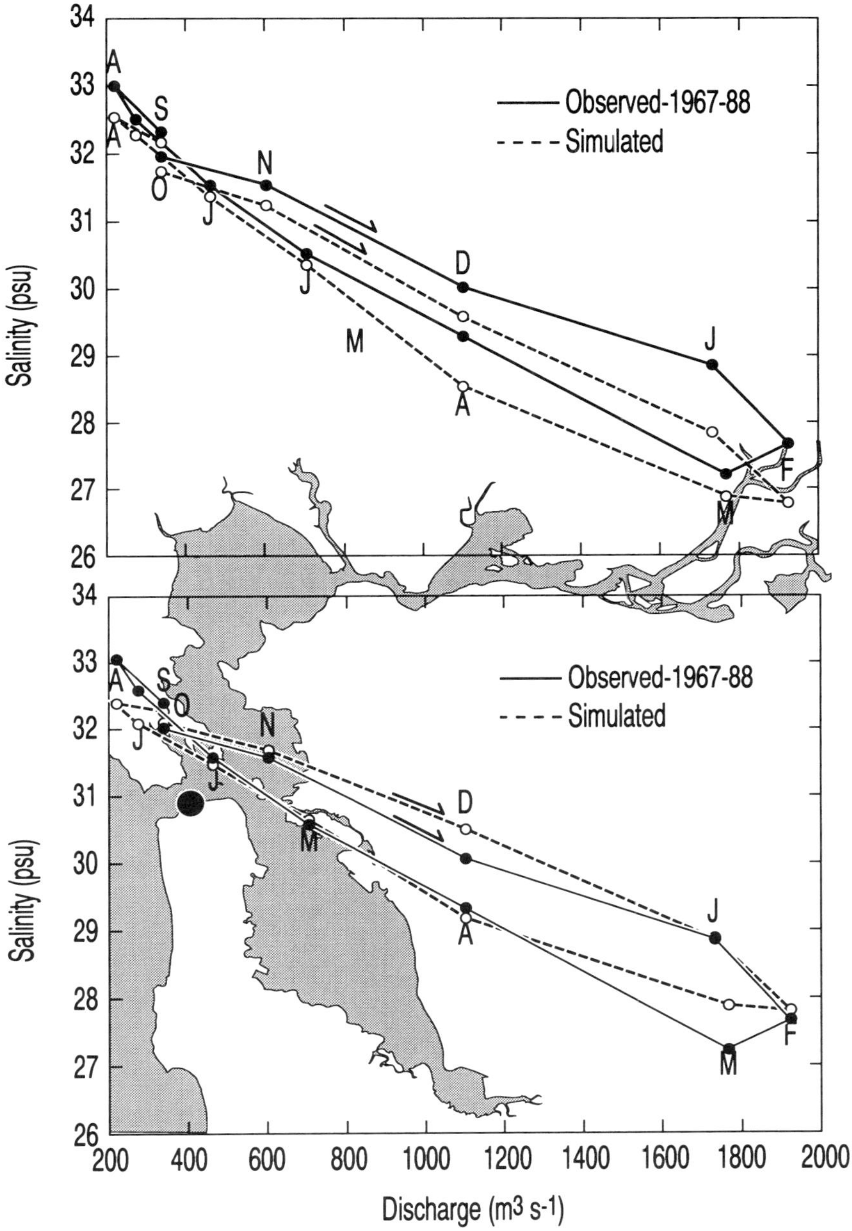

FIGURE 5. Upper Panel, mean annual cycle of salinity versus delta flow at Golden Gate from observations (1967-1988 at Fort Point) and numerical box model simulation (near surface segment # 49 in Fig. 1, 1967-1988). Lower Panel, mean annual cycle of salinity versus delta discharge at Golden Gate from observations (1967-1988) at Fort Point and from a statistical-dynamical model of salinity driven by delta flow using the observed salinity to estimate the response coefficients.

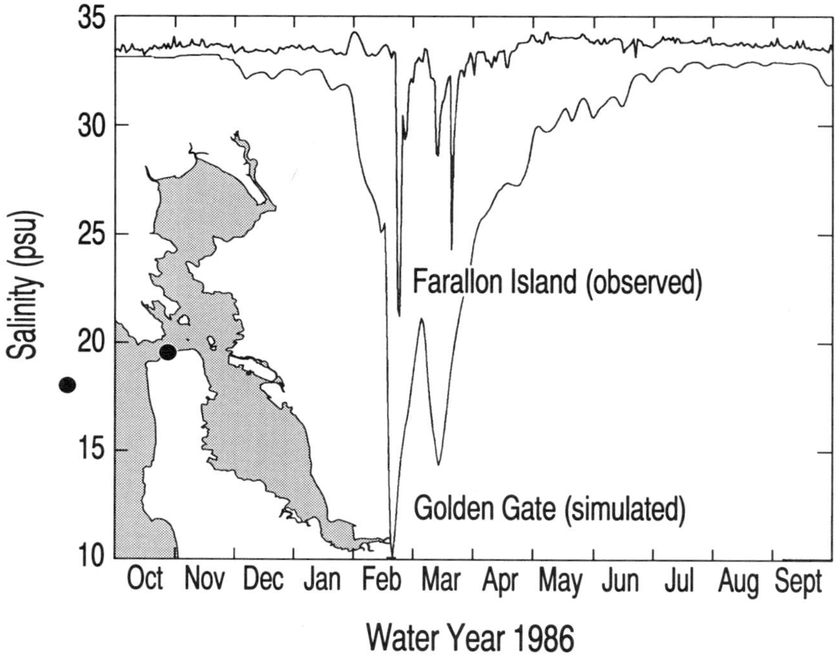

FIGURE 6. Daily near-surface salinity at Farallon Island and simulated daily salinity at Golden Gate, 1986. Simulated values are from the numerical box model.

tion. To illustrate observed and modeled hysteresis, we select results near the mouth (linear response) rather than near the head (nonlinear response).

The observed and modeled annual cycles illustrate a subtle difference in model behavior (Fig. 5). Both the NBM and statistical – dynamical models show the observed hysteresis in the salinity response to delta flow amplitudes. The width of the hysteresis loop is about the same as is measured, though the difference in salinity values just after a peak flow is less than measured. Also, the observed salinity appears more linear in response to falling delta flow. It is not yet clear how much these differences are due to a model parameter "tuning" problem rather than characteristics of the models themselves.

THE COASTAL OCEAN SALINITY

As discussed above, numerical simulations provide a reasonable description of the variations in salinity as a function of delta flow on mean-monthly time scales. However, in using mean-monthly anomalies, it is difficult to identify how changes in salinity in the coastal ocean affect salinity in the bay and vice versa (Peterson *et al.* 1989; Cayan & Peterson 1993).

To investigate this problem, we compare the NBM's daily estimates of salinity at Golden Gate with the observed daily salinity fluctuations at Farallon Island,

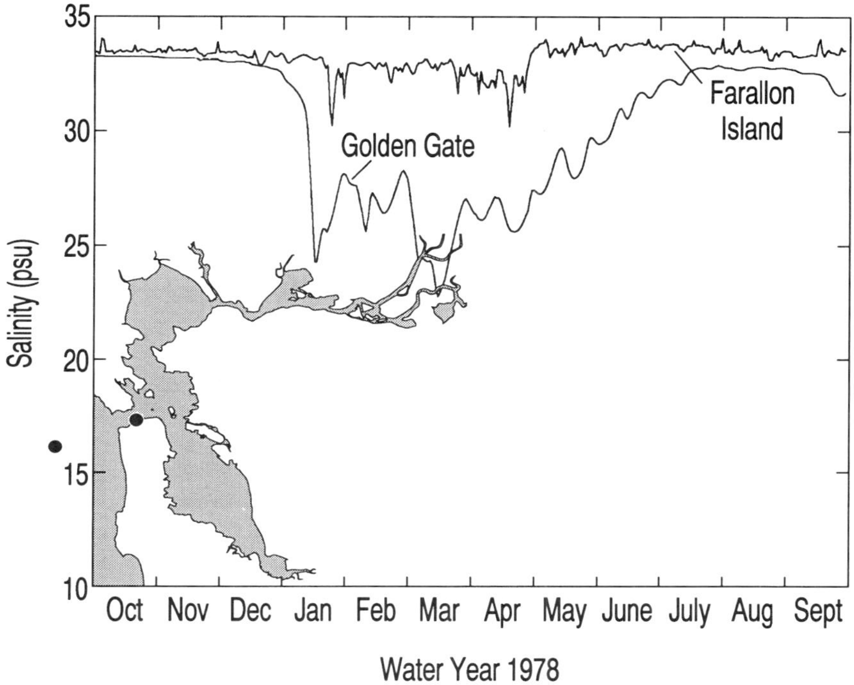

FIGURE 7. Daily near-surface salinity at Farallon Island and simulated daily salinity at Golden Gate winter 1978. Simulated values are from the numerical box model.

about 48 km offshore, and at Bodega Bay, 60 km to the north (Fig. 1). Gaps in daily Farallon Island salinity were filled using spline methods.

The typical annual cycle shows that salinities near the mouth of San Francisco Bay becomes much fresher in late winter early summer than at Farallon Islands (Fig. 6, 7). These large, fresh-water anomalies at the Golden Gate are largely due to an increase in delta flow (Uncles & Peterson, in press). On a monthly time scale, over 80% of the salinity variability at the Golden Gate can be explained solely by delta flow (Peterson *et al.* 1989). However, the coastal ocean clearly affects the bay. After freshening events, the bay gradually becomes more saline, usually well after the coastal ocean returns to a saline condition (Fig. 6, 7). This slow, delayed salinity increase suggests that the dominant processes that return salt to the bay are tidal mixing and diffusion.

Event-scale processes, such as storm driven exchanges are probably of less importance than tidal diffusion, but still can affect the salinity field in the bay. For example, upwelling events associated with the changing wind patterns in spring bring saltier water onto the shelf (Strub & James, 1988). The possible subtle role of these changes in spring coastal salinity on San Francisco Bay requires much more detailed near-surface and near- bottom observations near the bay's mouth than exist today (*cf.*, Uncles & Peterson, in press). In essence, the effects on the bay of relatively rapid changes in the coastal ocean as associated with a strong spring transition are unknown but may serve as a valuable tracer as the change propagates into the estuary.

Processes in the bay also affect the coastal ocean; low-salinity anomalies are seen in the Farallon Island records. The low-salinity spikes are stronger in wet years (1986 vs. 1978), are smaller than equivalent spikes at the Golden Gate, and typically appear after spikes appear at the Golden Gate (Fig. 6, 7). It is likely that low salinity plumes from the bay are important, but are not the sole source of this spring freshening. Other processes, such as local precipitation and runoff, upwelling, and advection also affect salinity in the coastal ocean. In particular, wind can mix the surface layers, causing a temporary increase in surface salinity.

Anecdotal accounts of the 1862 flood state that a brackish plume of water from San Francisco Bay reached the Farallon Island (Young 1929). We suggested that smaller Sacramento-San Joaquin River floods may also cause some of the low salinity spikes at Farallon Island. The oceanography of the Gulf of the Farallons, however, is complex and poorly understood (Noble, Ramp & Kinoshita 1992). Thus, at this stage, understanding the relationships between the bay's salinity and that of the coastal ocean may have more to do with interpreting variability in the Gulf of the Farallons than variability in the bay. As a simple conceptual model (Fig. 8), plume trajectories may tend towards offshore or northward pathways during pre-spring and post-fall transitions and offshore or southward during post-spring and pre-fall transitions (Conomos 1979; Breaker & Bratkovich 1993; Strub & James 1988; Noble, Ramp & Kinoshita 1992). These wind- responding transitions are stronger and more persistent in some years than others and, therefore, serve only as a generalized framework for interpreting coastal ocean variability.

Observations at Bodega Bay to the north were initiated in 1976, but it is hard to use these data to study the interaction between the bay and the coastal ocean. Major rainfall-runoff storms produce low salinity spikes at both locations (Fig. 9) because major storms are generally regional in size. Bodega Bay has its own local response to such storms, sometime freshening as much as San Francisco Bay, sometimes less. In addition, low salinity water can be advected between the bays by coastal currents. Thus, a relatively detailed knowledge of the precipitation and wind regimes is needed to help resolve possible interactions between the locations.

TIDAL VARIABILITY

Large differences in the time scales of salinity fluctuations are caused by changes in delta flow and tidal currents. To minimize the effects of delta flow at tidal frequencies we first examine tidal variations during low and constant delta flow. In this section, we focus on salinity variations in the mid-estuary reach.

Low Delta Flow

Tidal and spring-neap variations in salinity during low delta flow were simulated with a simple statistical-dynamical model. The model is:

$$y(t) = b_o x(t) + b_1 x(t-1) + b_2 x(t-2) + \ldots e(t) \qquad (2)$$

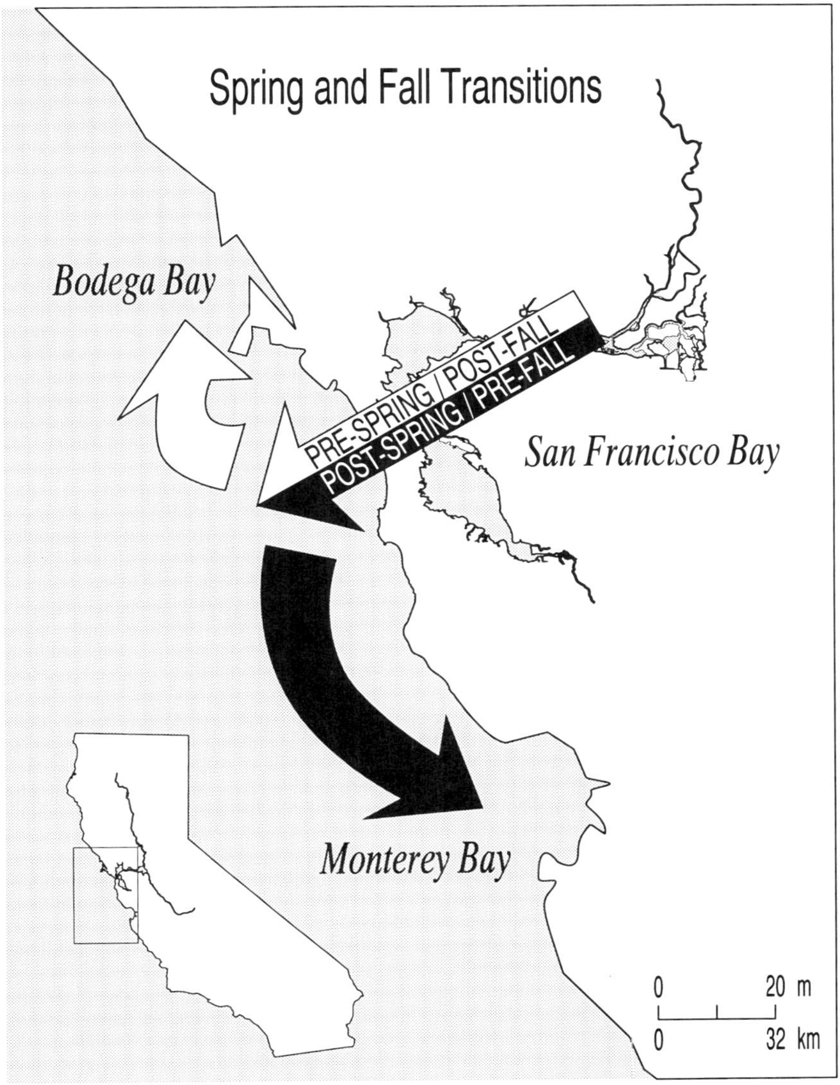

FIGURE 8. Sketch of plume pathways with wind from the north (post-spring and pre-fall transition and from the south (pre- spring and post-fall transition).

where present salinity, $y(t)$, is a linear combination of present time (t) and past forcing, $x(t\text{-}n)$, where n is a 1 hour backwards time step. For example, t-3 is the value of the forcing from 3 hours earlier; bn are the response coefficients (Table 4).

The forcing time series is estimated sea surface elevation (ESSE) at Golden Gate for calendar days 14 through 39, 1992 (Fig. 10 upper panel). Near-surface and near-bottom salinity at a station near western Carquinez Strait (Fig. 1, station # 4)

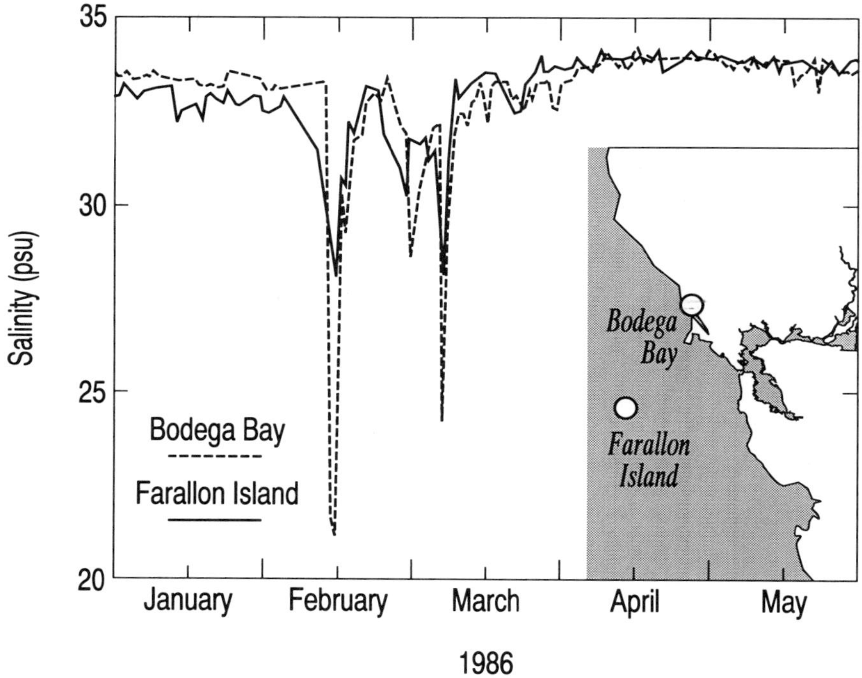

FIGURE 9. Daily near-surface salinity at Farallon Island and Bodega Bay winter 1986.

TABLE 4. Model parameters and simulation statistics.[1]

Observations	*Model Parameters*	*Estimate*	*% of Variance Simulated*	*Standard Deviation of Residual (PSU)*
Surface	b_0	0		
	b_1	0		
	b_2	0	77	± 0.99
	b_3	2.47		
	b_4	0.34		
Bottom	b_0	0		
	b_1	0		
	b_2	0		
	b_3	3.02	83	± 0.92
	b_4	0.26		

[1] Includes the 4 days 36 through 39 not used to estimate model parameters.

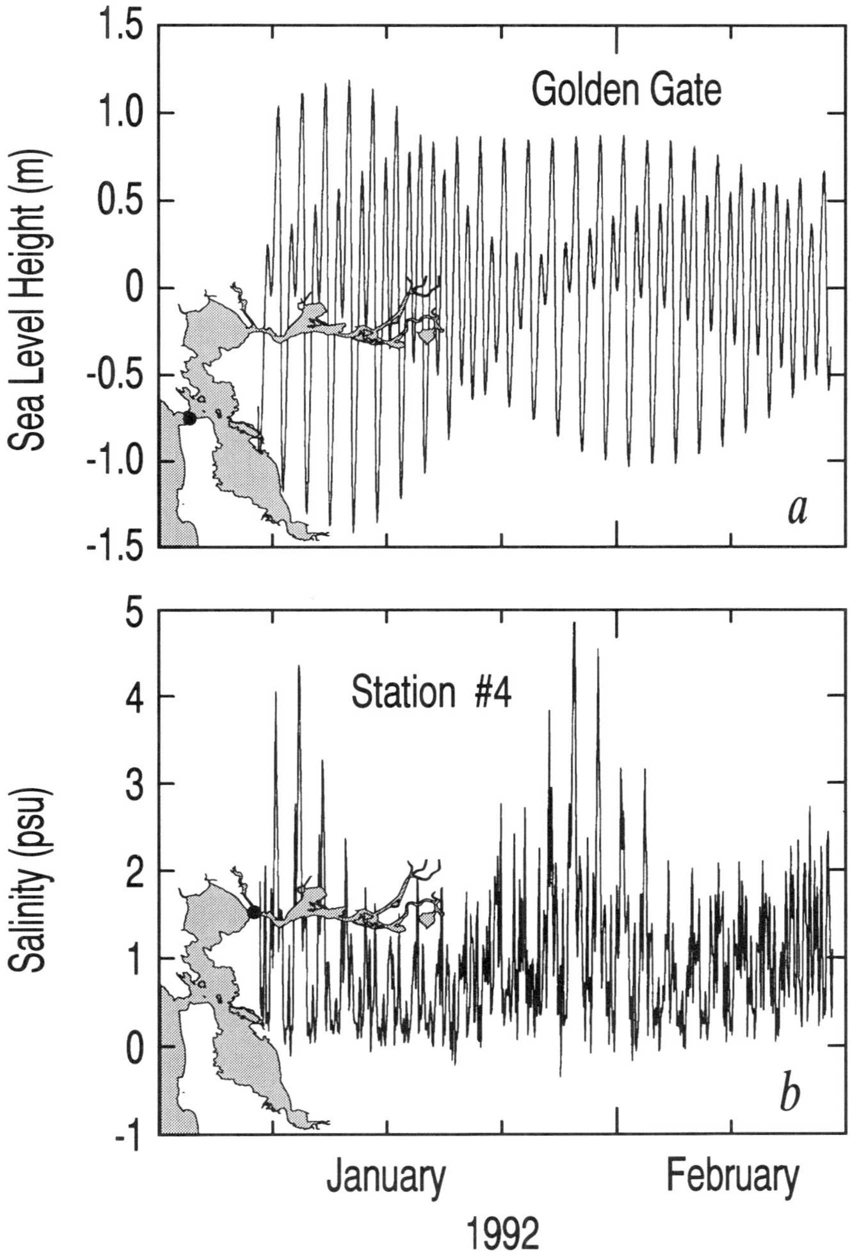

FIGURE 10. Upper panel: estimated estuarine surface elevation at Golden Gate calendar days 14 through 39, 1992. Lower panel: observed salinity stratification (bottom minus surface salinity) at station # 4 for the same period.

were simulated based solely on Golden Gate ESSE. The model structure was estimated from salinity observations of days 14 through 35. This model was also used to extend the simulation to include 4 more days beyond the model-building period (days 36 through 39).

There is a 3 hour lag in salinity response to ESSE forcing (e.g., $b_0 = b_1 = b_2 = 0$). This is partly because the station is inland from the location of the forcing in a kinematic sense (Golden Gate). Although simple, the model estimates about 80 percent of the hourly salinity variance (less for surface, more for bottom), suggesting changes in delta flow and wind regimes preceding and during the study period were small. Residual values are defined here as observed minus simulated salinity values. This instrumental variable method (Ljung 1987 and 1988) produced residuals with a Gaussian distribution (Fig. 11).

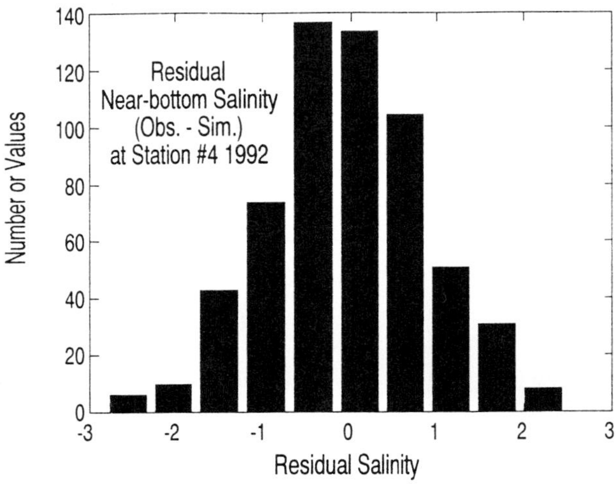

FIGURE 11. Histogram of residual salinity anomalies (observed minus simulated) for near-bottom salinity calendar days 14 through 39, 1992 at station # 4 San Francisco Bay. Simulated values are from a statistical-dynamical model driven by the estimated estuarine surface elevation at Golden Gate.

Both observed and time-series modeled data show the water column was usually stratified during the simulation time period (Fig. 10). Periods of stronger stratification were followed by periods of weaker stratification. The weaker stratification periods tended to occur when tidal amplitudes were highest (Fig. 12). The current speeds associated with changes in surface elevations are estimated from the rate of change in that elevation. In general, stratification develops and increases when currents are weak, and is destroyed when currents are strong (Fig. 13).

However, the stratification pattern is not simply related to the absolute tidal amplitudes. In general, in an elevation sequence of lower-high water followed by higher-low water, some stratification persists if the energy associated with vertical turbulent mixing is less than the potential energy associated with stratification. Apparently the next phase of the cycle is affected by this persistence in stratifica-

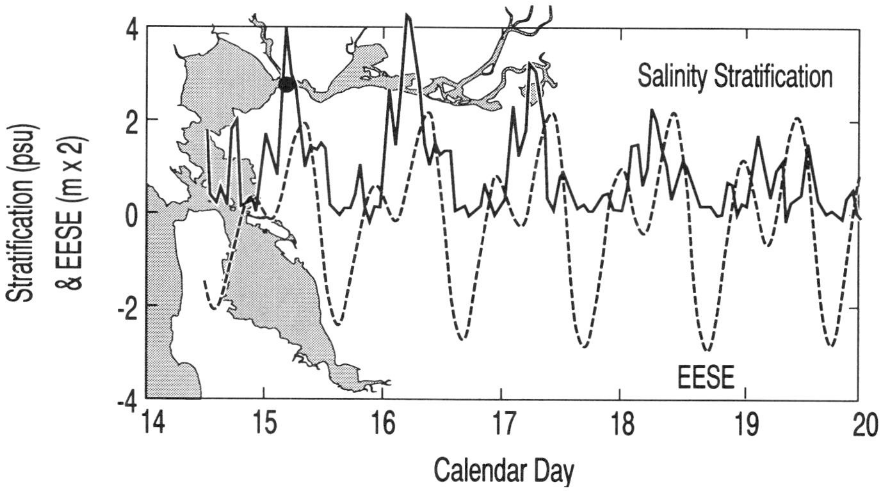

FIGURE 12. Salinity stratification at station # 4 and estimated estuarine surface elevation (EESE) at Golden Gate calendar days 15 through 20, 1992.

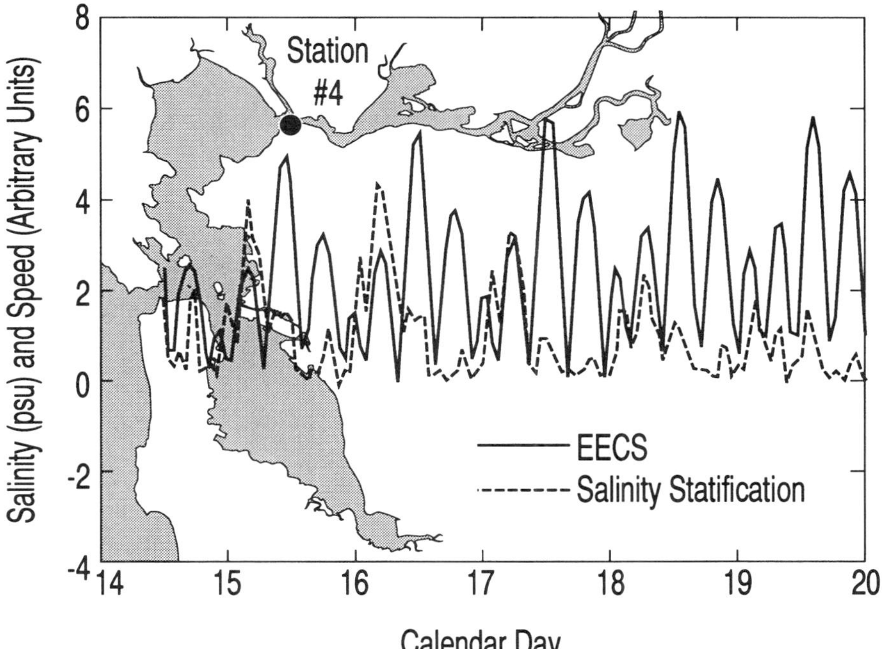

FIGURE 13. Salinity stratification at station # 4 and estimated relative estuarine current speed (EECS) at Golden Gate calendar days 15 through 20, 1992.

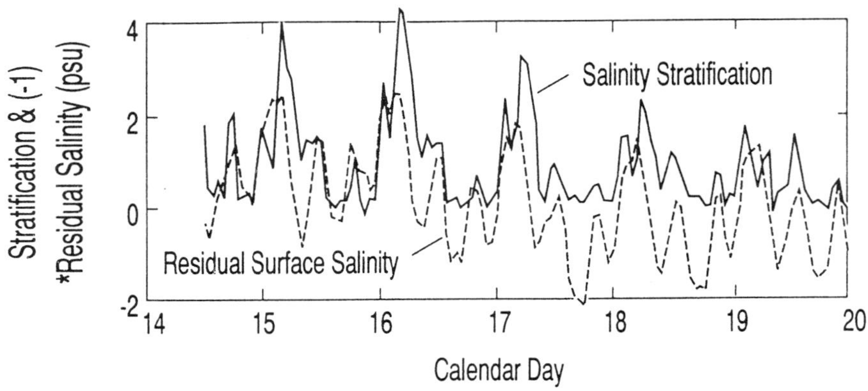

FIGURE 14. Salinity stratification at station # 4 and residual in near-surface salinity (observed minus simulated) at station # 4, calendar days 15 through 20, 1992. Simulated values are from a statistical-dynamical model driven by the estimated estuarine surface elevation at Golden Gate.

tion. It takes tidal cycles with enough energy to mix the water column before stratification breaks down (Fig. 13).

The coupling between stratification events and tidal variations of bottom stress and between feedback effects is complex (Simpson, Sharples & Rippeth 1991; Nunes Vaz & Simpson 1994). In this instance, internal dynamics not captured by our tidal amplitude-driven models appears to produce residual salinities on the order of 1 or 2 psu and time scales from hours to days (Fig 14).

High Delta Flow

The behavior of the salt field during periods of high or variable delta flow can be shown by extending the time series about 100 days to include 3 flood events (Fig. 15a). Tidal frequencies are removed from the time series by low-pass filtering. As expected, the magnitude of the drop in salinity is to some extent proportional to the magnitude of the flood (Fig. 15b). However, the timing of peak flow in relation to peak stratification varies. Presumably stratification follows the back side of peak delta flow as the salt field "reacts" to the flow. If the back side of the peak flow occurs during a neap-tide, stratification is greater than if it occurs during a spring tide.

CLIMATE AND THE HUMAN CONNECTION

The atmosphere is the major control on delta flow through the river basin, and, at subtidal frequencies, delta flow is the major control on San Francisco Bay salinity. The following is a brief overview to emphasize that the role of the atmosphere over the river basin-estuary-coastal ocean is all-encompassing.

Most of the precipitation in California occurs in winter and the relative wetness or dryness of winter precipitation is explained largely by a regional California pressure index (CPI, Cayan & Peterson 1989). When the regional atmospheric

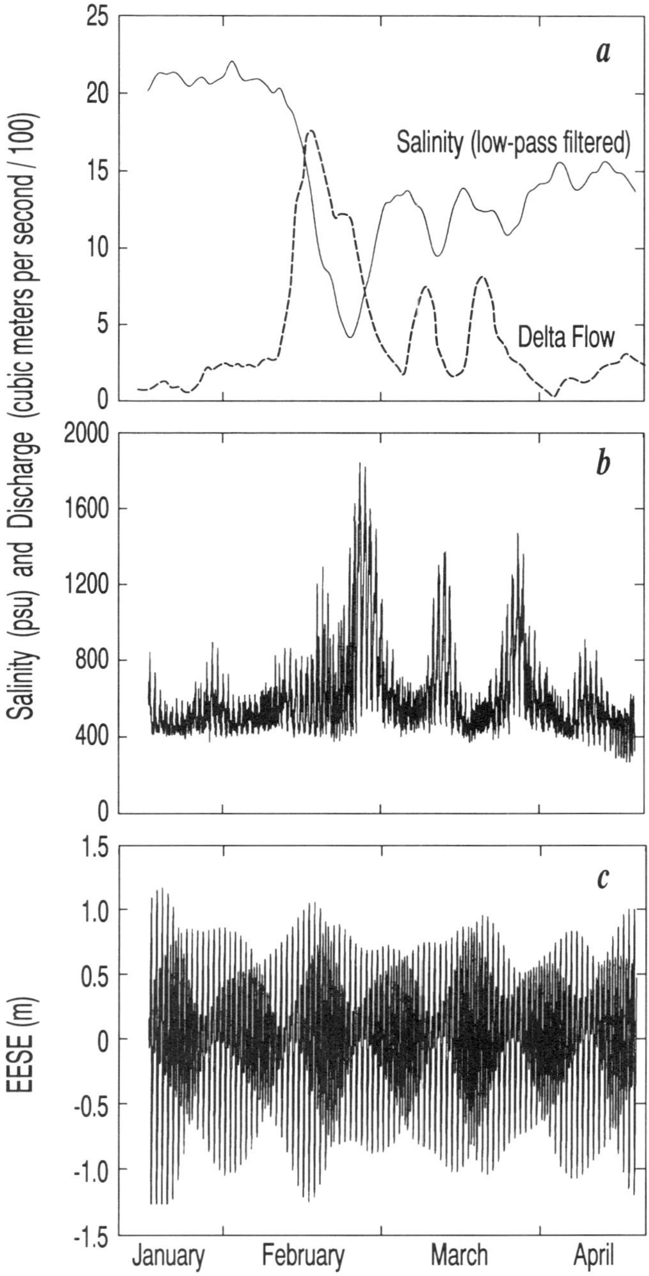

FIGURE 15. Panel A, Delta discharge winter 1992 and low-pass filtered near-surface salinity station # 4, 1992. Panel B, Salinity stratification at station # 4. Panel C, Estimated estuarine surface elevation (EESE) at Golden Gate.

pressure is high, forming a ridge, storms tend to be deflected to the north. When the pressure is low, offshore storms invade California. High or low precipitation can be further classified by air temperature: cool or warm. The four resulting winter-season categories: cool and wet, cool and dry, warm and wet, and warm and dry are associated with four distinct large-scale atmospheric circulation patterns (Cayan & Peterson 1993). Further, in winter, when the Aleutian-Alaskan low atmospheric pressure pattern over the north Pacific strengthens as is typical in El Niño years, California is very wet if that pattern is centered near the west coast of North America (*e.g.*, 1982-1983). If the pattern is farther offshore, a high pressure system tends to occupy the CPI region and California is dry (*e.g.*, 1976-1977).

Dettinger & Cayan (1995) have identified even longer time scales of influential atmospheric circulation patterns. Since the 1940s, north Pacific – western North American atmospheric circulation pressure and temperature fields in January, February, and March have pronounced long-term trends. The temperatures in these three winter months have been rising over the last five decades, resulting in less snowpack at mid-elevation Sierra Nevada mountain basins and an earlier spring snowmelt-driven runoff (the Roos effect – named after Maury Roos, California Department of Water Resources, who identified the trend in decreasing percentage of annual flow during spring). This natural effect accounts for an approximately 10 to 20 % reduction in spring flow as a fraction of total flow since the early 1940s.

This long-term reduction in spring flow as a percentage of the annual flow in the Sacramento – San Joaquin system is disconcerting because it is in the same direction as the artificial reduction in spring delta flow due to fresh water exports to the south. Reductions in spring delta flow is a critical issue in managing the bays fisheries (Jassby 1993).

These climate - delta export covariations are connected even on shorter inter-annual time scales. Because the human demand for water is as high or higher in dry vs. wet years, the percentage of delta flow that is exported is much higher in dry years.

Perhaps the above can best be summarized by showing the trend in spring (May) salinity at Pittsburg since the mid-1950s (Fig. 16). Following completion of the Shasta reservoir (early 1940s) an increasing fraction of delta flow is exported especially in spring. Secondary to this, is a natural-forced reduction in percent of annual Sacramento-San Joaquin river flow during spring (Dettinger & Cayan 1995). In effect spring delta flow has been declining and spring salinity rising. Furthermore, the large interannual swings in the percentage of delta flow exported are also coupled to wet and dry atmospheric circulation regimes. In a sense both the annual fluctuations and trend in salinity in Figure 16 are climate and human-caused, a striking example of why both scientific and management concerns of such a complex system as San Francisco Bay need a broad perspective (Peterson et al. 1995).

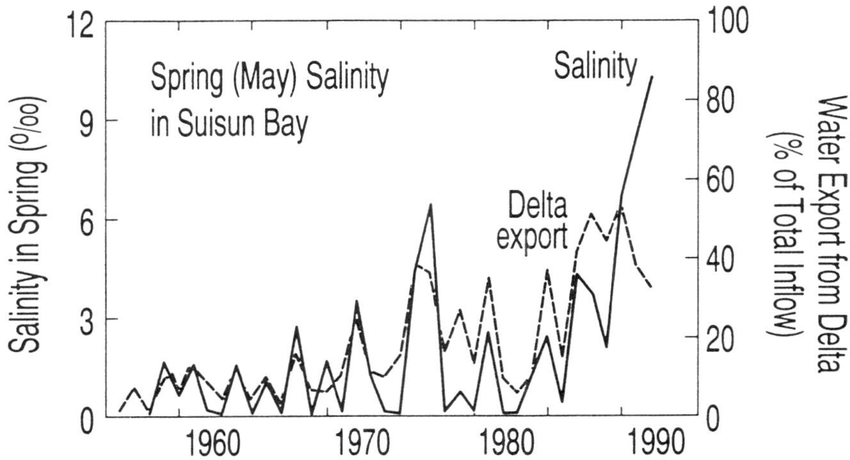

FIGURE 16. Mean-monthly (May) salinity at Pittsburg and water exported from the delta as a percentage of annual inflow.

ACKNOWLEDGMENTS

Reviews of W. Broenkow, E. McDonald and two anonymous reviewers and technical assistance of J. DiLeo-Stevens and L. Thomas are appreciated.

REFERENCES

Breaker, L. C., and A. Bratkovich. (1993). Coastal-Ocean processes and their influence on the oil spilled off San Francisco by the M/V Puerto Rican. *Marine Environ. Res.* 36:153-184.

Burau, J., S. G. Monismith, and M. Stacey. (1995). Hydrodynamic transport and mixing processes in Suisun Bay. (this volume)

California, State of. (1931). *Variation and Control of Salinity in the Sacramento-San Joaquin Delta and Upper San Francisco Bay*. Bull. 27, California State Printing Office, Sacramento. 440 pp.

California, State of. (1986). *Day Flow Program Documentation and Day Flow Data summary User's Guide*. California Dept. Water Resources, Central District, 3251 S Street, Sacramento, CA.

Carlton, J. T., J. K. Thompson, L. E. Schemel, and F. H. Nichols. (1990). Remarkable invasion of San Francisco Bay (California, USA) by the Asian clam *Potamocorbula amurensis*. I. Introduction and dispersal. *Marine Ecol. Progr. Ser.* 66:81-94.

Cayan, D. R., and D. H. Peterson. (1989). The influence of North Pacific atmospheric circulation on stream flow in the West. Pages 375-397 *in* D. H. Peterson, ed., *Aspects of Climatic Variability in the Pacific and the Western Americas*. Geophys. Monogr. Ser., vol. 55. American Geophysical Union, Washington, D.C.

Cayan, D. R., L. G. Riddle, and E. Aquado. (1993). The influence of precipitation and temperature on seasonal stream flow in California. *Water Resources Res.* 29:1127-1140.

Cayan, D. R., and D. H. Peterson. (1993). Spring climate and salinity in the San Francisco Bay estuary. *Water Resources Res.* 29:293-303.

Cheng, R. T., V. Casulli, and J. W. Gartner. (1993). Tidal, residual, intertidal mudflat (TRIM) model and its applications to San Francisco Bay, California. *Estuarine Coastal Shelf Sci.* 36:235-280.

Cheng, R. T., and J. W. Gartner. (1994). *Tides, Tidal and Residual Currents in San Francisco Bay, California – Results of Measurements, 1979-1980. Part I: Description of Data.* U.S. Geol. Surv., Water Resources Inv. Rept. 85-4339. 72 pp.

Cloern, J. E. (1984). Temporal dynamics and ecological significance of salinity stratification in an estuary (South San Francisco Bay, USA). *Oceanologica Acta* 7:137-141.

Cloern, J. E., and F. H. Nichols, (eds.). (1985). *Temporal Dynamics of an Estuary: San Francisco Bay.* Junk, Dordrecht, Netherlands. 237 pp.

Cloern, J. E. (1991). Tidal stirring and phytoplankton bloom dynamics in an estuary. *Jour. Marine Res.* 49:203-221.

Conomos, T. J. (1975). Movement of spilled oil as predicted by estuarine nontidal drift. *Limnol. Oceanogr.* 20:159-173.

Conomos, T. J. (1979). Properties and circulation of San Francisco Bay waters. Pages 47-84 *in* T. J. Conomos, ed., *San Francisco Bay: The Urbanized Estuary.* Pacific Division, American Association for the Advancement of Science, San Francisco, CA.

de Boor, C. (1990). *Spline Toolbox for Use with Matlab TM User's Guide.* The Mathworks, Inc., 21 Eliot St., South Natick, MA. 57 pp.

Denton, R. A. (1993a). *Accounting for Antecedent Conditions in Seawater Intrusion Modeling – Applications for the San Francisco Bay-Delta.* Hydraulic Engineering 93, vol. 1. Proc. ASCE National Conference on Hydraulic Engineering, San Francisco, July 1993, pp. 448-453.

Dettinger, M. D., and D. R. Cayan. (1995). Large-scale atmospheric forcing of recent trends toward early snowmelt runoff in California. *Jour. Climate* 8:606-623.

Ebbesmeyer, C. A., and W. Tangborn. (1992). Linkage of reservoir, coast, and strait dynamics, 1936-1990: Columbia River Basin, Washington Coast and Juan de Fuca Strait. Pages 288-299 *in* Helen I. Klose & Elizabeth J. Acomb, eds., *Interdisciplinary Approaches in Hydrology.* American Institute of Hydrology, Minneapolis, MN.

Festa, J. F., and D. V. Hansen. (1978). Turbidity maxim in partially mixed estuaries: A two-dimensional numerical model. *Estuarine Coastal Marine Sci.* 7:347-359.

Fischer, H. B., E. J. List, R. C. Koh, J. Imberger, and N. H. Brooks. (1979). *Mixing in Inland and Coastal Waters.* Academic Press, New York. 483 pp.

Ford, M., J. Wang, and R. T. Cheng. (1990). Predicting the vertical structure of tidal current and salinity in San Francisco Bay, California. *Water Resources Res.* 26:1027-1045.

Friedrichs, C. T., and D. G. Aubrey. (1994). Tidal propogation in strongly convergent channels. *Jour. Geophys. Res.* 99:3321-3336.

Glover, R. E. (1955). A new method for predicting transient states of salinity intrusion into the Sacramento-San Joaquin Delta. *Trans. American Geophys. Union* 36:641-648.

Hager, S. W., and L. E. Schemel. (1996). Dissolved inorganic nitrogen, phosphorus and silicon in South San Francisco Bay (parts I and II). Pages 189-215, 217-235 *in* James T. Hollibaugh, *San Francisco Bay: The Ecosystem.* Pacific Division, American Association for the Advancement of Science, San Francisco, CA.

Harder, J. A. (1977). Predicting estuarine salinity from river inflows. *Jour. Hydraulics Div., ASCE* 103:877-888.

Hill, C. L., and C. A. Kofoid. (1927). *Marine Borers and Their Relation to Marine Construction on the Pacific Coast.* San Francisco Bay Marine Piling Committee, San Francisco, CA. 357 pp.

Ingram, B. L., and D. Sloan. (1991). Strontium isotopic composition in estuarine sediments as paleosalinity and paleoclimate indicator. *Science* 255:68-72.

Ingram, B. L., and J. C. Ingle. (1996). Pre-historic record of environmental and climatic change in San Francisco Bay. Pages 35-61 *in* James T. Hollibaugh, *San Francisco Bay: The Ecosystem.* Pacific Division, American Association for the Advancement of Science, San Francisco, CA.

Jassby, A. D. (1993). Isohaline position as a habitat indicator for estuarine resources: San Francisco Bay Estuary. Appendix B, pages B1-B21 in *Managing Freshwater Discharge to the San Francisco Bay/Sacramento – San Joaquin Delta Estuary: The Scientific Basis for an Estuarine Standard.* Report to San Francisco Estuary Project, 75 Hawthorne Street, San Francisco, CA.

Kimmerer, W., and S. Monismith. (1993). An estimate of the historical 2 ppt salinity in the San Francisco Bay Estuary. Appendix A, pages A1-A29 in *Managing Freshwater Discharge to the San Francisco Bay/Sacramento – San Joaquin Delta Estuary: The Scientific Basis for an Estuarine Standard.* Report to San Francisco Estuary Project, 75 Hawthorne Street, San Francisco, CA.

Kinnetics Laboratory Inc. (1981). *In-Situ Field Data Gathering Stations San Francisco Bay-Delta Salinity Intrusion with Navigation Channels.* Final Report to U.S. Army Corps of Engineers, San Francisco District, KLI-81-1. 107 pp.

Koseff, J. R., J. K. Holen, S. G. Monismith, and J. E. Cloern. (1993). Coupled effects of vertical mixing and benthic grazing on phytoplankton populations in shallow, turbid estuaries. *Jour. Marine Res.* 51:843-868.

Largier, J. L. (1996). Hydrodynamic exchange between San Francisco Bay and the ocean: The role of ocean circulation and stratification. Pages 69-104 *in* J. T. Hollibaugh, ed., *San Francisco Bay: The Ecosystem.* Pacific Division, American Association for the Advancement of Science, San Francisco, CA.

Ljung, L. (1987). *System Identification: Theory for the User.* Prentice-Hall, Inc., Englewood Cliffs, NJ. 515 pp.

Ljung, L. (1988). *System Identification Toolbox User's Guide for use with Mathlab.* The Mathworks, Inc., 21 Eliot St., South Natick, MA. 59 pp.

McCulloch, D. S., D. H. Peterson, P. R. Carlson, and T. J. Conomos. (1970). *A Preliminary Study of the Effects of Water Circulation in the San Francisco Bay Estuary – Some Effects of Fresh-Water Inflow on the Flushing of South San Francisco Bay.* U. S. Geol., Surv. Circ. 637-A. 27 pp.

Nichols, F. H., J. K. Thompson, and L. E. Schemel. (1990). Remarkable invasion of San Francisco Bay (California, USA) by Asian clam *Potamocorbula amurensis.* II. Displacement of a former community. *Marine Ecol. Progr. Ser.* 66:95-101.

Nichols, F. H. (1985). Increased benthic grazing: An alternative explanation for low phytoplankton biomass in northern San Francisco Bay during 1976-1977 drought. *Estuarine Coastal Shelf Sci.* 21:379-388.

Noble, M., S. R. Ramp, and K. Kinoshita. (1992). *Current Patterns Over the Shelf and Slope Adjacent to the Gulf of the Farallones, Executive Summary.* U.S. Geol. Surv., Menlo Park, CA. Open-File Rept. 92-382. 26 pp.

Nunes Vaz, R. A., and J. H. Simpson. (1994). Turbulence closure modeling of estuarine stratification. *Jour. Geophys. Res.* 99:16143-16160.

Peterson, D. H., D. R. Cayan, J. F. Festa, F. H. Nichols, R. A. Walters, J. V. Slack, S. E. Hager, and L. E. Schemel, L.E. (1989). Climate variability in an estuary: Effects of river flow on San Francisco Bay. Pages 419-442 *in* D. H. Peterson, ed., *Aspects of Climate Variability in the Pacific and the Western Americas.* Geophys. Monogr. Ser., vol. 55. American Geophysical Union, Washington, DC.

Peterson, D. H., D. R. Cayan, J. DiLeo, M. A. Noble, and M. D. Dettinger. (1995). The role of climate in estuarine variability. *American Scientist* 83:58-67.

Pritchard, D. W. (1958). The equations of mass continuity and salt continuity in estuaries. *Jour. Marine Res.* 15:412-423.

Pritchard, D. W. (1957). Discussion of "On estimating stream flow into a tidal estuary" by David K. Todd and Lau Leung-Ku. *Trans. American Geophys. Union* 38(4):581-583.

Schwing, F. B. (1994). Long-term and seasonal patterns in coastal temperature and salinity along the North American west coast. Pages 129-143 *in* K. T. Redmond & V. L. Thary, eds., *Proceedings of the Tenth Annual Pacific Climate (PACLIM) Workshop, Asilomar California.* Tech. Rept. 36 of the Interagency Ecological Studies Program for the Sacramento-San Joaquin Estuary.

Simpson, J. H., J. Sharples, and T. P. Rippeth. (1991). A prescriptive model of stratification induced by freshwater runoff. *Estuarine Coastal Shelf Sci.* 33:23-35.

Smith, L. H. (1992). Unpublish software for filtering tidal variability from time series.

Smith, L. H., and R. T. Cheng. (1987). Tidal and tidally averaged circulation characteristics of Suisun Bay, California. *Water Resources Res.* 23:143-155.

Smith, L. H. (1987). *A Review of Circulation and Mixing Studies of San Francisco Bay California.* U.S. Geol. Surv., Circ. 1015. 38 pp.

Smith, P. E. and R.T. Cheng. (1991). Gravitational circulation in a tidal strait. Pages 429-434 in *Hydraulic Engineering 1991, Conference Proceedings, Nashville, TN, July 29-August 2, 1991.* WW Division of the American Society of Civil Engineers.

Strub, P. T., and C. James. (1988). Atmospheric conditions during the spring and fall transitions in the coastal ocean off western United States. *Jour. Geophys. Res.* 93:15561-15584.

Todd, D. K., and Lau Leung-Ku. (1956). On estimating stream flow into a tidal estuary. *Trans. American Geophys. Union* 37:468-473.

Todd, D. K. and Lau Leung-Ku. (1957). A reply to Donald W. Pritchard's discussion of "On estimating stream flow into a tidal estuary." *Trans. American Geophys. Union* 38(4):583-584.

Uncles, R. J. and M. B. Jordan. (1980). A one-dimensional representation of residual currents in the Seven estuary and associated observations. *Estuarine Coastal Marine Sci.* 10:39-60.

Uncles, R. J., and D. H. Peterson. (1995). A Computer model of long-term salinity in San Francisco Bay: Sensitivity to mixing and inflows. Environment International 21:647-656.

Uncles, R. J., and D. H. Peterson. (in press). Modeling the long-term salinity field in San Francisco Bay. *Continental Shelf Res.*

U.S. Bureau of Reclamation. (1987). *Water Development: Central Valley/Delta.* Document for State of California Water Resources Control Board's Bay/Delta Hearings, Sacramento. 45 pp.

Walter, R. A., and J. W. Gartner. (1985). Sub-tidal sea level and current variations in the northern reach of San Francisco Bay. *Estuarine Coastal Shelf Sci.* 21:17-32.

Washburn, L., M. S. Swenson, J. L. Largier, P. M. Kosro, and S. R. Ramp. (1993). Cross-shelf sediment transport by an anticyclonic eddy off Northern California. *Science* 261:1560-1564.

Winkler, E. D. (1985). *Preliminary Development of a Statistical Approach to Salinity Modeling in the Western Sacramento-San Joaquin Delta and Suisun Bay.* California Dept. Water Resources, Tech. Inf. Rec., Study Code No. 1463-CD-04.

Young, W. R. (1929). *Report on Salt Water Barrier Below Confluence of Sacramento and San Joaquin Rivers, California.* State of California, Dept. Public Works Bull. 22, vol. 1. 667 pp.

Zetler, B. D., and R. E. Flick. (1985). Predicted extreme high tides for mixed-tide regimes. *Jour. Phys. Oceanogr.* 15:357-359.

ISOTOPIC RECORDS OF PRE-HISTORIC SALINITY AND RIVER INFLOW IN SAN FRANCISCO BAY ESTUARY

B. LYNN INGRAM[1], JAMES C. INGLE[2], AND MARK E. CONRAD[3]
[1] Department of Geography, University of California, Berkeley CA 94720;
[2] Department of Geological and Environmental Sciences, Stanford University, Stanford, CA 94305; [3] MS 70A-3363, Berkeley Center for Isotope Geochemistry, Earth Sciences Division, Lawrence, Berkeley Laboratory, Berkeley, CA 94720

Isotopic compositions of strontium, carbon, and oxygen in San Francisco Bay estuarine waters vary systematically as a function of salinity. The isotopic composition, and thus salinity, of bay waters are recorded in the carbonate shells of mollusks living in the bay, which are incorporated into sediments accumulating beneath the Bay. Fossil mollusk shells contained in sediments cored beneath San Francisco Bay are used to derive a record of mean annual salinity over the past 5,900 years. The data indicate that average paleosalinity over this time period is substantially lower than the modern values corresponding to diversion-corrected streamflow. Because salinity in the Bay is primarily a function of river inflow, the paleosalinity values can be converted to paleodischarge values using a transfer function based on historic salinity and delta flow data. The data suggest that the annual average freshwater inflow to the Bay over much of the past 5,900 years was ≥ 1100 m^3/sec, the modern diversion-corrected delta flow value. The period between 3,700 yr BP and 5,100 yr BP appears significantly wetter than over the past 2,400 years, consistent with lake level and treeline records.

Sediments accumulating beneath San Francisco Bay contain a detailed, but little studied, paleo-environmental record of the estuary. The isotopic compositions of the sediments contain unique information about paleosalinity and freshwater inflow to the Bay. Long-term inflow records derived from bay sediment isotopic characteristics could be a valuable data source for water resource management. Instrumental stream-flow records are no longer than 100 years in drought-prone California. Pre-historic salinity and streamflow records can also serve as a baseline against which man-made modifications of the San Francisco Bay-Delta system can be assessed, and as a monitor of climate change in California. The discharge of the Sacramento and San Joaquin rivers integrates precipitation over an area equal to 40% of California, and therefore is a particularly good measure of regional climate (Fig. 1; Cayan & Peterson 1989).

Here, we present a record of paleosalinity, and inferences about paleo-streamflow, spanning much of the past 5,900 years. Paleosalinity is estimated using strontium, oxygen, and carbon isotopic measurements in carbonate fossils separated from estuarine sediments cored beneath the Bay. Isotopic measurements in modern river and estuarine waters, and of modern mollusks collected live in the Bay, are used to calibrate the relationships between salinity and isotopic compositions in the modern environment.

Ninety percent of the freshwater that enters San Francisco Bay comes from the Sacramento-San Joaquin watershed (Fig. 1). The waters of the Bay are mixtures of saline ocean water that enters through the Golden Gate (salinity ≈ 33), and fresh

FIGURE 1. Map of California showing drainage area and major rivers in San Francisco Bay watershed.

water that enters through the Delta (salinity ≈ 0.1; Fig. 2; Conomos 1979). The volume of river inflow is the primary control on salinity in the Bay (Peterson *et al.* 1989). The salinity varies systematically from high values in the Golden Gate to low values in the Delta. The salinity isopleths move upstream as river discharge decreases, and downstream as river inflow increases. At any one location in the Bay, salinity varies in response to variations in freshwater inflow. Freshwater inflow and salinity vary seasonally, with higher flows in the winter causing lower salinity in the bay, and lower flows in the summer resulting in higher salinity in the bay (Conomos *et al.* 1979). Salinity in the Bay also varies on longer timescales (hundreds to thousands of years) with changing sea level, and with changing climate. It is the longer-term variations that can be reconstructed from the estuarine sedimentary record (Ingram & Sloan 1992).

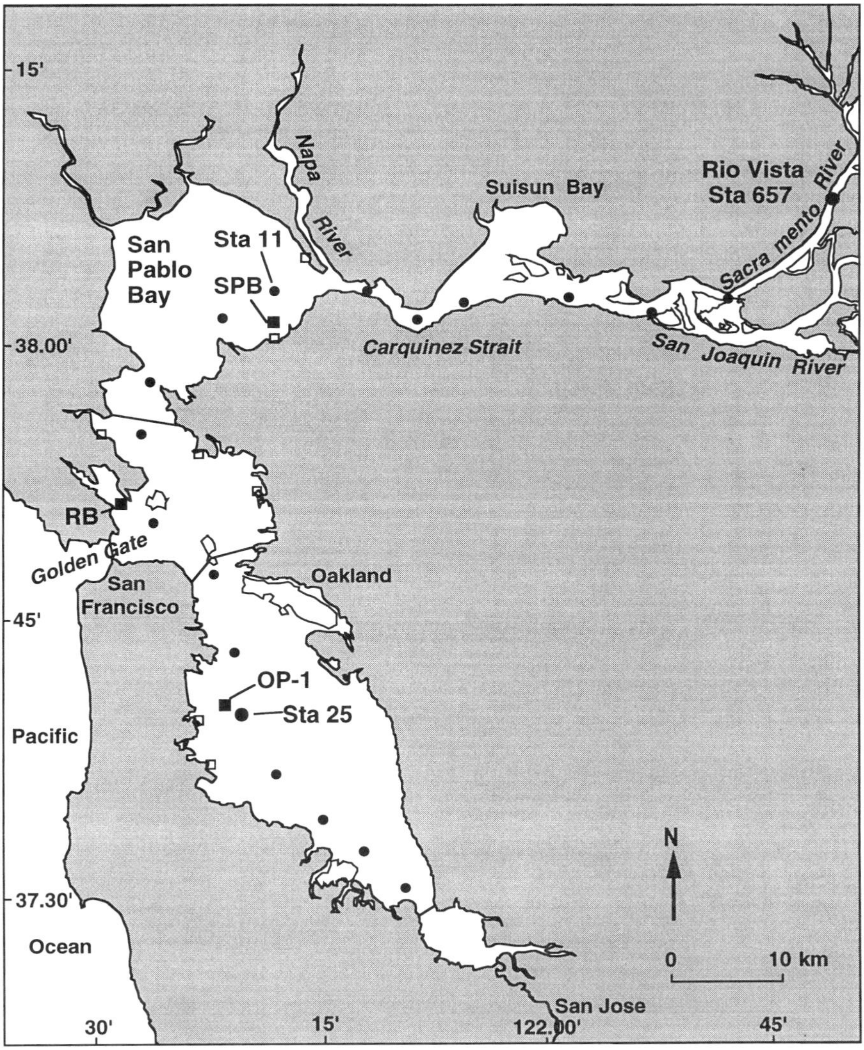

FIGURE 2. Map of San Francisco Bay, showing surface water sampling sites (closed cirles), modern mussel shell sampling localities (open squares), and coring localities (closed squares).

The sediments beneath San Francisco Bay consist of alternating estuarine and fluvial deposits over the past 450,000 years or more, reflecting sea level changes associated with fluctuations in the volume of continental ice (Atwater *et al.* 1977, 1979, 1981). Modern San Francisco Bay began forming at the end of the last glacial period (10,000 years ago), as recorded in estuarine sediments cored beneath the Bay (Atwater *et al.* 1977, 1979). Sea level reconstructed in San Francisco Bay sediments corresponds closely to the global sea level curve from Barbados (Fig. 3;

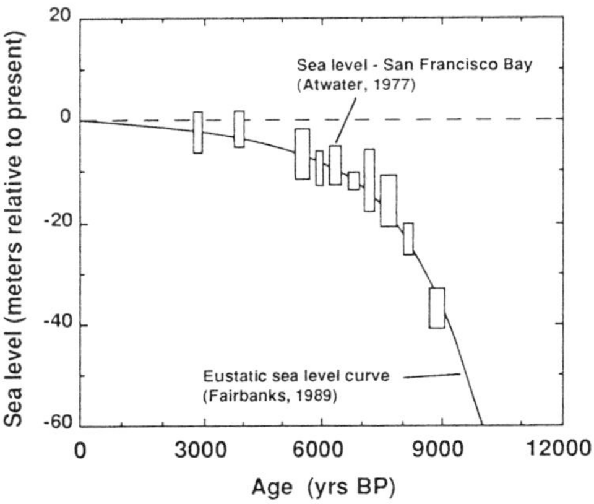

FIGURE 3. Sea level over the past 10,000 years, showing both the eustatic sea level recorded in Barbados coral reefs (Fairbanks 1989), and sea level recorded San Francisco Bay sediments (Atwater *et al.* 1979).

Atwater *et al.* 1977; Fairbanks, 1989). The sea level data indicates that the volume of San Francisco Bay has been controlled almost entirely by eustatic sea level and intra-estuarine sedimentation rather than by tectonic uplift or subsidence over the past 6000 years.

Paleosalinity Methods

Isotopic ratios of Sr ($^{87}Sr/^{86}Sr$, or $\Delta^{87}Sr$), oxygen ($^{18}O/^{16}O$ or $\delta^{18}O$), and carbon ($^{13}C/^{12}C$ or $\delta^{13}C$) can be related to salinity because the saline ocean water and the freshwater entering the Bay are markedly different in isotopic composition. The mixed waters of the Bay have intermediate salinity and isotopic compositions, and for each element, a one-to-one relationship between salinity and isotopic ratio can be established. The isotopic ratio of the water is recorded in the calcium carbonate shells of organisms growing in the Bay, and it is from the shell material that the paleosalinity is determined.

Strontium Isotopes. To relate the Sr isotopic variations measured in fossil materials to estuarine salinity in San Francisco Bay, it is necessary to know the strontium isotopic ratio and Sr concentration of both seawater and river water entering the Bay. The values are well known for seawater, which has had a constant $^{87}Sr/^{86}Sr$ ratio of 0.70919 and Sr concentration of 8 ppm for at least the past 200,000 years (DePaolo & Ingram 1985; Capo & DePaolo 1990). The Sr isotopic composition of river water, however, is somewhat variable, being controlled by the lithologic composition of rocks weathering in the catchment basin. The San Joaquin River drains the southern Sierra Nevada batholith (granitic rocks), with $^{87}Sr/^{86}Sr$ ratios of 0.70694 to 0.70785 (Ingram & Sloan 1992). The Sacramento River drains the northern Sierra Nevada and the southernmost Cascade Range, yielding an average $^{87}Sr/^{86}Sr$ ratio of 0.70485 to 0.70626 (Ingram & Sloan 1992). The average $^{87}Sr/^{86}Sr$ ratio and Sr concentration of river water entering the Bay is 0.70655 ($\Delta^{87}Sr$ value of -265), and 130 ppb. The effects of the variability of $^{87}Sr/^{86}Sr$ in the freshwater inflow are discussed by Ingram & DePaolo (1993), who determined that within reasonable bounds, the isotopic characteristics of average river water entering San Francisco Bay can be considered unchanging over the time period of interest here.

Oxygen Isotopes. The oxygen isotopic compositions of the river water entering San Francisco Bay depends on the $\delta^{18}O$ values of precipitation and groundwater in the Bay's watershed. The $\delta^{18}O$ values of precipitation in California vary widely, but systematically (Ingraham & Taylor 1991) from ≈ -5.5‰ on the Pacific Coast of California to ≈ -16.0‰ in the high elevations of the Sierra Nevada. River water entering San Francisco Bay is a mixture of surface runoff (from precipitation and snowmelt) and groundwater integrated over the entire drainage basin, and would therefore be expected to have a $\delta^{18}O$ value near the middle of this range. The actual value has been determined by measurements of estuarine waters (see below).

The salinity and stable isotopic compositions of coastal Pacific waters along the northern California margin are modified by freshwater from the continents and coastal upwelling (Peterson *et al.* 1989). Increases in both freshwater inflow and upwelling are expected to produce small decreases in the $\delta^{18}O$ values of coastal surface waters (by about 0.5 to 1.0‰). Because the variations in the stable isotope compositions of coastal waters are relatively small compared with the difference between their isotopic compositions and those of Sacramento River water, we assume the coastal end-member remains constant.

Due to the large difference between the oxygen isotope compositions of seawater and freshwater, the $\delta^{18}O$ variations in mollusk shell calcium carbonate are primarily determined by the salinity of the water in which it grew. However, because water temperature also causes oxygen isotope fractionation during carbonate precipitation, temperature effects in San Francisco Bay must be considered. The temperature of estuarine waters is a function of local weather conditions (ambient air temperature) and the amount and temperature of seawater and freshwater entering the estuary. The longer residence time in the summer months caused by decreased river inflow is also associated with an increase in water temperature of shallow areas of the Bay (particularly the south bay; Conomos 1979).

While large seasonal temperature variations have been observed throughout most of the bay, year to year variations in the average annual water temperatures are relatively small (Conomos 1979; Conomos *et al.* 1979). To correct for temperature-induced fractionation of oxygen isotopes in mollusk carbonate shells, we used the following equation (modified from Epstein *et al.* 1953):

$$\delta^{18}O_{carbonate\ (PDB)} = \{(T°C - 16.9)/-4.2\} + \delta^{18}O_{water(SMOW)}, \quad (1)$$

where $\delta^{18}O$ carbonate is expressed relative to the PDB standard and $\delta^{18}O$ water is expressed relative to the SMOW standard (*cf.* Faure 1986). Note that at 16.9°C, the value of $\delta^{18}O_{carbonate\ (PDB)}$ is equivalent to $\delta^{18}O_{water(SMOW)}$.

DESCRIPTION OF SAMPLES

The results reported here derive from three types of samples: estuarine waters, modern mollusk shells, and fossil shells from sediment cores. Estuarine waters were sampled in March, 1991, along a transect from the Delta (east of Suisun Bay) to the Golden Gate and South Bay (Fig. 2) for Sr isotopic measurements (Ingram & Sloan 1992). For oxygen isotopic measurements, estuarine waters were sampled from the same transect in March, 1992. In addition, estuarine waters were collected every two months over a three-year period (1991-1993) from three stations (11, 25, and 657; Fig. 2) from south-central San Francisco Bay, San Pablo Bay, and Rio Vista (by A. van Geen, U.S. Geological Survey).

To verify the relation between $\delta^{18}O$ values in carbonate shells and salinity in estuarine waters, modern mussel shells (*Mytilus edulis*) were collected live at eight bay locations during June-July, 1994 (Fig. 2) for isotopic analyses. The average $\delta^{18}O$ values for the whole shells, representing 1.0 to 1.5 years of growth, were compared with the $\delta^{18}O$ values of water samples collected with the mussels. In addition, shells from several of the sites were subsampled along growth bands in order to determine the extent to which seasonal isotopic variations were recorded in the shells.

The sediment cores used for this study were from Richardson Bay, San Pablo Bay, and the south-central part of San Francisco Bay near Oyster Point (Fig. 2). The cores from Richardson Bay included a 2.5 meter-long, 15 cm-diameter gravity core, and an 11 meter-long, 10 cm-diameter core (taken in 1991 by the U.S. Geological Survey). The cores from San Pablo Bay and south-central Bay (near Oyster Point) were taken by the authors during 1992 and 1993. Coring was accomplished using a modified Osterberg vacuum piston coring sampler, with a piston core barrel 1.15 m in length and inside diameter of 10 cm. The Oyster Point and San Pablo Bay cores are 5.72 m- and 6.7 m-long, respectively (Fig. 4). All of the cores were x-rayed for examination of sedimentary structures prior to sampling.

In the Richardson Bay gravity core, the sediments consist of layered brown-gray silt and clay with distinct layers rich in sand, mollusks (fragments and whole shells), and organic material (wood and seeds). Identifiable mollusk shells included clams and oysters (*Macoma nasuta* and *Ostrea lurida*), as well as foraminifers, fecal pellets, fish remains, and diatoms. The sediments from the Richardson Bay long core were similar in composition to those of the gravity core, but contained less abundant fossil mollusk remains (Ingram & DePaolo 1993).

The sediments from the San Pablo Bay core were composed of laminated clay and silt, fine sand, organic remains (plant fragments, wood, charcoal, and marsh plant seeds), calcareous and agglutinated foraminifers, diatoms, ostracods, mollusks (whole shells and fragments of bivalves), and fish remains. The dominant bivalve in the upper part of the core (0 to 120 cm) was *Macoma nasuta*. Below this level, *Mytilus edulis* was the dominant bivalve present. The fine-grained sand and silt was composed of mica, quartz, feldspar, and chrysotile (serpentine).

Sediments cored at Oyster Point (OP-1) consist of laminated and layered clay, silt, fine sand, organic remains (plant fragments, wood, seeds from marsh plants), calcareous and agglutinated foraminifers, diatoms, mollusks (whole shells and fragments of mussels, clams, oysters, and gastropods), barnacle shells, and fish remains. The upper 3 meters of the core at OP-1 is composed mainly of shell layers in a silt/sand matrix. The lower 2.5 meters consists of faintly bedded-to-laminated mud, with a few shelly layers; the mud is also burrowed and bioturbated in some horizons. The shell layers in OP-1 contain both abundant broken shells and whole shells, some with articulated valves in growth position. The dominant mollusks that could be identified were clams, oysters, and mussels (*Macoma nasuta, Ostrea*

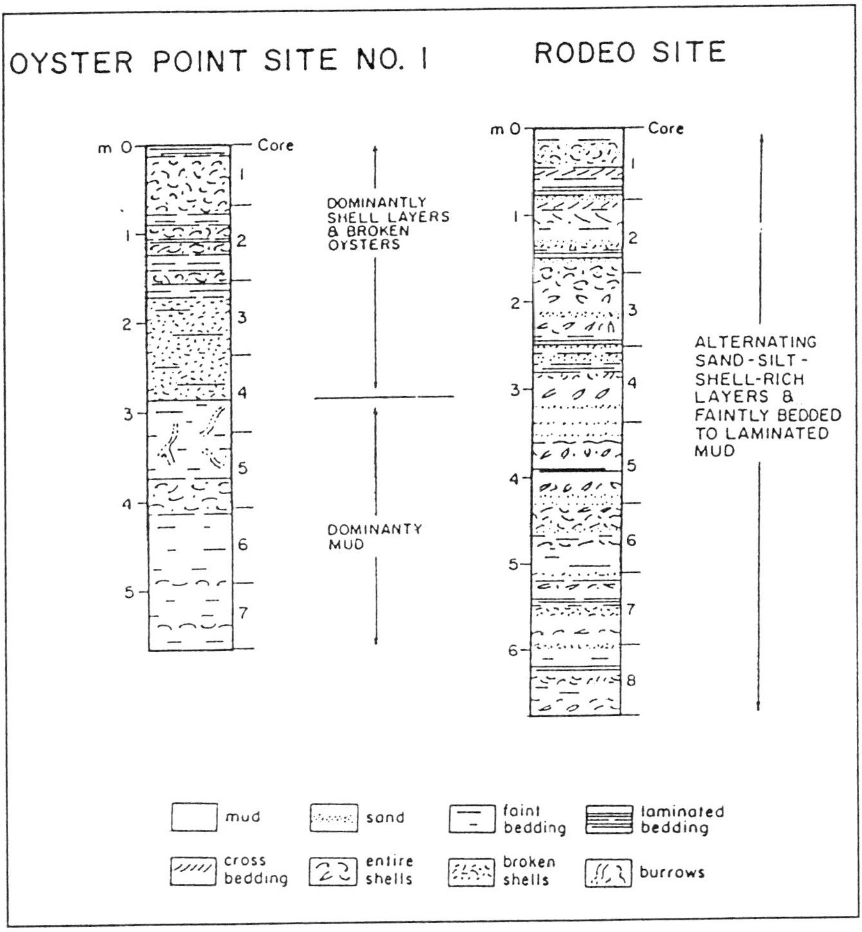

FIGURE 4. Lithologic columns of the Oyster Point and San Pablo Bay cores, showing sedimentary structures and sediment type. Lithologies and sedimentary structure based on soft x-ray analysis and visual inspection of split cores.

lurida, and *Mytilus edulis*). Although many of the mollusk shells in the cores were fragmented, in most cases the fragments were large and angular with little or no evidence of abrasion, suggesting minimal transport within the estuary. The sedimentary structures and sediment-type cored at Oyster Point and San Pablo Bay are shown in Fig. 4.

The Oyster Point and San Pablo Bay cores were subsampled sequentially every 2.0 cm, consuming one-half of the core, yielding approximately 50 grams of wet sediment per sample. Because the density of shell material was highest in the upper 350 cm of the Oyster Point core, the sampling resolution is highest over that interval (every 2 cm sample contained shell material). In the lower portion of the core (446 to 578 cm), the sediments were predominantly laminated mud with interbedded shell layers, allowing only intermittent sampling of shell material. The interval from 350 to 446 cm contained no shell material at all; it was the most heavily bioturbated.

ANALYTICAL METHODS

Prior to separation of the fossil samples, the sediments were soaked in distilled water overnight and then wet sieved using a 250-mesh screen (63 μm) and a coarser screen (> 841 μm) to separate macrofossil shell material from the microfossils (foraminifers). Mollusk shells (whole and fragments) from the coarse fractions of each sample were separated and sorted for isotopic analysis and radiocarbon dating.

Radiocarbon Analyses

Radiocarbon measurements were made at the Center for Accelerator Mass Spectrometry (CAMS), Lawrence Livermore National Laboratory (Davis *et al.* 1990). Prior to analysis, the shell surfaces were etched with 0.1N hydrochloric acid and rinsed with deionized water to remove adhered organic acids and secondary carbonate. The entire shell specimen was powdered to homogenize any variability in radiocarbon content along growth layers. In samples containing adequate numbers of foraminifers, radiocarbon analyses were also made of these to compare with the molluscan data.

Carbonate samples (ca. 8 mg) were placed in a 10 ml vacutainer (disposable blood sample vial) and evacuated through the rubber stopper using a hypodermic needle. After evacuation to below 20 mtorr, 0.5 ml of phosphoric acid was added to the vacutainer with a syringe, and the sample hydrolyzed for 30 to 60 minutes at 90°C to generate CO_2; the CO_2 was reduced to graphite using hydrogen with a cobalt catalyst (Vogel *et al.* 1987). The $^{14}C/^{13}C$ ratios were measured by AMS, and radiocarbon ages determined following standard conventions (Stuiver & Polach 1977; Donahue 1992).

Radiocarbon ages of mollusk shells were corrected for secular variations in the carbon cycle and production rate of ^{14}C in the atmosphere, as well as for upwelling of ^{14}C-depleted coastal water (Stuiver *et al.* 1986). Because of exchange between surface and intermediate water, and coastal upwelling, the seawater endmember in San Francisco Bay estuary has an average "reservoir age" of about 625 years (Stuiver & Braziunas 1993), with a corresponding ΔR of 225 years. Radiocarbon measurements of modern Sacramento river water indicate that it is similarly depleted in ^{14}C, so no correction for salinity was applied. We used the ΔR value of 225 and the marine ^{14}C-calibration curve (Stuiver & Braziunas 1993) to correct the radiocarbon ages. Because of the relatively large corrections applied to the data, we considered the age uncertainties to be no better than ± 100 years.

Strontium Isotopic Analyses

Mollusk shells were cleaned ultrasonically in deionized water, and oven-dried. Whole shells or several shell fragments were powdered in order to homogenize any seasonal bias that may exist a single shell fragment. A 1 to 5 mg split was taken from the powder and dissolved in weak acetic acid for isotopic analysis. The acetic acid solution was evaporated under a heat lamp and the sample redissolved in 1.5N HCl. One aliquot of the carbonate sample was used to determine Sr and Rb concentrations by isotope dilution using a spike solution of ^{84}Sr and ^{85}Rb. Another aliquot was passed through an ion-exchange column to separate purified Sr for isotopic analysis.

Estuarine and river waters were centrifuged to remove suspended sediment and organic material. One hundred to 500 ml of river water, and 10-20 ml of estuarine water, were evaporated, followed by the same treatment as the carbonate samples.

The Sr isotopic ratio measurements were made on a Fisons multicollector mass spectrometer, as described in Ingram & DePaolo (1993). The $\Delta^{87}Sr$ notation used is from DePaolo & Ingram (1985): $\Delta^{87}Sr = \{(^{87}Sr/^{86}Sr_{sample}) - (^{87}Sr/^{86}Sr_{standard})\} \times 100{,}000$. The EN-1 seawater standard has a $^{87}Sr/^{86}Sr$ ratio of 0.70919 ± 1.

Oxygen and Carbon Isotopes

The $\delta^{18}O$ values of water samples were measured using a variation of the method outlined by Socki *et al.* (1992). Two-ml water samples were equilibrated at 25°C (± 0.2°C) for at least 12 hours with approximately 800 mbars of CO_2 in 7 ml Vacutainers (blood sample vial). The CO_2 was then extracted from the Vacutainers, purified and analyzed with a Fisons Instruments Prism Series II isotope ratio mass spectrometer. Raw values from the mass spectrometer were corrected for the fractionation of oxygen isotopes between the CO_2 and H_2O using a CO_2-H_2O value of 1.0412. Duplicate analyses were within ± 0.1.

The modern mussel shells were cleaned to remove all tissue and organic material and dried prior to subsampling. The shells were sampled across their full length to determine the average oxygen and carbon isotopic compositions for the entire growth period. Growth bands in the shells were subsampled with a hand-held engraving drill with a 0.5 mm bit. In both cases, only the outer calcitic prismatic layer of the shell was sampled.

Fossil mollusk shells (whole and fragments) were separated and powdered for stable isotopic analyses. CO_2 for isotopic analyses was produced by reacting 1-2 mg of the shell samples with phosphoric acid (following the methods of McCrea 1950) using an Isocarb automated carbonate device on line with the Prism. The raw isotopic data from the mass spectrometer were corrected relative to the average values of 6-9 analyses of a carbonate standard, CM-1 ($\delta^{13}C = 2.05‰$, $\delta^{18}O = -1.94‰$), per sample run of 24 to 36 unknowns. Oxygen and carbon isotopic data for carbonate samples are both reported relative to the PDB Belemnite standard. The precision for these analyses was ± 0.05 for carbon and ± 0.1 for oxygen. All stable isotope analyses for this study were done at the Center for Isotope Geochemistry (CIG) at Lawrence Berkeley Laboratory.

RESULTS

Sediment Ages and Sedimentation Rates

The corrected basal age of the Oyster Point Core is 5,900 yrs BP at a depth of 578 cm, giving an average sedimentation rate of 0.98 mm/yr over the entire core. Radiocarbon dates from the middle of the core indicate that sedimentation rates were higher in the upper 350 cm of the core (1.7 mm/yr) compared with the lower 278 cm of the core (0.6 mm/yr; Fig. 5). The corresponding time interval represented by a 2-cm thick sample from the upper 350 cm of core is about 12 years, compared with 34 years in the lower 278 cm. The extent of bioturbation in the sediments by burrowing organisms also determines the time resolution in the sediments. Most of the core was not significantly bioturbated, although portions of the core showed

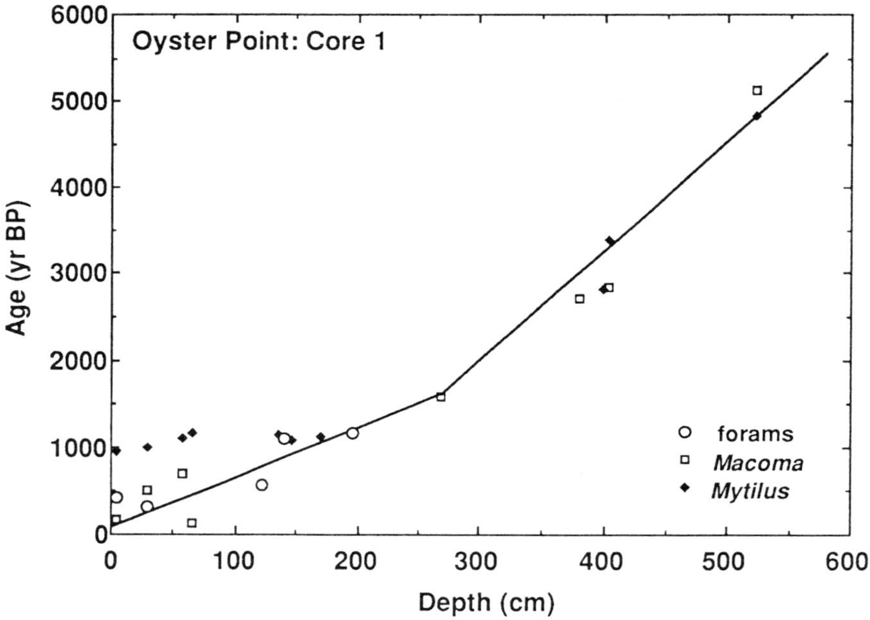

FIGURE 5. Sediment thickness (cm) versus corrected radiocarbon age (yr BP) for the Oyster Point core.

bioturbated intervals extending 5-10 cm. In the Oyster Point core, the most heavily bioturbated interval was 350 to 450 cm (Fig. 4). In the bioturbated intervals, the time resolution is limited to ≥ 50 years. The age of the sedimentary discontinuity separating the upper shell-rich unit from the lower laminated mud unit is approximately 2,100 yrs BP. The bioturbated interval (350 to 446 cm) occurred between 2,400 and 3,700 yrs BP.

In the San Pablo Bay core, the calibrated radiocarbon age of the base of the section is 2130 yrs, yielding an average sedimentation rate over the entire core of 3.14 mm/yr. However, sedimentation is more rapid 1600 to 2130 years ago relative to the past 700 years (Fig. 6). Between 105 and 127 cm, there is a 900-yr period of

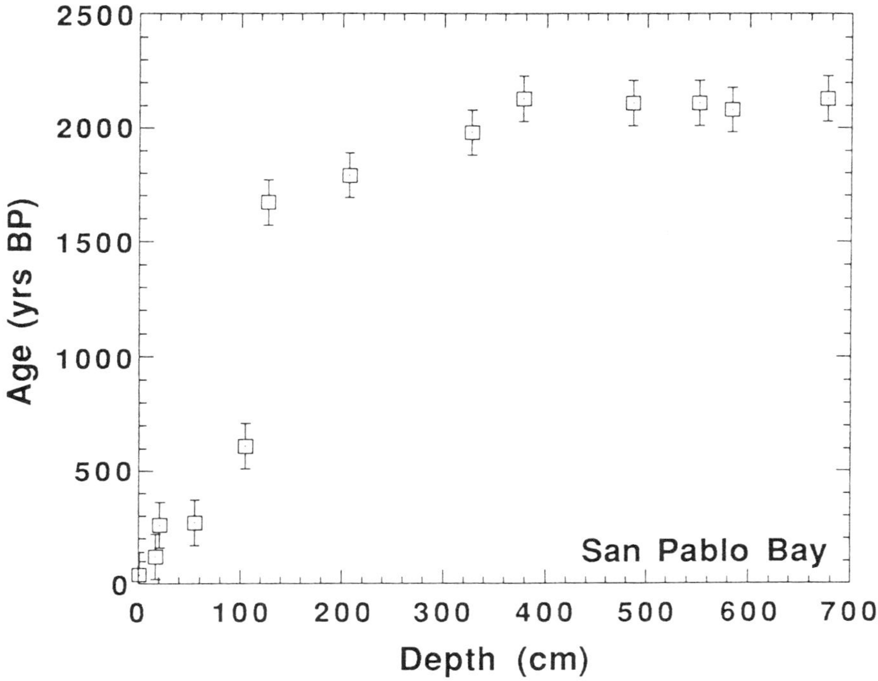

FIGURE 6. Sediment thickness (cm) versus corrected radiocarbon age (yr BP) for the San Pablo Bay core.

little of no deposition (700 to 1600 years BP; Fig. 6). The cause for the hiatus is not known, but it may be related to an extreme hydrological event during that time.

Calibration of Methods in Modern San Francisco Bay

Estuarine Waters

Sr Isotopes. The Sr concentration in estuarine waters exhibits a linear relationship with salinity (Fig. 7A), indicating that strontium is conservative in San Francisco Bay (Ingram and Sloan, 1992). The relationship between $^{87}Sr/^{86}Sr$ and salinity in estuarine water is calculated using the Sr concentrations and Sr isotopic ratios of the river water and seawater entering the bay. The $^{87}Sr/^{86}Sr$ ratio of the mixture (R_{mix}) is a function of the $^{87}Sr/^{86}Sr$ ratio and concentration of Sr in the river water (R_r and C_r) and ocean water (R_o and C_o) endmembers:

$$R_{mix} = [C_r R_r f + C_o R_o(1\text{-}f)]/C_{mix} \qquad (2a)$$

$$C_{mix} = C_r f + C_o(1\text{-}f) \approx C_o(1\text{-}f) \qquad (2b)$$

where C_{mix} is the concentration of Sr in the estuarine water, and f is the fraction of river water. To a good approximation, salinity (Smix) is given by: $S_{mix} \approx (1\text{-}f)\, S_o$, where So is the salinity of ocean water.

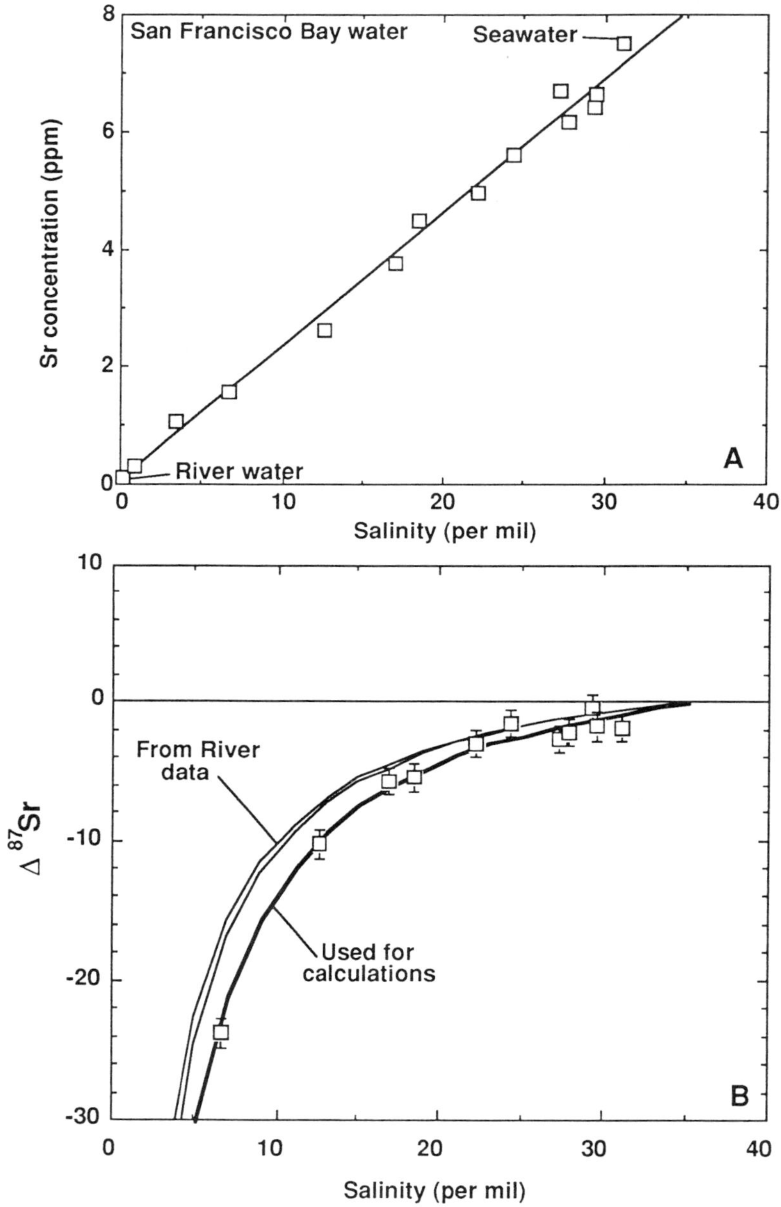

FIGURE 7. (A) Sr concentration (ppm) against salinity (per mil) for San Francisco Bay estuarine waters. (B) Relationship between Δ^{87}Sr and salinity (‰) in estuarine water, showing calculated mixing curves for addition of measured San Joaquin and Sacramento river water endmembers (narrow lines), and for addition of a model river water component with properties selected to fit the estuarine water data (wide line).

The Sr isotopic data are plotted against salinity in Fig. 7B. The narrow lines show the relationship calculated from the measured seawater and riverine endmember compositions. The calculated lines do not exactly fit the estuarine data. The wide line is a fit to the estuarine water data, which require a riverine Sr concentration of 160 ppb, somewhat higher than the average of the measured values in the streams in the delta (130 ppb).

Oxygen Isotopes. The salinities and $\delta^{18}O$ values of the estuarine waters sampled in March 12, 1991, are shown in Fig. 8a. The sample with the highest $\delta^{18}O$ (-1.2‰) and salinity (29.5‰) was collected near the Golden Gate. The samples with the lowest salinities and $\delta^{18}O$ values were collected at station 657 at Rio Vista, from the Sacramento River just before its confluence with the San Joaquin River. The rest of the samples from the estuary with intermediate salinities define a linear mixing relation between these river water and seawater end-members (Fig. 8a). The best fit line to the 1991-1993 data yield the following relation between $\delta^{18}O$ and salinity:

$$\delta^{18}O_{water} = 0.34(\text{salinity}) - 11.6 \qquad (3)$$

A 3‰ change in salinity corresponds to a 1‰ change in $\delta^{18}O$ of the water. The average deviation of the points from the best-fit line is $\pm$ 0.3 units of $\delta^{18}O$ which corresponds to a salinity uncertainty of $\pm$ 1‰.

The $\delta^{18}O$ values and salinities of the water samples collected during the period from 1991 to 1993 are plotted versus collection date (Fig. 9). The Sacramento River waters from station 657 (near Rio Vista; Fig. 2) had very consistent $\delta^{18}O$ values during 1991 and 1992, averaging -10.7‰. During 1993, however, the average $\delta^{18}O$ values of the river water dropped to -12.0‰. Both 1991 and 1992 were very dry years, occurring at the end of a six-year drought, whereas 1993 was a relatively wet year. The decrease in $\delta^{18}O$ values between wet and dry years may reflect an increased proportion of surface water (precipitation and snowmelt) relative to groundwater in the river during the wet year as compared with the two drought years. Beyond this yearly variation, no seasonal variation in the $\delta^{18}O$ values of the Sacramento River waters, such as a decrease in the $\delta^{18}O$ value during winter or spring, was observed. We do not know if the lack of seasonal variation represents the natural situation, or due to mixing and averaging of seasonal $\delta^{18}O$ caused by the mixing of waters stored in upstream reservoirs.

Both the salinity and $\delta^{18}O$ values of the estuarine waters from stations 11 and 25 show large seasonal fluctuations in response to changes in freshwater flow into the Bay (Fig. 9). The $\delta^{18}O$ values were lowest during the winter and early spring of each year and increased during the late spring and fall. Further, the magnitude of this seasonal variability was roughly twice as large during 1993 (wet year) than during 1991-1992 (dry years).

Previous studies have shown that although the isotopic composition of dissolved inorganic carbon is not conservative in the estuary due to modification by biological processes (Spiker 1980), $\delta^{13}C$ varies systematically with salinity in San Francisco Bay. This is because dissolved inorganic carbon in river water entering the bay has

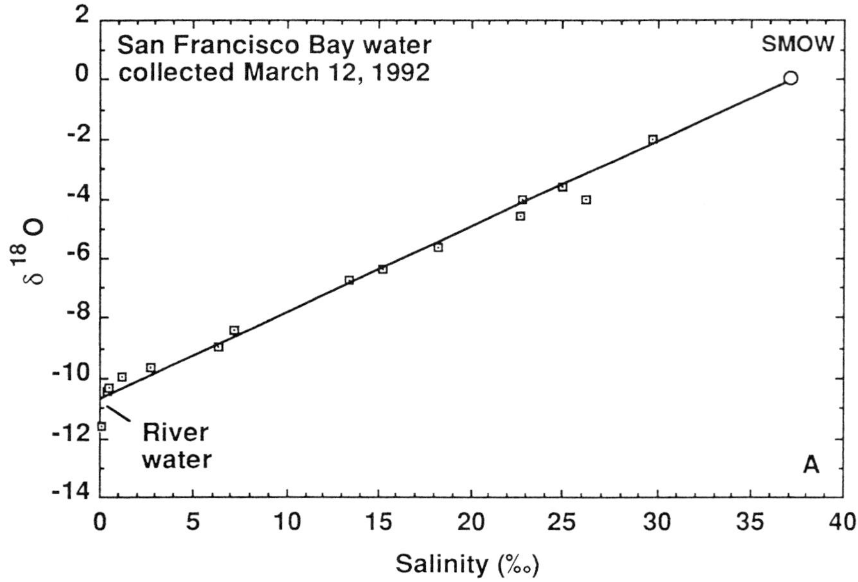

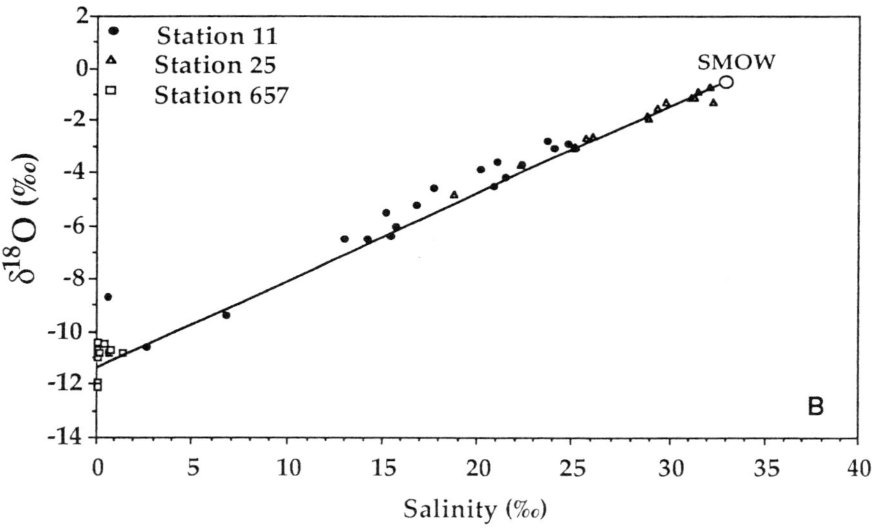

FIGURE 8. $\delta^{18}O$ values of San Francisco Bay waters plotted against salinity for samples collected (A) from salinity gradient on March 12, 1992, and (B) at stations 657 (Rio Vista), 11 (San Pablo Bay), and 25 (south-central bay) between 1991 and 1993.

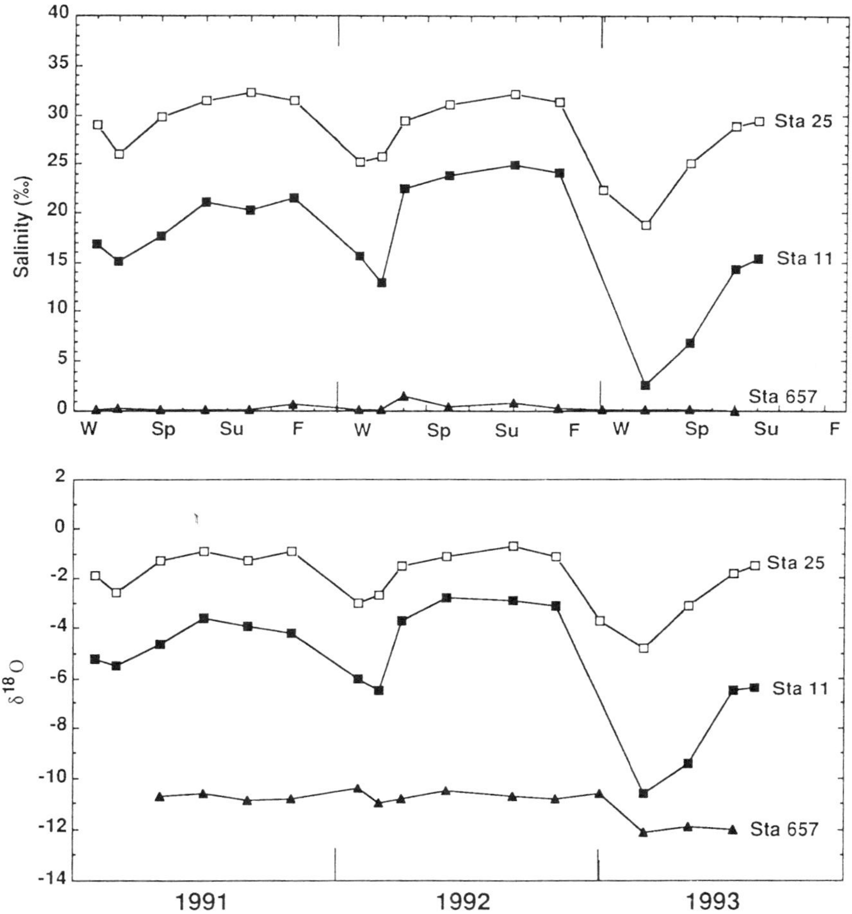

FIGURE 9. $\delta^{18}O$ and salinity of Bay waters from Fig. 11 plotted against collection date for 1991 to 1993.

a very low $\delta^{13}C$ value (~ -9.2‰), relative to seawater (~ 0‰). The relation between $\delta^{13}C$ and salinity is the following (from Spiker 1980):

$$\text{Salinity} = 2.94\ (\delta^{13}C + 9.2) \qquad (4)$$

Modern Shell Samples

The oxygen isotope ratios ($\delta^{18}O$ relative to PDB) for the whole mussel shells (*Mytilus edulis*) are plotted against the $\delta^{18}O$ values (relative to SMOW) of the waters (Fig. 10). The $\delta^{18}O$ of shell carbonate is the same as the $\delta^{18}O$ of water, indicating that the average water temperature must be close to 16.9°C, the temperature at which $\delta^{18}O_{\text{shell}} = \delta^{18}O_{\text{water in the Bay}}$. At 15°C, the relationship between $\delta^{18}O$ in shells and water would be shifted slightly to the right (Fig. 10). The error

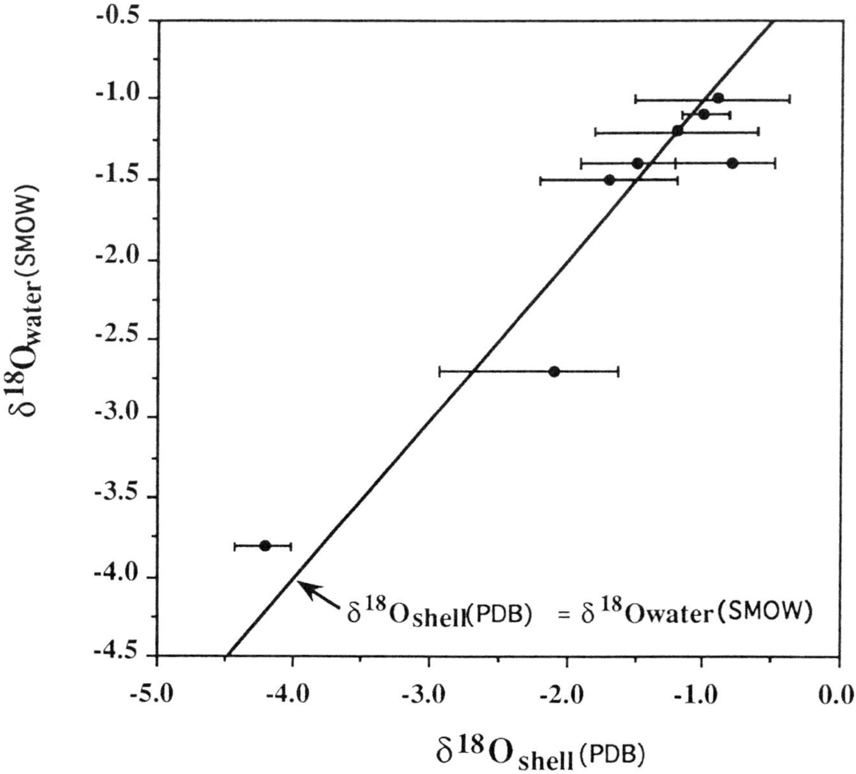

FIGURE 10. $\delta^{18}O_{PDB}$ values of modern *Mytilus edulis* shells versus $\delta^{18}O_{SMOW}$ of ambient water.

bars on Fig. 10 show the range the oxygen isotope values of the shells, and may be due to differences in the ages of the mussels collected at each of the sites. Also, some mussels were collected from tide pools, while other were attached to open surfaces of rocks. Despite these potential sources of scatter, the data conform well to the expected relationship.

In addition to measuring the average oxygen isotopic compositions of the modern mussel shells, we also measured the isotopic compositions of growth bands within several shells from two sites, to determine if seasonal salinity and oxygen isotopic changes in the water were recorded by the shells. Three individual shells, with an average length of 3.5 cm, were measured from the Vallejo Marina Site, and two shells, with an average length of 4.5 cm, were measured from Oyster Point (Fig. 2). The oxygen isotope data for the subsampled shells are plotted against the distance (mm) from the outer edge of the shell (Fig. 11). Although we know only the date of collection, the data clearly show one complete seasonal cycle, representing 1.0 to 1.5 years of growth. The oxygen isotopic variations in the shells at both sites are well-correlated (Fig. 11). The shells from the Vallejo Marina have

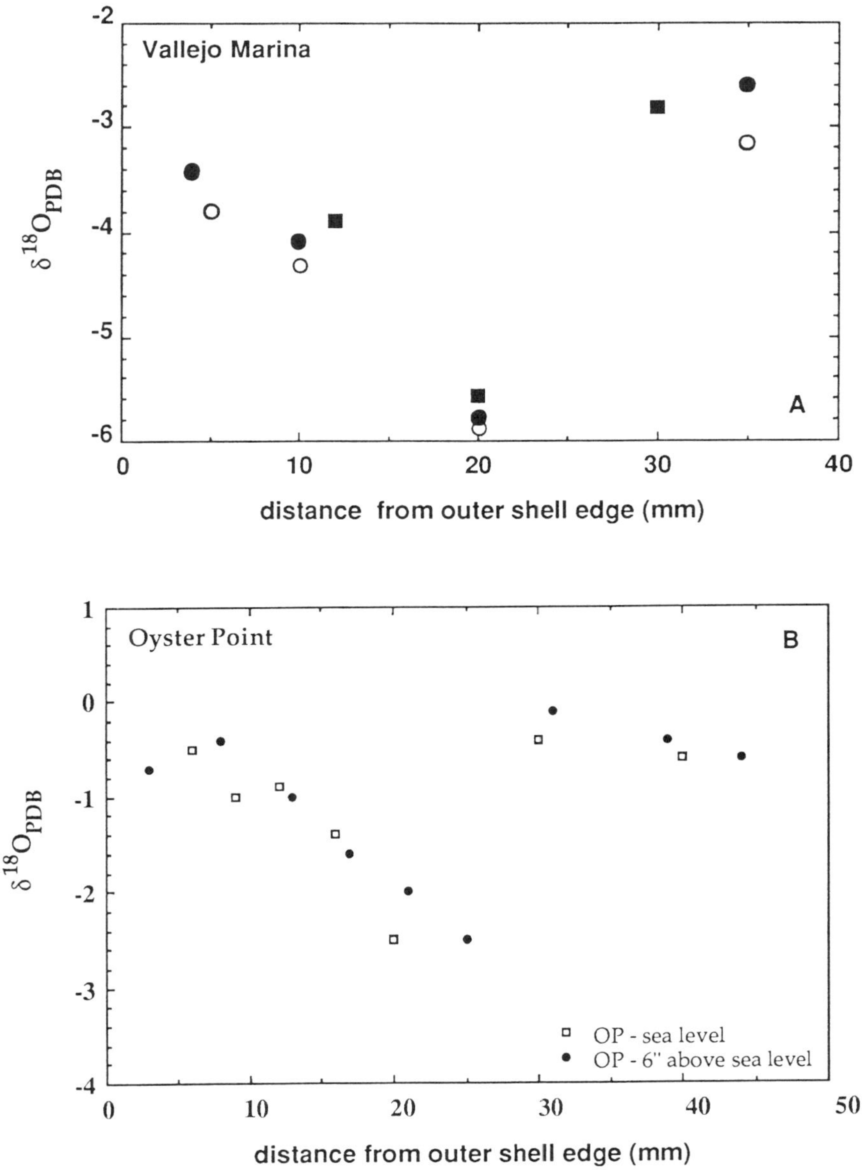

FIGURE 11. $\delta^{18}O$ PDB in growth layers of modern *Mytilus edulis* shells from Vallejo Marina (a) and Oyster Point (b), versus distance from edge along outer surface of shell (mm).

$\delta^{18}O$ values of -2.6‰ to -5.9‰, with a seasonal isotopic variation of about 2.4‰. The oxygen data for the shells from Oyster Point (close to station 25) ranged between -0.1‰ to -2.5‰, with a seasonal isotopic variation of 2.4‰. These results indicate that fossil shells may be used to determine natural seasonal salinity variations in San Francisco Bay over the past several thousand years.

Paleosalinity from Measurements of Fossil Shell Samples

Sr Isotopes: Richardson Bay. Strontium isotopic compositions and age (in yrs BP) for Richardson Bay are shown in Fig. 12A. The $\Delta^{87}Sr$ values are between 0 and about -5. In the shallow core, there is a discernible trend with higher values in the 50-120 yrs BP age range than in the 200-600 yrs BP age range. In the deeper core, five of the analyses yield $\Delta^{87}Sr$ values similar to those expected for growth in the modern environment (-2 to -3), whereas two analyses at 3400 yrs BP, and one from 4600 yrs BP give lower $\Delta^{87}Sr$ values.

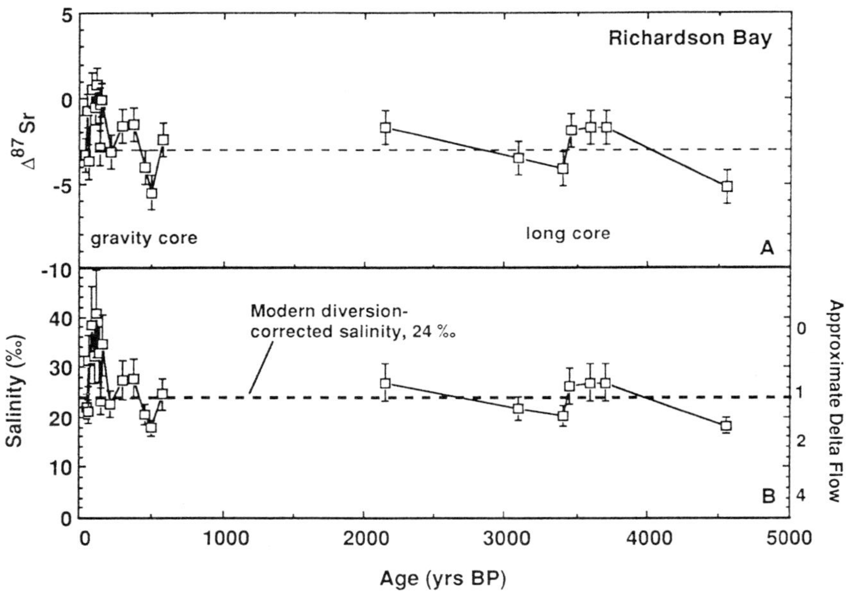

FIGURE 12. (A) $\Delta^{87}Sr$ values measured in molluscan carbonate plotted against age (yrs BP) for Richardson Bay gravity core and long core. (B) Calculated salinity plotted versus age for Richardson Bay gravity core and long core. The salinity corresponding to the "diversion-corrected" inflow (24) is plotted with dashed line for comparison. Approximate delta flow values (1000 m^3/sec) are shown on the right-hand scale.

Paleosalinities, calculated from $\Delta^{87}Sr$ using equation 4, are shown as a function of time in Fig. 12B. The data indicate relatively high salinity at about 40, 140-210, 270, 2100, and 3450-3700 yrs BP, and relatively low salinity at about 80, 220, 310, 440-500, 3100-3400, and 4600 yrs BP.

Oxygen and Carbon Isotopes: Oyster Point. The $\delta^{18}O$ and $\delta^{13}C$ values of the one hundred and sixty-seven *Mytilus edulis* fossil shells are plotted versus sample depth (Fig. 13). The $\delta^{18}O$ values of the mussel fossils in the upper 350 cm of the core are between -1.20‰ and -3.94‰, with an average value of -2.4‰. Below this depth, the $\delta^{18}O$ values in the intervals containing *Mytilus* shell layers (446 and 528 cm) are -2.69‰ to -5.58‰, averaging -3.9‰. The lowermost sample (at 578 cm; 5,900 yrs BP), has a $\delta^{18}O$ value of -2.68‰. The $\delta^{13}C$ values in the 0 to 350 cm interval range between 0.03‰ and -2.15‰, with an average value of -0.9‰. In the 446 to 528 cm interval, the $\delta^{13}C$ values are -0.62‰ to -3.26‰, averaging -1.9‰. The lowermost sample (528 cm, at 5,900 yr BP) has a $\delta^{13}C$ value of -0.94‰.

The average $\delta^{18}O$ value was 1.5‰ lower in the 446 to 578 cm interval (from about 3,800 to 5,100 yrs BP), than in the upper 350 cm of the core (0 to 2,400 yrs BP). The average $\delta^{13}C$ value was also lower in the older part of the core, by 1‰. Although more data is needed to document the trends in the lowest part of the core (older than 5,100 yrs BP), there is a suggestion that salinity at 5,900 yrs BP was similar to that over the past 2,400 years.

The co-variation between the oxygen and carbon isotopic compositions within the core (Fig. 13), suggests that $\delta^{18}O$ variations in the mussel carbonate primarily reflect salinity variations in the ambient water. However, a small correction was made for temperature using an average annual temperature at Oyster Point of 15°C (equation 4). The salinity range corresponding to the $\delta^{18}O$ carbonate values in the upper part of the core (0-350 cm) was 21.5‰ to 28.8‰, with an average annual salinity of 25.9‰ (Fig. 14). In the lower part of the core (446 to 578 cm), the salinity range was 15.2 to 24.4‰, with an average salinity of 23.3‰.

For the sake of comparison, salinity was also determined using the salinity – $\delta^{13}C$ relation from Spiker (1980; equation 3), as a check on those determined by the oxygen isotopic values. The average salinity in the upper 350 cm of the core was 24.4‰, and it was 21.5‰ in the lower portion of the core. These salinity values are offset by 1 to 2‰ from those obtained using the $\delta^{18}O$ values, but show the same variation with depth in the core.

The annual average salinity at Oyster Point measured between 1969 and 1977 was 28.0‰ (Conomos *et al.* 1979). The data presented here indicate that natural average salinity at Oyster Point was 2.1‰ lower than modern salinity averaged over the past 2,400 years. Between 3,800 and 5,100 yrs BP (Fig. 14), the average salinity at OP-1 was 4.7‰ lower than the 1969-1977 average.

San Pablo Bay. Ninety-four samples of *Macoma nasuta* were analyzed for stable isotopic compositions from the San Pablo Bay core, from depths of 0 to 279 cm (0 to 1900 yrs BP). The $\delta^{18}O$ and $\delta^{13}C$ values are plotted versus depth (Fig. 15). The average $\delta^{13}C$ and $\delta^{18}O$ values of *Macoma nasuta* shells for the upper 124 cm (0 to 700 yrs BP), were -2.24‰ and -3.66‰, respectively. In the interval from 124 to 279 cm (1600 to 1900 yrs BP), the average $\delta^{13}C$ and $\delta^{18}O$ values were -4.37‰ and -5.20‰, respectively.

Oxygen and carbon isotopic compositions are also well-correlated within the San Pablo Bay core (Fig. 16), suggesting that $\delta^{18}O$ variations in the carbonate of *Macoma nasuta* primarily reflects salinity variations in the ambient water. The

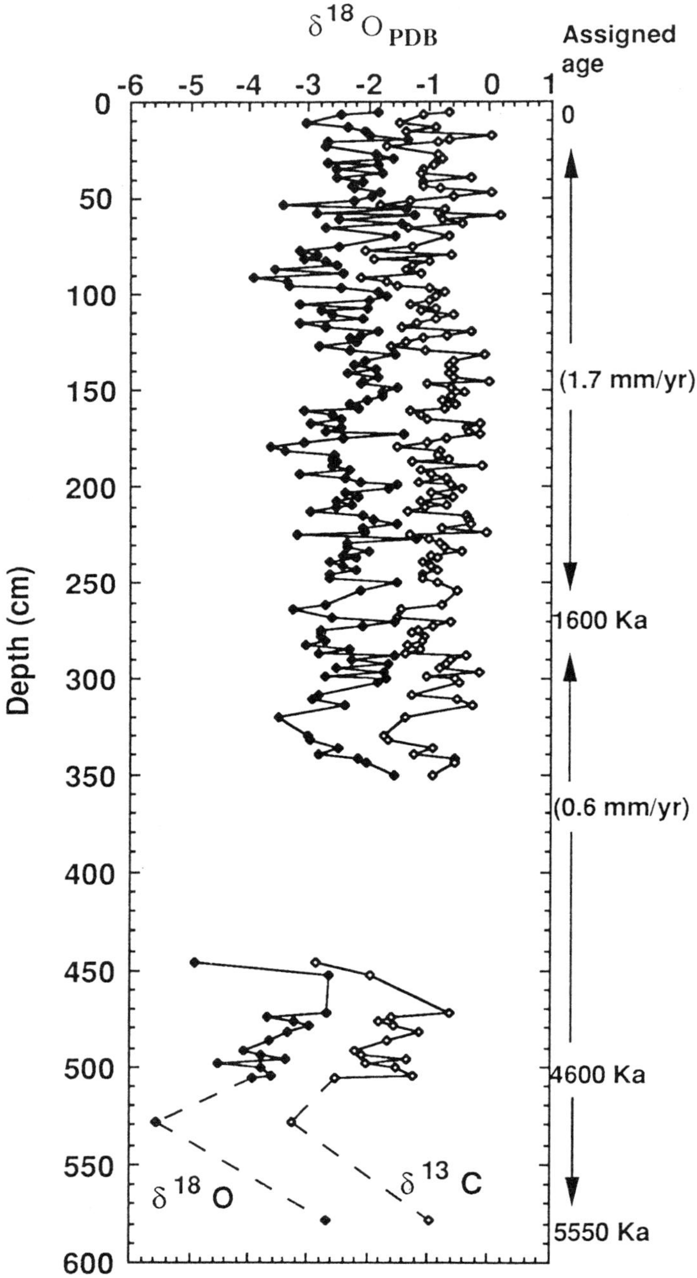

FIGURE 13. $\delta^{18}O$ and $\delta^{13}C$ (relative to PDB) values plotted against depth (cm) for Oyster Point core. Calibrated radiocarbon ages in core and average sedimentation rates are shown on right of graph.

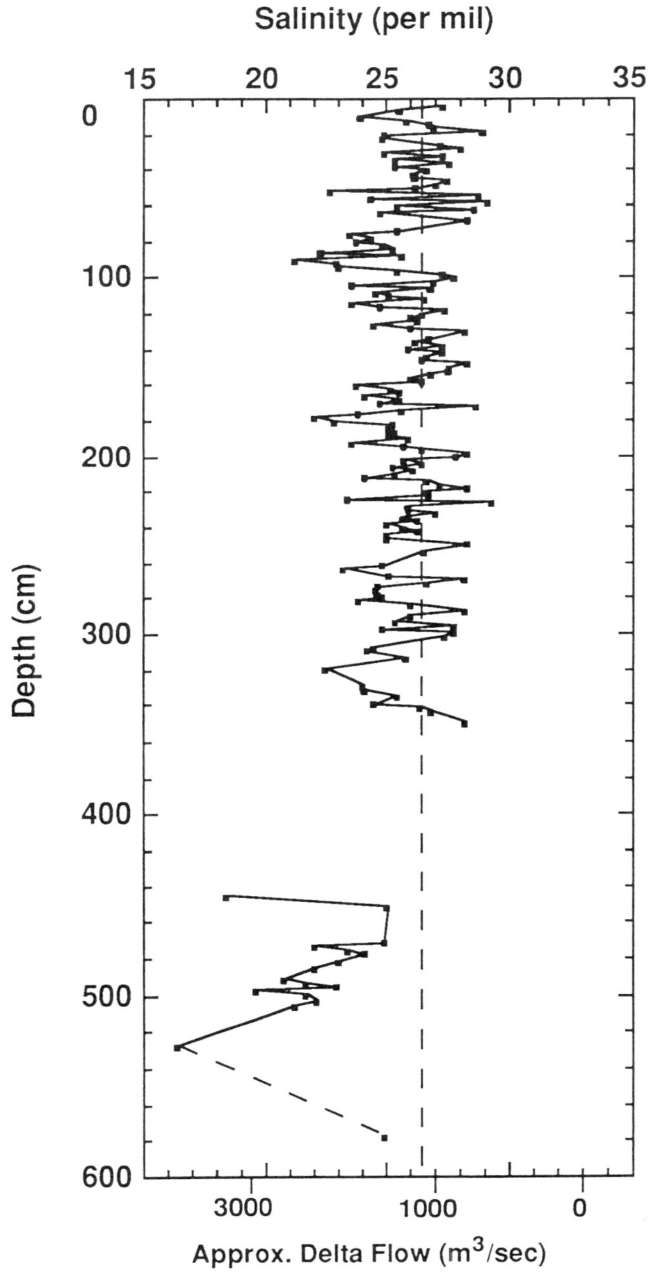

FIGURE 14. Calculated salinity (‰) plotted versus depth for Oyster Point core. The modern diversion-corrected salinity of 26‰ is plotted with dashed line for comparison, and approximate delta flow values (m^3/sec) are shown on bottom scale.

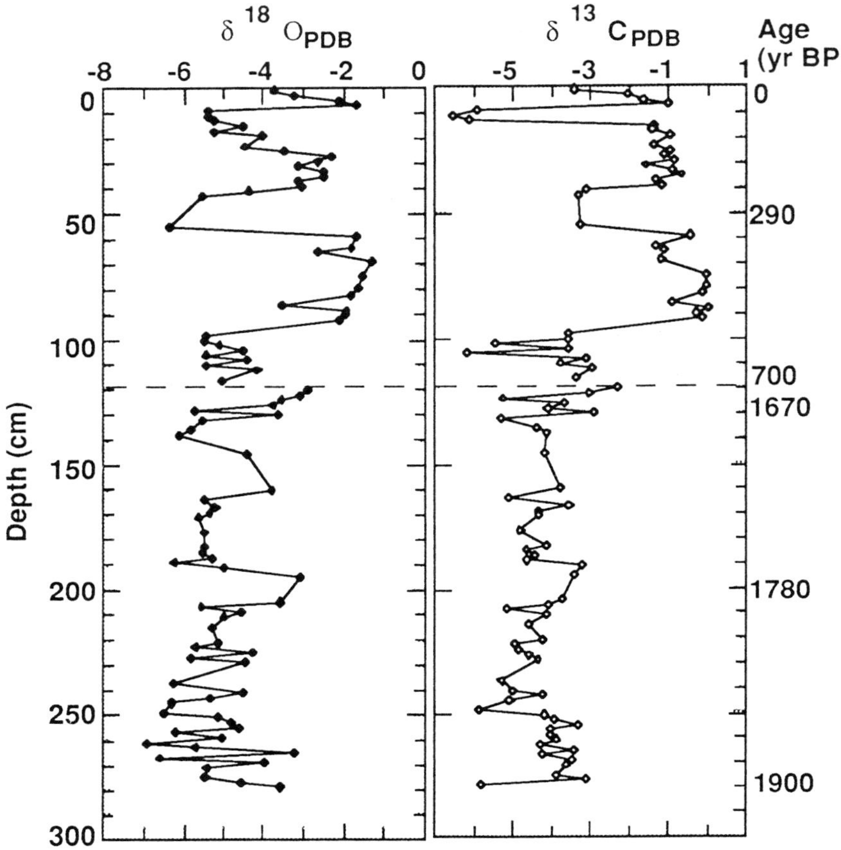

FIGURE 15. $\delta^{18}O$ and $\delta^{13}C$ (relative to PDB) values plotted against depth (cm) for upper 300 cm of San Pablo Bay core. Corrected ^{14}C ages (yr BP) are shown in the right-hand column.

salinity range corresponding to the $\delta^{18}O_{carbonate}$ values in the upper part of the core (0 to 124cm; 0-700 yrs BP) was 16.3‰ to 27.4‰, with an average annual salinity of 22.4‰ (Fig. 17). In the lower part of the core (124 to 279 cm; 1600 to 1900 yrs BP), the salinity range is 12.1‰ to 23.4‰, with an average salinity of 18.2‰.

The annual average salinity in San Pablo Bay near Rodeo measured between 1969 and 1977 is 24.0‰ (Conomos *et al.* 1979). The data presented here indicate that natural average salinity in San Pablo Bay over the past 700 years was 1.5‰ lower than modern salinity, and between about 1600 and 1900 yrs BP, it was 6.1‰ lower than modern.

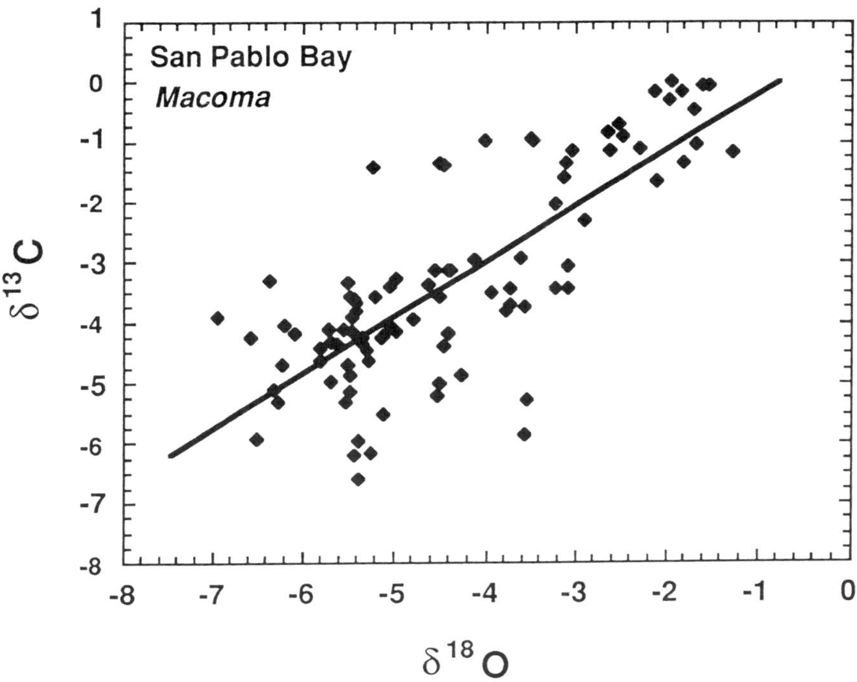

FIGURE 16. $\delta^{18}O$ values plotted against $\delta^{13}C$ values (PDB) for *Macoma nasuta* shells from the San Pablo Bay core.

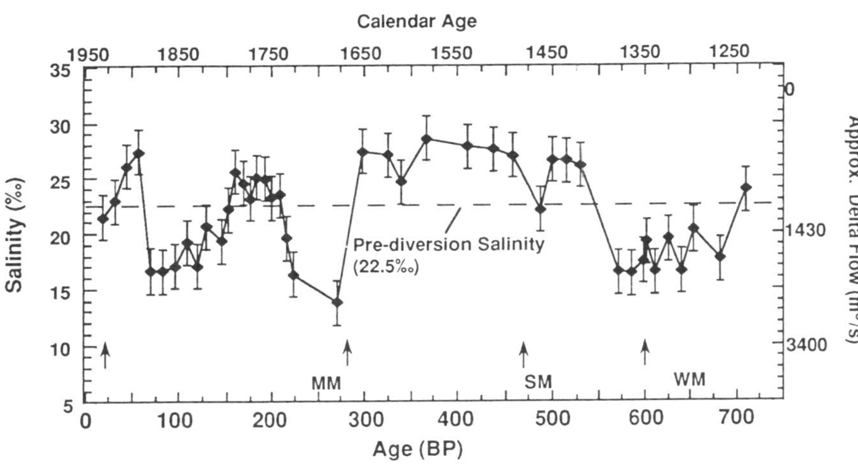

FIGURE 17. Salinity plotted against age for upper 120 cm (0 to 700 yr BP) of the San Pablo Bay core. Calendar year ages also shown. Periods of high levels of Mono Lake are indicated with up arrows (Stine 1990). Known sunspot minima are also shown: MM = Maunder Minimum, SM = Sporer Minimum, and WM = Wolf Minimum.

Paleo-Discharge into San Francisco Bay

Paleosalinity variations in San Francisco Bay may have two causes: (1) Changes in the bay volume, primarily associated with changing sea level, (2) Changes in the amount of freshwater inflow to the Bay. Over the past 6,000 years, sea level rose by 5 - 10 meters, approximately the amount of sediment that has accumulated in the south-central bay. This suggests that there has been little net change in bay volume over the past 6,000 years.

The relation between salinity and discharge (or delta flow) in San Francisco Bay is a function of the distance upstream from the mouth of the bay (Peterson *et al.* 1989). The best fit relation between surface water salinity and delta flow for the site closest to our core location south-central Bay is the following:

$$S(Q) = S_{oe}{}^{-0.00019Q},$$

where S(Q) is the salinity for a given river discharge, S_o is the salinity at zero river discharge (Peterson *et al.* 1989). For San Pablo Bay, we use the relation

$$S(Q) = S_{oe}{}^{-0.00035Q},$$

and for Richardson Bay, the relationship used here is

$$S(Q) = S_{oe}{}^{-0.00025Q}.$$

Paleosalinity was converted to mean river discharge, as shown on the right vertical axis of Figures 12B, 14, and 17. The absolute values of the calculated pale-inflow data will be subject to some revision, but the relative highs and lows will not change.

The modern average "diversion-corrected" delta flow value of 1100 m^3/s is used for comparison. This hypothetical value is determined by adding back all the water diverted from the estuary for agricultural and other uses (Nichols *et al.* 1986).

In the Richardson Bay core, the average annual river inflow (or delta flow) value is 1220 m^3/sec (calculated from salinities shown in Fig. 12). The periods of lowest salinity correspond to average annual delta flow values of 1500-2000 m^3/sec (Fig. 12). In the Oyster Point core, the calculated average delta flow value in the upper 350 cm of core (0 to 2,400 yrs BP) is 1260 m^3/sec; two times the modern average of 600 m^3/sec (close to the estimated pre-diversion inflow of 1100 m^3/sec; Fig. 14). In the lower portion of the core (446 to 528 cm, 2,400 to 3,700 yr BP), the average delta inflow is about 1850 m^3/sec; three times the modern average of 600 m^3/sec and about 1.5 times the estimated pre-diversion inflow of 1100 m^3/sec. In the San Pablo Bay core, the average delta flow value in the upper 124 cm (0 to 700 yrs BP) is about 1250 m^3/s, whereas in the interval from 124 to 279 cm (1600 to 1900 yrs BP), the average delta flow was substantially higher (1,840 m^3/s). These values are very close to the estimated pre-diversion value (1100 m^3/s).

Paleoclimatic Implications

The calculated paleodischarge values determined from our isotopic data are consistent with other paleoclimatic data in California. For example, high river discharge values from 3,800 and 5,100 yr BP indicated by our stable isotope data, and from 3,400 to about 4,300 from Sr isotope data (Figs. 12 and 14) corroborate paleobotanical evidence for cooler and wetter conditions in the White Mountains of eastern California, between 4,200 years and 2,000 yrs BP (Lamarche 1973), and with a high stand of Mono Lake at ca. 3,800 years BP (Stine 1990). In addition, a large body of water also covered the now-dry Silver Lake Playa in the Mojave Desert at approximately 3,600 yr BP (Enzel *et al.* 1989), as well as in the Searles Basin (Smith 1979). Thus, all existing climate records in California indicate that from 3,600 to at least 4,200 yr BP, average climate in California was wetter than during the modern century, although causes for the increase are not currently known.

The San Francisco Bay isotopic record for the past 2000 years is also compatible with other paleoclimatic records. Treeline variations in the White Mountains of eastern California indicate generally cooler and drier conditions between 2,000 years BP and the present (Lamarche 1974), in agreement with the lower average river inflow to San Francisco Bay during that period. The most recent period of low river inflow to San Francisco Bay at 210 to 160 yrs BP, coincides with a period of extended drought recorded in tree-rings from the western U.S. (Fritts *et al.* 1979). Other dry periods include 300 to 550 yr BP, and prior to 700 yr BP (Fig. 17). The low paleo-discharge between 600-900 yr BP (1100 to 1350 AD) recorded in San Francisco Bay (Oyster point core) coincides with a dry period indicated by relict stumps rooted in presently existing lakes and rivers in California's Sierra Nevada (Stine 1994).

Our isotope record indicates periods of relatively low salinity (high river inflow) in the present century, at *ca.* 200-300 and 550 to 700 BP (Fig. 17). These periods of high river inflow occur during times of high Mono Lake level (Stine 1990; Fig. 17). A particularly wet interval over the past 1000 years occurred at 300 yr BP; San Francisco Bay had its lowest salinity, Mono Lake level was highest, and an unusually wet episode in the Mojave Desert occurred (Enzel *et al.* 1989).

A Fourier analysis of the $\delta^{18}O$ values in the 0 to 150 cm interval of the Oyster Point core suggests that salinity in San Francisco Bay varied periodically over the past 1000 years. The identifiable periods in the spectra are at 443, 220, 148, 93, and 60 yrs, although relatively large uncertainties as associated with these periods due to the large age uncertainties. In addition, a clear 200-yr cycle is present in the San Pablo Bay core (Fig. 17), with periods of high river inflow in San Francisco Bay corresponding to low solar activity (and high Mono Lake level, Stine 1990). Climatic variations with 88- and 200-yr periods, such as those found in cosmogenic ^{10}Be variations in ice cores (Beer *et al.* 1988) and varved lake sediments (Halfman & Johnson 1988), have also been attributed to fluctuations in solar activity (Stuiver & Quay 1980), which are also thought to be responsible for the periods observed in the radiocarbon production rate spectrum (at 512, 416, 215, 147, and 88 years; Stuiver & Braziunas 1992). Further research is clearly needed to confirm these variations in San Francisco Bay and elsewhere.

ACKNOWLEDGMENTS

This work was supported by the Interagency Ecological Program, California Department of Water Resources Grant #B-58859 (to JCI, Stanford University). BLI was supported in part by an appointment to the Global Change Distinguished Postdoctoral Fellowships, sponsored by the U.S. Department of Energy, Office of Health and Environmental Research, and administered by the Oak Ridge Institute for Science and Education. MEC was supported by a grant from the Director, Office of Energy Research, Office of Basic Energy Sciences, Engineering and Geosciences Division of the U.S. Department of Energy under Contract No. DE-AC03-76SF00098. We thank A. van Geen for providing estuarine water samples, and Tracy Washington for collection and analysis of mussel shells as part of a summer internship at the Lawrence Berkeley Laboratory.

REFERENCES

Atwater, B. F., C. W. Hedel, and E. J. Helley. (1977). *Late Quaternary Depositional History, Holocene Sea-level Changes, and Vertical Crustal Movement, Southern San Francisco Bay, California*. U.S. Geol. Surv. Prof. Pap. 1014. 15 pp.

Atwater, B. F., S. G. Conard, J. N. Dowden, C. W. Hedel, R. L. MacDonald, and W. Savage. (1979). History, landforms and vegetation of the estuary's tidal marshes. Pages 347-385 *in* T. J. Conomos, ed., *San Francisco Bay: The Urbanized Estuary*. Pacific Division, American Association for the Advancement of Science, San Francisco, CA

Atwater, B. F., B. E. Ross, and J. F. Wehmiller. (1981). Stratigraphy of late Quaternary estuarine deposits and amino acid stereochemistry of oyster shells beneath San Francisco Bay, California. *Quat. Res.* 16:181-200.

Beer, J., U. Siegenthaler, G. Bonani, R. C. Finkel, H. Oesheger, M. Suter, and W. Wolfli. (1988). Information on past solar activity and geomagnetism from 10Be in the Camp Century ice core. *Nature* 331:675-679.

Capo, R. C., and D. J. DePaolo. (1990). Seawater strontium isotopic variations from 2.5 million years ago to the present. *Science* 249:51-55.

Cayan, D. R., and D. H. Peterson. (1989). The influence of North Pacific Atmospheric Circulation on streamflow in the West. Pages 375-398 *in* D. H. Peterson, ed., *Aspects of Climate Variability in the Pacific and the Western Americas*. Geophys. Monogr. 55.

Conomos, T. J. (1979). Properties and circulation of San Francisco Bay waters, Pages 47-84 *in* T. J. Conomos, ed., *San Francisco Bay: The Urbanized Estuary*. Pacific Division, American Association for the Advancement of Science, San Francisco, CA.

Conomos, T. J., R. E. Smith, D. H. Peterson, S. W. Hager, L. E. and Schemel. (1979). Processes affecting seasonal distributions of water processes in the San Francisco Bay estuarine system. Pages 115-142 *in* T. J. Conomos, ed., *San Francisco Bay: The Urbanized Estuary*. Pacific Division, American Association for the Advancement of Science, San Francisco, CA

Davis, J. C., *et al.* (1990). LLNL/UC AMS facility and research program. Pages 269-272 in *Proceedings of the 5th International Conference on Accelerator Mass Spectrometry, Nuclear Instruments and Methods* B52(3,4).

DePaolo, D. J., and B. L. Ingram. (1985). High resolution stratigraphy with strontium isotopes. *Science* 227:938-941.

Donohue, D. J., T. W. Linick, and A. J. T. Jull. (1992). Isotope-ratio and background corrections for AMS radiocarbon measurements. *Radiocarbon* 32:135-142.

Enzel, Y., D. R. Cayan, R. Y. Anderson, and R. G. Wells. (1989). Atmospheric circulation during Holocene lake stands in the Mojave Desert: Evidence of regional climate change. *Nature* 341:44-47.

Epstein, S., R. Buchsbaum, H. A. Lowenstam, and H. C. Urey. (1953). Revised carbonate-water isotopic temperature scale. *Geol. Soc. America Bull.* 64:1315-1326.

Fairbanks, R. G. (1989). A 17,000-year glacio-eustatic sea level record: Influence of glacial melting rates on the Younger Dryas event and deep-ocean circulation. *Nature* 342:637-642.

Faure, G. (1986). *Principles of Isotope Geology*. John Wiley & Sons, New York. 464 pp.

Fritts, H. C., G. R. Lofgren, and G. A. Gordon. (1979). Variations in climate since 1602 as reconstructed from tree rings. *Quatern. Res.* 12:18-46.

Halfman, J. D., and T. C. Johnson. (1988). High-resolution record of cyclic climatic change during the past 4 ka from Lake Turkana, Kenya. *Geology* 16:196-500.

Ingram, B.L, and DePaolo, D.J. (1993) A 4,300 year strontium isotope record of estuarine paleosalinity in San Francisco Bay, California. *Earth and Planet. Sci. Lett.* 119: 103-119.

Ingram, B. L., and D. Sloan. (1992). Strontium isotopic record of estuarine sediments as paleosalinity-paleoclimate record indicator. *Science* 255:68-72.

Ingraham, N. L., and B. E. Taylor. (1991). Light stable isotope systematics of large-scale hydrologic regimes in California and Nevada. *Water Resour. Res.* 27:77-90.

Lamarche, V. C. (1973). Holocene climate variations inferred from treeline fluctuations in the White Mountains, California. *Quatern. Res.* 3:632-660.

McCrea, J. M. (1950). On the isotope chemistry of carbonates and a paleotemperature scale. *Jour. Chem. Physics* 18:849-857.

Nichols, F. H., J. E. Cloern, S. N. Luoma, and D. H. Peterson. (1986). The modification of an estuary. *Science* 231:525-648.

Peterson, D. H., D. R. Cayan, J. F. Festa, F. H. Nichols, R. A. Walters, J. V. Slack, S. E. Hager, and L. E. Schemel. (1989). Climate variability in an estuary: Effect of river flow on San Francisco Bay. Pages 419-442 *in* D. H. Peterson, ed., *Aspects of Climate Variability in the Pacific and the Western Americas*. American Geophys. Union Monogr. 55. American Geophysical Union, Washington DC.

Smith, G. I. (1979). *Subsurface Stratigraphy and Geochemistry of Late Quaternary Evaporites, Searles Lake, California*. U.S. Geol. Surv. Prof. Pap. 1043. 130 pp.

Socki, R. A., H. R. Karlsson, and E. K. Gibson. (1992). Extraction technique for the determination of oxygen-18 in water using preevacuated glass vials. *Analytical Chem.* 64:829-831.

Spiker, E. C. (1980). The behavior of ^{14}C and ^{13}C in estuarine water: Effects of *in situ* CO_2 production and atmospheric exchange. *Radiocarbon* 22(3): 647-654.

Stine, S. (1990). Late Holocene fluctuations of Mono Lake, eastern California. *Palaeogeog. Palaeaclim. Palaeoecol.* 78:333-381.

Stine, S. (1994). Extreme and persistent drought in California and Patagonia during Medievel time. *Nature* 369:546-549.

Stuiver, M., and T. F. Braziunas. (1993). Modeling atmospheric ^{14}C influences and ^{14}C ages of marine samples to 10,000 BC. *Radiocarbon* 35(1):137-189.

Stuiver, M., Pearson, G. W., T. and Braziunas, T. F. (1986), Radiocarbon age calibration of marine samples back to 9000 cal yr BP. *Radiocarbon* 28(2B):980-1021.

Stuiver, M., and Quay, P. D. (1980). Changes in atmospheric carbon-14 attributed to a variable sun. *Science* 207:11-19.

Stuiver, M., and T. F. Braziunas. (1992). Evidence of solar activity variations. Pages 593-605 *in* R. S. Bradley & P. D. Jones, eds., *Climate Since A.D. 1500*. Routledge, London.

Stuiver, M., and H. A. Polach. (1977). Reporting of ^{14}C data. *Radiocarbon* 19:355-363.

Vogel, J. S., D. E. Nelson, and J. R. Southon. (1987). ^{14}C background levels in an accelerator mass spectrometry system. *Radiocarbon* 29(3):323-333.

RECENT SEDIMENTATION IN THE SAN FRANCISCO BAY SYSTEM

RAY B. KRONE
Department of Civil & Environmental Engineering
University of California at Davis, Davis, CA 95616

BACKGROUND

Historical sediment accumulations in the various parts of the San Francisco Bay system prior to 1950 were described in a previous symposium volume (Krone 1979). The distributions of sediment accumulations in the various bays were based on comparisons of successive hydrographic surveys (Smith 1965) and were adjusted for sea level rise. They showed that Grizzly, Suisun, and San Pablo Bays received large amounts of sediment during the period of hydraulic mining in the Sierra, 1853 to 1884. Suspension of the large deposits in Grizzly and Suisun Bays by wind-generated waves and transport by tidal currents during subsequent years moved some of the sediment deposited in these bays downstream to San Pablo Bay, and accretion continued in that bay to the end of the study period, 1950. As water depths in San Pablo Bay became shallow enough for wave action to suspend deposits there, increasing amounts accumulated in North San Francisco Bay. There was a significant increase in the rate of deposition in North Bay between 1923 and 1950, indicating that inflows of suspended sediment continued after the mining era and that sediment worked its way down through the upper bays. Agricultural activities, including logging, have become significant sources of suspended sediment.

The hydrographic survey comparisons showed that South San Francisco Bay deepened throughout the period 1870 to 1950, indicating a net loss of sediment from that bay to the ocean.

Water inflows and diversions in the Delta and outflows from the Delta to the Bay system were used to calculate sediment inflow to the Bay system. These flows were determined by the U.S. Bureau of Reclamation by application of the 1960 and the then planned 1990 and 2020 water water projects' operating rules to records of monthly flows for the years 1921 to 1971. The U.S. Geological Survey had measured suspended sediment loads in the San Joaquin and Sacramento Rivers during the years 1957 to 1966, and a correlation between the water year annual flows and annual sediment production was found that enabled a calculation of a statistical distribution of the sediment discharge to the Bay system from the Bureau data. This distribution indicated an annual median discharge of 2.6 million tons under 1960 conditions and a 20-fold range between the 2 and 98 percentile discharges. The projections by the Bureau included diversion of water from the Sacramento River by the Peripheral Canal. The projected median sediment discharges were 0.7 M tons in 1990 and 0.4 M tons in 2020. Forty percent of the years would have had discharges less than 0.2 M tons under the 2020 scenario. Sediment inflows from local streams tributary to the bays were calculated by Porterfield, *et. al*. (1980).

The sediment inflows, accumulations in the bays, and data on land disposal enabled an estimate of the annual loss of sediment to the ocean. A total of 10.5 M cubic yards (Mcy) entered the system, about 1 Mcy of dredged sediment was disposed on land, about 5.5 Mcy accumulated in the bays, and 4.0 Mcy exited the Golden Gate.

A new hydrographic survey of the bays was completed in 1990. This survey provided an opportunity to update sediment accumulation rates and to review sediment inflows during the period since the previous hydrographic survey in 1955. An update and review was prepared for the San Francisco District, U.S. Army Corps of Engineers, by Ogden Beeman & Associates and Ray B. Krone & Associates (1992). That update is the primary source of the material presented below.

SEDIMENT INFLOWS

Suspended solids data for the years 1980 to 1988 were added to the 1957 to 1966 data to improve the correlation between water flows and sediment production, and this correlation was used to calculate the sediment discharge to the bays from Delta water discharges. There has been little modification of discharges from local catchments around the bays since the Geological Survey study, and those sediment discharges were added to the calculated discharge from the Delta.

Annual water inflows to the Delta and outflows from the Delta to the bays for the years 1956 to 1990 were provided by the California Department of Water Resources for use in calculating a sediment budget. The average annual discharge to the system, calculated from these flows with the use of the above correlation, was 7.88 Mcy.

The Department also provided flows to the Bay system under 1990 and projected 2035 water project operating conditions by applying the operating rules applicable to those years to the 1921 to 1991 recorded runoffs. The calculated distribution of water discharge to the bays is plotted in Fig. 1 together with the previously determined distribution for 1960 operation. This figure shows the wide range of fresh water flows to the bays under present operation of the water diversion facilities and the increase that was planned in 1992 for the year 2035. This plot shows that under present operating procedures the annual fresh water flow to the bays will be only three million acre feet ten percent of the years, and that half of the years the flow will be less than eight million acre feet — about half of the flows under 1960 operating procedures.

Distributions of sediment discharges to the bay system are plotted in Fig. 2. This plot shows median annual sediment discharges of 2.6 Mcy under 1990 conditions (1.2 million tons more than the 0.7 million tons that had been projected in 1966) and about 2.1 Mcy under 2035 conditions. Average annual discharges are more useful for estimating impacts on dredging requirements than are medians. The average annual discharge under 1990 conditions is 5.9 Mcy and under 2035 conditions is 5.5 Mcy. These figures are much smaller than the 10.5 Mcy estimated for 1960 operating conditions.

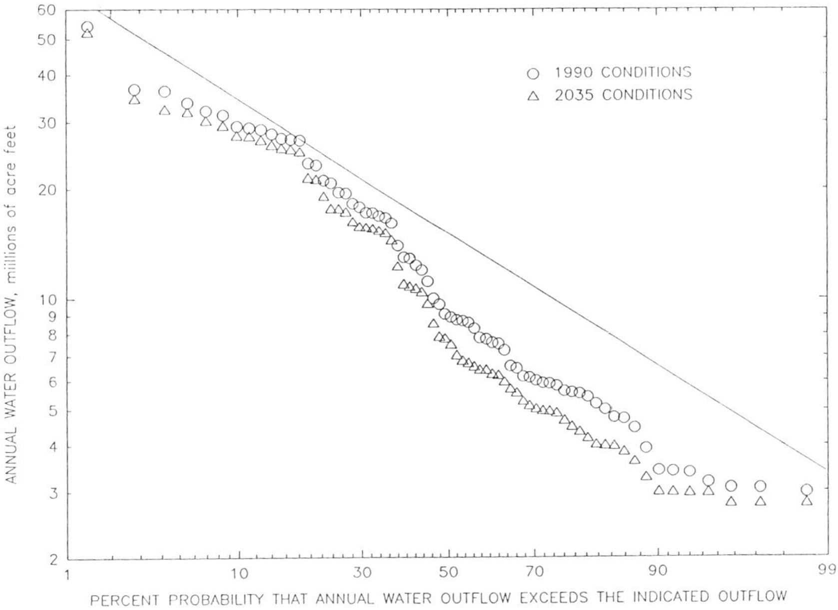

FIGURE 1. Probability distribution of annual fresh water outflows from the Central Valley.

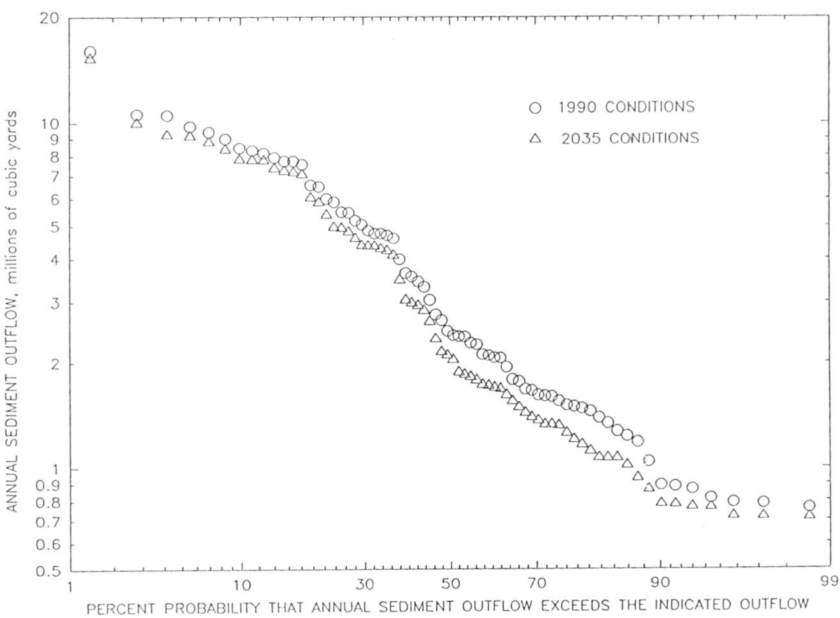

FIGURE 2. Probability distribution of sediment outflows from the delta under 1990 and projected 2035 water project operating procedures.

SEDIMENT ACCUMULATIONS IN THE BAYS

Charts of the Bay system were subdivided into South Bay, North Bay, San Pablo Bay, and Suisun Bay, and each of these was further subdivided into three to six sections for digitizing. The southern end of South Bay was further subdivided to facilitate correction for subsidence. The digitized data were then gridded into quadrangles of approximately 400 feet on a side, and changes in depths referenced to mean lower low water were determined for each subsection. The 1990 shoreline was used as the basis for comparison.

Volumes of accretion and erosion were calculated from the grid size and changes in bed elevation. A correction of 0.16 feet for sea level rise between 1955 and 1990 was applied to the entire system for this calculation, and corrections for subsidence due to ground-water pumping in south Bay were interpolated from the data presented in Poland and Ireland (1988).

Contour plots of accretion and erosion were prepared from these data. These plots show that accretion occurred in the shallow areas of northeast San Pablo Bay, Grizzly Bay, Honker Bay, and the southernmost portion of South San Francisco Bay. A natural levee appeared along the margins of the channel across San Pablo Bay.

Annual average volumes of sediment accumulation in the system during the 35 years between 1955 and 1990 are shown in Table 1. These data show that the accumulation rates were very low in the upper bays, appreciable in North San Francisco Bay, and small in South San Francisco Bay. Sedimentation is maintaining water depths in the upper bays as sea level rises and is creating a deposit that is advancing in North Bay. The low but positive deposition rate in South Bay indicates that sediments from the Central Valley are now accumulating there.

The total area of the 1990 hydrographic surveys was 11 square miles less than the areas covered in the mid-1950s. Advancing mud flats and marsh along the shores, particularly in the upper bays, caused this reduction.

TABLE 1. Changes in sediment volumes.

Area	*Vol. Change, Mcy*
Suisun Bay	0.30
San Pablo Bay	0.20
North Bay	2.71
South Bay	0.89
TOTAL	4.10

SEDIMENT BUDGET

The estimates of sediment accumulation, the sediment inflows during the years between the surveys, and the data for the local catchments enabled an estimate of the loss of suspended sediment to the ocean. The data are summarized in Table 2. The sediment outflow to the ocean, 3.37 Mcy, is 43 percent of the inflow. This

TABLE 2. Estimate of annual average sediment loss to the ocean, 1955-1990.

	Sediment Volume, Mcy
Sediment Inflow	7.88
Upland Disposal	-0.41
Accumulation in Bays	-4.10
Sediment Loss	3.37

figure is larger than the previous estimate, 38 percent, probably because the upper bays are not trapping sediment as effectively as they were before and consequently greater amounts of suspended sediment are carried to the Golden Gate. The increase in sedimentation rate in the South Bay, compared to the previous estimate, is very likely also due to larger amounts of suspended sediment being carried down into the lower bays.

FUTURE CHANGES

The drastic reductions of fresh water and sediment inflows to the San Francisco Bay system that was projected in 1966 for 1990 operating conditions did not materialize. The reduction that did occur during the intervening period is large, however, and was made without evaluation of the impacts of such changes on water quality of the system for fish. Increasing concerns for the health of the bays may encourage greater efforts on managing its water quality than have prevailed to date, and there may be significant changes in fresh water inflows. The distributions of water and sediment inflows to the bays presented above should be viewed with these reservations.

It appears that sediment discharges to the Bay system will continue to be sufficient to compensate sea level rise in the upper bays, that is to maintain present water depths there, and to continue the advance of deposition in North San Francisco Bay. As that bay accumulates sediment, greater loss to the ocean can be expected, as well as more rapid accumulation of sediment in South San Francisco Bay.

REFERENCES

Krone, R. B. (1979). Sedimentation in the San Francisco Bay System. Pages 85-96 *in* T. J. Conomos, ed., *San Francisco Bay, The Urbanized Estuary*. Pacific Division, American Association for the Advancement of Science, San Francisco, CA.

Ogden Beeman & Associates and Ray B. Krone & Associates. (1992). *Sediment Budget Study for San Francisco Bay*. Final Report to the San Francisco Distr., U.S. Army Corps of Engrs. 101 pp.

Poland, J. F., and R. L. Ireland. (1988). *Land Subsidence in the Santa Clara Valley, California, as of 1982*. U.S. Geol. Surv. Prof. Pap. 497-F. 61 pp.

Porterfield, G. (1980). *Sediment Transport of Streams Tributary to San Francisco, San Pablo, and Suisun Bays, California 1909-1966.* U.S. Geol. Surv. Water Resour. Inv. 80-64, Aug. 1980. 70 pp.

Smith, B. J. (1965). *Sedimentation in the San Francisco Bay System*. Proc. Fed. Interagency Sedimentation Conf. 1963. U.S. Dept. Agr. Misc. Publ. 970. 79 pp.

HYDRODYNAMIC EXCHANGE BETWEEN SAN FRANCISCO BAY AND THE OCEAN: THE ROLE OF OCEAN CIRCULATION AND STRATIFICATION

JOHN L. LARGIER
Center for Coastal Studies Scripps Institution of Oceanography University of California, San Diego La Jolla, CA 92093-0209

Tidal forcing dominates the mass exchange between San Francisco Bay and the coastal ocean, but the renewal of water in the Bay also depends on the "tidal exchange ratio" – the proportion of the flood tide volume that is new ocean water. The tidal exchange ratio is related to stratification and to cross-flows in the ambient ocean waters. Based on recent ideas about coastal circulation, the tidal exchange ratio is investigated during the "upwelling" and "relaxation" seasons that are characterized by very weak inflow of freshwater to the Bay. Results are obtained from historical velocity, temperature, salinity, sealevel and wind data.

It is shown that tidal cross-flows are not very effective in enhancing tidal exchange ratio values above their base level which is determined by the interaction of deterministic tidal flows with fixed topography. This base value is low, varying between about 20% and 40% over a spring-neap cycle. It is argued, however, that both the barotropic and baroclinic components of wind-driven upwelling play an important role in enhancing the tidal exchange ratio for San Francisco Bay. The dense upwelled water is selectively drawn into the Bay, intruding as a near-bottom density current once over the offshore bar.

San Francisco Bay consists of a northern bay through which the Sacramento, San Joaquin, and several smaller rivers enter the ocean, and a southern bay which is fed only by small rivers, such as Coyote Creek, and wastewater run-off (Conomos 1979). The morphology of the Bay is determined by that of the Pleistocene stream valley which it occupies. In spite of some deep channels, it has an average depth of 6 m (excluding mudflats) owing to extensive shallow areas. The maximum depth of 100 m occurs in the mouth of the Bay. The basin is almost closed as this connecting channel between the Bay and the ocean, known as Golden Gate, is very narrow relative to the surface area of the Bay. Although this channel is deeper than both the Bay and the immediate offshore waters, seaward of Golden Gate a shallow sandbar is found (Fig. 1). A narrow shipping channel has been dredged to 20-m depth through this bar that otherwise forms a 10-m sill between the Bay and ocean. Offshore of the bar, tidal flows become less important and wind-driven circulation becomes more important. The coastal ocean off northern California is typical of the midlatitude eastern boundaries of the ocean, in that it is characterized by strong and seasonal wind-driven upwelling. In the Gulf of the Farallones (Fig. 1), both the ocean and the atmospheric boundary layer exhibit complex flow fields due to the interaction of current and wind with topographic features such as Point Reyes and the Bolinas Ridge.

Flow through Golden Gate is due to the tidal rise and fall of sea level, the throughflow of river water, the rise and fall of sea level due to propagating shelf waves, and the surface stress applied by local winds. In addition, density-driven

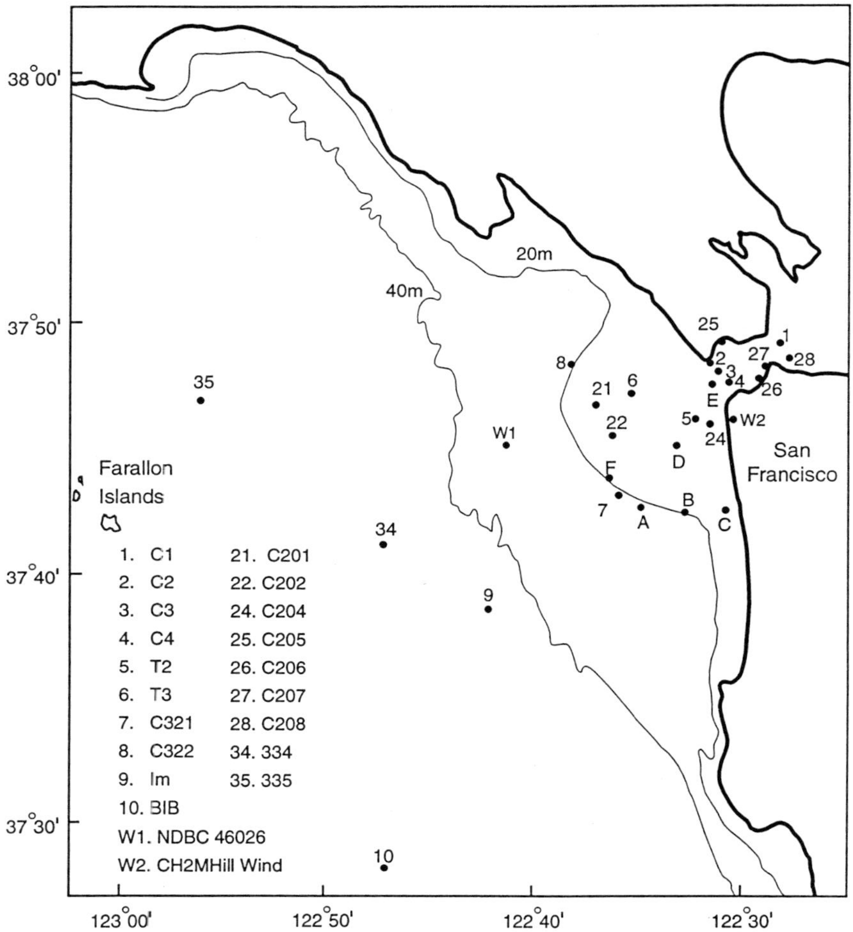

FIGURE 1. The study area, showing the outer part of Central Bay, the Golden Gate (narrowest section of mouth channel), and the Gulf of the Farallones. The 20-m isobath indicates the outer edge of the ebb-tide bar (Note that isobaths inshore of the bar are not shown). The locations of instruments deployed by Cheng & Gartner (1984), CH_2MHill (1989), and Noble & Gelfenbaum (1990) are indicated by dots. Wind records are available from NDBC Buoy 46026 (1981-90) and from a coastal site during the CH_2MHill 1987-88 study.

exchange flows may be established owing to the less saline (in winter) or warmer (in summer) nature of bay waters. The tide dominates current velocities and the integrated mass transport through the Gate – this is discussed in section 3. However, the primary concern is with the *net* exchange of water that occurs. The concern is not so much with the amount of water moving through the Gate, which can be well-described by tidal models (*e.g.*, Cheng *et al.*, in press) or observations of the rise and fall of sea level in the bay, but rather with the origin of this water. If an outflow of Bay water is simply returned to the Bay on the subsequent flood tide, the nearfield ocean (inner Gulf of the Farallones) is effectively an extension of the Bay and there is no net flushing of water from the Bay. Similarly, if the inflow of oceanic water on the flood tide is simply returned to the ocean on the subsequent ebb tide, the outer Bay is effectively an extension of the ocean and water is not flushed from locations farther into the Bay.

The net exchange is a product of the mass exchange and a (tidal) exchange ratio. The simplest form of this tidal exchange ratio (*TER*) can be expressed in terms of salinity, a conservative tracer of differences between Bay and ocean water: $TER = (S_{in} - S_{out}) / (S_{oc} - S_{out})$ where S_{out} is the average salinity of outflow through mouth, S_{in} is the average salinity of the inflow and S_{oc} is the ocean salinity (Parker *et al.* 1972; Fischer *et al.* 1979). The basis for this is a comparison of the inward and outward salt fluxes, thus the averages are velocity-weighted averages. The salt flux (transport) through the Golden Gate is given by AuS, where A is cross-sectional area and u is velocity. This can be decomposed by decomposing the velocity and salinity variables (Jay 1991):

$$u(y,z,t) = u_o + u_t(t) + u_y(y) + u_z(z) + u'(y,z,t)$$
$$S(y,z,t) = S_o + S_t(t) + S_y(y) + S_z(z) + S'(y,z,t)$$

where u_o and S_o are averages over the cross-sectional area and over the tidal cycle; u_t and S_t are time-dependent cross-sectional averages; u_y and S_y are averages over depth and over the tidal cycle, but dependent on cross-channel (lateral) position; u_z and S_z are depth-dependent cross-channel and tidal averages; u' and S' contain the residual variability. In addition to temporal changes in u and S, in shallow waters the cross-sectional area A also varies significantly in time due to the tidal rise and fall in water level. In expanding the salt flux AuS, one obtains several terms which can be grouped into five types of exchange. The first type of exchange is the mean-flow flux term, or "throughflow," which is associated with u_o and S_o. It relates to a net cross-sectional average outflow or inflow over a tidal cycle – this is due to river inflow to the Bay or long-period changes in sea level and filling/emptying of the Bay. Second are the time-dependent flux terms or "reversing flow with mixing," associated with u_t and S_t. It relates to temporal changes in the cross-sectional salinity of the reversing tidal flow– this is generally explained by mixing of the outflow with the oceanic ambient so that the inflow salinity is different from the outflow salinity (section 4). The third exchange type is due to the lateral-dependent flux terms or "lateral circulation" which relates to cross-channel differences usually associated with tidal jets, flow-topography interaction and the Coriolis effect (section 3). It is associated with u_y and S_y. Fourth, the depth-dependent exchange or "vertical circulation," associated with u_z and S_z, relates to vertical differences in velocity and salinity explained by stratification and by straining due to surface or bottom stresses. Typically, one finds a tidal-average outflow near-surface and a tidal-average inflow near-bottom (section 5). Finally, the residual flux term due to smaller scale effects associated with u' and S' can be referred to as "diffusion." While exchange is presented here in terms of salinity, the above

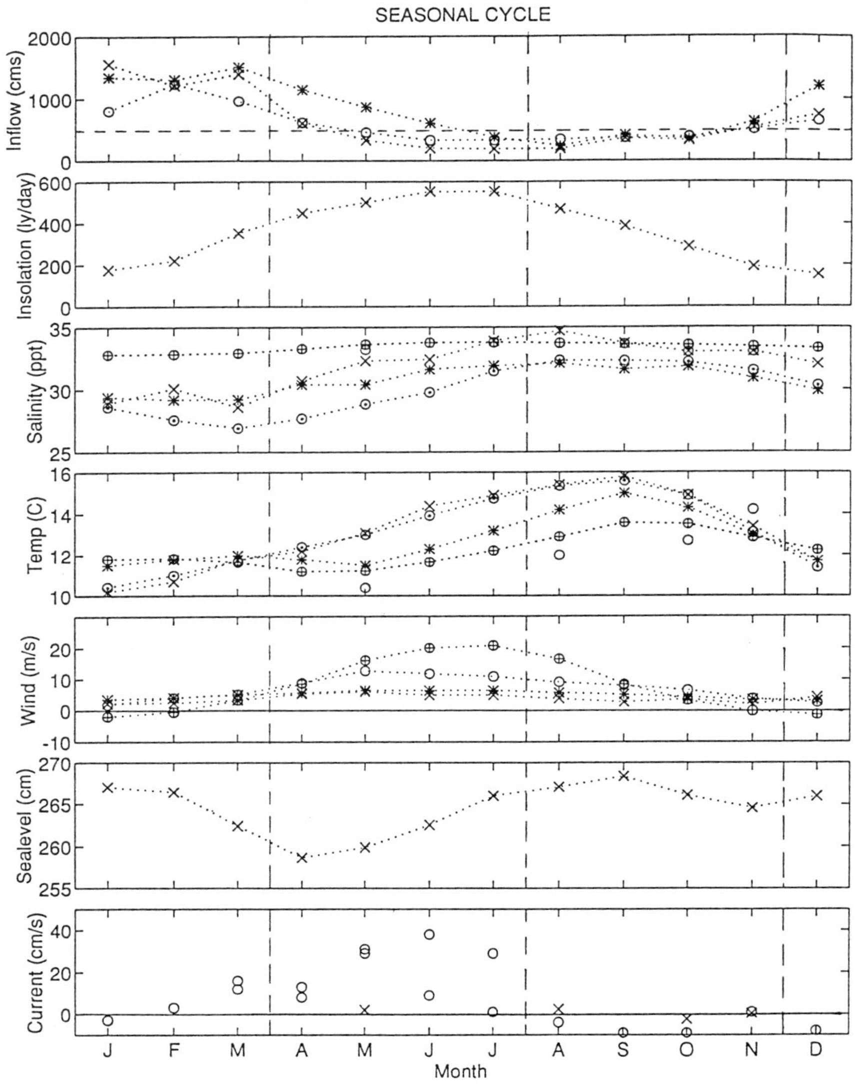

FIGURE 2. Seasonal cycles in monthly means of forcing and water type variables in the vicinity of San Francisco Bay. The seasons used by Largier *et al.* (1993) are indicated by dashed lines – upwelling (April-July), relaxation (August-November), winter (December-March). (a) Delta inflow (m^3s^{-1}) shows wet winter and dry summer/fall (dashed line at 500 $m^{-3}s^{-1}$): × = 1969-77 (Conomos *et al.* 1979), O = 1960-70 (Conomos & Peterson 1977), ✳ = 1978-85 (CH2MHill 1989). (b) Insolation (ly/day) shows summer warming: × = (Conomos *et al.* 1979). (c) Salinity (ppt) shows minimum bay values and maximum offshore-Bay difference in spring (systematic offset between different records may be sampling bias): × = Golden Gate for 1969-77 (Conomos 1979), O = Mooring A for 1987-88 (this paper), ✳ = Golden Gate for 1960-70 (Conomos & Peterson 1977), ⊙ = San Francisco for 1921-87 (Cayan *et al.* 1991), ⊕ = Farallon Islands for 1925-86 (Cayan *et al.* 1991).

discussion applies equally well to the fluxes of other dissolved or suspended matter that can be expressed as a concentration.

The hydrodynamic exchange through the Golden Gate appears to be dominated by the time- and depth-dependent flux terms (*cf.* Columbia River, Jay & Smith 1990). This exchange through the Golden Gate has an effect on the exchange between the ocean and all saline waters landward of the Gate. However, the importance of Golden Gate hydrodynamics becomes less important to residence and flushing times as one moves farther into the bay. Local longitudinal dispersion dynamics become more important. Ultimately, in the Delta, the exchange is dominated by throughflow due to the river.

The aim of this work is to point out the influence of oceanic circulation and density differences on hydrodynamic exchange between San Francisco Bay and the coastal ocean. Much of this study is relevant to other bays found in mediterranean climates (*e.g.*, San Diego Bay, California – Chadwick *et al.*, in press; Saldanha Bay, South Africa – Monteiro & Brundrit, in press). The results presented here are based on the analysis of a variety of historical data from the Golden Gate and the Gulf of the Farallones (Fig. 1).

In section 2, seasonal and large-scale structures are discussed. Since stratification, coastal circulation and freshwater inflow to the Bay vary seasonally, a seasonal variation in bay-ocean exchange is expected. The bulk of this work, however, is on tidal flow (section 3) and the dynamics that control the tidal exchange ratio *TER* (sections 4 and 5) as the tide dominates flow velocities and mass transport in the vicinity of the Golden Gate. Particular attention is given to the following and their effect on the tidal exchange ratio: (i) the three-dimensional structure of tidal flow through the Golden Gate; (ii) the phase-locked tidal cross-flow in the ocean; (iii) the wind-driven cross-flow; (iv) the sea level and isopycnal fluctuations associated with upwelling; and (v) the stratification observed in the vicinity of the mouth. Section 6 notes the primary contributions to nontidal mass transport through the Golden Gate.

FIGURE 2 (continued). (d) Temperature (°C) shows a large bay-offshore difference during the upwelling season and warmest water in fall: × = Golden Gate for 1969-77 (Conomos 1979), O = Mooring A for 1987-88 (this paper), ✳ = NDBC Buoy 46026 for 1981- 90 (Dorman and Winant, 1994), ⊙ = San Francisco for 1855-1987 (Cayan *et al.* 1991), ⊕ = Farallon Islands for 1925-1986 (Cayan *et al.* 1991). (e) Wind forcing shows a maximum in the upwelling season, particularly north of the Gulf of the Farallones: Alongshore wind speed (ms^{-1}, southward positive) at × = NDBC Buoy 46026 for 1981-90 (Dorman & Winant 1994), and O = NDBC Buoy 46013 for 1981-90 (Dorman & Winant 1994), wind speed (ms^{-1}) at ✳ = San Francisco Airport for 1959-70 (Conomos 1979), ⊕ = Upwelling Index ($m^{-3}s^{-1}$ per km of coastline) for 1972-83. (f) Sea level (cm) shows a minimum early in the upwelling season, when water is still cold: × = Golden Gate (Fort Point) for 1854-1986 (Cayan *et al.* 1991). (g) Alongshore current speed (cms^{-1}, southward positive) shows a maximum in the upwelling season, particularly north of the Gulf of the Farallones: × = Mooring A for 1987-88 (this paper), O = Mooring on 90-m isobath north of Point Reyes 1988-89 (Largier *et al.* 1993).

2. SEASONAL VIEW OF THE BAY AND ADJACENT OCEANIC REGIONS

Northern California experiences a mediterranean climate with hot dry summers and cool wet winters (Fig. 2), although the maritime influence at the coast results in relatively warmer winters and cooler, foggy summers. Seasonal influences are communicated via the ocean (*e.g.*, wind forcing and upwelling), the river (*e.g.*, volume of freshwater inflow), and directly to the bay (*e.g.*, surface heating). Riverflow is strongest during winter, as a result of the episodic rainfall associated with mid-latitude cold fronts. Following winter, the larger rivers continue to flow strongly as the snowpack melts in the upper catchment. In summer, riverflow weakens. Evaporation exceeds precipitation during the summer and bays/estuaries with little freshwater inflow may experience a net loss of freshwater, leading to hypersalinity (*e.g.*, Tomales Bay, Largier *et al.* 1994). Southern San Francisco Bay appears to be on the margin of this estuarine type whereas northern San Francisco Bay receives significant freshwater inflow all year (Conomos 1979). In addition to this cycle of freshwater fluxes, there is a distinct seasonal cycle in atmosphere-ocean heat flux, resulting in the warmer water temperatures observed in the bay during summer (Fig. 2).

The seasonality of coastal circulation off northern California has been described by Lentz & Chapman (1989) and by Largier *et al.* (1993) for the region immediately north of Point Reyes. The three-season year is equally relevant in discussing the Gulf of the Farallones (Figs. 2 and 3), although mean current and wind speeds in the Gulf are much weaker than on the open coast north of Point Reyes. In early spring (March-April), winter storms with westerly winds and rain give way to strong and persistent northerly winds which drive southward currents and upwelling of cold water along the coast. The "spring transition" to upwelling conditions is in response to the poleward movement of the North Pacific High Pressure cell (Lentz & Chapman 1989). The "upwelling season," characterized by northerly winds, upwelling circulation, cold water, coastal mists, and low sea level, persists until mid-summer, roughly August.

In response to the weakening of large-scale atmospheric pressure gradients, the wind weakens, the surface water warms, the sea level rises, the mists clear, and the near-coastal currents turn to northward (Largier *et al.* 1993). This "relaxation season" is terminated by the arrival of the winter storms in November or December. The winter, or "storm season," is characterized by cold water, strong but nonpersistent winds, freshwater run-off, and higher sea levels due to westerly and southerly wind forcing. In response to the seasonal cycles of freshwater supply, wind stress, sea level, seawater temperature, and surface heating, one can observe a seasonal cycle in the temperature and salinity of San Francisco Bay water (Conomos 1979). In the wet winter months, the Bay salinity is lower on average (Fig. 2) and greater vertical stratification is observed (Fig. 4). In spring the salinity of bay water increases. At the same time, however, water temperatures in the Gulf decrease owing to upwelling while water temperatures in the bay increase owing to surface heating (Fig. 2). Combined with the residual salinity difference, this

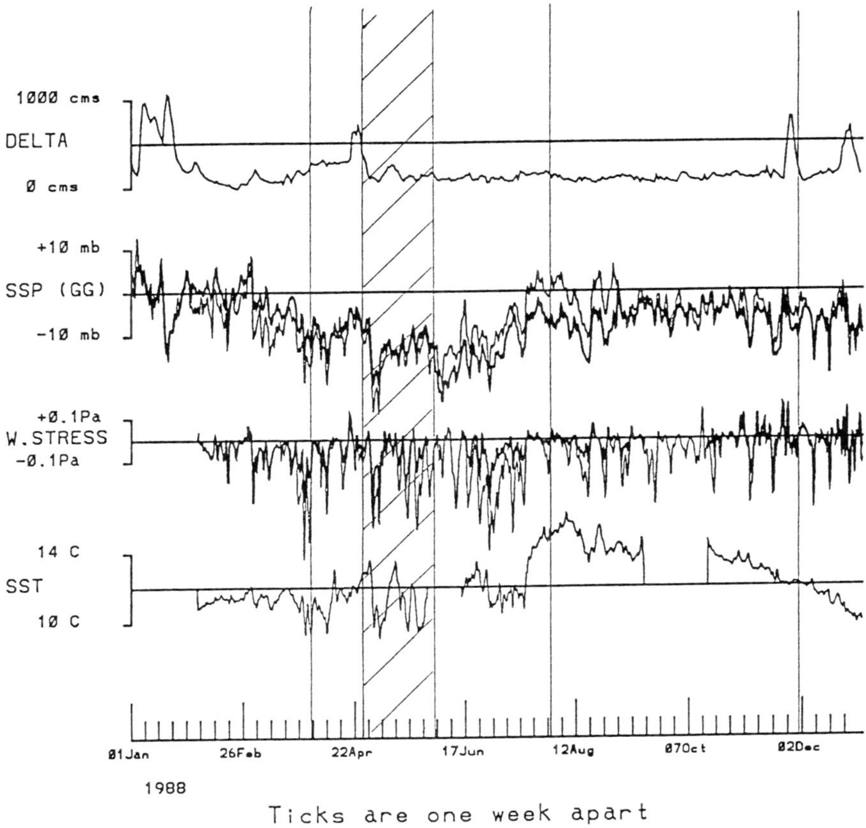

FIGURE 3. Delta inflow, synthetic subsurface pressure (sea level corrected for fluctuations in atmospheric pressure) at the Golden Gate (Fort Point), wind stress at NDBC Buoy 46026 (Fig. 1), and sea surface temperature at NDBC Buoy 46026. Fluctuations with periods shorter than 38 hours have been removed from these data (*e.g.*, diurnal and tidal fluctuations). Note the delta inflow event in late April followed by strong wind forcing, sea level set-down (about 15 cm) and the appearance of cold water. The period during which moorings were deployed offshore (CH_2MHill 1989), is shaded. The three seasons are indicated by lines: upwelling (April-July), relaxation (August-November), winter (December-March).

temperature difference maintains some stratification in the outer bay until late summer (Fig. 4). Water in the coastal ocean warms up during the relaxation season, at the same time as river inflow to the bay is a minimum. During this period (September- October), bay-ocean density differences are very small and stratification does not appear to be important (Fig. 4).

From the data collected by Cheng & Gartner (1984), one can plot mean vertical shear at mooring C1, in the mouth of the bay (Fig. 1), against mean "Delta Outflow" over the same period (Fig. 5). Significant vertical shear is only observed for Delta outflows that exceed about 400 m^3s^{-1}, typical of winter conditions (see months December to March in Fig. 2). If the shear throughout the cross-section is compa-

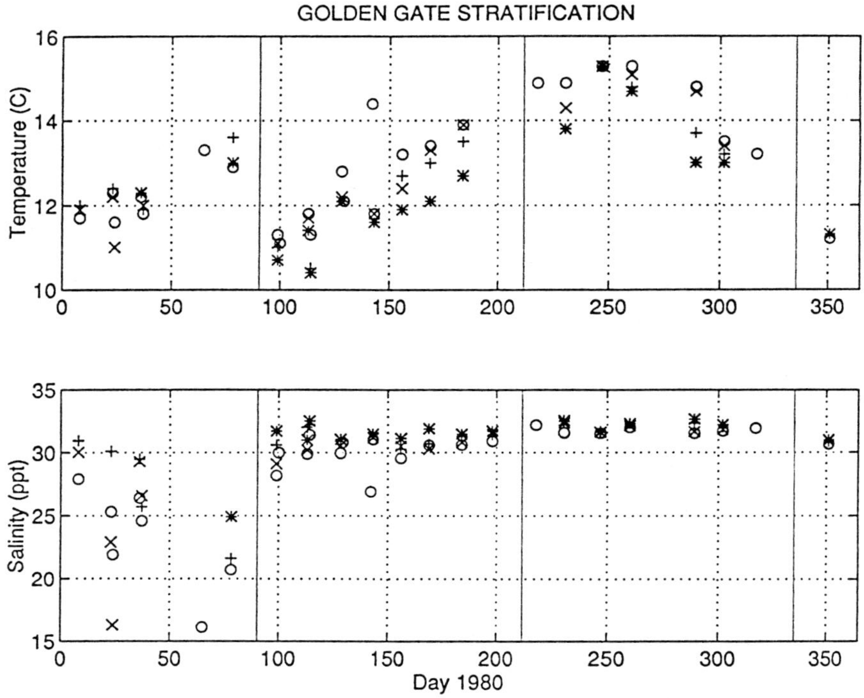

FIGURE 4. Variation in temperature and salinity at multiple depths at station 19, as occupied by Ota *et al.* (1989) during 1980 (nominal depths: O = 2 m, × = 10 m, + = 20 m, ✳ = 40 m). Tidal variation is aliased by the monthly sampling, leading to substantial inter-month variability that may not be real. However, the general trend of low-salinity, cold water in winter and cool water during the upwelling season is apparent. The significant vertical differences in winter decrease through the upwelling season. There is negligible stratification in the relaxation season. Solid vertical lines indicate the seasons.

rable to that at C1, then for Delta outflows of order 500 m^3s^{-1} one could expect the vertical circulation (due to a shear of 0.2 cm $s^{-1}m^{-1}$ through a 20-m surface layer) to be of the order of 680 m^3s^{-1}. In 115 days, this would exchange a volume of water equal to that of the bay. In the upwelling and relaxation seasons (and some dry winter periods), the delta flow is much less than 500 m^3s^{-1} and freshwater-related vertical circulation is clearly inadequate to maintain a well-flushed bay. Having identified the essential characteristics of these three seasons (winter, spring-summer and summer-fall – or storm, upwelling and relaxation, following Largier *et al.* 1993), one can consider qualitative Bay-ocean exchange scenarios for each season:

(1) In winter, one expects more rapid exchange owing to the larger freshwater inflow (Fig. 2) and vertical stratification (Fig. 4). This freshwater inflow enhances the bay-ocean exchange through the mean-flow and depth-dependent flux terms. Vertical estuarine circulation is evident in Fig. 5 and in the tracks of drifters released by Conomos & Peterson (1977) during 1970-73. Expecting

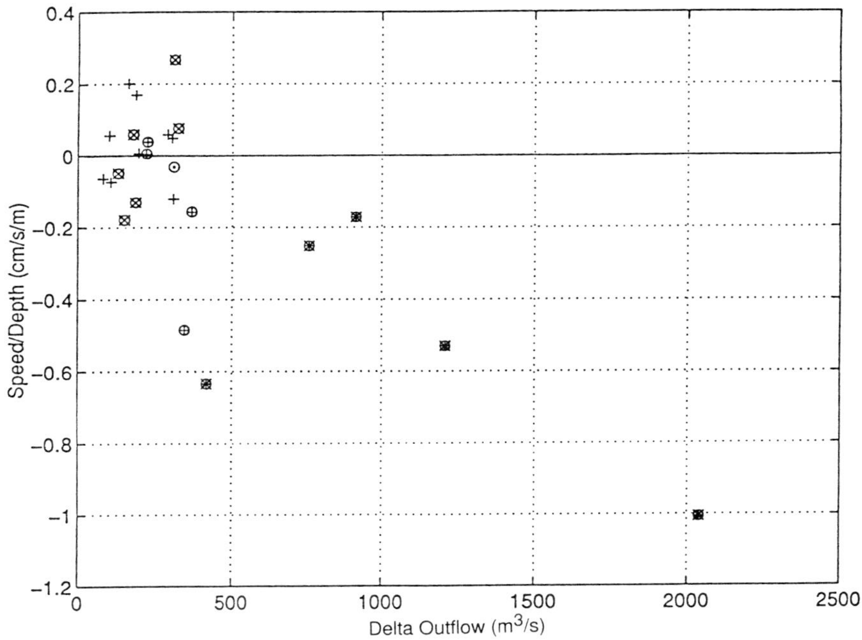

FIGURE 5. Average vertical shear in along-channel velocity at mooring C1 as a function of Delta outflow during 1979-80 (Cheng & Gartner, 1984). Both velocity difference and Delta outflow averages are taken over 12 M_2 tidal cycles. Negative values indicate stronger bay outflow nearer to the surface (estuarine circulation). Symbols are as follows: ✳ = one instrument in 0-10 m, other in 10-20 m; ⊕ = both instruments in 10-20 m; ⊗ = one instrument in 10-20 m, other in 20-30 m; ⊙ = one instrument in 0-10 m, other in 50-60 m; + = one instrument in 20-30 m, other in 50-60 m (total depth 92 m). For delta flow less than about 400 m^3s^{-1}, vertical shear is not well-resolved at this location. For larger delta flow, near-surface vertical shear is consistent with estuarine circulation and it increases with increasing delta inflow.

strong bay-ocean exchange due to freshwater inflow, the exchange during this season is not of concern in this paper.

(2) In spring-summer (the upwelling season), river inflow is weaker but the bay-ocean density difference may still be significant (Figs. 2 and 4), enhancing the depth-dependent flux term. Further, wind-forced upwelling results in strong currents in the Gulf (Fig. 3) and multi-day fluctuations in sea level which may be important in enhancing the time-dependent and the mean-flow flux terms. In spite of their importance, these upwelling-related processes have not been well described in previous work. Their role in exchange between San Francisco Bay and the ocean is the focus of this paper.

(3) In summer-fall (the relaxation season), the Bay-ocean density difference, river inflow and coastal circulation are at their weakest (Figs. 2 and 3). It is during this season that one expects the weakest exchange as it would appear that none of the flux terms are enhanced above their year-round base values which are due

to the tidal flow of homogeneous water to and from a quiescent ocean. The weak exchange processes during the relaxation season are explored further in this paper.

3. TIDAL CURRENTS

With a large surface area (10^9 m^2) relative to the cross-sectional area of the channel through the Golden Gate (10^5 m^2), mesotidal fluctuations in sea level lead to strong currents of over 2 ms^{-1} at times. The average tidal prism volume ($1.6 \times 10^9 m^3$) is similar to the average monthly river discharge to the bay, about a quarter of the total bay volume and about 56% of the Central Bay volume (Conomos 1979; Cheng & Gartner 1984). From the simple tidal prism method, a flushing time of 1.8 tidal cycles is thus obtained for the Central Bay. Water velocities and transport through the Golden Gate are dominated by the tidal flows due to the barotropic pressure gradients set up by tidal fluctuations in sea level in the Gulf of the Farallones. These tides propagate into the bay as shallow water waves and they evolve as described by Walters *et al.* (1985). In particular, as the wave propagates into the southern bay it becomes more like a standing wave with currents leading water levels. In contrast, as the reflected tidal wave is not as important in the longer and more dissipative northern bay, the tide remains more like a progressive wave and currents are approximately in-phase with the water levels. Near the Golden Gate, this north-south difference in tidal character results in a distinct phase difference and significant tidal pressure gradients between the northern and southern branches of San Francisco Bay and across the Golden Gate (Walters *et al.*, 1985).

The tide is a mixture of semidiurnal and diurnal components: major semidiurnal components are M_2, S_2, and N_2 with Golden Gate amplitudes of 0.58, 0.14, and 0.12 m, and Golden Gate velocities of 1.06, 0.29, and 0.25 ms^{-1}, respectively; major diurnal components are K_1 and O_1 with amplitudes of 0.37 and 0.23 m and velocities of 0.22 and 0.18 ms^{-1} (Walters *et al.* 1985; Petzrick 1993). This mixed tide displays two high waters and two low waters every day. The higher high water (HHW) is followed by the lower low water (LLW) so that the strongest velocities are typically observed on the large ebb tide. This ebb dominance explains the presence of the crescentic ebb-tide bar due to sediment deposition where the outflow velocities decrease offshore of the Golden Gate (Fig. 1). The bar is found about 9 km offshore, roughly a tidal excursion outward from the mouth of the bay.

The M2 tidal flow within and offshore of Golden Gate is summarized in Fig. 6, which is a plot of the orientation and magnitude of the major axis of current variance. These data are collated from a number of previous studies. During 1979/80, Cheng & Gartner (1984) deployed current meters at locations within the mouth and offshore, in addition to many moorings throughout San Francisco Bay. The data of interest to this study are from moorings at locations C1, C2, C3, C4, T2, T3, C321, C322, C201, C202, C204, C205, C206, C207, and C208 (Fig. 1). While these moorings were not all deployed at the same time, moorings were deployed at C1 (near the narrowest section of the mouth, *i.e.*, Golden Gate), C2, C3, C4 (across the westernmost end of the mouth, between Point Bonita and Seal

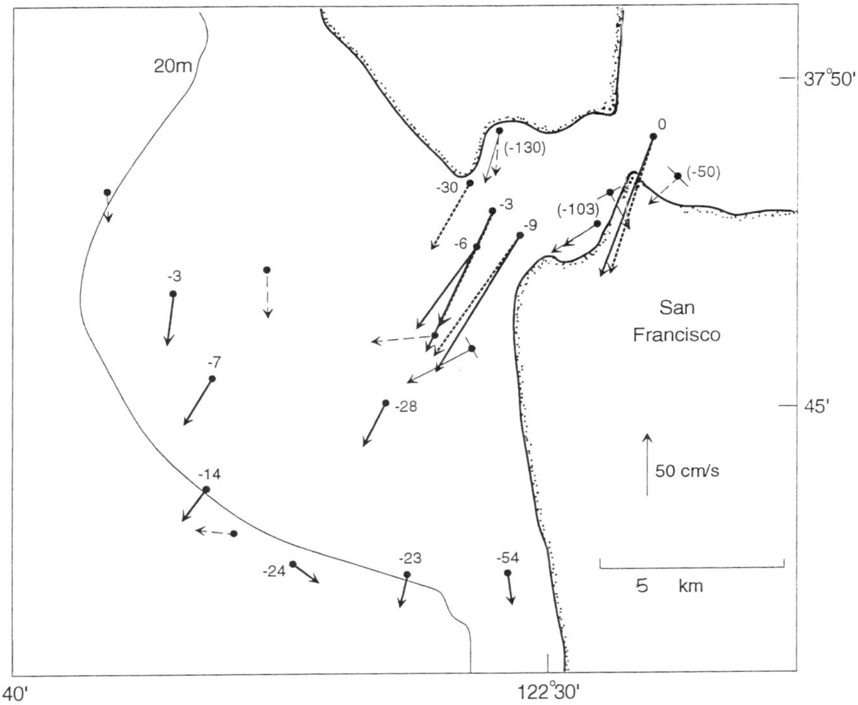

FIGURE 6. The orientation and magnitude of the major axis of current velocity at the M_2 tidal frequency in the vicinity of the mouth of San Francisco Bay. Solid arrows indicate velocity data in the 5 to 10-m depth bin and dashed arrows are for the 10 to 15-m depth bin. Finer solid or dashed arrows indicate shorter records for which there is marginal confidence in the result. The minor axis is only displayed when it is significant in this plot. The phase of the M_2 tidal currents (in degrees relative to that at the Golden Gate) are noted alongside the mooring positions.

Rock), and C202 (at the offshore bar) simultaneously during September 1979. These data are discussed in detail in this paper. In addition, a mooring was deployed at C1 many times during 1979/80 and this provides some seasonal coverage of tidal and mean currents in the mouth of the bay (*e.g.*, Fig. 5). A second valuable data set is from current meters moored at the mouth of the bay and offshore of the city of San Francisco during 1987/88 (CH$_2$MHill 1989). The one offshore mooring (A, Fig. 1) was maintained for a full year, while the other moorings (B, C, D, E, F) were deployed for a month at a time – in October 1987 and May 1988. The May 1988 data are discussed in detail in this paper. In addition, Noble & Gelfenbaum (1990) deployed current meters in the outer Gulf of the Farallones during 1988/89 (Im, BIB, 334 and 335, Fig. 1). While these data are not analyzed in detail, they provide valuable information on the tidal and wind-driven currents offshore, in the far-field of the San Francisco Bay outflow.

While tidal flow through Golden Gate is strong and its direction is controlled by the narrow well-defined channel, this is not true farther offshore. The direction of the major axis of the M_2 tide is not well defined by the limited amount of data available from offshore, particularly at locations removed from the tidal jet. It appears, however, that these offshore tidal currents resemble those elsewhere along the northern California coast, far removed from San Francisco Bay. These currents are explained as a Kelvin wave propagating northward over the continental shelf (Munk *et al.* 1970; Rosenfeld & Beardsley 1987; Noble & Gelfenbaum 1990). In contrast, between the Gate and the offshore bar, tidal currents are dominated by the

FIGURE 7. A photograph from Skylab (26 January 1976) of an unusually large turbid outflow from San Francisco Bay during strong freshwater inflow to the Bay (about 15,000 m^3s^{-1}). The mooring positions from Fig. 1 are plotted over the photograph. Note the nearshore areas of clear water immediately north and south of the outflow jet. The outer edge of the low-salinity outflow is defined by a marked color change.

jet outflow and the potential sink inflow structures which are characteristic of the mouth of a semi-enclosed bay/estuary (Fischer *et al.* 1979). Following Wilkinson (1978), the length scale of such a periodic outflow jet is scaled by $\ell = (\pi^2 P^2/Ah)^{1/3}$ where P is the tidal prism volume, A is the channel cross-sectional area, and h is the mean depth of flow in the channel (30 m). This suggests that the San Francisco Bay tidal outflow jet would extend about 15 km offshore (in the absence of stratification effects, and neglecting the effect of the ebb-tide bar). In Fig. 7, the turbid outflow from the Bay contrasts with the clear nearshore waters immediately north and south of this jet. Similarly, in Fig. 8, the warm Bay waters demarcate a jet outflow from the Gate into the cooler Gulf waters during the upwelling season.

The tidal mass transport to and from San Francisco Bay is well-described by vertically-averaged numerical models. In particular, Cheng *et al.* (in press) have developed the model TRIM to describe tidal and tidal residual circulation in San Francisco Bay. While the two-dimensional TRIM model resolves horizontal residual flow reliably, it is still difficult to accurately represent the three-dimensional detail observed in bay and estuarine flows. In addition to the vertically sheared density-driven flows discussed in section 5, other complex three-dimensional structures are observed in the tidal flow at the mouth of San Francisco Bay. In order to investigate these mechanisms one would need a three-dimensional model that can reproduce such secondary circulations in the vertical-lateral plane. Further, the complexity of nontidal flow on the shelf can only be adequately represented with

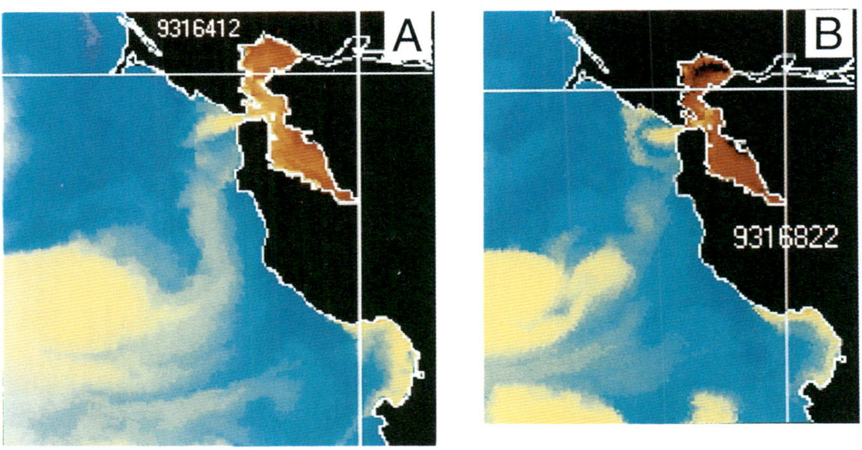

FIGURE 8. (a) AVHRR image of sea surface temperature in San Francisco Bay and the Gulf of the Farallones (redder colors are warmer and bluer colors are cooler), 13 June 1993. This is the typical temperature pattern during upwelling winds. Warmer Bay water is transported southward from the mouth. Inshore of this warmer water, cold water is upwelled along the coast as Ekman transport moves the warmer surface water offshore. In particular, note the local patch of cold water nearshore, immediately south of the warm outflow. (b) AVHRR image of sea surface temperature in San Francisco Bay and the Gulf of the Farallones (redder colors are warmer and bluer colors are cooler), 18 June 1993. Between the active outflow jet and the warm water arc from the previous outflow, a cold arc of water is observed. This is understood as being due to upwelling at the bar during the flood tide.

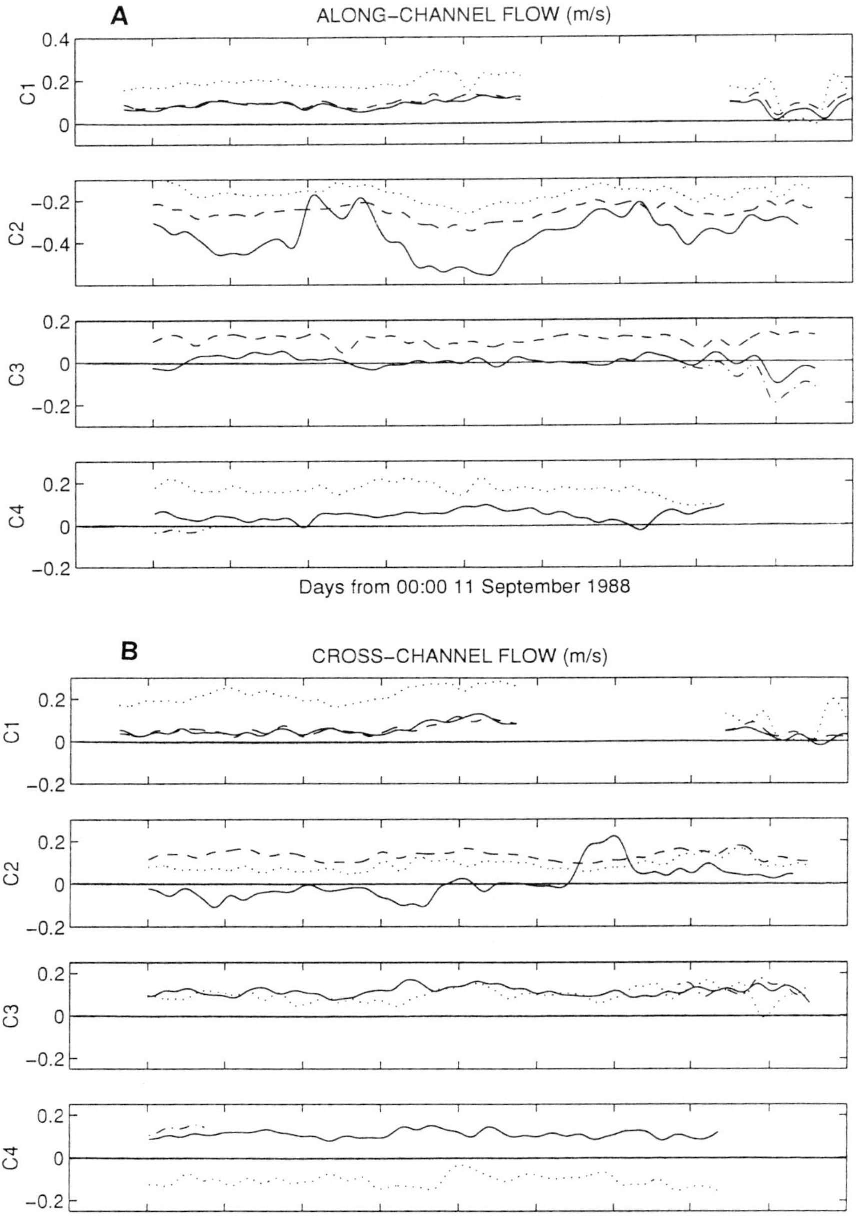

FIGURE 9. (a) Subtidal along-channel velocity (ms^{-1})at moorings C1, C2, C3, and C4 in the mouth of San Francisco Bay (Fig. 1). The along-channel orientation is defined by that of the M_2 axes in Fig. 6 and indicated in the y-axis label for each plot (positive values are into the Bay). All variance at tidal and diurnal frequencies has been removed by way of a

a bay model that is coupled to a shelf circulation model that adequately resolves flows like wind-driven upwelling, coastal-trapped waves, and deep ocean forcing (*cf.*, the observations of Largier *et al.* 1993, from north of Point Reyes). The effect of such shelf circulation dynamics on bay-ocean exchange is discussed in sections 4 and 6.

Current velocity data from moorings at C1, C2, C3, and C4 were low-pass filtered (PL64 filter) to remove all variability at higher frequencies, such as that due to tidal and diurnal cycles. Only variability with periods longer than 1.5 days is retained in the subtidal records of along-channel and cross-channel current at these moorings (Fig. 9). The "along-channel" orientation is chosen to be in the direction of the tidal currents, as plotted in Figure 6, with positive values indicating inflow. "Cross-channel" is at right angles, positive being toward San Francisco City (southeastward). In addition to the horizontal differences in subtidal current (*e.g.*, mean inflows at C4 and C1 versus mean outflow at C2), there are consistent vertical differences in subtidal currents at each mooring. These differences are also apparent in the mean currents plotted in Fig. 10. The vertical shear in along-channel flow is primarily a symptom of the vertical circulation that occurs in response to the density difference between ocean and bay water. The mean along-channel current at each instrument is given as a function of depth in Fig. 11. Data are also given for the current meters deployed offshore during May 1988. While vertical shear does vary from location to location and with depth and time, in this study no values larger than about 0.01 s^{-1} are observed in the vicinity of the mouth. This corresponds to a vertical circulation no larger than 10^4 m^3s^{-1}, which would circulate a volume of water equal to that in the Bay in about 8 days. This estimate is typical of periods with a strong Delta outflow (Fig. 5). During late summer and fall, when ocean-bay salinity/temperature/density differences are weakest, the vertical circulation may be two orders of magnitude smaller than this estimate – and the associated time scales, two orders of magnitude longer.

In addition to vertical shear in the along-channel flow, there is a strong subtidal shear in the cross-channel flow, with the near-surface and near-bottom mean velocities in opposite directions (Fig. 9b). This shear is best developed at C4, although it is apparent at all of the other moorings as well. Exploring the tidal structure of this cross-channel flow at C4 (Fig. 12), one can see it is rectified with both flood and ebb velocity pulses in the same direction – toward the shore near-surface and toward midchannel near-bottom. While the strength of this shear at C4 may be partially due to an ebb tide eddy occurring in the vicinity of C206 and C207 (note large differences in tidal phase, Figure 7, and TRIM model prediction, Burau and Monismith, pers. commun.), the observation of mean cross-

FIGURE 9 (continued). PL-64 low-pass filter with half-power point at 38 hours. Dot-dash lines are for data from top 10 m (C3: 6.7 m; C4: 6.1 m); solid lines for data from 10 to 20-m bin (C1: 11.9 m, C2: 12.2 m, C3: 12.2 m, C4: 12.2 m); dashed lines are for data from the 20 to 30-m in (C1: 21.0 m, C2: 28.1 m, C3: 25.3 m) and dotted lines are for deep data (C1: 89.7 m, C2: 33.5 m, C4: 40.6 m). (b) Subtidal cross-channel velocity (ms^{-1}) at moorings C1, C2, C3, and C4 in the mouth of San Francisco Bay (Fig. 1). The cross-channel orientation is at 90° to the right of the along-channel orientation.

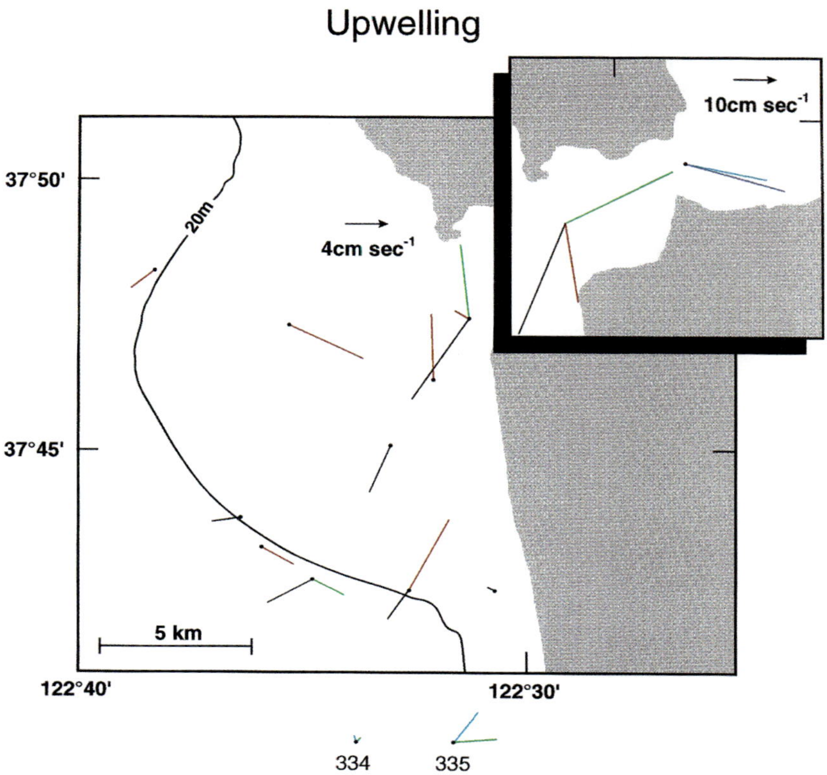

FIGURE 10a. Mean current velocities for instruments deployed by Cheng & Gartner (1984), CH2MHill (1989), and Noble & Gelfenbaum (1990) at various locations and various depths. Results (in Figs. 10a-c) are grouped according to season: (a) upwelling season (April-July). Black: data from top 10 m, Red: data from 10 to 20 m, Green: 20 to 40 m, Turquoise: 40 to 60 m, Blue: 80 to 100 m. (Figures 10b-c continued next page.)

channel shear at C1, C2, C3, and E as well, suggests that this secondary lateral circulation fills the channel. Lateral circulation is important as it spreads out the narrow band of momentum – the jet that emerges from the constriction at the Golden Gate. As a result, the tidal outflow jet from San Francisco Bay is not scaled by the width of the Gate, but by the width of the wider channel at the westernmost end of the mouth. This is consistent with jet theory as the lateral boundaries of the channel do not expand rapidly enough for a jet to form between the Golden Gate and the westernmost end of the mouth – the outflow interacts with the boundaries and the outflow momentum is spread across the full channel width, as observed.

In addition to the strong lateral circulation showing up in the tidal mean, there are strong fluctuations in tidal velocity at frequencies higher than that of the tidal forcing (*i.e.*, within a single in/outflow). For example, at the 40-m instrument on mooring C4 (Fig. 12), the flood tide starts with an inflow to the left of the channel axis and then it rapidly swings 60° to the right – the inflow speed levels off at about

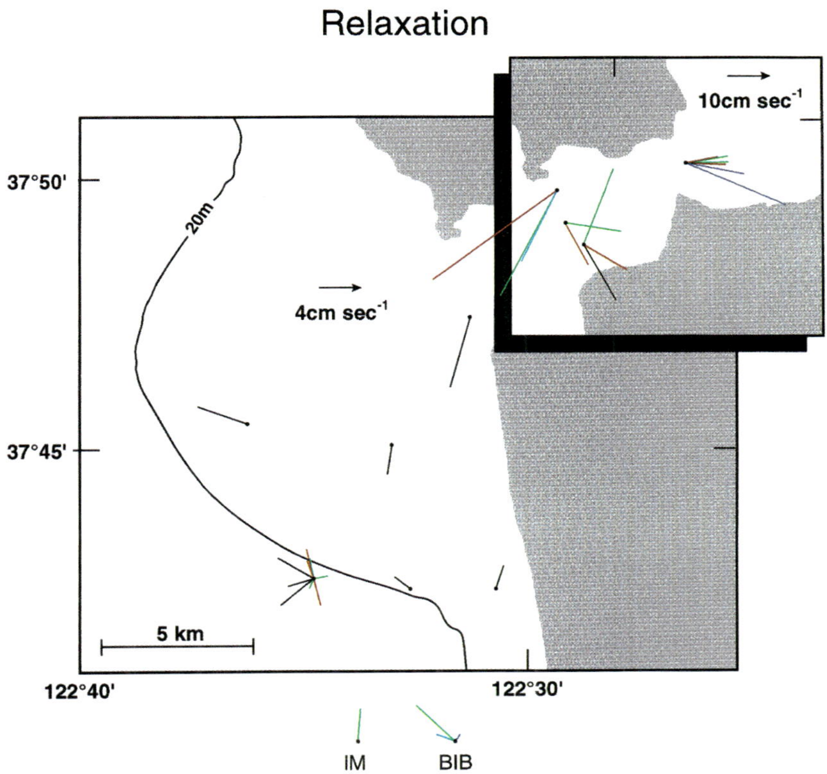

FIGURE 10b. Mean current velocities for instruments deployed by Cheng & Gartner (1984), CH_2MHill (1989), and Noble & Gelfenbaum (1990) at various locations and various depths. Results (in Figs. 10a-c) are grouped according to season: (b) relaxation season (August-November). Black: data from top 10 m, Red: data from 10 to 20 m, Green: 20 to 40 m, Turquoise: 40 to 60 m, Blue: 80 to 100 m. (Figure 10c continued next page.)

0.75 ms^{-1} and in a direction about 30° right of the channel axis for about 2 hours. Rather suddenly, then, in the middle of the flood tide, the 40-m velocity swings back to the left and accelerates – this 1.0 ms^{-1} flow, at about 45° to the left of the channel axis, continues for about 4 hours before decelerating and turning to flow out. The high-frequency structure of this deeper lateral flow is matched by flow in the opposite direction at the near-surface instrument (Fig. 12). Thus, the secondary lateral circulation reverses during midflood tide. This pattern of lateral circulation is also observed at mooring C3, with opposed flows at 12 m and 26 m. Similar supertidal fluctuations in speed and direction are also observed at C2 and C1, suggesting that these high-frequency fluctuations are not due to local topographic separation. Rather one may expect that these cross- channel fluctuations are related to flow curvature and changes in flow curvature with tidal phase (*e.g.*, the phase difference between tidal flows in the southern and northern arms of San Francisco

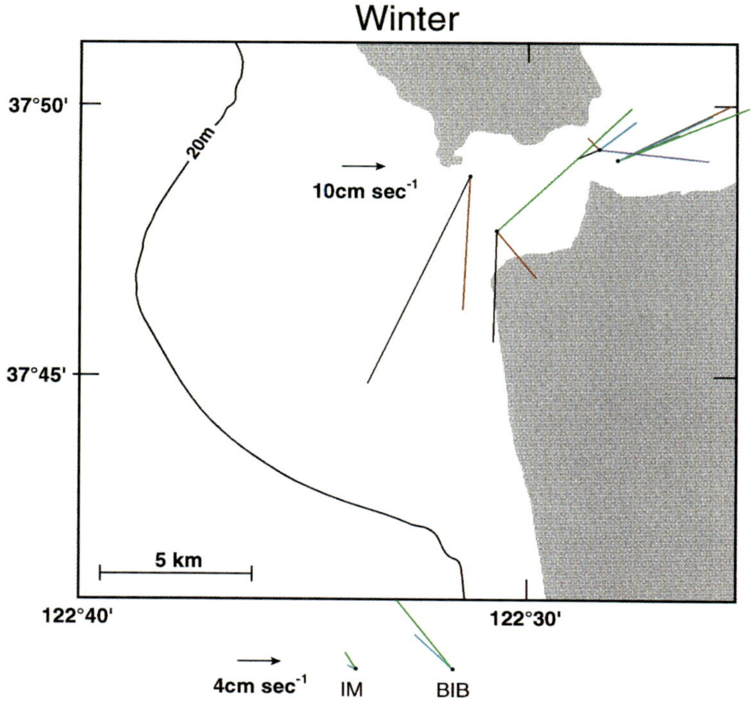

FIGURE 10c. Mean current velocities for instruments deployed by Cheng & Gartner (1984), CH2MHill (1989), and Noble & Gelfenbaum (1990) at various locations and various depths. Results (in Figs. 10a-c) are grouped according to season: (c) winter or storm season (December-March). Note that a different scale is used in the inset plot for the large tidal residual vectors observed in the mouth channel. Black: data from top 10 m, Red: data from 10 to 20 m, Green: 20 to 40 m, Turquoise: 40 to 60 m, Blue: 80 to 100 m.

Bay and in the coastal ocean). Additional lateral circulation may be driven by lateral density gradients created by lateral shear in the along-channel transport of density.

The importance of secondary circulation in the mouth of San Francisco Bay to the tidal exchange ratio (TER), is related to its role in producing a broader, slower outflow jet. The temporal fluctuations in TER are controlled primarily by the nature of the circulation in the ocean and the bay/ocean density difference and these are explored further in the next two sections. In the absence of these other influences, Ozsoy (1977) scaled the *TER* for a simple jet-and-sink "tidal pumping" scenario. This is a way of estimating the lower bound of the *TER* value, since any stratification or ambient cross-flow can be expected to improve exchange and increase *TER*. The tidal exchange ratio *TER* is related to $\tau_f = u_o t_f / b_o$ as in Fig. 13, where u_o is the velocity at the origin of the jet (mouth of Bay), b_o is the width, and t_f is the tidal period. Using the slower and broader values from the westernmost end of the mouth channel, values of 0.55 and 0.14 are estimated for the tidal pumping *TER* for typical spring and neap tidal ranges, respectively. These values are much less than 0.69

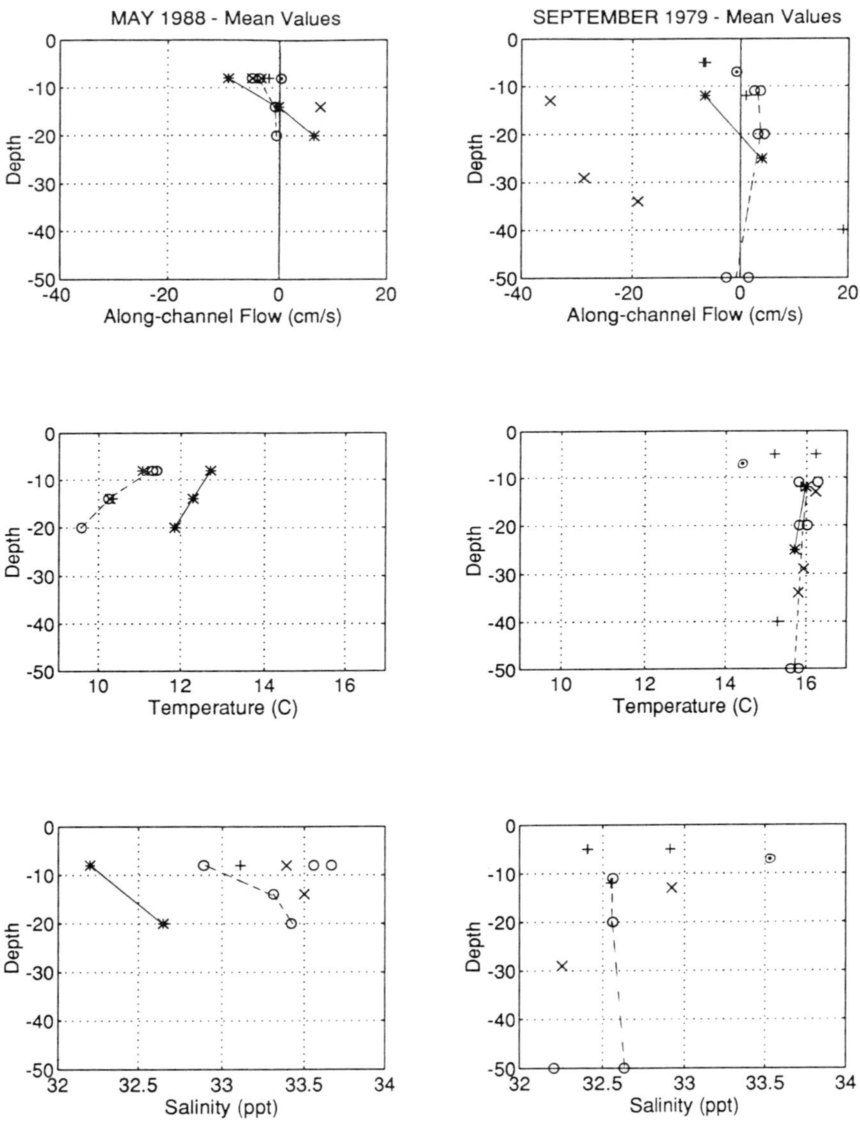

FIGURE 11. Mean along-channel velocity, temperature, and salinity as a function of depth, at moorings deployed during May 1988 (left panels) and September 1979 (right panels). In the left panels, data are from moorings A (O, dashed), B (×), C(O), D (⊗), E (✳, solid), F (+) and in the right panels, from C1 (O, dashed), C2 (×), C3 (✳, solid), C4 (+), C202 (⊙). Mooring positions are given in Figure 1.

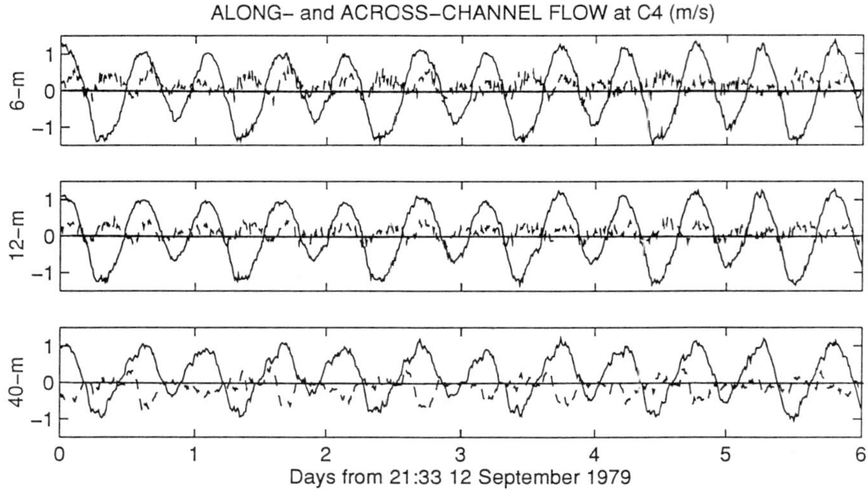

FIGURE 12. Variation in along-channel (54°, solid line) and cross-channel (144°, dashed line) velocities at mooring C4. Data is from instruments at depths of 6 m (top panel), 12 m (middle panel) and 40 m (bottom panel).

and 0.42, estimates of *TER* for a jet that originated from the narrow Gate itself. The *TER* can be calculated directly from observations, as Parker *et al.* (1972) did in September 1970, in a season of minimum stratification and minimum oceanic circulation (*TER* values 0.18 and 0.31, Fig. 13). However, the errors in the method itself, particularly with older low-resolution instruments, may be large because the *TER* is obtained as a small difference between two large numbers (Jay *et al.* 1996). Nevertheless, even with large error, these low *TER* values (Fig. 13) provide a qualitative confirmation that tidal pumping at the mouth of San Francisco Bay is not very effective. This is consistent with the suggestion that the initial conditions of the outflow jet are set by the broader westernmost entrance to the bay.

4. AMBIENT OCEANIC CURRENTS

The exchange between San Francisco Bay and the coastal ocean can be increased by the action of currents in the receiving waters – the ambient ocean. These coastal currents are strongly polarized alongshore and they act to remove the outflow water from the vicinity of the mouth so that the subsequent inflow is composed of new oceanic water. Thus, these currents increase exchange by increasing the exchange ratio rather than by increasing the volume of water exchanged. However, the role of ambient oceanic currents in determining the tidal exchange ratio at the mouth of San Francisco Bay is poorly known and poorly understood. These cross-flows may be forced by local winds, remote winds (*i.e.*, by way of shelf waves), tides, deep ocean circulation, density differences, or large-scale pressure gradients (Largier *et al.* 1993). In particular, observed currents in the Gulf of the Farallones are due to tides, wind, and shelf waves (Noble &

Gelfenbaum 1990). The wind-driven ambient currents are independent of the tide and they are generally of lower frequency than the tidal flow to and from the bay so that they may persist through a number of tidal outflow cycles. Further, the wind-driven currents vary with the seasons (see section 2). In contrast, tidal ambient currents are composed of the same frequency components as the tidal flow to and from the bay and these ambient tidal currents bear a constant phase relationship to the tidal outflow. In the absence of a lower frequency ambient flow, the tidal cross-flow reverses predictably and their role in bay-ocean exchange should be predictable (*e.g.*, with TRIM, Cheng *et al.*, in press). The tidal currents do not change with the seasons, although they are modulated by longer-period tidal components (*e.g.*, the spring-neap cycle). In the spring-summer-fall period when river inflow and density-driven exchange is weaker, the ambient currents in the Gulf of the Farallones are important in enhancing bay-ocean exchange. In this section, analyses of current velocity data from the Gulf of the Farallones will be presented and discussed in the light of this hypothesis.

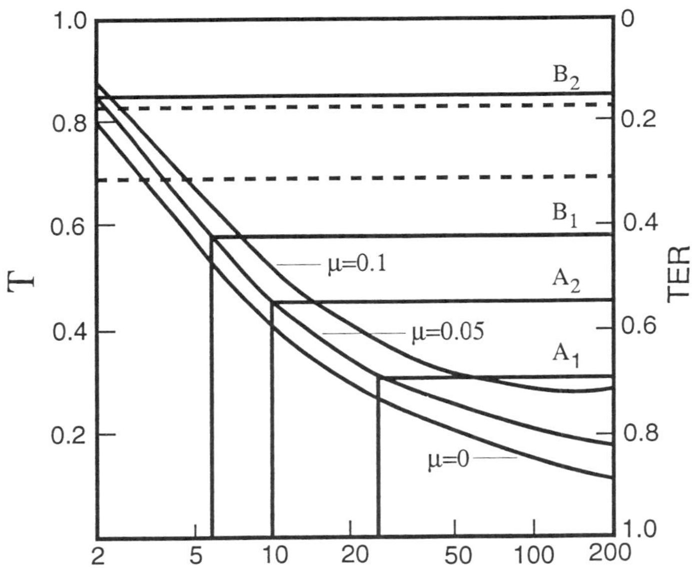

FIGURE 13. The dependence of the tidal exchange ratio *TER* on a dimensionless parameter $\tau_f = u_o t_f / b_o$ where u_o is the outflow velocity at the mouth of the bay, b_o is the width of the mouth and t_f is the period of the tidal flow. The three results shown in this figure correspond to different bottom friction values μ – they are from Ozsoy (1977), based on a geometrical calculation of the effect of tidal pumping (jet outflow and potential-sink inflow, with no stratification and no cross-flows). Guidelines labelled A1 and B1 represent *TER* values estimated for an outflow jet starting from Golden Gate and lines A2 and B2 represent *TER* values estimated for an outflow jet starting from the westernmost end of the mouth of the bay. The two values are representative of spring (A1, A2) and neap (B1, B2) tide flow velocities. The dashed lines represent the *TER* values obtained by Parker *et al.* (1972) during September 1970, with tidal ranges of 1.01 and 1.74 m, larger range yielding larger *TER*.

As tides are predictable, periodic motions which can be related to Greenwich phase, one can compare the tidal components of data collected at different times. All of the current velocity data collected at locations shown in Fig. 1, and briefly described in Section 3, have been subjected to tidal harmonic analysis. Tidal currents in San Francisco Bay, the Golden Gate, and the Gulf of the Farallones are dominated by the M_2 component. In Fig. 6, the results of these tidal analyses are plotted for the M_2 tide. The orientation and amplitude of the major axis is shown for near-surface velocities. M_2 currents are strongest in mid-channel within the mouth. Immediately offshore of the mouth, there is evidence of jet-related enhanced amplitudes and a southwestward orientation at moorings E, D, C202, and even at F and B. The minor axis is an order of magnitude smaller than the major axis and it is only plotted where it is large enough to be resolved in this figure. The phases of tidal currents, plotted as numbers in Fig. 6, are not precisely determined by these short records, owing to the contaminating presence of nontidal currents. Nevertheless, it is clear that the tide in the region of the offshore moorings leads that in the Golden Gate by about an hour (*i.e.*, about 29° phase difference). At the moorings farther offshore (Fig. 1), the oceanic tidal currents appear to lead by a larger amount (up to 2 hours). This means that a southward tidal cross-flow occurs during outflow from the Bay, consistent with the southwestward orientation of major axes near to the mouth of the Bay. During inflow to the Bay, a northward ambient cross-flow is found. The southward cross-flow starts about an hour before the Bay outflow and it turns northward about an hour before the inflow starts. The tidal currents at all moorings immediately offshore of the mouth exhibit a clockwise rotation, consistent with outflow persisting about an hour longer than the southward cross-flow. In the absence of other ambient currents the M_2 tidal outflow from the Bay is advected southwestward from the mouth, under the combined action of jet and ambient tidal currents, and then offshore before it is advected northward during the flood phase. Although one may expect that a significant fraction of the Bay outflow water will be beyond the radius of the subsequent potential-sink inflow to the Bay, the low *TER* values obtained by Parker *et al.* (1972), in spite of the tidal ambient flows which are always present, suggest that these tidal cross-flows are not very effective in enhancing bay-ocean exchange. This is not entirely surprising if one considers that the amplitudes of the major axes offshore of the mouth are less than 0.5 m s^{-1} and, thus, the transient southward tidal displacement is less than 7 km, which is comparable with the radius of the potential sink.

In contrast to tidal cross-flows, which have the same frequency as the Bay inflow/outflow, alongshore wind-driven currents typically have longer time scales (Largier *et al.* 1993). Alongshore-polarized southward winds force southward currents, which are also polarized alongshore. The surface wind stress drives offshore Ekman transport and the set-up of a cross-shore sea level slope that is in geostrophic balance with the alongshore current. In a similar way, northward currents occur during northward wind, although they also are observed during calm periods following southward wind (Wing *et al.* 1995). This wind-current correlation is obvious in Figure 14, as it is in the data of Largier *et al.* (1993). The upwelling season, by definition, is characterized by these southward upwelling winds and southward near-surface, wind-driven currents (Fig. 14). Although no stronger than tidal flows near the mouth of the bay, if wind-driven flow is southward during the tidal outflow, it probably remains southward during the inflow as wind forcing exhibits maximum energy in a 3- to 6-day band. Thus, ebbing Bay water that is advected south will not be returned northward during the flood tide phase. However, typical wind-driven velocities in the Gulf of the Farallones are weak (Figs. 2 and 14), less than 0.2 ms^{-1}, so that much of the outflow is not removed beyond the radius of the flood-tide potential sink. Water exiting the Bay on the early ebb will be transported 4.5 km by a 0.2 ms^{-1} cross-flow, but water that exits late on the ebb tide will not be removed at all.

The wind-driven and tidal cross-flows occur simultaneously and it is the combination of the southward tidal flow during outflows and southward wind-driven currents during the spring and early summer upwelling season that is

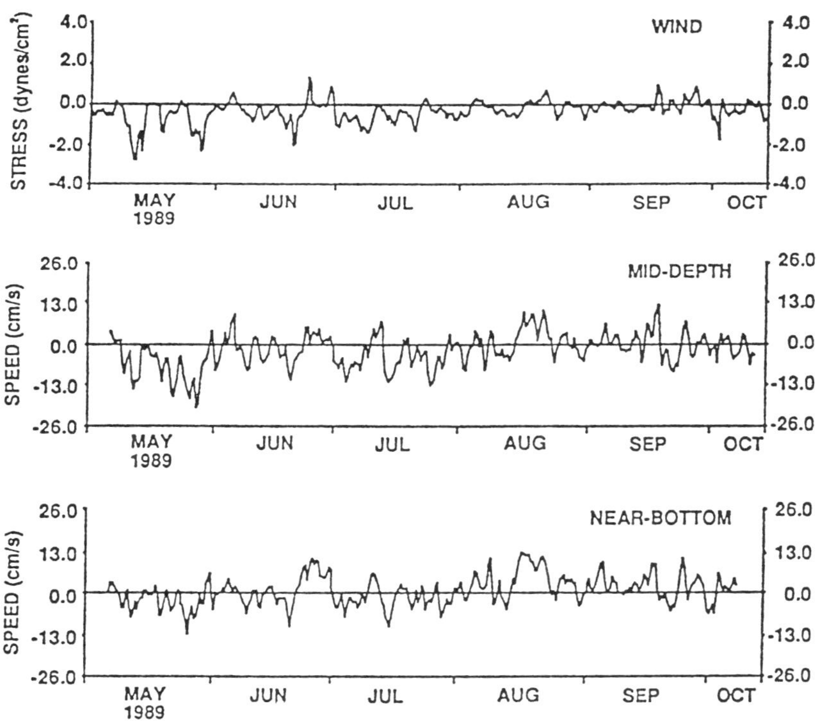

FIGURE 14. Subtidal alongshore wind stress and alongshore current speeds at mooring 315, Fig. 1 (from Noble & Gelfenbaum 1990). Positive flow is toward the northwest. Mid-depth and near-bottom currents correlate with the local wind stress, explaining 78% and 80% of the subtidal current variance, respectively. Mid-depth currents lag wind by 6 hours while near-bottom currents lag wind by 12 hours. Mid-depth and near-bottom currents respond with 6.2 and 5.6 cm s^{-1} speed for every 1 dyne cm^{-2} of wind forcing. With zero wind forcing, currents of -0.2 and 2.6 cm s^{-1}, respectively, are predicted. All results are from Noble & Gelfenbaum (1990), used here as an illustration of wind-driven subtidal flow in the Gulf of the Farallones at a location well-removed from the Bay outflow.

effective in increasing the *TER*. The southward tidal flow enhances the weak wind-driven flow during the ebb phase, thus advecting a far greater proportion of the outflow beyond the radius of the flood-tide potential sink. Further, if the wind-driven flow is greater or equal to the amplitude of the tidal flow, then the ambient current does not reverse during Bay inflow. As one expects near-surface, wind-driven flow to strengthen with distance offshore (Largier *et al.* 1993), the offshore component of tidal flow is also important as it transports Bay outflow into stronger cross-flow. As the old Bay water moves southward away from the mouth region, it is transported farther offshore by the wind-driven Ekman transport – cold, newly upwelled water is often observed inshore (*e.g.*, Fig. 2a).

The subtidal currents during fall in the Gulf of the Farallones are not well-known, but through reference to Fig. 14 and Largier *et al.* (1993), one would expect a weak northward mean, with occasional southward flow events. During northward subtidal flow, one would expect poor removal of Bay outflow and a decreased *TER* (*e.g.*, Parker *et al.* 1972). Water which escapes the Bay will be removed northward. Local northward flow could also occur at the mouth of San Francisco Bay as a result of a large-scale recirculation of the southward flow past Point Reyes. Associated with the upwelling center at Point Reyes (see Fig. 8) and/or the mesoscale filament offshore of Point Reyes (Strub *et al.* 1991), this cyclonic recirculation has been suggested previously by Gezgin (1990) and Wing *et al.*, (1995). Under these conditions, the tidal outflow would be advected northward and entrained into this recirculation eddy during upwelling conditions. Northward cross- flows may also be a result of buoyancy forcing over the shelf (Wing *et al.* 1995) or large-scale pressure gradients (Hickey & Pola 1983; Largier *et al.* 1993). Remote wind forcing of shelf waves can produce both southward and northward cross-flows.

5. DENSITY STRATIFICATION

The difference in the density of Bay and ocean water results in vertical stratification at the mouth of the Bay, where the two water types exchange. The stronger the vertical difference in density, the stronger the separation of dense inflow from less dense outflow and the greater the net exchange over a tidal cycle. As seen in Section 3 (Figs. 5 and 11), vertical shear in tidal mean currents through the Golden Gate persists throughout the year. However, it is only when the freshwater inflow to the Bay is large (Delta outflow greater than about 500 m^3s^{-1}) that the vertical shear is significant. Large freshwater inflow typically occurs during the winter (or storm season) months of December to March (Fig. 2). Of concern in this paper is what happens during drier periods. How important is density stratification in bay-ocean exchange during spring-summer-fall, and what form does it take? Are density differences primarily due to salinity or temperature differences, and how do these vary with time?

During spring, the salinity of San Francisco Bay water increases as the fresh-water inflow decreases (Fig. 2). While the salinity difference decreases, however, the temperature difference increases. Bay water warms rapidly during spring while coastal waters remain cold as a result of wind-driven upwelling (Fig. 2; Largier *et al.* 1993; Largier *et al.* 1996). It is only in late summer, when upwelling weakens and coastal waters warm, that bay-ocean densities may equalize. From historical observations, it appears that the shallower, semi-enclosed Bay remains warmer than the coastal waters even through fall (Fig. 2). This explains the persistence of weak vertical shear in the mouth of the bay throughout the year.

On 25 April 1988, a few days after a 600 m^3s^{-1} Delta outflow (Fig. 3), vertical profiles of salinity and temperature were collected along a line southwestward from the mouth of the Bay (Fig. 15). Although the data are sparse, the shape of the contoured isohalines suggests that hydraulic control may occur over the ebb tide bar during outflow (Largier, 1992). The lift- off of 33 ppt water over 34 ppt ambient water (density difference of 0.9 kg m^{-3}) and tightening of vertical stratification over the bar, is consistent with a supercritical Froude number associated with outflow velocities of the order of $0.4ms^{-1}$, as observed over the bar at mooring D. This lift-off of the low density bay outflow could explain the mean inflow at depth on the B mooring outside the bar (Fig. 10): as the outflow lifts over the denser ocean water, it exerts an offshore interfacial stress and entrains part of the ocean water so that seaward of the bar an onshore compensation flow is required at depth. From observations of tidal differences in density and maximum tidal velocity at mooring

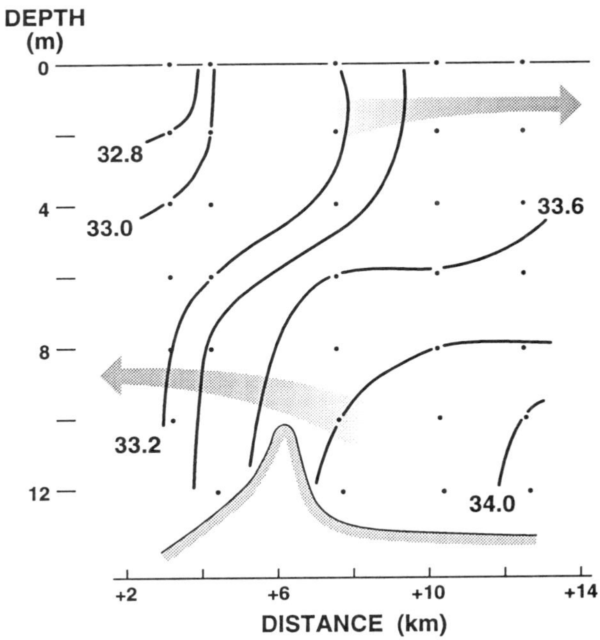

FIGURE 15. Contour plot of salinity in a vertical section along the 12-m isobath south from the mouth of San Francisco Bay. The left of the plot is near mooring E and the right of the plot is near mooring C (Fig. 1). Distance is from an arbitrary origin at the mouth of the bay. The bottom is sketched roughly, indicating the position of the ebb-tide bar. The broad arrows indicate the direction of density-driven flow.

D, it appears that supercritical Froude numbers are still found in early May after Delta outflow and Bay-ocean salinity differences have decreased. The persistence of this hydraulic control appears to be due to the large temperature difference resulting from coastal upwelling. Further evidence of the control of stratified exchange occurring at the bar is provided by the data in Fig. 16. At mooring F, outside the bar, tidal stratification is observed. Between the 8-m and 14-m instruments, a 2°C temperature and 1 ppt salinity difference is observed during large outflows. The stratification is due to an overflow of warm, low-salinity water. In contrast, at mooring E, within the bar, tidal stratification is also observed, but here maximum stratification occurs during the inflow of a cold and salty basal layer. In both cases, the onset of stratification can be sudden (e.g., 4 and 5 May, Fig. 16), as one would expect due to the passage of the front of a density-driven plume (Simpson 1982). These results suggest that during spring, the Bay outflow typically lifts off and forms a stratified flow at the bar. Conversely, the inflow of cold upwelled water forms a stratified basal flow. This vertical separation of tidal outflow and inflow significantly enhances the *TER*. Satellite images which show warm outflow water separated from the mouth of the Bay by a crescentic band of cold water (*e.g.*, Fig. 8b), further support this hypothesis.

During parts of May 1988, the density differences near the mouth of San Francisco Bay are largely due to upwelling-related temperature-salinity patterns. Upwelling in the Gulf of the Farallones is driven by a combination of the divergence in surface Ekman transport due to the coastal boundary and the curl in the wind field as it curves southward past Point Reyes and then eastward through the Golden Gate. While Bay water is warm and salty, these southerly winds upwell cold, saltier water. Once upwelled above the crest of the bar (10 to 15 m below surface), and if within the stratified tidal sink radius of the bar, this cold and salty water will be selectively drawn over the bar during tidal inflow. The combined action of the barotropic pressure gradient owing to the inward sloping sea level and the baroclinic pressure gradient owing to the inward sloping isopycnals exerts strongest inward forcing on the deeper layers of the water column. Once over the bar, the dense water can intrude into the bay as a density-driven "drainage" flow, even during the ebb tide (as seen at mooring E, Fig. 16).

The interaction of these wind-forced barotropic and baroclinic pressure gradients with flood and ebb tidal flows is well illustrated by observations during an upwelling event at moorings B, C, D, E, and F (Fig. 1). On 29 April 1988, the local wind strengthened to 10 ms^{-1} (Fig. 17a) and upwelling occurred offshore of San Francisco at moorings C and B. A 0.2 m drop in sea level was observed. Over a 2-day period, the near- bottom (14-m) temperature at B decreased from 11.7°C to 10.5°C and then, during a subsequent wind pulse on 3 and 4 May, the temperature dropped further to 10°C (Fig. 17b). A concurrent increase in salinity was observed (Fig. 17c). While this drop in near-bottom temperature indicates the onset of upwelling (Lentz & Chapman 1989), it is the temperature at depths above the crest of the bar that are relevant to Bay inflow (*e.g.*, that at 8 m on the B and C moorings). After 6 days of wind (5 May), the maximum water densities at 8 m were 1.0 kg m^{-3} greater than that prior to the onset of wind (Fig. 17d). This corresponds to a 0.7 ppt

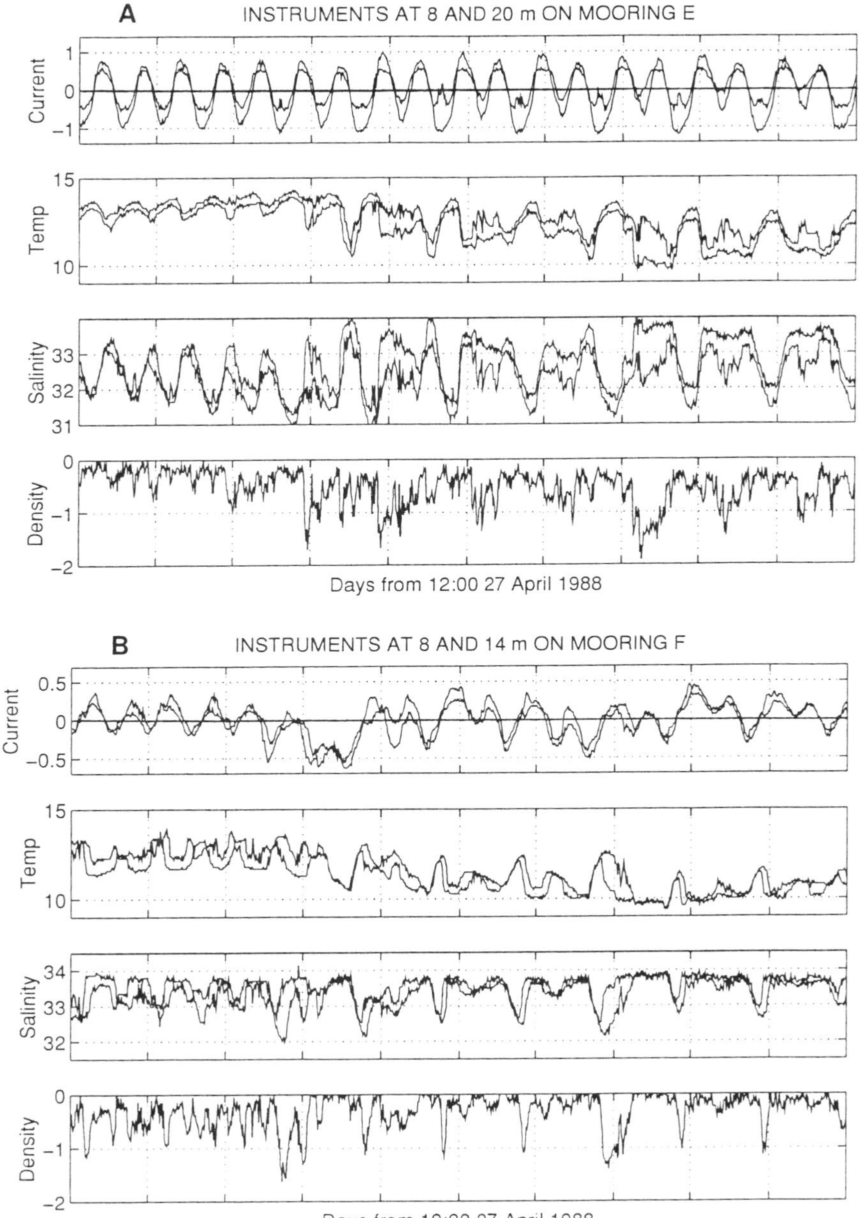

FIGURE 16. (a) Along-channel velocity (ms^{-1}), temperature (°C), salinity (ppt), and vertical density difference (kg m^{-3}) for 8-m and 20-m instruments at mooring E on the 24-m isobath, inside of the crest of the ebb-tide bar. (b) Along-channel velocity (ms^{-1}), temperature (°C), salinity (ppt), and vertical density difference (kg m^{-3}) for 8-m and 14-m instruments at mooring F on the 18-m isobath, outside of the crest of the ebb-tide bar.

increase in salinity and a 3°C drop in temperature. Close to the mouth of San Francisco Bay, the upwelling period is also characterized by salinity minima and temperature maxima as Bay water moves past these moorings on the late ebb tide (Fig. 17bc). In the absence of upwelling, the Bay outflow jets offshore and it does not directly impact mooring C (Figs. 7 and 17bc). On ebb tides during upwelling, however, it appears that the Bay outflow is drawn in along the coast to fill in the wind-driven coastal divergence. Conversely, on the flood tide during upwelling, tidal divergence may enhance the upwelling of cold water. The small ebb tide on 30 April, as the upwelling wind stress increased, resulted in very low salinities at C whereas the large ebb tide on 1 May, as the upwelling wind stress weakens, resulted in no drop in salinity (Fig. 17c). Corresponding to this increase and decrease in upwelling and associated change in sealevel, the ebb-tide velocities on 30 April are enhanced and the flood tide velocities are enhanced on 1 May (Fig. 17e). Satellite imagery displays warm Bay water nearshore off the San Francisco Peninsula and coastal upwelling of cold water south of Pillar Point (Fig. 18). Also shown is the southward removal of Bay outflow, typical of circulation during upwelling periods (see previous section).

While the wind forced component dominates the record at moorings B and C, at moorings E, D, and F the wind acts to modulate the dominant tidal forcing. With the onset of wind, ebb-tide salinity and density minima are observed at all instruments on E, within the bar, and at the 8-m instruments at D and F, outside the bar (Fig. 17cd). As discussed above, significant stratification is observed at F during these large outflow events (*e.g.*, 0.3 kg m^{-3} early on 30 April). Equally important are the increase in flood-tide salinity and density maxima and the decrease in near-bottom temperature minima at mooring E during the upwelling on 30 April and 1 May. Near-bottom, flood-tide density increases 1 kg m^{-3} through the 30 April wind event and another 0.5 kg m^{-3} through the 4 May wind event (Fig. 17d). Not only is this evidence of increased density forcing, but it is a direct observation of undiluted oceanic water intruding into the bay. While cold, salty upwelled water appears at B and C with the onset of upwelling, it is during the relaxation of upwelling that the most significant intrusions of upwelled water are observed at mooring E. As the wind relaxes, the coastal sea level rises and a large volume of water must enter the Bay – the inflow of dense upwelled water is enhanced by baroclinic pressure gradients. This baroclinic forcing is due to a density difference of as much as 3 kg m^{-3}, observed as the difference between inflow and outflow water at mooring E. The clear semi-diurnal intrusions at mooring E prior to the onset of upwelling (*e.g.*, 28 April) is replaced by aperiodic fluctuations during changes in wind forcing (*e.g.*, 3 May) and eventually diurnal fluctuations during the asymmetrical tides following upwelling (*e.g.*, 5-7 May). The weak currents between lower high water and higher high water are too weak to overcome the vertical stratification at E (Fig. 17bcd).

Wind-driven coastal upwelling impacts bay-ocean exchange through both a barotropic and a baroclinic term: set-down of the sealevel and upwelling of isopycnals during increased wind forcing and relaxation of these during weakening of the wind. While the mass exchange due to fluctuations in coastal sea level are

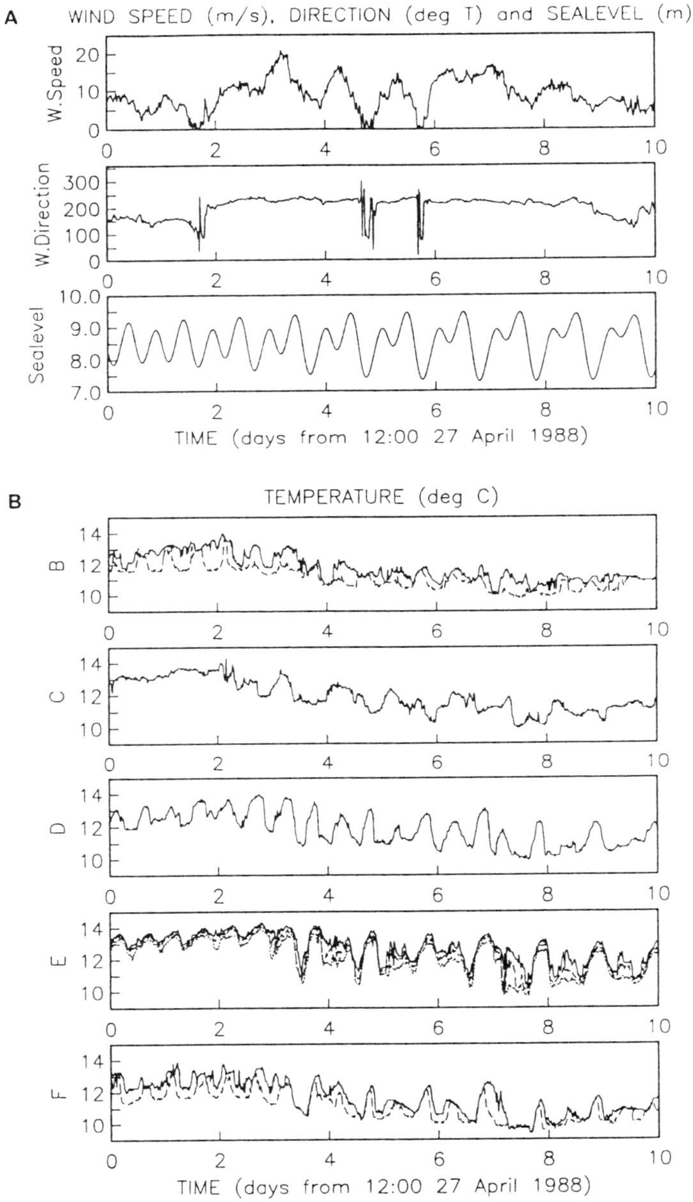

FIGURE 17. (a) Wind speed and direction at the coast south of the mouth to San Francisco Bay and water level recorded on mooring C (Fig. 1). Data presented for 10 days starting midday 27 April and ending midday 7 May 1988, for a period in which two strong wind forcing events were recorded and following a period of enhanced freshwater inflow to San Francisco Bay (see Fig. 4). (b) Temperature at moorings B, C, D, E, and F over the same period (solid lines: 8-m data; dashed lines: 14-m data; dotted lines: 20-m data).

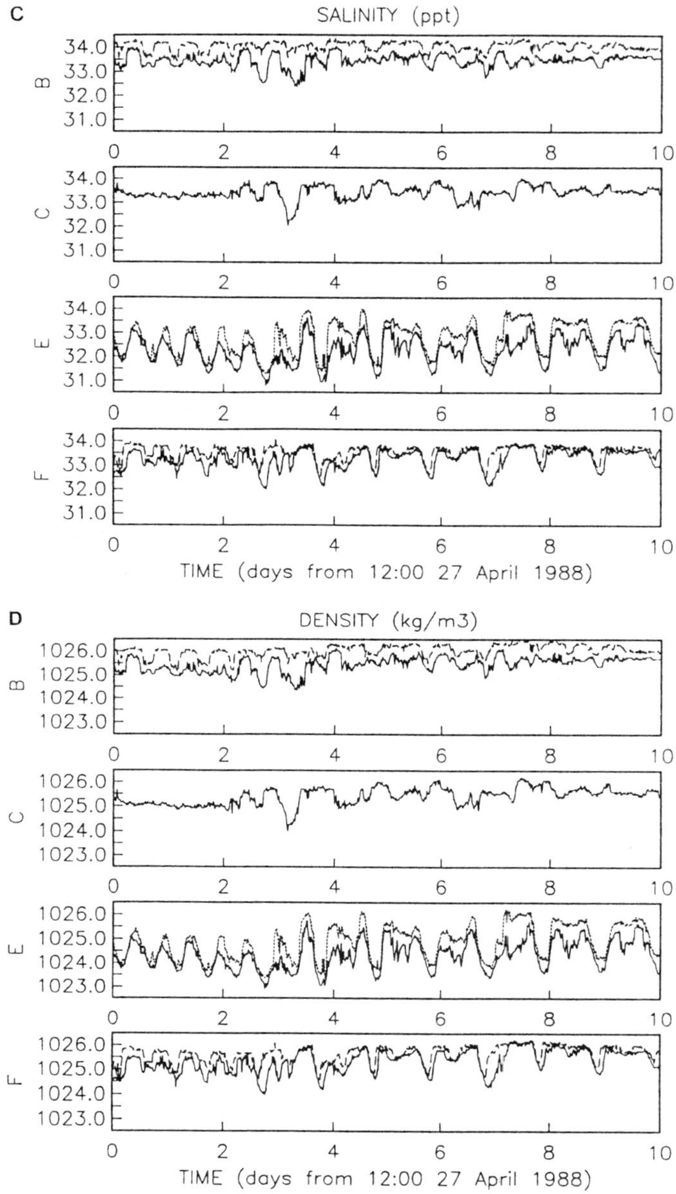

FIGURE 17 (continued). (c) Salinity data over the same period. (d) Density data over the same period.

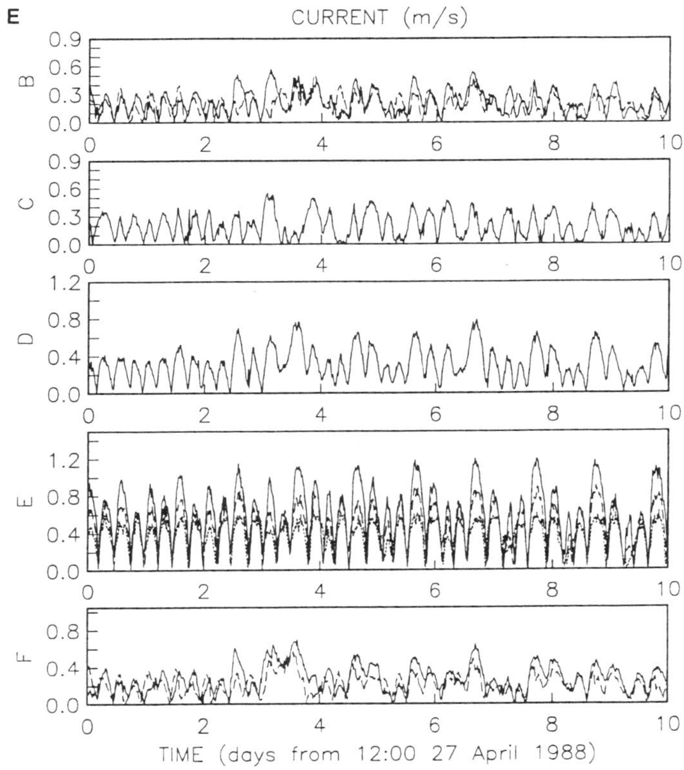

FIGURE 17 (continued). (e) Current speed over the same period.

small and well appreciated, the role of upwelling in enhancing the *TER* through increased density-driven exchange is poorly appreciated. Density-driven flushing of San Francisco Bay during the spring upwelling season is as much due to increased density in the ocean as it is due to decreased density in the bay. Cannon (1990) has suggested a similar importance of upwelling to the renewal of bottom water to Puget Sound. In the case of tributary estuaries, the estuary-ambient exchange may also be dominated by density changes in the ambient waters. This density-driven exchange fluctuates with the tide, synoptic fluctuations in wind forcing (supply of upwelled water), and seasonal fluctuations in rainfall (supply of river water). As the supply of dense oceanic water is due to upwelling, but the intrusion of it into the bay appears to occur as it relaxes, this suggests that intermittent upwelling may be more effective than steady upwelling in enhancing bay-ocean exchange.

6. DIRECT WIND FORCING

In addition to transport of mass through the Golden Gate due to tidal forcing, transport occurs in response to the net freshwater inflow to the Bay, nontidal fluctuations in sea level at the mouth of the Bay, and direct wind stress forcing on the surface in the Gate. Under normal circumstances, this nontidal transport is much weaker than that due to the tide, but it may result in better net exchange as nontidal transport may be non-reversing (*e.g.*, freshwater throughflow, afternoon sea breezes), vertically sheared (*e.g.*, freshwater throughflow, surface wind stress) or low frequency (*e.g.*, wind-driven fluctuations in sea level). As the focus of this paper is on the drier seasons, periods of large freshwater throughflow will not receive further attention here. Subtidal sealevel fluctuations were addressed briefly in the previous sections. In this section, direct wind forcing is discussed.

The stresses exerted by the wind on the surface and by the drag of the bottom boundary are particularly important in shallow water flows. But, wind stress is particularly difficult to analyze as it exhibits a complex spatial structure that fluctuates temporally with changes in the large-scale wind field and the thickness and density of the marine boundary layer in the atmosphere.

While the small-scale wind patterns in the inner Gulf of the Farallones are not well described, one expects the wind to curve eastward from the general southward winds further offshore. The cold marine air moves eastward through the gap in the coastal relief that is the Golden Gate; it does this as a density-driven flow known as a "gap wind" (Overland & Walter 1981). Wind data at San Francisco International and at Oakland airports (Conomos 1979) exhibit the strong diurnal cycle that occurs in spring and summer. Maximum eastward flow occurs in the afternoon when ocean-land temperature differences are maximum. As the density-driven flow through the Golden Gate constriction is expected to be supercritical (Smith 1979; Armi 1986), one expects significantly larger velocities there than at the airports where the diurnal range is about 6 ms^{-1} (Conomos 1979). The wind velocities through the Golden Gate should also exceed those observed west of the city of San Francisco (Fig. 1) where the diurnal fluctuation in wind strength has a range of 5 to 10 ms^{-1} (Fig. 17a). This strong eastward wind stress can be expected to retard the outflow and to enhance the inflow of near-surface water, contrary to the effect of density forcing. In the absence of multiple current meters near the surface in the Golden Gate one can only make estimates of wind-driven velocities as the frequency of the K_1 tide is so close to that of the diurnal wind forcing. In fact, much of the K_1 tidal energy in the mouth of the bay is probably due to wind forcing (as Monterey Bay HF Radar Consortium, 1994, have found for Monterey Bay). Very roughly, if one expects a surface velocity that is of the order of 3% of the wind speed, then with a 10 ms^{-1} wind one expects a 0.3 ms^{-1} current. It is unknown how this current extends vertically and, thus, the associated transport that must be balanced either by outflow at depth or by a rise in water level within the bay. While data are not available to make a calculation, it is expected that this diurnal effect is not a significant term in the Bay-ocean exchange. Nevertheless, in any future study of the importance of diurnal winds through the Golden Gate, attention should be given to diurnal-tidal interactions and to the effect of surface stress on any hydraulic control that may occur at the offshore bar (see section 5).

Bottom stresses may also be important to Bay-ocean exchange as they also exhibit complex spatial patterns and temporal fluctuations. Temporal fluctuations

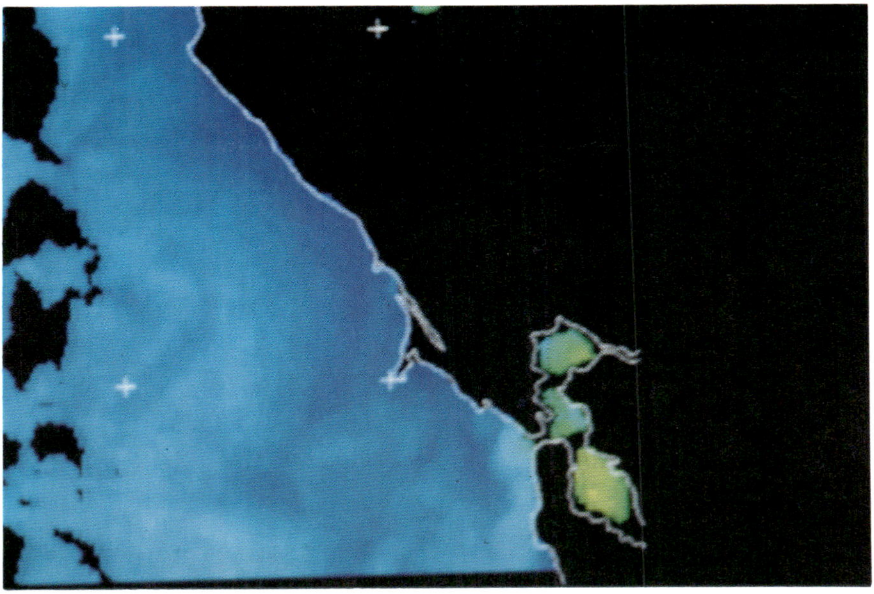

FIGURE 18. AVHRR image of sea surface temperature on 1 May 1988 (compare with Figures 8 and 17).

in bottom stress are correlated with the near-bottom tidal velocities. Further, fluctuations in vertical stratification modulate this bottom stress as well as its effect on total transport in the water column. This may result in partial rectification of tidal currents. Bottom stress is also an important contributor to the vertical mixing that reduces stratification and vertical exchange flows.

7. CONCLUSION

The exchange of water between San Francisco Bay and the coastal ocean is dominated by tidal transport. However, it is not immediately clear how much of the outflow is returned on the subsequent inflow. Expressed via the tidal exchange ratio, *TER*, this issue has been discussed extensively here. The bay is "well flushed" if most of the outflowing bay water is removed and then replaced with new ocean water. This is achieved most effectively when there is a strong freshwater inflow into San Francisco Bay so that significant vertical exchange circulation is maintained at the mouth of the bay. This is typical of the winter months of December to March. In spring, summer, and fall, the freshwater inflow is very small and the *TER* is expected to decrease. This paper has addressed the hydrodynamics which may enhance the *TER* during the drier "upwelling" and "relaxation" seasons. Attention has been given to the non-seasonal three-dimensional structure of tidal flow through Golden Gate (section 3) and the phase- locked tidal cross-flow in the ambient ocean waters (section 4) – these mechanisms are responsible for setting a base level for the *TER* in the absence of stratification and wind-driven circulation.

Seasonally varying mechanisms that received attention are the wind-driven cross-flow (section 4), the sea level and isopycnal fluctuations associated with wind-driven coastal upwelling (sections 5 and 6) and the stratification observed in the vicinity of the mouth (section 5). These mechanisms act to enhance the value of the *TER* above its base value.

During the upwelling season (typically April to July), the combination of wind-driven cross-flow and wind-driven upwelling of dense water at the mouth of the Bay appear to enhance *TER* values. To explore this further, one needs long-term data from the Central Bay which can be correlated with long-term records of wind forcing. It is hypothesized that flushing of the Bay increases during periods of upwelling.

During the relaxation season that follows the upwelling season, wind-driven cross-flow is weak or northward (counter to ebb tide cross-flow) and the Bay-ocean density difference is a minimum. It is in this season that the *TER* is likely to be a minimum and flushing of the bay at its weakest. Excessively long residence times may be observed at the back of the bay, particularly South Bay (*cf.*, those observed in Tomales Bay under similar circumstances, Largier *et al.* 1996).

In conclusion, then, freshwater inflow during winter and wind- driven upwelling during spring and early summer are expected to maintain effective tidal exchange between San Francisco Bay and the ocean. In late summer and fall, however, the effectiveness of tidal exchange is reduced and the residence of water in the bay is expected to attain a maximum.

ACKNOWLEDGMENTS

This study was made possible by the support of the Department of Water Resources through Interagency Agreement Numbers B-58371 and B-59296. I thank Ralph Cheng, Jeff Gartner, Larry Smith and Marlene Noble (USGS), and Steve Costa (CH_2MHill) for making their data available for use in this work.

REFERENCES

Armi, L. (1986). The hydraulics of two flowing layers with different densities. *Jour. Fluid Mech.* 163:27-58.

Cayan, D. R., D. R. McLain, W. D. Nichols, and J. S. DiLeo-Stevens. (1991). *Monthly Climatic Time Series Data for the Pacific Ocean and Western Americas.* U.S. Geol. Surv. Open- File Rept. 91-92 (USGS, Denver, Co.). 380 pp.

CH_2MHill. (1989). *Wastefield Transport and Bacteriological Compliance Studies of the San Francisco Ocean Outfall.* City and County of San Francisco, CA.

Chadwick, D. B., J. L. Largier, and R. T. Cheng. (in press). The role of thermal stratification in tidal exchange at the mouth of San Diego Bay. *In* D. Aubrey, ed., *Proceedings of the 7th Conference on the Physics of Estuaries and Coastal Seas.* American Geophys. Union.

Cheng, R. T., and J. W. Gartner. (1984). *Tides, Tidal and Residual Currents in San Francisco Bay, California – Results of Measurements, 1979-1980, Part IV. Results of Measurements in Central Bay Region.* U.S. Geol. Surv. Water Resources Inv. Rept. 84-4339, 757 pp.

Cheng, R. T., V. Casulli, and J. W. Gartner. (in press). Tidal, residual, intertidal mudflat

(TRIM) model and its applications to San Francisco Bay, California. *Estuarine Coastal Shelf Sci.*

Conomos, T. J. (1979). Properties and circulation of San Francisco Bay waters. Pages 47-84 *in* T. J. Conomos, ed., *San Francisco Bay: The Urbanized Estuary*, Pacific Division, American Association for the Advancement of Science, San Francisco, CA.

Conomos, T. J., and D. H. Peterson. (1977). Suspended-particle transport and circulation in San Francisco Bay: An overview. Pages 82-97 *in* Dave Peterson, ed., *Estuarine Processes, Vol. II*. Academic Press, New York.

Conomos, T. J., R. E. Smith, D. H. Peterson, S. W. Hager, and L. E. Schemel. (1979). Processes affecting seasonal distributions of water properties in the San Francisco Bay estuarine system. Pages 115-142 *in* T. J. Conomos, ed., *San Francisco Bay: The Urbanized Estuary*. Pacific Division, American Association for the Advancement of Science, San Francisco, CA.

Dorman, C. E., and C. D. Winant. (1995.) Buoy observations the atmosphere along the West Coast of the U.S., 1981-1990. *Jour. Geophys. Res.* 100(C8):16029-16044.

Fischer, H. B., E. J. List, R. C. Y. Koh, J. Imberger, and N. H. Brooks. (1979). *Mixing in Inland and Coastal Waters*. Academic Press, San Diego, CA. 483 pp.

Gezgin, E. (1990). *A study on hydrographic conditions and salt budget calculations for the Gulf of the Farallones with the data collected in August 1990*. M.S. Thesis. Naval Postgraduate School, Monterey, CA.

Hickey, B. M., and N. E. Pola. (1983). The seasonal alongshore pressure gradient on the west coast of the United States. *Jour. Geophys. Res.* 88(C12):7623-7633.1

Jay, D. A. (1991). Estuarine salt conservation: A Lagrangian approach. *Estuarine Coastal Shelf Sci.* 32:547-565.

Jay, D. A. and J. D. Smith. (1990). Circulation, density distribution and neap-spring transitions in the Columbia River Estuary. *Prog. Oceanogr.* 25:81-112.

Jay, D. A., R. J. Uncles, W. R. Geyer, J. L. Largier, J. Vallino, and W. R. Boynton, (1996). A review of recent developments in estuarine scalar flux estimation. *Estuaries* (in press).

Largier, J. L. (1992). Tidal intrusion fronts, *Estuaries* 15(1):26-39.

Largier, J. L., J. T. Hollibaugh, and S. V. Smith. (1996). Seasonally hypersaline estuaries in Mediterranean climate regions. *Estuarine Coastal Shelf Sci.* (accepted)

Largier, J. L., B. A. Magnell, and C. D. Winant. (1993). Subtidal circulation over the northern California shelf, *Jour. Geophys. Res.* 98(C10):18,147-18,179.

Lentz, S. J., and D. C. Chapman. (1989). Seasonal differences in the current and temperature variability over the northern California shelf during the Coastal Ocean Dynamics Experiment, *Jour. Geophys. Res.* 94(C9):12,571-12,592.

Monteiro, P. M. S., and G. B. Brundrit. (1996.) Exchange between Saldanha Bay and the coastal ocean in the southern Benguela Upwelling System. *Estuarine Coastal Shelf Sci.* (submitted)

Monterey Bay HF Radar Consortium. (1994). *Science and Engineering Plan 1994-95*.

Munk, W., F. Snodgrass, and M. Wimbush. (1970). Tides offshore: Transition from California coastal to deep-sea waters. *Geophys. Fluid Dyn.* 1:161-235.

Noble, M., and G. Gelfenbaum. (1990). *A Pilot Study of Currents and Suspended Sediment in the Gulf of the Farallones*. U.S. Geol. Surv. Open-File Rept. 90-476. 90 pp.

Ota, A. Y., L. E. Schemel, and S. W. Hager. (1989). *Physical and Chemical Properties of San Francisco Bay, California, 1980*. U.S. Geol. Surv. Open-File Rept. 89-41. 251 pp.

Overland, J. E., and B. A. Walter (1981). Gap winds in the Strait of Juan de Fuca. *Mon. Wea. Rev.* 109:2221-2233.

Ozsoy, E. (1977). *Flow and Mass Transport in the Vicinity of Tidal Inlets, UFL/COEL/TR-636*. Coastal & Oceanographic Eng. Dept., University of Florida, Gainesville, FL. 200 pp.

Parker, D. S., D. P. Morris, and A. W. Nelson. (1972). Tidal exchange at Golden Gate. *Proc. American Soc. Civil Eng., Jour. Sanitary Eng. Div.* 98:305-323.

Petzrick, E. P. (1993). *Currents through the Golden Gate.* M.S. Thesis. Naval Postgraduate School, Monterey, CA.
Rosenfeld, L. K., and R. C. Beardsley. (1987). Barotropic semidiurnal tidal currents off northern California during the Coastal Ocean Dynamics Experiment (CODE). *Jour. Geophys. Res.* 92(C2):1721-1732.
Simpson, J. E. (1982). Gravity currents in the laboratory, atmosphere, and ocean. *Ann. Rev. Fluid Mech.* 14:213-234.
Smith, R. B. (1979). The influence of mountains on the atmosphere. *Adv. Geophys.* 21:187-230.
Strub, P. T., P. M. Kosro, and A. Huyer. (1991). The nature of the cold filaments in the California current system. *Jour. Geophys. Res.* 96(C8):14,743-14,768.
Walters, R. A. (1982). Low-frequency variations in sea level and currents in south San Francisco Bay. *Jour. Phys. Oceanogr.* 12(7):658-668.
Walters, R. A., and J. W. Gartner. (1985). Subtidal sea level and current variations in the northern reach of San Francisco Bay. *Estuarine Coastal Shelf Sci.* 21:17-32.
Walters, R. A., R. T. Cheng, and T. J. Conomos. (1985). Time scales of circulation and mixing processes of San Francisco Bay waters. *Hydrobiologia* 129:13-36.
Wilkinson, D. L. (1978). Periodic flows from tidal inlets. *Proc. Coastal Eng. Conf., ASCE* 16(2):1336-1346.
Wing, S. R., J. L. Largier, L. W. Botsford, and J. F. Quinn. (1995). Settlement and transport of benthic invertebrates in an intermittent upwelling region. *Limnol. Oceanogr.* 40:309-321.
Wing, S. R., L. W. Botsford, J. L. Largier, and L. E. Morgan. (1996). Spatial structure of relaxation events and crab settlement in the northern California upwelling system. *Marine Ecol. Progr. Ser.* 128:199-211.

CURRENTS THROUGH THE GOLDEN GATE

ERNEST P. PETZRICK[1], CURTIS A. COLLINS, AND WILLIAM C. BOICOURT[2]
Department of Oceanography
Naval Postgraduate School
Monterey, CA 93943-5111

In late fall 1992, an upward-looking acoustic Doppler current profiler (ADCP) and a conductivity-temperature-pressure (CTD) recorder were moored in the Golden Gate. The 36-day record revealed a remarkably consistent subtidal shear associated with the estuarine circulation of San Francisco Bay. This shear, modulated only by neap-spring variations in the tide, was 0.02 s^{-1} during neaps, increasing slightly during springs. Wind-driven variability did not appear to significantly change the shear. Vertically integrated flow at the mooring site was 23 cm/s into the Bay, corresponding to a 26×10^3 m^3/s inward transport if this flow were constant across the Golden Gate section. Such a transport would flood San Francisco, indicating that there was across-channel structure of the mean flow compensating for this inward transport above the mooring. This increase in influx on the southern side of the Golden Gate is in agreement with earlier observations. The strong tidal currents (M2 amplitude was 100 cm/s) moving through the marked topographic gradients of the Golden Gate produced vertical structure in the shallow-water overtides and compound tides with rapid acceleration of flow occurring shortly after the beginning of flood tide. In addition, hodographs show that these interactions resulted in a repeatable pattern of intratidal development in the across-channel velocity. Mean energy fluxes at the site were 3×10^4 W for kinetic energy advection and 8.5×10^6 W for the work done by pressure.

On November 9, 1992, we deployed a bottom mounted instrument package consisting of an acoustic Doppler current profiler (ADCP) and a conductivity-temperature-depth (CTD) instrument just to the east of the Golden Gate Bridge (see Fig. 1). Our ADCP collected a vertical profile of currents from a distance of 8 m above bottom up to a depth of 12 m below the ocean surface while the CTD provided a time series of salinity, temperature and pressure at a distance of 2 m above the bottom. The purpose of this paper is to report the results of these measurements and to provide an improved description of Golden Gate circulation.

The Golden Gate is the narrow, deep channel that connects San Francisco Bay to the Pacific Ocean. The Golden Gate and the sill that marks its western boundary control the exchange between the Pacific Ocean and San Francisco Bay waters. The Golden Gate is relatively narrow and deep, 1 km and 110 m, with an along-channel length of 5 km. It is considerably deeper than the sill depth, which is 7 to 10 meters. A navigational channel aligned with the natural channel is maintained at about 0.6 km width and 20 m depth through the sill. Immediately to the east of the Golden Gate, the tidal channel diverges into northerly and southerly branches. The 20 m isobath associated with these channels extends south to Yerba Buena Island and north to Carquinez Strait.

---

[1] Current address: NOCC/JTWC, PSC 489 Box 2, FPO AP 96536-0051.
[2] Afflliation: University of Maryland, Center for Environmental and Estuarine Studies, Horn Point Environmental Laboratory, P.O. Box 775, Cambridge, MD 21613.

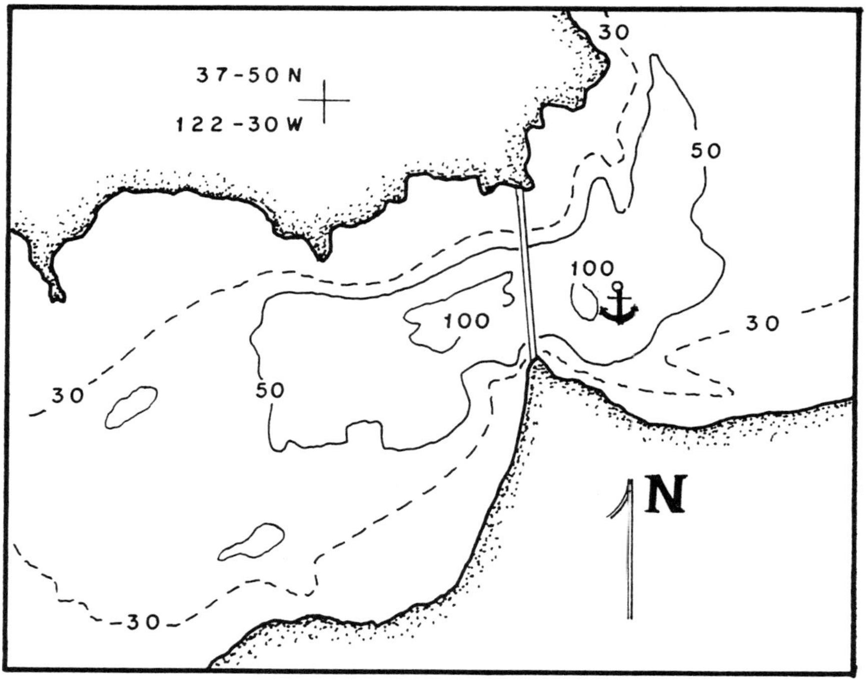

FIGURE 1. Measurement location (indicated by an anchor symbol) in the Golden Gate. Isobath depths are given in meters and are based upon a chart prepared by T. F. Chase, J. D. Young, C. H. Degman and E. J. Maple, USGS Menlo Park (1991). The double line connecting San Francisco and Marin represents the Golden Gate bridge.

Previous studies of flow in the Golden Gate (Conomos *et al.* 1985; Conomos 1979; Edmonston & Matthew 1931) have noted a large tidal exchange forced by mixed semidiurnal tides and winds, topographic steering, and a complex subtidal and residual flow. Strong winds were found to significantly influence the character of the current structure. Despite the narrow channel, the southern side of the Gate tends to have more influx than the north. During flood tides, a jet is formed. Largier (1996) points out the importance of across-channel flow in spreading the ebb flow across the entire western end of the Golden Gate. Some seiche-related forcing from the Bay is also felt in the Gate.

The primary motivation for these measurements was to obtain an improved description of the tidal stream and circulation through the Golden Gate. During summer 1992 the National Ocean Survey (NOS) surveyed currents in San Francisco Bay using acoustic Doppler current profilers (ADCP) in order to improve tidal current predictions for the Bay, but failed to obtain data for the Golden Gate (Williams *et al.* 1992). Our data provide a detailed view of the vertical structure of the flow and the relationship between pressure changes and the flow. They also reinforce the conclusions of others (Largier 1996; Conomos *et al.* 1985; Walters *et al.* 1985; Conomos 1979; Edmonston & Matthew 1931) regarding the complexity of the flow in the Golden Gate.

Observations

An RD Instruments self-contained ADCP (operating at 307 kHz, sampling at 1 Hz, using thirty 4-m bins, and averaging over 3 minutes) and a portable Seabird conductivity-temperature-depth (CTD) recorder (sampling rate 2 minutes) were deployed in a steel protective cage on the ocean bottom just inside the Golden Gate bridge from November 9 to December 15, 1992. The ADCP transducers were 2 m above the bottom and the CTD was 5 cm below the ADCP transducers. The ADCP was set to record in beam coordinates and with an extended speed range and these choices limited velocity resolution to 0.25 cm/s. This package was deployed just south of the main navigational channel in 100 m of water. On 11 and 12 November, the package was dragged by a fishing vessel southward and eastward into 91 m of water, remaining very near the original placement.

On November 9, during deployment, and on December 4, during a slack-to-flood period, single CTD profiles were obtained at the deployment site. The Golden Gate Bridge Authority provided wind data from an anemometer located 85 m above the water on the central span of the Golden Gate Bridge. Hourly water level data were obtained from NOS for San Francisco, Alameda, Point Reyes and Port Chicago. Naval Air Station Alameda provided meteorological data. Finally, USGS provided hourly surface temperature and salinity data for Fort Point, approximately 914 m to the southwest of our ADCP deployment site (located in San Francisco at the foot of the Golden Gate Bridge).

In presenting ADCP results in this paper, we have rotated currents clockwise 5.5 degrees. This was done to minimize the variance of the vertically-averaged across-channel flow and to define the direction of the along-channel flow, namely 84.5°T and is referred to as a "principal axis" rotation. An alternative coordinate system, used by Largier (1996), is defined by the major axis for the M2 tidal flow. For our vertically-averaged currents, the orientation of the major axis of the M2 tidal flow was 84.9°T, essentially the same as that yielded by the principal axis rotation based upon the variance of the vertically-averaged flow.

Results

Water Properties. Our two CTD casts provide limited information on the stratification during ADCP deployment. On November 9 a midlayer halocline and thermocline result in density differences of about 0.5 kg/m^3 between upper and lower layers. On December 4, the pycnocline is near surface, associated with fresher waters. These vertical profiles suggest that the estuary was well mixed.

The changes in the water properties between the two casts are large, corresponding to cooling of about 1°C and a freshening of 1 S. The moored CTD provided additional information about the temporal variability. Temperature decreased steadily during the first 27 days of deployment from 15°C on November 11 to 13°C on December 8. Temperature then warmed a few tenths of a degree during the next few days before resuming a cooling trend. Salinity remained relatively constant between 31 and 32 S while water cooled during the first 27 days of deployment, then decreased about 1 S during the remainder of the record. As a result of these changes, density increased 0.1 kg/m^3 per day during the first 27 days, with subsequent density changes due to cooling balanced by density changes due to freshening. Note that water property changes occurred in two distinct phases: cooling followed by freshening.

Figure 2 shows bottom temperature and pressure measurements during spring tides on November 23. Temperature decreases (increases) about an hour after low (high) time. Temperature changes are out of phase with the sea level so that cooling is associated with flood, *i.e.*, the ocean waters that are advected into the Bay are colder than those waters found in the Bay. During the period of 7-8 December, temperature did not change and the Bay and the ocean were nearly the same temperature. Subsequently the Bay became cooler so that warmer ocean waters were advected into the Bay on floods. Salinity (not shown) also lagged high or low water by an hour but changed in the same manner as the pressure so that higher salinity water was advected into the Bay on floods. Higher frequency temperature and salinity variability was greatest during flood, and during ebb increases were nearly monotonic.

A least squares analysis (Foreman 1977) was used to estimate the amplitude and phase of salinity and pressure variability at diurnal and semidiurnal frequencies. The amplitude of the M2 components were 0.34 S and 0.59 kPa with salinity lagging pressure by 58 minutes. The amplitude of the K1 components were 0.24 S and 0.47 kPa, with salinity lagging pressure by 80 minutes.

We had hoped that the surface temperature and salinity measurements at Fort Point, when combined with our bottom measurements, would provide a measure of vertical stability. Unfortunately, surface salinity measurements failed on No-

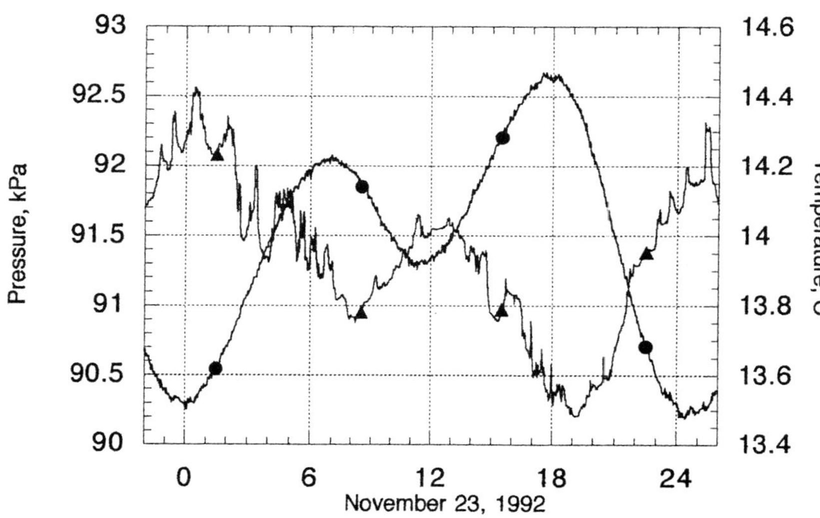

FIGURE 2. Temperature (triangle symbol) and pressure (circle symbol) at a distance of 2 m above the bottom. Measurements were recorded at two minute intervals.

vember 25 and did not start again until the day our instruments were retrieved. Surface (Fort Point) minus subsurface (bottom moored CTD) differences were about -0.5 S and 0.5°C. Salinity differences decreased to zero on November 21 and temperature differences became positive on December 9. Because of a lack of profile measurements, we are unsure as to how horizontal and vertical gradients contribute to these differences.

Pressure. Strong tidal forcing was considered the primary energy source for the Golden Gate. Bottom pressure (Fig. 2) resembles surface tide gauge data for San Francisco: a mixed semidiurnal tide, with neaps on November 19 and December 3 and springs on November 12 and 24 and December 12. As expected, amplitudes for semidiurnal and diurnal constituents were the same as those for the San Francisco Bay tide gauge. The variance of pressure, 35 kPa^2, is proportional to the mean potential energy, 1800 J/m^2.

Figure 3 shows the relationship between the amplitude of semidiurnal variability for pressure, salinity, density, temperature and the along-channel component of the vertically integrated velocity. The amplitude of the pressure variability in this frequency band varied between 0.4 and 0.8 kPa, and the pattern of neaps and springs in the pressure data was similar to the variability of salinity, density and along-channel flow. The amplitude of semidiurnal temperature variability behaved differently. The amplitude remained nearly constant, 0.25°C, during the first ten days. It then decreased steadily to zero on December 7. On December 8 the amplitude of the M2 temperature variability increased again, and had an amplitude of 0.1°C at the end of our measurement period.

Velocity. The character of the flow and the nature of tidal variability is illustrated

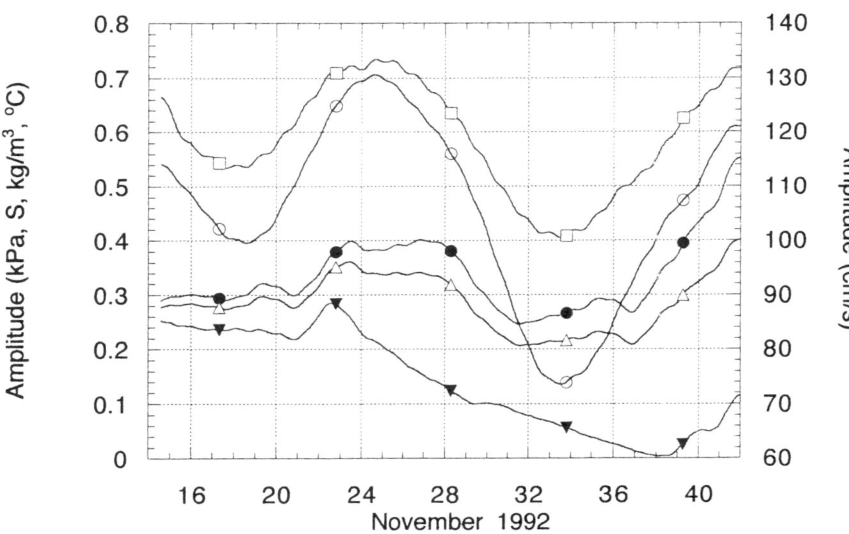

FIGURE 3. Demodulation of M2 (12.42 hour) amplitudes. Pressure is labeled with an open square, the vertically-averaged velocity with an open circle, salinity with a closed circle, density with an open triangle and temperature with an inverted closed triangle.

for 23 November (Fig. 4) during a period of spring tides. Flood and ebb flows dominated the along-channel flow and ranged from 213 cm/s to -178 cm/s, respectively, during 23 November. During floods, velocities at a given time were similar at all depths; so isotachs appear to be parallel to the depth axis, although maximum velocities occurred at mid-depth in the water column. Ebbs were weaker than floods, and during ebb, strongest velocities were found near the surface. Maximum accelerations occurred about 1 hour after the beginning of the flood. The mean along-channel flow was 23 cm/s.

Across-channel velocities, constrained by the sides of the Golden Gate, were almost an order of magnitude less than the along-channel velocities. The isotachs appeared noisier, few extended from top to bottom, the flow was mostly negative (the mean is -11 cm/s), and a tidal cycle was more difficult to discern. One to two hours prior to the beginning of ebb flow, southward flow in the lower half of the

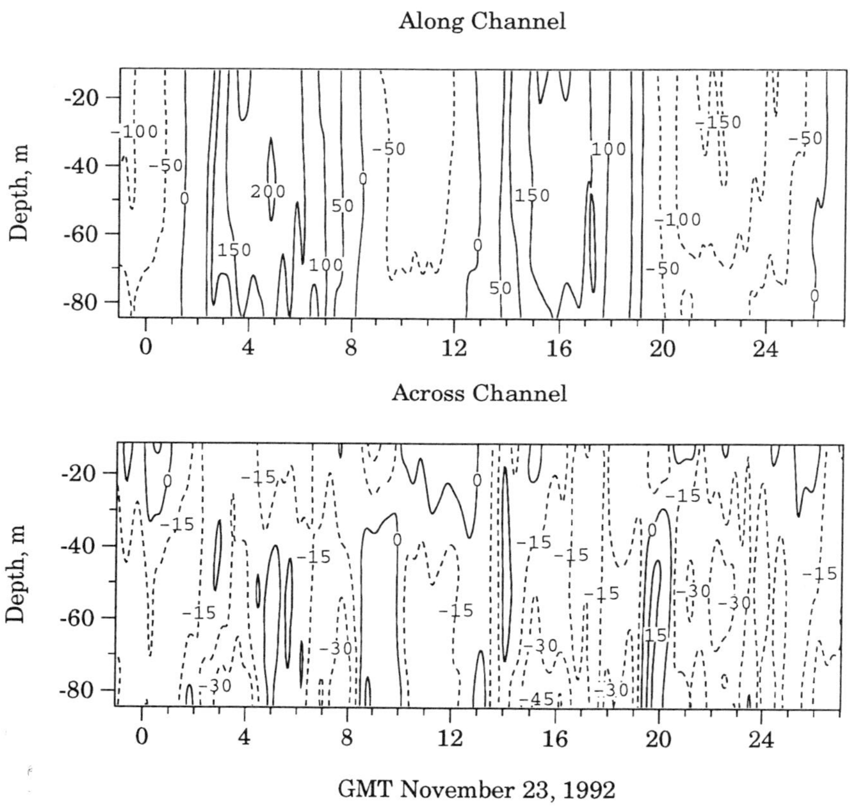

FIGURE 4. Along and across-channel velocities during spring tides on November 23, 1992. Negative velocities are indicated by dashed isotachs. The contour interval for the along-channel flow is 50 cm/s and for the across-channel flow is 15 cm/s. For the across-channel flow, unlabeled solid isotachs represent zero cm/s and unlabeled dashed isotachs are -15 cm/s except for a few isotachs near the upper and boundaries that exceed these values. Bottom depth is 91 m. These charts were constructed from 3-minute velocity samples.

water column intensified, and at the onset of the ebb, the across-channel flow in the lower layer reversed and was positive for 1-2 hours before becoming negative again. During this cycle, flow at the surface tended to be opposite the flow at depth. The other discernible tidal feature in the across channel flow was a negative pulse in the lower layer that occurred during the period of time before the flood reached its maximum velocity.

To illustrate the relationships between the two components of the flow in Fig. 4, a scattergram (Fig. 5) was formed by placing dots at the head of the vectors of the vertically-integrated velocity. The scale of the ordinate of Fig. 5 is only one quarter that of the abscissa, so that the relationships between the along- and across-channel flows can be seen easily, but the direction of the flow is distorted. The cluster of points resembles the letter "z" lying on its side. At the start of the flood the across-channel flow was negative, and as the flood progressed the flow

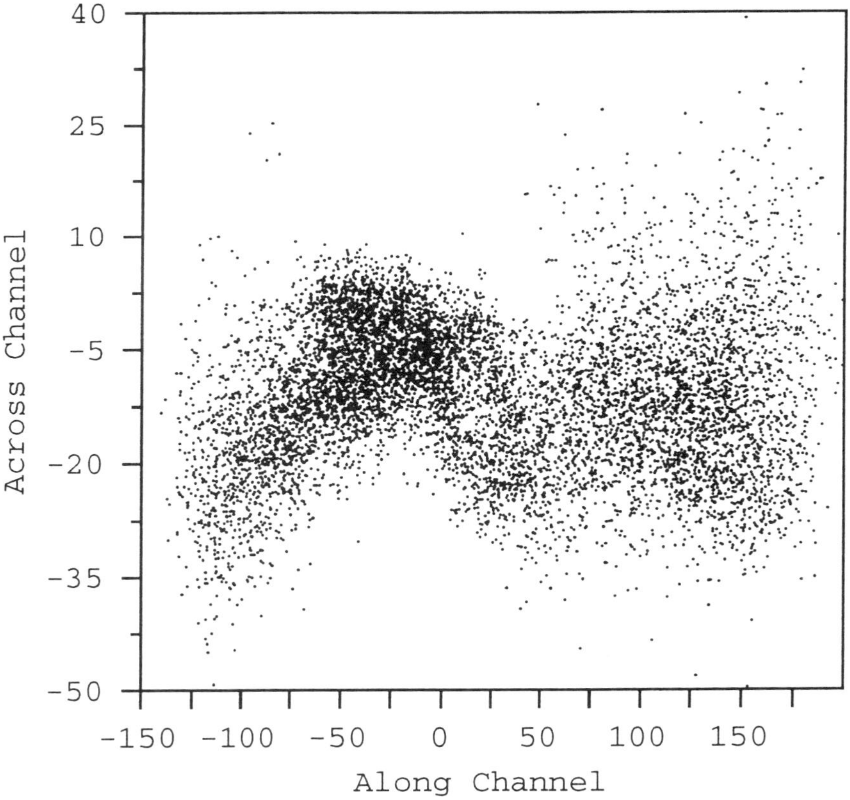

FIGURE 5. Scattergram for the vertically integrated flow. The coordinate system has been rotated clockwise 5.5° so the abcissa represents along-channel flow and is oriented along 84.5 - 264.5°*T*. Different scales have been used for the ordinate and abcissa so that across-channel flow is exaggerated. Even 3-minute samples were used so the total number of data points included in this figure is 8,172.

rotated counterclockwise, increasing the across-channel flow. As the flood waned, the across-channel flow decreased. At the start of the ebb flow the across-channel flow increased, becoming positive on some occasions. The flow rotated counter-clockwise as the ebb waxed, and the across channel flow decreased, becoming stronger. As the ebb waned, the magnitude of the across-channel velocity decreased but remains negative. The strongest flood and ebb currents, associated with spring tides, exhibited the strongest positive and negative across-channel flows, respectively.

The shape of the scattergram formed by the current vectors changes with depth. At the surface, the sideways "z" pattern was least evident, with strengthened ebb flow and little dependence between the range of across-channel flow on the magnitude of along-channel flow. With depth, the "z" became more pronounced, with a negative to positive change of across-channel flow associated with the switch from flood to ebb currents and positive to negative change of across-channel flow associated with the switch from ebb to flood curreis.

Figure 6 is a histogram of speed and direction for the 3-minute samples of the vertically integrated flow. Speeds have been classified as negative for ebb and positive for flood based upon the sign of the along-channel flow, but directions are given as °*T* and have not been rotated to the principal axis. Modes of speed and direction were associated with both the flood and ebb flow. For flood, the modes were 91°*T* and 140 cm/s, and for ebb, 254°*T* and -50 cm/s. Few northward velocities were observed, and a distinct gap in the speed histogram occurred when speeds were near zero, giving rise to secondary speed modes at ± 20 cm/s. The lack of observations for speeds close to zero is attributed to the rotary nature of the flow and not the observational technique.

Ebb flows (Fig. 6) had greater variability of direction and lower speeds than floods. With respect to vertical changes of these modes, the direction of the flood remains the same with depth, but speeds were greater at mid-depth. For ebbs, modal direction rotated counterclockwise with depth, from 275°*T* at 12m to 235°*T* at 84 m . The modal speed also decreased with depth for the ebb flow, from 60 cm/s at 12 m to 35 cm/s at 84 m.

The standard deviation of the along-channel flow with depth increased from a minimum of 67 cm/s at 84 m to a maximum of 92 cm/s at 44 m, and then decreases to 88 cm/s near the surface (12 m). The standard deviation of the across-channel flow was less than 20 cm/s, decreasing slightly from 84 m to a minimum at 26 m. The direction of the principal axis rotated counter clockwise from 87°*T* near surface (12 m) and near bottom (84 m) to 82°*T* at 56 m. The vertical profile for kinetic energy was similar to that for the standard deviation of the flow along the principal axis, varying from 300 J/m^2 at 84 m to a maximum of 480 J/m^2 at 44 m. The mean kinetic energy for the water column was 37,000 J/m^2.

Tidal Currents. Tidal currents were estimated using a least-squares analysis (Foreman 1978). From 33 days of current observations, 39 tidal constituents were derived (two by inference) and agree with those obtained in previous studies (Walters *et al.* 1985). The largest constituents were the five primary constituents

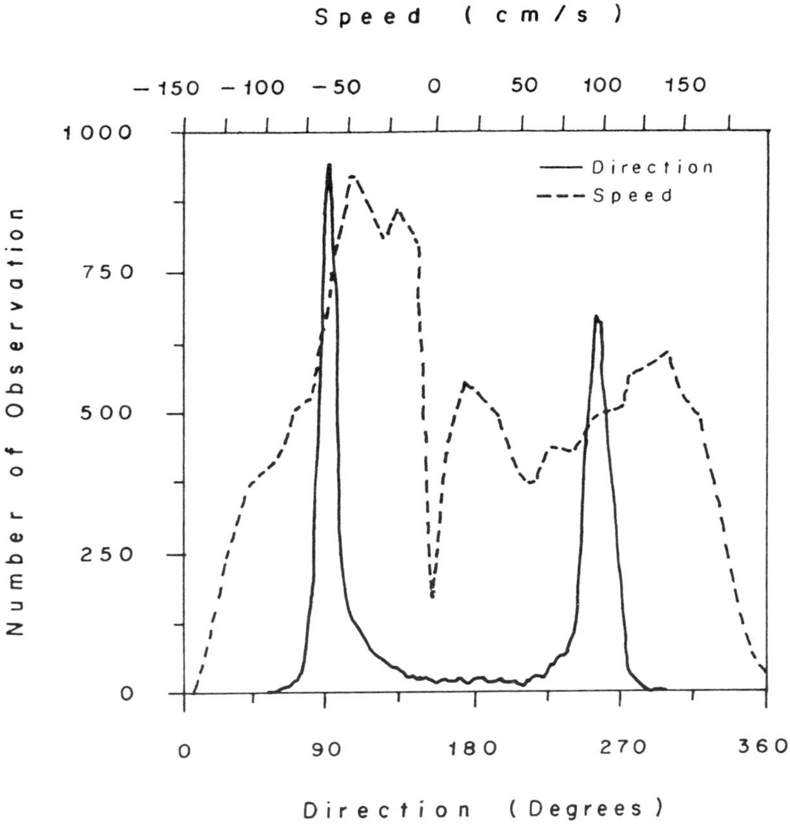

FIGURE 6. Histogram of speed and direction (°*T*) for vertically integrated flow. Negative speeds indicate ebb flow. The integration interval was 10 cm/s for speed and 1.8° for direction. A total of 16,344 3-minute observations were used.

(O1, K1, N2, M2, S2), three compound tides (2MK5, MK3, 3MK7), and two overtides (M4, M6). Results are presented in Fig. 7 for the lunar semidiurnal tide (M2) and the two overtides. (Complete results for all constituents, including standard deviations for the constituents, can be found in Petzrick, 1993.) The ellipse orientation (or inclination) is the angular measure of the clockwise rotation of the semi-major axis from north (0°*T*). A negative semi-minor axis indicates clockwise rotation of velocity. Since the Foreman analysis uses hourly samples, our three-minute samples allowed 20 estimates of each constituent, so that standard deviations could be calculated directly.

The M2 tidal currents had an amplitude three times greater than the next largest constituent, K1. The semi-major axis was 108 cm/s near the surface, increased to a maximum of 115 cm/s at 44 m, and then decreased to 82 cm/s at 84 m (Fig. 7, top left). The semi-minor axis was less than 4% of the major axis, so that motion at this frequency was nearly rectilinear. For M2 tidal currents the semi-major axis

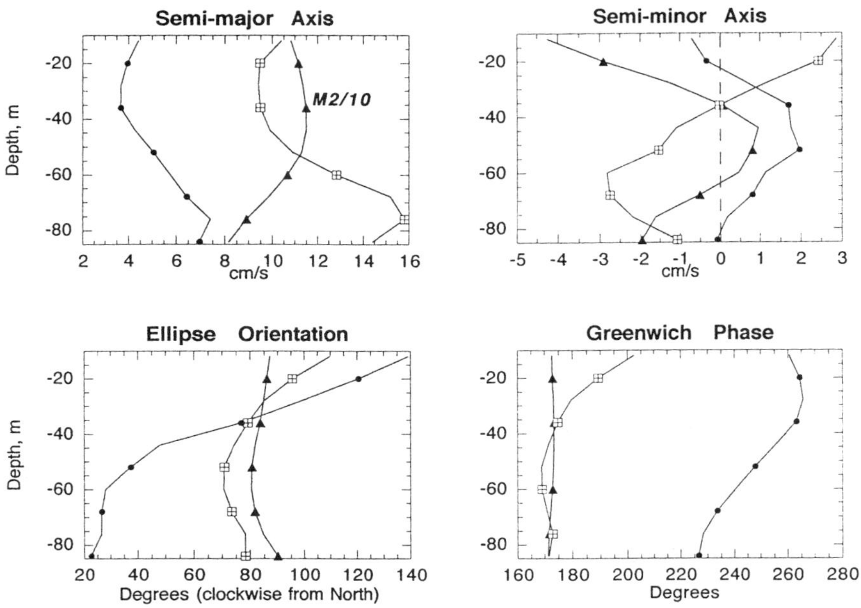

FIGURE 7. Depth variation of flow characteristics determined by harmonic analysis for M2 (solid triangle), M4 (open box) and M6 (solid circle) tides. The amplitude of the M2 semi-major axis is ten times as large as that given by the ordinate in the upper left panel and this is indicated by the "M2/10" label.

was aligned with along-channel flow (84.5° *T*) except at 84 m, where the axis rotated clockwise to 90° *T*. The phase shift from Greenwich varied little with depth and was 172°. Standard deviations (not shown) were about 0.5 cm/s and 0.5° for amplitudes and phases, respectively.

The remaining primary constituents, O1, K1, S2, and N2, were of the same order of magnitude, ranging from about 30 cm/s for K1 to 10 cm/s for N2. For each of these constituents the semi-major axes were nearly uniform in the upper 40-60 m, and then decreased to about 50% of these values at 84 m. As for M2, the semi-minor axes were small, so that flow was rectilinear for these constituents.

The M4 (6.21-hour period) and M6 (4.14-hour period) overtides served to modify the M2 tide, and their characteristics are also presented in Fig. 7. The M4 overtide was the strongest of the secondary constituents, with a maximum amplitude of 15 cm/s. Both the M4 and M6 semi-major axes were largest in the lower portion of the water column (76 m), with an amplitude twice that in the upper layer. Below 40 m, the M4 constituent rotated clockwise and the M6 constituent rotated counterclockwise. The M4 ellipse was oriented along 70-80° *T*, while the M6 ellipse rotates counterclockwise with depth from 140° *T* at 12 m to 20° *T* at 84 m. The Greenwich phase was the same for M2 and M4, about 170°, but 50-90° greater for M6. Thus the M4 tide was in phase with the M2 tide and increased the accelerations and the magnitude of the flow associated with flood and ebb currents.

The M6 currents were maximum 1, 5, and 9 hours after maximum M2 flow, with a pattern of northward and southward flow at depth (and opposing flow in the surface layer) similar to that shown in Fig. 4.

The compound tidal constituents had amplitudes which ranged from 13.5 cm/s to 2 cm/s. The MK3 (8.18-hour period) constituent had the largest semi-major axis, exceeding 10 cm/s at depths shallower than 84 m, with an ellipse orientation and Greenwich phase similar to that for M2. The 3MK7 (3.53-hour period) constituent appeared to be surface and bottom intensified, and the 2MK5 (4.93-hour period) changes its sense of rotation from the surface to mid-layers. The ellipse orientation for these constituents was north-south, so that currents at this frequency were associated with across-channel flow.

Low-frequency Circulation. Results of tidal analysis (above) were used to construct tidal currents for the period of measurement, and these currents were subtracted from the observed data record. The detided data were then smoothed using a 72-hour Butterworth filter. Data from the moored CTD, Fort Point surface measurements, winds on Golden Gate Bridge, NAS Alameda and NOAA tide gauges were also smoothed using a 72-hour Butterworth filter. Stream flow during the period of our measurements was negligible until a major event started on December 5, peaking on December 13 at 800 m^3/s and ending on December 22. These subtidal flows and related environmental data were analyzed qualitatively using stacked plots of smoothed variables and cross correlation functions.

Subsurface pressure and salinity were positively correlated with along-channel transport with a lag of one day (*e.g.*, salinity in the Golden Gate increased with flow into the Bay). Temperature was negatively correlated with along-channel transport, with a minimum correlation at a lag of -0.5 day.

Low frequency winds were persistently southeastward from November 11 through 17. During this period, the vertically-averaged through-channel current was into the Bay, opposite to Ekman transport. Pulses of southeastward wind on November 27, December 1, and December 6 were also correlated with flow into the Bay. The strongest meteorological forcing was northwestward winds which occurred on December 9. This event also appeared to coincide with increased flow from the ocean into the Bay through the Golden Gate.

High-frequency Variability. Despite tidal domination of the flow, the structure of the across-channel flow appears to be often dominated by flows with periods of 1 - 2 hours. These flows also tend to have opposing surface and deep flow or mid-level maximum or minimum. Some evidence of these shorter period fluctuations can also be seen in the along-channel flow, *i.e.*, the waves on the 150 cm/sec isotach that occur between 0500-0600 on November 23 (Fig. 4). In a false-color image of the along-channel flow, striations associated with these ~1-hour period along-channel and across-channel flows are readily apparent. High frequency variability was also clearly visible in bottom temperature and salinity records during flood (Fig. 2).

First we analyzed the detided current data for variability at higher frequencies. The detided current data were dominated by an along-channel flow with a period of 5 hours. This was much longer than the ~1-hour striations discussed above, but is close to the period of the 2MK5 tidal constituent, the stationary part of which should have been removed by Fourier analysis. The covariance of 12 m along-channel flow with flow at deeper depths had a sharp peak at zero lag, changed sign below 64 m, and had coincident zero crossings at +/- 1.3 hours. The covariance functions for across-channel flow were similar in shape and magnitude, although there was more scatter in the first zero crossings. This implies opposing flows in upper and lower layers with remarkably little temporal distortion.

Spectral analysis of detided data also revealed a possible quarter-wave seiche of the along-channel high frequency flow. Energy was present at 1.67 hours. Since the quarter-wave seiche response, T_c, is

$$T_c = \frac{4L_c}{\sqrt{gh}}$$

reasonable estimates of T_c range from 0.5 to 2 hours for the central Bay depending on the values of depth, h, and length, L_c (Chuang & Boicourt 1989).

Different high pass filters were used with the original time series study the structure of variability of flow for the frequency range 6 cph to 0.5 cph. For along-channel flow, variance was greatest at 76 m, minimum at 40 m, and increased to 12 m. Covariances for along-channel flow were positive except for those between the bottom two or three depths and the two or three observations closest to the surface. The vertical structure of the variance of the across-channel flow was similar to that for the along-channel flow except that the minimum was shallower, 32 m. Covariances for along-channel flow changed sign in the middle of the water column. Reynolds stresses (u′v′) were largest at 80 m, with the minimum at 60 m, and a maximum at 44 m, and were negative at 12 m. Covariances between salinity and along-channel flow were positive.

DISCUSSION

The apparent flooding of San Francisco Bay implied by large mean currents associated primarily with floods that exceeded ebbs was due to the two dimensional structure of the flow and the Eulerian nature of our measurements. Some indication of this in our measurements is given by southward deflection of waning floods and northward deflection of waxing ebbs. If the gravitational circulation is primarily associated with Delta outflow through Carquinez Strait, it should enter the Golden Gate from the north, perhaps reversing flow there first, and forcing inflow in the southern part of the Golden Gate to the south. If the distribution of ebb and flood currents are not symmetric mirror images across the Golden Gate, a bias will be introduced to the measurements at a fixed location. Conomos *et al.* (1985) noted that tidal flood currents followed the depths along the south shore and that ebb currents were more concentrated toward the northern shore and shoal areas. They also noted a tidally driven residual flow directed up-estuary in the center of the Golden Gate and toward the ocean next to shore.

This pattern of flow also agrees with theoretical and numerical analysis of Park & Wang (1991). They studied the vorticity balance of tidal flow over a hollow in a straight rotating channel, a situation resembling the topography of the Golden Gate. They found a pattern of residual currents with inflow to the south and outflow to the north. The vorticity distribution in Cheng *et. al.* (1993) also shows this asymmetry.

Analysis of across-channel flow supports the exiting of fresher surface waters near the surface on the north side of the Golden Gate and the entering of denser waters following depths on the south side (Conomos *et al.* 1985). The near-bottom across-channel currents showed pulses of 15 cm/s northward, which were partially balanced by southward flow between 12 and 36 m. The bimodal structure of speed and lack of reciprocity for ebb and flow direction indicate effects of variable topographic steering at different depths.

During the period of observations, waters of the Golden Gate clearly cooled and freshened. Vertical profiles showed a deepening and cooling of the surface layer, which extended throughout the water column. Time series data also showed a general cooling trend which exceeded 0.02°C per day from mid-November until early December. In early December, a series of winter storms reversed the cooling trend for about three days, after which it resumed. During the November cooling phase, salinity remained fairly constant, with tidal oscillations about a mean salinity of 31.5. During the December storms, the salinity freshened rapidly, reaching a low of 29.6 by the end of the deployment. Salinity variability retained a tidal character throughout the measurement, freshening almost 1 S on each ebb. However, the normal cooling of 1°C that occurred with each flood disappeared completely as the waters of the Bay became cooler than ocean waters.

These data clearly span the change from the "fall transition" to the winter season (see Largier 1996 for a definition of seasons). The point in time when Bay waters become colder than ocean waters is easily detected by the change in phase of temperature with respect to either pressure or current that occurs in the Golden Gate, and hence is a useful index for the change of seasons in the Bay. In 1992, this change occurred at about the same time as the first major winter storm and associated fresh water inflow to the Bay.

The effect of these changes in water properties increased the amplitude of the semidiurnal density variations after December 8 (Fig. 3). Note that prior to December 8, the semidiurnal density variations did not appear to completely follow the spring-neap cycle so that during the period from November 24-28, the amplitude of the density variations remained relatively constant, 0.35 kg/m^3, while tides and tidal currents increased. The amplitude of the semidiurnal density variability depends upon the horizontal stratification. If the vertical stratification of the column of water that passes through the Golden Gate during a tidal cycle remains uniform during that cycle, then the horizontal density gradient, ρ_x, can be estimated from the semidiurnal amplitudes of the vertically-integrated flow, $A_{\bar{u}^z}$, and density, A_ρ, as

$$\hat{\rho}_x \approx \frac{14 A_\rho}{A_{\bar{u}^z}} \text{ kg/m}^3\text{-km.}$$

The time series for $\hat{\rho}_x$ ranged from 0.034 to 0.04 kg/m^3-km during November 16 to

December 6, with maximum values associated with neap conditions as well as points about halfway through the spring-neap cycle. This suggests that the horizontal density structure changed during the spring-neap cycle, possibly due to the character of the ocean inflow or differing exchanges between the southern, central and northern Bay. During the period when temperatures were the same in the Bay and Pacific Ocean, $\hat{\rho}_x$ was minimum, 0.031 kg/m^3-km. Subsequently, as fresh water associated with winter storms flowed into the Bay $\hat{\rho}_x$ increased to 0.046 kg/m^3-km.

The vertical shear of the subtidal along-channel flow can be used to estimate the vertical circulation associated with the gravitational flow in the estuary. The mean vertical shear of the across-channel flow was 0.01 s^{-1}, about 2.5 times greater than predicted from the thermal wind equation using the horizontal density gradient estimated above. Largier (1996) estimated vertical shear of 0.01 s^{-1} in the Golden Gate from historical current meter data and notes that this requires a vertical transport of 10^3 m^3/s in the estuary. Figure 8 shows the baroclinic structure of the subtidal flow for our measurements. For along-channel flow, the velocity increased with depth to 70-75 m . The magnitude of the shear was 0.02 s^{-1}, and it increased slightly during spring tides, both in upper and lower layers. The effects of bottom stress and mixing were also evident in the along-channel flow, and are associated with decreasing velocities below 75 m.

The vertical structure of the across-channel flow (Fig. 8) is such that the flow was directed toward mid-channel in the upper layer and toward shore at mid-depth. At depth, the direction of the across-channel flow appeared to alternate, with the strongest pulses toward shore on November 20 and December 14. The shear was weakest during periods of spring tides, but was always less than that for along-channel flow. To the west of the Golden Gate Bridge, Largier (1996) noted a stronger pattern of opposite flow in surface and bottom layers for across-channel flow and attributed this to spreading of the velocity jet associated with ebb tides.

Tidal forcing dominated the kinetic energy of the currents and the potential

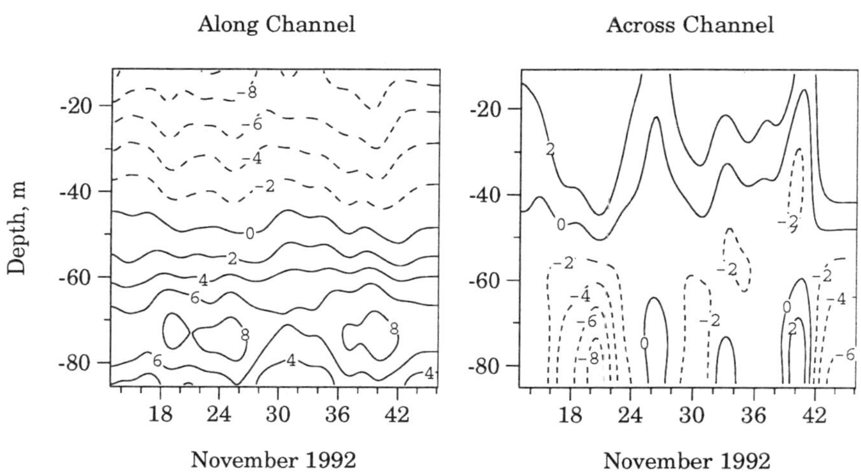

FIGURE 8. Subtidal baroclinic flow $(u(z,t) - \bar{u}^z(t))$. Bottom depth is 91 m.

energy of the sea level. The standing wave character of the tidal motion is indicated by the phase between the vertically-integrated velocity and sea level and the fact that the kinetic energy was about twenty times greater than the potential energy (this ratio should be closer to unity if we could integrate horizontally over a wavelength). The mixed character of the tide is reflected in strong, nearly rectilinear motion associated with primary tidal constituents. Shear (about 20% for the M2 tide) was observed with a subsurface maximum in the upper layer. Overtide and compound tides had amplitudes of 5-10 cm/s, with more complex vertical structures.

A common index for nonlinearity of tides is the ratio of the M4 and M2 constituents. In the Golden Gate, ratios of 0.04 were found for sea level variability and 0.09 for kinetic energy at these frequencies. These results are far below those for regimes considered to be highly nonlinear due to frictional effects. On the other hand, the strong accelerations which occur after the start of flood tides indicate the nonlinear character of the flow and were associated with the M4 tide. The M6 tide was directed across channel in the upper and lower layers so that negative flow in the upper layer was opposed by positive flow in the lower layer.

As expected, there was a strong correlation between the amplitude of the semidiurnal variations of pressure, A_p, and vertically integrated flow, $A_{\bar{u}^z}$ (Fig. 3). Linear regression yielded $A_{\bar{u}^z} = 10 + 162A_p$ with a correlation of 0.98, and this relationship can be used to estimate the amplitude of the tidal flow at our observation site from the amplitude of sea level or pressure variability. The regression coefficient, 162 cm/s-kPa, yields an estimate of the dependence of tidal prism on the amplitude of the tide.

ADCP and bottom CTD data were decimated to six-minute samples and used to estimate volume and energy flows at the mooring site. The volume flux was estimated by vertically summing the along-channel currents. The velocity of the bottom layer was assumed to be one half that of the deepest ADCP bin, while the velocity of the surface layer was set equal to the velocity of the upper ADCP bin. This yielded a mean volume flux along a 1-m wide column at our measurement site of 21 m^3/s. If this rate of inflow occurs over half the channel, the volume flux into the Bay would be 13,000 m^3/s. A better estimate for this inflow can be obtained by using the observed mean shear, assuming lateral homogeneity, and then adjusting the net transport to match the mean freshwater inflow during the period (150 m^3/s). The resulting rate of inflow is 2,200 m^3/s. Using this latter estimate of inflow, the flushing rate (volume of the Bay divided by inflow rate) for San Francisco Bay would be 35 days.

Similarly, we can make rough estimates of terms in the energy equation. The divergence of the potential energy in the Bay should be balanced by the kinetic energy flux through Golden Gate and the work done by pressure. Kinetic energy flux for a 1-m wide column at our measurement site was estimated using the same approximations for the surface and bottom layer as were used for the volume flux. For this calculation, we also assumed that the water column was homogeneous and used the density measured by the bottom moored CTD, and evaluated

$$\int_{h}^{0} \tfrac{1}{2}\rho v^{2} u\,dz.$$

This yielded a mean kinetic energy flux of 2.8×10^4 W. The energy flux due to the work done by pressure was estimated by assuming hydrostatic equilibrium and neglecting the effects of the changes of density within the water column and evaluating

$$\int_{h}^{0} \rho u\,dz.$$

The mean energy flux for the 1-m wide column at our measurement site was 8.5×10^6W. If these energy fluxes were the same across the entire Golden Gate, the total energy flux through the Golden Gate would be 8.5×10^9W.

ACKNOWLEDGMENTS

We would like to thank Marla Stone for making sure that our instruments worked, Andy Anderson for helping with deployment tasks, Tarry Rago for checking calculations, and the U.S. Coast Guard for providing ship support for deployment and recovery of our instruments. Larry Smith graciously provided Fort Point data, Marlene Noble provided recent charts for the Golden Gate, Reginaldo Durazo helped with processing ADCP data, and Bob Williams provided encouragement for us to undertake these measurements. Tom Herbers helped with wave dynamics and Gilberto Fuentes drafted Figures 1 and 6. John Largier graciously provided an advance copy of his work for our use and comments from reviewers were helpful in revising this manuscript. The research was sponsored by the Commander, Naval Oceanography Command, and the Naval Postgraduate School.

REFERENCES

Cheng, R. T., V. Casulli,and J. W. Gartner. (1993). Tidal, Residual, Intertidal Mudflat (TRIM) Model and its Applications to San Francisco Bay, California. *Estuarine Coastal Shelf Sci.* 36(3):235-280.

Chuang, W. S., and W. C. Boicourt. (1989). Resonant Seiche Motion in the Chesapeake Bay. *Jour. Geophys. Res.* 94(C2):2105-2110.

Conomos, T. J. (1979). Properties and circulation of San Francisco Bay waters. Pages 47-84 *in* T. J. Conomos, ed., *San Francisco Bay: The Urbanized Estuary*. Pacific Division, American Association for the Advancement of Science, San Francisco, CA.

Conomos, T. J., R. E. Smith, and J. W. Gartner. (1985). Environmental setting of San Francisco Bay, 1985. Pages 1-12 *in* J. E. Cloern & F. H. Nichols, eds., *Temporal Dynamics of an Estuary: San Francisco Bay*. Dr. W. Junk Publishers, Dordrecht, Netherlands.

Edmonston, A. D., and R. Matthew. (1931). *Variation and control of salinity in Sacramento-San Joaquin Delta and upper San Francisco Bay*. California Dept. Public Works Bull. 27. 440 pp.

Foreman, M. G. (1977). *Manual for Tidal Heights Analysis and Prediction*. Pacific Marine Sci. Rept. 77-10. Institute of Ocean Sciences, Patricia Bay, Victoria, BC. 9 pp.

Foreman, M. G. (1978.) *Manual for Tidal Currents Analysis and Prediction*. Pacific Marine Sci. Rep. 78-6. Institute of Ocean Sciences, Patricia Bay, Victoria, BC. 9 pp.

Fornshell, J. A., and J. C. Reed. (1982). *Tidal Current Atlas for Long Island Sound and San Francisco Bay*. Oceanographic Unit Tech. Rept. 82-4, U.S. Coast Guard, Washington, DC. 39 pp.

Largier, J. L. (1996). Hydrodynamic Exchange between San Francisco Bay and the Ocean: The Role of Ocean Circulation and Stratification. Pages 69-103 *in* J. T. Hollibaugh, ed., *San Francisco Bay: The Ecosystem*. Pacific Division, American Association for the Advancement of Science, San Francisco, CA. (this volume.)

Park, M., and D. Wang. (1991). Transient Tidal Vorticity over a Hollow. Pages 419-434 *in* B. Parker, ed., *Tidal Hydrodynamics*. John Wiley & Sons, Inc., NY.

Petzrick, E. P. (1993). *Currents through the Golden Gate*. M.S. Thesis. Naval Postgraduate School, Monterey, CA. 119 pp.

Stommel, H. (1951). *Recent Developments in the Study of Tidal Estuaries*. WHOI Tech. Rept. 51-33. Woods Hole Oceanographic Institution, Woods Hole, MA. 202 pp.

U.S. Department of Commerce. (1950). *Manual of Current Observations*, pp. 78-83, U.S. Govt. Printing Office, Washington, DC.

Walters, R. A., R. T. Cheng, and T. J. Conomos. (1985). Time scales of circulation and mixing processes of San Francisco Bay waters. Pages 13-36 *in* J. E. Cloern & F. H. Nichols, eds., *Temporal Dynamics of an Estuary: San Francisco Bay*. Dr. W. Junk Publishers, Dordrecht, Netherlands.

Williams, R. G., W. L. Wilmot, and H. R. Frey. (1992). *San Francisco Bay Current Prediction Quality Assurance Miniproject*. NOAA Tech. Memorandum NOS OES 003, Rockville, MD. 40 pp.

STRATIFICATION DYNAMICS AND GRAVITATIONAL CIRCULATION IN NORTHERN SAN FRANCISCO BAY

STEPHEN MONISMITH, JON R. BURAU AND MARK STACEY
Environmental Fluid Mechanics Laboratory
Stanford University, Stanford CA 94305-4020

In this paper we discuss field observations and modeling results examining the connection between flow-induced stratification and gravitational circulation. We present simultaneous ADCP and top and bottom salinity records collected by the USGS showing that density-driven flows are highly intermittent, with strong gravitational circulation accompanying stratification episodes. Tidal variations in gravitational circulation appear to be linked to tide-induced vertical mixing and to the strain-induced periodic stratification (SIPS) that is typically found in Northern San Francisco Bay. Results obtained using a 1D water column model suggest that there exists a critical condition where tidal mixing is unable to overcome the creation of stratification by advection. Variability of observed stratification is better explained by a parameter which explicitly considers SIPS, and thus which depends explicitly on the longitudinal salinity gradient, than it is by a similar parameter based on flow.

The tides and tidal currents in San Francisco Bay are extremely energetic. At the Golden Gate, sea level can change as much as 2.6 m in ~ 6 hours on a spring tide. Because of the geometry of the Bay, these elevations can drive maximal tidal currents in the channels of San Francisco Bay in the range of 50 to 200 cm/s (Cheng & Gartner 1984). The magnitude of the energy input by the tides sets the stage for most of the transport and mixing processes that provide the dominant linkage between physical forcing, *e.g.*, river flow, and biological response, *e.g.*, annual primary production by phytoplankton.

Biologically relevant transport and mixing processes include those associated with vertical turbulent mixing (*i.e.*, over depth), horizontal advection (*e.g.*, landward or seaward) or dispersion by some combination of mixing and advection (Fischer *et al.* 1979). Reviews of horizontal dispersion mechanisms associated with tides, *e.g.*, depth-averaged Eulerian residual currents or by wave-induced transport in San Francisco Bay can be found in Walters *et al.* (1985). In this paper we focus on stratification dynamics and gravitational circulation, with particular attention to northern San Francisco Bay (see Fig. 1). We focus on stratification because rates of vertical mixing can be much smaller in the presence of stratification than in an homogeneous water column (see *e.g.*, Abraham 1988 or Ivey & Imberger 1992). Hence, biological processes that are affected by vertical exchanges should be dependent on water column stability. We examine gravitational circulation, flows driven by baroclinic pressure gradients, because they are thought to be a primary influence on the longitudinal transport of organisms (Herbold *et al.* 1992), and on the formation of the Estuarine Turbidity Maximum (ETM – see below). The pairing of gravitational circulation with stratification is not an arbitrary choice: as we will discuss below, the two are strongly connected. Moreover, both are likely

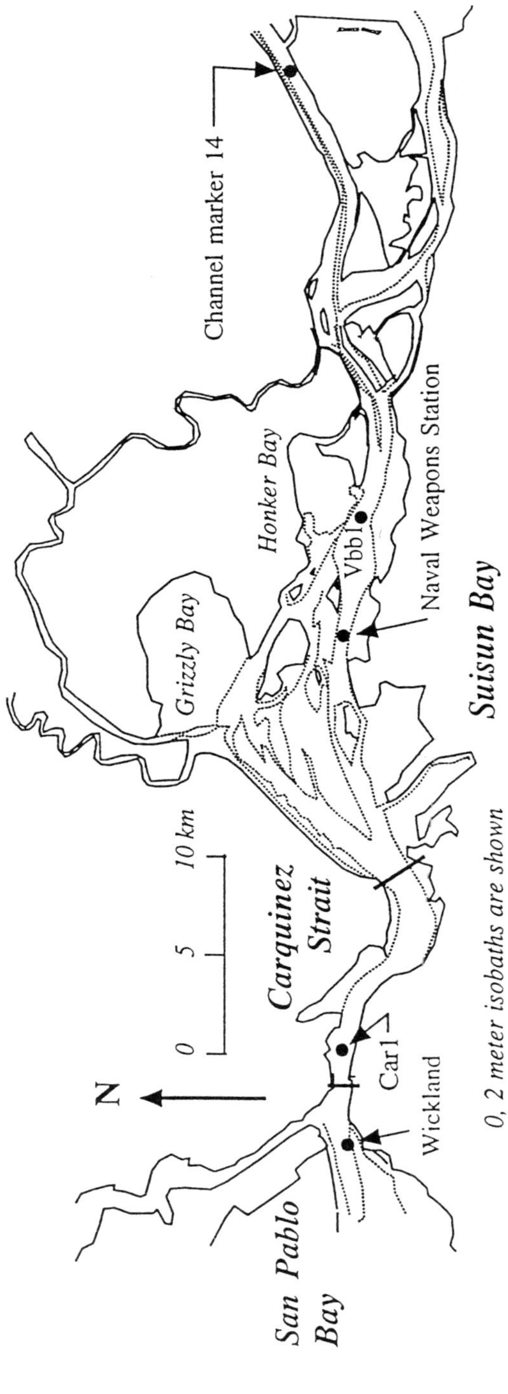

FIGURE 1. The northern reach of the San Francisco Bay estuary.

to be directly influenced by the buoyancy input provided by Delta inflow to the Bay, and thus to represent a means by which inflows to the Bay might influence ecological processes.

Before presenting our observations, in §2 we attempt to make the case that dynamics of density stratification are relevant to ecological processes in San Francisco Bay, both directly, and through their influence on gravitational circulation. In §3 we use simultaneous observations of salinity and velocity structure to show that gravitational circulation and stratification are, in fact, strongly linked to each other and to the tides, with flow playing a somewhat secondary role. We pursue this coupling using a one-dimensional model of salinity and velocity structure in §4, finding that there appears to be a critical condition, which can be described using a parameter, Ri_X, determined by the tides and by the longitudinal structure of the salinity field, which separates episodes of strong, persistent stratification, intense gravitational circulation and substantial upstream salt flux, from periods in which all of these are relatively weak. We summarize and discuss our findings in §5.

2. DENSITY STRATIFICATION IN SAN FRANCISCO BAY

2.1. Linkage of stratification and primary production by phytoplankton

In seeking to explain his observations of spring phytoplankton blooms in South San Francisco Bay, where sufficiently strong benthic grazing appeared to be available to consume any and all primary production, and equally, where there was no obvious nutrient limitation to suppress blooms, Cloern (1982) hypothesized that density stratification was the key. When stratification was present, as is usually the case in the spring, the benthos and the photic zone were de-coupled, thus permitting phytoplankton growth unhindered by benthic grazing. In the summer and autumn, when the water column is usually un-stratified, the abundant population of benthic filter-feeders could consume whatever phytoplankton biomass that might be produced.

Examining a 10 year record of phytoplankton biomass, Cloern(1992) found that most blooms coincided with the period of weakest tides in the spring. To examine this connection he developed a one-dimensional (vertical) phytoplankton model that incorporated tidally-derived vertical using an eddy mixing coefficient that varied with the spring-neap tidal cycle. When this coefficient was small (mixing was weak), blooms could be produced.

Koseff *et al.* (1993) refined this model, building in direct effects of stratification on vertical mixing (albeit in an ad-hoc fashion), and including tidal as well as neap-spring variations in mixing that were directly coupled to tidal currents (using formulae given in, *e.g.*, Fischer *et al.* 1979). They also presented a scaling analysis that shows how bloom formation in homogeneous water columns depended on the relationship between the time scale for growth, the time scale for consumption and the time scale for vertical mixing of the water column. As suggested by scaling, they were never able to produce blooms when they used values of tidal currents and rates of benthic grazing typical for South SF Bay. However, reductions in vertical mixing due to stratification were sufficient to permit bloom formation, both because of de-coupling the water column and the benthos, and because of reduction of respiratory losses of phytoplankton biomass due to retention of biomass in the photic zone. However, once stratification was removed (reflecting the persistent effects of vertical mixing), and mixing coefficients were reset to their homogeneous values, the blooms disappeared, thus reinforcing the view that blooms and stratification episodes were intimately connected.

Videgar *et al.* (1993) extended the Koseff *et al.* (1993) model to include explicit modeling of turbulence (using the model described in §4) and to include separate shoal and channel water columns that obeyed local water column balances but that also exchange fluid horizontally. Again stratification was found to be crucial to bloom formation, although horizontal exchange was also important (see Jassby *et al.*, this volume).

More generally, one can surmise that any biological or geo-chemical process can be altered by variations in vertical mixing, *e.g.*, by altering coupling between the water column and the bed, or by altering the light climate seen by phytoplankton cells, will be influenced by stratification.

2.2. The Estuarine Turbidity Maximum (ETM)

A second case where buoyancy effects are important to biologically relevant transport is that of the estuarine turbidity maximum (ETM). Considerable ecological value has been attached to the ETM that often is found in Suisun Bay in the spring (Arthur & Ball 1979; Kimmerer 1992). As it often described, the ETM is the transition zone between Eulerian residuals that are directed seawards at all depths (in the channel), and a region of "estuarine circulation" in which flow at depth is directed landwards and flow at the surface is seawards ((Hansen & Rattray 1966; Arthur & Ball 1979). As Jay & Musiak (1994b) point out, this Eulerian picture must be treated with some caution because Stokes drift (wave) transport can be important, implying that Lagragian (particle) motions can be significantly different from Eulerian ones.

From the mean Eulerian flow field picture, it can be inferred that in this transition region, mean vertical velocities can be sufficiently large to counter the sinking of some sediment or phytoplankton particles thus causing them to be retained in the resulting turbidity maximum (Schubel 1968, Peterson *et al.* 1975; Arthur & Ball 1979). Along with elevated levels of turbidity and relatively high concentrations of phytoplankton biomass (which are to be expected according to this conceptual model), zooplankton and larval fishes are often abundant there (see *e.g.*, Kimmerer 1992).

However, the ETM is typically also the boundary between stratified and unstratified parts of the estuary. Geyer (1993b) carried out calculations suggesting that the effect of stratification in the downstream part of the ETM on turbulent mixing was much more important in maintaining a turbidity maximum than was the mean flow pattern. Thus, the structure and function of the ETM also depends on stratification as well as on baroclinic pressure gradients.

2.3. Stratification, gravitational circulation and SIPS

The linkage between stratification and gravitational circulation is quite strong. Gravitational circulation represents a balance between $O(\beta\Gamma gH)$ baroclinic pressure gradients and $O(\nu_t U H^{-2})$ frictional forces. Here β is the coefficient of saline expansivity, Γ is the longitudinal salinity gradient (which is assumed to be constant), H is the fluid depth, ν_t is an eddy viscosity, and U is the induced residual velocity. Equating these two we find that

$$U \sim \frac{\beta\Gamma g H^3}{\nu_t} \tag{1}$$

If the flow is not stratified, $\nu_t \sim C_D^{1/2}\, U_{rms} H$ (Prandle 1985), and so

$$\frac{U}{U_{rms}} \sim \frac{\beta\Gamma g H^3}{C_D^{1/2} U_{rms}^2} \tag{2}$$

In the absence of stratification, we should expect a factor of 4 variation in gravitational circulation between the weakest and strongest observed currents due to a factor of 2 variation in U associated with the spring-neap cycle (Walters & Gartner 1985; Smith *et al.* 1991), and due to a factor of 2 variation in G caused by flow variations (Jassby *et al.* 1995). Variations in stratification can induce much larger variations in U: For example a factor of 10 reduction in ν_t^{-1} from its neutral value would not be unusual (see Ivey & Imberger 1991), giving a factor of 10 variation in U. Because salinity stratification in the northern reach is typically stronger in the winter and spring when Delta outflow is larger (Conomos 1979), it seems reasonable to expect that gravitational circulation should be stronger in the winter than in the summer. Indeed, this is the basis for the assumption that spring river flows are important to the dispersal of larval fish in the northern reach (Herbold *et al.* 1992).

The strong coupling between stratification and gravitational circulation can be enhanced through the process known as Strain Induced Periodic Stratification "SIPS" (Simpson *et al.* 1990; Nunes-Vaz & Lennon 1991; Nunes-Vaz & Simpson 1994; Sharples *et al.* 1994). The underlying physics are simple in concept: Tidal flows will be sheared because of bottom friction; in the presence of an horizontal salinity gradient this shear will alternately develop and remove stratification. To see how this works in its simplest fashion, we consider the 1D balance for the local salinity, $S(z,t)$:

$$\frac{\partial S}{\partial t} = -\,U(z,t)\Gamma \tag{3}$$

If the horizontal salinity gradient, Γ, is independent of depth, the time rate of change of the vertical salinity gradient will be

$$\frac{\partial}{\partial t}\left(\frac{\partial S}{\partial z}\right) = -\left(\frac{\partial U}{\partial z}\right)\Gamma \tag{4}$$

Given that Γ is negative (salinity decreases with x = distance from the Golden

Gate), stable salinity stratification $\partial S/\partial z < 0$ will develop whenever $\partial U/\partial z < 0$. In an homogeneous boundary layer flow this will occur on the ebb; in a stratified flow it can happen on the ebb or on the flood if, as is often observed, the lower part of the water column turns before the upper layer. Moreover, if the flow stratifies, the upper layer will accelerate relative to the lower layer, enhancing the stratification that develops (Jay & Musiak 1994a; Monismith & Fong 1995).

During the flood in an homogeneous flow, $\partial U/\partial z > 0$ and salinity stratification will decrease or be eliminated since a water column with $\partial S/\partial z > 0$ is unstable. Gravitational circulation also tends to increase stratification since it transports salty water upstream at depth and fresher surface water downstream. Turbulent mixing can significantly influence the development of stratification during SIPS. Stratification can only form on the ebb if the stratifying effects of advection can overcome the effects of mixing associated with bottom and internally-generated turbulence (Abraham 1988). Therefore, variations in stratification naturally arise through neap-spring variations in the rate of production of kinetic energy by barotropic tidal currents (Simpson *et al.* 1990; Nunes Vaz *et al.* 1989).

Based on this discussion, we expect salinity stratification and gravitational circulation to depend on Delta outflow (which alters the longitudinal salinity gradient) and on the tides. The dependence on Delta outflow and tides has been parameterized using the Estuarine Richardson number (Fischer *et al.* 1979)

$$Ri_E = \frac{\beta S_0 g Q}{W U_{rms}^3} \tag{5}$$

where W = the estuary width at a given section (which is constant), Q is the flow of freshwater, and S_0 is ocean salinity. Ri_E is thought to be a good bulk parameter, irrespective of the detailed physical processes involved, for measuring salinity intrusion; it is intended to quantify the stratifying effect of the input buoyancy flux on salt transport (Abraham 1988). As such, it is also a candidate for predicting flow-induced stratification. As we will discuss below, in fact, it appears that baroclinic currents and tidal variations in stratification are more directly linked to variations in the things that affect SIPS, *i.e.*, tidal velocities, and the strength of the longitudinal salinity gradient, than to flow *per se*.

In summary, if we are interested in stratification (because of its effects on vertical mixing) or in gravitational circulation (because of its importance to horizontal transport), it appears we must focus on their coupling. We can hypothesize that stratification and gravitational circulation intensity are linked through vertical mixing-induced changes in stratification and through periodic straining of the salinity field. It is this coupling we pursue in §3 and §4 below.

3. OBSERVATIONS OF STRATIFICATION AND SHEAR IN CARQUINEZ STRAIT

3.1. Background

The geometry of the region of interest, the Northern Reach of San Francisco Bay is indicated in figure 1. It consists of a several sub-embayments, San Pablo

Bay and Suisun Bay that consist of narrow, deep (~ 10 m) channels incised through a series of relatively broad, shallow (~ 2 m) shoals, and between several islands (Fig. 1). San Pablo Bay and Suisun Bay are connected by a relatively deep (20 m), channel, Carquinez Strait that curves significantly and has relatively little shoal.

General reviews describing the how tides and the seasonal variations in flow affect salinity in the Northern Reach have been presented in Walters & Gartner (1985) and Denton & Hunt (1985). As would be expected, high salinities are seen in summer and early fall when flows from the Delta into the Bay are low and low salinities are seen the northern reach when flows can be high, typically in winter and spring. One notable feature of the dependence of the salinity field on flow is that it appears that depth-averaged salinities and top-bottom salinity differences are related to the ratio of the distance from the Golden Gate to X_2, the distance to the point in the channel where the bottom salinity is 2 ppt (Fig. 2a - taken from Jassby *et al.* 1995). Upstream of X_2 there is little salinity gradient and little stratification, whereas downstream the opposite is true (Fig. 2b). Because X_2 depends on flow (magnitude and history), it can be used as a general indicator of the response of the salinity field to flow[1].

3.2 Salinity, Flow and ADCP data

The observations we discuss are a set of 15 minute top and bottom salinity records from the USGS station at Wickland Oil at the western end of Carquinez Strait from the 13th of February to the 14th of March, 1991. This data, which also included Dayflow (inflow of water from the Delta into the Bay) estimates, was provided for us by the USGS California District Office in Sacramento[2]. The salinity data are taken at the USGS Wickland station (Fig. 1) which has been in operation, almost continuously, since 1988. Estimates of X_2 were taken from the time series used by Jassby *et al.* (1995) that is based on Bureau of Reclamation shore station salinity data.

Along with salinity data, we will also discuss velocity data from an Acoustic Doppler Current Profiler (ADCP) deployment in the western end of Carquinez Strait near Wickland Oil from December 1990 to March 1991. These data were acquired using a 1.2 MHz narrow band RDI ADCP taking an average velocity profile with 1m resolution (Gartner *et al.* 1995). In discussing these data it must be borne in mind that the ADCP cannot measure the whole of the velocity profile; it misses the region extending from the bottom to about 1 m above the transducer, as well as the top 3 m of the water column (Burau *et al.* 1993). In all of the data shown below, negative streamwise velocities are seaward and positive velocities are landward.

Overall, this data set shows the response of the salinity and velocity fields to two spring-neap tidal cycles and to a single flow event in which the flow increased from a baseline value of approximately 100 m3/s to a peak of about 1200 m^3/s (Fig. 3a). In response to these tidal and flow conditions, mean salinities varied from 11

---

[1] This does not hold for flow events like the flood of February 1986 which resulted in salinities considerably less than those that would be inferred from figure 2a (R.T. Cheng, pers. commun. 1992).

[2] This data set was generously assembled for us by Rick Oltmann of the USGS California District Office..

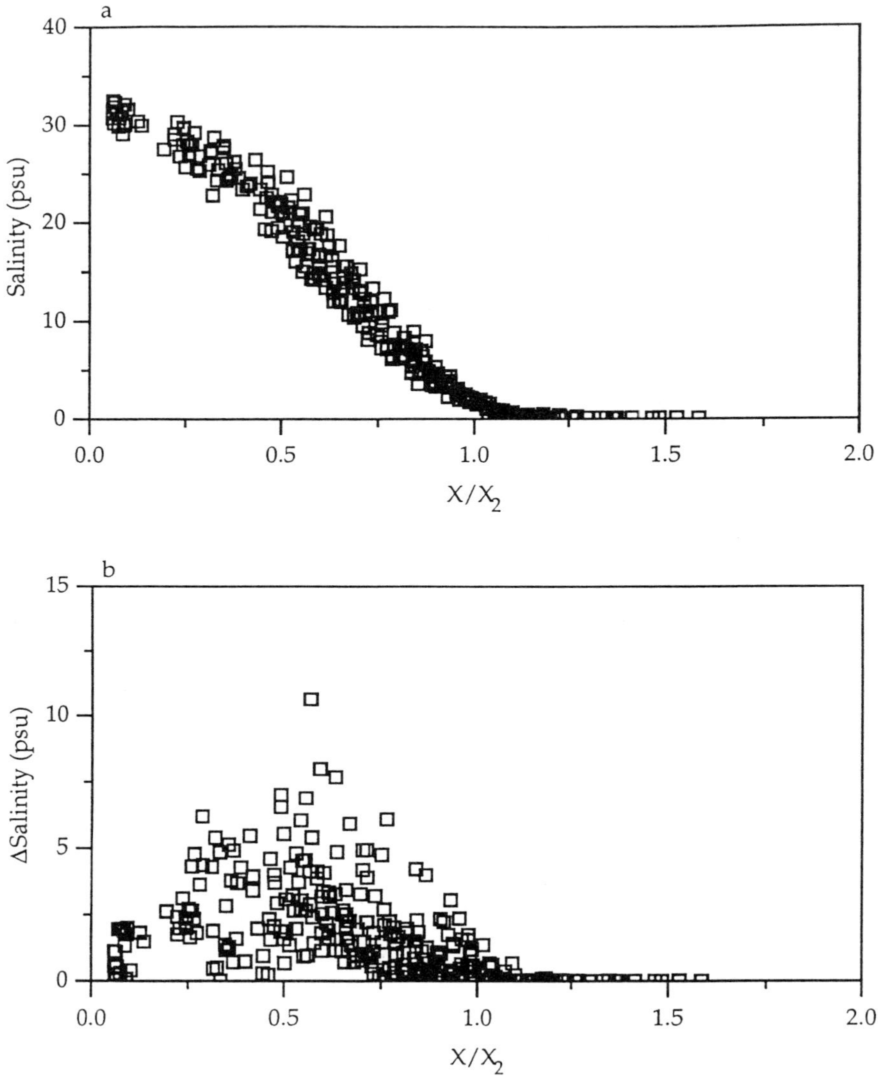

FIGURE 2. (a) Depth-averaged salinity as a function of X/X_2 for 1990-1992, where X is the distance along the channel from the Golden Gate, and X_2 is the estimated distance to where the bottom salinity is 2 psu (after Jassby *et al.* 1995); (b) Top-bottom salinity difference as a function of X/X_2 for 1990-1992 (after Jassby *et al.* 1995).

to 25 psu (psu = practical salinity unit ≈ ppt), top bottom salinity differences varied from 0 to 9 psu, and X_2 varied from 76 to 93 km (Fig. 3b).

At the start of this period, flows were low, and tidally averaged salinities increase slightly, probably reflecting the latter phases of the recovery of the salinity field after a small flow event of 600 m^3/s between 800 and 1000 hours (prior to the period

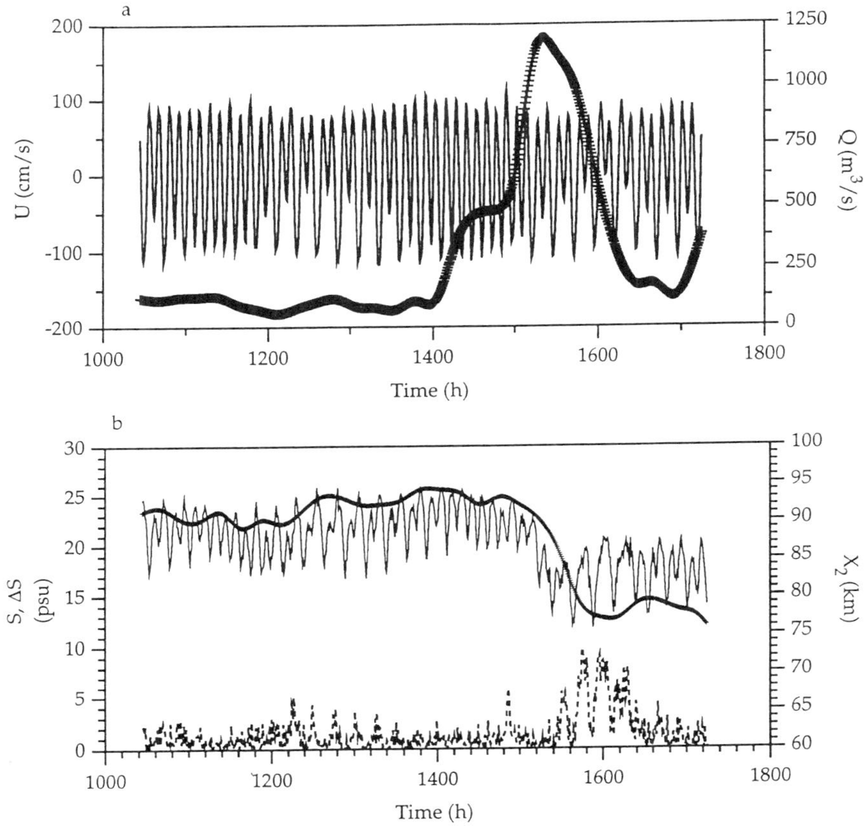

FIGURE 3. (a) Depth-averaged velocity at Wickland as measured by the ADCP (—), and Dayflow estimate of Delta Outflow (+); (b) depth-averaged salinity at Wickland (—), top-bottom salinity difference (- - -), and estimated X_2 (+). In these plots, time is given in hours elapsed since midnight January 1st.

of interest). During this recovery, X_2 moves slightly upstream. As expected, the salinity field responds directly to flow, with both the top and bottom salinities dropping in response to the flow event that starts at 1400 hours. This appears to reflect the downstream migration of the salinity field as measured by a reduction in X_2.

Large tidal variations in salinity are evident throughout the record. For example, the 6 psu fluctuation seen circa 1200 hours are the result of 15 km tidal displacements in the presence of a mean longitudinal salinity gradient we would estimate to be 0.4 psu/km. As noted by Smith & Cheng (1987), salinity and velocity tend to be nearly 90 degrees out of phase in Suisun Bay, reflecting the dominant influence of advection on salinity. Stratification does not directly respond to flow (Fig. 3b). Periodic stratification develops between 1250 and 1350 hours despite the fact that flow remains constant; what changes is the tides. The strongest stratification appears 25 hours after the peak in Dayflow at 1530 hours, and again appears to coincide with a neap-spring change in character of the tides from semi-diurnal to nearly diurnal.

The connection between tidal motions and stratification can be more clearly seen in Figs. 4-6, a sequence of plots of top-bottom salinity difference and the square of the depth-averaged velocity (Figs. 4a to 6a), and of the streamwise velocity component for 3 depth cells, z_1, z_7, and z_{13}, that are 3 m, 9 m, and 15 m above the bottom respectively (Fig. 4b to 6b). In these figures we have also marked periods in which the top-bottom shear is negative; it is during these times that shear should act to create stratification (see eq. 4). In all of these plots, it appears that negative shear is not sufficient for stratification to develop; instead low tidal energy, indicative of weak turbulent mixing, and negative shear are both important.

For example, the sequence of 5 diurnal stratification events that start at 1220 hours (Fig. 4a) all commence in periods when mixing is weak and shear is

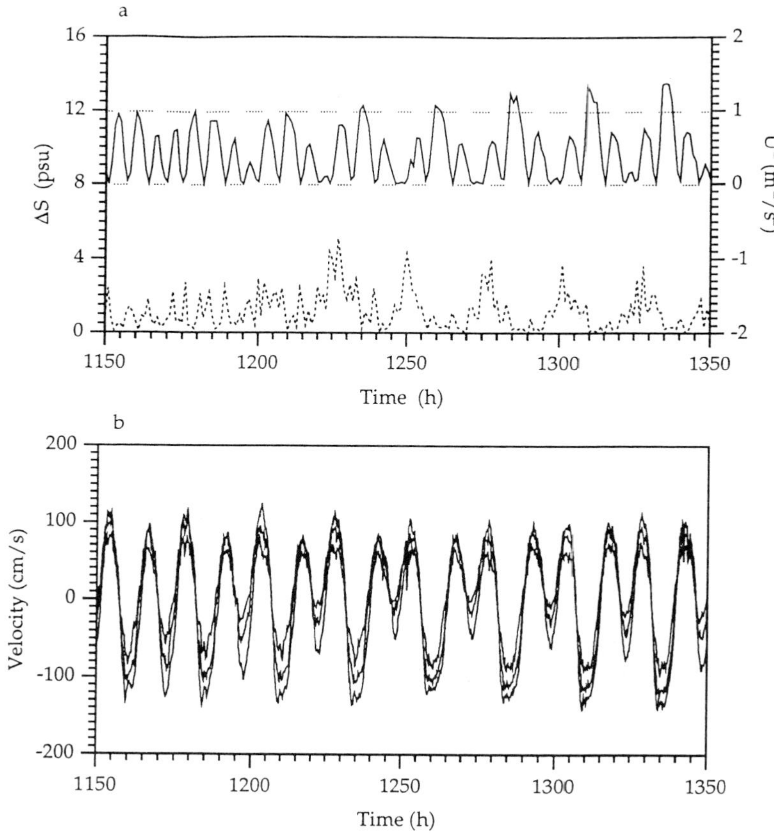

FIGURE 4. Conditions observed between 1150 and 1350 hours in Carquinez Strait: (a) Top-bottom salinity difference (- - -), depth-averaged velocity squared (—), and sign of top bottom shear (dots) - when shear is negative sign is set to 1, otherwise zero; (b) Velocities measured by ADCP 3 m, 9 m, and 15 m above the bed.

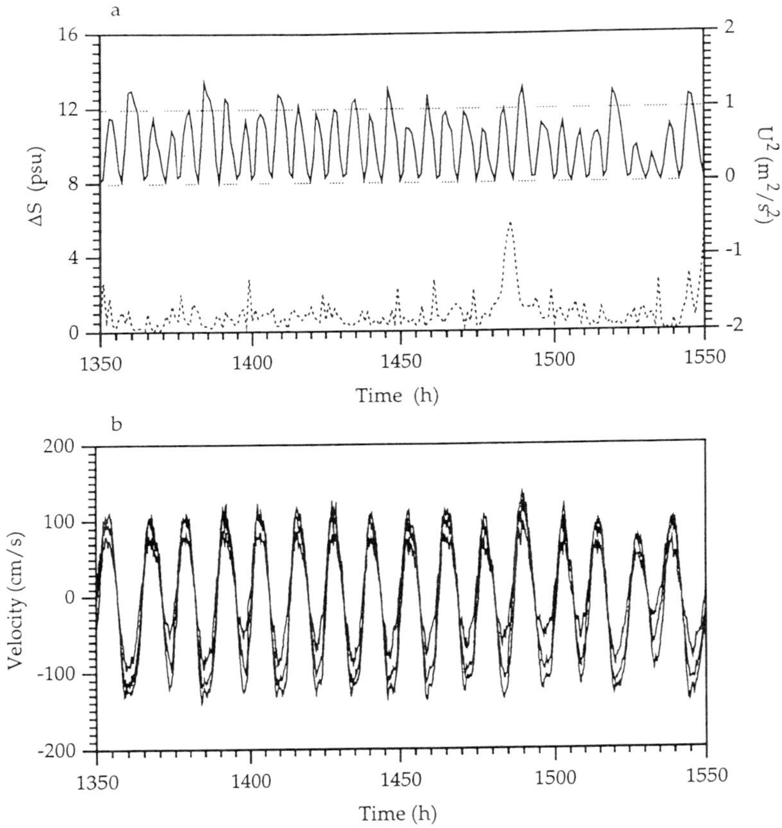

FIGURE 5. Same as 4 but for 1350-1550 hours.

stabilizing. This coincides with alternate weak ebbs. Showing the importance of vertical mixing, the stratification disappears during each of the subsequent strong ebbs despite the fact that these ebbs could, via the SIPS mechanism, lead to stratification formation rather than destruction. This pattern in tidal mixing led to a predominantly diurnal stratification cycle. During this period of time there was little change in Q or in X_2, hence all of the modulation in stratification was associated with variations in the barotropic tide.

Figure 4b shows the development of large tidally varying shears, mainly on the ebbs, with flood shears generally much weaker. This is what Jay & Musiak (1994a) would refer to as ebb-flood asymmetry. Even the weakest ebb shears offer considerable potential for producing stratification. If we assume an M_2 varying sinusoidal shear with a peak value of 0.5 m/s and a longitudinal gradient of 0.4 psu/km, we can calculate that in the 6 hours starting at 1220 hours, tidal straining could create a salinity difference of about 3 psu, slightly less than the 4 psu observed. It seems likely that the difference here is the upper 3 m of the water column the ADCP does not see. This region could contain significant shear, and since it is further removed

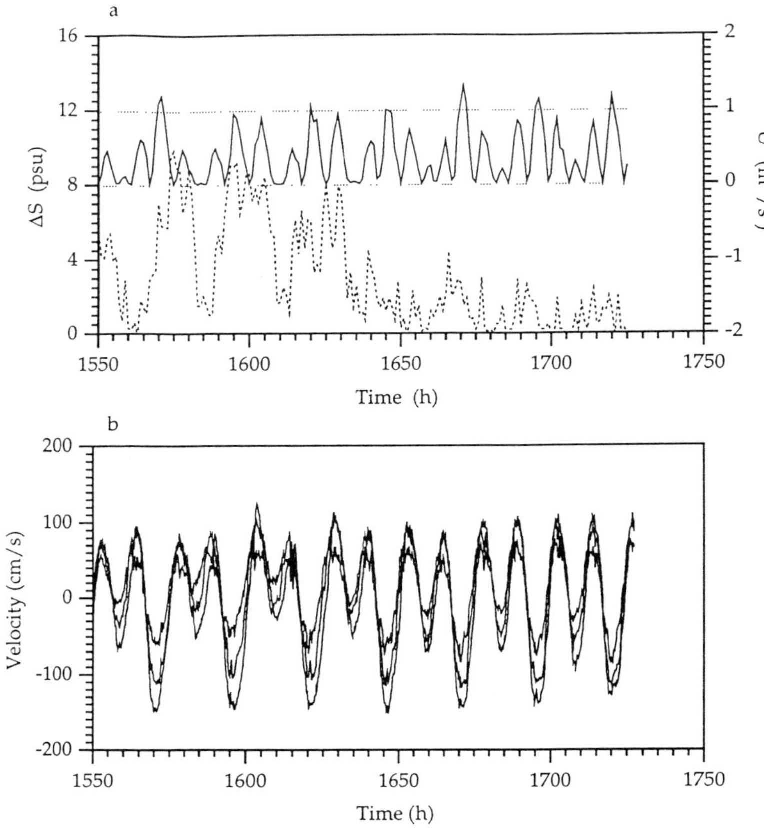

FIGURE 6. Same as 4 but for 1550-1750 hours.

from bottom-produced turbulence could contribute to the strength of the observed stratification.

As the tides strengthened from 1250 onwards, and the two daily tides became more equal, the periodic diurnal stratification weakened, nearly disappearing between 1350 and 1550 (Fig. 5a), excepting an inexplicable stratification event near 1485. During this period, the pattern in the ADCP data of strongly sheared ebbs and weakly sheared floods persisted. It is possible that the Wickland salinity station is not recording stratification, that may have been forming in the channel. Indeed, the observed shear is consistent with stratification formation and destruction.

Finally, in the last period of the record, the flow event from 1400 to 1600 overlaps with a neap tide near 1580 (Fig. 6a), producing the strongest stratification (9 psu) and shear (100 cm/s). Strong shears are even seen throughout the flood starting at 1600. Again, as tidal energy increased from 1580 onwards, stratification and shear weaken despite the lack of any change in X_2 and hence in baroclinic pressure gradient. Even during the period of strongest baroclinic response, the stratification retains its diurnal character. By 1730, the stratification returns to background levels.

3.3. Observations in the Sacramento River

To illustrate that the observations discussed in §3 are not unique to Carquinez Strait, we present in Fig. 7 a data set acquired in the Sacramento River near channel marker 14 (below Decker Island) in the Sacramento River (see Fig. 1) by the Corps of Engineers. This data set shows top and bottom currents and salinities for several days in 1991. Note that the "chopped off" appearance of the salinity data is because the maximum reading of the salinity sensors was set to 4 ppt.

As in the Carquinez data, tidal stratification and de-stratification, along with the concomitant development of sizeable top-bottom shear, are evident. As before, stratification and shear are associated mainly with ebbs. Given the relatively low salinities seen at this site, *i.e.*, bottom salinities in the range of 1 to 4 ppt, the small degree of stratification that forms (a few tenths of a ppt) is consistent with the relatively weak salinity gradient that appears to exist near X_2 (see Fig. 2a). It is impressive that even this degree of stratification is sufficient to dramatically modify vertical momentum transport so as to produce significant shear.

In one regard, this data seems to differ from that measured farther downstream: strong shears and strong stratification occur during strong rather than weak ebbs. This may be the result of the fact that the longitudinal salinity gradient should be

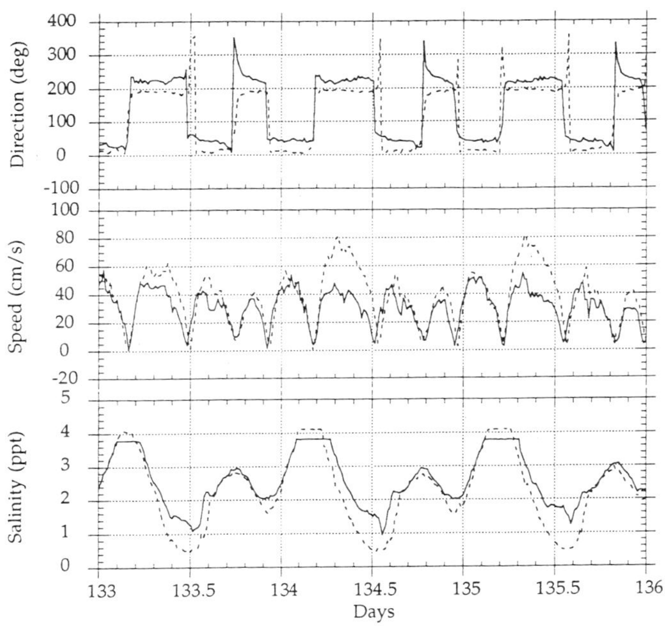

FIGURE 7. Surface/near-bed salinities and tidal currents on the Sacramento River at channel marker 14. Surface sensor (- - -) and near-bed sensor (—). At this station, the channel alignment is such that ebb flows are directed at roughly 235 degrees true north while floods are at 45 degrees.

much stronger during the strong ebb at this location because salinities are higher at the end of flood, and hence so are horizontal salinity gradients.

3.4. Analysis of Carquinez ADCP data using Principal Components Analysis

The dynamic nature of the vertical velocity shear is well illustrated through the use of principle components analysis (PCA). PCA is a multivariate statistical method that separates a set of (observed) time series into a number of orthogonal (independent) spatial functions, or principle components, each of which has an associated amplitude, varying in time (see Kundu & Allen 1976; Walters & Gartner 1985). It has been widely used in oceanography to extract the behavior of different modes of motion or fluctuation from observations of currents, water levels, salinities, etc.

The relative importance of each of the principal components is expressed as a percent of the variance of the original data set explained by each component. The original data time series can be reconstructed by a linear combination of the amplitude of each principle component at that depth, multiplied by the appropriate temporally-varying amplitude. PCA is most useful when a physically relevant mode can be identified, since then it can be used to define the time behavior of that mode (Kundu & Allen's 1976). Thus, as we will see below, the separation of baroclinic (gravitational circulation) and barotropic (tides) flows need not be defined by temporal filtering. As we will demonstrate below this is especially important when examining gravitational circulation, which appears to fluctuate in strength diurnally and semi-diurnally, and thus is not, strictly speaking, as is commonly assumed, a subtidal phenomenon.

To analyze ADCP data, complex PCA (Preisendorfer 1988) must be used: The horizontal velocity vector in a given bin is split into a real part, the streamwise velocity, U, and an imaginary part, the transverse velocity, V, *i.e.*, we from the complex velocity Q as:

$$Q(zi\ ,t) = U(z_i\ ,t) + i\ V(z_i\ ,t), \qquad (6)$$

where z_i is the velocity in the ith bin. The rest of the analysis proceeds as with a single set of time series, albeit with complex covariance matrices, rather than real ones, yielding complex structures for the principal components, *i.e.*, structures that have both transverse and streamwise variability. Without detailing the specific mathematical manipulations, the result is that we write Q as:

$$Q\ (z_i,t) = \sum_{m=1}^{\infty} A_m(t)\ \Phi_{\ m}(z_i) \qquad (7)$$

where $A_{\rm m}$ the complex amplitude of the mth mode and $\Phi_{\rm m}$, the complex vertical structure function for the mth mode, are calculated from the $n \times n$ covariance matrix formed with Q.

The Carquinez Strait ADCP data set yielded suitable time series of velocity in 14 bins, accordingly, PCA gave 14 eigenfunctions ($n = 14$). According to the "scree" test which examines the way in which the variance explained depends on mode number (Preisendorfer 1988), the first three of these are statistically signifi-

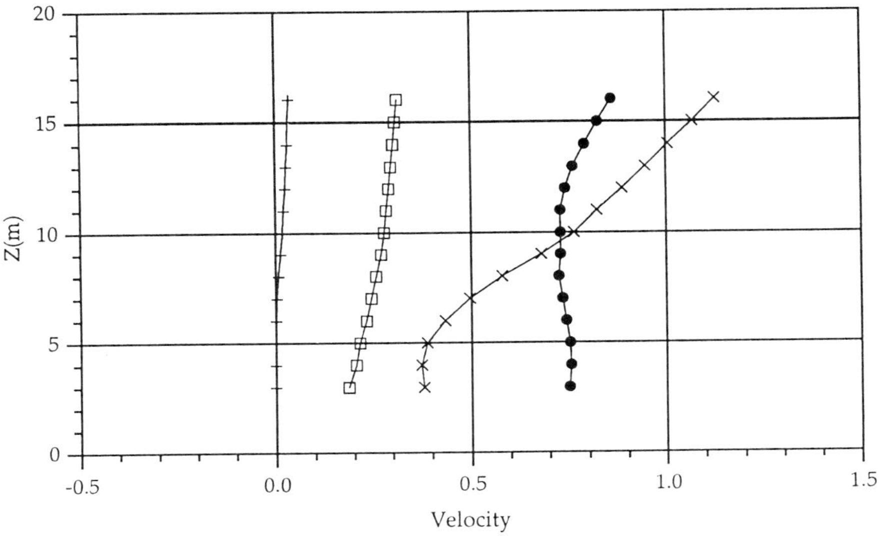

FIGURE 8. Shapes of the first two principal components of ADCP velocity data from Carquinez Strait, February/March 1991: PC_1 ( ) U(z), (+) V(z); PC_2 (•) U(z), (×) V(z). U and V for PC_2 have been offset by 0.75 units for clarity.

cant. The vertical structure of the first two principal components which result are displayed in Fig. 8. The first principal component, PC_1 has the boundary-layer appearance one would expect for the barotropic tide and rotates a little with depth, presumably due to curvature in Carquinez Strait (Geyer 1993). The velocity shear seen in PC_1 is somewhat larger than would be expected for an homogeneous boundary layer, suggesting that this mode is influenced by stratification. Although it too turns with depth, the second principal component, PC_2, has the classic profile of gravitational circulation, with two counterflowing layers. The first component accounts for 97.5% of the variance in the ADCP data while the second component accounts for nearly 80% of the remaining variance. Clearly, these two components define the flow.

The time dependence for PC_1 and PC_2 are given in Fig. 9 where it can be seen that both modes oscillate semi-diurnally. A comparison of the time series of PC_1 amplitude with harmonic predictions made for nearby station C24 shows that the amplitude of PC_1 is predictable (a linear relation can be fit with $R^2 = 0.95$) and thus that PC_1 is the barotropic tide.

While the vertical structure of PC_2 appears to be that which would be expected for gravitational circulation, its time variation is different in that it fluctuates tidally rather than slowly responding to changes in longitudinal salinity gradient or tidal energy. However, confirming the fact that it accounts for the non-boundary-layer portion of the top-bottom shear, the low frequency variability of PC_2 tracks that of $\Delta U = U(z_{13}) - U(z_1)$. This is illustrated in Fig. 10 where a 30 h moving-average smoothing has been applied to both these variables to remove tidal variations. The correspondence of the smoothed ΔU and PC_2 time series offers strong support for

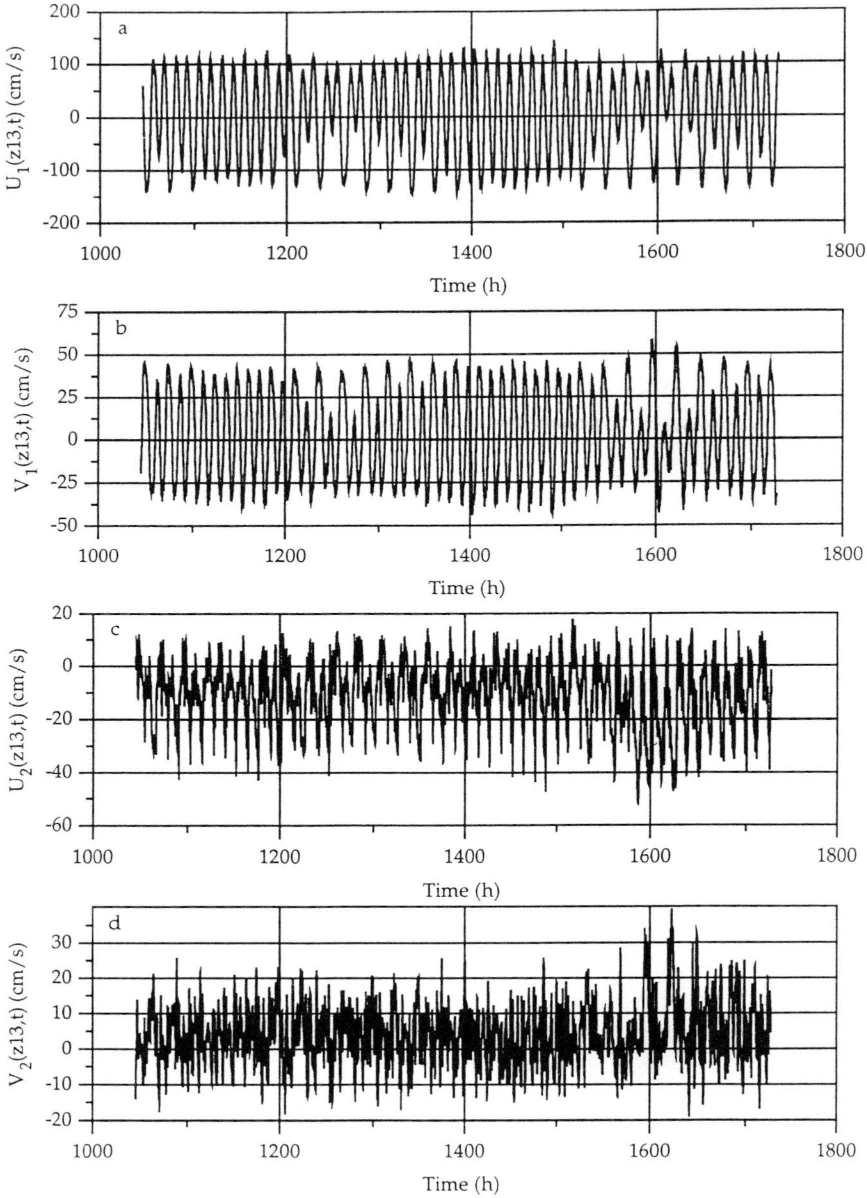

FIGURE 9. Time series of PC_1 and PC_2 amplitudes for ADCP data from Carquinez Strait, February/March 1991: (a) PC_1:$U(z_{13}, t)$; (b) PC_1:$V(z_{13}, t)$; (c) PC_2:$U(z_{13}, t)$; (d) PC_2: $V(z_{13}, t)$.

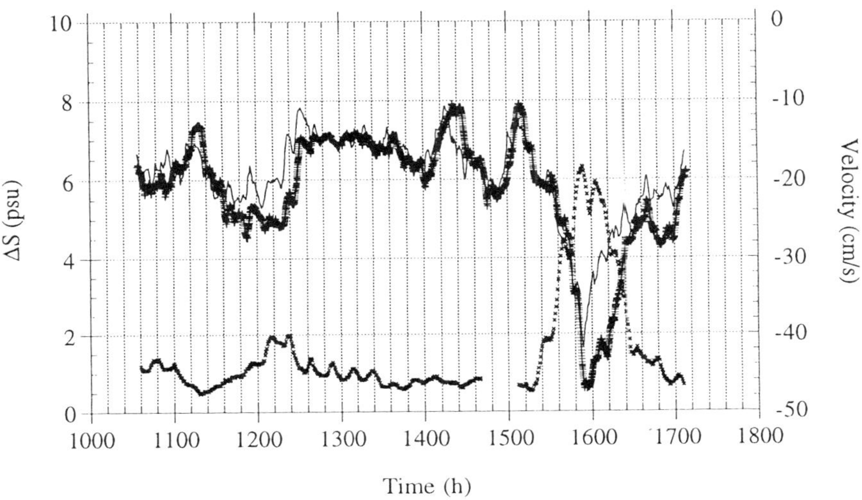

FIGURE 10. Subtidal behavior of: ΔS (×); $U2(z_{13},t)$ multiplied by 1.92 to show correspondence to ΔU (+); and PC_2 (—).

our interpretation of PC_2 as a gravitational circulation mode. Most importantly, as we discussed in §1, the connection between ΔS, also smoothed to remove tidal variations, and PC_2, is striking. When ΔS is large, PC_2, *i.e.*, gravitational circulation is strong; conversely, when ΔS is small, gravitational circulation is weak.

To examine how the barotropic and baroclinic flows behave at the tidal time scale, we focus separately on three periods during the flow: the first neap tide from hours 1195-1295 (neap 1), the spring tide from hours 1370-1470 (spring) and the second neap tide from hours 1545-1645 (neap 2). These are given as Fig. 11a, 11b, and 11c respectively.

Neap 1 (1195-1295)

During this period, the first principal component displays the strong diurnal inequality that characterizes neap barotropic tides in this part of San Francisco Bay at this time of year. (Cheng & Gartner 1984) At the same time, the baroclinic flow consists of a series of "pulses" of exchange flows (in the streamwise direction only – the same pulsing does not appear in the cross-channel direction). PC_2 tends to "turn on" for periods ranging from 1 or 2 hours to as many as 5 hours. During these pulses, the surface current is strongly down-estuary, at as much as 35 cm/s, implying similarly intense up-estuary flows near the bottom. These pulses seem to extend from near the maximum ebb tide to just after the start of the flood tide.

In terms of SIPS, we would expect that when turbulent mixing is weak, a stratified water column will be produced during ebbs that then will destratify during the ensuing flood. The strong pulses of baroclinic flow which begin during the weak ebbs (at hours 1220, 1245, and 1270) thus should be coincident with stratification. During the strong ebb tides (at hours 1210, 1235, 1260, and 1285), the behavior of the baroclinic flow alters and displays a "double-pulse" behavior.

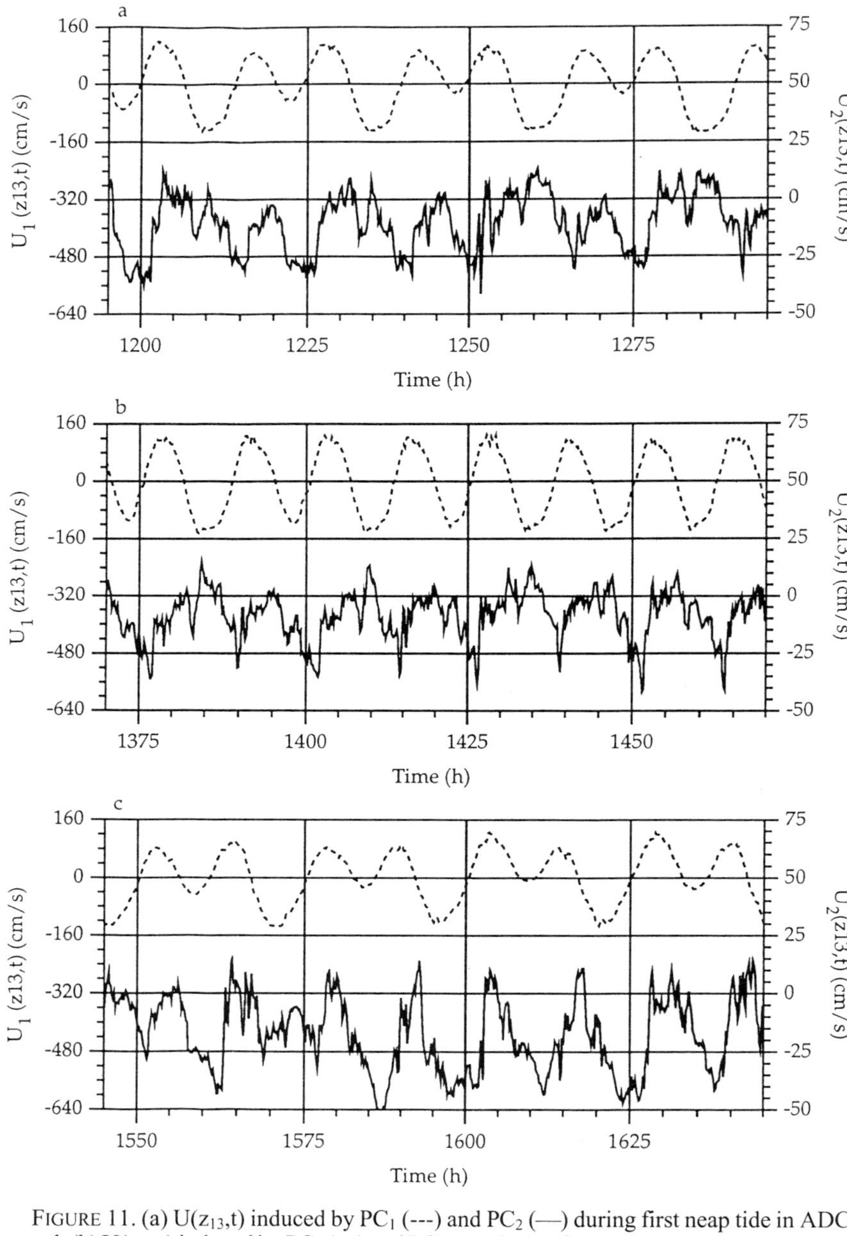

FIGURE 11. (a) $U(z_{13},t)$ induced by PC_1 (---) and PC_2 (—) during first neap tide in ADCP record; (b) $U(z_{13},t)$ induced by PC_1 (---) and PC_2 (—) during first spring tide in ADCP record; (c) $U(z_{13},t)$ induced by PC_1 (---) and PC_2 (—) during second neap tide in ADCP record. Note that this neap tide followed a flow event of 1200 m^3/s.

When the barotropic tide accelerates, the baroclinic flow turns on briefly, then turns off as the barotropic tide nears its maximum, and the water column becomes well-mixed.

During the ensuing barotropic deceleration, the baroclinic flow again turns on and continues to act until the following flood tide. Just as above, we can hypothesize that tidal straining of the density field plays an integral role in establishing this behavior. As the flow begins to ebb, tidal straining begins to stratify the water column; as in the weak ebb case, the resulting stratification allows the baroclinic exchange flow to develop. In this case, however, the flow continues to accelerate, producing increased mixing, until the stratification is broken down, thus shutting off the baroclinic flow. Turbulent mixing prevents stratification from forming through the maximum ebb. Then, during the deceleration, turbulent mixing decreases, and tidal straining again stratifies the water column, resulting in a reappearance of the baroclinic flow. The net result is the double-pulse behavior which is seen near hours 1210, 1235, 1260, and 1285.

Spring (1370-1470)

During the spring tide we show the tides have uniform strength on ebb and flood (about 130/s) and there is no diurnal inequality evident (except slightly weak ebbs at hours 1370 and 1395). The result of this stronger, sustained barotropic forcing is a marked change in the behavior of the second principal component. Instead of sustained pulses of baroclinic flow, the time behavior of the exchange flow consists of a series of "spikes" (particularly after hour 1410 when the barotropic tide becomes truly symmetric). The spikes of baroclinic flow differ from the pulses seen during the neap tide both in timing and duration. While the baroclinic flow pulses lasted several hours during the neap tide, during the spring tide they tend to appear and disappear in less than an hour. Further, the spikes tend to occur as the barotropic tide turns from ebb to flood, rather than just after the maximum ebb.

Again the interaction of the density stratification with turbulent mixing is central: During spring tides, energetic barotropic flows keep the water column from becoming significantly stratified and thus gravitational circulation remains weak. The observed positive and negative spiking of PC_2 is the result of a difference in phase between different parts of the water column, largely due to the differential effects of bottom friction acting on a stratified water column (Monismith & Fong 1995).

Neap 2 (1545-1645)

As during the first neap tide, PC_1 displays a strong diurnal inequality during the second neap period. The time series of the PC_2 is characterized by a series of pulses that are much stronger than those seen during the first neap. In fact, the strongest pulse (just after hour 1585) results in a surface velocity of about 50 cm/s. A surface velocity of this magnitude would correspond to a bottom velocity of about 41 cm/s, or a top-bottom velocity difference of 91 cm/s. Clearly, the effect of stratification that follows the freshwater flow event in this period has been to intensify and prolong the pulses of baroclinic flow.

This event emphasizes the nonlinearity inherent to SIPS: the strong pulse of exchange flow near hour 1585 also acts to rapidly stratify the water column. Taking a value of 90 cm/s for the top-bottom velocity difference, and a typical horizontal salinity gradient of 0.5 psu/km, the strong pulse of exchange flow at hour 1585 would stratify the water column at a rate of 1.6 psu/hr. The pulse lasted for several hours, resulting in a top-bottom salinity difference which was probably greater than 5 psu. In fact according to Fig. 6, the maximum stratification was approximately 8 psu at 1590 hours. The ensuing flood did not break down this stratification and the level of stratification at the beginning of the next ebb was not zero; thus the double pulse behavior seen in the first neap was replaced by a single pulse of longer duration. The increased intensity and persistence of the exchange flow is what leads to the strong shear evident even after low-pass filtering.

3.5. The Creation of Residual Gravitational Circulation

The mean flow recorded by the ADCP is shown in Fig. 12 along with a cubic fit to the mean flow described by the theoretical shape given (for example) in Prandle (1991) (*i.e.*, eq. 2) and calculated assuming that $\Gamma = 0.46$ psu/km, a net specific discharge of 0.3 m^2/s, and a vertical eddy viscosity $\nu_t = 3.7 \times 10\text{-}3$ m^2/s.

While the fit to theory appears good, both the discharge and nt were adjusted to give a best fit (the latter as in Lung & O'Connor 1984). The specific discharge is roughly twice as large as what would be calculated by dividing the average flowrate (300 m^3/s) by the local width (2000 m). This is not too serious a discrepancy given that this purely Eulerian analysis doesn't account for the inwards Stokes drift flow (Jay 1991), and even the Eulerian mean flow need not be distributed uniformly across the cross section (Fischer 1972).

However, the eddy viscosity appears to be roughly 10 times smaller than would

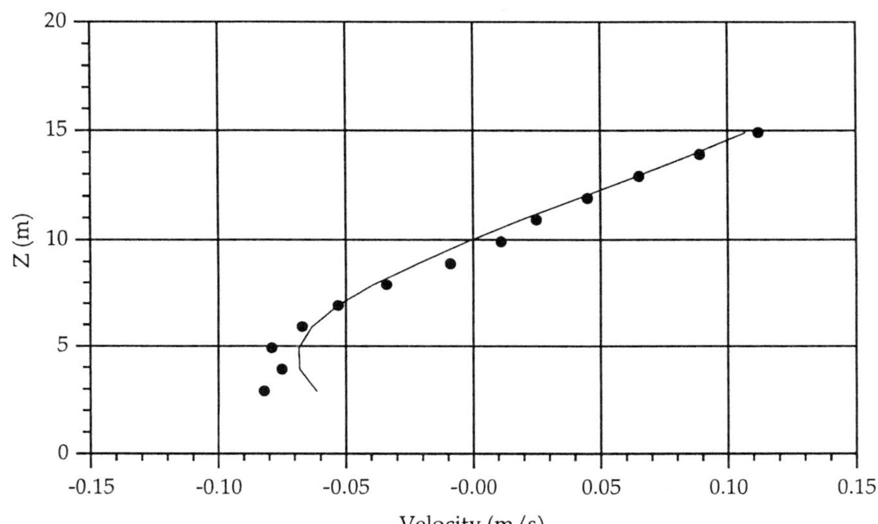

FIGURE 12. Time-averaged velocity profile measured by ADCP (•) in Carquinez Strait for February/March 1991 compared with theoretical profile for gravitational circulation (—).

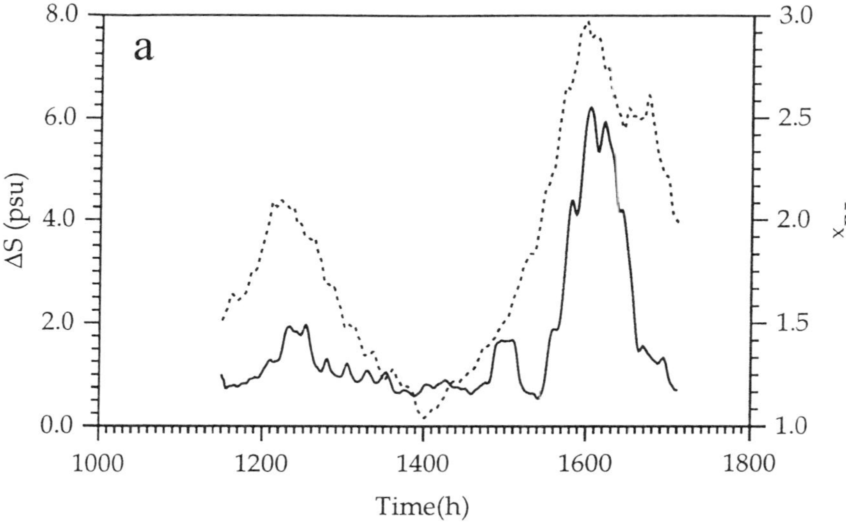

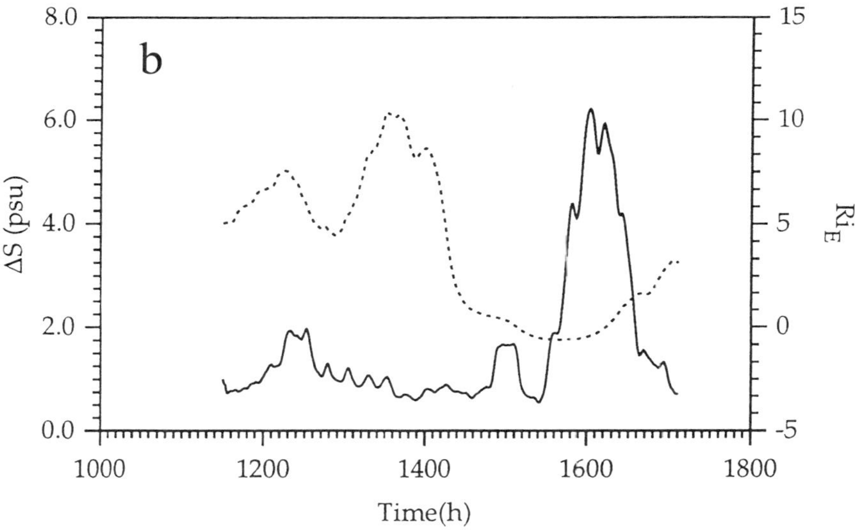

FIGURE 13. Time variations of stratification and stability parameters: (a) SIPS Richardson number, Ri_x (----), and ΔS (—); (b) Estuarine Richardson number, Ri_E (---), and ΔS (—).

have been predicted using the standard formula for the depth-averaged viscosity (Prandle 1991) based on the fluid depth and on the rms tidal current. It is possible that this reduction is the averaged effect of the time-varying stratification, however, the stratification correction Uncles *et al.* (1990) used for modeling flows in the Tamar only gives a factor of two reduction in mixing rates when applied to this data set, leaving a substantial discrepancy.

Alternatively, the ADCP data itself makes the case that the gravitational flow is highly unsteady, behavior that is entirely consistent with the underlying physics. Thus, as we have seen above, variations in salinity stratification, vertical shear and gravitational circulation are all intertwined. Two key pieces of this puzzle are the stabilizing effect of stratification on turbulence and the creation and destruction of stratification by differential advection. The net effect is periodic stratification that is similar to but not quite identical to that which would be expected from the SIPS model advanced by Simpson and his colleagues. Accompanying this periodic stratification are periods of intense gravitational circulation. Looking at subtidal variability, when several strong stratification events occur sequentially, the concomitant gravitational circulation pulses appear as a sustained period of gravitational circulation.

4. MODELING STRATIFICATION DYNAMICS

4.1. Model Philosophy

To model the stratification response of the northern reach to flow and tide variations one can either proceed by applying a full three-dimensional hydrodynamic model (*i.e.*, Casulli & Cheng 1992) or by simplifying the flow in some way. The flow to be calculated can be simplified by averaging over the width of the estuary; Ford *et al.* (1990) presented such model which they applied with some success to the northern reach.

A third alternative is one in which streamwise variations in all flow properties (excepting salinity) are eliminated and the dynamics of a complex three-dimensional estuary are reduced to that appropriate to a single hypothetical water column (Nunes-Vaz & Simpson 1994). This approach, which we pursue, represents a dynamic version of the steady state analytical theories that have traditionally been used to describe these flows.

4.2. The Model

We used Blumberg *et al.* (1992)'s model, referred to hereafter as the BMY model3, to calculate water column structure. Rather than use parametric representation of turbulent mixing processes like billowing, *etc.* (Monismith & Fong 1995), this model uses formal turbulence closure to obtain eddy mixing coefficients and so close the mean momentum and salt equations. As in other 1D models (*e.g.*, Simpson & Sharples 1992, Nunes-Vaz & Simpson 1994), the flow is assumed to be homogenous in both horizontal directions, thus eliminating gradients of everything except pressure and salinity in the x directions. If we neglect rotation, we have two equations for the two unknowns:

$$\frac{\partial U}{\partial t} = \frac{-1}{\partial_0}\frac{\partial P}{\partial x} + \frac{\partial}{\partial z}\left(K_m \frac{\partial U}{\partial z}\right) \tag{8a}$$

$$\frac{\partial S}{\partial t} = -\Gamma\, U + \frac{\partial}{\partial z}\left(K_s \frac{\partial S}{\partial z}\right) \tag{8b}$$

To model SIPS, we have included streamwise advection of the salt field in the

[3] This code was graciously provided to us by Dr. Alan Blumberg of Hydroqual, Inc.

salt balance using a value of Γ that is constant. In contrast, Sharples *et al.* (1994) found it necessary to use time and depth-dependent values of Γ to qualitatively reproduce stratification in the York River Estuary. The eddy mixing coefficients K_m and K_s are solved for using the Mellor-Yamada level 2.5 turbulence closure (Mellor & Yamada 1982) which defines them in terms of the turbulence length scale, L, and turbulence velocity scale, q, and two empirical constants, C_m and C_s as

$$K_m = C_m q L \tag{9a}$$
$$K_s = C_s q L \tag{9b}$$

Two more transport equations must be solved for q and L; these are based on approximation of exact equations for transport of the Reynolds stresses.

The Mellor-Yamada level 2.5 closure is widely used in modeling coastal and estuarine flows; it is thought to be reasonably effective for describing effects of stratification on turbulence (Blumberg *et al.* 1992). However, in the form we use, there is an additional significant buoyancy modification. The master length scale L is subject to the stratification limit prescribed by Galperin *et al.* (1988):

$$\frac{q}{NL} \leq 0.53 \tag{10}$$

where N is the buoyancy frequency. This incorporates the well-known tendency of stratified turbulence to limit the vertical length scale of turbulence to be no larger than the Ozmidov scale

$$L_o = \left(\frac{\varepsilon}{N^3}\right)^{1/2} \tag{11}$$

where e is the rate of turbulence dissipation ~ q^3/L (see Turner 1973).

The one-dimensional model requires that the pressure gradient be specified. While in principle this could be done using pressure gradients derived from a 2D circulation model like that described in Smith & Cheng (1987), we split the pressure gradient into two components. The first oscillates with period of 12 hours (approximately the M_2 tide) and is the same as used by Monismith & Fong (1995). For reference sake this is:

$$\frac{-1}{\rho_0}\frac{\partial P_{M_2}}{\partial x} = \frac{2\pi U_{\max}}{T_{M_2}} \cos\left(\frac{2\pi t}{T_{M_2}}\right) \tag{12}$$

The second component, the baroclinic pressure gradient, drives gravitational circulation; it is derived from the longitudinal salinity gradient. It has two components, one associated with the direct effect of the longitudinal salinity gradient, the other with the surface expression of the baroclinic pressure gradient. The combination of is:

$$\frac{-1}{\rho_0}\frac{\partial P_B}{\partial x} = g\beta\Gamma\left[z - H + \alpha\left(\frac{H}{2}\right)\right] \tag{13}$$

where $z = 0$ is the bottom, $z = H$ is the free surface. The free parameter α is used to adjust the tidally averaged flow in the model (Simpson & Sharples 1991).

The full set of equations are advanced in time using central differences in space and time, with a smoothing function used to eliminate temporal oscillations (see Blumberg & Mellor 1987). To model the Carquinez Strait data, we used a water column of 20 m discretized in 10 cm vertical increments. We found by experimentation that a time step of approximately 3 minutes was needed to give results that were independent of time step. The bottom roughness, z_0, which must be specified, was set to 1 cm. Adjustment of the flow to the given initial conditions were taken care of by allowing the flow to "spin-up" for one tidal cycle. This meant running the model so as to allow turbulence fields to develop and for the mean velocity profiles to adjust to the applied pressure gradients and mixing coefficient distributions before we allowed the salt field to evolve. Once the flow was initialized, we ran the code to simulate 30 tidal cycles, or 15 days.

4.3. Model Results

Initially (and rather naively), we chose $U_{max} = 0.75$ m/s (although as seen in the ADCP data it could easily be more during spring tides) and $\Gamma = 0.5$ ppt/km. With these parameters specified we ran the code for different values of α. Depending on the choice of α, the water column went into one of three states: When a was set for a strong river flow, the water column would become completely fresh and the resulting flow would show a net outward flow superposed on the tidal oscillations. When a was set for a strong "negative" river flow (salt water moving upstream), the water column would become completely saline (match the ocean salinity). The third state, one of zero net flow, a condition that must be determined by repeatedly running the model, was a completely stratified regime, with completely fresh water flowing out at the surface and ocean water flowing in at the bottom. Apparently, the tidally-produced mixing energy was not sufficient to break down the stratification induced by the baroclinic motions.

However, by repeatedly running the model with different values of U_{max} , Γ, and H, it became evident that it was possible for the tidally produced turbulent mixing to balance the stratifying effects of Γ and create periodic stratification. Empirically we found that boundary between periodic stratification and the "two-layer" estuary was determined by the expression

$$U_{max} = 0.12 H \Gamma^{1/2} \tag{14}$$

which has a simple physical interpretation: At the threshold velocity, we expect that the tidal generation of turbulent kinetic energy (TKE), say E_{tides}, exactly balances the buoyancy flux, B_{strat}, associated with the stratifying effects of baroclinic forcing (Abraham 1988):

$$E_{tides} \sim B_{strat} \tag{15}$$

The rate of creation of turbulent kinetic energy (TKE) by the barotropic tides will vary as:

$$E_{tides} \sim C_D U_{max}^3 \tag{16}$$

where C_D is the bottom drag coefficient. Note that in the Blumberg *et al.* (1992) implementation of the Mellor-Yamada closure, C_D is not specified, instead a bottom roughness z_0 is given. In general, C_D depends on the height to which it is referenced. For example in our calculations if u_* is related to the depth averaged tidal velocity, we find $C_D \approx 0.002$.

The horizontal buoyancy flux that drives stratification formation is

$$B_{strat} \sim g\beta\Gamma U_{max} H^2 \tag{17}$$

Equating these two estimates (assuming a constant efficiency of conversion of TKE to potential energy – see Ivey & Imberger 1991) we expect that the non-dimensional stratification parameter

$$Ri_x = \frac{g\beta\Gamma H^2}{C_D U_{max}^2} \tag{18}$$

determines the boundary between flow states. Using our model runs, and assuming that $C_D \approx 0.002$, we find a critical value of $Ri_x \approx 0.25$. When Ri_x is greater than this O(1) constant, tidal energy is not sufficient to overcome the stratifying effects of baroclinic pressure gradients, tidal straining and frictional shears; accordingly, the water column will stratify. On the other hand, when it is subcritical, our model predicts that tidal mixing will be able to eliminate stratification caused by tidal straining and by the baroclinically forced mean flow.

4.4. Comparison of Model Results to Observations

The model runs and scaling analysis we discuss above suggest that during particularly strong tides, tidally produced turbulence should be able to overcome the stratifying effects of longitudinal variations in density and weak periodic stratification should appear. Conversely, when the tides are "weak" (in the sense defined by eq. 18), the tendency to create stratification will win out, possibly creating stratification that is not broken down during flood tides, and thus which strengthens during successive ebbs.

In our idealized model, this is the route to a "river-over-ocean" estuary. In reality, this scenario is not completed for several reasons:

(1) Γ is a function of mean salinity (see Fig. 2a) – hence the assumption that Γ is does not depend on position in the water column or on time breaks down;

(2) We omit the effects of cross-sectional, *i.e.*, shoal-channel, variability that could provide, in effect, additional vertical mixing (Huzzey & Brubaker 1988);

(3) More generally, dispersion due to horizontal variability of tidal currents (Zimmerman 1986; Fischer *et al.* 1979) will also tend to act to maintain salinities at intermediate values – *i.e.*, the 1D model can not represent the full salt balance that determines Γ.

Nonetheless, we can hypothesize that Ri_x is a good parameter for delineating stratification behavior. We tested this hypothesis using the 1991 Carquinez Strait field data discussed in §3. We first interpolated the salinity, velocity, flow and X_2 data to a common timebase with 1 hour increments. X_2 is required in order to calculate Ri_x as a function of time since using figure 2b,

$$\Gamma \approx 37.5/X_2. \tag{17}$$

Note that this representation of the longitudinal salinity gradient is a coarse approximation to reality as in reality, Γ may vary somewhat in time and space. Using this approximation, we find that

$$Ri_x \sim \frac{g\beta S_0 H^2}{C_D U_{\max}^2 X_2} \tag{18}$$

which depends explicitly on X_2. For the sake of comparison, we also calculated the Estuarine Richardson number, E, defined by eq. (5). Note that Ri_E is dependent on Q as opposed to Ri_x, which depends on X_2 . In essence these two parameters differ only in their estimates of the horizontal buoyancy flux, *i.e.*, one, Ri_E, that assumes that the major salinity flux is by mean advection as opposed to one, Ri_x, that assumes that the major salinity flux is purely tidal.

Figure 13a shows the resulting time series of Ri_x and ΔS in Carquinez Strait for the Feb-March 1991 period plotted using 30h smoothed values of U^2 and ΔS. Remarkably, ΔS tracks Ri_x, with the two major peaks in ΔS corresponding to the two major peaks in Ri_x. In contrast, Fig. 13b, a similar plot of Ri_E and ΔS , shows that Ri_E does not describe the flow dynamics that result in the peak in ΔS that occurred near 1240 hours, nor the timing of the response to the peak in Q near 1550h. Our interpretation of the behavior seen in Fig. 13a is that twice during this period the flow passed through the stability boundary between runaway stratification and purely periodic stratification. From the sudden rise in ΔS near 1540 hours we deduce that the critical value of Ri_x, at least in Carquinez Strait, is 2, approximately 8 times larger than what we calculate.

The difference in the critical value of Ri_x deduced from the observations and from modeling may be due to several causes:

(1) For the sake of comparison, the filtered U^2 should be multiplied by two factors: (a) a factor of 2 which corresponds to the average of the square of a sinusoid (our model tide); (b) a factor of 2 which represents the difference between the square of the calculated surface currents for these conditions and the effective current which represents the tidal pressure gradient. Since we have used the observed surface currents, and not the pressure gradient, in calculating Rix, this suggests that we are off by a factor of two (= 8/(2*2)) in our estimate of the critical value of Ri_x.

(2) It could be effect of transverse variations in the flow properties. As seen in our PCA analysis, and in plots of ADCP data from Carquinez Strait given in Agostini *et al.* (1993), there appear to be relatively strong O(10 cm/s) transverse flows organized in several vertically circulating cells. These could be the result of flow curvature in Carquinez Strait, Coriolis forces, and spanwise variations in depth. These could serve to increase vertical mixing by increasing vertical shear and hence turbulence, and by advective changes in the salinity field, especially those associated with motions driven by the transverse salinity gradients like those observed by Huzzey & Brubaker (1988). The major potential source of extra mixing is breaking by internal waves, a feature of estuarine flows that is only

beginning to attract attention. For example recent experiments in the Rotterdam waterway suggest that internal wave breaking may contribute as much mixing as bottom-generated turbulence, and potentially at locations remote from their source (J. Imberger, pers. commun. 1993).

In any event, it appears that Ri_x, which is based on the horizontal salinity gradient, which itself depends on X_2, appears to be much better than E, a flow-based parameter, for parametrizing the tendency for the water column to stratify, and thus, for strong gravitational circulation to develop. Moreover, given that $Ri_x \propto (H/U)^2$, and that $U \propto H$ (Walters & Gartner 1985), we would expect that the transition between sustained stratification and periodic stratification should be a global rather than local event. Salinity data collected in Suisun Bay in the spring of 1993 seems to confirm this view (Gartner & Burau 1995).

5. SUMMARY AND CONCLUSIONS

Excepting large flow events, the Northern Reach is the prototype of a partially mixed (or weakly stratified – *c.f.* Jay & Smith 1990) estuary: it has generally weak and highly ephemeral vertical density differences, and persistent horizontal density gradients. As we have discussed above, the horizontal density gradient has important implications, most notably the generation of tidally as well as subtidally varying stratification and gravitational circulation. Thus, we can conclude that for moderate flows, the important effect of river flow is to strengthen the longitudinal salinity gradient by compressing the longitudinally varying sainity field, *i.e.*, by decreasing X_2.

The linkage between stratification and gravitational circulation is such that during neap tides, which, for winter and early spring, typically involve substantial diurnal inequality in the tides, diurnal episodes of stratification and strong gravitational circulation develop. From a physical standpoint, the pulsing of stratification and gravitational circulation is easily understood in terms of the significant reduction in vertical mixing that accompanies stratification. This reduction in mixing allows the baroclinic pressure gradient (arising from the horizontal salinity gradient) to drive stronger flows because off the weakened effect of bottom friction. Moreover, time-varying stratification affects the momentum balance such that barotropic pressure gradients can also produce gravitational circulation when the retarding effects of bottom friction are confined to the lower part of the water column. by the stratification.

Our modeling of an idealized dynamic one-dimensional water column roughly replicated the observed periodicity in stratification and gravitational circulation. This modeling indicated that there exists a critical condition parametrized by a Richardson number, Ri_x, based on tidal conditions and on the longitudinal salinity gradient, Γ. Below the critical value of Ri_x, periodic flow conditions, *i.e.*, SIPS, prevail. When Ri_x increases beyond the critical value, runaway stratification develops.

While our modeling results are in qualitative agreement with field observation, it appears that our one-dimensional water column undermixes, that is the field seems to be more strongly mixed than our model would lead us to expect. This could be the result of ignoring the mixing effects of internal waves, or, given that turbulence closures like that which we use tend to overmix, it the most likely source of the discrepancy arises through our neglect of transverse flow variations like those observed by Huzzey & Brubaker (1988) or Simpson & Turrell (1986).

This mismatch of observations and modeling is not new. In their original modeling of Liverpool Bay, Simpson & Sharples (1991) were forced to use stronger currents in their model than they had observed in order to reproduce observed salinity stratifications. Applying several turbulence closures of varying sophistication and finding results that depended on the closure used, Nunes-Vaz & Simpson (1994) make the case that SIPS provides a tough test for any turbulence modeling. Thus, if three-dimensional hydrodynamic models of the type described by Casulli and Cheng (1992) are to be employed with any predictive skill in northern San Francisco Bay, it seems likely that turbulence models capable of accurately representing the effects of stratification on vertical mixing will be required.

What are the implications of the behavior discussed above for the transport of sediment or organisms? Given that sediment concentrations, for example, fluctuate tidally, it is conceivable that the actual rate of sediment transport is very much different from what one would calculate using mean velocities and mean concentrations. For example, if sediment concentrations were in phase with tidal currents, *i.e.*, peak concentrations occur at or near peak velocities, little sediment would be transported in the fashion that is assumed to maintain the ETM, since gravitational circulation and sediment concentrations would be nearly out of phase.

More generally, if we attempt to postulate circulation patterns that might exchange fluid particles (or organisms) between the shoals and channels of the Northern Reach, we must bear in mind the fact that the gravitational circulation pattern might only be operating for a part of the tidal cycle. One can imagine that a parcel of water leaving the shoals might be carried into a stratified, ebbing water column in the channel, and thus be transported relatively far downstream during the gravitational circulation pulse. At the end of the stratified period, this parcel of water would be mixed with salty water flowing landward underneath it. Regions of the shoals that exchanged fluid in this manner might be expected to have shorter residence times than those for which water parcels entered the channel during phases of the tide when the water column in the channel is not stratified.

Finally, when thinking about any process that depends on rates of vertical mixing, it is important to consider tidal and subtidal variability of stratification. Unlike the simple models of decaying stratification considered in Koseff *et al.* (1993), it may be necessary to consider a SIPS state as the basic flow configuration for, *e.g.*, phytoplankton modeling (J. Koseff, pers. commun. 1995).

ACKNOWLEDGEMENTS

This work was supported by the EPA San Francisco Estuary Project and by the Interagency Ecological Studies Program. We acknowledge the support provided by NSF through grant CTS-8958314 to SGM and support from the US Geological

Survey to JRB. We also recognize the generous gift of computer equipment from Silicon Graphics. George Nichol of the Corps of Engineers provided the Sacramento River data at channel marker 14. The authors thank Ralph Cheng, Jeff Gartner, Pete Smith and Jeffrey Koseff for their help and advice. Finally, SGM thanks Vincenzo Casulli for hosting him in excellent style in Trento, Italy while he finished this paper.

REFERENCES

Abraham, G. (1988). Turbulence and mixing in stratified tidal flow. Pages 149-180 *in* J. Dronkers and W. van Leussen, eds., *Physical Processes in Estuaries*. Springer-Verlag., New York, NY.

Agostini, D., J. Evans, F. Wen, and P. Smith. (1993). Visualization and analysis of multi-dimensional velocity measurements in Carquinez Strait, California. Pages 1096-1101*in* H. W. Shen, S. T. Su, & F. Wen, eds., *Hydraulic Engineering '93*, vol. 2. American Soc. Civil Eng., New York, NY.

Armor, C. A., and P. L. Herrgesell. (1985). Distribution and abundance of fishes in the San Francisco Bay Estuary between 1980 and 1982. Pages 211-227 *in* J. E. Cloern & F. H. Nichols, eds., *Temporal Dynamics of an Estuary: San Francisco Bay*. Junkers, Hingam, MA.

Arthur, J. F., and M. D. Ball. (1979). Factors influencing the entrapment of suspended material in the San Francisco Bay Estuary. Pages 143-174 *in* T. J. Conomos, ed., *San Francisco Bay: The Urbanized Estuary*. Pacific Division, American Association for the Advancement of Science, San Francisco, CA.

Blumberg, A. F., B. Galperin, and D. J. O'Connor. (1992). Modeling vertical structure of open channel flows. *Jour. Hydrol. Div., American Soc. Civil Eng.* 118:1119-1134.

Blumberg, A. F., and G. L. Mellor. (1987). A description of a three-dimensional coastal ocean circulation model. Pages 1-16 *in* N. S. Heaps, ed., *Three Dimensional Coastal Ocean Models*. American Geophysical Union, Washington, DC.

Burau, J. R., M. R. Simpson, and R. T. Cheng. (1993). *Tidal and Residual Currents Measured by an Acoustic Doppler Current Profiler at the West End of Carquinez Strait, San Francisco Bay, California, March to November 1988*. U.S. Geol. Surv. Water Resour. Inv. Rept. 92-4064. 76 pp.

Casulli, V., and R. T. Cheng. (1992). Semi-implicit finite difference methods for three-dimensional shallow water flow. *Internat. Jour. Num. Methods in Fluids* 15:629-648.

Cheng, R. T., and J. W. Gartner. (1984). *Tides, tidal and residual currents in San Francisco Bay, California – Results of measurements, Part 1*. U.S. Geol. Surv. Water Resour. Inv. Rept. 84-4339. 72 pp.

Cloern, J. E. (1982). Does benthos control phytoplankton biomass in south San Francisco Bay? *Marine Ecol. Progr. Ser.* 9:191-202.

Cloern, J. E. (1991). Tidal stirring and phytoplankton bloom dynamics in an estuary. *Jour. Marine Res.* 49:203-221.

Conomos, T. J. (1979). Properties and circulation of San Francisco Bay waters. Pages 143-174 *in* T. J. Conomos, ed., *San Francisco Bay: The Urbanized Estuary*. Pacific Division, American Association for the Advancement of Science, San Francisco, CA.

Denton, R. A., and J. R. Hunt. (1986). *Currents in San Francisco Bay – Final Report*. Hydrol. and Coastal Eng., Dept. Civil Eng., Univ. California, Berkeley. Tech. Rept. UCB/HEL-86/01.

Fischer, H. B. (1972). Mass transport mechanisms in partially stratified estuaries. *Jour. Fluid. Mech.* 53:671-687.

Fischer, H. B., E. J. List, J. Imberger, R. C. Y. Koh, and N. H. Brooks. (1979). *Mixing in Inland and Coastal Waters*. Academic Press, New York, NY. 483 pp.

Ford, M., J. Wang, and R. T. Cheng. (1990). Predicting the vertical structure of tidal current and salinity in San Francisco Bay, California. *Water Resour. Res.*26:1027-1045

Galperin, B., and L. H. Kantha, S. Hassid, and A. Rosati. (1988). A quasi-equilibrium turbulent energy model for geophysical flows. *Jour. Atmos. Sci.* 45:55-62.

Gartner J. W., M. R. Simpson, and R. N. Oltmann. (1995). *Velocity Measurements by Acoustic Doppler Current Profiler and Conventional Current Meters in Suisun Bay, San Pablo Bay, and Carquinez Strait, California, 1990-1991.* U.S. Geol. Surv. Water Resour. Inv. Rept. (in press)

Gartner J. W. and J. R. Burau. (1995). *Report of field measurements during a salt balance study an Suisun Bay, California, 1992-1993.* U.S. Geol. Surv. Water Resour. Inv. Rept. (in press)

Geyer, W. R. (1993a). Three-dimensional tidal flow around headlands. *Jour. Geophys. Res.* (Oceans) 98:955-966.

Geyer, W. R. (1993b). The importance of suppression of turbulence by stratification on the estuarine turbidity maximum. *Estuaries* 16:113-125.

Hansen, D. V., and M. Rattray, Jr. (1966). New dimensions in estuary classification. *Limnol. Oceanogr.* 11:319-326.

Herbold, B., A. D. Jassby, and P. B. Moyle. (1992). *Status and Trends Report on Aquatic Resources in the San Francisco Estuary.* Report to the EPA San Francisco Estuary Project. 257 pp.

Huzzey, L. M. and J. M. Brubaker. (1988). The formation of longitudinal fronts in a coastal plain estuary. *Jour. Geophys. Res.* (Oceans) 93:1329-1334.

Ivey, G. N. and J. Imberger. (1991). On the nature of turbulence in a stratified fluid. Part 1: The energetics of mixing. *Jour. Phys. Oceanogr.* 21:650-658.

Jassby, A. D., W. M. Kimmerer, S. G. Monismith, C. Armor, J. E. Cloern, T. M. Powell, J. Schubel, and T. Vendlinski. (1995). Isohaline position as a habitat indicator for estuarine resources: San Francisco Bay-Delta, California, U.S.A. *Ecol.Applications.* 5:272-289.

Jassby, A. D., J. R. Koseff, and S. G. Monismith. (1995 [1996]). Processes underlying phytoplankton variability in San Francisco Bay. Pages 325-349 *in* T. Hollibaugh, ed., *San Francisco Bay: The Ecosystem.* Pacific Division, American Association for the Advancement of Science, San Francisco, CA.

Jay, D. A. (1991). Estuarine salt conservation: A Lagrangian approach. *Estuarine Coastal Shelf Sci.* 32:547-565.

Jay, D. A., and J. D. Musiak. (1995a). Internal tidal symmetry in channel flows: Origins and consequences. *In*: C. Pattiaratchi, ed., *Mixing Processes in Estuaries and Coastal Seas.* American Geophys. Union. (in press).

Jay, D. A., and J. D. Musiak. (1995b). Particle trapping in estuarine tidal flows. *Jour. Geophys. Res.* (Oceans) (in press).

Jay, D. A., and J. D. Smith. (1990). Residual circulation in shallow estuaries 2.Weakly stratified and partially mixed, narrow estuaries. *Jour. Geophys. Res.* (Oceans) 95(c1): 773-748.

Kimmerer, W. M. (1992). *An Evaluation of Existing Data in the Entrapment Zone of the San Francisco Bay Estuary.* Interagency Ecol. Studies Program for the Sacramento-San Joaquin Estuary Tech. Rept. 33. 49 pp.

Koseff, J. R., J. K. Holen, S. G.Monismith, and J. E. Cloern. (1993). The effects of vertical mixing and benthic grazing on phytoplankton populations in shallow turbid estuaries. *Jour. Marine Res.* 51:1-26.

Kundu, P. K., and J. S. Allen. (1976). Some three-dimensional characteristics of low-frequency current fluctuations near the Oregon coast. *Jour. Phys. Oceanogr.* 6:181-199.

Lung, W. S., and D. J. O'Connor. (1984). Two-dimensional mass transport in estuaries. *Jour. Hydrol. Eng.* 110:1340-1357.

Mellor, G. L., and T. Yamada. (1982). Development of a turbulence closure model for geophysical fluid problems. *Rev. Geophys. Space Phys.* 20:851-875.

Monismith, S. G, and Fong, D. A. (1995). A simple model of mixing in stratified tidal flows. *Jour. Geophys. Res.* (Oceans) (in press).

Nunes Vaz R. A. and G. W. Lennon. (1991). Modulation of estuarine stratification and mass flux at tidal frequencies. Pages 505-520 *in* B. B. Parker, ed., *Tidal Hydrodynamics*. Wiley Interscience, New York, NY.

Nunes Vaz R. A. and J. R. Simpson. (1994). Turbulence closure modeling of estuarine stratification. *Jour. Geophys. Res.* (Oceans) 99:16143-16160.

Nunes Vaz, R. A., G. W. Lennon and J. R. de Silva Samarasinghe. (1989). The negative role of turbulence in estuarine mass transport. *Estuarine Coastal Shelf Sci.* 28:361-377.

Prandle, D. (1985). On salinity regimes and the vertical structure of residual flows in narrow tidal estuaries. *Estuarine Coastal Shelf Sci.* 20:615-635.

Prandle, D. (1991). Tides in estuaries and embayments (review). Pages 125-152 *in* B. B. Parker, ed., *Tidal Hydrodynamics*. Wiley Interscience, New York, NY.

Preisendorfer, R. W. (1988). *Principal Component Analysis in Meteorology and Oceanography*. Elsevier, New York, NY. 425 pp.

Schubel, J. R. (1968). Turbidity maximum of northern Chesapeake Bay. *Science* 161:1013-1015.

Sharples, J., J. H. Simpson, and J. M. Brubaker. (1994). Observations and modelling of periodic stratification in the Upper York River Estuary, Virginia. *Estuarine Coastal Shelf Sci.* 38:301-312.

Simpson, J. H., and J. Sharples. (1991). Dynamically-active models in the prediction of estuarine stratification. Pages 101-113 *in* D. Prandle, ed., *Dynamics and Exchanges in Estuaries and the Coastal Zone*. Springer-Verlag, New York, NY.

Simpson J. H., J. Brown, J. Matthews, and G. Allen. (1990). Tidal straining, density currents, and stirring in the control of estuarine stratification. *Estuaries* 13:125-132.

Simpson, J. H., and W. R. Turrell. (1986). Convergent fronts in the circulation of tidal estuaries. Pages 139-152 *in* D. A. Wolfe, ed., *Estuarine Variability*. Academic Press, New York, NY.

Smith L. H., and R. T. Cheng. (1987). Tidal and tidally averaged circulation characteristics of Suisun Bay, California. *Water Resour. Res.* 23:143-155.

Smith, P. E., R. T. Cheng, J. R. Burau, and M. R. Simpson. (1991). Gravitational circulation in a tidal strait. Pages 429-434 *in* R. M. Shane, ed., *Hydraulic Engineering 1991*. American Soc. Civil Eng., New York, NY.

Turner, J. S. (1973). *Buoyancy Effects in Fluids*. Cambridge Univ. Press, Cambridge, UK. 368 pp.

Uncles, R. J., J. A. Stephens, and M. L. Barton. (1991). Observations of fine-sediment concentrations and transport in the turbidity maximum region of an estuary. Pages 101-113 *in* D. Prandle, ed., *Dynamics and Exchanges in Estuaries and the Coastal Zone*. Springer Verlag, New York, NY.

Vidergar, L. L., J. R. Koseff, and S. G. Monismith. (1993). Numerical models of phytoplankton dynamics for shallow estuaries. Pages 1628-1634 *in* H. W. Shen, S. T. Su, & F. Wen, eds., *Hydraulic Engineering '93*, vol. 2. American Soc. Civil Eng., New York, NY.

Walters R. A., and J. W. Gartner. (1985). Subtidal sea level and current variations in the northern reach of San Francisco Bay. *Estuarine Coast. Shelf Sci.* 21:17-32.

Zimmerman, J. T. F. (1986). The tidal whirlpool: A review of horizontal dispersion by tidal and residual currents. *Netherlands Jour. Sea Res.* 20:133-154.

WATER PROPERTIES AND QUALITY

IMPORTANCE OF DISSOLVED SULFIDES AND ORGANIC SUBSTANCES IN CONTROLLING THE CHEMICAL SPECIATION OF HEAVY METALS IN SAN FRANCISCO BAY

JAMES S. KUWABARA, CECILY C. Y. CHANG,
ANDALUSIA I. KHECHFE, AND YVONNE R. HUNTER
U.S. Geological Survey
345 Middlefield Road, MS 439
Menlo Park, CA 94025

In an urbanized estuary like San Francisco Bay, a significant source of trace inorganics is anthropogenic, and yet the processes that control the biological availability of those solutes in the estuarine system have only recently been investigated. One process described in our previous work is the complexation of metals with dissolved organic substances. Given the variety of sources for these organic constituents, developing a thermodynamic data base for such reactions may indeed be a formidable task, although initial thermodynamic data on copper interactions with dissolved organics have been recently published by others. We also have reported that the presence of metastable dissolved-sulfide species in the oxic water column of the bay could affect trace metal speciation over ecologically significant concentration ranges. However, metal speciation control by sulfides is likely to be more episodic or transient than control by dissolved organic complexation, based on results showing only 1 out of 23 cases where the dissolved sulfide concentration was greater than the concentration for dissolved copper (a metal with a high affinity to complex with sulfides). In contrast to longitudinal gradients for dissolved organic carbon (DOC) concentrations (from 92 to 367 μmoles-CL^{-1}) in both the northern and southern components of the Bay, dissolved sulfides (from <1 to 162 nM) typically increased with depth, suggesting a primary benthic source. This hypothesis was supported by subsequent benthic flux measurements from incubated cores and from in-situ benthic flux chambers. Dissolved sulfide fluxes were consistently positive (*i.e.*, out of the sediment) while DOC fluxes showed seasonal shifts in direction. Initial direct measurements of the benthic flux for dissolved copper and cadmium are presented that, similar to DOC fluxes, indicated a temporal variability in direction across the sediment-water interface.

Major urbanized estuaries like San Francisco Bay serve as receiving waters for a variety of trace contaminants from numerous point and nonpoint sources. Once in the estuarine environment, these solutes may be repartitioned or altered in chemical speciation. The ecological importance of understanding the processes controlling chemical speciation has been consistently demonstrated (Anderson *et al.* 1978; Gavis *et al.* 1981; Kuwabara *et al.* 1986). Kuwabara *et al.* (1989) noted two potentially important mechanisms for controls on trace-metal speciation during their studies of the South San Francisco Bay water column. First, dissolved-copper and zinc concentrations were strongly correlated with concentrations of dissolved organic substances, suggesting the importance of metal-organic complexation reactions. This idea was supported by their chemical speciation calculations for copper, but available thermodynamic data used in these calculations were not specific to San Francisco Bay. Recent electrochemical and resin-partitioning studies by Donat *et al.* (1994) provided initial thermodynamic data directly describing the complexation of dissolved copper and nickel with naturally occurring organic ligands from the South Bay, and also demonstrated the importance of these reactions. Secondly, the distribution of dissolved sulfides in the oxic water column of both the northern and southern components of San Francisco Bay (hereafter referred to as the North Bay and South Bay) led Kuwabara & Luther (1993) to conclude that sulfide reactions in the water column should be considered as a controlling mechanism for trace-metal speciation in estuarine waters. Although the water column of the Bay is consistently well-oxygenated and typically unstratified with respect to dissolved oxygen (Smith & Herndon 1980), the kinetics of sulfide oxidation suggest that sulfide could exert at least transient controls on metal speciation. Though one might question how sulfides could be an important factor in this oxidizing environment, thermodynamic data for metal-inorganic ligand complexes (Hogfeldt 1983; Dyrssen & Wedborg 1989) indicated that submicromolar levels of dissolved sulfides could impose a pronounced effect on metal speciation. This paper: (1) synthesizes recent information on the distribution of dissolved-sulfide and dissolved organic carbon concentrations in the North Bay and South Bay (Fig. 1), (2) discusses the implications of recent benthic flux measurements for dissolved sulfides and dissolved organic carbon, and (3) presents initial benthic flux measurements for dissolved copper (Cu) and cadmium (Cd) to compare with previously modelled estimates of internal fluxes.

MATERIALS AND METHODS

This study involved the collection of water-column and benthic-flux-core samples. Initial water-column samples were collected from the main channel of the North Bay and the South Bay (Fig. 1) in 1991 during a multiyear drought. Prior to the drought, the average delta outflow during the wet season between October 1, 1984 and May 31, 1985 was 440.1 m^3s^{-1}. In contrast, during the 1991 wet season between October 1, 1990 and May 31, 1991 within the drought period, the average delta outflow was only 185.6 m^3s^{-1}. The post-drought condition between October 1, 1992 and May 31, 1993 also contrasted the 1991 water year with an average delta outflow of 769.7 m^3s^{-1}. The North Bay was sampled during a high discharge event that occurred during the spring of 1991 (March 12-15) when the average daily delta outflow for the period was 261.6 m^3s^{-1}. The water-column samples were collected at three U.S. Geological Survey (USGS)- designated stations (Stations 4, 8, and 18 with depths of 14, 15, and 32 m, respectively; see Fig. 1). Station numbers increase in the North Bay toward the Golden Gate.

In 1991, samples were collected from the South Bay at Stations 27, 32 and 36 (Fig. 1) during neap tides before, during, and after the spring phytoplankton bloom in April. The South Bay bloom periods during this study are defined here as the time interval during which Chl-a concentrations exceeded 10 $\mu g\ L^{-1}$(Weinke *et al.* 1992, 1993; Caffrey *et al.* 1994). Water column depths at each station varied with sampling date due to tidal effects; depths at Stations 27, 32 and 36 ranged from 11 to 13 m, 10 to 12 m, and 7 to 9 m, respectively. Station numbers increase in the South Bay with distance from the Golden Gate.

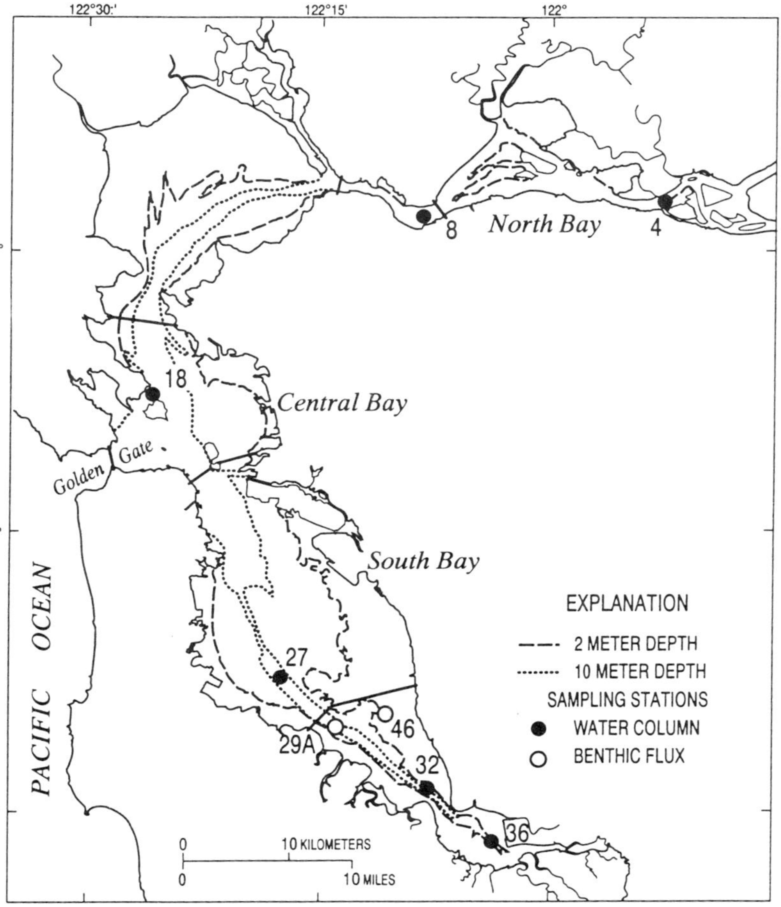

FIGURE 1. Study site showing water column and benthic flux core sampling stations in San Francisco Bay, California.

In 1993, sample collection was similar in the South Bay except that Station 29A rather than 27 represented the northernmost station to correspond with the benthic flux core sampling location. Sampling depths for 1993 are listed in Table 1.

At each station, samples were collected from the Research Vessel Polaris by means of 10-L fluoroethylene polymer (FEP)-lined sampling bottles designed for trace metal sampling at three depths: 1 m from the surface, at approximately 50 percent of maximum depth, and 2 m from the bottom (hereafter designated shallow, mid-depth, and deep samples, respectively). The exception to this procedure was made at the shallowest sampling station 36, where no mid-depth sample was collected. Water column samples were also collected for measurement of pH and concentrations of dissolved oxygen (DO by colorimetry and membrane electrode),

TABLE 1. 1993 Water column data for South San Francisco Bay. Dissolved-oxygen (DO) saturation is given as percent (% DO SAT), salinity (SAL) in practical salinity units (PSU). Tidal conditions (TC[1]) at each station are given in parentheses after the station number. Symbol "–" indicates that the sample was not analyzed because of cracks in the polyethylene sample bottle. The 1991 water column data was previously reported (Kuwabara & Luther 1993). Confidence intervals (95%) are presented for n ≥ 3 analytical replicates.

Date	*Station # (TC)*	*Depth (m)*	*% DO SAT*	*SAL (PSU)*	*DOC (μM)*	*Dissolved Cd (nM)*	*Dissolved Cu (nM)*	*SULF (nM)*
2/16	29A(EF)	1	97	20.3	198±10	0.640±0.048	31±2	3.6± 3.6
		6	98	20.9	206± 9	0.695±0.082	31±1	2.0± 0.8
		13	101	21.1	202± 7	0.980±0.037	31±2	12.8± 5.9
	32(BE)	1	108	19.7	229± 4	0.739±0.037	34±2	2.0± 1.4
		6	108	19.9	223± 7	0.750±0.029	34±2	31.4± 9.0
		12	108	20.0	220± 4	0.728±0.034	32±2	137±40.0
	36(E)	1	96	19.0	238± 7	0.934±0.060	46±4	0.8± 1.0
		7	93	19.4	233± 7	0.744±0.032	42±3	0.8± 1.2
3/25	29A(SL)	1	151	20.6	198±14	0.874±0.086	35±1	6.5± 4.0
		4	141	20.9	197±21	0.354±0.016	33±3	7.6±16.3
		8	137	21.2	212±16	0.897±0.056	31±1	8.7± 4.8
	32(BE)	1	201	18.6	210±18	0.284±0.028	39±2	0.4± 0.6
		7	143	19.8	213±13	–	–	9.8± 0.2
		12	139	20.0	230±13	0.284±0.009	39±2	19.6± 8.8
	36(SH)	1	132	16.9	262±26	0.398±0.032	46±3	29.9± 4.9
		7	127	17.6	233±13	0.398±0.057	40±2	13.3± 2.1
4/22	29A(F)	2	104	20.3	175±11	0.710±0.097	63±4	1.8± 3.4
		6	96	20.9	168± 8	0.747±0.144	56±6	5.0± 2.6
		11	92	21.1	169±11	0.684±0.084	53±4	10.2± 5.6
	32(SL)	2	101	19.7	255±16	0.419±0.041	28±3	1.1± 0.5
		5	100	19.9	256±15	0.492±0.063	31±2	5.8± 0.5
		8	100	20.0	260±26	0.510±0.080	28±3	10.9± 4.1
	36(BF)	2	99	19.0	310±12	0.509±0.067	58±3	19.8± 0.9
		5	98	19.4	324±13	0.666±0.081	67±5	11.3± 1.8

[1] Tidal conditions (TC) are abbreviated as follows: EF = End of flood; BE = Beginning of ebb; SL = Slack low tide; SH = Slack high tide; F = Flood; BF = Beginning of flood.

and dissolved organic carbon. Depth profiles for salinity (in PSU [Practical Salinity Units]), DO and temperature were obtained using a data-acquisition system with an arrangement of multiple probes to monitor various water-quality properties. Barometric pressure was determined on site for DO-concentration corrections and verified against measurements made at San Francisco International Airport (National Weather Service, oral commun. 1991). DO concentrations were also corrected for salinity and temperature (Franson 1985).

Benthic flux measurements of dissolved (0.2 μm-filtered) sulfide and DOC were made before, during and after the spring (*i.e.*, late March through early April) phytoplankton bloom of 1992 from flux core and in-situ flux chambers at two sites in the South Bay: a shoal station 46 and a channel station 29A (Fig. 1). In 1993, similar benthic flux measurements were also made except: (1) measurements of benthic flux for dissolved copper and cadmium were made in addition to those for dissolved sulfide and DOC, (2) only incubated cores from main channel station 29A were used to allow for higher sampling frequency during incubation, and (3) two sampling dates were selected during the bloom period, one near the peak of the bloom (April 7) and the other near the end of the bloom (April 23). This phase of the study was done in collaboration with macronutrient flux work being done by J. M. Caffrey of the U.S. Geological Survey, Menlo Park, and D. E. Hammond and W. M. Berelson at the University of Southern California. Methods for flux-core sampling were similar to those described by Kemp *et al.* (1990), except that in 1992, samples were collected only at the beginning and end of the flux-core incubation period because of sample volume constraints. Therefore, those flux determinations assumed a linear change in solute concentration over the incubation period. This assumption was tested in 1993 experiments with additional flux cores taken to accommodate increased sample volume removal (4 sampling times per core) at the expense of sampling only the main channel station. Details of the chamber design and method of operation are presented by Caffrey *et al.* (this volume).

Dissolved-sulfide concentrations were determined on site electrochemically by square-wave voltammetry (Kuwabara & Luther 1993). Total dissolved sulfides were measured by the method of addition with sulfide species depositing on a hanging mercury drop electrode at -0.4 V. A cathodic scan to -1.0 V was use to strip the deposited sulfides. Sulfide-speciation measurements were not performed in order to avoid sample contamination at low anticipated sulfide concentrations and to complete the replicate sample analysis within 2 hours of sampling.

Dissolved (0.2 μ filtered) organic carbon (DOC) concentrations were processed and analyzed by low-temperature (<100°C), ultraviolet/oxygen/persulfate oxidation within 24 h of sample collection (Hunter & Kuwabara 1994). With the exception of the silver membrane filter, all wetted surfaces in the DOC sample processing were glass, or fluoroethylene polymer.

Dissolved (0.2 μm filtered) trace metals from 1993 samples were analyzed by a modification of a cobalt co-precipitation method (van Geen, pers. commun.), hereafter referred to as the mini-method. Samples were filtered using acid-washed polycarbonate membranes, then immediately acidified with 12 N quartz-redistilled HCl (2 mL concentrated HCl to 1-liter sample). The mini-method was used because of sample volume constraints in the benthic flux experiments. For Cd determinations, 1.2 mL of sample was dispensed into 1.5 mL unpigmented, polyethylene, centrifuge tubes (Eppendorf 22-36320-4), followed by the addition of 30 μL of a 33.6 mM cobalt chloride solution and 30 μL of 4% ammonium pyrrolidinedithiocarbamate (APDC). The procedure for Cu analysis was similar, except that 40 μL of the cobalt and APDC solutions were used. Samples were shaken using a vortex mixer after the addition of each reagent. The sample tubes, then containing a green precipitate, were centrifuged at 15,000 g for 14 min after which the supernatant was siphoned off with a 300 μL needle-nosed pipette tip under vacuum. The pellet was resuspended with 1 mL of deionized water (1 mL), shaken, and then centrifuged again. The supernatant was siphoned off and the remaining pellet was dried under a heat lamp until most of the liquid had evaporated. Complete drying of the pellet was avoided to facilitate subsequent dissolution and sample recovery. After the addition of 20 μL of 8N HNO_3 to the pellet, the centrifuge tube was capped and then reexposed to the heat lamp until the contents clarified. Samples were then evaporated to dryness to form a pink cobalt bead that was redissolved with 400 μL of 0.16 N HNO_3, resulting in a 3-fold concentration. The samples were then capped, exposed for 5-min to the heat lamp, cooled, and shaken before analyzing by atomic absorption spectrometry. Deionized water (18 Mohm) blanks, certified seawater standards (Slew-1, National Research Board of Canada), and spiked standards were analyzed with each set of three samples to determine and correct for recovery efficiency (n = 6 for each sample type). Table 2 provides quality control parameters for this low-volume technique. Standard clean techniques were used to avoid sample contamination (Chang *et al.* 1992). All wetted parts of hardware used in the benthic flux experiments (sampling and incubation) were non-metallic, acid-washed, and tested for metal contamination before use.

TABLE 2. Quality control parameters associated with the trace metal analysis (concentrations given in nanomolar units).

Metal	*Detection limit*[1]	*Blank*	*Seawater Standard*[2]	*Measured*	*n*	*CV (%)*
Cd	0.015	0.115	0.160±0.027	0.187±0.027	14	20
Cu	20	0	28.00±0.05	24±1	14	9

[1] Estuarine and seawater samples were preconcentrated by a factor of 3 into a 0.1N HNO_3 matrix.
[2] Error terms represent 95% confidence intervals for *n* sample sets.

RESULTS AND DISCUSSION

Water-Column Distributions

More than 80 percent of the DO measurements were within 90 to 110 percent of saturation, indicating that the water column at our sampling sites was an oxic environment (Table 1) in which the presence of "free" sulfides (H_2S, HS^-) was not thermodynamically favored. Measured pH values were fairly uniform throughout the sampling period, with more than 60 percent of the values between 7.9 and 8.0 (7.92 ± 0.05).

During the drought period of 1991, DOC concentrations in the North Bay decreased with increasing distance from freshwater end members (Fig. 2), and generally decreased with increasing depth (Kuwabara & Luther 1993). In the South Bay, the DOC concentrations also increased with distance from the mouth of the estuary. However, the horizontal DOC gradient was more pronounced than in the

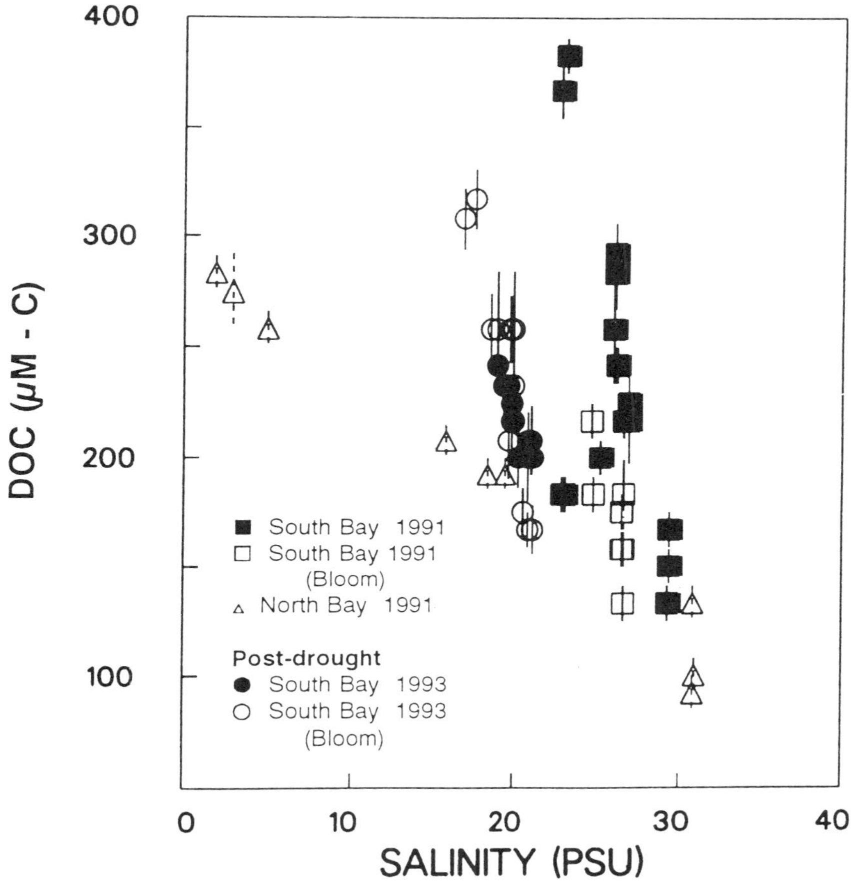

FIGURE 2. Distributions of DOC concentrations along longitudinal salinity gradients in the North Bay and South Bay. The plot synthesizes measurements made during and after a 6-year drought period.

North Bay (Fig. 2), and, as with salinity, no vertical DOC gradient was observed in the South Bay, which receives minimal freshwater input relative to the North Bay (Conomos *et al.* 1979). The data suggest a higher freshwater end-member concentration in the South Bay or higher rates of evaporation in the shallower South Bay (*i.e.*, a linearly extrapolated DOC concentration at a salinity of 0 PSU of 730 $\pm$ 30 μmoles-C L^{-1}) than in the North Bay (290 $\pm$ 20 μmoles-C L^{-1}). This difference is not surprising, because of the predominance of industrial and municipal low-salinity point sources in the South Bay especially during drought periods (Conomos *et al.* 1979). The extrapolated zero-salinity DOC intercepts for the North Bay and South Bay in 1991 are consistent in trend with measurements made by Hunter & Kuwabara (1994) for major freshwater sources to the South Bay (*i.e.*, municipal and riverine sources ranging from 500 to 1000 μmoles-C L^{-1}), and the

North Bay (about 200 μmoles-C L^{-1} for the Sacramento River). In 1993, similar horizontal trends in DOC concentration were observed, though the extrapolated zero-salinity DOC intercept for 1993 (610 ± 20 μmoles-C L^{-1}) was significantly lower than in 1991 (730 ± 30 μmoles-C L^{-1}, as previously indicated). We presume that this temporal difference in freshwater end-member concentrations resulted from increased dilution of DOC point sources during the post-drought period. For example, the mean daily discharge for the wet season measured along the Guadalupe River at San Jose was 0.69 m^3 s^{-1} between October 1, 1990 and May 31, 1991 compared with 3.32 m^3s^{-1} between October 1, 1992 and May 31, 1993 (Trujillo *et al.* 1992 and 1994). Weinke *et al.* (1992) observed a much lower maximum Chl-*a* concentrations (14.7 μg L^{-1}) during the spring phytoplankton bloom of 1991 than reported by Caffrey *et al.* (1994) for the 1993 bloom (maximum Chl-*a* of 76.9 μg L^{-1}). Coincidently, DOC concentrations were not elevated during the spring phytoplankton bloom of 1991, but did reach maximum concentrations in 1993 during the post-drought bloom. This may have been due to the release of cell exudates and lysis of senescent cells during the more productive 1993 bloom.

In contrast to DOC distributions, dissolved-sulfide concentrations did not co-vary with the longitudinal salinity gradient in the North or South Bay, and were elevated at depth. A significant positive correlation between relative depth (*i.e.*, the ratio of sampling depth to water column depth) and measured sulfide concentration ($r^2 = 0.244$, $n = 57$) suggests a benthic source of sulfide (Fig. 3). Kuwabara & Luther (1993) reported that the spring phytoplankton bloom of 1991 did not result in elevated water column concentrations of dissolved sulfide. Even with greater planktonic biomass in the 1993 bloom relative to 1991, there was no indication of a water-column-sulfide source.

Dissolved trace metal data for Cd and Cu showed no apparent trends with depth (r^2 for relative depth with Cd and Cu were 0.028 and 0.003, respectively, $n = 23$; Table 1). Dissolved Cd increased with salinity ($r^2 = 0.226$, $n = 23$). In contrast, dissolved-Cu distributions before the phytoplankton bloom showed a significant decrease with increasing salinity ($r^2 = 0.674$, $n = 15$). This trend for Cu has been previously reported (Kuwabara *et al.* 1989; Flegal *et al.* 1991). As with DOC concentrations, elevated Cu concentrations were observed during the bloom sampling of April 22, 1993 (Table 1).

Chemical-speciation calculations of Kuwabara & Luther (1993) demonstrated the potential importance of dissolved sulfides as ligands in the oxic San Francisco Bay water column. Their analysis indicated that, as the dissolved-sulfide concentration declined below the dissolved-Cu concentration (modeled at 30 nM, Table 1), Cu complexation with dissolved organic matter (DOM) became increasingly important. The stability constants associated with Cu-DOM complexes are lower than Cu sulfide complexes by orders of magnitude (Dyrssen & Wedborg 1989). Hence, the concentration of uncomplexed, and presumably more bioavailable Cu increased as complexation by organic matter exerted greater controls over Cu complexation. Concentrations of inorganically complexed Cu (other than sulfide species) would also increase. The range (Fig. 4) of calculated Cu ion activity was

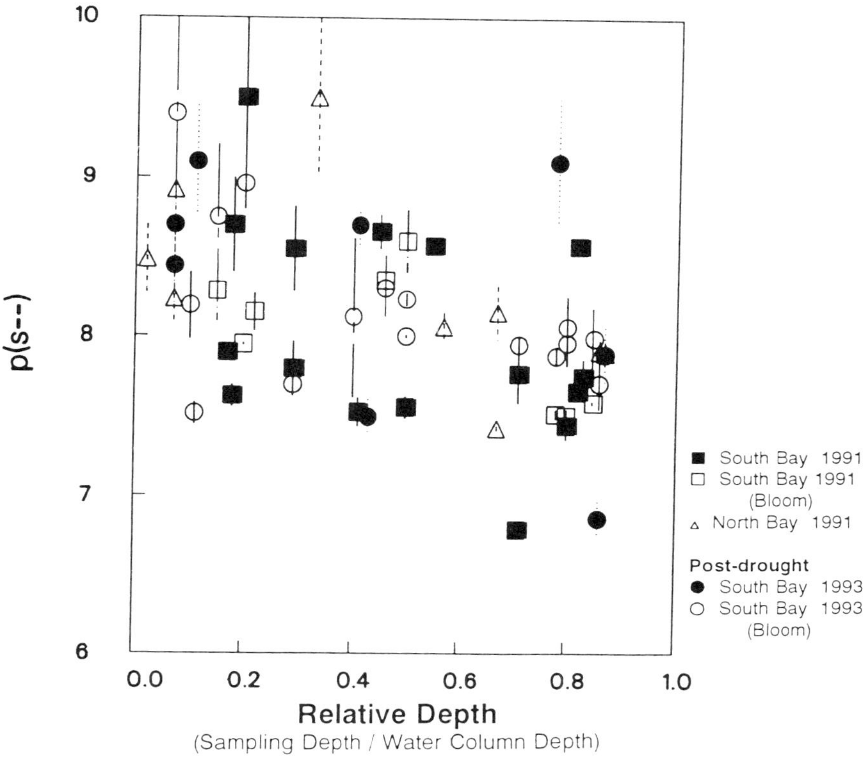

FIGURE 3. Water column measurements of dissolved sulfides (in p or -log units) both during and after a 6-year drought period indicating elevated concentrations at depth. Relative depth is depicted as the ratio of sampling depth over water column depth.

ecologically relevant because it bracketed toxic Cu ion activities for phytoplankton groups (*e.g.*, negative log of the Cu ion activity (pCu) of 8.5 to 10.9 for diatom species; Gavis *et al.* 1981). Similar to Cu, an analogous sulfide effect on Cd speciation was also reported although the dominant, nonsulfide species for Cd were chloro-complexes (Kuwabara & Luther 1993).

Recently, direct measurements of thermodynamic data involving Cu complexation with organic ligands in the South Bay have been reported (Donat *et al.* 1994). Using electrochemical and resin repartitioning techniques, Cu complexation with two organic ligand types were described: the first ligand type, L_1, had a lower concentration (about 13 nM) and higher binding affinity ($\log K \geq 13.5$) relative to a second ligand type, L_2, of higher concentration (about 70 nM) but lower binding affinity ($\log K$ between 9.0 and 9.6). By comparison, Kuwabara & Luther's (1993) calculations assumed only one organic ligand type with a $\log K$ of 9.5, and a ligand concentration of 2 μM (*i.e.*, 10^{-6} moles binding sites per mg-C; McKnight *et al.* 1983). Figure 4 represents a comparison between speciation effects using Donat *et al.*'s directly determined thermodynamic data (assuming a $\log K$ for CuL_1 and CuL_2

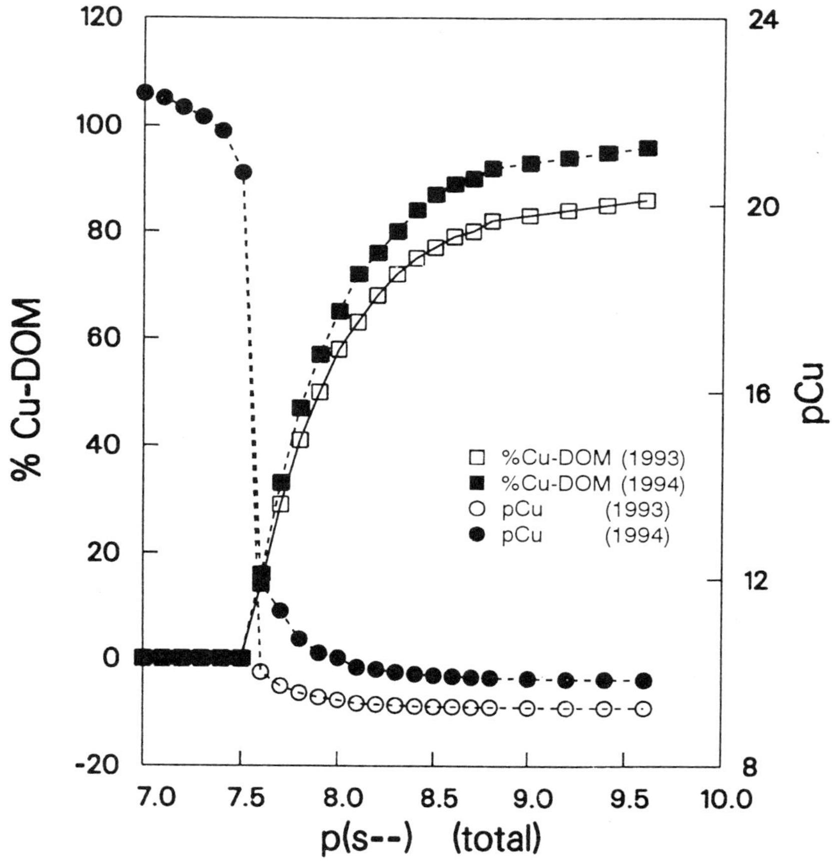

FIGURE 4. The implications of recent thermodynamic data for the complexation of metals with dissolved organic substances determined for the South Bay. Open symbols represent speciation calculations based on estimated thermodynamic data (Kuwabara & Luther 1993). Filled-in symbols represent calculations based on measured thermodynamic data for copper complexation specific to the South Bay (Donat *et al.* 1994). Note the shift in speciation controls when dissolved sulfide concentrations (in p or -log units) decrease below the dissolved Cu concentration of 30 nM (*i.e.*, $pS^{2-}>7.5$).

of 13.8 and 9.3, respectively) with earlier speciation estimates by Kuwabara & Luther (1993). The comparison is made over the observed dissolved-sulfide concentration range with dissolved Cu fixed at 30 nM and organic ligand sites as specified above. The results indicate that as dissolved sulfide decreases below the total Cu concentration (30 nM as mentioned above), the added presence of the high affinity ligand, L_1, results in: (1) a lower asymptotic Cu ion activity (*i.e.*, pCu of 9.8) relative to earlier estimates (9.4), and (2) ultimately a greater predominance of Cu association with dissolved organics (96% organically complexed using Donat *et al.*'s data versus 86% by earlier estimates.) Clearly, Donat *et al.*'s (1994) results provide strong support for the importance of organic complexation as a process regulating Cu speciation over the entire concentration range of DOC, L_1 and L_2 reported for the South Bay (Fig. 2). In contrast to DOC concentration ranges (typically within 1 order of magnitude), sulfide concentrations varying over 3 orders of magnitude have been observed in the oxic water column (Fig. 3). Although the observed dissolved-sulfide-concentration range in these oxic waters indicate a potential for ecologically significant effects on trace-metal speciation (Fig. 4), only 14% of the dissolved-sulfide measurements were greater than 30 nM (*i.e.*, similar to dissolved-Cu concentrations), and only 1 out of 23 were greater than the measured Cu concentration. This suggests that sulfide control over metal speciation is likely to be more episodic or transient than control by dissolved organics. In cases where the dissolved sulfide concentration exceeds dissolved Cu, sulfide might then exert similar speciation controls on metals like Cd with lower association constants than Cu.

Benthic Fluxes

Sulfide benthic fluxes were always out of the sediment (positive in sign) over the course of our sampling period in 1992 and 1993 (Tables 3 and 4; Fig. 5). This was consistent with our hypothesis of a primary benthic source, on the basis of initial concentration distributions in the water column. Fluxes reached maximum observed values during the spring at both the shoal and channel stations in 1992, although the peak at the channel station lagged by approximately a month. A post bloom peak like that observed in 1992 was not determined from our April 23, 1993 core incubation experiment (Fig. 5). This may have been due to the fact that Chl-*a* concentrations remained elevated 29 days after peak Chl-*a* concentrations were observed at channel Station 30 (5.15 km from our channel sampling Station 29A). In contrast to sulfide fluxes, DOC fluxes seasonally changed direction over the period of study, going from negative (into the sediment) in the winter to positive values during the spring and summer (Caffrey *et al.*, this volume).

Initial dissolved-metal flux measurements from benthic flux core experiments in 1993 are presented in Table 4 (To our knowledge the first direct measurements made for the Bay). The data are believed to be significant in two respects. With an understanding of the temporal and spatial limitations of this initial data set, if one extrapolates these metal fluxes over the South Bay area (554 km^2; Cheng & Gartner 1985), the estimated daily internal flux of Cu and Cd to the South Bay ranges between -179 and 30.7 kg d^{-1} for Cu and between -16.3 and 28.1 kg d^{-1} for Cd. Given those temporal and spatial limitations, one might expect that the actual range for these values would be even greater. By comparison, point source inputs to the South Bay have been estimated at 28 to 46 kg d^{-1} for Cu and 4 kg d^{-1} for Cd (Flegal *et al.* 1991; Wood *et al.* 1995). The relative magnitude of these two transport processes support the notion that benthic fluxes play an important role in regulating trace metal distributions in the South Bay. Second, data indicate temporal shifts in the direction of benthic flux for Cu and Cd, although the magnitude of Cu fluxes was generally orders of magnitude greater than that for Cd. Because sulfide fluxes were consistently positive (Fig. 3), these temporal shifts in the direction of the

TABLE 3. 1992 benthic fluxes from Shallow Station 46 with 95 % confidence intervals (n samples) for dissolved organic carbon (DOC, in mmoles $m^{-2}h^{-1}$), sulfides (in µmoles $m^{-2}h^{-1}$). Note that flux measurements in 1993 were made only from the Main Channel Station 29A due to increased core sampling in 1993 with the addition of trace metal flux determinations. Positive flux values indicate a flux out of the sediment and negative values into the sediment. Data for samples taken at the beginning and end of the incubation period (see Methods Section) were regressed against elapsed incubation time to calculate fluxes (r^2 values indicate fit for the regression). Paired flux measurements indicate determinations made on replicate cores. Symbol "*" indicates that flux measurement was determined from Lander (*in situ*) samples as opposed to flux core experiments.

Sampling Date		*DOC Flux*	*Sulfide Flux*
3/17	Flux	-0.94±0.43	0.14±0.03
	n, r^2	5, 0.90	6, 0.96
	Flux	-0.80±0.33	0.16±0.03
	n, r^2	6, 0.82	6, 0.99
4/13	Flux	0.08±0.11 (Unstirred)	0.37±0.09
	n, r^2	12, 0.04	6, 0.92
	Flux	1.92±0.15	0.42±0.09
	n, r^2	8, 0.97	6, 0.94
	Flux	1.34±0.10*	0.42±0.08
	n, r^2	7, 0.99	6, 0.95
	Flux	0.95±0.10*	0.71±0.02
	n, r^2	7, 0.97	4, 0.99
5/11	Flux	0.45±0.08	0.42±0.03
	n, r^2	15, 0.54	5, 0.99
9/9	Flux	0.06±0.64	0.50±0.04
	n, r^2	16, < 0.01	6, 0.98

benthic flux suggest that the metal speciation near the sediment-water interface is not consistently controlled by sulfide complexation. The ecological implication is that internal sources of dissolved Cd and Cu to the water column are transported at a lower pCd and pCu (*i.e.*, greater biological availability) than sulfide thermodynamic data would predict. Although there is a tendency for metals like Cu and Cd to complex with sulfides to a much greater extent than with dissolved organics, the presence of sulfides can also impose metal-solubility limitations (Hogfeldt 1983). The formation of insoluble metal-sulfide precipitates could explain the lack of correlation between dissolved-sulfide and dissolved-metal fluxes. Although the observed metal fluxes are better correlated with the DOC flux, the precision of the DOC flux measurements (in mmoles $m^{-2}h^{-1}$) do not provide adequate resolution relative to the metal fluxes (in nmoles $m^{-2}h^{-1}$) to be definitive about organic complexation controls.

Although point sources of trace metals to the South Bay can be readily calculated on the basis of municipal treatment plant data, non-point (including internal) sources are presently much more difficult to estimate. The importance of non-point

TABLE 4. 1992 and 1993 benthic fluxes from Main Channel Station 29A with 95 % confidence intervals (n time series concentrations) for dissolved organic carbon (DOC, in mmoles $m^{-2}h^{-1}$), sulfides (in μmoles $m^{-2}h^{-1}$), and dissolved Cd and Cu (in nmoles $m^{-2}h^{-1}$, determined in 1993 only) in South San Francisco Bay. Note that flux measurements in 1993 were made only from the Main Channel Station 29A due to increased core sampling in 1993 with the addition of trace metal flux determinations. Positive flux values indicate a flux out of the sediment and negative values into the sediment. Time series concentration data from the cores was regressed against elapsed incubation time to calculate fluxes (r^2 values indicate fit to the linear model). In 1992, samples were only taken at the beginning and end of the core incubation period (see Methods Section). Paired flux measurements indicate determinations made on replicate cores. Symbol "*" indicates that flux measurements were determined from Lander (in situ) samples as opposed to flux core experiments. Symbol "–" denotes that trace metal fluxes were not determined in 1992 experiments.

Sampling Date		*DOC Flux*	*Sulfide Flux*	*Cd Flux*	*Cu Flux*
2/25/92	Flux	-0.83±0.17	0.25±0.30	–	–
	n, r^2	18, 0.37	6, 0.98		
3/17/92	Flux	-0.11±0.55	0.17±0.01	–	–
	n, r^2	6, 0.03	6, 0.99		
	Flux	-0.64±0.71	0.09±0.03	–	–
	n, r^2	6, 0.39	5, 0.97		
4/13/92	Flux	-0.16±0.21	0.36±0.02	–	–
	n, r^2	7, 0.18	6, 0.99		
	Flux	-0.13±0.15	0.57±0.02	–	–
	n, r^2	7, 0.25	6, 0.99		
	Flux	0.53±0.16*	1.05±0.02*	–	–
	n, r^2	5, 0.96	4, 0.99		
	Flux	0.72±0.16*	0.90±0.02*	–	–
	n, r^2	5, 0.97	4, 0.99		
5/11/92	Flux	0.55±0.10	1.02±0.07	–	–
	n, r^2	15, 0.55	6, 0.99		
9/9/92	Flux	1.67±0.43	0.79±0.09	–	–
	n, r^2	15, 0.54	6, 0.93		
4/7/93	Flux	0.65±0.65	0.57±0.09	23.8±5.2	-175±9
	n, r^2	12, 0.33	12, 0.80	18, 0.56	24, 0.84
4/23/93	Flux	0.13±0.34	0.54±0.10	-5.7±2.2	46±32
	n, r^2	12, 0.02	12, 0.74	20, 0.25	21, 0.03
10/23/93	Flux	-0.64±0.06	0.40±0.06	-13.8±1.2	-269±82
	n, r^2	16, 0.99	12, 0.84	29, 0.45	23, 0.34

sources within the South Bay as a major process affecting the distribution and transport of trace metals has been suggested in recent studies. Flegal *et al.* (1991) correlated data for cadmium, zinc, cobalt, and iron with previous benthic flux results for dissolved silica (Hammond *et al.* 1985) to hypothesize the importance of diagenetic remobilization of these metals from benthic sediments. A similar analogy to dissolved-silica fluxes was used to suggest the predominance of benthic flux (*e.g.*, ≈ 0.7 nmoles-Ag $m^{-2}h^{-1}$ in the South Bay) relative to fluvial transport

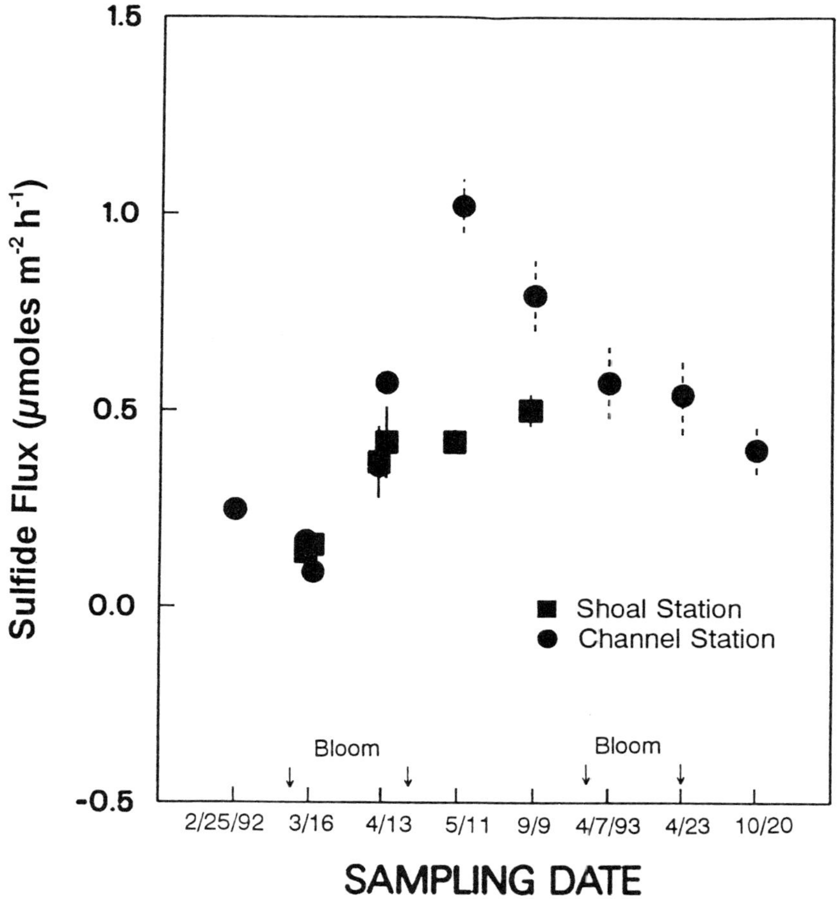

FIGURE 5. Measured benthic flux of dissolved sulfides from a main channel and shoal station in the South Bay. The approximate duration of the phytoplankton blooms in 1992 and 1993 are bracketted along the horizontal axis. Note that the horizontal axis indicates sequential dates of flux core experiments and not equivalent time increments. Positive values indicate fluxes out of the sediment, negative values into the sediment.

in controlling dissolved-silver distributions in the entire Bay (Smith & Flegal 1993). Rivera-Duarte & Flegal (1994) espoused the importance of benthic fluxes for dissolved lead and silver in regulating water column distributions based on pore water distributions and diffusive flux calculations. For copper and nickel, Flegal & Sanudo-Wilhelmy (1993) suggested the importance of urban surface runoff as a major non-point source. Wood *et al.* (1995) used a diagnostic computer model, integrating chemical and physical processes in the South Bay, to estimate non-point sources for Cu, Cd and Zn under the assumption that they were dominated by internal sources (376, 6 and 345 nmoles $m^{-2}h^{-1}$, respectively). These flux estimates are similar in magnitude to the direct measurements reported here and suggest that the loading due to non-point sources, that was estimated using their model, could well be dominated by benthic fluxes. However, the direction of that measured flux was temporally variable. In this case, the data (Table 4) represent periods during and after the post-drought phytoplankton bloom of April 1993 at only one main channel site in the South Bay. In light of the distinct seasonal patterns in freshwater inflow, primary productivity and fluxes of other constituents (*e.g.*, macronutrients) in the South Bay, there is a need to directly consider the temporal variability of benthic fluxes for trace metals. In processes-integrated, water quality models, internal sources cannot otherwise be decoupled from other non-point sources to the South Bay. Clearly, the spatial variability of benthic fluxes should also be quantified.

ACKNOWLEDGMENTS

We thank J. M. Caffrey of the U.S. Geological Survey, and D. E. Hammond and W. Berelson of the University of Southern California for their collaboration in executing the benthic flux experiments. Hydrologic and biological information regarding the South Bay between 1991 and 1993 from J. R. Burau, B. E. Cole, J. W. Gartner and L. H. Smith of the U.S. Geological Survey are acknowledged. We are also grateful for the valuable help of the crew of the Research Vessel Polaris. Colleague reviews by C. C. Fuller, T. M. Wood and two anonymous reviewers are much appreciated. We acknowledge support from the U.S. Geological Survey Toxic Substances Hydrology Program. Use of trade names herein are for identification purposes only and do not constitute endorsement by the U.S. Geological Survey.

REFERENCES

Anderson, M. A., F. M. M. Morel, and R. L. L. Guillard. (1978). Growth limitation of a coastal diatom by low zinc ion activity. *Nature* 276:70-71.

Caffrey, J. M., B. E. Cole, J. E. Cloern, J. R. Rudek, A. C. Tyler, and A. D. Jassby. (1994). *Studies of the San Francisco Bay, California, Estuarine Ecosystem: Pilot Regional Monitoring Results, 1993*. U.S. Geol. Surv., Open-File Rept. 94-82. 411 pp.

Chang, C. C. Y., J. S. Kuwabara, and S. P. Pasilis. (1992). Phosphate and iron limitation of phytoplankton biomass in Lake Tahoe. *Canadian Jour. Fish. Aquatic Sci.* 49:1206-1215.

Cheng, R. T., and J. W. Gartner. (1985). Harmonic analysis of tides and tidal current in South San Francisco Bay, California. *Estuarine Coastal Shelf Sci.* 21:57-74.

Conomos, T. J., R. E. Smith, D. H. Peterson, S. W. Hager, and L. E. Schemel. (1979). Processes affecting seasonal distributions of water properties in the San Francisco Bay estuarine system. Pages 115-142 *in* T. J. Conomos, ed., *San Francisco Bay:Tthe Urbanized Estuary*. Pacific Division, American Association for the Advancement of Science, San Francisco, CA.

Donat, J. R., K. A. Lao, and K. W. Bruland. (1994). Speciation of dissolved copper and nickel in South San Francisco Bay: A multi-method approach. *Analytica Chimica Acta* 284:547-571.

Dyrssen, D., and M. Wedborg. (1989). The state of dissolved trace sulphide in seawater. *Marine Chem.* 26:289-293.

Flegal, A. R., G. J. Smith, G. A. Gill, S. Sanudo-Wilhelmy, G. Scelfo, and L. C. D. Anderson.

(1991). Dissolved trace element cycles in the San Francisco Bay estuary. *Marine Chem.* 36:329-363.

Flegal, A. R., and S. Sanudo-Wilhelmy. (1993). Comparable levels of trace metal contamination in two semienclosed embayments: San Diego Bay and South San Francisco Bay. *Environ. Sci. Technol.* 27:1934-1936.

Franson, M. A. H. (1985). *Standard Methods for the Examination of Water and Wastewater*, 16th ed. Section 421, American Public Health Assoc., Washington, DC. 1268 pp.

Gavis, J., R. L. L. Guillard, and B. L. Woodward. (1981). Cupric ion activity and the growth of phytoplankton clones from different marine environments. *Jour. Marine Res.* 39:315-333.

Hammond, D. E., C. C. Fuller, D. Harmon, B. Hartman, M. Korosec, L. G. Miller, R. Rea, S. Warren, W. Berelson, and S. W. Hager. (1985). Benthic fluxes in San Francisco Bay. *Hydrobiologia* 129:69-90.

Hogfeldt, E. (1983). *Stability constants of Metal-Ion Complexes. Part A, Inorganic Ligands.* Pergamon Press, Oxford, UK. 310 pp.

Hunter, Y. R., and J. S. Kuwabara. (1994). Ionic strength and DOC determinations from various freshwater sources to the San Francisco Bay. *Bull. Environ. Contamin. Toxicol.* 52:311-318.

Kemp, W. M., P. Sampou, J. Caffrey, M. Mayer, K. Henriksen, and W. R. Boynton. (1990). Ammonium recyling versus denitrification in chesapeake Bay sediments. *Limnol. Oceanogr.* 35:1545-1563.

Kuwabara, J. S., J. A. Davis, and C. C. Y. Chang. (1986). Algal growth response to particle-bound orthophosphate and zinc. *Limnol. Oceanogr.* 31:503-511.

Kuwabara, J. S., C. C. Y. Chang, J. E. Cloern, T. L. Fries, J. A. Davis, and S. N. Luoma. (1989). Trace metal associations in the water column of South San Francisco Bay, California. *Estuarine Coastal Shelf Sci.* 28:307-325.

Kuwabara, J. S., and G. W. Luther, III. (1993). Dissolved sulfides in the oxic water column of San Francisco Bay, California. *Estuaries* 16:567-573.

McKnight, D. M., G. L. Feder, E. M. Thurman, R. L. Wershaw, and J. C. Westall. (1983). Complexation of copper by aquatic humic substances from different environments. *Sci. Total Environ.* 8:65-76.

Rivera-Duarte, I., and A. R. Flegal. (1994). Benthic lead fluxes in San Francisco Bay. *Geochimica et Cosmochimica Acta* 58:3307-3313.

Smith, G. J., and A. R. Flegal. (1993). Silver in San Francisco Bay estuarine waters. *Estuaries* 16:547-558.

Smith, R. E., and R. E. Herndon. (1980). *Physical and Chemical Properties of San Francisco Bay Waters, 1977.* U.S. Geol. Surv., Open-File Rept. 80-1191. 51 pp.

Trujillo, L. F., K. L. Markham, J. R. Palmer, and M. F. Friebel. (1992). *Water Resources Data California Water Year 1991*, Vol. 2. U.S. Geol. Surv.-Data Rept. CA-91-2. 315 pp.

Trujillo, L. F., K. L. Markham, J. R. Palmer, and M. F. Friebel. (1994). *Water Resources Data California Water Year 1993*, Vol. 2. U.S. Geol. Surv.- Data Rept. CA-93-2. (in press)

Weinke, S. M., B. E. Cole, J. E. Cloern, and A. E. Alpine. (1992). *Plankton Studies in the San Francisco Bay, California. XIII. Chlorophyll Distributions and Hydrographic Properties of San Francisco Bay, 1992.* U.S. Geol. Surv., Open-File Rept. 92-158. 116 pp.

Weinke, S. M., B. E. Cole, and J. E. Cloern. (1993). *Plankton Studies in the San Francisco Bay, California. XIV. Chlorophyll Distributions and Hydrographic Properties of San Francisco Bay, 1992.* U.S. Geol. Surv., Open-File Rept. 93-423. 175 pp.

Wood, T. M., A. M. Baptista, J. S. Kuwabara, and A. R. Flegal. (1995). Diagnostic modeling of trace metal partitioning in South San Francisco Bay. *Limnol. Oceanogr.* 40:345-358.

METAL CONTAMINATION IN SAN FRANCISCO BAY WATERS: HISTORIC PERTURBATIONS, CONTEMPORARY CONCENTRATIONS, AND FUTURE CONSIDERATIONS

A. R. FLEGAL[1,2], I. RIVERA-DUARTE[1], P. I. RITSON[1], G. M. SCELFO[2], G. J. SMITH[2], M. R. GORDON[3] AND S. A. SAÑUDO-WILHELMY[1,4]

[1] **Earth Science Board, WIGS, University of California, Santa Cruz, CA 95064**

[2] **Institute of Marine Science, WIGS, University of California, Santa Cruz, CA 95064**

[3] **Moss Landing Marine Laboratories, Moss Landing, CA 95039**

[4] **Marine Sciences Research Center, State University of New York, Stony Brook, NY 11794-5000**

Contemporary concentrations of trace elements (Ag, Cd, Cu, Ni, Pb, and Zn) in surface waters of the San Francisco Bay estuary appear both to reflect historic perturbations of the system and to presage future levels of contamination during protracted periods of limited hydraulic flushing. The elemental distributions corroborate established concepts on the importance of suspended particulate scavenging on the biogeochemical cycles of particle reactive elements in estuarine waters, and they also attest to new perceptions of the importance of diagenetic remobilization from benthic sediments on some elemental cycles within the estuary. That remobilization appears to be most pronounced in the southern reach of the estuary, where anthropogenic inputs have been most concentrated historically. This suggests that benthic fluxes of elements from sediments will become a more important source of contamination in the future, as anthropogenic fluxes from point source discharges, surface runoff, and aeolian deposition to bay waters are reduced. It also suggests that some elemental concentrations in the southern reach may remain elevated for decades in spite of the projected decreases in those anthropogenic inputs. Moreover, the potentially adverse impact of benthic fluxes may be exacerbated by future reductions in hydraulic flushing of the estuary, which may be caused both by reduction in freshwater inflow because of diversions from the estuary and by climatological decreases in surface runoff in its drainage basin.

San Francisco Bay (Fig. 1) is a physically and hydrologically complex estuarine system, which has been dramatically altered by anthropogenic activities within the past two centuries (Nichols *et al.* 1986). That complexity and those alterations have contributed to existent seasonal and spatial perturbations in trace element distributions within its surface waters. Factors influencing those distributions have been discussed in recent reports (Abu-Saba & Flegal 1994; Flegal *et al.* 1991; 1995; Sañudo-Wilhelmy & Flegal 1994; Smith & Flegal 1993; Wood *et al.* 1994). Current levels of elemental contamination above natural levels in San Francisco Bay have not been quantified, because there are relatively few historic data for comparisons and most of those data are circumspect (Flegal *et al.* 1991; Flegal & Coale 1989).

Consequently, this inchoate review of temporal variations in trace metal distributions in the estuary is based on a limited number of comparisons between recent and selected historic data. Those qualified comparisons suggest that benthic fluxes of metals from previously contaminated sediments have contributed to the elevated concentrations of some trace metals in contemporary estuarine surface waters. The

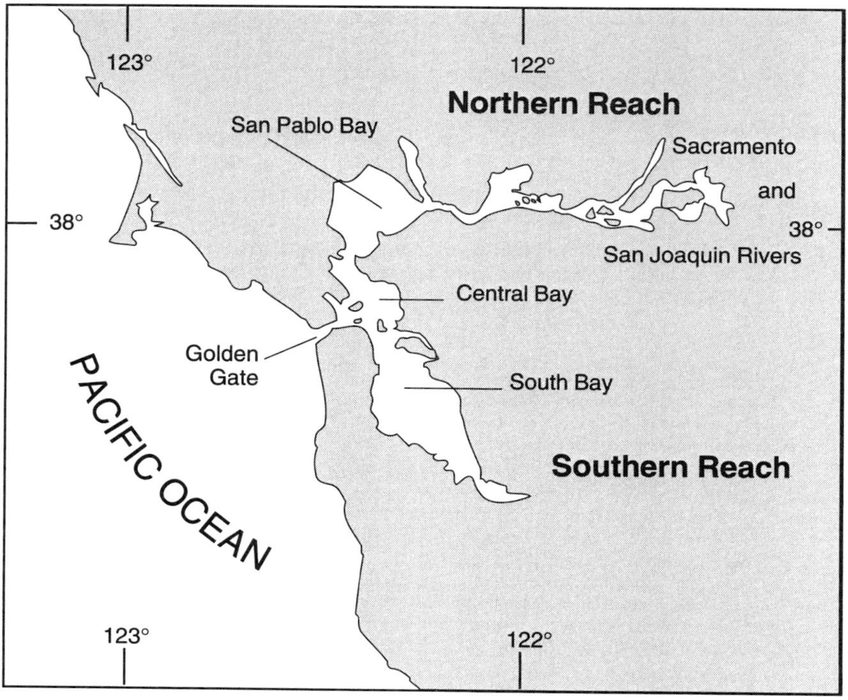

FIGURE 1. The San Francisco Bay estuary.

comparisons also suggest, that contamination will persist for decades, in spite of recent and projected reductions in anthropogenic loadings of trace elements to the estuary from point source discharges, surface runoff and aeolian deposition. The comparisons suggest that levels of contamination may be exacerbated by additional decreases in fluvial discharges to San Francisco Bay that will further attenuate hydraulic flushing in the estuary, especially in its southern reach or South Bay.

BACKGROUND

Factors Governing Trace Element Distributions in the Estuary

Distributions of trace metals in San Francisco Bay (Abu-Saba & Flegal 1994; Flegal *et al.* 1991, 1994; Smith & Flegal 1993) exhibit pronounced spatial and temporal concentration gradients (Fig. 2). The distributions are often consistent with those of metalloid (*e.g.*, Cutter 1989) and organic (*e.g.*, Spies *et al.* 1987) contaminants. Many of those contaminant distributions are consistent with nutrient distributions in the estuary (*e.g.*, Cloern *et al.* 1985; Peterson *et al.* 1985, 1986, 1989). *In toto*, the distributions of dissolved constituents validate established concepts on the primary factors governing their biogeochemical cycles in the estuary, as explained below.

Briefly, trace metal distributions in the estuary appear to be regulated by the following factors. (1) San Francisco Bay is a physically modified, urban estuary

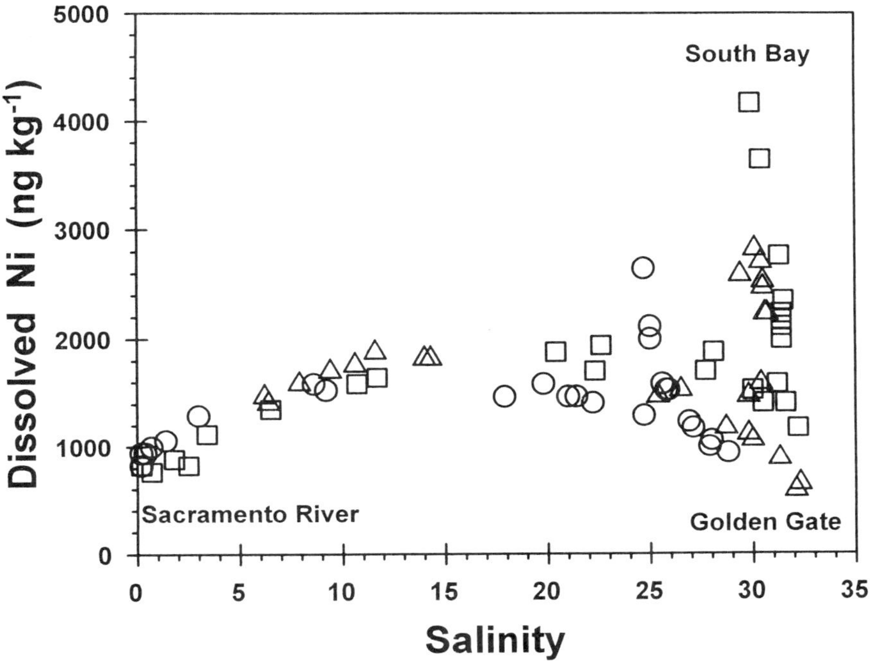

FIGURE 2. Temporal and spatial variations of dissolved (0.45m) nickel concentrations (ng kg^{-1}) in San Francisco Bay in 1989. Samples collected in April (○), August (□), and December (Δ). Concentrations are plotted versus salinity, as reported in Flegal *et al.* (1991).

with two components that are becoming increasingly disparate with time (Fig. 1): (i) the northern reach, which is a positive estuary surrounded by about 775,000 acres of relatively natural marshlands with some urban (2 million people) and industrial areas and (ii) the southern reach, which is a seasonally negative estuary surrounded by a megalopolis (4 million people) and some industrial tidal ponds (Flegal *et al.* 1991). There are about 10 million people within the drainage basin of the estuary (State of California 1993). (2) Anthropogenic inputs of many elemental contaminants to the estuary have increased with time in parallel with the population increase. Disproportionately large inputs from point source discharges and urban surface runoff occur in the central and southern reaches of the estuary where the population increase is centered (Davis *et al.* 1991, 1992). (3) Particle reactive contaminants tend to be recycled within the estuary, between dissolved and particulate phases in the water column and surface sediments (Flegal & Sañudo-Wilhelmy 1993). (4) The relative importance of those cyclic phase transformations has recently been evidenced by studies showing that seasonal releases of some trace elements from sediments within the estuary are order(s) of magnitude greater than fluvial fluxes of those elements to the estuary and comparable to their anthropogenic fluxes (Rivera-Duarte & Flegal 1994, 1995). (5) Essentially all (> 95 %) fluvial discharges (natural average flow $\approx 1.3 \times 10^{10}$ m^3yr^{-1} or 10.6 million acre-feet year[1]) to the estuary are located in the northern reach, which limits hydraulic flushing in the southern reach to seasonal periods of high fresh water discharges (Smith, 1987). (6) Fluvial discharges to the estuary have been substantially reduced ($\approx$ 60%) by fresh water diversions from the drainage basin that have diminished hydraulic flushing in the estuary, especially the southern reach (Peterson *et al.* 1989). (7) Reductions in hydraulic flushing have exacerbated problems of elevated concentrations of some contaminants in the South Bay (Nichols *et al.* 1986).

Recent Reductions in Trace Element Loadings to the Estuary

The supposition of those factors accounts for elevated concentrations of some contaminants in San Francisco Bay, and that has catalyzed efforts to decrease pollutant loadings to the estuary (Condit 1987; Davis *et al.* 1991; U.S. Environmental Protection Agency 1992; U.S. General Accounting Office 1991). This has included numerous projects to reduce the amount of anthropogenic discharges to the estuary through source control, to upgrade the treatment of effluents discharged from publicly operated treatment works (POTW) into the bay from no treatment to advanced treatment, and to redirect point source discharges from the hydrologically restricted extremities of the estuary to the Central Bay and Pacific Ocean. Approximately 63% of POTW effluent now receives secondary treatment and 37% receives tertiary treatment. Approximately 64% of the POTW effluent to the estuary is released into deep waters in the Central Bay, 28% is released to shallow waters in the southern reach, and 9% is released to shallow waters in the northern reach. Additionally, the Southwest Ocean Outfall Project (SWOOP) has consolidated and upgraded the treatment of waste waters from the San Francisco Peninsula and redirected discharges of those wastes (60-75 mg d^{-1}) from the bay to an ocean outfall (6.5 km offshore). [Approximately 155 mg d^{-1} are discharged from POTW's into the South Bay (Steve Moore, San Francisco Regional Water Quality Control Board, personal communication).] That project cost $\approx$ \$2 billion, and over \$3 billion has been invested on waste water treatment facilities to improve the water quality of the estuary during the past three decades (Condit 1987).

Those efforts have dramatically decreased the amount of trace element discharges to the estuary. While there are insufficient reliable data to substantiate public media reports that some metal loadings to the estuary decreased by 90% (Davis *et al.* 1992; Batts 1993; S. Moore, pers. commun.), the media reports are partially supported by recent estimates of the reductions in some waste water loadings to the estuary (S. Moore, pers. commun.).

Some of the decreases in trace element discharges are illustrated in Fig. 3. It shows the systematic decrease in reported trace metal loadings (Cd, Cu, Ni, and Zn) from some of the major POTWs in the San Francisco Bay estuary between 1975 and 1985. These include reported decreases in discharges of copper (from 343 to 52 kg d^{-1}), zinc (from 653 to 117 kg d^{-1}), cadmium (from 16 to 7 kg d^{-1}), and nickel (from 87 to 56 kg d^{-1}) during that period. With subsequent improve-

[1] One $km^3 = 10^9 m^3 = 10^{12} L \approx 2.64 \times 10^{12}$ gallons $\approx$ 810,190 acre-feet.

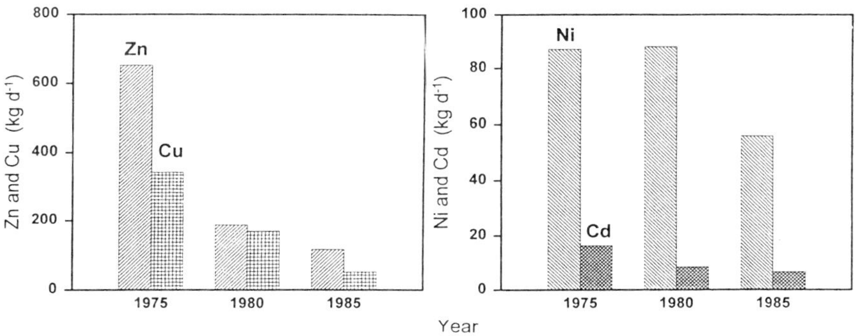

FIGURE 3. Systematic decrease in reported trace metal loadings (kg d^{-1}) from publicly owned treatment works (POTWs) to San Francisco Bay (from S. Moore, San Francisco Bay Regional Water Quality Control Board).

ments in waste water treatment since then, it is reasonable to assume that point source loadings of heavy metals to the estuary truly have decreased by at least one order of magnitude within the past two decades.

Measurements of Trace Element Contamination in the Estuary

Contingent with elemental decreases in point source loadings, decreases in heavy metal concentrations within the estuary have not been quantified for several reasons. Historic comparisons of elemental concentrations in indicator species (*e.g.*, *Mytilus californianus* and *M. edulis*) in the estuary are of limited value, because the analyses commonly reflect the amount of sediment in undepurated organisms rather than the concentrations of their tissues (Smith *et al.* 1986). Historic comparisons of elemental concentrations of surficial sediments are also of limited value, because those sediments integrate contaminant inputs over decades (Ritson & Flegal 1994). Finally, even simple comparisons of temporal changes in elemental concentrations in estuarine waters are restricted, because most previous measurements were limited in scope and did not utilize currently accepted techniques that are required for accurate trace element measurements (Flegal *et al.* 1991; Flegal & Coale 1989).

Fortunately, the techniques used in recent San Francisco Bay studies are comparable with those utilized by Gordon (1980) for a preliminary study of elemental distributions in the northern reach of the estuary in 1979-1980. Similar trace metal clean techniques were employed to collect the samples and measure their trace element concentrations in both studies. Consequently, those historic data may be compared with contemporary data to assess recent temporal variations in the levels of trace element concentrations in the northern reach of the estuary over that period.

HISTORIC VARIATIONS IN TRACE ELEMENT CONCENTRATIONS IN THE BAY

Lead

With the initiation of accurate measurements of trace element concentrations in the oceans more than two decades ago, it has been possible to document temporal changes in oceanic concentrations associated with subsequent changes in anthropogenic inputs of those elements (Bruland 1983). The most notable of those changes has been the systematic decrease in lead concentrations in oceanic surface waters that corresponds with the concurrent decrease in aeolian depositions of industrial lead aerosols to those waters, which is primarily due to the reduction in lead additives to gasolines in the last two decades (National Research Council 1993). This was initially shown in the North Atlantic (Boyle *et al.* 1986), where lead concentrations above the main thermocline have decreased by 30-40% between 1979 and 1989 in response to the five to eight-fold decrease in aeolian lead deposits during that period (Veron *et al.* 1993). Similarly, the three-fold decrease in lead in the Southern California Bight has paralleled the three-fold reduction in industrial lead inputs to those coastal waters over the past two decades (Sañudo-Wilhelmy & Flegal 1994).

While there has been an even greater decrease (>100-fold) in atmospheric concentrations of lead in the San Francisco Bay area, there has not been a corresponding decrease in lead concentrations in its estuarine waters (Fig. 4). Urban atmospheric concentrations of lead in San Francisco declined from 2.4 $\mu g\ m^{-3}$ in 1970 to 0.1 $\mu g\ m^{-3}$ in 1985 (Flegal & Smith 1987); and continued to decline with the systematic elimination of lead additives in gasolines in 1992 (National Research Council 1993). The decrease in aerosol lead was reflected in reported decreases in loadings of lead from the principal POTWs to the estuary from 116 kg d^{-1} in 1975 to 25 kg d^{-1} in 1985 (S. Moore, pers. commun.). However, there was no apparent decline in lead concentrations throughout the northern reach of the estuary between 1980 (Gordon 1980) and 1989 (Flegal *et al.* 1995), as illustrated in Fig. 4.

Isotopic composition analyses show that essentially all of the lead in San Francisco Bay waters is still derived from industrial sources (Flegal *et al.* 1995). The isotopic ratios of lead in bay waters are not comparable with the isotopic ratios of deep sediments, which reflect natural inputs of lead to the estuary, nor with the ratios of contemporary aerosols, which reflect current aeolian inputs of industrial lead to San Francisco Bay (Ritson & Flegal 1994). The isotopic compositions of the estuary's surface waters are, though, comparable with the isotopic ratios of its surficial sediments, which reflect decadal accumulations of industrial lead inputs to the estuary (Ritson & Flegal 1994). That isotopic symmetry between surface sediments and bay waters in San Francisco Bay is consistent with lead concentration gradients in sediment pore waters that evidence the diagenetic cycling of contaminant lead between surficial sediments and overlaying waters in the estuary (Rivera-Duarte & Flegal 1994).

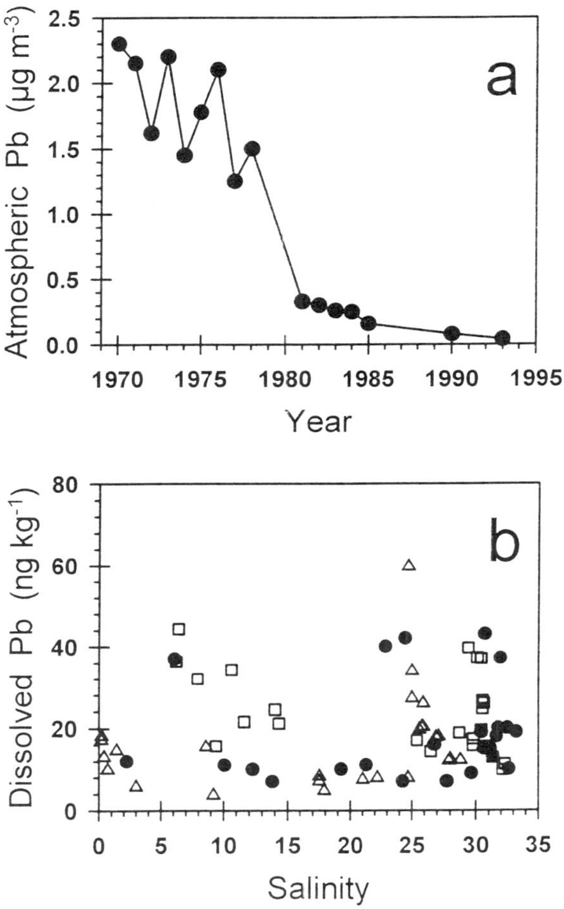

FIGURE 4. Temporal variations of (a) particulate lead concentrations in San Francisco aerosols (μg m^{-3}) and (b) dissolved (< 0.45 μm) lead (ng kg^{-1}) in the San Francisco Bay estuary. Particulate aerosol data provided by J. Sandberg (San Francisco Air Quality Management District), as reported in Flegal & Smith (1988). Dissolved lead concentrations are plotted versus salinity. The initial water samples were collected in August, 1979 (•) and March, 1980 (•) in the northern reach of the estuary from the Sacramento – San Joaquin Delta to the Golden Gate (Gordon (1980); and the latter samples were collected in August, 1989 (Δ) and April, 1989 (□) from throughout the estuary (Flegal *et al.*, manuscript in preparation).

Other Trace Elements

There are similar absences in overt decreases in other trace element concentrations (Cd, Cu, Ni, and Zn) in the northern reach of the San Francisco Bay estuary, in spite of observed order of magnitude reductions in waste water loadings of those elements to the estuary (Fig. 5). Those absences are also attributed to persistent inputs of trace elements from historically contaminated sediments, as well as to

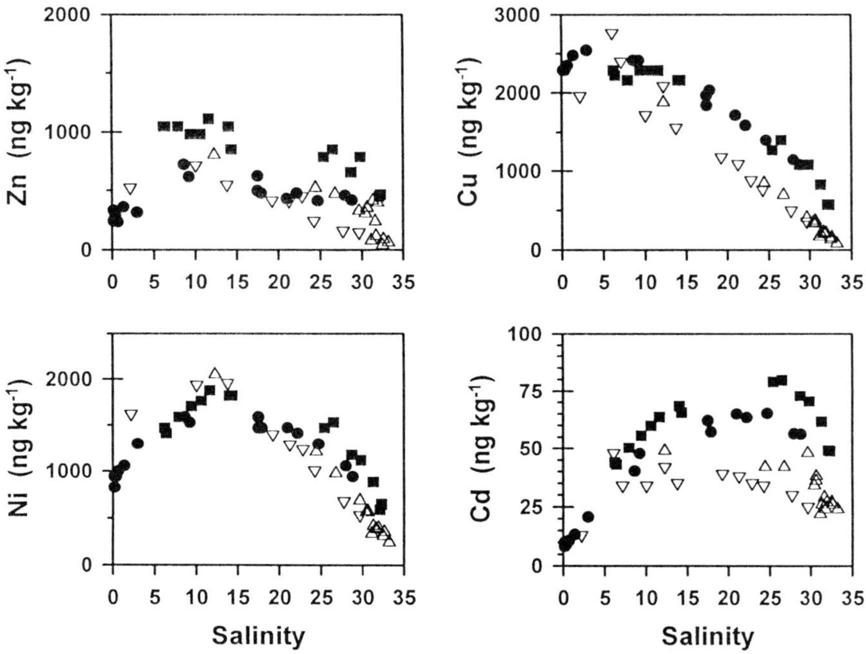

FIGURE 5. Temporal and spatial variations in zinc, nickel, copper, and cadmium concentrations (ng kg^{-1}) in the San Francisco Bay estuary. Concentrations are plotted versus salinity. The initial water samples were collected in August, 1979 (∇) and March, 1980 (Δ) in the northern reach of the estuary from the Sacramento – San Joaquin Delta to the Golden Gate (Gordon 1980); and the latter samples were collected in August, 1989 (■) and April, 1989 (•) from throughout the estuary (Flegal *et al.,* manuscript in preparation).

differences in fresh water outflow to the delta during the two sampling periods. The reported outflow (W. Dibben, California Department of Water Resources, pers. commun.) was relatively high in 1980 ($2.7 \times 10^{10} m^3 yr^{-1}$), and it was relatively low in 1989 ($8.2 \times 10^9 m^3 yr^{-1}$).

Those outflows are inversely correlated with hydraulic residence times. Consequently, the hydraulic residence times in 1989 were substantially longer than those in 1980. This presumably contributed to elevated elemental concentrations in the estuary to levels that were comparable to those in the preceding decade, in spite of the order of magnitude reduction in POTW inputs of those trace elements between the two sampling periods.

HYDRAULIC FLUSHING AND FRESH WATER DIVERSIONS

The preceding data indicate that elevated concentrations of trace elements in San Francisco Bay are partially due to limited hydraulic flushing. This is a natural seasonal and inter annual characteristic of the system. It is now being exacerbated by engineered diversions of freshwater discharges and by climatological reductions in those discharges. The latter, in turn, appear to be due to both natural drought cycles and global warming.

Current and Projected Fresh Water Diversions

The impact of fresh water diversions on trace element concentrations in the estuary has recently been evidenced by analyses of sediment cores from the bay. The cores reveal pronounced decreases in sedimentation rates and increases in the concentrations of several trace elements in the 1950's, when large scale diversions of fresh water from the Sacramento and San Joaquin Rivers were initiated (S. Luoma, United States Geological Survey, pers. commun.). Approximately 60% of the water in that drainage basin is now diverted to support agriculture, industry, and municipalities in the Central Valley and Southern California. Demands for additional diversions continue to increase as the population of California increases.

The potential impact of the expanding population is illustrated in Fig. 6. It shows projected water consumption levels in California, based on current consumption rates and population projections (State of California 1993), relative to annual

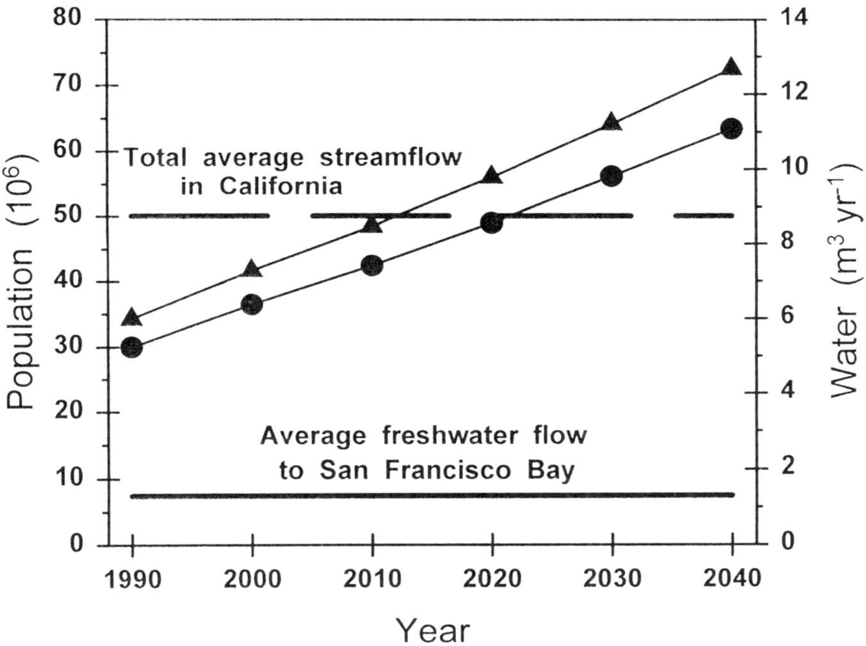

FIGURE 6. Current and projected population and water consumption rates (m^{-3} yr^{-1}) in California from 1990 to 2040. The population estimates (•) are from the Official Population Projections of the State of California (Anonymous 1993). The water consumption rates (▲) are from current levels (W. Dibben, California Dept. Water Resources, pers. commun.); and the projections based on existent per capita consumption and population projections. Average freshwater flows to San Francisco Bay through the Sacramento - San Joaquin Delta and total average stream flows in California are from the California Department of Water Resources (W. Dibben, pers. commun.).

average freshwater discharges in San Francisco Bay and in the entire state. The total annual water use in California in 1985 was 4.2×10^{10} (34,000,000 acre-feet) for a population of 26 million. This included 6.7×10^{9} m^3 (5,500,000 acre-feet) for municipal and industrial use and 3.5×10^{10} m^3 (28,500,000 acre-feet) for agricultural irrigation; and it constituted an average water consumption of 1,614$m^3 capita^{-1} yr^{-1}$ (1.3 acre-feet per capita per year). Based on that average consumption rate, the water demand in 2040, when the population of California is projected to be 63,343,000 (State of California 1993), will be 1.02×10^{11} $m^3 yr^{-1}$. That demand will be an order of magnitude greater than the entire fresh water discharge in the San Francisco Bay drainage basin (1.3×10^{10} $m^3 yr^{-1}$), and it will be 15 % greater than the average annual stream flow for all of California (8.76×10^{10} $m^3 yr^{-1}$). Therefore, water demands, based on current consumption rates, of the state's population are projected to exceed its capacity within forty years.

Historic and Current Droughts

The importance of hydraulic flushing on trace element concentrations in the San Francisco Bay was punctuated by drought conditions which existed during our initial surveys in 1989 (Flegal *et al.* 1991). Those surveys revealed anomalously elevated dissolved concentrations of several trace elements in the South Bay during seasonally low fresh water discharges, which were further reduced by the drought (Fig. 2). Low flow conditions (130 to 330 $m^3 s^{-1}$, with the exception of an ephemeral flow of 1,100 $m^3 s^{-1}$) persisted during the three survey periods (October 1988 to September 1989). This was naively referred to as the consequences of a "protracted drought," which extended from 1987 to 1992, in our initial reports on trace element cycles in the estuary.

Although the drought has now continued (with one brief interruption) for eight years (1987-1994), it is inconsequential compared to historic droughts. This is evidenced by recent paleoclimatological studies which have shown that previous drought conditions extended for ≥ 80 years within San Francisco Bay (Ingram *et al.* 1994) and for up to 250 years in its drainage basin (Stine 1994). The latter study has shown that the average season during those historic droughts were drier than any year in the 1987-1994 drought or the 1928-1934 drought. The paleoclimatological studies have also shown that precipitation during the past 150 years was atypically high, constituting the third wettest period in the past 3,500 years. Therefore, the current "average" fresh water discharge to the estuary, that is based on data collected within the past century, is anomalously high; and future freshwater discharges to the estuary may substantially decrease if climatological conditions revert toward more typical drought conditions.

Future Droughts

Future discharges to the estuary are also anticipated to be substantially reduced by global warming (Flegal *et al.* 1991; Smith & Flegal 1993). The projected reductions are based on models of the regional consequences of increases in anthropogenic emissions of CO_2 and other gases (Gleick 1987; Waggoner 1991).

The models indicate that an atmospheric increase of 1°C to 2°C (which is consistent with recent conservative projections of global warming) will actuate a 10% decrease in precipitation in California and a 40% to 70% decrease in surface runoff in the San Francisco Bay estuary drainage basin. Those theoretical models have been substantiated by decreases in surface runoff associated with increases in temperature in the Sierras (D. Peterson, U.S. Geological Survey, pers. commun.).

This suggests that recent reductions in fresh water flushing may appear trivial within the next few decades, if global warming increases atmospheric temperatures as projected. When coupled with population projections and associated estimates of future water demands, it also indicates that the amount of water per capita in California within the next fifty years will be an order of magnitude lower than it is today. This would substantially limit the amount of fresh water available to flush out the bay. Consequently, some elemental concentrations in the estuary may remain elevated for decades, in spite of order of magnitude decreases in anthropogenic inputs of those elements to the estuary.

DIAGENETIC REMOBILIZATION

Another factor that appears to have substantially contributed to relatively elevated concentrations of some trace elements in San Francisco Bay is diagenetic remobilization. The potential role of benthic sediments as a source of trace elements in estuarine waters has been recognized for decades (Turekian 1977), and that source has previously been proposed to account for the nonconservative excesses of some trace elements in the northern reach of the San Francisco Bay estuary (Gordon 1980). But benthic fluxes and associated excess concentrations in the water column have primarily been observed in the null zone near the head of the estuary, while trace element concentrations in the lower reaches of most other estuaries exhibit nonconservative losses due to biogeochemical scavenging processes (Turekian 1977). Consequently, the anomalously high concentrations of some dissolved trace elements in the high salinity waters in the southern reach of San Francisco Bay (Flegal *et al.* 1991; Flegal & Sañudo-Wilhelmy 1993) appear to contrast with the concept that estuaries are a sink for particle reactive elements.

The diagenetic remobilization of trace elements from sediments throughout San Francisco Bay, and especially in the South Bay, is consistent with previous reports on the diagenetic remobilization of nutrients in the estuary (Peterson *et al.* 1986; Hammond *et al.* 1985; Schemel *et al.* 1984). For example, measurements of silicate concentrations in benthic sediments and pore waters indicated that 65% of the silica and 90% of the carbon, were remobilized from sediments in the South Bay (Hammond *et al.* 1985). Those initial measurements have recently been substantiated by measurements of nutrient concentrations in sediment pore waters (Rivera-Duarte & Flegal 1994) and in incubated cores and benthic flux chambers (Caffrey *et al.* 1994; Kuwabara *et al.* 1996). They have also been corroborated by recent measurements of benthic nutrient fluxes in other estuaries (*e.g.*, Aller 1988; Boynton & Kemp 1985; Burdige *et al.* 1992; Kemp *et al.* 1990).

The relative magnitude of benthic fluxes of trace elements in San Francisco Bay has been derived from diffusion and mass balance calculations. Analyses of excess silver concentrations in the southern reach of the estuary indicated that the benthic flux of silver in the South Bay was ≈ 1.7 μg $m^{-2}d^{-1}$ and that the benthic flux of silver throughout the estuary was ≈ 1.2 μg $m^{-2}d^{-1}$ (Smith & Flegal 1993). Those estimates have been corroborated by two independent calculations of benthic silver fluxes in the estuary, using (i) pore water concentration gradients and (ii) a mass balance model (Rivera-Duarte & Flegal 1995). Both sets of calculations have indicated that the total benthic flux of silver (up to 8,800 kg yr^{-1}) in the South Bay during periods of intense benthic activity is an order of magnitude greater than the fluvial flux of dissolved silver to the estuary (12 kg yr^{-1}; Smith & Flegal 1993) and comparable to the total anthropogenic flux of silver from all municipal and industrial discharges to the estuary (2,700 to 7,200 kg yr^{-1}; Davis *et al.* 1991). The relative sizes of those fluxes is illustrated in Fig. 7.

The benthic fluxes of some other trace elements also appear to be relatively large. For example, the estimated benthic flux of dissolved lead is 400 to 14,100 kg yr^{-1} (Rivera-Duarte & Flegal 1994), using the irrigation benthic flux model of Hammond *et al.* (1985). This is one to three orders of magnitude greater than the fluvial flux of dissolved lead to the estuary (15 kg yr^{-1}), and it is within an order of magnitude of the total (dissolved and particulate) anthropogenic flux of lead to the estuary (72,500 to 650,000 kg yr^{-1}). Also, independent analyses of pore waters (Rivera-Duarte & Flegal 1994b) and benthic flux chambers (Kuwabara *et al.* 1996), indicate that benthic sediments in San Francisco Bay appear to act as a relatively

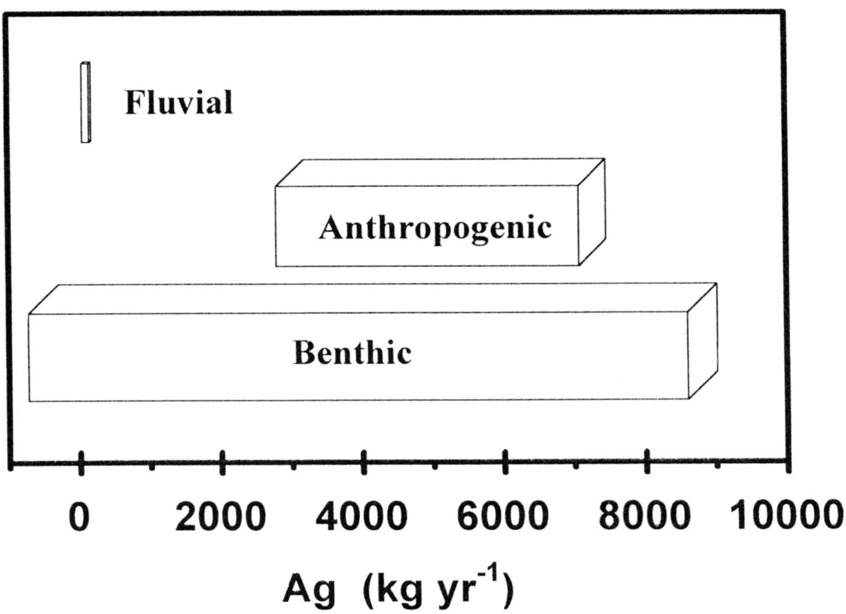

Figure 7. Silver inputs (kg yr^{-1}) to the San Francisco Bay estuary from fluvial, anthropogenic and benthic fluxes (from Rivera-Duarte & Flegal 1995).

large source of copper during some periods and as a sink for copper during other periods.

The preceding estimates of benthic fluxes of trace elements in San Francisco Bay have substantiated the proposal that contaminated sediments represent a persistent source of contamination to overlaying waters in semi-enclosed embayments (Flegal & Sañudo-Wilhelmy 1993). The potential impact of benthic fluxes on surface water concentrations in the San Francisco Bay estuary was indicated by the elevated concentrations of trace metals in San Diego Bay. The concentrations of several trace metal concentrations (Ag, Cd, Co, Cu, Ni, and Pb) in the southern reach of San Diego Bay were anomalously higher than those of adjacent surface waters in the Southern California Bight, and they were similar to those in the most contaminated waters in the southern reach of San Francisco Bay. The copper concentrations in San Diego Bay in 1989 (Flegal & Sañudo-Wilhelmy 1993) were also comparable to those previously measured in 1976 (Zirino *et al.* 1978), in spite of the elimination of all point source discharges to that system nearly three decades ago. This suggests, by analogy, that contaminated sediments in the South Bay represent a chronic source of contamination to the overlaying waters, as they seem to be in San Diego Bay.

SUMMARY

In summary, elevated concentrations of trace elements in the San Francisco Bay appear to be due to both historic and anthropogenic inputs. While most metals are sequestered by suspended particulates, many of those particulates are entrained within the estuary and the metals may then be periodically recycled back into the water column by diagenetic and physical mixing processes. This suggests that current anthropogenic inputs are additive to historic inputs that have been accumulated in benthic sediments and are periodically recycled into overlaying waters. Trace element concentrations in those waters may also be elevated by reductions in hydraulic flushing. This has been caused by both engineered diversions of fresh water discharges from the estuary and by episodic decreases in fresh water discharges within its drainage basin. It is expected that flushing will be further reduced by additional diversions required to sustain projected population growth and by suspected decreases in fresh water discharges caused by global warming and natural climatological cycles. Consequently, elevated concentrations of some trace elements in the estuary may persist in spite of current and projected decreases in anthropogenic loadings.

ACKNOWLEDGMENTS

This review is based on a preliminary study directed by John Martin at Moss Landing Marine Laboratories and subsequent studies conducted by the WIGS Laboratory at the University of California at Santa Cruz. These include complementary analyses by Khalil Abu-Saba, and Jonathan Crick. The studies have been supported by the National Science Foundation, Environmental Protection Agency, National Oceanographic and Atmospheric Administration, and the California State Water Resources Control Board.

REFERENCES

Abu-Saba, K., and A. R. Flegal. (1995). Chromium in San Francisco Bay: Superposition of geochemical processes causes complex spatial distributions of RLDOX species. *Marine Chem.* 49:189-199.

Aller, R. C., and J. Y. Aller. (1992). Meiofauna and solute transport in marine muds. *Limnol. Oceanogr.* 37:1018-1033.

Batts, C. W. (1993). Discussion of: Priority pollutant loads from effluent discharges to the San Francisco Estuary, J. A. Davis *et al.* 64:134 (1992). *Water Environ. Res.* 65:82-83.

Boyle, E. A., S. D. Chapnick, G. T. Shen, and M. P. Bacon. (1986). Temporal variability of lead in the Western North Atlantic Ocean. *Jour. Geophys. Res.* 91:8,573-8,593.

Bruland, K. W. (1983). Trace elements in sea-water. Pages 157-220 *in* J. P. Riley and R. Chester, eds., *Chemical Oceanography*, Vol. 8. Academic Press, New York, NY.

Caffrey, J. M., D. E. Hammond, J. S. Kuwabara, L. G. Miller, and R. R. Twilley. (1996). Benthic processes in San Francisco Bay: The role of organic inputs and bioturbation. Pages 425-442 *in* J. T. Hollibaugh, ed., *San Francisco Bay: The Ecosystem*. Pacific Division, American Association for the Advancement of Science, San Francisco, CA.

Cloern, J. E., B. E. Cole, R. L. Wong, and A. E. Alpine. (1985). Temporal dynamics of estuarine phytoplankton: A case study of San Francisco Bay. *Hydrobiologia* 129:153-176.

Cloern, J. E., and F. H. Nichols. (1985). Time scales and mechanisms of estuarine variability, a synthesis from studies of San Francisco Bay. *Hydrobiologia* 129:153-176.

Condit, R. J. (1987). Fiscal year 1985-1986 Municipal Division program activities. California Regional Water Quality Control Board, San Francisco Bay Region. Appendix A. in *BADA, 1987, Pollutants in the Bay-Delta Estuary*. Tech. Rept. for the State Water Resources Control Board, Bay-Delta Hearings. Bay Area Dischargers Association, Oakland, CA.

Conomos, T. J. (1979). Properties and circulation of San Francisco bay waters. Pages 47-84 *in* T. J. Conomos, ed., *San Francisco Bay: The Urbanized Estuary*. Pacific Division, American Association for the Advancement of Science, San Francisco, CA.

Cutter, G. A. (1989). The estuarine behavior of selenium in San Francisco Bay. *Estuarine Coastal Shelf Sci.* 28:13-34.

Davis, J. A., A. J. Gunther, B. J. Richardson, J. M.O'Conner, R. B. Spies, E. W. Wyatt, E. Larson, and E. C. Meiorin. (1991). *Status and Trends Report on Pollutants in the San Francisco Bay Estuary*. San Francisco Estuary Project, Oakland, CA. 240 pp.

Davis, J. A., A. J. Gunther, and J. M. O'Conner. (1992). Priority pollutant loads from effluent discharges to the San Francisco Estuary. *Water Environ. Res.* 64:134-140.

Donat, J. R., K. A. Lao, and K. Bruland. (1994). Speciation of dissolved copper and nickel in South San Francisco Bay: a multi-method approach. *Analytica Chimica Acta* 284:547-571.

Flegal, A. R., and K. H. Coale. (1989). Discussion: Trends in lead concentrations in major U.S. rivers and their relation to historical changes in gasoline-lead consumption. *Water Resources Bull.* 25:1275-1277.

Flegal, A. R., G. J. Smith, G. A. Gill, S. Sañudo-Wilhelmy, and L. C. D. Anderson. (1991). Dissolved trace element cycles in the San Francisco Bay estuary. *Marine Chem.* 22:36: 329-363.

Flegal, A. R., and S. Sañudo-Wilhelmy. (1993). Comparable levels of trace metal contamination in two semi-enclosed embayments: San Diego Bay and South San Francisco Bay. *Environ. Sci. Technol.* 27:1934-1936.

Flegal, A. R., S. Sañudo-Wilhelmy, P. Ritson, I. Rivera-Duarte, G. Smith, and G. Scelfo. (1995). Lead in the San Francisco Bay estuary. (Manuscript in preparation.)

Flegal, A. R., and D. R. Smith. (1988). Lead dispersion gradients in San Francisco Bay. Pages 468-483 *in* Proceedings of the Seventh International Ocean Disposal Symposium. Environment Canada.

Gleick, P. H. (1987). Regional hydrologic consequences of increases in atmospheric CO_2 and other trace gases. *Climatic Change* 10:137-167.

Gordon, R. M. (1980). *Trace Element Concentrations in Seawater and Suspended Particulate Matter from San Francisco Bay and Adjacent Coastal Waters*. MA Thesis. Dept. Biol. Sci., San Jose State Univ., San Jose, CA. 85 pp.

Hammond, D. E., C. Fuller, D. Harmon, B. Hartman, M. Korosec, L. G. Miller, R. Rea, S. Warren, W. Berelson, and S. W. Hager. (1985). Benthic fluxes in San Francisco Bay. *Hydrobiologia* 129:69-90.

Ingram, B. L., J. C. Ingle, and M. E. Conrad. (1994). [Abstract] Oxygen and carbon isotopic records of paleosalinity in San Francisco Bay. Page 150 *in* Eighth International Conference on Geochronology, Cosmoschonology and Isotope Geology. Berkeley, CA.

Kemp, W. M., P. Sampou, J. Caffrey, M. Mayer, K. Henriksen, and W. R. Boynton. (1990). Ammonium recycling versus denitrification in Chesapeake Bay sediments. *Limnol. Oceanogr.* 35:1545-1563.

Kuwabara, J. S., C. C. Y. Chang, A. I. Khechfe, and Y. R. Hunter. (1996). Implications of dissolved sulfides and organic substances on the chemical speciation of trace contaminants in the water column of San Francisco Bay, California. Pages 157-172 *in* J. T. Hollibaugh, ed., *San Francisco Bay: The Ecosystem*. Pacific Division, American Association for the Advancement of Science, San Francisco, CA.

Luoma, S. N., and D. J. H. Phillips. (1988). Distribution, variability, and impacts of trace elements in San Francisco Bay. *Marine Pollution Bull.* 19:413-425.

National Research Council. (1993). *Measuring Lead Exposure in Infants, Children, and Other Sensitive Populations*. National Academy Press, Washington, DC. 337 pp.

Nichols, F. H., J. E. Cloern, S. N. Luoma, and D. H. Peterson. (1986). The modification of an estuary. *Science* 231:567-573.

Peterson, D. H., D. R. Cayan, and J. F. Festa. (1986). Interannual variability in biogeochemistry of partially mixed estuaries: Dissolved silicate cycles in northern San Francisco Bay. Pages 123-138 *in* D. A. Wolfe, ed., *Estuarine Variability*. Academic Press, New York.

Peterson, D. H., H. R. Cayan, J. F. Festa, F. H. Nichols, R. A. Walters, J. V. Slack, S. E. Hager, and L. E. Schemel. (1989). Climate variability in an estuary: Effects of riverflow on San Francisco Bay. Pages 419-442 *in* D. H. Peterson, ed., *Aspects of Climate Variability in the Pacific and the Western Americas*. Geophys. Monogr. 55. American Geophysical Union, Washington, DC.

Peterson, D. H., R. E. Smith, S. W. Hager, D. F. Harmon, R. E. Herndon, and L. E. Schemel. (1985). Interannual variability in dissolved nutrients in northern San Francisco Bay estuary. Pages 37-58 *in* J. E. Cloern & F. H. Nichols, eds., *Temporal Dynamics of an Estuary: San Francisco Bay*. Dr. W. Junk Publ., Dordrecht, The Netherlands.

Ritson, P. I., and A. R. Flegal. (1994). [Abstract] Isotope systematics of contaminant lead fluxes in the San Francisco Bay estuary.Page 268 *in* Eighth International Conference on Geochronology, Cosmoschonology and Isotope Geology. Berkeley, CA.

Rivera-Duarte, I., and A. R. Flegal. (1994). Benthic lead fluxes in San Francisco Bay, California, USA. *Geochimica et Cosmochimica Acta* 58:3307-3313.

Rivera-Duarte, I., and A. R. Flegal. (1994b). [Abstract] Dissolved copper concentrations in pore waters of San Francisco Bay sediment. Page 91 *in* 75th Annual Meeting of the Pacific Division, American Association for the Advancement of Science, June 19-24, San Francisco, CA.

Rivera-Duarte, I., and A. R. Flegal. (1995). Porewater silver concentration gradients in contaminated sediments of San Francisco Bay. Marine Chemistry (submitted for publication).

Sañudo-Wilhelmy, S. A., and A. R. Flegal. (1994). Lead fluxes in the Southern California Bight. *Geochimica et Cosmochimica Acta* 58:3315-3320.

Schemel, L. E., D. D. Harmon, S. W. Hager, and D. H. Peterson. (1984). Response of northern San Francisco bay to riverine inputs of dissolved inorganic carbon, silicon, nitrogen, and phosphorous. Pages 221-240 *in* V. S. Kennedy, ed., *The Estuary as a Filter*. Academic Press, New York.

Smith, D. R., M. D. Stephenson, and A. R. Flegal. (1986). Trace metals in mussels transplanted to San Francisco bay. *Environ. Toxicol. Chem.* 5:129-138.

Smith, G. J., and A. R. Flegal. (1993). Silver in San Francisco Bay estuarine waters. *Estuaries* 16:547-558.

Smith, L. H. (1987). *A Review of Circulation and Mixing Studies of San Francisco Bay, California*. U. S. Geol. Surv. Circ. 1015. 38 pp.

Spies, R. B., J. S. Felton, and D. W. Rice. (1987). Benzthioles in estuarine sediments as indicators of street runoff. *Nature* 327:697-699.

State of California. (1993). *Official Population Projections*. Sacramento, California. 3 pp.

Stine, S. (1994). Extreme and persistent drought in California and Patagonia during mediaeval time. *Nature* 369:546-549.

Turekian, K. K. (1977). The fate of metals in the oceans. *Geochimica Cosmochimica Acta* 41:1139-1144.

U.S. General Accounting Office. (1991). *Water Pollution: Stronger Efforts Needed by EPA to Control Toxic Water Pollution*. GAO, Washington, DC. 58 pp.

U.S. Environmental Protection Agency. (1992). *National Water Quality Inventory-1990. Report to Congress*. EPA 503/9-92/006. EPA, Washington, DC. 174 pp.

Veron, A. T., M. Church, C. C. Patterson, Y. Erel, and A. R. Flegal. (1993). Response of lead cycling in the surface Sargasso Sea to changes in the tropospheric input. *Jour. Geophys. Res.* 98:18,269-18,276.

Waggoner, P. E. (1991). U.S. water resources versus announced but uncertain climate change. *Science* 251:1002.

Wood, T. M., A. M. Baptista, J. S. Kuwabara, and A. R. Flegal. (1994). Diagnostic modeling of trace metal partitioning in South San Francisco Bay. *Limnol. Oceanogr.* (in press).

Zirino, A., S. H. Lieberman, and C. Clavell. (1978). Measurements of Cu and Zn in San Diego Bay by automated anodic stripping voltammetry. *Environ. Sci. Technol.* 12:73-79.

DISSOLVED INORGANIC NITROGEN, PHOSPHORUS AND SILICON IN SOUTH SAN FRANCISCO BAY

I. MAJOR FACTORS AFFECTING DISTRIBUTIONS

STEPHEN W. HAGER AND LAURENCE E. SCHEMEL
U.S. Geological Survey, 345 Middlefield Road, Menlo Park, CA 94025

The distributions of the plant nutrients, dissolved inorganic nitrogen, dissolved reactive phosphorus and dissolved silica in the southern reach of San Francisco Bay (South SFB) are affected by many factors, including wastewater inputs, freshwater inputs from streams in the watershed, exchange with Central SFB, evaporation, and biological uptake and regeneration. Wastewaters, particularly those entering south of Dumbarton Bridge, dominate the distributions of nutrients in South SFB, because for much of each year, wastewaters are the only input of water to the head of the South SFB estuary. Dilution of these wastewaters by freshwaters from local streams is significant, on the average, only during the wettest month of the year. Except in very dry winters, increases in freshwater flow into northern SFB from the Sacramento – San Joaquin Delta rapidly change the composition of the waters of Central SFB, which brings about enhanced exchange with South SFB, and a seasonal dilution of nutrients in the waters of South SFB. If there is only a little freshwater dilution, ratios of nitrogen to silicon in the waters of South SFB can be higher than ratios of utilization by diatoms, and silicon depletion can occur during spring phytoplankton blooms, possibly affecting the species distribution of the phytoplankton population. The number of factors affecting this tributary estuary leads to complicated nutrient/salinity plots, suggesting that quantitative modelling is required for a more complete understanding of nutrient dynamics in this system.

Within the San Francisco Bay estuarine system, South San Francisco Bay (South SFB) was once considered to be the part most adversely impacted by wastewater discharges (Pearson 1958). This was due in part to the high loading with organic matter by wastewater treatment facilities, which amounted to 157,000 kg biological oxygen demand (BOD) day^{-1} in 1965 (California RWQCB 1987). Improved wastewater treatment practices reduced BOD loading to 16,000 kg day^{-1} in 1985, and as a result, autotrophic production is now the largest source of organic matter to the South SFB ecosystem (Jassby *et al.* 1993). This change from a net heterotrophic system to a net autotrophic one has renewed interest in the factors affecting the plant nutrients, nitrogen, phosphorus and silicon in South SFB. During recent spring phytoplankton blooms, we have observed reductions of the concentrations of ammonium, nitrate, nitrite and dissolved silica to levels that are likely to limit production by phytoplankton, and thus, potentially, the productivity of the system. In this paper, we review factors that affect concentrations of nutrients in South SFB, building on and updating the work of Conomos *et al.* (1979). Additionally, we present observations made in South SFB that may yield insights regarding the behavior of systems with high loadings of inorganic nutrients, contributing to the worldwide database on the problem of coastal eutrophication. Factors affecting nutrient concentrations in South SFB also are of interest because of similarities between the distributions of nutrients and other waste-derived substances in South

SFB, including cadmium, copper, and zinc (Kuwabara *et al.* 1989; Flegal *et al.* 1991; van Geen & Luoma 1993), nickel (Flegal *et al.* 1991; van Geen & Luoma 1993), cobalt and iron (Flegal *et al.* 1991), and selenium (Cutter 1989).

South SFB is a large embayment extending south-easterly from Central SFB, through which it is connected to the Golden Gate, the Gulf of the Farallones, and the Pacific Ocean (Fig. 1). Because Central SFB is the seaward end of the northern SFB estuary, the estuary of the Sacramento and San Joaquin Rivers flowing from the great Central Valley of California, South SFB is considered a tributary estuary of this main estuary. South SFB is relatively shallow, and has an axial channel, greater than 10 m deep, running most of its length. The San Bruno Shoal, an extensive shallow water area beginning about 19 km from the Golden Gate, through which a deep-water channel has been dredged, separates upper South SFB from middle South SFB. The San Mateo Bridge appears to separate the middle part into two hydrologically different basins (Powell *et al.* 1989). The narrow strait at Dumbarton Bridge about 51 km from the Golden Gate demarcates middle South SFB and lower South SFB. The mean depth of South SFB is substantially influ-

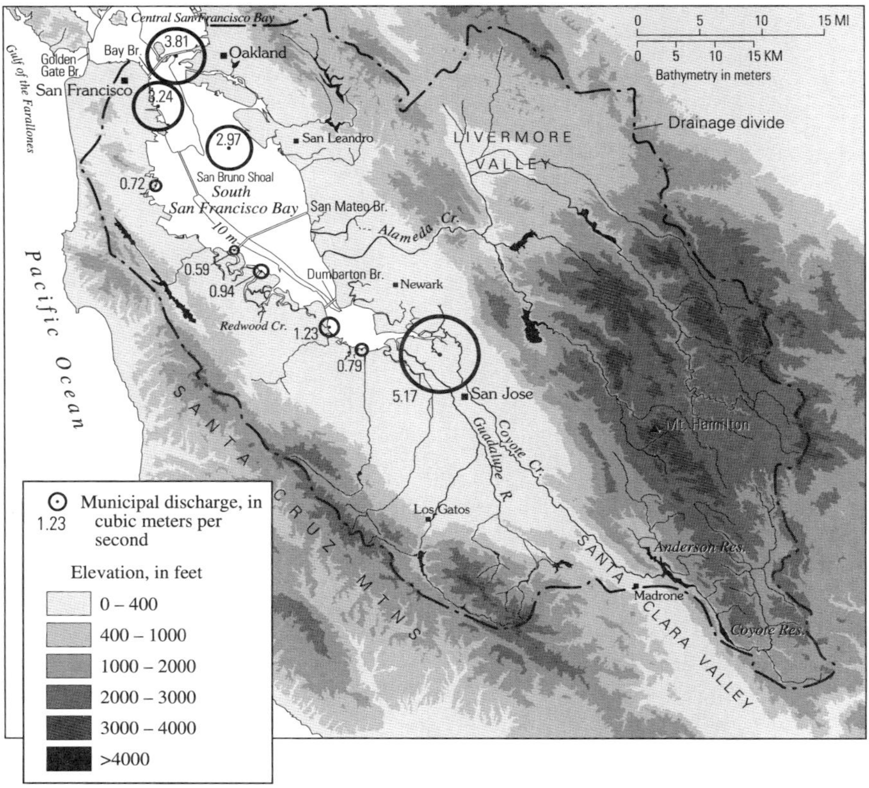

FIGURE 1. Map of South San Francisco Bay, showing the bathymetry, the extent of the local watershed, and the locations of municipal wastewater outfalls.

enced by the deep waters north of San Bruno Shoal (depth at mean tide level, 7.0 m), and mean depths for San Bruno Shoal to San Mateo Bridge, San Mateo Bridge to Dumbarton Bridge, and south of Dumbarton Bridge are much smaller (Table 1).

The climate varies substantially over the South SFB basin, in part because it is located on the boundary of the marine and continental air masses (Gilliam 1962). Rainfall varies from 35 cm yr^{-1} at San Jose on the valley floor (1955-93), to 54 cm yr^{-1} on Mt. Hamilton east of San Jose (1955-93), and 128 cm yr^{-1} at Los Gatos west of San Jose (1959-83) (U.S. Dept. Commerce 1956-94). Rainfall is highly seasonal, occurring for the most part during winter. Evaporation and evapotranspiration rates also show large seasonal variation, with highest values in spring and summer due to high temperatures and afternoon winds generated by the marine/continental airmass interaction (*cf.* Conomos 1979).

The watershed of South SFB is quite small in relation to the volume of the estuary. The total area of the watershed is only about 4800 square kilometers (San Francisco Bay Conservation and Development Commission *et al.* 1992), much of it lying south and east of the estuary (primarily the Santa Clara and Livermore

TABLE 1. Hydrography of South San Francisco Bay[1]

Segment	*Area* $10^9\ m^2$	*Volume* $10^9\ m^3$	*Average Depth* *m*
At Mean Lower Low Water:			
Upper			
Bay Bridge to San Bruno Shoal	0.225	1.40	6.2
Middle			
San Bruno Shoal to San Mateo Bridge	0.129	0.37	2.9
San Mateo Bridge to Dunbarton Bridge	0.068	0.26	3.8
Lower			
South of Dunbarton Bridge	0.019	0.058	3.1
At Mean Tide Level:			
Upper			
Bay Bridge to San Bruno Shoal	0.231	1.62	7.0
Middle			
San Bruno Shoal to San Mateo Bridge	0.138	0.50	3.6
San Mateo Bridge to Dunbarton Bridge	0.088	0.34	3.9
Lower			
South of Dunbarton Bridge	0.034	0.086	2.6
At Mean High Water:			
Upper			
Bay Bridge to San Bruno Shoal	0.235	1.82	7.7
Middle			
San Bruno Shoal to San Mateo Bridge	0.144	0.62	4.3
San Mateo Bridge to Dunbarton Bridge	0.102	0.42	3.5
Lower			
South of Dunbarton Bridge	0.046	0.117	2.5

[1] Data from Selleck *et al.*, 1966; areas from Table VI, and volumes calculated from coefficients in Table III, using tidal datums of 0, 0.97 and 1.78 meters. The Upper segments are their sections 3-5, the Middle segments are their section 6, and sections 7-9, and the Lower segment is sections 10-15.

Valleys, Fig. 1). The lowland area immediately surrounding the South SFB estuary is almost completely urbanized, containing most of the almost 4 million people now living in the watershed (Association of Bay Area Governments 1991). Because of this large number of people in the watershed, wastewater inputs to the estuary are substantial (California Regional Water Quality Control Board, San Francisco Bay Region 1987). Urban runoff and other nonpoint sources of nutrients may also be a significant factor in the ecology of South SFB (San Francisco Bay Conservation and Development Commission *et al.* 1992).

Our purpose here is to develop a general understanding of the factors that affect concentrations of plant nutrients in South SFB. Important factors include wastewater inputs, freshwater inputs, water exchanges brought about by variable conditions in Central SFB, evaporation, and biological uptake and regeneration. Here we focus on wastewater inputs, the influences of conditions in Central SFB, and biological effects. We will show that information about some of the factors is rather limited, and thus identify needs for future research. More detailed treatment of the effects of meteorological forcing, evaporation, and local stream inflows are in Schemel & Hager (this volume).

APPROACH AND SOURCES OF DATA

Data on concentrations of nutrients in the waters of South SFB from a variety of sources were consulted for this study (Table 2). The data presented here are primarily longitudinal transects of the surface waters of the main channel from our studies, extending from the Oakland-San Francisco Bay Bridge in the north to the limit of the deepwater channel in the south, about 8 km south of Dumbarton Bridge.

TABLE 2. Sources of nutrient data consulted for this study.

Location	*Year*	*Citation*
Bay	59-60	Harris *et al.*, 1961
	60-61	McCarty *et al.*, 1962
	61-62	Storrs *et al.*, 1963
	63-64	Storrs *et al.*, 1965
	64	Bain & McCarty, 1965
	68	Kramer *et al.*, undated
	69-70	Schemel, 1971
	69-76	Smith *et al.*, 1979
	76-77	Cole & Herndon, 1979
	77	Smith & Herndon, 1980
	78	Smith & Herndon, 1983
	79	Cloern & Cole, 1979
	80	Ota *et al.*, 1989
	81-93	City of San Jose, self-monitoring reports
	82	USGS, unpublished data
	82-85	Kinnetic Laboratories, Inc., . . . 1984, '85, '86
	84	USGS, unpublished data
	88-91	Hager, 1993
	89	Flegal *et al.*, 1991
	91-93	Hager, 1994
Streams	79-90	Earthinfo, 1993

We use plots of nutrient concentrations against salinity (nutrient/salinity diagrams) to describe the processes that affect the distributions of dissolved inorganic nutrients in the South SFB estuary. One advantage of using nutrient/salinity diagrams is that spatial variations in concentrations of the nutrients caused by the tidal excursion are minimized. Because of steep longitudinal concentration gradients within South SFB, this can be a substantial simplification. Another reason to use nutrient/salinity plots is that inclusion of a "conservative" variable often makes it easier to understand the physical processes involved. Instead of showing nutrient versus location and salinity versus location plots to explain an observation, the nutrient/salinity plot often suffices.

A time series of estuarine transects on a nutrient/salinity diagram often shows the relative importance of the processes that cause the changes over time, because these processes can be represented as unique vectors on the diagram (Fig. 2). A water parcel with characteristics represented at point W can be modified by mixing with freshwater (FW), wastewater (WW), and seawater (SW). Simultaneously it can also be modified by evaporation (E), precipitation (rainfall) (P), biological and

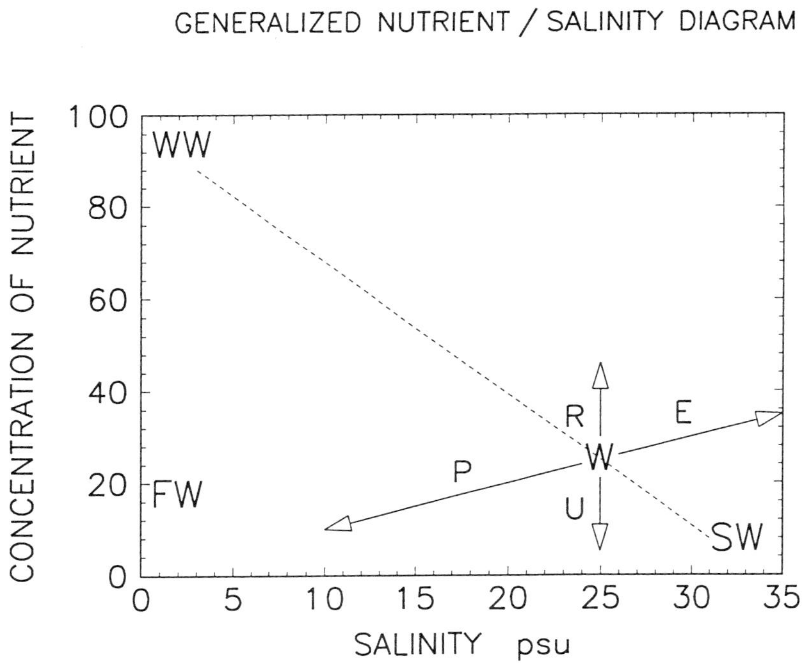

FIGURE 2. Generalized nutrient / salinity diagram showing vectors corresponding to processes acting on water parcels in South San Francisco Bay. WW, FW and SW are the wastewater, freshwater and seawater mixing members, respectively. E is evaporation, P is precipitation, U is uptake by either biological or abiological processes, and R is regeneration, either by organic decomposition or other release processes.

perhaps abiological uptake (U), and regeneration from organic matter or release from inorganic matter (R). The evaporation vector, if extended backward, would pass through the origin, because it is the inverse of the vector describing mixing with distilled water. The rainfall vector, however, does not quite pass through the origin, because the rainwater contains some nutrients, although usually at low concentrations. To avoid confusion, the term freshwater will refer to surface waters entering the bay in streamflow or runoff, and the FW and WW sources will be collectively referred to as non-saline waters.

In most of this study, we will use three terms for the nutrients which may warrant a brief introduction for some readers. First, dissolved inorganic nitrogen (DIN), the sum of nitrate, nitrite and ammonium, is used so that we do not need to account for nitrification of ammonium. Second, dissolved silica (DSi) is used, which is the same as silicate. Third, dissolved reactive phosphorus (DRP) is used, which makes reference to the possibility that the acidic conditions of the method of analysis convert some other forms of phosphorus to ortho-phosphate. It is approximately equal to ortho-phosphate under most natural conditions.

FACTORS AFFECTING NUTRIENT DISTRIBUTIONS

Wastewater Inputs

Wastewater inflows are, in most cases, the most important factor responsible for the distributions of nutrients observed in South SFB. The concentrations of DIN and DRP in wastewater are much higher than those in natural surface waters or in sea water. In areas where there is substantial municipal use of groundwater, even the concentration of DSi in wastewater can be as much as twice that of the surface waters entering the bay (see below).

About three-fourths of the wastewater inputs to lower and middle South SFB enter lower South SFB, and because of long residence times there, these sources mix and appear to be one large source. Thus, hypothetically, the simplest nutrient/salinity diagrams could look like mixing lines between this single, high-nutrient, non-saline water source landward, and a low-nutrient, saline source seaward (*cf.* dashed line in Fig. 2). This condition is virtually never achieved, however, because the residence time of the waters is long relative to the time scales of variability of the various factors affecting concentrations. None-the-less, recognition of the general condition of a strong longitudinal concentration gradient, with high concentration, low salinity waters toward the landward end, is a key to understanding the more complex nutrient/salinity diagrams which describe nutrients in South SFB.

Over the past few decades, there have been major changes in the number and location of outfalls into South SFB, and in the quantity and quality of the wastewaters entering the bay. For example, in the mid-1960s, there were 17 outfalls (13 from sewage treatment plants [STPs]) south of the San Mateo Bridge, 29 outfalls (12 STPs) between the San Mateo Bridge and Hunter's Point, and 40 outfalls (3 STPs and at least 18 sewers) between Hunters Point and the Oakland-San Francisco Bay Bridge (Pearson *et al.* 1969). Consolidations began as early as 1965, and by about 1982, the existing configuration was in place (Fig. 1). Probably the major change (excluding improvements in treatment) was the collection of all effluents from the southeast side of the bay into a deep water outfall north of the San Bruno Shoal (Fig. 1). The exception was the Milpitas plant, which joined with the San Jose/Santa Clara Water Pollution Control Plant (hereafter, SJ/SC WPC Plant), and opted for improved treatment and continued discharge into lower South SFB.

In this time period, there were many significant changes in wastewater treatment processes. Most dischargers had attained secondary treatment by 1979, and some began it as early as 1962 (Table 3). These improvements have produced major changes in the composition of the wastewaters entering South SFB. As an example, we show data from the SJ/SC WPC Plant, which generates about 72 percent of the wastewater volume to lower South SFB (Davis *et al.* 1991). The plant was modified in March of 1979 to tertiary treatment (removal of additional oxygen demand and

TABLE 3. Wastewater outfalls into South San Francisco Bay.

Receiving Water Outfall	*Volume*[1] $m^3 s^{-1}$	*Degree of Treatment*
Northern South SFB		
East Bay M.U.D.	3.81	secondary (1978)
San Francisco Southeast	3.24	secondary (1982)
East Bay Discharges	2.97	secondary
South San Francisco and others	0.72[2]	secondary (mid-late 60s)
Middle South SFB		
San Mateo	0.59	tertiary (1980)
South Bayside System	0.94	tertiary
Lower South SFB		
Palo Alto	1.23	tertiary
Sunnyvale	0.79	tertiary
San Jose / Santa Clara	5.17	tertiary (1979)

[1] Volume data for 1984 - 87, from Davis *et al.*, 1991, except as noted. Degree of treatment data from California Regional Water Quality Control Board, San Francisco Bay Region, 1987.

[2] Estimate from Brown & Caldwell *et al.*, 1975.

nitrification of the effluent). The time series of the composition of the effluent from the plant is shown in Fig. 3. The improved treatment clearly lowered organic nitrogen concentrations in the effluent, and also nitrified most of the ammonium in the wastewater to nitrate prior to its discharge to the bay.

The effect of the change in the plant effluent on the composition of the waters of South SFB was rapid (Fig. 4). In samples collected prior to the change (November, December, January and February), ammonium was a fairly large fraction of the dissolved inorganic nitrogen even at our sampling sites beginning over 13 km from the outfall. After the change to in-plant nitrification (May, June and July), ammonium was a much smaller fraction of the DIN at our sampling sites. In March, nearest in time to the change-over date, the waters furthest from the plant followed the pre-change pattern, whereas the waters nearer the plant had started to respond

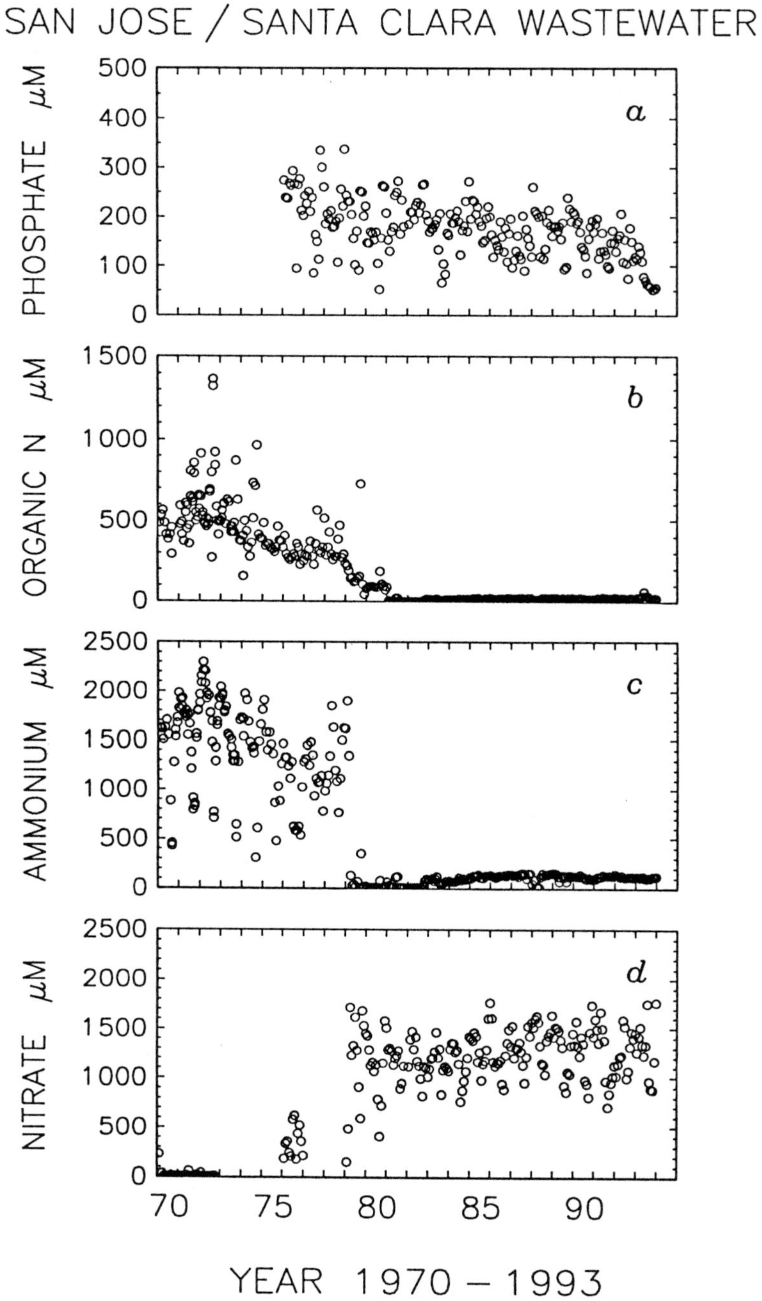

FIGURE 3. Changes in the composition of the wastewaters discharged by the San Jose / Santa Clara Water Pollution Control plant: (a) – phosphate; (b) – organic nitrogen; (c) – ammonium; (d) – nitrate. The plant was upgraded to tertiary treatment in March 1979.

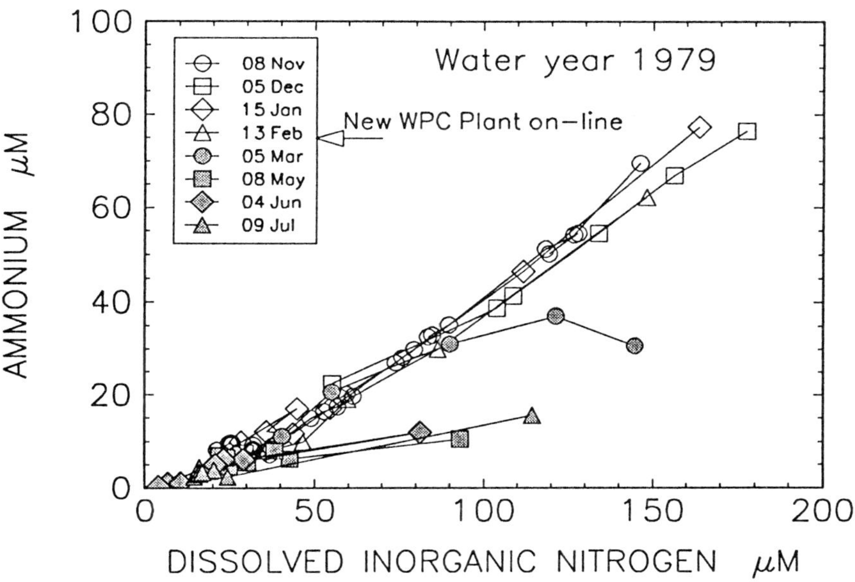

FIGURE 4. Concentrations of ammonium in the waters of South San Francisco Bay in relation to concentrations of dissolved inorganic nitrogen, showing the effect of the March 1979 change in the composition of the effluent of the San Jose/Santa Clara Water Pollution Control plant on the fraction of DIN that was ammonium.

to the lower ammonium in the effluent. Prior to the change, ammonium concentrations over 400 μM had been observed (Conomos *et al.* 1979), and now, ammonium concentrations rarely exceed 20 μM.

We analyzed the effluent data from the SJ/SC WPC Plant to see if there were any trends over time, using only the data since the conversion to advanced treatment. In the period 1980-93, there were slight decreases in volumes discharged and concentrations of organic nitrogen, and small increases in concentrations of ammonium and nitrate, and thus DIN. The rate of increase of concentrations of DIN was about 1.5 percent per year. However, because of the decreasing volumes discharged, there was no significant change in the loading of either DIN or total nitrogen (DIN plus organic nitrogen) over the period. There was a distinct downward trend (about 2 percent per year) in concentrations of phosphate, although the concentration data are more variable than those for nitrate.

An additional factor influencing nutrient distributions is the variation of the composition of the wastewater effluent on a seasonal time scale. There is undoubtedly variation on shorter time scales as well, but these variations are probably not important because of the large volume of lower South SFB (86 x 10^6 m^3 at MTL, Table 1) relative to the volume of daily discharge (0.62 x 10^6 m^3d^{-1}, Table 4). The observed annual cycle in the effluent of the SJ/SC WPC Plant is characterized by relatively little change in concentrations of nitrate and phosphate over most of the year, and lower concentrations of nitrate and phosphate in the months of July,

August and September, the minimum occurring reliably in August (Fig. 5). In these summer months, there is little variation in volume discharged, and no comparable variation in the concentrations of ammonium or organic nitrogen.

DSi is rarely measured in effluents from wastewater treatment plants, but DSi was measured by the Department of Water Resources in wastewaters entering South SFB in 1962 (Storrs *et al.* 1963). The flow weighted concentrations for the municipalities of northern Santa Clara County were 498 and 507 μM for February 19 - March 3 and April 23 - May 3, 1962, respectively. These are twice the typical surface water concentrations for this area (see below), probably because a large fraction of the municipal water used in this region in the Spring of 1962 was well

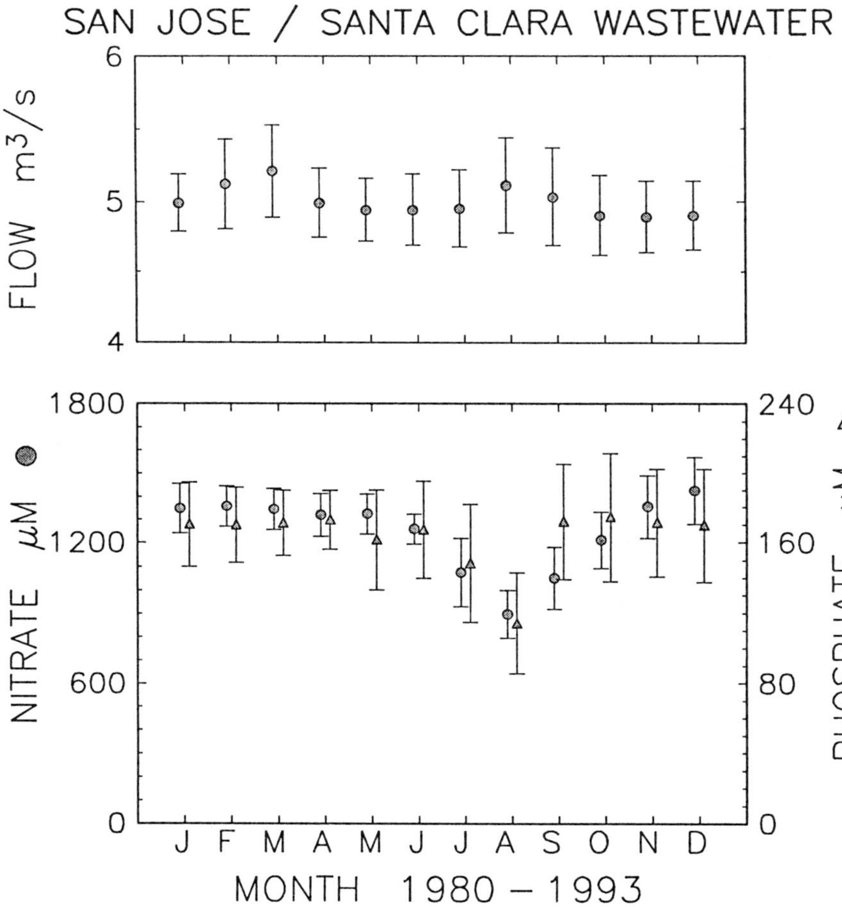

FIGURE 5. Variations in the composition of the wastewaters of the San Jose/Santa Clara Water Pollution Control plant on a annual time scale. Error bars shown are 95 percent confidence intervals on the monthly means.

TABLE 4. Flow characteristics of sources of non-saline waters (creeks, rivers, and wastewaters) entering South San Francisco Bay.

Receiving Water Sources	*Area of Watershed*	*Period of Record*	*Annual Mean*	*Maximum Monthly Mean*	*Flow Exceeded Percent of Time 10%*	*50%*
	km²	*Water Years*	m^3s^{-1}			
Upper South SFB						
San Lorenzo	97.1	1968-92	0.473	1.44	1.05	0.057
Colma	28	1964-92	0.216	0.53	0.37	0.048
Wastewater	—	1984-87	8.06	—	—	—
Middle South SFB						
Patterson[1,3]	1639	1959-92	2.31	8.38	3.26	0.000
Redwood	4.7	1960-92	0.031	0.098	0.042	0.003
Wasterwater	—	1984-86	2.33	—	—	—
Lower South SFB						
Guadalupe[3]	378	1959-92	1.29	4.02	1.56[2]	0.062[2]
Saratoga	23.9	1934-92	0.245	0.82	0.54	0.023
San Francisquito	96.9	1931-92	0.538	1.92	0.85	0.009
Coyote[3]	508	1959-87	1.65	2.48	2.58	1.22
Penitencia	55.7	1961-87	0.18	0.555	0.31	0.016
Wastewater	—	1984-87	7.23	—	—	—

Sources of data: Creeks and rivers, USGS, 1993, except Coyote and Guadalupe, our calculation from USGS data; Wastewaters, Davis *et al.* 1991, except for Middle South SFB, Gunther *et al.* 1987.

[1] Patterson Creek is a distributary of Alameda Creek.
[2] For 1930 - 92.
[3] At lower flows, most of this water may have been used for groundwater recharge.

water since imports totalled about 1 m^3s^{-1} while groundwater pumpage equalled almost 5 m^3s^{-1} (Poland & Ireland 1985). In general, groundwater has higher concentrations of DSi than surface waters. Concentrations in wastewaters entering upper South SFB were more similar to concentrations in surface waters, ranging for the most part from 210 μM to 420 μM, with a flow weighted average of 304 μM (Storrs *et al.* 1963).

Extrapolation of DSi/salinity plots to zero salinity, using data sets from lower South SFB that extended to low salinities, and were collected during times of low freshwater inflow and when uptake by diatoms was probably low, gave concentrations of 442 μM, 450 μM, and 474 μM. In general, we expect that concentrations of DSi in wastewaters entering lower South SFB will be similar to the reported measurements and to our extrapolations, or about 440 to 500 μM.

Freshwater Inputs

Effects of local freshwater inputs on distributions of nutrients in South SFB are usually difficult to quantify, in part because of the effects of other processes such as mixing with waters from Central SFB. The amount of local freshwater inflow is poorly known because only about 47 percent of the watershed is currently gaged, down from 62 percent in 1963. Some of the gaging is upstream from groundwater recharge facilities, so that even the gaged flows do not always accurately represent flows to the bay. For example, flows reported for Coyote Creek (station near Madrone, discontinued after 1987) were usually water released for recharge, and only the highest flows reached the bay. These flows exceed 3 m^3s^{-1} only 23 days per year on the average. Impoundments also play a role in the response of potential local freshwater sources to storm events. An extreme example of the influence impoundments occurred during the storm of January 11-13, 1952, which caused substantial damage in the Alameda Creek and Guadalupe River drainages, but virtually no flow in Coyote Creek at Madrone, because all of the runoff from the upper Coyote Creek watershed was stored in Coyote and Anderson reservoirs (capacity 142 x 10^6 m^3, USGS 1985) (Rantz 1956). There are also significant impoundments on the Alameda Creek (277 x 10^6 m^3) and Guadalupe River (52 x 10^6 m^3) systems. Note that the sum of the capacities of these reservoirs is greater than the volume of South SFB below San Mateo Bridge at mean tide level. Finally, urbanization of areas immediately surrounding the bay has made them more impervious, and thus direct sources of storm runoff (Waananen 1969; Crippin & Waananen 1969). Probably the best estimate of runoff available combines Rantz's (1974) data for natural runoff with land-use data and runoff coefficients, as done by the San Francisco Bay Conservation and Development Commission (1992) for an average precipitation year.

The response of South SFB to local freshwater inflows is relatively rapid, related to storm events or sequences of storms that increase inflows over a period of days to weeks (Schemel & Hager, this volume). This episodic nature is seen in the variability in the gaged inputs from the Guadalupe River (1949-93) and the Alameda Creek system (Patterson Creek distributary, 1959-93) (Fig. 6). During the period 1959-93, these major sources had flows less than 0.1 m^3s^{-1} for 212 days and 263 days per year, respectively. Statistics on these and other freshwater flows to South SFB, and for comparison, wastewater flows, are shown in Table 5. Comparisons should take into consideration the physical differences between the receiving areas of South SFB (Table 1). North of San Bruno Shoal, freshwater flows are much smaller than wastewater flows, but the relatively large volume of this area minimizes effects of these inflows. In middle South SFB, most of the freshwaters discharge onto the broad shallow area on the east side of the bay, while the wastewaters flow into the deep water channel on the west side, which tends to separate their effects. In lower South SFB, freshwaters and wastewaters are both discharged into an area of relatively small volume, where they mix together. However, freshwater inflows are similar in magnitude to the wastewater inputs only during the wettest month of an average year (Table 5).

Although nutrients in runoff do represent additional loading to South SFB, runoff probably lowers concentrations of DIN and DRP in the ambient waters of South SFB, because concentrations appear from limited data to be relatively low in runoff waters. Median concentrations in the Guadalupe River in the period 1979-90 were: ammonium, 4 μM (n = 71); nitrate plus nitrite, 79 μM (n = 54); and o-PO_4, 0.3 μM (n = 71, Earthinfo 1993). When these nutrient concentrations were plotted against streamflow rates, higher concentrations were found only at the

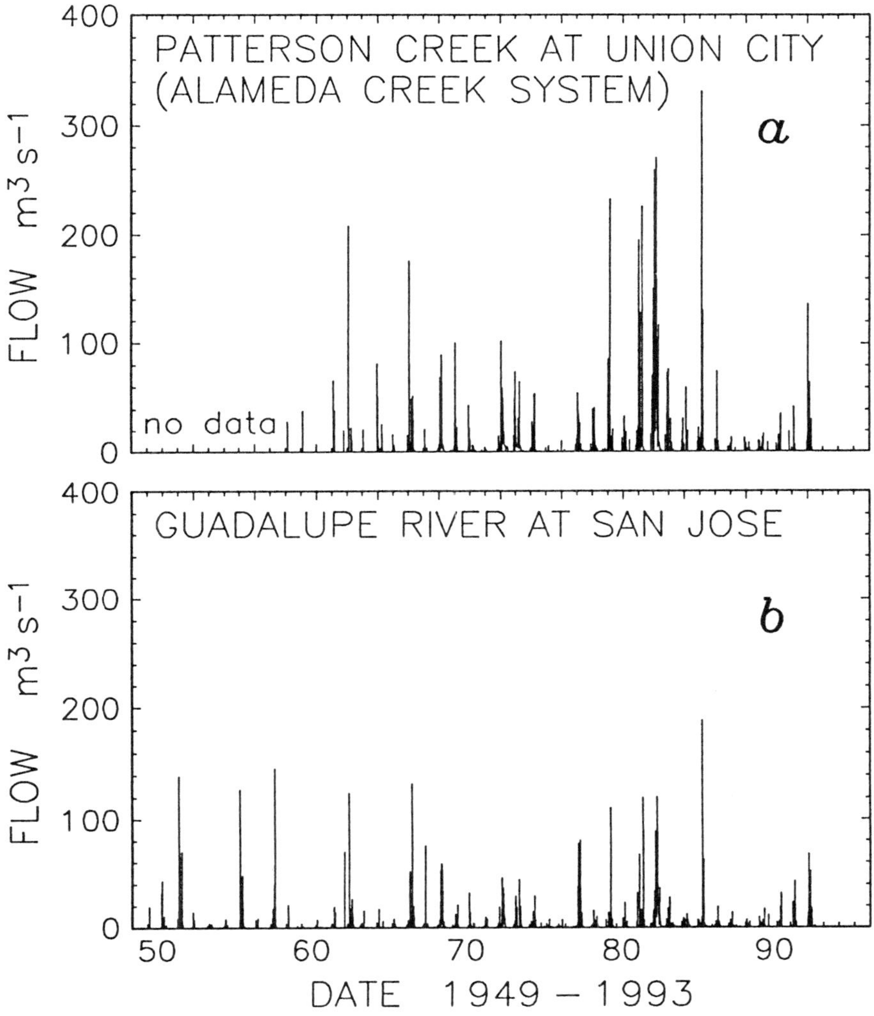

FIGURE 6. Gaged inflows to South San Francisco Bay from: (a) the Alameda Creek system; and (b) the Guadalupe River at San Jose. Data from U.S. Geological Survey Water-Data Reports.

lowest flow rates. This would tend to moderate the effects of high flow rates on the transport of dissolved nutrients into the estuary.

The only N and P data available for the Alameda Creek system are for the period 1974-80, when the creek was used as a conduit to carry wastewater from the Livermore Valley, and thus these data are not indicative of the current situation. This practice ceased in January of 1980.

TABLE 5. Average monthly precipitation rate measured at Newark, and pan evaporation rates measured at Newark and Redwood City, California. (Pan coefficients are those estimated for Newark, but were not applied to the pan evaporation rates in this table.)[1]

Month	*Precipitation*	*Pan Evaporation*			*Pan Coefficient*
	cm mo^{-1}				
	Newark			*Redwood City*	
	1949-93	*1964-77*	*1990-93*	*1990-93*	
January	7.5	4.61	5.03	3.74	0.80
February	5.7	6.35	6.76	5.70	0.70
March	5.4	12.0	11.0	9.26	0.70
April	2.6	17.0	18.2	16.2	0.75
May	0.81	22.3	22.7	19.9	0.75
June	0.23	23.7	24.6	22.3	0.75
July	0.07	25.1	24.8	22.9	0.80
August	0.16	22.2	22.7	20.5	0.80
September	0.42	18.5	17.9	16.6	0.80
October	2.0	12.7	13.4	13.2	0.85
November	4.7	6.86	8.03	7.12	0.85
December	6.0	4.64	5.50	3.73	0.85

[1] Precipitation data from U.S. Department of Commerce, Weather Bureau, annual summary (1993). Evaporation data from U.S. Department of Commerce, Weather Bureau, annual summaries (1964-1977), and Cargill Salt Company (pers. commun., 1994). Pan coefficients are those reported by Selleck and others (1966).

For dissolved silica, data for the Guadalupe River in the period 1979-90 (n = 73) and those for Alameda Creek (1952-64, n = 199) both give median values of 250 μM (Earthinfo 1993). Concentrations of DSi were not related to flow rate in either system, which probably indicates that the concentrations observed in Alameda Creek were not influenced by wastewater inputs of DSi.

Urban runoff can contribute significant amounts of nutrients to some estuarine systems, but does not seem to be very important to South SFB. The total annual transport of nutrients to South SFB by runoff, including urban runoff, in an average precipitation year has been estimated to be 1.1×10^6 kg Total Kjeldahl N yr^{-1}, 0.3×10^6 kg Nitrate-N yr^{-1}, and 0.18×10^6 kg Phosphate-P yr^{-1} (San Francisco BCDC 1992). We estimate that these inputs would amount to 12 percent of the total N inputs and 5 percent of the total P inputs by the point sources to lower South SFB.

Seasonal Flushing due to Mixing with Central SFB

South SFB is a tributary estuary (or sub-estuary) of the northern SFB estuarine system. This means that South SFB responds to changes in the main estuary, and particularly, to changes in Central SFB. Changes in salinity and nutrient concentrations in Central SFB correspond with increases or decreases in Delta outflow (Peterson *et al.*, this volume). Mixing between Central SFB and South SFB is enhanced during large increases in Delta outflow and by gravitational flows during neap tides (Walters *et al.* 1985). Mixing with waters from Central SFB removes nutrients from South SFB. This aspect of the interaction between South SFB and the main estuary has been recognized for over two decades (McCulloch *et al.* 1970).

Effects of mixing with Central SFB are greatest in the upper portion of South SFB, but can extend far down South SFB. The most important factor affecting the degree of penetration of Central Bay water into South SFB is the magnitude of Delta outflow (see Conomos 1979, Table 5). Conditions in Central SFB respond rapidly (within 7 days on average; Conomos 1979, Fig. 17) to changes in Delta outflow. The rapidity of this response can be seen in Fig. 7, where a large increase in Delta outflow beginning 9 days prior to the sampling date lowered salinities by as much as 10 psu. The effects extended as far south as Redwood Creek.

This penetration of lower salinity water into South SFB lowered the concentrations of DRP and DIN in parts of South SFB (Figs. 7a and 7c). This is because the concentrations of nutrients in the waters of Central SFB (a mixture of freshwaters from the Delta with seawater) are much lower than those in the ambient waters in South SFB (a mixture primarily of wastewaters and seawater). The plots clearly show the different compositions of the two different waters of low salinity. If the two low salinity waters had similar nutrient concentrations, the transect would simply turn back on itself, as the DSi/salinity diagram nearly does (Fig. 7b), instead of making an angled turn. There is even some difference for DSi, because DSi concentrations in the wastewaters contributing to the waters of lower South SFB are generally higher than the concentrations in the Sacramento and San Joaquin Rivers (see above).

From Fig. 7, it can be seen that concentrations of DRP will be lowered more than those of the other nutrients by intrusion and mixing of water from the north, because concentrations in those waters are very low relative to the ambient waters of South SFB. Thus, to show the seasonal effect of interaction with Central Bay waters, we compare DRP/salinity transects in a moderately wet year, 1980, with those in a very dry year, 1977. In 1980 (Fig. 8a), intrusion and mixing of water from northern SFB was brought about by two high Delta outflow events in the early part of the year. The creeks entering South SFB also may have contributed to the lower salinities, as there was a record high flow (February 19) from the Alameda Creek system and near record flows from other South SFB creeks at the same time (Wahl *et al.* 1980). After local streamflow and Delta outflow receded, the nutrients and salinity began to increase back toward their previous, waste-influenced, nearly marine conditions. However, because the freshening of South SFB had been substantial, even October and November distributions did not attain the salinities and nutrient concentrations of the summer of 1977 (Fig 8b).

Evaporation

In estuaries with relatively low rates of freshwater input, like South SFB, net evaporation may play a role in the seasonal variation of salinity and concentrations of nutrients (*cf.* Smith *et al.* 1987). Precipitation and pan evaporation rates at locations adjacent to South SFB are given in Table 5. The differences between evaporation rates for Redwood City and Newark are indicative of the difficulty in predicting an evaporation rate for the surface of South SFB. The need to correct pan evaporation rates using pan coefficients introduces further uncertainty. Thus, these data are shown primarily to indicate the general trend. For much of the year,

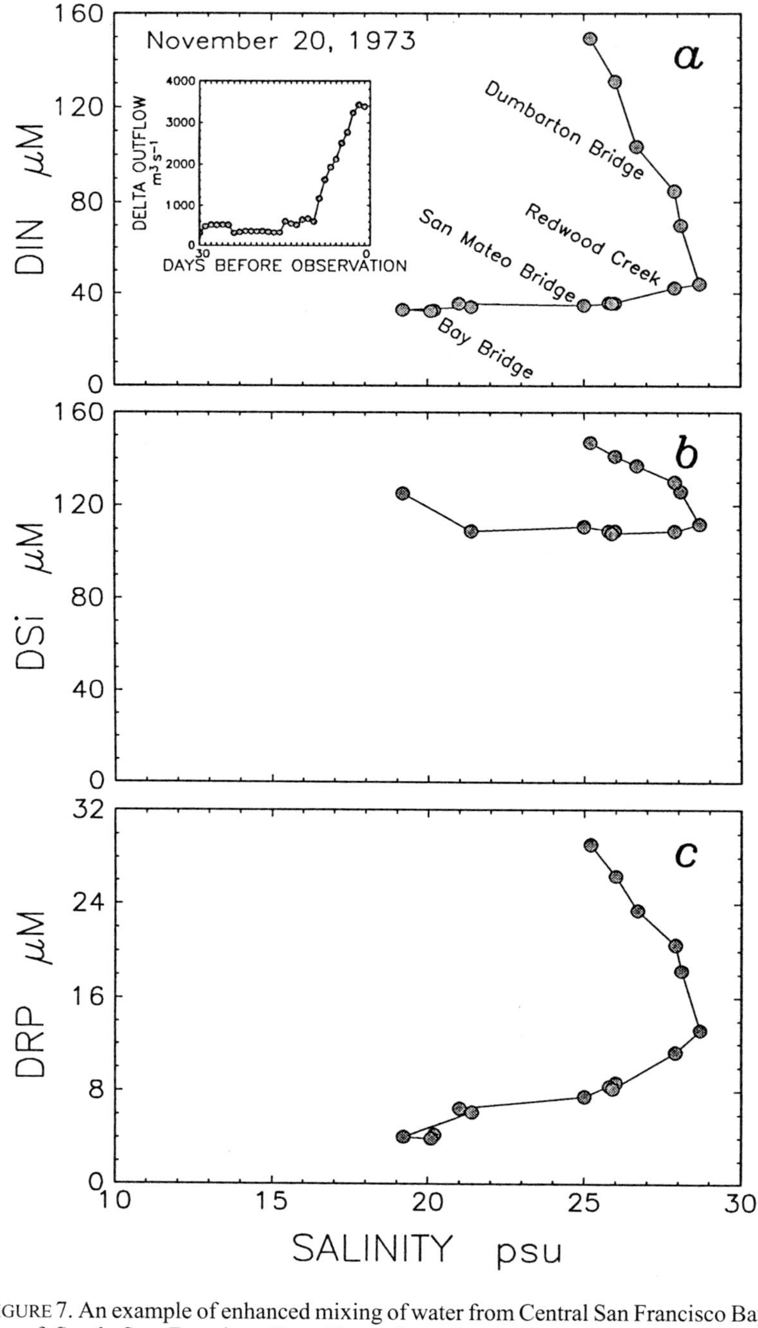

FIGURE 7. An example of enhanced mixing of water from Central San Francisco Bay with waters of South San Francisco Bay, November 20, 1973, shown on nutrient / salinity diagrams for: (a) DIN; (b) DSi; (c) DRP.

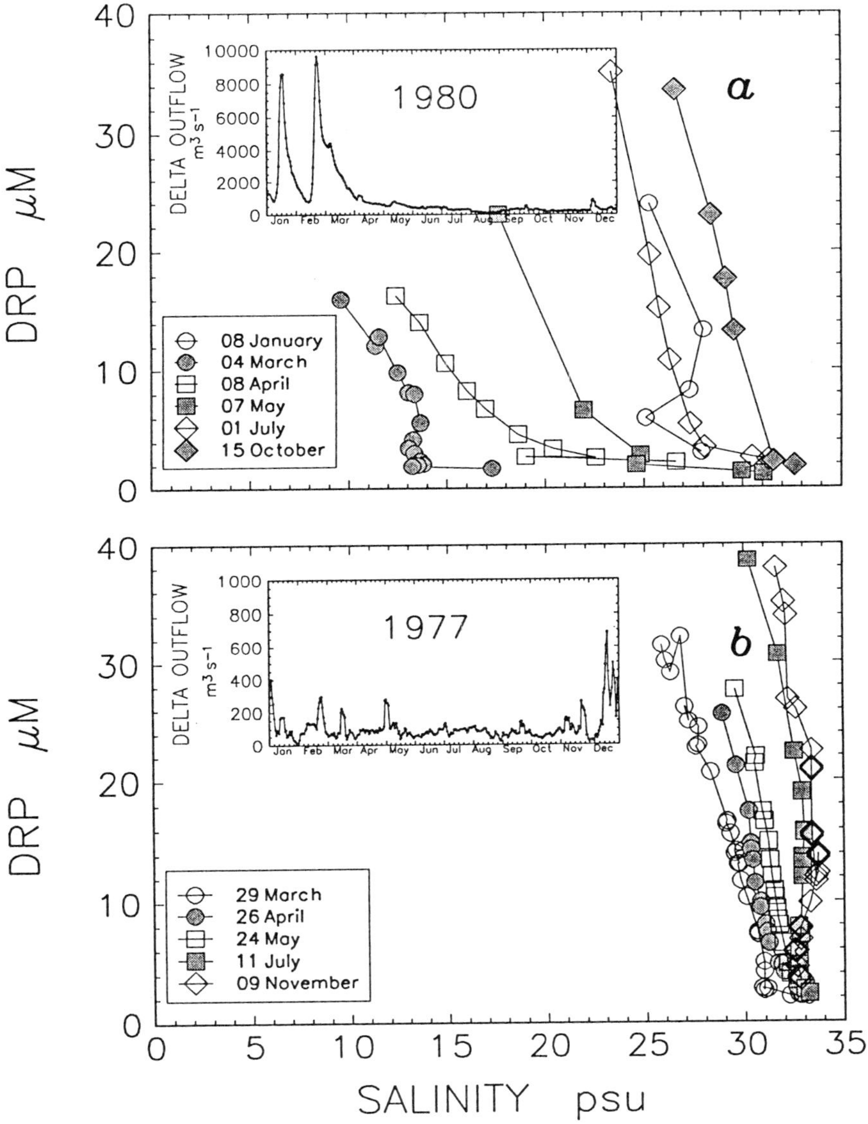

FIGURE 8. Comparison of DRP / Salinity plots for: (a) a moderately wet year, 1980; and (b) a very dry year, 1977. Note the factor of ten difference in the scales of the Delta Outflow insets.

there is net evaporation. Multiplying the average net rate (corrected using the pan coefficients) for the three months with the highest rates by the high tide surface area provides a measure of the potential of evaporation to affect nutrient and salinity values. In lower South SFB, almost 10 percent of the volume at mean tide level would be lost per month, with 5 percent and 3 percent lost from the middle and upper segments respectively. In this high evaporation rate period, the rate for lower and middle South SFB would be equivalent to a freshwater loss of about $20 m^3 sec^{-1}$, or twice the waste input rate to those parts of South SFB.

Evaporation concentrates solutes, including nutrients, and raises salinities. When the rate of water loss is greater than the rate of inflow of non-saline water, seawater flows in from Central SFB to replace the lost water, bringing with it additional salt. Evaporative concentration effects are sometimes seen on nutrient/salinity diagrams as mid-bay salinity maxima during dry months. This effect was seen in all nutrient salinity diagrams when mean June-September Delta outflows were lower than 200 $m^3 s^{-1}$ (*cf.* Fig. 8b), was sometimes seen when flows ranged from 200-600 $m^3 s^{-1}$, but was not seen when flows were higher. There is some uncertainty in assigning evaporation as the cause of salinity maxima in South SFB, because similar effects are created by fluctuating flows out of the Delta, as described above.

Biological Uptake and Regeneration from Organic Matter

The spring phytoplankton bloom is a recurrent feature in South SFB. The conditions under which blooms are likely to occur include low tidal energy and adequate freshwater buoyancy, which produce stratification in the water column. Then, because of a combination of reduced benthic grazing of phytoplankton production and increased light exposure in the surface layer, the phytoplankton increase their biomass rapidly, attaining chlorophyll a concentrations of over 40 mg m^{-3} (Cloern 1991). During these blooms, nutrient uptake rates exceed supply rates and concentrations of ammonium, nitrite, nitrate and dissolved silica are reduced to very low levels. Concentrations as low as 0.3 μM DIN and 0.7 μM DSi have been observed during the spring blooms of 1990 and 1993.

In Fig. 9, we can follow the effects of the bloom of the spring of 1993 as it developed and dissipated. In late January, mixing with freshened waters of Central SFB reduced salinities and DIN and DRP concentrations in upper South SFB, in part setting up the buoyancy conditions conducive to a bloom. On the 11th of March, nutrients in the highest salinity part of South SFB (middle South SFB) were present at concentrations of about 30 μM DIN, 80 μM DSi, and 3 μM DRP. By March 30th, DIN and DSi had been completely depleted, whereas DRP was still present at a little less than 2 μM. The bloom was still keeping concentrations of DIN and DSi low in mid-April, but by June, nutrient concentrations had increased.

The removal of nutrients by phytoplankton in March show the importance of recycling to the late stages of blooms, because the 65-80 μM of DSi consumed was accompanied by an apparent consumption of only 30-50 μM DIN. By comparing the stoichiometry of these nutrients with the Redfield ratios (N:Si ~ 1:1, see later), we can infer that maybe as much as half of the N was recycled to support that DSi uptake. Therefore, both N and Si reach rate limiting concentrations in the same bloom (*cf.* Levasseur *et al.* 1990), because recycling of N allows continued uptake of Si after N has been depleted.

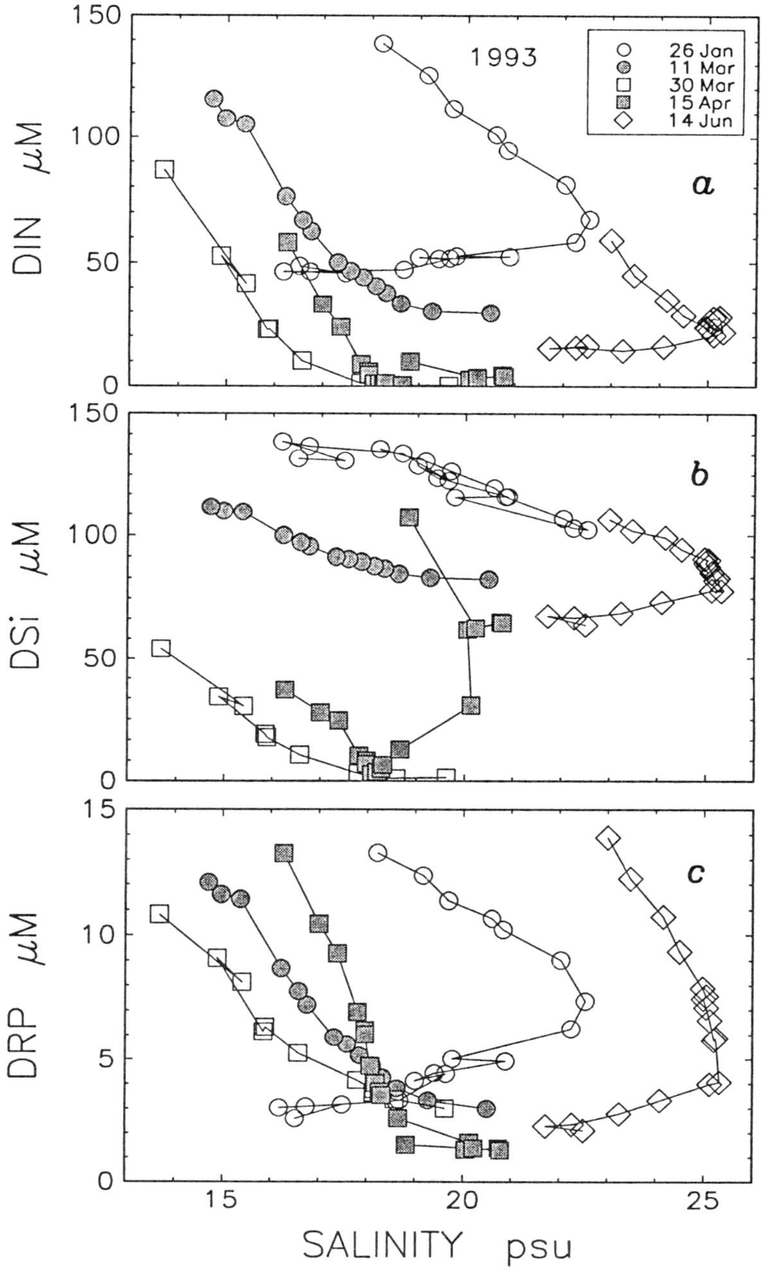

FIGURE 9. Details of the progression of the 1993 Spring phytoplankton bloom for: (a) DIN; (b) DSi; (c) DRP. Some editing has been done to the individual transects to simplify the figure.

Although spring blooms produce the most dramatic draw-downs of nutrients, it is likely that some uptake of nutrients by phytoplankton goes on all year in South SFB (Cole & Cloern 1984). Much of the time, however, rates are low and salinity gradients are small, such that the effects are not obvious on nutrient/salinity plots. In this case, nutrient/nutrient plots sometimes do show evidence, because of the differences in the stoichiometry of uptake relative to supply. Plots of DIN against DRP for 1990-93 show concavity in late summer (Fig. 10a), and are typical of late summer conditions in all 14 of the years for which there are records. This could be due to either losses of nitrogen or additions of phosphorus, or a combination of the two. With these data, we are unable to decide, although the minima in some DIN/DRP lines certainly suggest that nitrogen losses are probable. These losses of nitrogen may be either directly or indirectly due to uptake by plants. An indirect mechanism might be phytoplankton uptake → benthic grazing → fecal deposition → ammonification → nitrification → denitrification (see, for instance, Seitzinger 1988). This mechanism is suggested by high biomass of benthic filter feeders in late-summer (J. Thompson, USGS, pers. commun. 1994). The nitrogen loss could also be due to denitrification supported by diffusion of nitrate into the sediments, but the benthic flux data do not give much evidence that this is occurring (Caffrey *et al.*, this volume).

Increases in DRP also are possible. Late summer increases in water column DRP are seen in many estuaries (*cf.* Kemp & Boynton 1984; Jordan *et al.* 1991), including northern San Francisco Bay (Peterson *et al.* 1985), and nearby Tomales Bay (Smith *et al.* 1987). Many mechanisms have been proposed for the observed late summer increases (*cf.* Froelich 1988). On the other hand, measurements of benthic fluxes of phosphorus in South SFB show rates to be very low in most cases (Caffrey *et al.*, this volume).

During the winter, when we would expect biological processes to be at their minimum rates, the mixing of wastewater and seawater appears to predominate, producing linear relations (Fig. 10b). The slopes of the two curves from 1993 indicate a relation of DIN and DRP in the source different from that reported by the wastewater treatment plants. These slope differences suggest dilution of the wastewater with surface freshwaters (*cf.* Schemel & Hager, this volume), which have a very different composition (see Table 6). Of the years shown, 1993 was the wettest year.

Because wastewater inputs are an important factor in nutrient behavior in South SFB, we consider here an aspect of systems affected by wastewater inputs, first pointed out by Officer & Ryther (1980). Wastewaters ordinarily have high nitrogen to silicon and phosphorus to silicon ratios compared with natural waters. This, combined with the relatively slow regeneration of DSi from phytoplankton detritus (N ~ 0.2 day^{-1}, Si ~ 0.02 day^{-1}, unpublished data), sometimes leads to depletion of DSi and consequent shifts in phytoplankton community structure (*cf.* Cloern *et al.* 1994). This can have important implications to the consumers of primary production (*cf.* Officer & Ryther 1980; Ryther & Officer 1981).

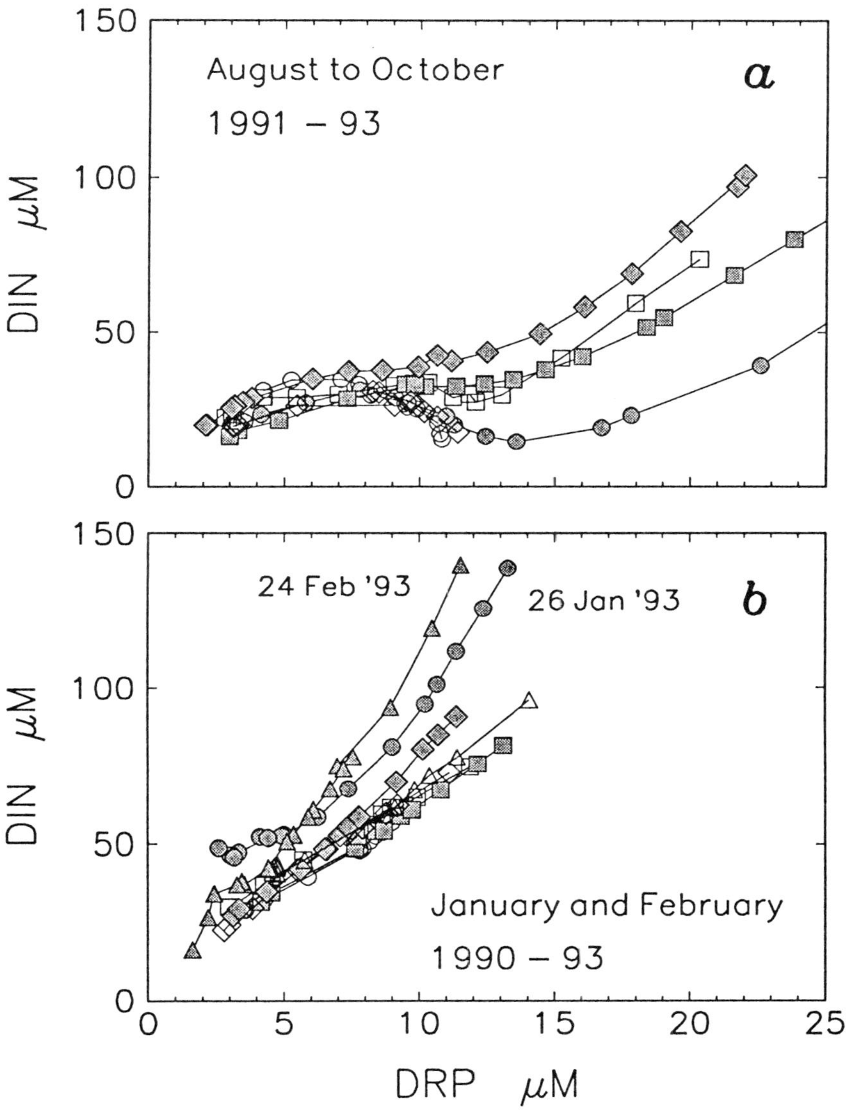

FIGURE 10. Dissolved inorganic nitrogen / dissolved reactive phosphorus for water year 1993, to show: (a) the non-linearities which develop in the late summer, (b) the effect of freshwater dilution in mid-winter.

The approximate stoichiometry of the sources of nutrients to South SFB is given in Table 6. "Redfield ratios" are touchstones in such considerations, because they are based on the observation that the proportions of carbon, nitrogen and phosphorus in phytoplankton (and therefore their net uptake) are relatively constant (C:N:P = 106:16:1, Redfield *et al.* 1963). When the phytoplankton population consists

TABLE 6. Approximate stoichiometry of nutrient sources to South San Francisco Bay.

Source	*Concentrations*			*Ratios*		
	DIN	*DSi*	*DRP*	*N/P*	*Si/P*	*N/Si*
	μM			*by atoms*		
Lower South SFB						
Wastewater[1]						
'80-Apr. '93	1336	450	169	7.9	2.7	3.0
Freshwater[2]	83	248	0.3	280		0.3
Upper South SFB						
Seawater[3]	15	32	2.0	7.5	16	0.5
River Water[4]						
(>1000 m^3s^{-1}	10	250	1.4	7	180	0.04
(~500 m^3s^{-1})	17	270	2.0	8	140	0.06
Wastewater[5]	1300	300	170	8	1.8	4.3
Redfield Ratios	—	—	—	16	~20[6]	~1

[1] N and P, San Jose/Santa Clara WPC plant self-monitoring reports, 1980-1993. Si, estimate from limited data (see text).
[2] Guadalupe River, 1979 to 1990 (N and P, n = 71; Si, n = 73). Earthinfo, 1993.
[3] Measurements of near-bottom waters of Central SFB, 1990-1993 (n=17), mean salinity, 31.49 psu (Hager 1993, 1994).
[4] N and P, Hager & Schemel 1992, Figure 5. Si, Schemel & Hager 1986, Table 1.
[5] N and P, estimated to be similar to wastewaters in note 1, above. Si, from Storrs & others 1963 (see text).
[6] Brzezinski 1985, and Park 1967 (see text).

primarily of diatoms, the approximate ratio of Si:P = 20:1 (from a Si:N of 1.2 for netplankton, Brzezinski 1985; see also Park 1967) can also be appropriate. The Redfield ratios can also be used in relation to the remineralization of the nutrients C, N and P, but are less useful in relation to the regeneration of DSi, because it cycles relatively slowly.

Only one source has an N/P ratio greater than 16 (the Redfield N/P), the freshwaters entering lower South SFB. Thus, the higher the levels of local streamflow, the more likely DRP is to become depleted by the spring bloom. However, depletion of DRP has not been observed.

The ratio of N to Si in the non-saline waters entering the south end of the bay will be about 3 (= wastewater) in a year with a relatively dry winter. In a wet winter, such as 1993, or when the input rate of freshwater amounts to approximately six times the wastewater input rate, the ratio may be 1 or lower, because the nitrogen will be greatly diluted by the surface waters, while the silica will be diluted only a little. Thus, it would seem that silica depletion would be more likely to occur, all other factors being equal, after drier winters. Silica depletion has only been seen in our surveys in the springs of 1990 and 1993, and while the winter of 1990 was dry, the winter of 1993 was fairly wet (*cf.* Fig 9). As discussed above, this depletion of DSi was probably the result of significant use of recycled nitrogen by the bloom. It may be that the temporal extent of the bloom was influenced by stratification related to the wet weather, and this allowed the DSi depletion to develop. Note that the wastewaters entering upper South SFB also have relatively high N/Si ratios, but that the presumed high dilution rate (large receiving volume) seems to prevent these effluents from having obvious impacts. Benthic fluxes of DSi, although often high enough to influence concentrations when residence times are long (Hammond *et al.* 1985), may not be a very important source of DSi at the time of the spring bloom. This is because flux rates of DSi are probably seasonal (Hammond *et al.* 1985), with highest rates in the summer as the bloom of the previous spring is partially redissolved, and lowest rates in the winter.

OVERVIEW — THE ANNUAL CYCLE

We end with a capsule description of the annual cycle exhibited by nutrient concentrations along the main channel of South SFB. This cycle is driven primarily by seasonal variations in rainfall and runoff. At the beginning of each water year (October), Delta outflow has usually just reached its annual minimum, salinity is high in most areas of South SFB, and nutrient concentrations show a distinctive gradient with highest concentrations landward of Dumbarton Bridge. During very dry years (*e.g.* 1977), there may be very little change in this gradient-dominated distribution over the year.

When the first major increase in Delta outflow event of the winter season occurs, both salinity and nutrient concentrations drop sharply, particularly in the upper part of South SFB. As additional pulses of Delta outflow enter the northern SFB system, salinity and concentrations of nutrients continue to decrease, but the effects of each successive pulse are diminished. This winter removal of salt and nutrients from South SFB also creates the hydrologic conditions that support the spring phytoplankton bloom. During the bloom, there is a rapid decrease in surface water concentrations of dissolved nutrients. If the bloom is not limited by physical factors, it may become limited by the availability of nutrients. Which of the nutrients is first to become limiting is substantially related to how much the waters of South SFB (largely a mixture of seawater and wastewater in drier years) have been diluted by mixing with waters of Central SFB, and also with the local freshwater sources from around the South Bay. This is because the various sources to South SFB contain nutrients in differing proportions. The likelihood of a nutrient becoming limiting can be estimated by comparing availability (approximated by water column concentrations) with need (Redfield ratios). Thus, dissolved silica is more likely to be limiting during a dry year, although recycling of nitrogen during a prolonged bloom has allowed depletion of DSi even in a relatively wet year. After the bloom and levels of delta outflow have receded, mixing with waters of Central SFB slowly increases salinity, and inputs of wastewater increase nutrient concentrations in South SFB. Here we have qualitatively explained much of the observed variation in nutrient concentrations over the annual cycle. Ultimately, however, numerical models of the relevant processes are essential to a quantitative assessment of the various sources and sinks.

ACKNOWLEDGMENTS

We would like to extend our thanks to all our USGS colleagues who so ably participated in our field programs over the last 25 years, particularly to Dana D. Harmon. We also thank Lawrence H. Smith who retrieved the local streamflow data for us. We also appreciate helpful reviews made by Richard E. Smith and Larry G. Miller in the early stages of the preparation of this manuscript, and later reviews by two anonymous reviewers.

REFERENCES

Association of Bay Area Governments. (1991). *Status and Trends Report on Land Use and Population. Final Report. The Geomorphology, Climate, Land Use, and Population Patterns in the San Francisco Bay, Delta, and Central Valley Drainage Basins*. Association of Bay Area Governments, Oakland, CA. 186 pp.

Bain, R. C., and J. C. McCarty. (1965). *Nutrient-productivity Studies in San Francisco Bay*. U.S. Public Health Service, Central Pacific Region Water Pollution Control Administration. 116 pp.

Brown and Caldwell and others. (1975). *Water Quality Control Plan, San Francisco Bay Basin* (2). State Water Resources Control Board, Regional Water Quality Control Board, San Francisco Bay Region, CA. (pagination varies)

Brzezinski, M. A. (1985). The Si:C:N ratio of marine diatoms: Interspecific variability and the effect of some environmental variables. *Jour. Phycol.* 21:347-357.

Caffrey, J. M., D. E. Hammond, J. S. Kuwabara, L. G. Miller, and R. R. Twilley. (1996). Benthic processes in San Francisco Bay: The role of organic inputs and bioturbation. Pages 425-442 *in* J. T. Hollibaugh, ed., *San Francisco Bay: The Ecosystem*. Pacific Division, American Association for the Advancement of Science, San Francisco, CA.

California Regional Water Quality Control Board, San Francisco Bay Region. (1987). *Historic Changes in Municipal Loadings to San Francisco Bay, 1955-1985*. California Regional Water Quality Control Board, San Francisco Bay Region, Oakland, CA. 6 pp., 2 appendices.

City of San Jose, California, San Jose / Santa Clara Water Pollution Control Plant, Environmental Services Department. (1970-1993). Self-monitoring reports.

Cloern, J. E. (1991). Annual variations in river flow and primary production in the South San Francisco Bay Estuary (USA). Pages 91-96 *in* M.Elliot & J. P. Ducrotoy, eds., *Estuaries and Coasts: Spatial and Temporal Intercomparisons*. Olsen and Olsen, Fredensborg, Denmark.

Cloern, J. E., and B. E. Cole. (1979). *Water Quality of South San Francisco Bay and Coyote Creek After Failure of the San Jose-Santa Clara Water Pollution Control Plant. I. 17 September - 10 October 1979*. U.S. Geol. Surv., Open-File Rept. 79-1600. 25 pp.

Cloern, J. E., B. E. Cole, and S. W. Hager. (1994). Notes on a *Mesodinium rubrum* Red Tide in San Francisco Bay (California, USA). *Jour. Plankt. Res.* 16(9):1269-1276.

Cole, B. E., and J. E. Cloern. (1984). Significance of biomass and light availability to phytoplankton productivity in San Francisco Bay. *Marine Ecol. Progr. Ser.* 17:15-24.

Cole, B. E., and R. E. Herndon. (1979). *Hydrographic Properties and Primary Productivity of San Francisco Bay Waters, March 1976 - July 1977*. U.S. Geol. Surv., Open-File Rept. 79-983. 97 pp.

Conomos, T. J. (1979). Properties and circulation of San Francisco Bay waters. Pages 47-84 *in* T. J. Conomos, ed., *San Francisco Bay: The Urbanized Estuary*. Pacific Division, American Association for the Advancement of Science, San Francisco.

Conomos, T. J., R. E. Smith, D. H. Peterson, S. W. Hager, and L. E. Schemel. (1979). Processes affecting seasonal distributions of water properties in the San Francisco Bay estuarine system. Pages 115-142 *in* T. J. Conomos, ed., *San Francisco Bay: The*

Urbanized Estuary. Pacific Division, American Association for the Advancement of Science, San Francisco.

Crippen, J. R., and A. O. Waananen. (1969). *Hydrologic Effects of Suburban Development near Palo Alto, California*. U.S. Geol. Surv., Open-File Rept. 126 pp., 1 appendix.

Cutter, G. A. (1989). The estuarine behavior of selenium in San Francisco Bay. *Estuarine Coastal Shelf Sci.* 28:13-34.

Davis, J. A., A. J. Gunther, B. J. Richardson, J. M. O'Connor, R. B. Spies, E. Wyatt, E. Larson, and E. C. Meiorin. (1991). *Status and Trends Report on Pollutants in the San Francisco Bay Estuary*. San Francisco Bay - Delta Aquatic Habitat Institute, Richmond, CA. 240 pp., 3 appendices.

Earthinfo, Inc. (1993). *Earthinfo CD , USGS Quality of Water*. Earthinfo Inc., Boulder, CO. (1 CD ROM)

Flegal, A. R., G. J. Smith, G. A. Gill, S. Sanudo-Wilhelmy, and L. C. D. Anderson. (1991). Dissolved trace element cycles in the San Francisco Bay estuary. *Marine Chem.* 36:329-363.

Froelich, P. N. (1988). Kinetic control of dissolved phosphate in natural rivers and estuaries: A primer on the phosphate buffer mechanism. *Limnol. Oceanogr.* 33(4, pt. 2):649-668.

Gilliam, Harold. (1962). *Weather of the San Francisco Bay Region*. University of California Press, Berkeley, CA. 72 pp.

Gunther, A. J., J. A. Davis, and D. J. H. Phillips. (1987). *An Assessment of the Loading of Toxic Contaminants to the San Francisco Bay-Delta*. Aquatic Habitat Institute, Richmond. 330 pp.

Hager, S. W. (1993). *Dissolved Nutrient and Suspended Matter Data for the San Francisco Bay Estuary, California, October 1988 through September 1991*. U.S. Geol. Surv., Open-File Rept. OFR 93-57. 57 pp.

Hager, S. W. (1994). *Dissolved Nutrient and Suspended Matter Data for the San Francisco Bay Estuary, California, October 1991 through September 1993*. U.S. Geol. Surv., Open-File Rept. 94-471. 53 pp.

Hager, S. W., and L. E. Schemel. (1992). Sources of nitrogen and phosphorus to northern San Francisco Bay. *Estuaries* 15(1):40-52.

Hammond, D. E., C. Fuller, D. Harmon, B. Hartman, M. Korosec, L. G. Miller, R. Rea, S. Warren, W. Berelson, and S. W. Hager. (1985). Benthic fluxes in San Francisco Bay. *Hydrobiologia* 129:69-90.

Harris, H. S., D. L. Feuerstein, and E. A. Pearson. (1961) *A Pilot Study of Physical, Chemical, and Biological Characteristics of Waters and Sediments of South San Francisco Bay (South of Dumbarton Bridge)*. Sanitary Engineering Research Laboratory, University of California, Berkeley, CA. 257 pp.

Jassby, A. D., J. E. Cloern, and T. M. Powell. (1993). Organic carbon sources and sinks in San Francisco Bay: Variability induced by river flow. *Marine Ecol. Progr. Ser.* 95:39-54.

Jordan, T. E., D. L. Correll, J. Miklas, and D. E. Weller. (1991). Nutrients and chlorophyll at the interface of a watershed and an estuary. *Limnol. Oceanogr.* 36(2):251-267.

Kemp, W. M., and W. R. Boynton. (1984). Spatial and temporal coupling of nutrient inputs to estuarine primary production: The role of particulate transport and decomposition. *Bull. Marine Sci.* 35:522-535.

Kinnetic Laboratories, Inc. and Larry Walker Associates. (1984). *South Bay Dischargers Authority Water Quality Monitoring Program. Second Year Monitoring Report, December 1982 - November 1983*. 235 pp., 3 appendices.

Kinnetic Laboratories, Inc. and Larry Walker Associates. (1985). *South Bay Dischargers Authority Water Quality Monitoring Program. Third Year Monitoring Report, December 1983 - November 1984*. 239 pp., 3 appendices.

Kinnetic Laboratories, Inc. and Larry Walker Associates. (1986). *South Bay Dischargers Authority Water Quality Monitoring Program. Fourth Year Monitoring Report, December 1984 - November 1985*. 260 pp., 4 appendices.

Kramer, J. R., W. F. Hunter, E. Callender, D. Peterson, D. S. McCulloch, and P. Carlson. (undated). *Geochemistry of San Francisco Bay*. (Unpublished manuscript, 65 pp.)

Kuwabara, J. S., C. C. Y. Chang, J. E.. Cloern, T. L. Fries, J. A. Davis, and S. N. Luoma. (1989). Trace metal associations in the water column of South San Francisco Bay, California. *Estuarine Coastal Shelf Sci*. 28:307-325.

Levasseur, M. E., P. J. Harrison, B. R. Heimdal, and J.-C. Therriault. (1990). Simultaneous nitrogen and silicate deficiency of a phytoplankton community in a coastal jet-front. *Marine Biol.* 104:329-338.

McCarty, J. C., R. A. Wagner, M. Macomber, H. S. Harris, M. Stephenson, and E. A. Pearson. (1962). *Appendixes to an Investigation of Water and Sediment Quality and Pollutional Characterisitics of Three Areas in San Francisco Bay, 1960-1961*. Sanitary Engineering Research Laboratory, University of California, Berkeley, CA. 285 pp.

McCulloch, D. S., D. H. Peterson, P. R. Carlson, and T. J. Conomos. (1970). *Some Effects of Fresh-water Inflow on the Flushing of South San Francisco Bay: A Preliminary Report*. U.S. Geol. Surv., Circ. 637A. 27 pp.

Officer, C. B., and J. H. Ryther. (1980). The possible importance of silicon in marine eutrophication. *Marine Ecol. Progr. Ser.* 3:83-91.

Ota, A. Y., L. E. Schemel, and S. W. Hager. (1989). *Physical and chemical Properties of San Francisco Bay, 1980*. U.S. Geol. Surv., Open-File Rept. 89-421. 251 pp.

Park, K. (1967). Nutrient regeneration and preformed nutrients off Oregon, *Limnol. Oceanogr.* 12:353-357.

Pearson, E. A. (1958). *"Reduced Area" Investigation of San Francisco Bay*. Report to the State Water Pollution Control Board, Lafayette, CA. 328 pp.

Pearson, E. A., P. N. Storrs, and R. E. Selleck. (1969). *A Comprehensive Study of San Francisco Bay. Final Report, volume III. Waste Discharges and Loadings*. SERL Rept. No. 67-3, Sanitary Engineering Research Laboratory, University of California, Berkeley, CA. 98 pp.

Peterson, D. H., D. R. Cayan, M. D. Dettinger, M. A. Noble, L. G. Riddle, L. E. Schemel, R. E. Smith, R. J. Uncles, and R. A. Walters. (1996). San Francisco Bay salinity: Observations, numerical simulations, and statistical models. Pages 9-34 *in* J. T. Hollibaugh, ed., *San Francisco Bay: The Ecosystem*. Pacific Division, American Association for the Advancement of Science, San Francisco, CA.

Peterson, D. H., R. E. Smith, S. W. Hager, D. D. Harmon, R. E. Herndon, and L. E. Schemel. (1985). Interannual variability in dissolved inorganic nutrients in Northern San Francisco Bay Estuary. *Hydrobiologia* 129:37-58.

Poland, J. F., and R. L. Ireland. (1985). *Land subsidence in the Santa Clara Valley, California, as of 1980*. U.S. Geol. Surv. Open-File Rept. 84-818. 93 pp.

Powell. T. M., J. E. Cloern, and L. M. Huzzey. (1989). Spatial and temporal variability in South San Francisco Bay (USA). I. Horizontal distributions of salinity, suspended sediments, and phytoplankton biomass and productivity. *Estuarine Coastal Shelf Sci.* 28:583-597.

Rantz, S. E. (1956). Flood of January 1952 in the South San Francisco Bay region. Pages 531-561 *in Floods of 1952 in California*. U.S. Geol. Surv., Water-Supply Pap. 1260-D.

Rantz, S. E. (1974). *Mean Annual Runoff in the San Francisco Bay Region, California, 1931-70*. U.S. Geol. Surv., Misc. Field Studies, Map MF-613. 24 pp. and 2 sheets.

Redfield, A. C., B. H. Ketchum, and F. A. Richards. (1963). The influence of organisms on the composition of sea-water. Pages 26-77 *in* M. N. Hill, ed., *The Sea*, Vol. 2. Wiley-Interscience, New York.

Ryther, J. H., and C. B. Officer. (1981). Impact of nutrient enrichment on water uses. Pages 247-261 *in* B. J. Neilson & L. E. Cronin, eds., *Estuaries and Nutrients*. Humana Press, Clifton, NJ. 643 pp.

San Francisco Bay Conservation and Development Commission and others. (1992). *The Effects of Land Use Change and Intensification on the San Francisco Estuary*. San Francisco Estuary Project, Oakland, CA. 157 pp.

Schemel, L. E. (1971). *Seasonal Distributions of Salinity, Phytoplankton Nutrients, and Dissolved Organic Carbon in South San Francisco Bay, 1969-1970.* M.S. Thesis. San Jose State College, San Jose, CA. 54 numbered leaves.

Schemel. L. E., and S. W. Hager. (1986). Chemical variability in the Sacramento River and in northern San Francisco Bay. *Estuaries* 9(4A):270-283.

Schemel. L. E., and S. W. Hager. (1996). Dissolved inorganic nitrogen, phosphorus, and silicon in South San Francisco Bay: II. A case study of effects of climate and weather. Pages 217-235 *in* J. T. Hollibaugh, ed., *San Francisco Bay: The Ecosystem.* Pacific Division, American Association for the Advancement of Science, San Francisco, CA.

Seitzinger, S. P. (1988). Denitrification in freshwater and coastal marine ecosystems: Ecological and geochemical significance. *Limnol. Oceanogr.* 33(4, part 2):702-724.

Selleck, R. E., E. A. Pearson, B. Glenne, and P. N. Storrs. (1966). *A Comprehensive Study of San Francisco Bay. Final Report, Volume IV. Physical and Hydrological Characteristics of San Francisco Bay.* SERL Rept. No. 65-10. Sanitary Engineering Research Laboratory, University of California, Berkeley, CA. 99 pp.

Smith, R. E., and R. E. Herndon. (1980). *Physical and Chemical Properties of San Francisco Bay Waters, 1977.* U.S. Geol. Surv., Open-File Rept. 80-1191. 28 pp.

Smith, R. E., and R. E. Herndon. (1983). *Physical and Chemical Properties of San Francisco Bay Waters, 1978.* U.S. Geol. Surv. Open-File Rept. 82-R-0273. 22 pp.

Smith, R. E., R. E. Herndon, and D. D. Harmon. (1979). *Physical and Chemical Properties of San Francisco Bay Waters, 1969-1976.* U.S. Geol. Surv., Open-File Rept. 79-511. 607 pp.

Smith, S. V., W. J. Wiebe, J. T. Hollibaugh, S. J. Dollar, S. W. Hager, B. E. Cole, G. W. Tribble, and P. A. Wheeler. (1987). Stoichiometry of C, N, P, and Si fluxes in a temperate-climate embayment. *Jour. Marine Res.* 45:427-460.

Storrs, P. N., R. E. Selleck, and E. A. Pearson. (1963). *A Comprehensive Study of San Francisco Bay, 1961-1962, South San Francisco Bay Area, Suisun Bay - Lower San Joaquin River Area, San Pablo Bay Area.* Sanitary Engineering Research Laboratory, University of California, Berkeley, CA. SERL Rept. No. 63-3. 221 pp.

Storrs, P. N., R. E. Selleck and E. A. Pearson. (1965). *A comprehensive Study of San Francisco Bay, 1963-1964, North San Francisco Bay Area, Central San Francisco Bay Area, Lower San Francisco Bay Area.* Sanitary Engineering Research Laboratory, University of California, Berkeley, CA. SERL Rept. No. 65-1. (pagination varies)

U.S. Department of Commerce, Weather Bureau. (1956-94). *Climatological Data, California, Annual Summaries.* U.S. Department of Commerce, Weather Bureau, Asheville.

U.S. Geological Survey. (1985). *Water Resources Data for California, Water Year 1984.* U.S. Geol. Surv., Water-Data Rept. CA-84-2. 315 pp.

U.S. Geological Survey. (1993). *Water Resources Data for California, Water Year 1992.* U.S. Geol. Surv., Water-Data Rept. CA-92-2. 405 pp.

van Geen, A., and S. N. Luoma. (1993). Trace metals (Cd, Cu, Ni, and Zn) and nutrients in coastal waters adjacent to San Francisco Bay, California. *Estuaries* 16(3A):559-566.

Waananen, A. O. (1969). Urban effects on water yield. Pages 169-182 *in* W. L. Moore & C. W. Morgan, eds., *Effects of Watershed Changes on Streamflow.* University of Texas Press, Austin, TX. 289 pp.

Wahl, K. L., J. R. Crippen, and J. M. Knott. (1980). *Floods of January and February 1980 in California.* U.S. Geol. Surv., Open-File Rept. 80-1005. 52 pp.

Walters, R. A., R. T. Cheng, and T. J. Conomos. (1985). Time scales of circulation and mixing processes of San Francisco Bay waters. *Hydrobiologia* 129:13-36.

DISSOLVED INORGANIC NITROGEN, PHOSPHORUS, AND SILICON IN SOUTH SAN FRANCISCO BAY

II. A CASE STUDY OF EFFECTS OF LOCAL CLIMATE, AND WEATHER

LAURENCE E. SCHEMEL AND STEPHEN W. HAGER
Water Resources Division, U.S. Geological Survey, Menlo Park, CA 94025

Salinity, temperature, and concentrations of dissolved nutrients, nitrate, (reactive) phosphate, and dissolved silica were measured in South San Francisco Bay (South SFB) from October 1990 through May 1991, a period with unusual weather patterns and low seasonal rainfall. Direct effects of inflows from North San Francisco Bay (North SFB) were limited to the reach seaward of San Mateo Bridge (seaward reach) during this study. Landward of San Mateo Bridge (landward reach), changes in physical and chemical properties were related to seasonal climate change and to local streamflow and winds associated with weather events. The seasonal decrease in evaporation over the autumn months was an apparent factor causing gradual decreases in salinity in the landward reach. Short periods of rainfall and local streamflow coincided with episodic changes at the most-landward station, Dumbarton Bridge, that directly affected relations between nutrient concentrations and salinity. During March, when rainfall and runoff was greatest, changes in nutrient-salinity gradients throughout the landward reach clearly indicated dilution of ambient waters by local streamflow, whereas the seaward reach was affected primarily by mixing with North SFB. Effects of short periods of strong winds were particularly evident at the San Mateo and Dumbarton bridges, but South SFB-wide changes in nutrient-salinity distributions were associated with an episode of unusually strong winds during early April.

South San Francisco Bay (South SFB) is a large tributary estuary of North San Francisco Bay (North SFB), the estuary of the Sacramento-San Joaquin River system (Fig. 1). Although South SFB was once the site of productive commercial fisheries, it has a long history of environmental degradation primarily associated with municipal and industrial wastes (Nichols *et al.* 1986). Improvements in waste treatment and discharge strategies have improved water quality over recent decades, but municipal waste treatment plants remain the largest sources of dissolved nutrients to South SFB (Hager & Schemel, this volume). Effects of municipal waste inflows on concentrations of nutrients and other waste-derived solutes are particularly evident in the landward reach, south of San Mateo Bridge (Fig. 1), which is less directly influenced by mixing and exchange with waters from North SFB. In this study we investigate factors affecting dissolved nutrients that are associated with local climate and weather, with particular emphasis on the landward reach.

Processes that disperse and transport nutrients and other waste-derived solutes operate over a wide range of time scales (Cloern & Nichols 1985), but effects of these processes are often identified by changes in fundamental properties, such as salinity and temperature. In this study we use daily measurements from three monitoring stations in South SFB to resolve variations in water-column properties related to seasonal climate change and short-term weather events. We link obser-

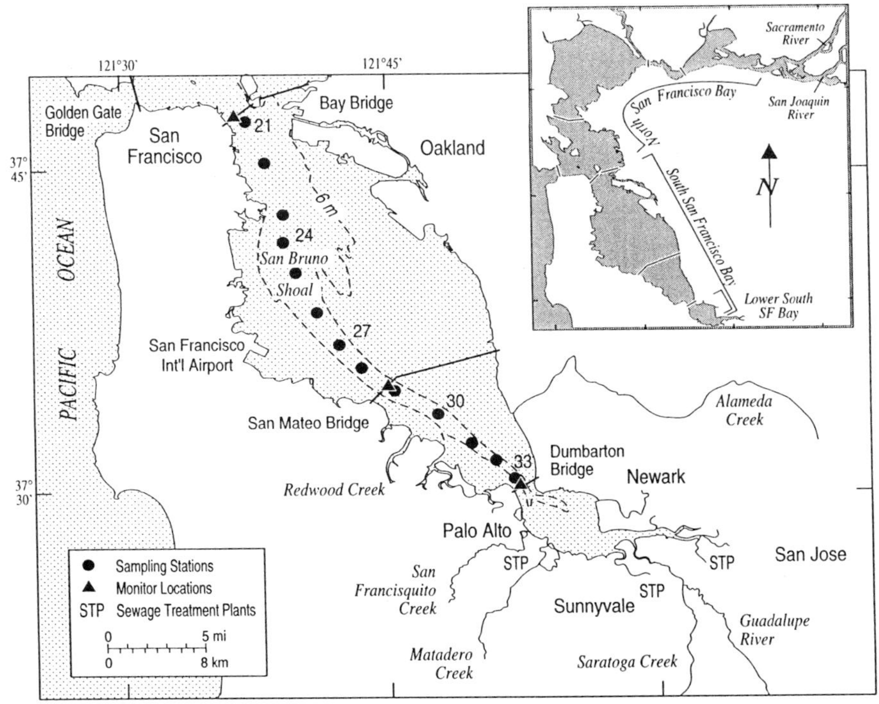

FIGURE 1. Map of San Francisco Bay showing locations of sampling and monitoring stations in South San Francisco Bay, local streams, and sewage treatment plants in Lower South SFB.

vations at the monitoring stations to distributions of nutrients measured monthly along the longitudinal channel of South SFB, then to concentrations of nutrients measured more frequently in the landward reach. A principle objective is to show effects of short-term events, such as inflows from local streams and winds, in the landward reach, particularly those that might not be resolved by observations made at fortnightly and longer intervals.

Longitudinal gradients in nutrients and other waste-derived solutes have been described in previous studies of South SFB (e.g., McCulloch *et al.* 1970; Conomos *et al.* 1979; Cutter 1989; Kuwabara *et al.* 1989; Flegal *et al.* 1991). Factors that create and maintain these gradients include the shape and bathymetry of South SFB, the locations of waste and stream inflows, and other factors that are reviewed by Hager & Schemel (this volume). Approximately 80% of the volume of South SFB lies in the area bounded by the San Francisco Bay Bridge (Bay Bridge) and the San Mateo Bridge (Selleck *et al.* 1966, Fig. 1), defined here as the seaward reach. Most municipal waste is discharged into the seaward reach near the Bay Bridge or into Lower South SFB, the small basin landward of Dumbarton Bridge. Wastes discharged into the seaward reach are rapidly dispersed into a large volume of water, whereas in Lower South SFB, wastes are discharged into a small volume and must mix seaward through the strait at Dumbarton Bridge, a comparatively slow process that contributes to high concentrations and large longitudinal gradients in nutrient concentrations in the landward reach.

An important aspect of the seasonal hydrology of South SFB is the extent to which it is influenced by mixing and exchange with the main estuary, North SFB (McCulloch *et al.* 1970; Imberger *et al.* 1977, Walters *et al.* 1985). In most years with normal or greater levels of precipitation, this can dominate the hydrology in large areas of South SFB. Relatively little emphasis has been placed on recognizing effects of local streams on South SFB, primarily because these exhibit little flow except during storm events, and, even then, discharges are much lower than freshwater flow to North SFB (primarily Delta outflow). Effects of local streamflow are potentially greatest in the landward reach, where streams drain 85% of the South SFB watershed, and particularly in Lower South Bay, where several large streams discharge. Low levels of Delta outflow as well as unusual weather patterns during this study helped us to discriminate between effects of inflow from North SFB and changes related to climate, weather events, and local sources of freshwater.

The 1991 Water Year

During the 1991 water year, unusual weather was an important factor that created less complex hydrologic conditions than in typical years. Rainfall from October 1990 through May 1991 at San Francisco International Airport (SFO) was about two-thirds of normal, and the distribution of rainfall over this period did not follow the typical pattern (Fig. 2A). October and November were very dry, and rainfall during the wettest months of the year, December, January, and February, was about one-third of normal. Rainfall during the 28-day period ending on March 26, however, was equivalent to three times the normal precipitation for March.

Intensity and frequency of rainfall affects discharge from local streams (Fig. 2B). The total amount of local streamflow to South SFB is unknown, in part because some flow is impounded and much of the runoff from the surrounding urban area is not measured (Hager & Schemel, this volume). Flow in streams that discharge into South SFB increases rapidly with the onset of rain, then usually decreases sharply after the rain stops. This pattern is typical when surface runoff is a major component of the flow (Sylvester & Brown 1977). In this study, gaged local streamflow was greater than municipal waste flow only during periods of heavy rainfall (Fig. 2A,B).

Levels of Delta outflow to North SFB result from precipitation and snowmelt in a large drainage basin after diversions for various uses and storage (Fig. 2C). The hydrodynamic influence of North SFB on South SFB depends on the level of Delta outflow, and this in turn affects distributions of nutrients in South SFB (Hager and Schemel, this volume). During the 1991 water year, Delta outflow from October to January remained low, and the major storm in early February resulted in only a moderate increase. Delta outflow increased during the heavy rains in early and late March, but only to levels that would directly affect waters in South SFB seaward of San Bruno shoal (Fig. 1, Imberger *et al.* 1977; Conomos 1979).

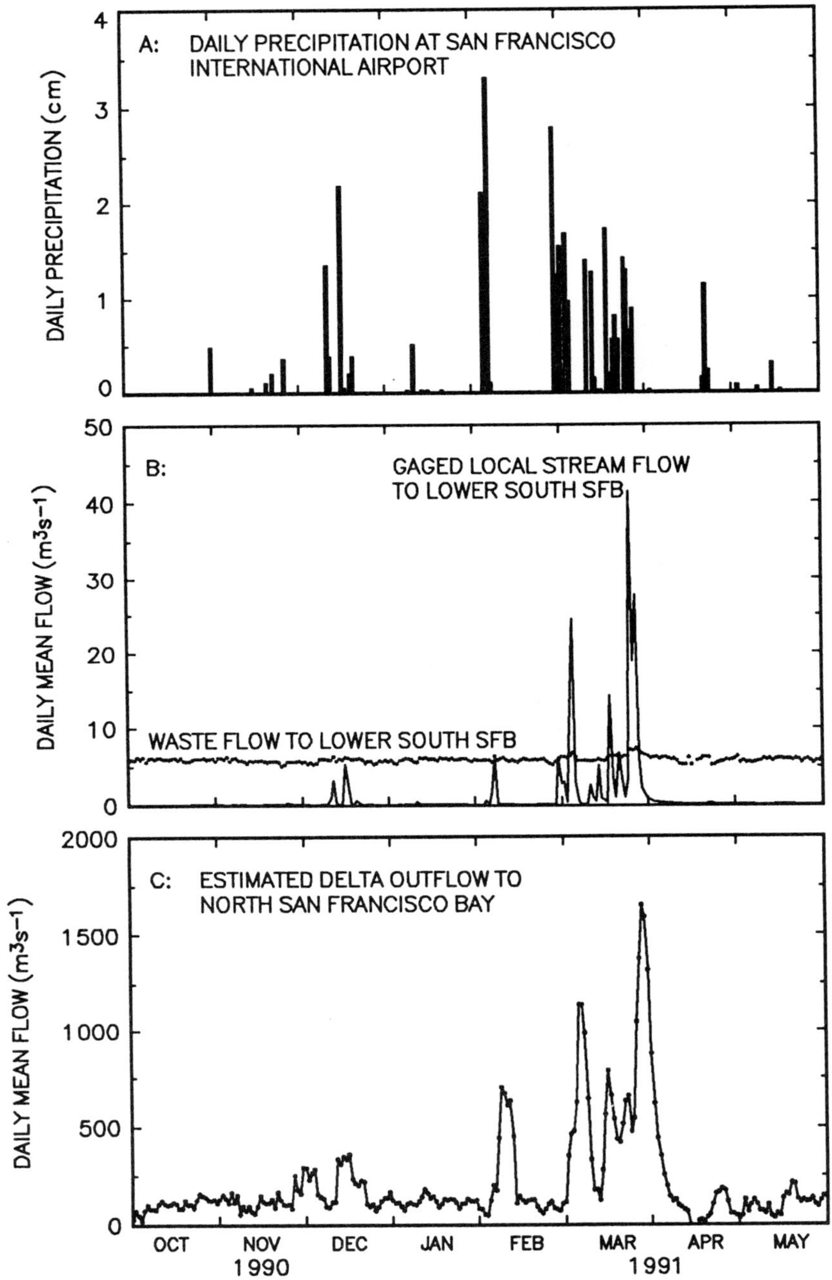

FIGURE 2. Daily data from October 1990 through May 1991 for (A) precipitation at San Francisco International Airport (SFO), (B) gaged local streamflow and waste flow to Lower South SFB, and (C) Delta outflow to North San Francisco Bay.

Concentrations of nutrients in South SFB can show large depletions associated with phytoplankton blooms that develop during early spring in most years (Hager & Schemel, this volume). During early April 1991, however, the initial stage of a bloom was interrupted by an unusual, 6-day-long period of strong winds (high wind speeds) that resuspended bottom sediments and severely reduced the photic depth (Wienke *et al.* 1992). Consequently, there was little evidence of uptake of nutrients by phytoplankton (Hager 1993), which provided the opportunity to observe changes in distributions of nutrients during this spring without the complicating effects of a phytoplankton bloom.

Periods of strong winds are expected to produce a surface current in the same direction as the wind and a return flow in the deep longitudinal channel of South SFB (Cheng & Casulli 1982). Wind-driven currents and wind-enhanced vertical mixing have been associated with changes in water-column properties in the seaward reach of South SFB (Powell *et al.* 1989; Huzzey *et al.* 1990).

METHODS

Field data in this study were obtained from monitoring stations located on three bridges in South SFB, water samples collected at the Dumbarton Bridge monitoring station, and water samples collected from the R/V Polaris at stations located along the deep-water channel (Fig. 1).

Salinity and temperature were recorded at 15-minute intervals at the three bridge monitoring stations between October 1990 and May 1991, although there are substantial gaps in these records. Sensors were mounted at two depths at both Bay Bridge and San Mateo Bridge, approx. 2-3m and 10-15 m below mean lower-low water. At Dumbarton Bridge, only one set of sensors was operated during most of this study. It was mounted 1m above the bottom in approximately 6 m of water at mean tide level. During March and April, salinity was also measured by a sensor that floated 1m below the surface of the water at Dumbarton Bridge. Salinity is reported in units of the Practical Salinity Scale (psu, Lewis 1980) and temperature in degrees Celsius. Other details are provided in the data reports (USGS 1991; Schemel 1995).

All daily mean values computed from the monitoring station data were based on a 24 h day, even though variations in these measurements were modulated by tides with a period of 24.8 h. Daily mean values for water-column temperature were computed with all available data from both upper and lower sensors, except at Dumbarton Bridge, where only the lower sensor values were used. Predicted values for daily root-mean-square tidal amplitudes (RMS tides) were computed (for Dumbarton Bridge) using data and methods from Cheng & Gartner (1984).

Surface water was collected with a plastic bucket at Dumbarton Bridge two or more times each week from January through April 1991, but less frequently in November and December 1990. Similarly, samples of near-surface water (1.5 m) were collected by pump from the R/V Polaris during regularly-scheduled cruises along the deep channel of South SFB (Fig. 1). All samples were analyzed for salinity and the dissolved nutrients, nitrate, nitrite, ammonium, (reactive) phosphate, and silicate (hereafter referred to as dissolved silica). Results from only 5 of the 12 cruises between December and April are described here, and we limit our presentation of the nutrients to phosphate, nitrate, and dissolved silica. All results from samples collected during all cruises and at Dumbarton Bridge are reported by Hager (1993) along with details of the sample processing and analytical methods.

Rates of freshwater flow from the Delta of the Sacramento and San Joaquin rivers to North SFB (Delta outflow) were estimated by the California Department of Water Resources (CDWR) DAYFLOW Program (CDWR 1986). Flow for gaged streams to Lower South SFB was based on data from Guadalupe River, Matadero Creek, Saratoga Creek, and San Francisquito Creek (USGS 1991). Flow for the Patterson Creek distributary of Alameda Creek and Redwood Creek were added to the flow to Lower South SFB to estimate total local streamflow.

Wind and rainfall data were obtained from monthly reports for SFO (National Oceanic and Atmospheric Administration, 1990-1991). Flow-weighted mean concentrations of dissolved inorganic nitrogen species and phosphate and daily total flows of municipal waste to Lower South Bay were computed from data contained in unpublished monthly reports from the San Jose, Palo Alto, and Sunnyvale waste treatment facilities.

The level of significance for all parametric statistical tests was 0.05.

RESULTS AND DISCUSSION

Results of this study are presented and discussed in three stages. First, we describe variations in the water-column properties, temperature and salinity, at the three bridge monitoring stations over the 8-month period and relate these to seasonal climate changes and weather-related events. Next, we link the major variations in salinity at the bridges to changes in salinity and concentrations of nutrients observed along the longitudinal axis of South SFB. Finally, we examine changes in salinity and concentrations of nutrients in the landward reach of South SFB at Dumbarton Bridge that were associated with episodes of rainfall, runoff, and strong winds (high wind speeds).

Temperature and Salinity in South SFB

Water-column temperature varies seasonally with the annual solar cycle and can be affected by short-term events associated with local weather. The general seasonal pattern of water-column temperatures in San Francisco Bay has been described by Conomos (1979). In recent years, monitoring stations at the Bay, San Mateo, and Dumbarton bridges have provided detailed data that show the seasonal pattern and also resolve variations over much shorter time scales. Water-column temperatures (daily mean values) showed large decreases at all three bridges over the autumn months of 1990, coinciding with the seasonal decrease in air temperature (Fig. 3A,B). Values were lowest near the winter solstice, when air temperatures reached unusually low values for the area. Both water-column and air temperatures increased over January and February, but were lower during the passage of many storm fronts in March. The response of water-column temperature to storms during March was least at Bay Bridge and greatest at Dumbarton Bridge in the shallower

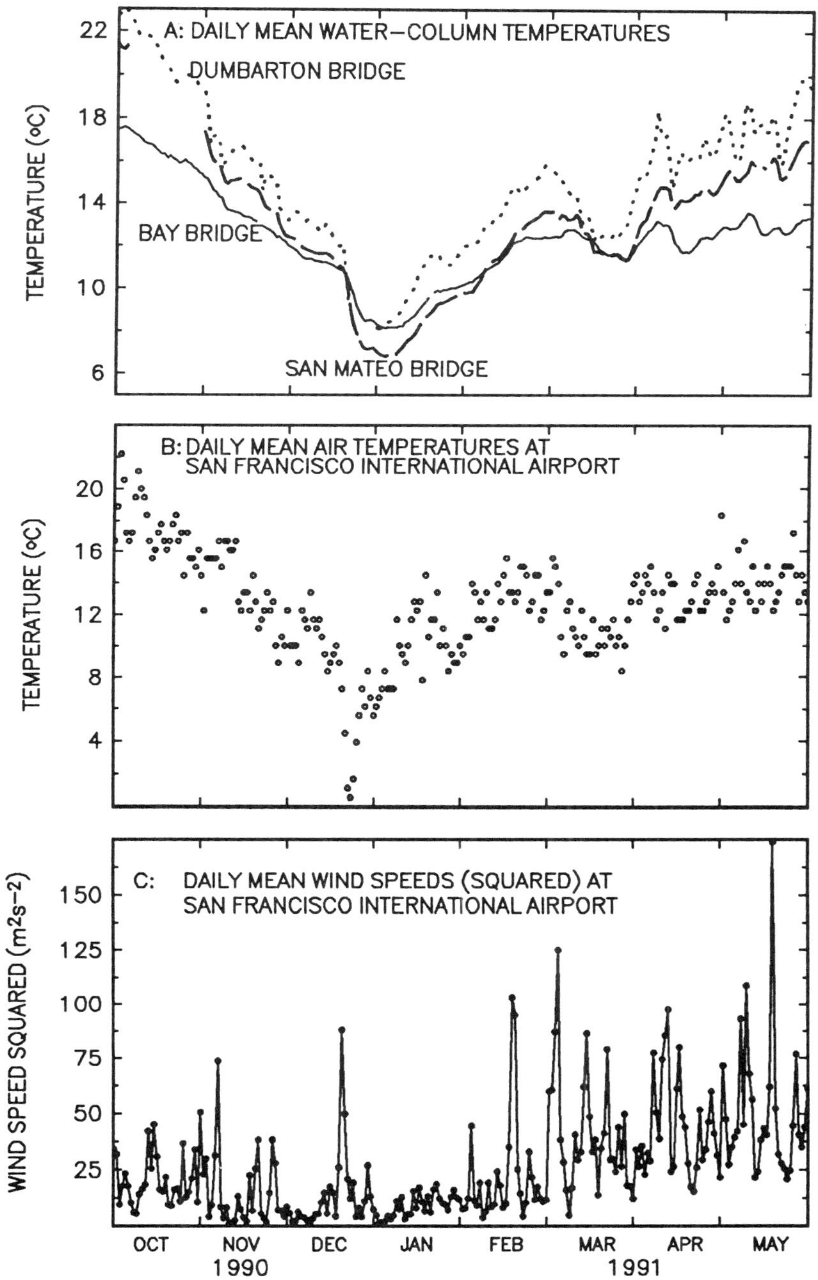

FIGURE 3. Daily mean values from October 1990 through May 1991 for (A) water-column temperature at the Bay, San Mateo, and Dumbarton bridges, (B) air temperature at San Francisco International Airport (SFO), and (C) daily mean wind speed (squared) at SFO.

landward reach of South SFB. In addition to cooler air temperatures during March, winds, which mix the water-column, were stronger than during the previous months (Fig. 3C). Variations in water-column temperatures during April and May also corresponded to periods of stronger winds, and effects were greatest at Dumbarton Bridge.

Salinity near San Francisco typically follows an annual pattern related to relatively large changes in Delta outflow to North SFB (see Peterson *et al.* this volume). During the dry autumn and early winter of this study, Delta outflow was low relative to most years and salinity values were high at Bay Bridge (Fig. 4A). A small increase in Delta outflow in late November corresponded with a distinct but small decrease in salinity. Until mid-December, salinity at San Mateo Bridge was higher than at the other two bridges (Fig. 4B). Salinity at both San Mateo and Dumbarton bridges exhibited gradual decreases until mid-December, when a small but abrupt decrease at Dumbarton Bridge coincided with a storm that produced a small amount of inflow from local streams. A few days later, a rapid decrease in salinity at San Mateo Bridge coincided with strong winds. After these two weather-related events, salinity continued to decrease gradually through January at both the San Mateo and Dumbarton bridges.

Moderate to large increases in Delta outflow during March were followed within a week by appreciable decreases in salinity at Bay Bridge (Fig. 4C). Greatest differences in salinity between upper and lower sensors, salinity stratification, followed the highest levels of Delta outflow and coincided with neap (weaker) tides, when gravitational circulation can be strongest in North SFB and the landward reach of South SFB (Walters *et al.* 1985). After Delta outflow decreased to low levels in early April, salinity rapidly increased at Bay Bridge, then increased more gradually through May.

Salinity variations over winter and spring at both San Mateo and Dumbarton bridges were distinctly different from those at Bay Bridge. Salinity at San Mateo Bridge decreased gradually over winter, reaching minimum values in mid-April shortly after a period of strong winds (Fig. 4D). Decreases in salinity at Dumbarton Bridge during winter were more episodic, and associated with storm events that produced inflow from local streams. The greatest salinity stratification at Dumbarton Bridge coincided with highest local streamflows, but stratification did not persist during this period of strong tides and did not develop during the following neap tides. Mixing by tides is enhanced by obstructions and topographic variations, such as those near Dumbarton Bridge (Geyer & Signell 1992). In addition to constriction through the strait, numerous pilings under two highway bridges, an aqueduct, and utility towers enhance turbulent mixing. Therefore, significant levels of stratification might not persist through periods of strong tides, except perhaps when inflows from local streams are very high.

Increases in salinity over April and May were small at Dumbarton Bridge, but they occurred primarily during episodes of strong winds (Fig. 4D). Because most of the bay is shallow, strong winds can rapidly mix the water column as well as redistribute water masses (Powell *et al.* 1989; Huzzey *et al.* 1990). During the six-day period of strong winds in early April, salinity decreased at San Mateo

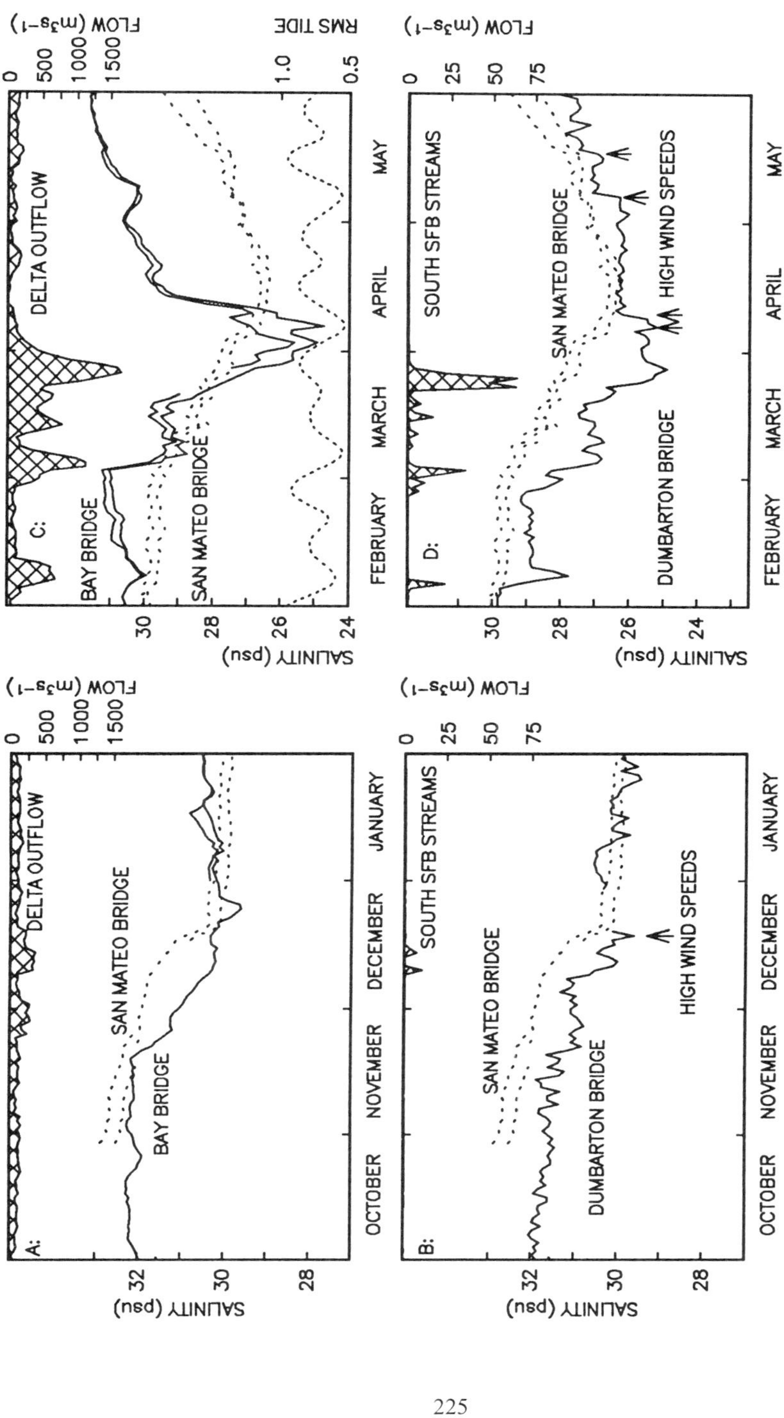

FIGURE 4. Daily mean values for Delta outflow and local streamflow (both on inverted axes), daily mean values for salinity (upper and lower sensors, when available) at the Bay, San Mateo, and Dumbarton bridges from October 1990 through May 1991, and RMS tides for February through May 1991. Arrows showing periods of high wind speeds refer to data shown in Fig. 3.

Bridge and increased at Dumbarton Bridge. During the last two days of the event, winds were unusually strong and from the north, conditions that produce a landward-moving, surface-water current (Walters 1982). Landward transport of less saline waters to San Mateo Bridge and more saline waters to Dumbarton Bridge along with mixing of the water column would account for the observed changes at these two bridges.

Some changes in temperature and salinity at the three bridge monitoring stations occurred within a few days (event time scale), whereas others occurred over periods of weeks to months (seasonal time scale). The large decrease in water-column temperature over the autumn months coincided with the seasonal decrease in solar radiation and changes in other factors that directly affect evaporation (Linsley *et al.* 1975). Effects of evaporation on concentrations of dissolved substances in South SFB are potentially greatest in the landward reach where water depths are relatively shallow. However, effects of evaporation in South SFB are difficult to identify, because distributions of dissolved substances are affected by many simultaneous processes (Hager & Schemel, this volume). It does appear, though, that the decrease in evaporation over autumn could have contributed to the decrease in salinity observed in the landward reach, particularly at Dumbarton Bridge.

Because of unusually dry weather, municipal waste accounted for nearly all the freshwater inflow to the landward reach of South SFB during autumn and early winter of this study (Fig. 2B). For example, from October through January the mean flow of waste water to Lower South SFB was about 6 m^3s^{-1}, whereas the mean flow from gaged streams to Lower South Bay was only 0.2 m^3s^{-1} (Fig. 2B). During the summer months, evaporation from Lower South SFB alone is equivalent to half the inflow of waste water, but by January the amount lost by evaporation is relatively small (Hager & Schemel, this volume). Consequently, inflow from local streams was small compared to the decrease in the amount of fresh water lost by evaporation over the autumn months. Less evaporation resulted in greater dilution by waste inflow. Although our results from Dumbarton Bridge showed that small decreases in salinity coincided with inflows from local streams, it is likely that the decrease in evaporation contributed to the more-gradual decrease in salinity over the many weeks with little or no precipitation during this dry period. A simple calculation treating Lower South SFB as a closed system suggested that the decrease in salinity due to decreased evaporation would have been greater than the amount we observed if there had been no mixing and exchange of waters through the strait at Dumbarton Bridge.

A similar influence of evaporation rates on salinity has been observed in the landward reach of a long, narrow embayment, Tomales Bay, that is located on the coast about 50 km north of San Francisco. Like South SFB, the landward reach is shallow (1-4 m), and inflow from local streams is highly seasonal (Smith *et al.* 1987). During summer, high evaporation rates increased salinity in the landward reach to levels that exceeded that of adjacent coastal waters. However, salinity decreased in the landward reach over the fall months. This decrease in salinity was attributed to the large decrease in evaporation rates, because no increases in freshwater inflow were observed (Smith *et al.* 1989).

Distributions of Dissolved Nutrients in South SFB

Concentrations of phosphate, nitrate, and dissolved silica in surface waters of South SFB during five cruises from December 1990 through April 1991 are plotted with respect to salinity in Fig. 5. A common feature in all these nutrient-salinity distributions was that concentrations of nutrients were higher in the landward reach with the highest concentrations measured at the most-landward station sampled on each cruise, typically station 33 (stn. 30 in December, stn. 32 in February). Although changes in the distributions over time were related to changes in both salinity and concentrations, changes in salinity provide useful information concerning mixing and circulation processes that affect concentrations of nutrients, but might not be easily seen in nutrient data alone (Hager and Schemel, this volume). Distributions during December showed the influence of a small increase in Delta outflow that was largely limited to the reach seaward of San Bruno shoal. In early February, salinity was lower bay-wide, and decreased landward by only a small amount, which was consistent with results from the bridge monitoring stations (Fig. 4). Distributions in early March clearly showed decreased salinity values in the landward reach following some inflow from local streams, but the seaward reach was just beginning to show effects from increasing Delta outflow. Delta outflow and flows from local streams were highest in late March, and lowest values for salinity were observed at both the Bay and Dumbarton bridges by early April (Figs. 4 and 5). The complex nutrient-salinity distributions in early April all exhibited salinity maxima and reversals or discontinuities in the relations between salinity and concentrations of nutrients near San Mateo Bridge. Distributions for all three nutrients indicated that the influence of waters from North SFB increased seaward from San Mateo Bridge, and that the influence of inflow from local streams increased landward. In addition, results from the four stations seaward of San Bruno shoal were grouped together in these distributions, showing that a lens of water from North SFB had penetrated South SFB and that a mixing zone extended to San Mateo Bridge.

Distributions during late April were greatly different from those earlier in the month, but discontinuities (changes in slope of relations between concentrations and salinity) were still apparent near San Mateo Bridge. Data from other cruises (not shown, see Hager 1994) indicated that most changes in distributions mid-bay had occurred relatively rapidly between April 5 and 11, which corresponded with the period of strong winds (Fig 4D). In contrast, the greatest increase in salinity at Bay Bridge occurred after April 11, when Delta outflow was low and tidal energy was greater (Fig. 4C).

It is difficult to compare absolute differences in concentrations of nutrients at the stations over time because of differences in stages of the tide among cruises. In the seaward reach, however, changes in concentrations of dissolved silica were large compared to changes in concentrations of phosphate and nitrate. This observation agreed with previous work in North SFB showing that ocean to river gradients in concentrations of phosphate and nitrate are small compared to that for dissolved silica, which is much more concentrated in Delta outflow than in seawater

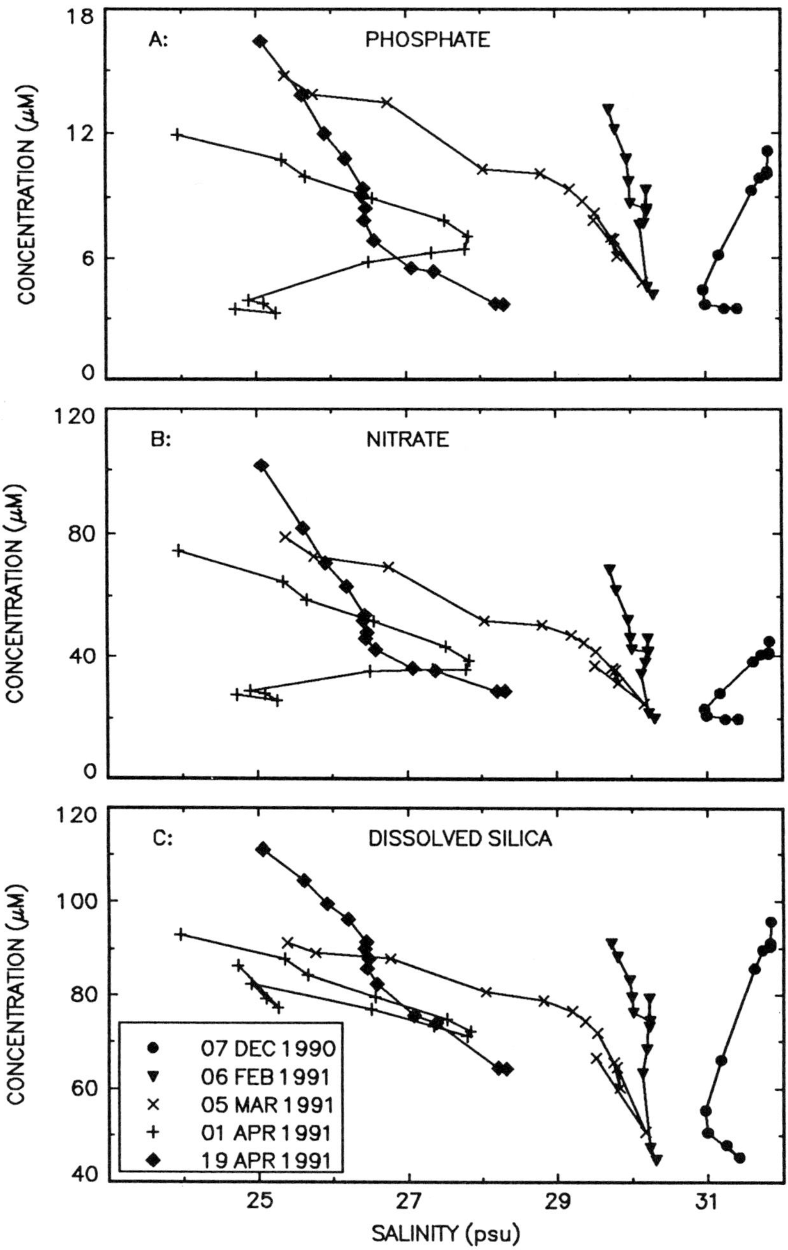

FIGURE 5. Distributions of dissolved inorganic nutrients, (A) phosphate, (B) nitrate, and (C) dissolved silica, with respect to salinity in South SFB during five cruises from December 1990 through April 1991. Symbols represent values at sampling stations shown in Fig. 1.

(Peterson *et al.* 1985). The distribution of dissolved silica in early April was different from those for phosphate and nitrate in that concentrations of dissolved silica increased both landward and seaward of San Mateo Bridge, and the differences between the two dissolved silica-salinity gradients were small. In contrast, waters in the landward reach of South SFB were clearly identifiable by much higher concentrations of nitrate and phosphate. Therefore, nitrate and phosphate could be used to differentiate between waters diluted by local streamflow in the landward reach and low-salinity waters affected by North SFB.

In the landward reach, changes in salinity and in concentrations of nutrients from February through April resulted in large changes in the longitudinal gradients, which can be represented by the slopes of the generally linear relations between these variables (stns. 27-33, Fig. 5). In the section below we examine this and other variations in salinity and concentrations of nutrients at the most-landward station, Dumbarton Bridge.

Dissolved Nutrients at Dumbarton Bridge

Variations in temperature and salinity shown above for Dumbarton Bridge demonstrated the sensitivity of the landward reach to seasonal climate change and weather-related events. During this study, changes in water-column properties were associated with even small inflows from local streams and other phenomena and might not be apparent during years with phytoplankton blooms or greater levels of Delta outflow. In this section we show effects of inflows from local streams and event-scale mixing processes on nutrient concentrations and their relations to salinity at Dumbarton Bridge.

Data in Fig. 6 were grouped to represent time periods between the local streamflow and other weather-related events identified above. The seasonal-scale trend of decreasing salinity in the landward reach was evident in the observations before mid-December. Within the subsequent time periods, most of the variability in values resulted from sampling at different stages of the tide, because longitudinal gradients in both salinity and concentrations of nutrients were large. From mid-December to early February, concentrations of each of the three nutrients were inversely related to salinity in a consistent, linear manner. Similarly, after the rainfall and runoff in early February, concentrations of all three nutrients were linearly related to salinity for the next three weeks. Most data from these two periods (18 Dec-02 Feb and 05 Feb-26 Feb) were distributed in two distinctly different groups, and data from the latter period appeared shifted toward lower salinity. However, slope and intercept values from regressions of these two data sets were not statistically different (Table 1). The consistency of relations between salinity and concentrations of nutrients over several weeks indicated a balance between of supply and mixing processes that was perturbed by rainfall-runoff events.

Highest rainfalls and local streamflows occurred in early and late March (Figs. 2 and 4). Although data for March to mid-April showed a trend toward lower values for salinity and concentrations of nutrients, concentrations were linearly related to salinity during mid-March (March 9-22, encircled plus symbols in Fig. 6), a period

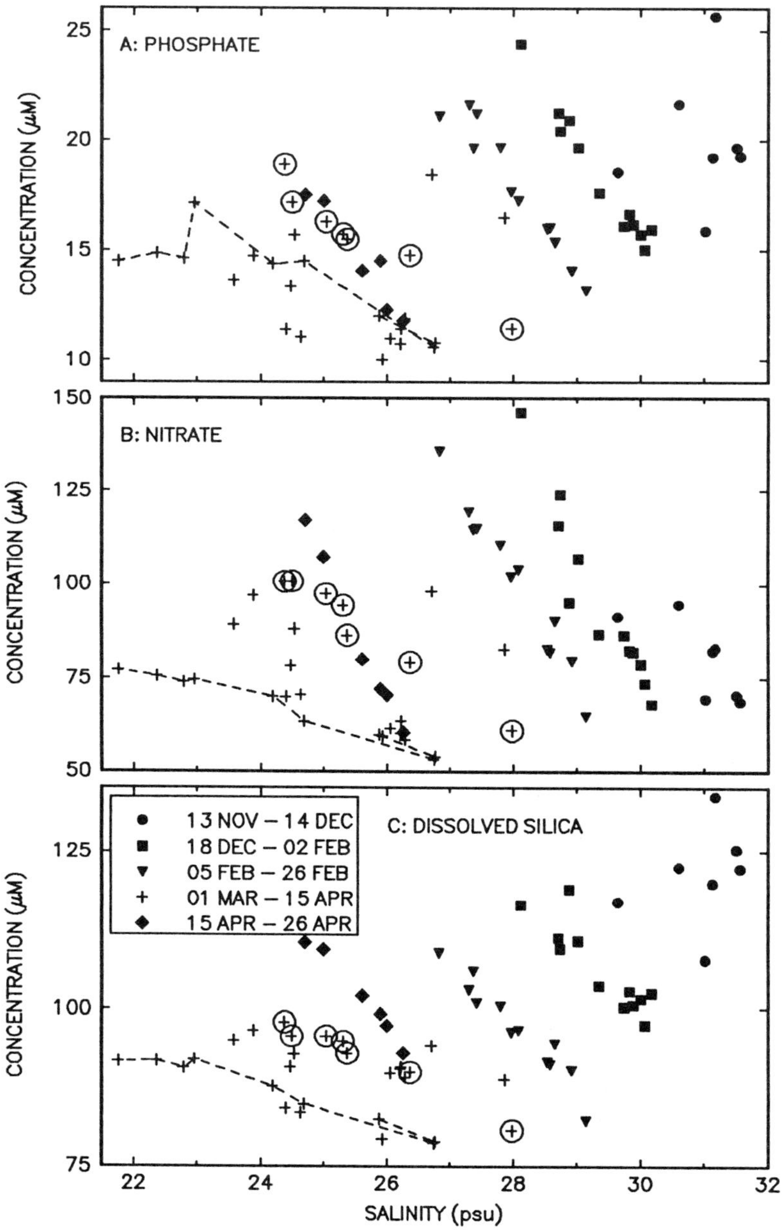

FIGURE 6. Concentrations of dissolved inorganic nutrients, (A) phosphate, (B) nitrate, and (C) dissolved silica, plotted against salinity at Dumbarton Bridge for selected time periods during the 1991 water year. Encircled plus symbols represent samples collected between March 9 and March 22, 1991. Plus symbols connected by dashes represent samples collected on the same day, March 26, 1991, from high to low tide.

with less local streamflow. In addition to a distinct shift from the distribution of values in mid-February toward lower salinity by mid-March, slopes of regressions for all three nutrients in mid-March were significantly less negative and intercept values were significantly lower compared to those for January and February (Table 1).

Storms in late March produced the highest local streamflows during this study, coinciding with large changes in salinity and concentrations of nutrients. Gaged local streamflow to Lower South SFB over the three days of highest flow (March 24-26) was equivalent to about 10-20% of the mean tide volume of Lower South SFB. Plus symbols connected by dashed lines in Figure 6 represent measurements made hourly between higher-high and lower-low tides on March 26 (tidal range = 2.4 m). Concentrations of nutrients were linearly related to salinity over this single excursion of the tide with the exception of concentrations of phosphate at the lowest salinities (and lowest tidal heights), which were much lower than might be expected from values at higher salinity. These lower values presumably resulted from the most-recent local streamflow that had not been dispersed by tidal mixing in Lower South SFB. Linear regressions of data from this day yielded the lowest values for intercepts and the least negative values for slopes (Table 1).

TABLE 1. Coefficients for linear regressions of concentrations of phosphate, nitrate, and dissolved silica with respect to salinity, as the independent variable, for surface samples at Dumbarton Bridge during selected time periods of the 1991 water year. 95% confidence intervals are shown for intercepts and slopes.

		Phosphate		
Date (month/day)	*Intercept*	*Slope*	r^2	*n*
12/18 to 02/02	145 ± 17.6	- 4.30 ± 0.60	0.96	12
02/05 to 02/26	125 ± 20.1	- 3.82 ± 0.71	0.93	12
03/09 to 03/22	60.8 ± 14.4	- 1.77 ± 0.57	0.93	7
03/26	36.6 ± 13.0	- 0.95 ± 0.54	0.71	9
04/15 to 04/26	111 ± 40.2	- 3.78 ± 1.58	0.92	6

		Nitrate		
Date (month/day)	*Intercept*	*Slope*	r^2	*n*
12/18 to 02/02	1072 ± 223	- 33.2 ± 7.6	0.90	12
02/05 to 02/26	876 ± 109	- 27.7 ± 3.8	0.96	12
03/09 to 03/22	381 ± 59	- 11.4 ± 2.3	0.97	7
03/26	186 ± 17	- 4.9 ± 0.7	0.97	9
04/15 to 04/26	1034 ± 108	- 37.1 ± 4.2	0.99	6

		Dissolved Silica		
Date (month/day)	*Intercept*	*Slope*	r^2	*n*
12/18 to 02/02	373 ± 100	- 9.06 ± 3.39	0.78	12
02/05 to 02/26	374 ± 56	- 9.88 ± 1.98	0.92	12
03/09 to 03/22	209 ± 25	- 4.55 ± 1.00	0.97	7
03/26	155 ± 12	- 2.81 ± 0.50	0.96	9
04/15 to 04/26	387 ± 60	- 11.16 ± 2.36	0.98	6

Concentrations of nutrients were again linearly related to salinity in late-April, after the period of strong winds. Linear regressions of these data yielded intercept values that were greatly different from those on March 26, but were not significantly different from those during January (Table 1). Therefore, the gradient in nutrient concentrations had returned to one primarily reflecting mixing between waste inflow and ambient South SFB waters.

Intercept values for linear regressions of nutrients at Dumbarton Bridge appeared to represent time-averaged estimates of concentrations in the composite freshwater mixing member (waste plus local streamflow) for the preceding inflow event (Table 1). During January and February, intercept values were similar in magnitude to but lower than mean concentrations for waste discharged at the plants, 1480 μM for nitrate and 180 μM for phosphate. Dissolved silica was not measured by the sewage treatment plants, but the mean concentration probably was between 400 and 500 μM (Hager and Schemel, this volume), a range that includes intercept values from January and February. Concentrations in three local streams on three dates during early, middle, and late March ranged from 24 to 91 μM for nitrate, 0.6 to 4.2 μM for phosphate, and 58 to 267 μM for dissolved silica (USGS 1991). Therefore, differences between waste and local stream concentrations of nitrate and phosphate were greater than a factor of ten, whereas the difference for dissolved silica was perhaps only a factor of two. This is reflected in decreases in the intercept values for the three nutrients after the major storms in March. Lowest intercept values for dissolved silica were only about one-half the value during January, but differences for nitrate (one-fifth) and phosphate (one-fourth) were much greater. Consequently, nitrate and phosphate were sensitive indicators of dilution of waste waters by local stream inflow, but dissolved silica was less sensitive because concentrations were not greatly different.

SUMMARY

Results of this study add some details to the general description of seasonal variations in nutrient-salinity distributions presented by Hager & Schemel (this volume). The seasonal decrease in evaporation from summer to winter appeared to be a factor causing small decreases in salinity and concentrations of nutrients, but these effects were limited to the landward reach and might not be evident during years with normal rainfall and runoff. The sensitivity of South SFB to local, short-term events, such as temperature variations, winds, and local streamflow, was greatest in the landward reach, and in particular at Dumbarton Bridge. During winter of 1991, Delta outflow and local streamflow reached only moderate levels, and San Mateo Bridge was an apparent boundary between waters directly affected by mixing and exchange with North SFB in the seaward direction and waters influenced by local streamflow in the landward direction. Changes over time in the slopes of the nutrient-salinity relations in the landward reach and at Dumbarton Bridge were clear indications of dilution by local streamflow. At Dumbarton Bridge, local streamflow resulted in episodic changes that persisted long after these short-term events. Salinity at San Mateo Bridge and distributions of nutrients mid-bay appeared less directly influenced by Delta outflow and local streamflow.

Effects of strong winds were particularly evident at the San Mateo and Dumbarton bridges, where effects sometimes persisted long after the events. In addition to effects at these two bridges, South SFB-wide changes in nutrient-salinity distributions were also associated with the early-April period of unusually strong winds. Our results generally demonstrate that seasonal-scale variations in local climate and short-term weather events are additional factors that influence the annual cycles exhibited by salinity and nutrient concentrations in South SFB.

ACKNOWLEDGEMENTS

We are greatly indebted to many USGS volunteers who participated in sample collection and processing and other duties during this study. These volunteers include Lars and Amy Olander, Norton Bell, Stan Brown, and Cindy Lee, who were particularly dedicated and helpful. We thank Randy Brown, Sheila Greene, and others with the California Department of Water Resources for their support of the monitoring stations in South Bay and the U.S. Fish and Wildlife Service for cooperation and assistance at the Dumbarton Bridge monitoring site. We also thank D.H. Schoellhamer and J.S. Kuwabara for helpful reviews of the manuscript and our colleagues at USGS for their assistance.

LITERATURE CITED

California Department of Water Resources. (1986). *DAYFLOW Program Documentation and Users Guide*. Central District, Sacramento, CA.

Cheng, R. T. and Vincenzo Casulli. (1982). On Lagrangian residual currents with applications in South San Francisco Bay, California. *Water Resources Res.* 18:1652-1662.

Cheng, R. T. and J. W. Gartner. (1984). *Tides, tidal and residual currents in San Francisco Bay, California – Results of Measurements, 1979-1980: Part 1. Description of Data*. U.S. Geol. Surv., Open-File Rept. 84-4339. 72 pp.

Cloern, J. E. and F. H. Nichols. (1985). Time scales and mechanisms of estuarine variability, a synthesis of studies from San Francisco Bay. *Hydrobiologia* 129:229-237.

Conomos, T. J. (1979). Properties and circulation of San Francisco Bay waters. 1979. Pages 47-84 *in* T. J. Conomos, ed., *San Francisco Bay: The Urbanized Estuary*. Pacific Division, Association for the Advancement of Science, San Francisco, CA.

Conomos, T. J., R. E. Smith, D. H. Peterson, S. W. Hager, and L. E. Schemel. (1979). Processes affecting seasonal distributions of water properties in the San Francisco Bay Estuarine System. Pages 115-142 *in* T. J. Conomos, ed., *San Francisco Bay: The Urbanized Estuary*. Pacific Division, Association for the Advancement of Science, San Francisco, CA.

Cutter, G. A. (1989). The estuarine behavior of selenium in San Francisco Bay. *Estuarine Coastal Shelf Sci.* 28:13-34.

Flegal, A. R., G. J. Smith, S. Sanudo-Wilhelmy, and L. C. D. Anderson. (1991). Dissolved trace element cycles in the San Francisco Bay estuary. *Marine Chem.* 36:329-363.

Geyer, W. R., and R. P. Signell. (1992). A reassessment of the role of tidal dispersion in estuaries and bays. *Estuaries* 15:97-108.

Hager, S. W. (1993). *Dissolved Nutrient and Suspended Particulate Matter Data for the San Francisco Bay Estuary, October 1988 through September 1991*. U.S. Geol. Surv., Open File Rept. 93-57. 52 pp.

Hager, S. W., and L. E. Schemel. (1996). Dissolved inorganic nitrogen, phosphorus, and silicon in South San Francisco Bay: I. Major factors affecting distributions. Pages 189-215 *in* J. T. Hollibaugh, ed., *San Francisco Bay: The Ecosystem*. Pacific Division, American Association for the Advancement of Science, San Francisco, CA.

Huzzy, L. M., J. E. Cloern, and T. M. Powell. (1990). Episodic changes in lateral transport and phytoplankton distributions in South San Francisco Bay. *Limnol. Oceanogr*. 35:472-478.

Imberger, J., W. B. Kirkland, and H. B. Fischer. (1977). *The Effect of Delta Outflow on the Density Stratification in San Francisco Bay*. Report to the Association of Bay Area Governments. Oakland, California. 109 pp.

Kuwabara, J. S., C. C. Chang, J. E. Cloern, T. L. Fries, J. A. Davis, and S. N. Luoma. (1989). Trace metal associations in the water column of South San Francisco Bay, California. *Estuarine Coastal Shelf Sci*. 28:307-325.

Lewis, E. L. (1980). The Practical Salinity Scale 1978 and its antecedents. IEEE *Jour. Ocean Engr*. OE-5:3-8.

Linsley Jr., R. K., M. A. Kohler, and J. L. H. Paulhus. (1975). *Hydrology for Engineers*, 2nd ed. McGraw-Hill, Inc., New York. 482 pp.

McCulloch, D. S., D. H. Peterson, P. R. Carlson, and T. J. Conomos. (1970). *A Preliminary Study of the Effects of Water Circulation in the San Francisco Bay Estuary: Some Effects of Fresh-water Inflow on the Flushing of San Francisco Bay*. U.S. Geol. Surv., Circ. 637-A. 27 pp.

National Oceanographic and Atmospheric Administration. (1990-1991). *Monthly Summary of Local Climatological Data for San Francisco International Airport*. Asheville, NC.

Nichols, F. H., J. E. Cloern, S. N. Luoma, and D. H. Peterson. (1986). The modification of an estuary. *Science* 231:567-573.

Peterson, D. H., R. E. Smith, S. W. Hager, D. D. Harmon, R. E. Herndon, and L. E. Schemel. (1985). Interannual variability in dissolved inorganic nutrients in Northern San Francisco Bay Estuary. *Hydrobiologia* 129:37-58.

Peterson, D. H., D. R. Cayan, M. D. Dettinger, M. A. Noble, L. G. Riddle, L. E. Schemel, R. E. Smith, R. J. Uncles and R. A. Walters. (1996). San Francisco Bay salinity: Observations, numerical simulation, and statistical models. Pages 9-34 *in* J. T. Hollibaugh, ed., *San Francisco Bay: The Ecosystem*. Pacific Division, American Association for the Advancement of Science, San Francisco, CA.

Powell, T. M., J. E. Cloern, and L. M. Huzzey. (1989). Spatial and temporal variability in South San Francisco Bay (USA). I. Horizontal distributions of salinity, suspended sediments, and phytoplankton biomass and productivity. *Estuarine Coastal Shelf Sci*. 28:583-597.

Schemel, L. E. (1995). *Measurements of Salinity, Temperature, and Tides in Southern San Francisco Bay, California, at Dumbarton Bridge: Water Years 1990-1993*. U.S. Geol. Surv., Open-File Rept. 95-326. 24 pp plus appendix.

Selleck, R. E., E. A. Pearson, B. Glenne, and P. N. Storrs. (1966). *A Comprehensive Study of San Francisco Bay, Volume IV. Physical and Hydrological Characteristics of San Francisco Bay*. Sanitary Engr. Res. Lab. Rept. No. 65-10, University of California, Berkeley. 99 pp.

Smith, S. V., W. J. Wiebe, J. T. Hollibaugh, S. J. Dollar, S. W. Haver, B. E. Cole, G. W. Tribble, and P. A. Wheeler. (1987). Stoichiometry of C, N, P, and Si fluxes in a temperate embayment. *Jour. Marine Res*. 45:427-460.

Smith, S. V., J. T. Hollibaugh, S. J. Dollar, and S. Vink. (1989). Tomales Bay, California: A case for carbon-controlled nitrogen cycling. *Limnol. Oceanogr*. 34:37-52.

Sylvester, M. A., and W. M. Brown. (1977). *Relation of Urban Land-use and Land-surface Characteristics to Quantity and Quality of Storm Runoff in Two Basins in California*. U.S. Geol. Surv., Water-Supply Pap. 2051. 49 pp.

USGS. (1991). *Water Resources Data for California, Water Year 1991*. U.S. Geol. Surv., Water Data Rept. CA-91-4, Sacramento, CA. 451 pp.

Walters, R. A. (1982). Low-frequency variations in sea level and currents in South San Francisco Bay. *Jour. Physical Oceanogr.* 12:658-668.

Walters, R. A., R. T. Cheng, and T. J. Conomos. (1985). Time scales of circulation and mixing processes of San Francisco Bay waters. *Hydrobiologia* 129:13-36.

Wienke, S. M., B. E. Cole, J. E. Cloern, and A. E. Alpine. (1992). *Plankton Studies in San Francisco Bay, XIII. Chlorophyll Distributions and Hydrographic Properties of San Francisco Bay, 1991.* U.S. Geol. Surv., Open-File Rept. 92-158. 116 pp.

THE SUPPLY AND CARBON CONTENT OF SUSPENDED SEDIMENT FROM THE SACRAMENTO RIVER TO SAN FRANCISCO BAY

LAURENCE E. SCHEMEL[1], STEPHEN W. HAGER[1] AND DALLAS CHILDERS JR.[2]
[1]**Water Resources Division, U.S. Geological Survey, Menlo Park, CA 94025**
[2]**Water Resources Division, U.S. Geological Survey, Vancouver, WA 98661**

The Sacramento River is the largest source of freshwater and suspended sediment to the Delta and to San Francisco Bay. Freshwater flow in this river is controlled most of the year, but winter storms typically produce high discharges that must be diverted from the river to the Yolo Bypass before flowing to the Delta and the bay. Concentrations of suspended sediments were measured in the Sacramento River and the Delta during 1979-1984. The river transported relatively small amounts of suspended sediment to the Delta over dry years and during periods of low flow, and because of water exports from the Delta, about one-half of this reached the bay. Major winter storms caused large increases in flow and in concentrations of suspended sediment. Most of the annual load of suspended sediment was transported during these flood events, when half or more of the suspended sediment entered the Delta and bay via the Yolo Bypass. During periods of unusually high flow, suspended sediment loads exceeded the annual average by factors of two or more within a few weeks. The particulate carbon content of the suspended sediment was measured during 1983-1984. During flood events, particulate carbon was 1-2% of the suspended sediment, but the fraction was higher (mostly 2-4%) during periods of low flow. Therefore, the Sacramento River supply of particulate carbon probably exceeds annual phytoplankton production in northern San Francisco Bay during years with average or higher sediment transports. Large event-scale, seasonal and interannual variability in this river system indicated that suspended sediments and particulate carbon are supplied to San Francisco Bay primarily in large, short-term pulses that might be separated by months to years of very low transport.

San Francisco Bay is the estuary of the Sacramento-San Joaquin river system, the river system that drains most of the central valley and adjacent mountain areas of California (Fig. 1). The importance of freshwater flow and the supply of dissolved and particulate substances from this river system to the Bay ecosystem is well recognized (*e.g.*, Cloern & Nichols 1985). This source of suspended particles (sediments) directly affects characteristics of suspended and bottom sediments in the Bay (Knebel *et al.* 1977; Krone 1979; Peterson *et al.* 1993), which in turn influence major ecosystem processes such as photosynthesis (*e.g.*, Cole & Cloern 1984). Organic matter in or on riverborne suspended particles is a potentially-large source of energy to the Bay ecosystem (Jassby *et al.* 1993), but past measurements of particulate organic carbon (Schemel & Dedini 1979; Spiker & Schemel 1979; Schemel 1983) have not adequately characterized variability in the supply. Suspended sediments also transport industrial and agricultural chemicals from the central valley to the Bay (Domagalski & Kuivila 1993), some of which are potentially toxic to estuarine organisms (Phillips & Spies 1988).

San Francisco Bay and the rivers that flow into the Bay have been extensively modified and greatly perturbed by anthropogenic activities over the past 150 years (Nichols *et al.* 1986). Consequently, freshwater flows and rates of suspended sediment transport in the rivers and into the Bay and the factors that control them have changed and continue to change. Our focus here is on the Sacramento River system, the largest source of freshwater and suspended sediment to the Bay. First, we review factors affecting freshwater flow and historical information on the supply of suspended sediment from this river system to the Bay. Then, we present and examine measurements of suspended sediment and particulate carbon collected during the most recent episode of high river flow (1979-1984), emphasizing variability in concentrations and transports on interannual, seasonal, and event time scales, and identifying areas where information is incomplete. Finally, we use relations based on these recent data to estimate suspended sediment and particulate carbon transports during the record-breaking flood of 1986.

Flows in the Major Rivers, to the Delta, and to the Bay

The Sacramento River and other rivers that drain the central valley of California flow to San Francisco Bay through the Delta, a network of channels at the confluence of the two largest rivers, the Sacramento and the San Joaquin (Fig. 1). Monthly mean values for flow of the Sacramento River to the Delta for water years 1956-1993 (a water year is October through September) show distinct high-flow and low-flow periods during most years (Fig. 2A), reflecting the typical seasonal climate pattern of cool, wet winters and warm, dry summers (Conomos *et al.* 1985). However, interannual variability in the climate in northern California is extremely large (Cayan & Peterson 1989), which is evidenced by large differences in flows among winter seasons and years. Such large variability in flow typically results in large differences in suspended sediment transport among seasons and years because highest concentrations and transports in most rivers coincide with times of highest flows (*e.g.*, Meade 1988).

Although highly variable on monthly and seasonal time scales, the Sacramento River was the largest source of freshwater to the Delta (Fig. 2B), supplying 83% of the total over the last 38 years (since completion of most major reservoirs on the Sacramento River). The second largest source of freshwater to the Delta was the San Joaquin River, which accounted for a third or more of the monthly flow during springs of some years. Much of the freshwater flow to the Delta was exported for agricultural and other uses or was consumed in the Delta during the irrigation season (defined here as May to October), and some was exported to off-stream storage at other times (Fig. 2C). In recent years, export pumping during summer withdrew most of the freshwater from the San Joaquin River in the southern Delta, and required diversion of about half the freshwater from Sacramento River upstream of Rio Vista. These diversions account for most of the differences between river flows to the Delta and freshwater flows (Delta outflows) to the Bay, and likewise affect the transports of dissolved and particulate substances.

When winter storms produce high (uncontrolled) discharges from the Sacramento River basin, much of this flow is diverted into flood bypasses and basins

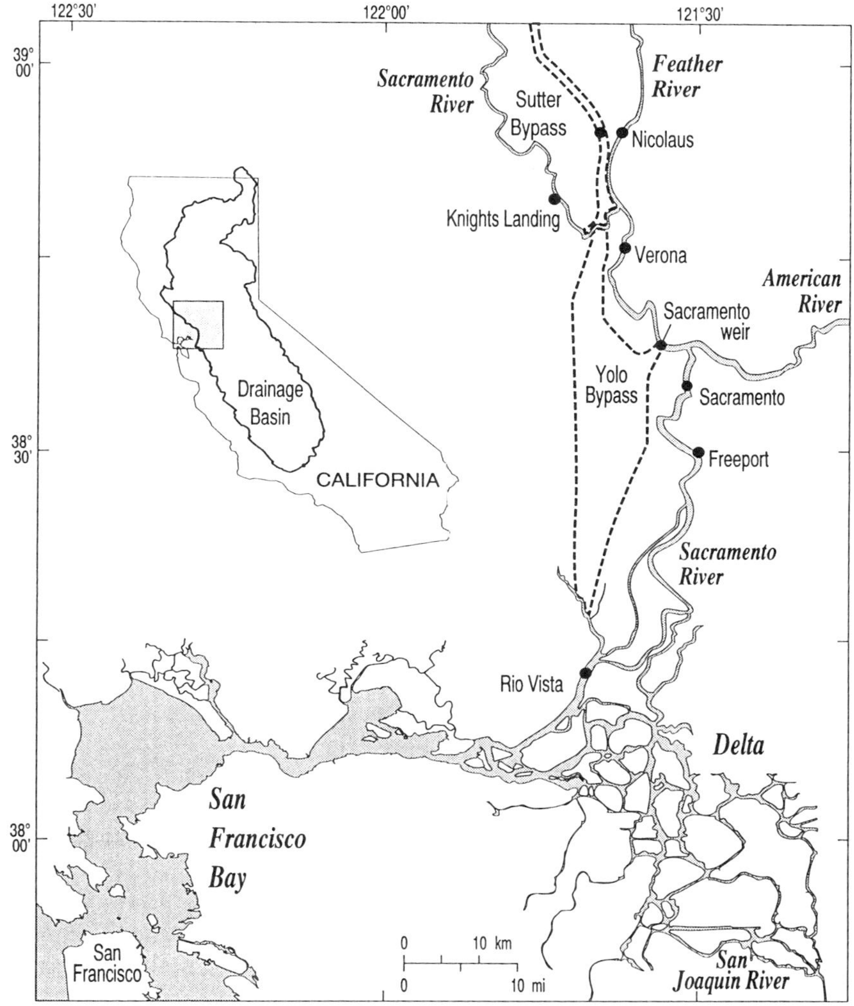

FIGURE 1. Map showing San Francisco Bay, the drainage basin of the Sacramento and San Joaquin rivers, flood bypasses, and locations on the Sacramento River and tributary rivers.

along the river as it traverses the valley floor (Karhl 1978). The Yolo Bypass routes this additional flow around the Sacramento metropolitan area to the western Delta upstream of Rio Vista, where it can flow directly to the Bay (Fig. 1). Consequently, during major floods, flow rates in the Yolo Bypass and at Rio Vista can greatly exceed those in the Sacramento River at Freeport. This is illustrated in Fig. 3, which shows daily mean flows measured during water years 1980 through 1986 at Freeport and in the Yolo Bypass relative to flow downstream at Rio Vista. There is little flow through the Yolo Bypass when flows at Freeport are less than about

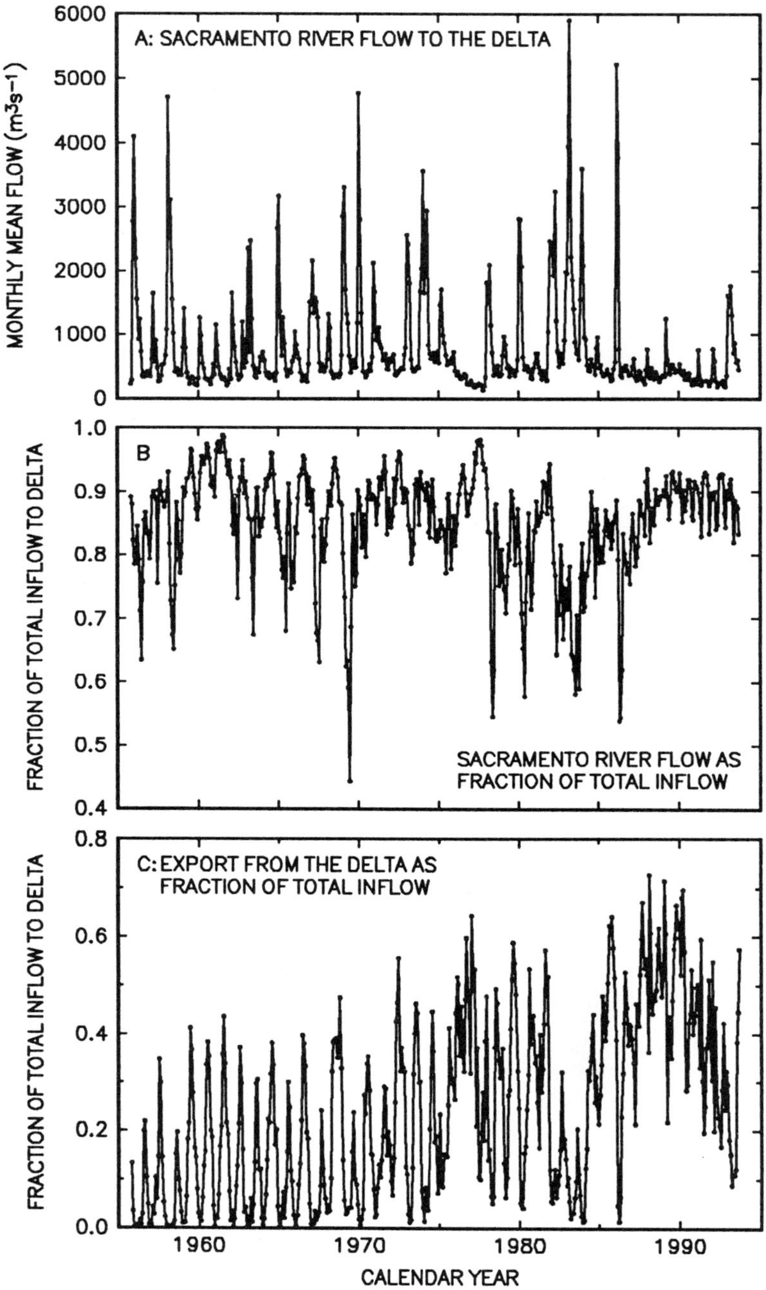

FIGURE 2. Flow of the Sacramento River to the Delta as monthly mean flow (A) and fraction of total flow to the Delta (B), and monthly mean water export from the Delta (C) during 1956-1993.

2000 m^3s^{-1}. However, when flows exceed 2000 m^3s^{-1} at Rio Vista, nearly all of this additional flow passes through the Yolo Bypass, and there is very little increase in flow at Freeport. Over the last 38 years, flow in the Yolo Bypass exceeded that at Freeport an average of one week each year. Flows in the Yolo Bypass exceeding 1000 m^3s^{-1} occurred relatively frequently, averaging two weeks per year. In contrast, flows in the Yolo Bypass exceeding 6000 m^3s^{-1} were relatively rare, occurring only 21 days in the 38-year period; the recent 7-year period shown in Fig. 3 included 11 of these 21 days.

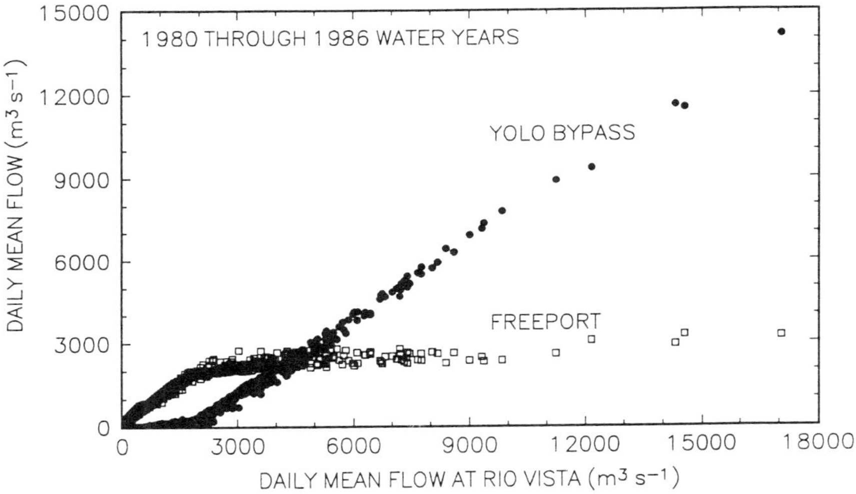

FIGURE 3. Graph showing relations between daily mean flow at Rio Vista and daily mean flows at Freeport and in the Yolo Bypass, 1980-1986 water years.

Suspended Sediment Transport to the Delta

The weathering of rocks and erosion of fine particles are natural processes that are accelerated by anthropogenic activities in river basins (*e.g.*, Meade 1988). Before the California gold rush, it is estimated that the annual transport of suspended sediment to the Delta from all rivers averaged 1.3×10^6t (Gilbert 1917; 1t = 10^3kg = 1.10 short tons). This value increased to an estimated $12 \times 10^6 ty^{-1}$ by the late 19th century, primarily because of hydraulic mining, which was then banned in 1884. By the middle of the 20th century, suspended sediment transport to the Delta had decreased to about $4.2 \times 10^6 ty^{-1}$ (Porterfield 1980), and it is expected to decrease to about $3.1 \times 10^6 ty^{-1}$ by the end of this century (Krone 1979), in part because of increasing impoundment and diversion upstream of the Delta.

Estimates of suspended sediment transports in studies during the first half of the 20th century were based on very few measurements (Porterfield 1972). The first systematic survey of sediment transport to San Francisco Bay began in 1957. This study included a large field program to measure suspended sediment transports in most of the influent rivers and streams, and collected the first measurements to assess the importance of suspended sediment transport through the Yolo Bypass.

Mean values for annual transports of suspended sediments from the major rivers over 1957-1959 probably were representative of long-term (annual) averages (Porterfield *et al.* 1961). The annual average value for transport of suspended sediment from the Sacramento River to the Delta was 3.5×10^6t (revised value from Porterfield 1980). This value was equivalent to 85% of the total transport of suspended sediment to the Delta, greatly exceeding that from the San Joaquin River (10%). The Yolo Bypass accounted for 27% of the transport from the Sacramento River, or 23% of the transport to the Delta. In terms of the supply to the Bay, suspended sediment transported from the Delta was 79% of the total, indicating that the Sacramento River supplied about two-thirds of the total to San Francisco Bay. These results clearly show the dominance of the Sacramento River and the importance of the Yolo Bypass in supplying suspended sediment to the Delta and the Bay. By compensating for effects of a major impoundment since this study (Porterfield *et al.* 1978), we estimated that the long-term annual average for suspended sediment transport from the Sacramento River to the Delta was about 3.1×10^6t during our study. This value is used in a general way as an "average" value for "normal" years in comparisons with our measurements below.

DATA SETS AND METHODS

Data sets and descriptions of methods used in sample collection and analyses have been published in reports that are identified below. Our overview of the methods here is brief, and we refer the reader to the appropriate reports for details.

Measured and estimated flow rates are used in this report. The primary sources for measured flows for 1979-1986 are annual reports by the U.S. Geological Survey (USGS 1979-1986). Other measured flows were obtained from the California Department of Water Resources (CDWR), Central Division, data base for the DAYFLOW Program. Flows at Rio Vista and from the Delta are not measured because of the strong influence of tides. Mean daily flows for these locations were estimated by the DAYFLOW Program (CDWR 1986 and 1987).

Suspended sediments are measured daily at Freeport (at Sacramento before 1980) as part of a long-term (since 1948) and continuing monitoring program. Suspended sediments were measured periodically at Rio Vista during 1979-1984. Measurements of suspended sediments during the 1980 water year comprise the most-complete and most-recent data set for the other locations shown in Fig. 4; measurements were discontinued at these locations after 1980. All measurements of suspended sediment were based on samples obtained with surface-to-bottom integrating samplers. Methods used in collecting samples and in computing daily mean values for suspended sediment transports and concentrations at all locations except Rio Vista are described in the USGS annual reports cited above. Daily values for suspended sediment transport at Rio Vista were computed from the estimated daily mean flow (from DAYFLOW) and a concentration (instantaneous) measured on that day. Suspended sediment concentrations at Rio Vista for 1979-

1982 are reported in the USGS annual reports, and those for 1983-1984 are reported by Harmon *et al.* (1986) and Ota *et al.* (1986).

The carbon content of the suspended sediment, primarily particulate organic carbon, was measured during two periods, January - May 1983 and November 1983 - November 1984 at Rio Vista. Although the analytical method did not discriminate between inorganic and organic carbon, subsequent laboratory studies indicated that the inorganic carbon fraction might have been about 5% of the total carbon. Data presented here are mean values for duplicate analyses. Details of the methods and the complete data set are reported by Hager (1994).

To supplement measured values for flows and suspended sediment transports, we computed estimates for some sites on a daily basis using flow and sediment transport budgets (eq.1-4). Similarly, two estimates for suspended sediment transports at Rio Vista (Est. Rio Vista) were computed (eq. 5-6), each using eq. 4 to estimate (daily) the fraction of the flow from Freeport that reached Rio Vista (PQ). The rationale for these budgets is illustrated by Fig. 4.

Sources = Knights Landing + Sutter Bypass + Nicolaus (eq. 1)
Est. Yolo Bypass = Sources - Verona (eg. 2)
Est. Freeport = Verona + American River - Sacramento Weir (eq. 3)
PQ = [Rio Vista - (Yolo Bypass + Sacramento Weir)] / Freeport (eq. 4)
Est. Rio Vista = Est. Yolo Bypass + Sacramento Weir+ (PQ)(Freeport)(eq. 5)
Est. Rio Vista = Yolo Bypass + Sacramento Weir+ (PQ)(Freeport) (eq. 6)

RESULTS AND DISCUSSION

Annual and seasonal flows from the Sacramento River varied greatly among the years of this study. Water years 1979 and 1981 were dry years with lower-than-normal flows at Freeport and Rio Vista and with no flow through the Yolo Bypass during winter (Fig. 5). Water years 1980 and 1982-1984 were years with normal or higher-than-normal flows. Most of the differences between flows at Freeport and Rio Vista were due to diversions from the Delta primarily during the irrigation season and flow through the Yolo Bypass during the winter storm season (see Introduction).

Both 1982 and 1983 were exceptional because river flows were unusually high for most months of these water years. Consequently, flows and concentrations of suspended sediments during the irrigation seasons of these two years might not be representative of typical years. During the other summers, most concentrations of suspended sediments were in the low range of about 15 to 40 mg l^{-1} at both Freeport and Rio Vista (Fig. 6), and differences between these locations were generally small. During winter, highest concentrations of suspended sediment at Freeport often coincided with the first major storm event of the season, a phenomenon that has been observed at Sacramento in previous years (*e.g.*, Goodwin & Denton 1991). During the two dry winters, 1979 and 1981, concentrations of suspended sediments at Freeport typically were higher than those at Rio Vista when runoff from storms caused moderate increases in flow (generally to less than 2000 m^3s^{-1}). However, during the other winters, concentrations at Rio Vista typically were higher than

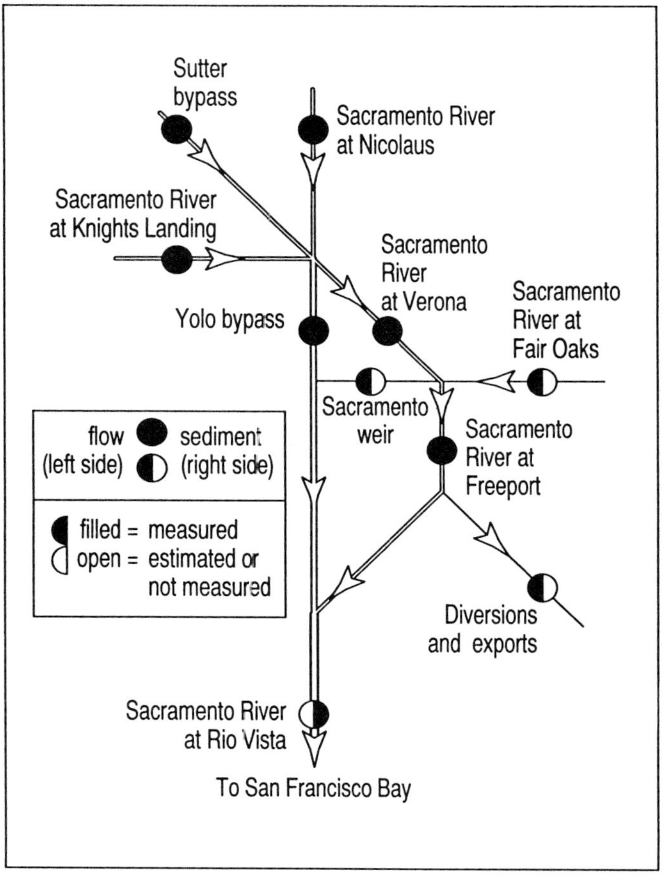

FIGURE 4. Diagram showing relations between locations where flow or suspended sediment or both were measured during 1980.

those at Freeport, and highest concentrations at Rio Vista coincided with periods of high flow from the Yolo Bypass.

Plots of suspended sediment concentration versus flow for Rio Vista and Freeport show wide ranges in concentration at most levels of flow (Fig. 7), indicating the potential for large errors in estimates of concentrations (and transports) based on flow. Goodwin & Denton (1991) found that concentration (and transport) varied by nearly two orders of magnitude for a given flow over a longer record from Sacramento (1953-1976), but achieved some success in improving correlations with flow by separating the data by season. Our approach was to base the analysis on levels of flow and knowledge of the ways freshwater can flow through the river system and the Delta. At low flows (to about 500-600 m^3s^{-1}), the level of flow was typically higher at Freeport, but the range of flow was greater at Rio Vista because a (variable) fraction of the river flow was diverted between

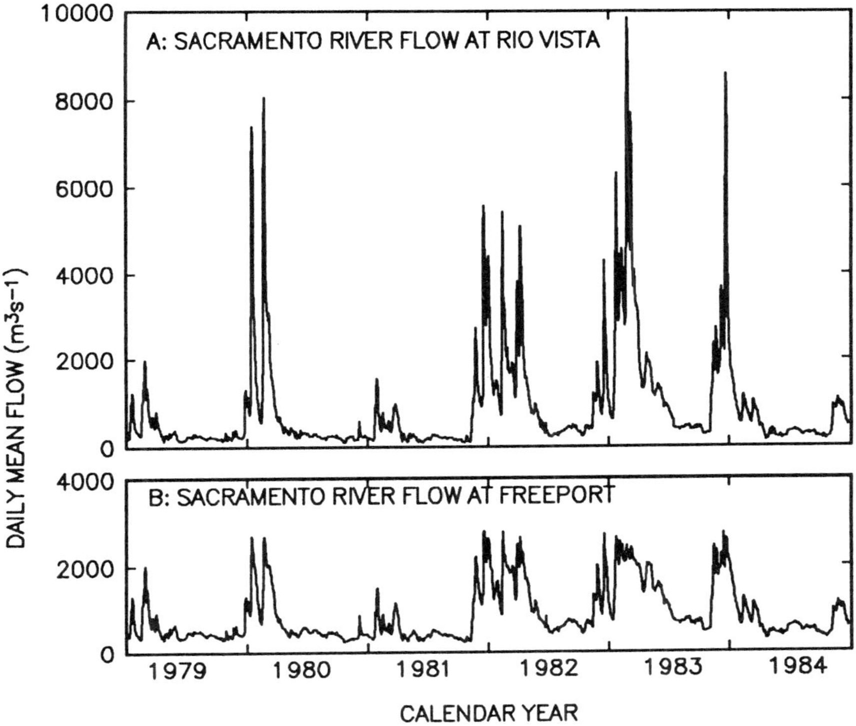

FIGURE 5. Daily mean flow of the Sacramento River at Rio Vista (A) and at Freeport (B) during 1979-1984.

Freeport and Rio Vista (Fig. 4). As a result, meaningful correlations of flows with concentrations of suspended sediments were not identified at Rio Vista during the irrigation season. Similarly, concentrations at Freeport showed little correlation with flows when river flow was low and controlled by releases from reservoirs, which was consistent with previous observations during summers at Sacramento (Goodwin & Denton 1991). Consequently, flow was not a good predictor of concentration at either location when flows were low.

Concentrations of suspended sediment increased with increasing flows over the range of roughly 500 to 1500 m^3s^{-1} at both locations, although the increase appeared somewhat greater at Freeport (Fig.7). This range in flow included runoffs from small- to moderate-sized storms, including many first or early storms of the season, when there was little or no flow through the Yolo Bypass. At greater flows, however, water was diverted upstream of Freeport, and although concentrations of suspended sediment remained in the high range at Freeport, they showed little dependence on the small range in flow (Fig. 7B). In contrast, discharge from the Yolo Bypass increased flows and concentrations of suspended sediments at Rio Vista, both to greater levels than those observed at Freeport. Although the relation

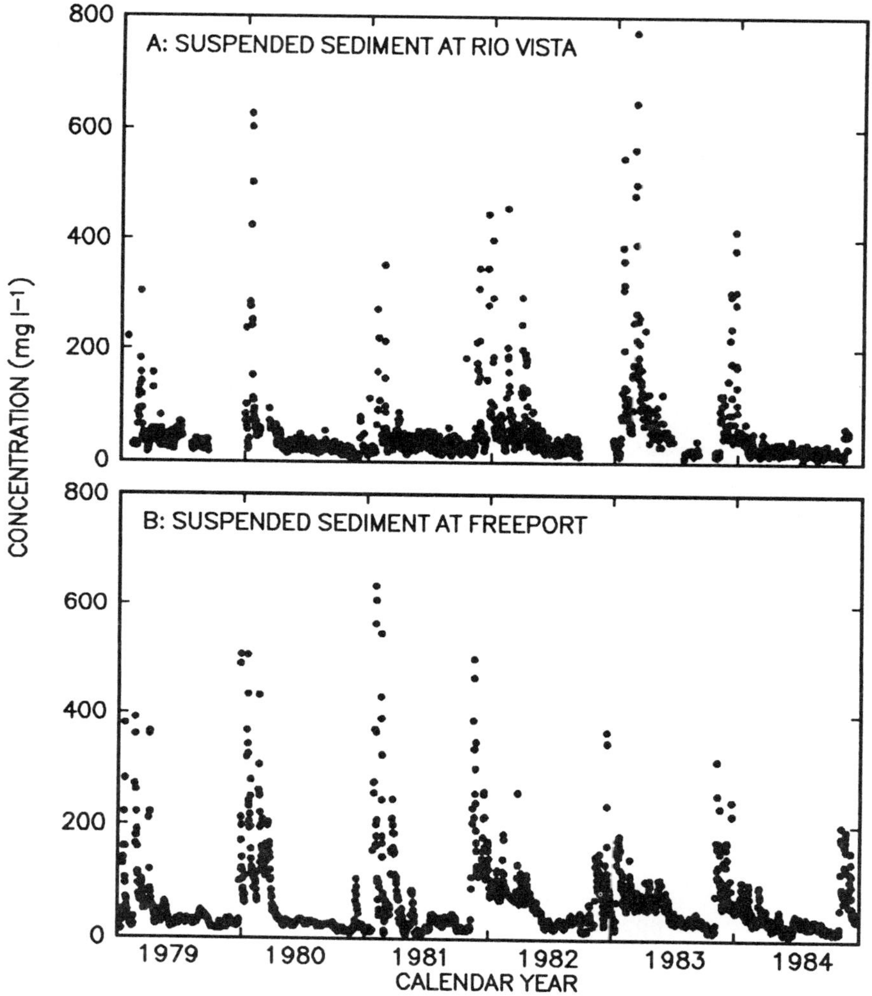

FIGURE 6. Instantaneous concentrations of suspended sediment at Rio Vista (A) and daily mean concentrations of suspended sediment at Freeport (B) during 1979-1984.

between flow and concentration was variable over the range of high flows, mean concentrations of suspended sediments at Rio Vista could be estimated from daily mean flows by the following equation (see Fig. 7A), that was computed by the group method described by Glysson (1987).

$$[\text{Suspended Sediment}] = 0.00185 \times \text{Flow}^{1.38} \qquad \text{(eq. 7)}$$

Individual (daily mean) estimates from this equation were greatly different from measured values in some cases. However, effects of these differences were minimized when transports were averaged over flood events or longer periods of time

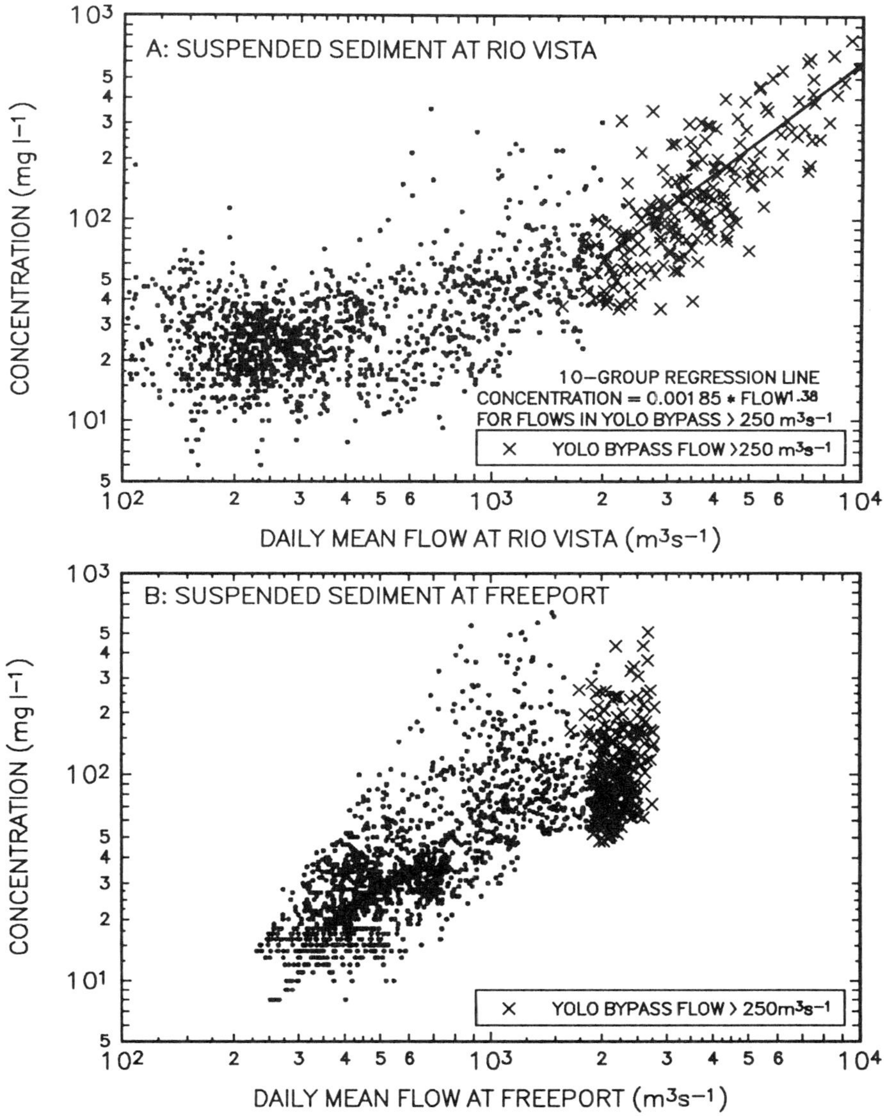

FIGURE 7. Log-log plots of suspended sediment concentrations and daily mean flows at Rio Vista (A) and at Freeport (B) during 1979-1984.

(Walling 1977). In addition to natural short-term variability, other factors might have influenced concentrations of suspended sediments during periods of high flow. This is examined in more detail below.

Water Year 1983 was remarkable because winter and annual flows (Delta outflow and Sacramento River) were the highest in the historical record dating from

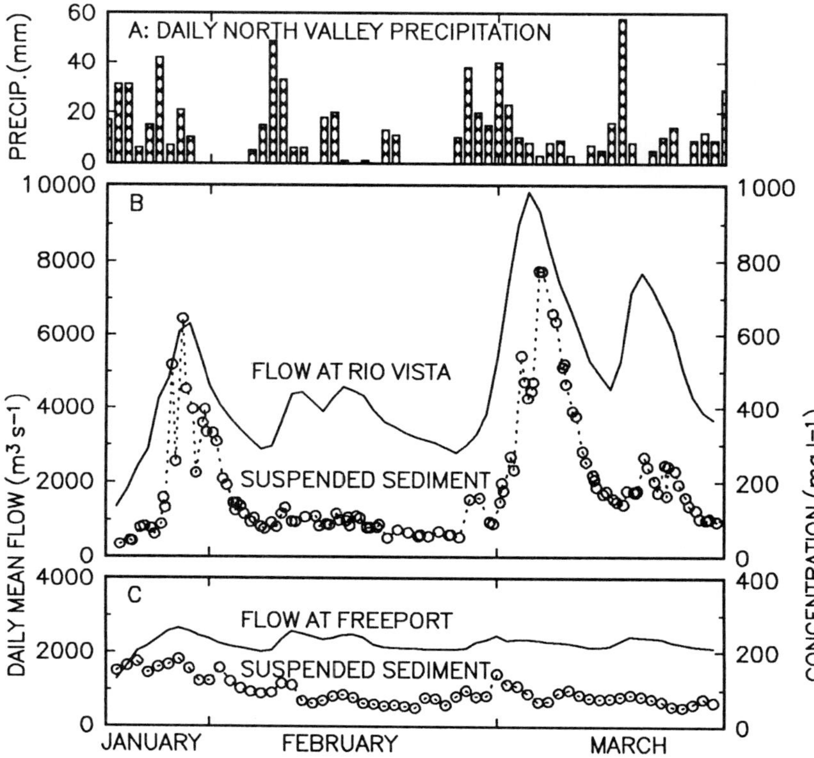

FIGURE 8. Daily precipitation in the north central valley (A) and daily mean flows and concentrations of suspended sediment at Rio Vista (B), and Freeport (C) during major storm events of January-March 1983.

1930. Links have been suggested between unusual periods of river flow (both high and low) to San Francisco Bay and large-scale atmospheric and oceanic circulation patterns (Cayan & Peterson 1989; Peterson *et al.* 1995). Water year 1983 exhibited the greatest sea-surface temperature anomalies along the California coast of the 1982-1983 El Niño-Southern Oscillation (ENSO) episode (Breaker 1989), which was perhaps the most intense of this century (Barnett *et al.* 1988).

During winter 1983, flows were high through the Yolo Bypass for several weeks in response to a sequence of frequent and intense storms (Fig. 8A). Highest flows at Rio Vista (early March) coincided with the highest concentrations of suspended sediments (Fig. 8B), both of which were the highest observed during the 1979-1984 period. During this major flood event, concentrations increased with increasing flow, reached highest concentrations near the highest flow, then decreased with decreasing flow. During the subsequent peak in flow, however, concentrations were much lower than might be expected from the levels of flow and concentrations during the previous event. A similar phenomenon was observed earlier in January, when a large peak in flow was followed by two smaller peaks. Examination of sequences of flood events at Rio Vista during other years revealed similar patterns of comparatively low concentrations of suspended sediment during high-flow events that closely followed high-flow events. This and other hysteresis effects have been attributed to natural processes in many rivers (*e.g.*, Meade 1988).

In addition to natural processes, flood-water management practices, such as the release of water from reservoirs as they approach their flood-control capacity, appeared to affect the relation between flow and concentrations of suspended sediment in the Sacramento River. Releases from large reservoirs typically have lower concentrations of suspended sediment than runoff from rainfall events (*e.g.*, Petts 1984). During winter 1983, flow released from large reservoirs on the tributary rivers constituted a large fraction of the total flow during peaks in flow that followed high-flow events, such as those described above, and during periods of sustained high flow. In general, concentrations of suspended sediments were highest at Rio Vista when large increases in flow were primarily the result of runoff from precipitation.

Concentrations of suspended sediments as well as flows at Rio Vista were higher than those at Freeport during the major flood events in January and March 1983 (Fig. 8c). Differences between these two locations during periods of high flow could be attributable to many factors. These include discharges from reservoirs on the American River, which tend to reduce concentrations at Freeport, and erosion or deposition in the Yolo Bypass, which affect concentrations at Rio Vista but not at Freeport. In general, concentrations measured at Freeport were poor indicators of the levels of concentration and the variability in concentrations at Rio Vista during major flood events.

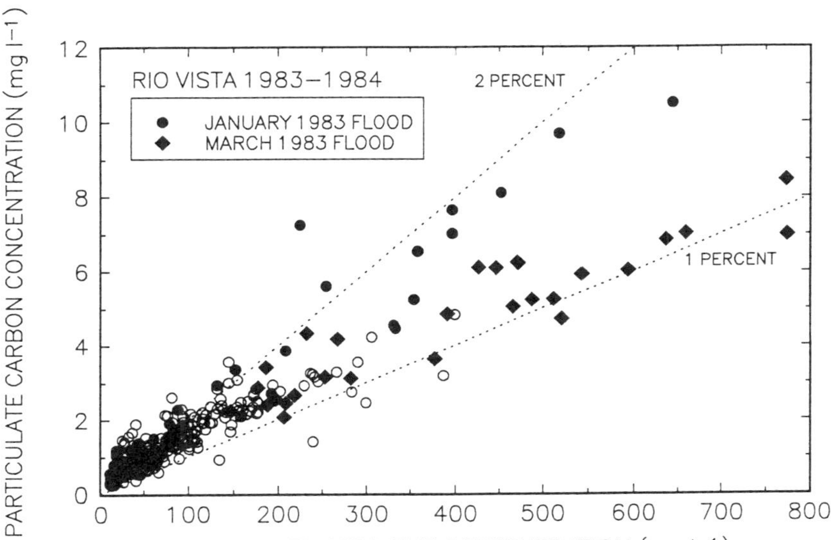

FIGURE 9. Graph showing relation between concentrations of particulate carbon and suspended sediment at Rio Vista, 1983-1984.

The two flood events in winter 1983 that yielded the highest concentrations of suspended sediments at Rio Vista (Fig. 8B) also produced the highest concentrations of particulate carbon measured over the 1983-1984 study periods (Fig. 9). Suspended sediments during the January 1983 flood were more enriched in carbon (nearly 2%) compared to the March flood (about 1%). During winter of 1984, concentrations of suspended sediments reached only about 400 mg l^{-1}, and the carbon content was near 1%. Generally, the carbon content for most samples collected during this study was between 1 and 2% (mean value of 1.48%) when concentrations of suspended sediment were greater than 100 mg l^{-1}. At lower concentrations of suspended sediment, the carbon content tended to be greater (mean value of 2.80% for suspended sediment concentrations less than 50 mg l^{-1}); however, few samples yielded contents exceeding 4%. Similar values for carbon contents (percent carbon) and inverse relations between carbon contents and concentrations of suspended sediments have been reported for large river systems (Meybeck 1982; Ittekkot & Laane 1991; Trefry *et al.* 1994). Algal production in the river and the Delta probably accounted for some of the high carbon contents during summer when concentrations of suspended sediments were in the low range (see Jassby *et al.*, this volume). Carbon contents during the first flood of 1983 (January) might have been high because the first major flood can mobilize organic matter produced over the previous summer. High carbon contents were not observed in early winter 1984, which followed unusually high river flows throughout the 1983 water year. Canuel and Cloern (this volume) report values in the 2-3% range during winter of 1993, which followed several years of unusually low flow in the Sacramento River (see Fig. 2A).

Although flows and concentrations of suspended sediment varied by over two orders of magnitude at Rio Vista, the carbon content of the suspended sediment varied over a much smaller range. In fact, during winter the carbon content varied by little more than a factor of two, indicating that suspended sediment concentrations might be reasonably good predictors of particulate carbon concentrations under moderate to high flow conditions. The following linear relation is based on suspended sediment concentrations greater than 50 mg l^{-1} ($n = 219$; $r^2 = 0.83$):

$$[\text{particulate carbon}] = 0.55 + 0.0114\,[\text{suspended sediment}]. \qquad \text{(eq. 8)}$$

Alternatively, a probable range of values can be estimated by assuming 1 to 2% of the suspended sediment is carbon.

Suspended Sediment Transports

Flow and concentration measurements presented above indicated large variability in the rate of sunply (transport) of suspended sediment to the Delta and to the Bay. Although day-to-day differences in suspended sediment transport can be large, transports over major flood events or sequences of events and seasonal and interannual variations in transports are probably more important in terms of effects on the San Francisco Bay ecosystem. Our objectives in this section are to quantify suspended sediment transports at Freeport and Rio Vista over periods of time ranging from flood events to water years, and to compare transports at these two locations, emphasizing influences of flow from the Yolo Bypass and water diversion from the Delta on suspended sediment transports to the Bay.

Flows and suspended sediment transports were measured at key locations on the Sacramento River and in major bypasses during two major flood events in January and February of 1980 (Fig. 4). This data set is particularly valuable because it contains measurements of suspended sediment in the Yolo Bypass, one of several locations on this river system where suspended sediments have not been measured since that year. Measurements and estimates based on flow and mass balances (see Methods) were used to calculate mean values for flows and total transports over each of these two major flood events (defined here and below as periods when flow at Rio Vista exceeded flow at Freeport).

Measured and estimated values for flow were similar at Freeport and in the Yolo Bypass, giving some assurance that flow and transport budgets produced reasonable estimates (Table 1). Flow budget calculations (eq. 4) indicated that about 90% of the flow from Freeport reached Rio Vista during each flood event. Most of the remaining flow from Freeport probably reached the Bay via channels in the Delta, but this was a small amount compared to flow at Rio Vista. Although suspended sediment transport at Freeport during January was greater than that during February, differences in the other measured and estimated quantities between the two periods were relatively small (Table 1). During January, suspended sediment transport at Freeport was between the measured and estimated values in the Yolo Bypass, and during February, both values for transport in the Yolo Bypass were greater than that at Freeport. Consequently, the suspended sediment transport at Freeport probably accounted for half or less of the transport from the Sacramento River to the Bay.

Estimated values for total (34-day) transport at Rio Vista over these two events ranged from 3.1 to 3.7 × 10^6t, which encompassed the value of 3.2 × 10^6t estimated from eq. 7. These results indicated that an amount equivalent to the average annual transport for the Sacramento River (approx. 3.1 × 10^6t, see Introduction) was

TABLE 1. Measured and Estimated flows and suspended sediment (SS) transports during major winter flood events of 1980.

Location	*January 1980 (16 days)* Mean Flow ($m^3 s^{-1}$)	SS Transport ($10^6 t$)	February 1980 (18 days) Mean Flow ($m^3 s^{-1}$)	SS Transport ($10^6 t$)
At Freeport				
Measured	2120	0.86	2180	0.69
Estimated	2330	—	2150	—
In Yolo Bypass				
Measured	1820	0.64	2320	0.76
Estimated	1860	0.91	2100	1.0
At Rio Vista				
Measured	—	1.9	—	—
Estimated	4020	1.4 to 1.7	4380	1.4 to 1.6

1t = 10^3kg = 1.10 short tons

supplied to San Francisco Bay in one month. In addition, estimates using eq. 7 and eq. 8 indicated that this sediment contained 43×10^9g carbon, which was equivalent to an estimate of the amount of phytoplankton carbon produced in northern San Francisco Bay over the entire 1980 calendar year (45×10^9g C, Jassby *et al.* 1993). Our value also indicated that the estimated riverine input of particulate organic carbon to the Bay given by Jassby *et al.* (16×10^9g C) was too low, probably because measurements of particulate organic carbon were not available for winter 1980.

Concentrations of suspended sediments were measured daily at Rio Vista during the January 1980 event, allowing calculation of transports from these instantaneous values and estimated daily mean flows. The transport from these measurements at Rio Vista exceeded the range of values estimated from measurements at Freeport and in the Yolo Bypass (Table 1). This comparison suggested that our measurements for 1982-1984 winter storm season transports at Rio Vista (presented in Table 2) might be somewhat high. Some error due to cross-channel variability at Rio Vista was expected, especially when flows were high from the Yolo Bypass; however, the sampling bias shown by this comparison was not unusually large (Walling 1977).

Values for daily mean flows and suspended sediment transports were summarized in Table 2 for selected periods when measurements were available for both Freeport and Rio Vista. The winter storm seasons of water years 1982-1984 were periods when substantial flows were diverted to the Yolo Bypass, as indicated by differences between flows at Rio Vista and Freeport (Fig. 5, Table 2). Flows at Rio Vista averaged over periods of 1 to 2 months were as great as twice the values at Freeport; however, total transports of suspended sediment at Rio Vista over these periods were 2 to 5 times greater than those at Freeport. Therefore, it was not

TABLE 2. Flows and measured transports of suspended sediment (SS) at Freeport and Rio Vista during selected periods of 1979-1984.

Water Year	*Days in Record*	*Mean Flows (m^3s^{-1})* Freeport	Rio Vista	Delta	*SS Transports (10^6t)* Freeport	Rio Vista
Winter Storm Season						
1982	66	2260	3290	4150	1.5	3.1
1983	78	2190	4330	5950	1.3	6.4
1984	33	2210	3620	4840	0.58	1.8
Irrrigation Season						
1979	107	436	224	195	0.14	0.07
1980	147	467	270	347	0.16	0.09
1981	152	382	189	151	0.15	0.07
1984	139	502	278	299	0.17	0.07
Esitmates of Annual Mean Values						
1981	350	442	285	300	1.1	0.41
1982	328	1210	1270	1660	3.1	3.9
1984a	348	859	842	1160	1.7	2.4

$1t = 10^3kg$ = 1.10 short tons
1984a = November 1983 through October 1984

uncommon for values measured at Freeport to represent much less than half of the suspended sediment transport to the Bay during these winter floods. Mean values for delta outflow indicated that flow from Rio Vista constituted at least three-quarters of the flow to the Bay (Delta outflow)and most likely was the dominant source of suspended sediments. During these periods of 1982-1984, most of this suspended sediment was transported through the Yolo Bypass. Estimates of suspended sediment transports based on daily mean flows at Rio Vista (eq. 7) were close to the measured values (Table 2) over these three periods (1982 = 2.9×10^6t; 1983 = 7.3×10^6t; 1984 = 2.2×10^6t).

The prolonged period (78 days) of high flow during 1983 resulted in suspended sediment transport at Rio Vista that was double the average annual transport for the Sacramento River. Similarly, total transport over 66 days in 1982 was equivalent to the annual average. However, in these cases and in others, a large fraction of the transport occurred over just a few days when flows were very high through the Yolo Bypass during and immediately following major rainfall events. For example, amounts equivalent to half the annual average were transported at Rio Vista in 2 weeks in 1980 (Table 1) and in only 3 days in March 1983. This situation is not unusual in rivers that exhibit great variability in flow, and shows the need to measure suspended sediment transports during high flow (Nordin & Meade 1981).

Measurements during the irrigation seasons of 4 years showed how the diversion of water from the Delta influenced flow and suspended sediment transport to the Bay. In general, about half the flow at Freeport was diverted or consumed in the Delta, and flows from the Delta to the Bay (Delta outflow) were similar in magnitude to those at Rio Vista. Transports of suspended sediment were consistent with differences in flow in that transports at Freeport were about twice those at Rio Vista. Similarly, we expect that suspended sediment transports to the Bay were also about half the values at Freeport. The 4- to 5-month-long records from irrigation seasons in Table 2 represent conditions that can persist for 8 or more months each year, depending on interannual variability in weather patterns. Assuming an average of approx. 3.1×10^6t of suspended sediment are transported to the delta annually, our data indicated that only a small fraction of this (probably less than 5%) occurs during the irrigation season in most years. In addition, suspended sediment transports to the Bay during even extended periods of such low flow are barely significant compared to transports during winter flood events, such as those during 1980 (see Tables 1 and 2).

Measurements over 3 water years were nearly complete at Rio Vista, which provided estimates of annual transports (Table 2). During 1981, a dry year, mean values for flow were in the range of those for the irrigation season, and the total transport of suspended sediment at Freeport was only one-third the annual average for the Sacramento River. The total transport of suspended sediment at Rio Vista was less than half that at Freeport, which also indicated that the supply of suspended sediment from the Sacramento River to San Francisco Bay during 1981 was only 10-15% of the value for an average year. Flows and suspended sediment transports over 1982 and 1984 were higher than those during 1981. However, three-quarters or more of these annual totals at Rio Vista were transported in just 1 or 2 months during winter.

Variability in climate and weather patterns (*e.g.*, the number of storms, the amount of precipitation, etc.) was ultimately responsible for most differences in the flow of the Sacramento River among seasons and years. Furthermore, the level of flow generated by a storm or a sequence of storms was a key factor influencing concentrations and transports of suspended sediment in the Sacramento River. These relationships suggest that San Francisco Bay could receive loads of suspended sediment equivalent to many times the annual average supply during major or record-breaking floods, such as the remarkable flood that occurred near the beginning of the 1986-1987 ENSO episode (Barnett *et al.* 1988).

The Flood of 1986

Major storms during winter 1986 caused extensive flooding and record-breaking flows in the Sacramento River system (Fig. 10). In February, records for flow were broken at Verona, Sacramento Weir, Yolo Bypass, and Sacramento, records that dated back as far as 1918 (Blodgett & Lucas 1988). River stage at Sacramento was the highest on record since the great flood of January 1862 (for a description of that flood see Peterson *et al.* 1985). Effects of the winter 1986 flood on the Bay ecosystem probably were many and diverse (*e.g.*, Peterson *et al.* 1989), ranging from major changes in distributions of dissolved and particulate substances baywide to sedimentation and disruption of benthic communities in the upper estuary. The successful invasion of the upper estuary by an Asian clam later that year and perhaps other major changes in the Bay ecosystem were influenced by this enormous flow and sediment-transport event (Nichols *et al.* 1990).

At Rio Vista, the highest flow in February 1986 (17100 m^3s^{-1}) greatly exceeded the highest flow during 1983 (9840 m^3s^{-1}; Fig. 11A), and flow reached high levels again in March 1986 (6450 m^3s^{-1}). Concentrations of suspended sediments were measured at Freeport during February and March, but not at Rio Vista or in the Yolo Bypass (Fig. 11B). Highest concentrations at Freeport were observed during the highest flows in February, reaching concentrations that were higher than any measured at Freeport during 1979-1984 (see Fig. 6). Concentrations of suspended sediments at Rio Vista were estimated from daily mean flows using eq.7. Uncertainties in these estimates might be very large, in part because flow in 1986 reached higher levels than in 1983.

Suspended sediment transports were computed over the February flood (the first 21 days, Fig. 11) and over the entire period (total of 40 days) from measurements at Freeport and estimates for Rio Vista. Total (40-day) transport at Freeport (1.6×10^6t) was only 20% of our estimated value for Rio Vista (8.1×10^6t). At Freeport, 71% (1.2×10^6t) of the 40-day total was transported over the first 21 days, and our estimates indicated 84% (6.8×10^6t) of the 40-day total was transported over the first 21 days at Rio Vista. However, it was clear that most of the suspended sediment was transported over much shorter periods of time in February. Over the 3 days that included record-breaking flows, suspended sediment transport at Freeport was

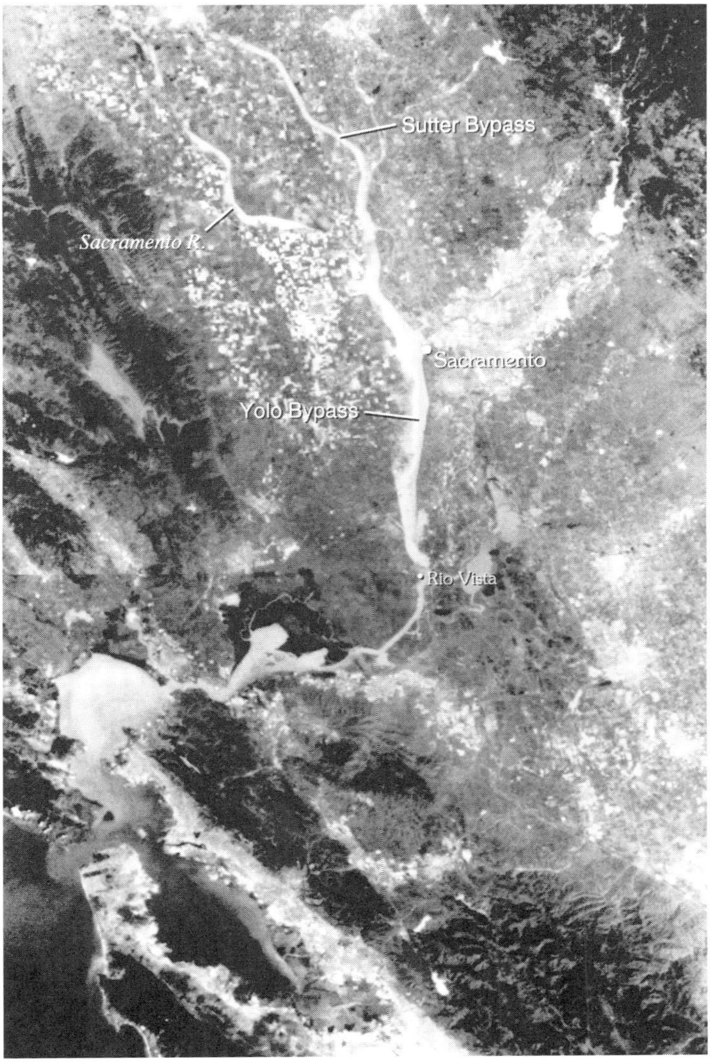

FIGURE 10. LANDSAT image on March 21, 1986, showing evidence of flooding in the Sutter and Yolo Bypasses over the previous month and sediment plumes in San Francisco Bay and the adjacent coastal Pacific Ocean.

0.49×10^6t, nearly one-third the measured value for the entire 40 days. Similarly, estimated transport at Rio Vista over these 3 days was 3.8×10^6t, half the value over the 40-day period and nearly a factor of 8 times greater than the transport measured at Freeport.

The estimated suspended sediment transport at Rio Vista over just 3 days in February 1986 exceeded the amount supplied by the Sacramento River over an

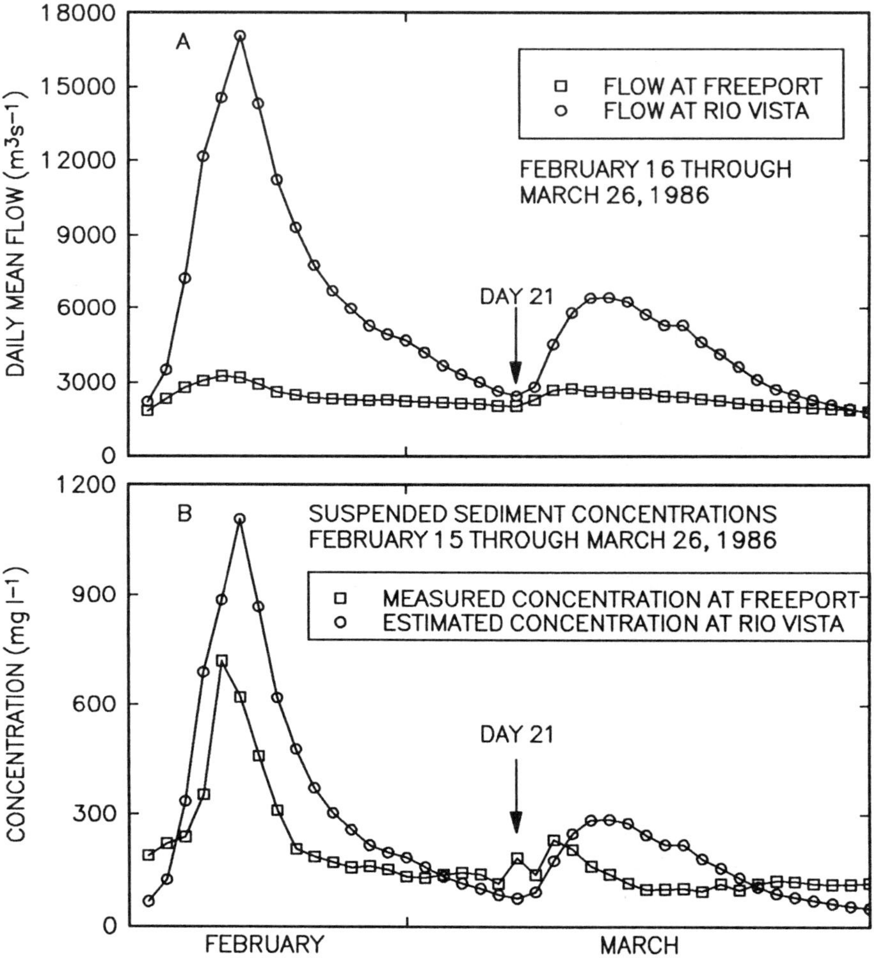

FIGURE 11. Daily mean flows at Rio Vista and Freeport (A) and estimated and measured concentrations of suspended sediment at Rio Vista and Freeport (B) during major flood events of February and March 1986.

average year. Although we would expect events of this scale to be infrequent, transports over a few weeks also exceeded the annual average during 1980, 1982, and 1983. In contrast, transports during dry seasons or years were much less than the annual average. Low flows over the recent episode of dry years (1987-1992) suggest that the Bay received little suspended sediment from the Sacramento River for an extended period of time (Fig. 2). Consequently, even though this river system has been highly modified from its natural state, variability in weather patterns and climate remains a key factor affecting the supply of suspended sediment from the Sacramento River to San Francisco Bay.

SUMMARY AND CONCLUSIONS

The Sacramento River is the largest source of freshwater and suspended sediment to the San Francisco Bay estuarine system. From 1979 through 1984, large variations in flows and in concentrations and transports of suspended sediments were observed over time scales of days to years. Seasonal and interannual variability in suspended sediment transport was particularly great. Over seasons with low flows and dry years, the supply of suspended sediment to the Delta and the Bay was a small fraction of what would be expected over an average year. In contrast, winters with high flows produced amounts equivalent to an average year's supply in just a few weeks.

During normal to wet years, a large fraction of the annual supply of suspended sediment was transported through the Yolo Bypass. Although we have provided a means of accounting for the supply from the Yolo Bypass by estimating suspended sediment transports at Rio Vista (eq. 7), the lack of measurements is likely to result in great uncertainties in estimates of suspended sediment transports to the Delta and Bay in future years. Quantitative measurements of supplies of dissolved and particulate substances from the river systems are fundamental information in studies of the San Francisco Bay ecosystem (*e.g.*, Jassby *et al.* 1993).

Suspended sediment from the Sacramento River system is a large source of organic matter to the Bay during normal to wet years, and might be an important source during summers and drought years if supplies of organic matter from other sources are similarly limited. Assuming the carbon content of the suspended sediment is 1%, in the low range of observed values, transport to the Bay during an average year (roughly 3×10^{10}g) will probably exceed the amount produced by phytoplankton in northern San Francisco Bay, particularly because production has decreased in recent years due to grazing pressure (Alpine & Cloern 1992). However, the nutritional value of river-supplied organic matter is highly variable and might be lowest during the times of highest transports (Canuel & Cloern, this volume). Even with this factor in mind, the supply of particulate carbon from the Sacramento River is a major component of the carbon budget for the San Francisco Bay ecosystem.

ACKNOWLEDGMENTS

We are grateful for the hard work and dedication of the many people involved in the collection of flow and sediment measurements. In particular, we wish to acknowledge the dedication of C.T. Henderson in collecting samples at Rio Vista. We also thank S. Greene for providing estimated flows from the DAYFLOW program, and J. Thompson, P. Lehman, and our colleagues at the U.S. Geological Survey for helpful comments and reviews. Special thanks is extended to D. Jones, A. Acosta, and L. Gaydos for help with the LANDSAT data and image (Fig. 10), and J. Dileo-Stevens for preparing other illustrations.

LITERATURE CITED

Alpine, A. E. and J. E. Cloern. (1992). Trophic interactions and direct physical effects control phytoplankton biomass and production in an estuary. *Limnol. Oceanogr.* 37(5):946-955.

Barnett, T., N. Graham, M. Cane, S. Zebiak, S. Dolan, J. O'Brien, and D. Legler. (1988). On the Prediction of the El Niño of 1986-1987. *Science* 222:1203-1210.

Breaker, L. C. (1989). El Niño and related variability in sea-surface temperature along the central California Coast. Pages 133-140 *in* D. H. Peterson, ed., *Aspects of Climate Variability in the Pacific and the Western Americas*. Geophysical Monogr. 55. American Geophysical Union, Washington, DC.

Blodgett, J. C. and J. B. Lucas. (1988). *Profile of Sacramento River, Freeport to Verona, California, Flood of February 1986*. U.S. Geol. Surv., Open-File Rept. 88-82. 16 pp. plus map.

California Department of Water Resources. (1986). *DAYFLOW Program Documentation and Users Guide*. Central District, Sacramento, CA.

California Department of Water Resources. (1987). *DAYFLOW 1930-1955: Historic Daily Flows (Water Years 1930-1955)*. Central District, Sacramento, CA.

Canuel, E. A. and J. E. Cloern. (this volume). Regional differences in the origins of organic matter in the San Francisco Bay ecosystem: Evidence from lipid biomarkers.

Cayan, D.R. and D.H. Peterson. (1989). The influence of North Pacific atmospheric circulation of streamflow in the west. Pages 375-198. *in* D. H. Peterson, ed., *Aspects of Climate Variability in the Pacific and the Western Americas*. Geophysical Monogr. 55. American Geophysical Union, Washington, DC.

Cloern, J. E. and F. H. Nichols. (1985). Time scales and mechanisms of estuarine variability, a synthesis from studies of San Francisco Bay. *Hydrobiologia* 129:229-237.

Cole, B.E . and J. E. Cloern. (1984). Significance of biomass and light availability to phytoplankton productivity in San Francisco Bay. *Marine Ecol. Progr. Ser.* 17:15-24.

Conomos, T. J., R. E. Smith, and J. W. Gartner. (1985). Environmental setting of San Francisco Bay. *Hydrobiologia* 129:1-12.

Domagalski, J. L. and K. M. Kuivila. (1993). Distributions of pesticides and organic contaminants between water and suspended sediment, San Francisco Bay, California. *Estuaries* 16:416-426.

Glysson, G. D. (1987). *Sediment-transport Curves*. U.S. Geol. Surv., Open-File Rept. 87-218. 47pp.

Gilbert, G. K. (1917). *Hydraulic Mining Debris in the Sierra Nevada*. U.S. Geol. Surv., Prof. Pap. 105. 154 pp.

Goodwin, P. and R. A. Denton. (1991). Seasonal influences on the sediment transport characteristics of the Sacramento River. *Proc. Instn. Civ. Engrs.* Part 2, 91:163-172.

Hager, S. W. (1994). *Particulate Carbon and Nitrogen and Suspended Particulate Matter in the Sacramento River at Rio Vista, California, January 3 - April 9, 1983, and October 31, 1983 - November 29, 1984*. U.S. Geol. Surv., Open-File Rept. 94-470. 35pp.

Ittekkot, V. and R. W. P. M. Laane. (1991). Fate of riverine particulate organic matter. Pages 231-243 *in* E. T. Degens, S. Kempe & J. E. Richey, eds., *Biogeochemistry of Major World Rivers*. John Wiley & Sons, New York.

Harmon, D. D., L. E. Schemel, S. W. Hager, and A. Y. Ota. (1986). P*hysical and Chemical Data for the Sacramento River at Rio Vista, California, January through May, 1983*. U.S. Geol. Surv. Open-File Rept. 86-230. 26 pp.

Jassby, A. D., J. E. Cloern, and T. M. Powell. (1993). Organic carbon sources and sinks in San Francisco Bay: variability induced by river flow. *Marine Ecol. Progr. Ser.* 95:39-54.

Jassby, A. D., J. Koseff, and S. Monismith. (1996). Processes underlying phytoplankton variability in San Francisco Bay. Pages 325-349 *in* J. T. Hollibaugh, ed., *San Francisco Bay: The Ecosystem*. Pacific Division, American Association for the Advancement of Science, San Francisco, CA.

Kahrl, W. L. (1978). *The California Water Atlas*. State of California, the Governor's Office of Planning and Research and the Department of Water Resources, Sacramento, Calif. 118 pp.

Knebel, H. J., T. J. Conomos, and J. A. Commeau. (1977). Clay-mineral variability in the suspended sediments of the San Francisco Bay system, California. *Jour. Sediment. Petrol.* 47(1):229-236.

Krone, R. B. (1979). Sedimentation in the San Francisco Bay system. Pages 85-96 *in* T. J. Conomos, ed., *San Francisco Bay: The Urbanized Estuary*. Pacific Division, American Association for the Advancement of Science., San Francisco, Calif.

Meade, R. H. (1988). Movement and storage of sediment in river systems. Pages 165-179 *in* A. Lerman.& M. Meybeck, eds., *Physical and Chemical Weathering in Geochemical Cycles*. Kluwer Academic Publishers, Boston, MA.

Meybeck, M. (1982). Carbon, nitrogen, and phosphorus transport by world rivers. *American Jour. Sci.* 282:401-450.

Nichols, F. H., J. E. Cloern, S. N. Luoma, and D. H. Peterson. (1986). The modification of an Estuary. *Science* 231:567-573.

Nichols, F. H., J. K. Thompson, and L. E. Schemel. (1990). Remarkable invasion of San Francisco Bay (California, USA) by the Asian clam *Potamocorbula amurensis*, II. Displacement of a former community. *Marine Ecol. Progr. Ser.* 66:95-101.

Nordin, C. F. and R. H. Meade. (1981). The flux of organic carbon to the oceans: Some hydrological considerations. Pages 173-218 *in* G. E. Likens, *et al.*, eds., *Flux of Organic Carbon by Rivers to the Oceans*. U.S. Department of Energy, Washington DC.

Ota, A. Y., L. E. Schemel, S. W. Hager, and D. D. Harmon. (1986). *Physical and chemical data for the Sacramento River at Rio Vista, California, November 1983 through November 1984*. U.S. Geol. Surv., Open-File Rept. 86-229. 33 pp.

Peterson, D. H., D. Cayan, J. DiLeo, M. Noble, and M. Dettinger. (1995). The role of climate in estuarine variability. *American Scientist* 83:58-67.

Peterson, D. H., R. E. Smith, S. W. Hager, D. D. Harmon, R. E. Herndon, and L. E. Schemel. (1985). Interannual variability in dissolved inorganic nutrients in Northern San Francisco Bay Estuary. *Hydrobiologia* 129:37-58.

Peterson, D. H., D. R. Cayan, J. F. Festa, F. H. Nichols, R. A. Walters, J. V. Slack, S. W. Hager, and L. E. Schemel. (1989). Climate variability in an estuary: Effects of riverflow on San Francisco Bay. Pages 419-442 *in* D. H. Peterson, ed., *Aspects of Climate Variability in the Pacific and the Western Americas*. Geophysical Monogr. 55. American Geophysical Union, Washington, DC.

Peterson, D. H., M. Noble, and R. E. Smith. (1993). *Suspended Sediment in San Francisco Bay Estuary, California – Recent History and Available Data Sets*. U.S. Geol. Surv., Water-Resources Inv. Rept. 93-4128. 35 pp.

Petts, G. E. (1984). *Impounded Rivers*. John Wiley & Sons. New York. 326 pp.

Phillips, D. J. H. and R. B. Spies. (1988). Chlorinated hydrocarbons in the San Francisco Estuarine System. *Marine Pollution Bull.* 19 (9):445-453.

Porterfield, G., N. L. Hawley, and C. A. Dunnam. (1961). *Fluvial sediments transported by streams tributary to San Francisco Bay Area*. U.S. Geol. Surv. Open-File Rept. 70 pp.

Porterfield, G. (1972). *An Inventory of Published and Unpublished Fluvial-sediment Data for California, 1956-1970*. U.S. Geol. Surv., Open-File Rept. 26 pp.

Porterfield, G., R. D. Busch, and A. O. Waananen. (1978). *Sediment transport in the Feather River, Lake Oroville to Yuba City, California*. U.S. Geol. Surv., Water-Resources Inv. Rept. 78-20. 73 pp.

Porterfield, G. (1980). *Sediment Transport of Streams Tributary to San Francisco, San Pablo, and Suisun Bays, California, 1909-1966*. U.S. Geol. Surv., Open-File Rept. 80-64. 92 pp.

Schemel, L. E. and L. A. Dedini. (1979). *Particulate Organic Carbon in San Francisco Bay, California, 1971-1977*. U.S. Geol. Surv., Open-File Rept. 79-512. 31 pp.

Schemel, L. E. (1983). *Salinity, Alkalinity, and Dissolved and Particulate Organic Carbon in the Sacramento River at Rio Vista, California, and at Other Locations in the Sacramento-San Joaquin Delta*. U.S. Geol. Surv., Water Resources Inv. Rept. 83-4059. 45 pp.

Spiker, E. C. and L. E. Schemel. (1979). Distribution and stable-isotope composition of carbon in San Francisco Bay. Pages 195-212 *in* T. J. Conomos, ed., *San Francisco Bay: The Urbanized Estuary*. Pacific Division, American Association for the Advancement of Science, San Francisco, CA.

Trefry, J. H., S. Metz, T. A. Nelsen, R. P. Trocine, and B. J. Eadie. (1994). Transport of particulate organic carbon by the Mississippi River and its fate in the Gulf of Mexico. *Estuaries* 17(4):839-849.

U.S. Geological Survey. annually, 1979-1986. *Water Resources Data for California*, Volume 4. Water Data Repts. CA-79-4 through 86-4.

Walling, D. E. (1977). Assessing the accuracy of suspended sediment rating curves for a small basin. *Water Resources Res*. 13:531-538.

ECOSYSTEM FUNCTION

DISTRIBUTION AND ACTIVITY OF BACTERIOPLANKTON IN SAN FRANCISCO BAY

JAMES T. HOLLIBAUGH AND PATRICIA S. WONG
Tiburon Center, San Francisco State University
P.O. Box 855, Tiburon CA 94920

We have examined the distribution of bacterioplankton abundance and L-leucine incorporation rate throughout San Francisco Bay over a 3 year period. Bacterioplankton were distributed fairly evenly through the bay and showed little seasonal variation, with the exception of a pronounced response to the spring phytoplankton bloom in South San Francisco Bay. L-leucine incorporation rate was maximal during the spring/summer period and also responded strongly to the spring bloom. Rates were highest at either end of the bay.

Bacterioplankton biomass calculated from abundance estimates averaged 30 mgC/m^3. L-leucine incorporation rates were converted to production estimates using empirical conversion factors. Bacterioplankton production integrated over the one year period from July 1988 to July 1989 averaged 9.3 gC/m^3. The doubling time for the bacterial population averaged 1.2 days and ranged from 0.28 to 6.04 days.

Bacterioplankton production was weakly correlated with phytoplankton biomass in the South Bay, primarily due to the magnitude of the impact of the spring bloom. During the rest of the year it was not related to chlorophyll. A gradient of biomass and production increasing toward the head of the South Bay appeared to be driven by a source of organic matter on the shoals.

In the North Bay, bacterioplankton variables were related to the salinity gradient. Abundance and activity were highest in the Sacramento River. There was an indication of a grazing impact by the introduced Asian clam, *Potamocorbula amurensis*, on bacterioplankton in Suisun Bay.

With the advent of methods for examining their distribution and activity, aerobic, heterotrophic bacteria have been shown to be important components of plankton foodwebs (Azam & Ammerman 1984; Azam *et al.* 1983; Ducklow 1983). Bacterioplankton depend on a supply of reduced carbon for growth and their abundance and production have been shown to be correlated with phytoplankton biomass and primary production over a broad range of marine and freshwater ecosystems (Cole *et al.* 1988; White *et al.* 1991) On an areal basis, bacterial production averaged 30% of primary production over the systems examined (Cole *et al.* 1988).

Although significant effort has been devoted to the study of other foodweb components of San Francisco Bay, bacterioplankton have received little attention. Harvey & Young (1980a, b) studied the contribution of marsh drainage to bacterioplankton biomass in South San Francisco Bay. Hollibaugh (1988) reported on anomalous thymidine metabolism in bacterioplankton from South San Francisco Bay. Hollibaugh (1994) expanded this analysis to include a comparison of San Francisco Bay populations with those from other marine ecosystems and derived generalities concerning the factors controlling thymidine metabolism in bacterioplankton. However, to date there has been no systematic analysis of spatial and temporal variation in biomass and production of San Francisco Bay bacterioplankton.

Primary production in San Francisco Bay is lower than that of other estuaries as a consequence of (1) high turbidity and the resultant shallow photic depth (Cloern *et al.* 1985; Cole & Cloern 1984; Cloern 1987; compare data in Cloern *et al.* 1985 with Nixon 1988) and (2) grazing by benthic filter feeders (Cloern 1982; Nichols 1985; Alpine & Cloern 1992; Werner & Hollibaugh 1993). If San Francisco Bay receives significant organic loading, as suggested by mass balance analyses (Jassby *et al.* 1993), the ratio of bacterioplankton production to primary production could deviate significantly from the average of 30% reported by Cole *et al.* (1988). Specifically, bacterioplankton might assume a much more important role in the San Francisco Bay foodweb than in other systems. An accurate assessment would permit us to determine whether bacterioplankton biomass and production has the potential to be a significant source of particulate organic carbon for planktonic bacteriovores or benthic filter feeders capable of using bacterial C (Werner & Hollibaugh 1993). It would also provide an indirect means of assessing the importance of allochthonous sources of labile organic matter to San Francisco Bay.

Analysis of surveys of thymidine incorporation rates conducted in conjunction with the studies reported in Hollibaugh (1988) and Hollibaugh (1994) has been hampered by methodological constraints. The incorporation of methyl-[^{3}H]-thymidine (me-TdR) into DNA (Fuhrman & Azam 1980, 1982) is a commonly-used method for assessing bacterioplankton production. The method relies on the assumption that me-TdR is not catabolized, an assumption that has been shown to be violated in a number of systems (Robarts & Zohary 1993) and that was found to be particularly a problem in San Francisco Bay (Hollibaugh 1988, 1994). Because of this constraint, we have used L-leucine incorporation rates (Kirchman *et al.* 1985, 1986; Simon & Azam 1989) measured in the same samples to assess bacterioplankton production in San Francisco Bay. We have used an empirical calibration method, rather than relying on conversion factors calculated from first principles, to convert incorporation rates into production estimates.

The purpose of the study presented here was to examine the spatial and temporal distribution of bacterioplankton biomass and production in San Francisco Bay. We also examined the relationship between bacterial activity and suspended particulate material to assess the availability of bacterioplankton biomass to grazers.

METHODS

Sample collection. Most of the data presented here are from analysis of samples collected on longitudinal transects of the San Francisco Bay system (R/V *Polaris*, U.S. Geological Survey, Menlo Park). Station locations are shown in Fig. 1. Sample collection began in July of 1988, near the beginning of a 5-year drought, and continued through July of 1990. Transects were taken monthly or bimonthly during the first year of the study, then at less regular intervals after that with an emphasis on sampling during the South Bay spring bloom period. All samples were collected

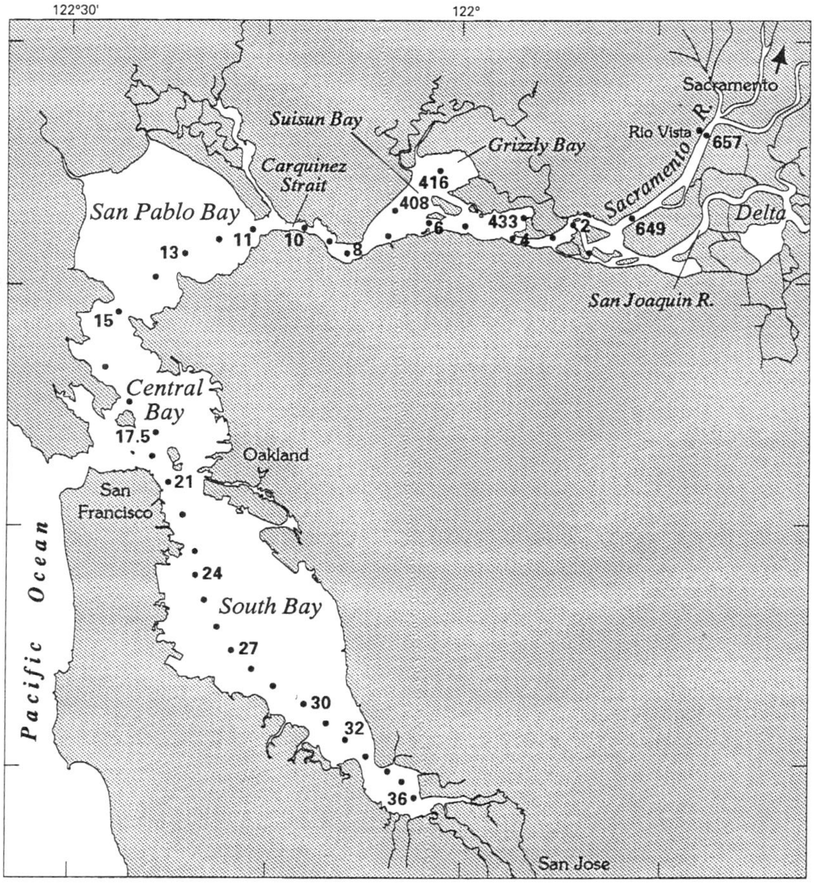

FIGURE 1. San Francisco Bay sampling stations.

during daylight hours. Most samples were taken from the bow, upstream of the hull, with a clean plastic bucket (nominal depth 0.1 m). Bottom water samples were collected occasionally with a 5-liter Niskin sampler. Samples were processed on board the R/V *Polaris* immediately after collection.

San Francisco Bay is generally well-mixed with little surface to bottom variation in properties (Conomos 1979; Walters *et al.* 1985). An exception to this generality occurs during brief periods of salinity stratification in South San Francisco Bay (Cole & Cloern 1984; Cloern 1982) and downstream of the estuarine turbidity maximum in northern San Francisco Bay (Peterson *et al.* 1975). We compared bacterioplankton abundance and production in surface water samples with bottom water samples (collected with a Niskin sampler) and found little difference, except briefly during the South Bay spring phytoplankton bloom when surface rates were less than 2-fold higher than those at a depth of 10 m.

Bacterioplankton Abundance and Biomass. A 20 milliliter subsample was removed from each sample and preserved with 2% (final concentration) filtered, borate buffered formaldehyde and refrigerated until counted, usually within 2 weeks. Bacterioplankton stained with DAPI were enumerated by epi-illuminated fluorescence microscopy at 1000 x magnification (Porter & Feig 1980). At least 10 fields containing at least 20 cells were counted for each sample. Typically only one slide was examined per sample; however, occasionally replicate slides were prepared and enumerated. The coefficient of variation for the count per field was typically 10-15%, and was greater than the coefficient of variation of abundance estimates from triplicate slides (5-10%). Bacterioplankton biomass was calculated from abundance data using a conversion factor of 20 fg C/cell (Lee & Fuhrman 1987). We then calculated mean annual standing crop at each station, weighted for the time between samplings, for the period July 1988 - July 1989.

L-leucine Incorporation Rates. L-leucine incorporation rates were measured as described in Hollibaugh & Wong (1992). Briefly, 4,5- or 3,4,5-[^{3}H]-L-leucine (L-leucine) was added to duplicate 5.0 mL subsamples to a final concentration of 100 nM. This concentration was chosen to minimize isotope dilution (Simon & Azam 1989) and after examining the relationship between incorporation rate and added substrate concentration for San Francisco Bay. Samples were incubated for 1-2 hr at approximately the *in situ* temperature (flowing surface sea water bath). Incorporation rates were constant over this time interval and all L-leucine was incorporated into protein (Hollibaugh & Wong 1992; unpublished data). Incubations were terminated by filtering the sample through a 25 mm dia., 0.45 μm pore-size membrane filter (Millipore HA), then rinsing sequentially with unlabeled sample water, iced 5% (w/v) trichloroacetic acid (TCA) and iced 80% ethanol. Blanks, which were not run for all samples, were either killed with formaldehyde prior to adding the radiolabeled substrate, or filtered immediately after adding substrate. Blanks were typically < 5% of total incorporation in this system, and were not subtracted.

Following the final ethanol rinse, filters were placed in scintillation vials, air dried overnight and dissolved in 1 mL of ethyl acetate. Four milliliters of Aquasol 2 (NEN) was added to the scintillation vial which was then radioassayed using a Beckman 3801 Liquid Scintillation Spectrometer. Counting efficiency was determined with an external standard and the quench correction program operated in the Compton edge shift ("H#") mode.

Empirical Conversion Factors. The relationship between L-leucine incorporation and bacterioplankton production was determined empirically at a number of stations on each cruise using dilution experiments similar to those described previously (Fuhrman & Azam 1982; Kirchman *et al.* 1982; Landry & Hassett 1982; Landry *et al.* 1984; Wright & Coffin 1984; Ducklow & Hill 1985a, b; Riemann *et al.* 1987, and others). We estimated the specific growth rate of cells in the original sample (of necessity all cells are assumed to have the same specific growth rate), then used this specific growth rate and the abundance of bacterioplankton in the original sample to calculate the number of cells produced during the incorporation

TABLE 1. Summary statistics of bacterioplankton variables for regions of San Francisco Bay, 1988-1991.

Variable and Region	*n*	*Range*	*Mean*	*S.E.*	*S.D.*
All Stations					
Abundancec (109 cells/L)	462	0.33-6.87	1.72	0.05	1.01
Biomass (μg C/L)	462	6.6-137	34.4	0.94	20.3
L-leu Incorp (pM/hr)	479	24-2689	314	14.3	314
Specific L-leu (amol/cell*hr)	448	0.02-145	0.21	0.01	0.19
Factor (cells/amol)	140	0.002-1.605	0.262	0.023	0.266
Production (μg C/L/d)	479	4-436	42.3	2.5	49.3
North Bay[1]					
Abundancec (109 cells/L)	169	0.47-3.77	1.68	0.06	0.83
Biomass (μg C/L)	169	9.4-75	33.6	1.28	16.6
L-leu Incorp (pM/hr)	168	64-1352	315	15.5	201
Specific L-leu (amol/cell*hr)	163	0.05-1.45	0.23	0.01	0.19
Factor (cells/amol)	61	0.018-0.923	0.172	0.018	0.141
Production (μg C/L/d)	168	8-111	27.7	1.3	17.0
Central Bay[2]					
Abundancec (109 cells/L)	153	0.46-4.14	1.36	0.05	0.60
Biomass (μg C/L)	153	9.2-83	27.2	0.98	12.0
L-leu Incorp (pM/hr)	150	38-919	212	14.5	178
Specific L-leu (amol/cell*hr)	130	0.04-0.76	0.15	0.01	0.12
Factor (cells/amol)	30	0.042-1.338	0.323	0.050	0.273
Production (μg C/L/d)	150	6-143	32.9	2.3	27.6
South Bay[3]					
Abundancec (109 cells/L)	159	0.33-6.87	2.08	0.10	1.32
Biomass (μg C/L)	159	6.6-137	41.6	2.08	26.3
L-leu Incorp (pM/hr)	180	24-2689	386	32.3	433
Specific L-leu (amol/cell*hr)	155	0.02-1.32	0.23	0.02	0.23
Factor (cells/amol)	49	0.002-1.61	0.338	0.049	0.342
Production (μg C/L/d)	180	4-436	62.5	5.2	70.2

[1] USGS station 8 to Rio Vista, > 46 km upriver of Alcatraz Is.
[2] USGS stations 9-21, < 46 km upriver to < 6 km south of Alcatraz Is.
[3] USGS stations 22-36, > 6 km south of Alcatraz Is.

rate measurement. Finally, we divided the number of cells produced by the amount of label incorporated by subsamples of the original, unaltered sample (performed as described above) to calculate a conversion factor: cells produced per mole incorporated.

Our approach differs somewhat from that used by other investigators (Kirchman *et al.* 1982; Ducklow & Hill 1985a, b; Riemann *et al.* 1987). These investigators measured incorporation of methyl-[^{3}H]-thymidine by the population in the dilution experiment and calculated conversion factors from the relationship between the number of cells produced per unit time and the amount of label incorporated during the same time interval. This approach assumes that thymidine metabolism in the bacterioplankton population does not change during the incubation and that it is the same as in the original sample. We tested this assumption and found that it did not hold for San Francisco Bay populations (Hollibaugh, unpublished data): the partitioning of ^{3}H from me-TdR between DNA, RNA and protein changed during the incubation. Because we measured L-leucine incorporation in the same samples for which thymidine incorporation was measured, we used the same calibration method for both substrates.

The diluent was prepared by filtering a subsample through a 47 mm dia., 0.45 μm pore size Millipore filter into a clean Millipore Sterifil unit at low vacuum (< 20 kPa). Two or three 25-50 mL volumes were filtered and used to rinse the filter, the filtration unit, a graduated cylinder and a clean 125 or 250 mL opaque HDPE screwcap bottle (Nalge). Additional sample was then filtered and used to prepare a 1:9 dilution of an unscreened, unfiltered subsample of the original sample. The dilution was placed in the plastic bottle and incubated at the *in situ* temperature for 24-48 hr. Samples were removed at approximately 6 hr intervals and preserved and enumerated as described above. The abundance estimate used for each time point was the mean of the counts of at least 10 replicate fields of one slide (coefficient of variation ca. 15%).

The specific growth rate, μ, of the bacterioplankton population in the dilution was calculated as the slope of the linear regression of the natural logarithm of abundance versus time. There were 4-6 time points per regression. This growth rate was assumed to represent the growth rate of the population *in situ* in the absence of predation. We rejected all regressions with $r^2 < 0.7$ to eliminate some of the variability due to erratic growth in dilution experiments, decreasing the number of estimates of μ from 216 to 140. However, by rejecting regressions with $r^2 < 0.7$ we have biased the data set of specific growth rate estimates used to calculate factors in favor of higher values of μ. Visual inspection of the plots revealed a few regressions that were linear with small positive slopes that were not statistically significant by the criterion of $r^2 > 0.7$ (*i.e.*, the slope could not be distinguished from 0 by this criterion). Based on inspection of the data, the lowest value of μ that could be detected reliably in the dilution experiments was ca. 0.001 hr^{-1}.

The empirically determined value of μ, the abundance of bacteria in the original, unaltered sample and the assumption of exponential growth were used to calculate the number of cells produced during a specified time interval (1 hr). Conversion factors were calculated by dividing this value by the amount of L-leucine incorporated during the same interval. Based on inspection of the data, conversion factors were averaged over the regions of San Francisco Bay given in Table 1.

Bacterioplankton Production. L-leucine incorporation rates, conversion factors averaged over the station groups given in Table 1 and a carbon conversion factor of 20 fg C/cell (Lee & Fuhrman 1987) were used to calculate bacterioplankton production. Annual production, integrated over the period July 1988 - July 1989, was calculated as follows. For all stations except 2 and 32, production was integrated over the time series using the trapezoidal approximation, then extrapolated to 1 year (365 days). The annual extrapolation for stations 2 and 32, which were missing samples from the summer of 1989, was calculated by assuming that rates during the summer of 1989 were the same as during the summer of 1988. The mean summer rate was substituted for the missing values, then the time series was integrated as above and extrapolated to 1 year. Vertical integrations are based on the station groupings and mean depths for channels and shoals used by Cloern *et al.* (1985) so that bacterioplankton production can be compared to their estimates of phytoplankton primary production for the same regions. Because production rates per unit volume at Suisun Bay shoal stations were similar to values in adjacent channel stations (Table 2), we calculated production on the San Pablo and South Bay shoals (where we have no measurements) using the volumetric rate for channel stations.

TABLE 2. Bacterioplankton Production and Weighted Mean Biomass at San Francisco Bay Stations. For each station: n - number of measurements at a given station during the time series; Days - length of the time series for a given station; Mean Biomass - biomass averaged over the time series and weighted by sampling interval, and; Integrated Production - production integrated over the time series. Columns under the general heading "Annual Mean Biomass and Production" are means for each station group of weighted mean biomass, of integrated production extrapolated to 1 year (365 days), of the mean production for each station group and of vertically integrated annual production for shoals and channels for each station group.

USGS Station	*n*	*Days*	*Weighted Mean Biomass* (mgC/m^3)	*Integrated Production* (gC/m^3)	*Annual Mean Biomass and Production* — *Biomass* (mgC/m^3)	*Production* — *Volume* ($gC/m^3/yr$)	*Production* — *Shoals* ($gC/m^2/yr$)	*Production* — *Channel* ($gC/m^2/yr$)
Suisun Bay Shoals					31	7.4	16	
433	15	356	32	7.6		7.8		
418	10	243	26	4.9		8.6		
416	12	343	37	6.2		6.6		
408.1	12	343	30	6.0		6.3		
Sacramento River/Suisun Bay Channel					30	9.7		107
657	14	355	37	11.2		11.5		
2	14	341	32	12.0		12.9		
4	16	355	32	7.9		8.1		
6	15	355	30	6.1		6.3		
8	15	355	26	9.5		9.8		
10	12	355	24	9.2		9.4		
San Pablo Bay					24	9.0	18	115
11	16	356	24	9.8		10.0		
13	16	356	24	8.8		9.0		
15	16	356	25	7.7		7.9		
Central Bay					26	8.6		
17.5	16	356	25	8.6	26	8.9		
21	21	357	27	8.2		8.4		
South Bay					43	11.8	32	150
24	22	357	30	7.4		7.6		
27	27	357	35	10.0		10.3		
32	18	330	44	15.4		17.7		

Size Fractionation Experiments. We also examined the degree to which actively metabolizing bacterioplankton, as indicated by their ability to incorporate L-leucine or thymidine into macromolecules, were associated with particles. Water samples were collected and incubated with radiolabeled substrates (4,5- [^{3}H]-L-leucine, 5.0 Ci/mmol; methyl-[^{3}H]-thymidine, 20 Ci/mmol; or 6-[^{3}H]-thymidine, 18.6 or 20.1 Ci/mmol) as described above except that 30 mL volumes were incubated in opaque plastic bottles. Incubations were terminated by withdrawing 5.0 mL subsamples with a transfer pipette then filtering the subsamples through 1.0 or 0.2 μm pore size Nuclepore filters. The filters were processed as described in Hollibaugh & Wong (1992) except that they were not dissolved in ethyl acetate. ^3H retained by 1.0 μm pore size filters is taken to be incorporation by large bacteria or by bacteria associated with flocs or other particles larger than 1 μm in diameter. ^3H retained by the 0.2 μm pore size filter is taken to be incorporation by both free-living and particle-associated bacteria.

Temperature, Salinity and Chlorophyll. Data on environmental variables including temperature, salinity and *in vivo* chlorophyll fluorescence were collected by USGS personnel on a continuous basis while underway and during hydrocasts at stations. Fluorescence measurements were converted to estimates of chlorophyll concentration using conversion factors calculated from regressions of fluorescence against the concentration of chlorophyll in discrete samples measured as described in Parsons *et al.* (1984).

Preliminary analysis indicated statistically significant relationships between L-leucine incorporation rate or cell-specific L-leucine incorporation rates and salinity for North Bay stations. However, these variables were also correlated with temperature, which covaries with salinity in this data set. Rates of microbial processes are strongly dependent on temperature with Q_{10} values in the range of 2-3 (Hollibaugh 1979; Li & Dickie 1984; White *et al.* 1991; Shiah & Ducklow 1994a, b). In order to remove the confounding influence of temperature, we followed Li & Dickie's (1984) approach and used the integrated form of the Arrhenius equation:

$$\ln (v_2/v_1) = (E_a/R)\ ((T_2\text{-}T_1)/T_2T_1)); \qquad (1)$$

to adjust rates measured at the *in situ* temperature to rates at 15°C. The apparent activation energy, Ea, was determined empirically by measuring L-leucine incorporation rate at a number of temperatures (range: 8-24°C) then fitting the data to the linearized form of the Arrhenius equation:

$$\ln (v) = (E_a/R) \times (1/T) + \log (A) \qquad (2)$$

as shown in Fig. 2. The experiment was performed a number of times with samples collected at different locations and at different times of the year. There was little seasonal or spatial variation in the apparent activation energy determined in this manner. Equation 1 and the average value of the apparent activation energy determined in this manner (8,032 cal/mole L-leucine incorporated) were then used to transform the data set. Finally, we subdivided the data set into two subsets that roughly correlate with periods of high and low freshwater inflow, and thus residence time, in the North Bay: winter (November through March) and summer (April through October).

RESULTS

Abundance and Biomass. Figure 3a presents the distribution of bacterioplankton through the San Francisco Bay ecosystem and Fig. 4a shows its temporal variation

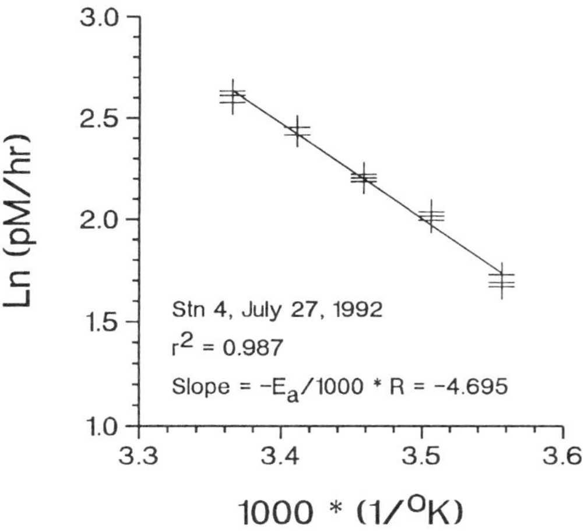

FIGURE 2. Example of regressions used to calculate coefficients for the Arrhenius equation used to adjust measured rates to rates at 15°C.

during the study period. Biomass calculated from abundance ranged between 6.6 and 137 g C/L (Table 1). Biomass was consistently lower in the Central Bay than in either South San Francisco Bay or in Suisun Bay and the Sacramento River (Fig. 3a, Table 1). Bacterioplankton abundance (and thus biomass) in samples from the Suisun Bay shoals (1.63 ± 0.11, $\overline{X}$ ± SE) was not significantly different ($P < 0.05$, 2-tailed Student's *t*-test) from the standing crop at adjacent channel stations (1.63 ± 0.09). Most of the temporal variation in standing crop (Fig. 4a) was due to variability at South San Francisco Bay stations (Fig. 3a). The influence of the South San Francisco Bay spring phytoplankton bloom on the bacterioplankton community is clearly evident in time courses for Stations 27 and 32 (Fig. 5d, e). There was also significant variation in the standing crop of bacterioplankton at North Bay stations (Fig. 3); however, it was not tied to the seasonal cycle (Fig. 5a, b).

The average of the weighted mean standing crops for all stations during the period July 1988-July 1989 was 30 mgC/m^3 (Table 2), similar to the unweighted mean of 34.4 for all stations and all samples given in Table 1. The weighted mean standing crop increased by a factor of 1.5 with distance into the South Bay and from Suisun Bay into the Sacramento River.

L-leucine Incorporation Rate. L-leucine incorporation rates varied over 2 orders of magnitude: from 24 to 2689 pM/hr (Table 1). Rates were highest and most variable in the South Bay and lowest and least variable in the Central Bay (Fig. 3b, Table 1). Temporal variation in L-leucine incorporation rate (Fig. 4b) was dominated by the response of the bacterioplankton community to the South Bay spring phytoplankton bloom (Fig. 3b, 5d, e). There is no obvious cause for the spike in L-leucine incorporation at Station 21 on 18 January, 1990 (deciyear 90.046, Fig. 5c). L-leucine incorporation rate was quite variable at North Bay stations, especially Station 2, but the variability did not reflect the seasonal cycle (Fig. 5a, b). The mean rate for all Suisun Bay shoal samples (288.5 ± 22.4; $\overline{X}$ ± SE) was not significantly different ($P < 0.05$, 2-tailed Student's *t*-test) from those measured in Suisun Bay channel samples (302.7 ± 22.5).

Empirical Conversion Factors. Our estimates of μ from the dilution experiments ranged from 0.0009 to 0.0842 hr^{-1} with a mean of 0.0298 hr^{-1}. There was a great deal of variability in the conversion factors we derived (Table 1, Fig. 6). The mean

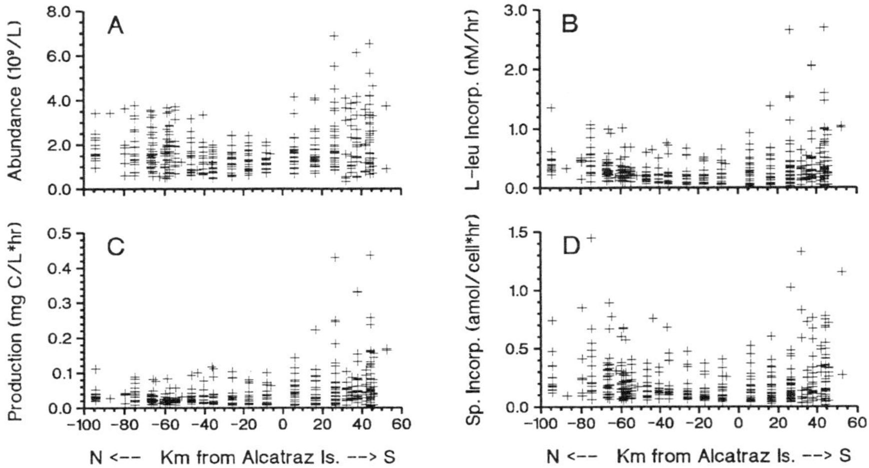

FIGURE 3. Distribution of bacterioplankton variables in San Francisco Bay by distance (in km) along the main shipping channel either north (negative values) or south (positive values) of Alcatraz Island, which is about 4 km due east of the Golden Gate. Panel A – abundance and biomass; panel B – L-leucine incorporation rate; panel C – production; panel; D – cell-specific L-leucine incorporation rate.

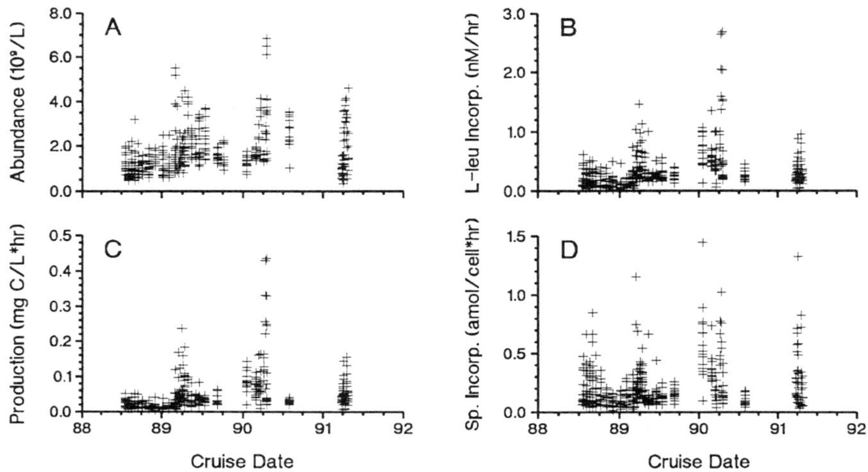

FIGURE 4. Distribution of bacterioplankton variables in San Francisco Bay by sampling date (given in decimal years). Panel A – abundance and biomass; panel B – L-leucine incorporation rate; panel C – production; panel D – cell-specific L-leucine incorporation rate.

factor for North Bay samples was lower than the mean factor for Central or South Bay samples by about 50% (Table 1). The mean value of the factors we determined (0.262 cells/amol L-leu incorporated) is greater than conversion factors derived from first principles (Student's *t*-test, $p < 0.05$, 0.11 and 0.154 cells/amol; Kirch-

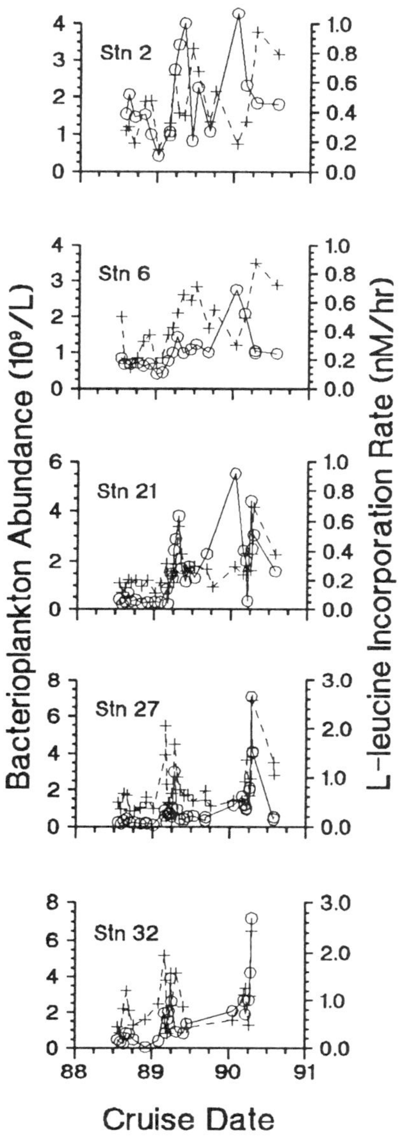

FIGURE 5. Time series of bacterioplankton abundance (+) and L-leucine incorporation rate (o) at 5 representative stations. Station 2 is at -74.91 on Fig. 3, Station 6 is at -57.84, Station 21 is at 6.00, Station 27 is at 26.5, and Station 32 is at 44.28 (or see Fig. 1).

man *et al.* 1986 and Simon & Azam 1989, respectively), but is within the range of empirical conversion factors reported by Kirchman *et al.* (1986) and is less than the median value (0.54 cells/amol) they recommend using.

Bacterioplankton Production. Bacterioplankton production ranged from 4 to 436 μg C/L/d, with a mean rate of 42.3 μg C/L/d (Table 1). Production was greatest and most variable at South Bay stations (Fig. 3c), in large part due to the response of the bacterioplankton community to the annual spring bloom (Fig 4c). Relatively lower rates of production at North Bay stations are due in part to the lower conversion factor determined for these stations and in part to a bias in the data set for sampling during the South Bay spring bloom.

The mean for all stations of annual production integrated over the period July 1988-July 1989 was 9.3 gC/m^3/yr (Table 2), with relatively little variability between stations. Production on Suisun Bay shoal stations was slightly lower than at adjacent channel stations. Production decreased from the Sacramento River into Suisun Bay, reaching a minimum at Station 6. Production increased (doubled) with distance towards the head of the South Bay.

Cell Specific Incorporation Rates. Specific incorporation rates ranged from 0.02 to 1.45 amol/cell × hr with a bay-wide mean value of 0.21. Specific incorporation rates were lowest in the Central Bay (Table 1, Fig 3d). There was pronounced seasonal variation in specific incorporation rate with lowest values seen during winter and highest values during the spring (South Bay) and summer (North Bay) (Fig 4d).

Particle-associated Bacteria. The substrate incorporated into macromolecules by bacteria associated with particles ranged from 10.3 to 67.1% of total sub-

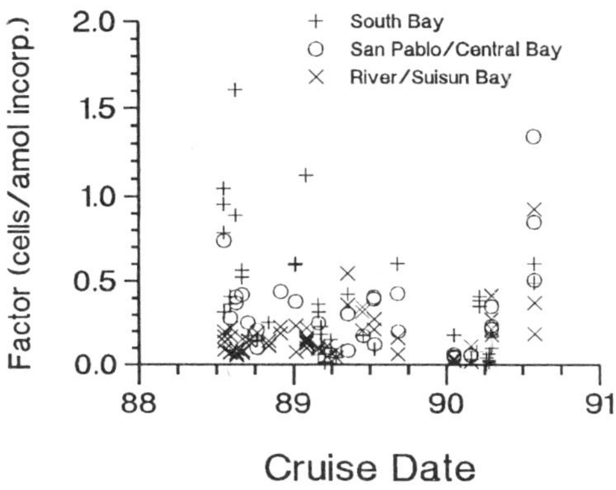

FIGURE 6. Empirical conversion factors calculated for San Francisco Bay bacterioplankton displayed by sampling date in decimal years. The general location of the station at which the sample was collected is also indicated, the stations in these location groups are given in Table 1.

strate incorporated by our samples (Table 3). More incorporation was associated with particles in the North Bay than in the South Bay or Central Bay. Particle-associated bacteria did not exhibit a clear preference for any of the substrates we tested. The percentage of L-leucine incorporation that was associated with particles increased somewhat with increasing L-leucine concentration.

Relationship to Chlorophyll and Salinity. The range of chlorophyll concentrations encountered in North Bay samples was small (mean: 1.6; range: 0.1-8.5) compared to the South Bay (mean: 6.6; range: 1-61.5). Only a limited range of salinities (23.5 to 32.5 PSU) was encountered in South Bay samples. Salinities sampled at North Bay stations ranged from 0 (river water) to 29.2 PSU.

We found no significant relationship between bacterioplankton variables and chlorophyll for samples from the North or Central Bay (Model 2 linear regressions, $p > 0.05$ data not shown). Poor, but significant relationships were found between bacterioplankton abundance or L-leucine incorporation rate and chlorophyll *a* concentration for South Bay samples (Fig. 7a-c). However, these correlations were driven by the spring bloom samples and the significance level of the test ($p < 0.05$) may be too low because of spatial or temporal autocorrelation of the variables. Chlorophyll concentration explained 43% of the variance in abundance and 38% of the variance in L-leucine incorporation rate (Fig 7). The relationship between cell-specific incorporation rate and chlorophyll concentration was not significant.

We found no significant relationships between the bacterioplankton variables (rates transformed to 15°C) and salinity for South Bay stations. We found statistically significant correlations (but see above) between all bacterioplankton variables and salinity for winter samples from the North Bay, with all variables inversely related to salinity (Fig 8). The relationship between abundance and salinity was linear (Fig 8a) and explained 30% of the variance in abundance. The relationships between L-leucine incorporation rate and cell-specific L-leucine incorporation rate and salinity were more complicated. Both distributions had lower thresholds that increased more or less linearly as salinity decreased (Fig 8b, c). Linear regressions of these data sets had slopes that were steeper than the lower threshold, indicating greater variability at the low salinity end of the distribution. The regressions explained 20% and 14% of the variance in L-leucine incorporation rate and cell-specific L-leucine incorporation rate, respectively.

The relationship between these variables and salinity was similar during the

TABLE 3. Substrate incorporation by particle-associated versus free-living bacterioplankton in San Francisco Bay. Station numbers are USGS stations, the "TCES" sample was collected off the dock at the Tiburon Center, SFSU. Substrates were 4,5-[^{3}H]-L-leucine (L-leu), methyl-[^{3}H]-thymidine (Me-TdR) and 6-[^{3}H]-thymidine (6-TdR). Mean (X) and standard deviation of the sample mean (SD) for three replicates (2 replicates where SD not shown) are given. The percentage of total incorporation attributed to particle-associated bacteria (Part Assoc) is the ratio of ^{3}H retained by the 1.0 µm pore size filter to ^{3}H retained by the 0.2 µm filter, expressed as a percentage. The chlorophyll *a* concentration (Chl*a*) and turbidity (expressed as an extinction coefficient, K_Z) of the water sampled are also given.

				Filter Pore Size (µm)						
				1.0		*0.2*				
Sampling Date	*Stn*	*Subst*	*Conc (nM)*	$\overline{X}$	*SD*	$\overline{X}$	*SD*	*Part Assoc (%)*	*Chla (µg/L)*	K_z (m^{-1})
Sept. 6, 1989	27	L-leu	100	30	2.0	76	2.6	39		
	2	L-leu	100	57	6.5	112	4.7	51		
Sept. 25, 1989	TCES	L-leu	100	29	1.0	150	6.0	19		
		L-leu	10	8.5	0.1	66	1.5	13		
		L-leu	1	1.2	0.1	12	0.8	10		
Oct. 4, 1989	27	L-leu	100	18	1.7	48	2.1	37	2.8	0.77
		L-leu	10	4.4		19		24		
		L-leu	1	0.6	0.06	3.7	0.10	17		
	2	L-leu	100	49	3.1	118	4.4	41	1.7	1.82
		L-leu	10	10		26		37		
		L-leu	1	1.0	0.08	2.6	0.06	36		
Jan. 17, 1990	27	Me-TdR	10	0.8		3.2	0.01	25	1.7	1.48
		6-TdR	10	0.6	0.06	2.7	0.01	23		
	11	Me-TdR	10	2.5	0.18	6.8		37	0.6	4.03
		6-TdR	10	2.0	0.07	4.8	0.06	42		
	4	Me-TdR	10	5.8		8.7	0.11	67	0.9	3.98
		6-TdR	10	3.4	0.16	5.4	0.29	63		
Mar. 20, 1990	32	Me-TdR	10	9.2	0.5	37	2.1	25	10.9	2.20
		6-TdR	10	5.7	0.2	29	1.5	19		
		L-leu	100	184	4	512	12	36		
Apr. 12, 1990	32	Me-TdR	10	3.2	0.15	22	2.8	15	17.4	2.58
		6-TdR	10	2.8	0.4	20	1.0	14		
		L-leu	100	201	7	810	37	25		
Apr. 18. 1990	32	Me-TdR	10	13	0.1	90	6.1	14	61.5	1.61
		6-TdR	10	0.2	0.06	0.5		50		
		L-leu	100	646	29	1672	90	39		

summer (Fig 9). There was more scatter in the abundance data (the relationship is no longer statistically significant, Fig 9a). There is less variability in the L-leucine incorporation rate distribution and the slopes of the regression line and of the lower threshold of the data are comparable (Fig 9b). Cell-specific incorporation rates are generally lower during the summer and the lower threshold for the distribution does not vary with salinity, although the linear regression of the data set is significant (Fig 9c).

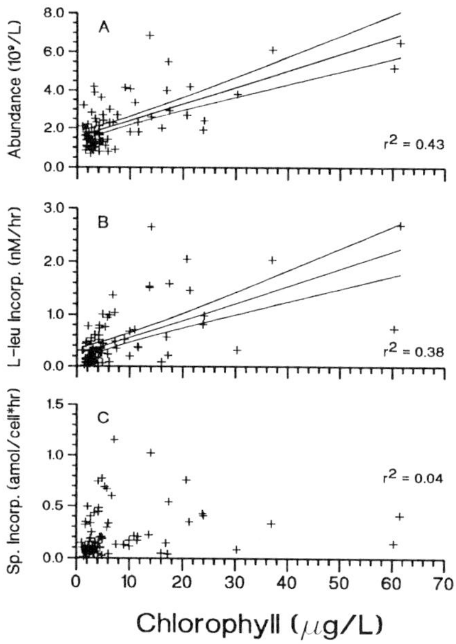

FIGURE 7. Relationship between bacterioplankton variables and chlorophyll concentration for South San Francisco Bay stations (see Table 1). Lines are regression line and 95% confidence interval. Panel A – abundance; panel B – L-leucine incorporation rate; panel C – cell-specific L-leucine incorporation rate.

DISCUSSION

Empirical Conversion Factors. Accurate and precise conversion factors are essential to interpreting activity measurements (*i.e.* L-leucine incorporation rate) in terms of bacterioplankton production. Unfortunately it is exceedingly difficult to verify that the factors are accurate, and they are rarely precise. We are forced to assume that the growth rates measured in dilution cultures reflect the growth rate of the undisturbed population *in situ*. If this assumption is not valid, it could generate a systematic bias in the factor (too large or too small). However, variability in the conversion factors, both natural and as an artifact of the experiment, is probably a more serious problem, and one that has been encountered by others (Kirchman *et al.* 1986; Kirchman & Hoch 1988; Coveney & Wetzel 1988; Cole *et al.* 1989; Simon & Azam 1989). The variability in conversion factors is in general one of the major shortcomings of current methods for estimating bacterial production (Robarts & Zohary 1993).

Some of the variability we encountered is due to inherent spatial and temporal variability in the physiology of San Francisco Bay bacterioplankton. This reflected in variability of the relative rates of L-leucine and methyl-[^{3}H]-thymidine incorporation (data not shown) or of methyl- and 6-[^{3}H]-thymidine incorporation (Hollibaugh 1994). We have also observed that the metabolism of San Francisco Bay bacterioplankton changes during incubation in dilution experiments, at least with regard to the metabolic fate of ^{3}H supplied as methyl-[^{3}H]-thymidine (unpublished data). Similar changes during prolonged incubations, including changes in mean cell volume, the percentage of "active" cells and species composition, have been reported by others (Ferguson *et al.* 1984; Lee & Fuhrman 1991) and may affect conversion factor calculation (Kirchman & Hoch 1988; Ducklow & Hill 1985b).

Another source of variability arises from experimental error, primarily from estimating specific growth rates (μ) with dilution cultures. One source of error was non-linear or erratic growth in the dilution experiments. We attempted to address this problem by eliminating regressions with low r^2 values from the data set used to calculate conversion factors. This improved the data set, but much variability

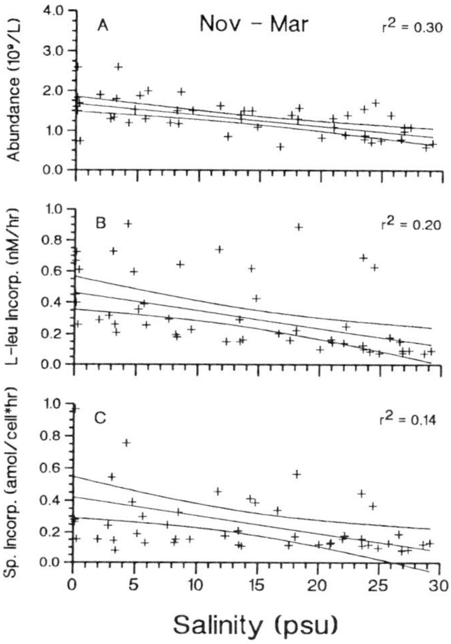

FIGURE 8. Relationship between bacterioplankton variables and salinity for Northern San Francisco Bay stations (see Table 1) during the winter. Lines are regression line and 95% confidence interval. Panel A – abundance; panel B – L-leucine incorporation rate adjusted to the rate at 15°C; panel C – cell-specific L-leucine incorporation rate adjusted to the rate at 15°C.

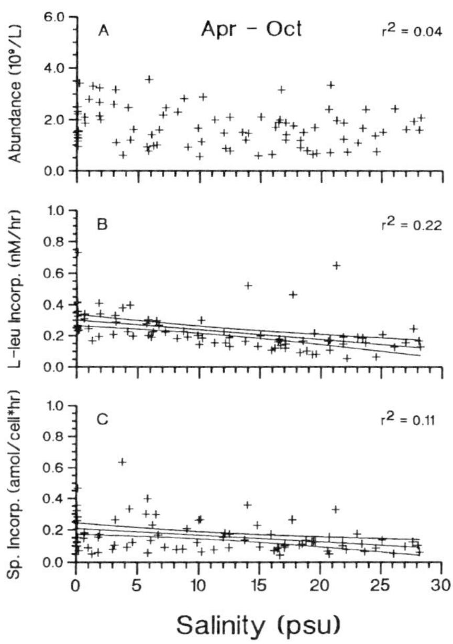

Figure 9. Relationship between bacterioplankton variables and salinity for Northern San Francisco Bay stations (see Table 1) during the summer. Lines are regression line and 95% confidence interval. Panel A – abundance; panel B – L-leucine incorporation rate adjusted to the rate at 15°C; panel C – cell-specific L-leucine incorporation rate adjusted to the rate at 15°C.

remained. We chose to estimate conversion factors by averaging over fairly large regions of San Francisco Bay. These regions were chosen primarily on the basis of environmental similarities; however, upon inspection the groupings were consistent with the distribution of conversion factors.

Because of the uncertainty in conversion factors, we have focused on primary variables (abundance, L-leucine incorporation rate, cell-specific L-leucine incorporation rate) in our attempts to understand the factors controlling bacterioplankton population dynamics in San Francisco Bay. We recognize the need to understand the role of bacterioplankton in the trophic structure of San Francisco Bay, so we have used the conversion factors to generate estimates of bacterioplankton production. These estimates should be interpreted with caution, given the problems with deriving conversion factors.

Annual Production. On an areal basis, bacterioplankton production in San Francisco Bay from July 1988 to July 1989 (Table 2) was comparable to the phytoplankton production rates for 1980 reported by Cloern *et al.* (1985). However, due to the impact of *Potamocorbula amurensis* on the phytoplankton population, bacterioplankton production in Suisun Bay is now approximately 5-fold greater than post-1986 rates for Suisun Bay reported by Alpine & Cloern (1992).

Our data suggest that production per unit volume was somewhat lower on the shoals of Suisun Bay than in the channel (Table 2). Furthermore, production reached a minimum in Suisun Bay at Station 6, located in the shipping channel amidst the sand flats of central Suisun Bay. This pattern of decreased production might result from grazing mortality caused by *Potamocorbula amurensis*, which were abundant on the shoals of Suisun Bay at this time. However standing crops do not appear to be depressed at these stations, which would be expected if clam grazing were having a significant impact (Alpine & Cloern 1992; Werner & Hollibaugh 1993). It is difficult to enumerate samples from this region of San Francisco Bay because of the high suspended particle load which increases counting error and might masked real differences in abundance. Alternatively, the production pattern may be a secondary effect of clam predation: clams may be out-competing bacteria for carbon by grazing phytoplankton or POC. Unfortunately, we do not have enough data from Suisun Bay prior to the *Potamocorbula amurensis* invasion for a direct analysis of the clam's impact on bacterioplankton.

We also observed a gradient of production in South San Francisco Bay, increasing toward the head of the bay. The production gradient was accompanied by a gradient in standing crop (Table 2). These observations suggest an important source of labile organic matter at the head of the South Bay. The gradient was still evident in the distribution of average biomass and productivity when samples from the spring bloom period (months of March and April) were excluded from the average (Table 4). The gradient of bacterioplankton productivity and biomass are clear, while it is equally obvious that there is no accompanying gradient of phytoplankton biomass. This suggests that bacterioplankton are growing on organic matter generated on the surrounding mudflats (phytoplankton, benthic microalgae or marshes), or on allochthonous C supplied by freshwater (primarily treated wastewater, Conomos 1979) discharged into the head of South San Francisco Bay. The weak salinity gradient suggests that the influence of freshwater discharge is minimal. The increase in the cell-specific L-leucine incorporation rate at Station 32 argues that the quality of growth substrates as well as their supply rate is elevated at the head of the South Bay. It also suggests that the distribution of productivity and biomass in the South Bay is not due to elevated grazing rates, dilution or tidal mixing further down the bay.

Cell Specific L-leucine Incorporation Rate. Bacterioplankton specific growth rates can be calculated from production and biomass data. We found that specific growth rates calculated this way were not correlated ($p > 0.05$) with the specific growth rates determined in dilution experiments. This is not surprising given the assumptions, variability and potential errors inherent in both measurements. As a consequence, we have used the cell-specific rate of L-leucine incorporation as a surrogate for specific growth rate to examine spatial and temporal variation in the activity of bacterioplankton in San Francisco Bay.

The cell-specific L-leucine incorporation rate should vary only as a function of the specific growth rate of the bacterioplankton population, the average protein content of the cells, and isotope dilution. The protein content of the cells is weakly dependent on cell size (Simon & Azam 1989), which is relatively constant for San Francisco Bay populations (*ca.* 0.7 μm diameter cocci). We have minimized variation in isotope dilution due to variation in the *in situ* concentration of L-leucine by adding ^{3}H-labeled L-leucine to concentrations (100 nM) that are much higher than *in situ* (typically 1-10 nM). These concentrations should also repress endogenous synthesis of L-leucine by bacterioplankton and thus minimize, or at least minimize the variability in, internal isotope dilution (Kirchman *et al.* 1986; Simon & Azam 1989). Thus the cell-specific rate of L-leucine incorporation should be the most direct indicator of the activity of bacterioplankton populations. Assuming the population is growing exponentially, the cell-specific incorporation rate is related to the specific growth rate of the bacterioplankton population by the equation:

$$\mu = \ln [1 + (I \times F)], \qquad (3)$$

where I is the cell-specific incorporation rate and F is the conversion factor.

Environmental Factors Controlling Bacterioplankton. To identify the relative importance of internal versus external control of the bacterioplankton community in San Francisco Bay, we examined the relationships between the primary bacterioplankton variables we measured (abundance, L-leucine incorporation and cell-specific L-leucine incorporation rate) and chlorophyll and salinity. Chlorophyll was used because Cole & Cloern (1984) have shown that primary production is directly related to phytoplankton standing crop in San Francisco Bay and Cole *et al.* (1988) and White *et al.* (1991) have shown that there is a strong correlation between bacterioplankton production and phytoplankton biomass in other systems. Salinity was selected because, given San Francisco Bay's current low primary production, we hypothesized a relationship between bacterioplankton variables and allochthonous carbon sources. An analysis of the carbon budget of Northern San Francisco Bay by Jassby *et al.* (1993) demonstrated that allochthonous C supplied by river inflow is a major component of the budget.

Regression analysis indicated that the South San Francisco Bay bacterioplankton community responded to changes in phytoplankton biomass, but the correlation was not strong (r^2 = 0.43; Fig 7). This weak correlation was obtained despite a very clear response of the bacterial community to the South Bay spring bloom (Figs 3, 4 and 5d, e) and is somewhat puzzling. A possible explanation is that the bacterial response is really driven by primary production, which is correlated indirectly with biomass. Alternatively, as suggested above, production on the South Bay shoals significantly influences the bacterioplankton community in the channel. The production cycle on the shoals and in the channel may be linked by physical factors so that they tend to covary but are not tightly linked. Chin-Leo & Benner (1992) evoke a similar explanation for the lack of correlation between chlorophyll and bacterioplankton variables in the Mississippi River plume, while Shiah & Ducklow (1994a) attributed the poor correlation to the control of bacterial (but not phytoplankton) process rates by the strong seasonal temperature variation in Chesapeake Bay and to lags imposed by the processing time of phytodetritus in the bottom water of highly stratified Chesapeake Bay.

In contrast to the South Bay, the North Bay displayed a clear, if complicated response to freshwater inflow. During winter the relationship between abundance and salinity showed no deviation from linearity, suggesting simple conservative mixing (Fig 8a). This implies that no net production or consumption of bacterioplankton biomass occurs in the North Bay during winter, possibly as a result of low temperatures and short water residence times. We saw no relationship between abundance and salinity during summer (Fig 9a) when the ratio of bacterioplankton turnover time to residence time was small.

We offer the following speculative interpretation of the complicated relationships between L-leucine incorporation rates or cell-specific L-leucine incorporation rates and salinity. These distributions may be decomposed into two distributions: a set of points along a lower threshold; and excursions above this threshold that become more frequent and pronounced at low salinities (Figs 8b, c; 9b, c). The lower threshold seems to fall along a conservative mixing line, suggesting that this relationship may be driven physically. The simplest explanation for the distribution of excursions above the lower threshold is variability in the low salinity source term propagating down the salinity gradient. Because these variables are rates and not concentrations, we infer that the bacterioplankton community is responding to periodic pulses of readily labile organic material delivered with freshwater, superimposed on a background of slower, more constant growth on relatively less readily available organic matter or on organic matter produced at a limited rate *in situ*, *i.e.* phytoplankton production. The slope and linearity of the lower threshold of the L-leucine incorporation rate distribution suggests that the refractory material is more abundant at the freshwater end of the gradient and that its concentration decreases more as a result of simple dilution than as a result of consumption by bacterioplankton. Chin-Leo & Benner (1992) offer a similar explanation for the distribution of bacterial production and growth rate with salinity in the Mississippi River plume.

The slope and linearity of the lower threshold of the cell-specific L-leucine incorporation rate (an index of specific growth rate) distribution during winter (Fig 9c) suggests that there is consistently more high quality, labile organic material in freshwater than in coastal salt water during this season. During the summer, the lower threshold of this distribution does not vary with salinity, suggesting that material fueling bacterial growth at this lower threshold is of equal quality and quantity throughout the North Bay. The generally lower cell-specific L-leucine incorporation rates during the summer suggest that there is less labile organic matter available to the North Bay bacterioplankton population during summer than during winter.

The distribution of bacterioplankton variables we saw in San Francisco Bay differs from distributions reported for other estuarine systems where there is typically a mid-estuary maximum of bacterioplankton abundance and productivity (Wright 1978; Valdes & Albright 1981; Albright 1983; Wright & Coffin 1984; Kirchman & Hoch 1988; Chin-Leo & Benner 1992). These estuarine systems also had mid-estuary maxima in chlorophyll concentration and, presumably, primary production, which provides a significant source of organic C that is more labile than allochthonous C transported into the estuary with freshwater.

Importance of Particle-Associated Bacteria. Although the biomass and production of bacterioplankton in San Francisco Bay are low compared to other systems (compare our data to the summaries in Ducklow & Shiah 1993), bacterioplankton may be of trophic significance due to the depressed phytoplankton production. A critical factor determining trophic significance of bacterioplankton is the ability of other organisms to consume them. If bacterioplankton biomass must be processed through a number of steps in the "microbial loop" (Azam *et al.* 1983), before it is available to higher trophic levels such as zooplankton, very little energy will be transferred to these trophic levels (Pomeroy 1974, 1979, 1980; Ducklow 1983; Ducklow *et al.* 1986). If higher trophic levels were able to graze directly on bacterioplankton, the efficiency of energy transfer would be much higher. It now appears that, with few exceptions (Peterson & Hobbie 1978, King *et al.* 1980, Wright *et al.* 1982), metazoan filter feeders are not able to retain free-living bacterioplankton effectively. However, bacteria associated with particles are available to these grazers. Thus, the partitioning of bacterial biomass between free-living, single cells and cells associated with aggregates or large particles has great ecological significance.

The results of size fraction experiments we conducted suggest that a significant fraction of the bacterioplankton in San Francisco Bay are associated with particles. However, in interpreting these data it must be remembered that we measured the distribution of bacterial activity, not of biomass. Our attempts to quantify directly by epifluorescence microscopy the percentage of bacterial biomass that is associated with particles were not successful due to interference by the particles themselves. The following discussion evaluates the constraints on inferring the partitioning of biomass between free-living and particle-associated populations from the partitioning of ^{3}H incorporation.

The partitioning of biomass will only be the same as the partitioning of ^{3}H incorporation if the biomass-specific incorporation rate of a substrate is the same for both particle-associated and free-living bacteria. For this to happen, the two populations must metabolize the substrate by similar metabolic pathways so that the net assimilation efficiency of the ^{3}H is similar for both populations. In addition, isotope dilution must be the same for both populations, *i.e.* the pools of naturally occurring substrate and the added, radiolabeled substrate must equilibrate rapidly with respect to the length of the incubation and achieve the same final specific activity.

We tested to see whether differences in isotope dilution influenced the partitioning of ^{3}H incorporation between particle-associated and free-living bacteria by determining how the partitioning changed as the concentration of added substrate increased. We found that the relative incorporation by particle- associated bacteria decreased (Table 3). This would be expected if the immediate environment of particle associated-bacteria is isolated from the bulk phase water by a diffusion barrier, or if natural substrate concentrations are elevated in the interior of particles (Alldredge 1979; Alldredge & Cox 1982; Logan & Hunt 1987). Both of these factors have been demonstrated to be important for marine aggregates. Alternatively, high substrate concentrations may enhance the relative importance of uptake and incorporation by eukaryots that are retained by the 1 μm pore size filter. However, the proportion of particle-associated incorporation in our experiments did not increase dramatically during the South Bay spring bloom when eukaryotic biomass (chlorophyll *a*) increased by over an order of magnitude (Table 3). Regardless of the cause, the difference in partitioning between free-living and particle-associated bacteria changed by a factor of no more than 2 when L-leucine concentrations were increased over 2 orders of magnitude. We conclude that isotope dilution effects are relatively minor.

The observation that the relative partitioning of the three substrates we used tends to covary in most samples suggests that the particle-associated and free-living bacteria are physiologically similar. This is somewhat surprising as there is clear evidence from work on marine aggregates that particle-associated bacteria are physiologically and taxonomically distinct from their free-living counterparts (Simon 1988; Smith *et al.* 1992; DeLong *et al.* 1994). While this conclusion needs to be verified with further experimentation, the data we have collected to date suggest that these populations are not physiologically distinct in San Francisco Bay.

Because we have not been able to obtain reliable counts of particle-associated bacteria in San Francisco Bay samples, we cannot directly evaluate whether or not the cell-specific incorporation rates are the same for particle-associated and free-living bacteria. Some reports indicate that particles are favorable environments for bacterial growth, while other reports conclude that particle-associated bacteria may be at a disadvantage compared to their free-living counterparts (Alldredge *et al.* 1986; Iriberri *et al.* 1987; Karl *et al.* 1988; Simon *et al.* 1990; Smith *et al.* 1992). The quality of particles changes depending on composition and age. San Francisco Bay particles are composed of clay particles held together by an organic matrix interspersed with bacteria. Bacteria of the free-living and particle-associated populations appear qualitatively the same — same size, same morphology, etc. — upon visual inspection by epifluorescence microscopy. The particles are not thickly colonized with bacteria and large cells, microcolonies or long filaments are rare, (comparable to their frequency in the free-living population, where they are also rare), suggesting that growth on San Francisco Bay particles is not dramatically enhanced over growth in the surrounding water. Thus, we expect the cell-specific incorporation rates to be approximately the same in both environments.

Finally, particles or flocs such as those found in San Francisco Bay are aggregates of smaller particles and organic matter. These may be disrupted and loosely associated bacteria may be dislodged by handling (pipette shear, mixing) while setting up the experiment. This would cause us to underestimate the fraction of the population that is associated with particles.

Although there are numerous caveats, as discussed above, we are reasonably confident that the partitioning of ^{3}H incorporation between particle-associated and free-living populations in San Francisco Bay reflects the partitioning of bacterial biomass between these environments. The most obvious feature of the distribution of particle-associated bacteria is that we found higher percentages of particle-associated ^3H incorporation in the more turbid waters of Northern San Francisco Bay (Table 2). This is consistent with the pattern reported for other systems (reviewed in Iriberri *et al.* 1987; Yoon & Rosson 1990).

Bacterial C Demand. One consequence of bacterioplankton production is a demand for labile, reduced carbon. This demand is inversely related to net growth efficiency. The net growth efficiency of bacterioplankton appears to vary between fairly wide limits, depending on carbon source. A net growth efficiency of 33%, frequently assumed for calculating bacterioplankton carbon demand, yields a carbon demand three times bacterioplankton production. For our samples, this demand would be about 30 gC/m^3/yr, or 60 gC/m^2/yr for a 2 m water column and 300 gC/m^2/yr for a 10 m water column. This calculation suggests that bacterioplankton are significant to San Francisco Bay carbon fluxes. Lower net growth efficiencies in the range of 10-40% have recently been reported for other systems (Kirchman *et al.* 1992, North Atlantic Spring Bloom; Chin-Leo & Benner 1992, Mississippi River plume; Biddanda *et al.* 1994, Louisiana shelf; Carlson & Ducklow, in press, Sargasso Sea). A lower net growth efficiency implies increased bacterioplankton C demand: with a net growth efficiency of 20%, the C demand of our samples is about 50 gC/m^3/yr, or 100 gC/m^2/yr and 500 gC/m^2/yr for 2 and 10 m water columns, respectively.

Comparison with Other Biota. To illustrate this point further, it is instructive to compare bacteria to three other, more well known, members of the San Francisco Bay biota. The first is "Humphrey the Whale," a young California Gray whale that visited San Francisco Bay briefly in 1985 and 1986. Humphrey was 45 ft. long and weighed approximately 40 metric tons at the time that he visited the Bay (Tokuda & Hall 1992). Assuming a dry weight to wet weight ratio of 0.2 and a carbon to dry weight ratio of 0.5, Humphrey's biomass was approximately 4,000 kg C.

The mean biomass of a bacterioplankton cell is about 20 fg C (Lee & Fuhrman 1987), and there are about $1.72 \times 10^{12}/m^3$ in San Francisco Bay (Table 1). Conomos *et al.* (1985) give the volume of San Francisco Bay at mean lower low water as 6.66×10^9 m^3 with a tidal prism of 1.6×10^9 m^3. The biomass of bacterioplankton in San Francisco Bay at high tide is thus about 230,000 kg C, or 57.5 Humphrey units.

Similar calculations for delta smelt (using an average wet weight of 20 g and a dry to wet weight conversion factor of 0.15) and phytoplankton (using a C to chlorophyll ratio of 55 from Weinke & Cloern 1987, and a mean chlorophyll concentration of 4 μg/L) indicates that the biomass of bacterioplankton in San Francisco Bay is the equivalent of 1.5×10^8 Delta Smelt and 16% of the phytoplankton biomass. And unlike either Humphrey or Delta Smelt, the mean doubling time of bacterioplankton in San Francisco Bay (calculated using Equation 3) is 1.2 days (range 0.28-6.04 days).

Bacterioplankton also play a significant role in the transformation of xenobiotic compounds, metals, etc. Unlike the detrital material that comprises the bulk of suspended particulate material in San Francisco Bay, the surface of bacterioplankton is biochemically active. The bacterial cell wall contains numerous sites for the adsorption of organic and inorganic solutes. The periplasmic space and plasma membrane contain enzymes that perform many biochemical functions including hydrolyzing polymers, facilitating redox reactions and concentrating solutes in the interior of bacterioplankton cells. The surface area of a typical coccoidal bacterioplankton cell 0.7 m in diameter is 1.54×10^{-12} m^2. The area of bacterioplankton membrane in one cubic meter of San Francisco Bay water is 2.65 m^2. At high tide, the total surface area of bacterioplankton membrane in San Francisco Bay is 18 times the surface area of San Francisco Bay, including mudflats.

ACKNOWLEDGEMENTS

Most of the financial support for this work came from the National Science Foundation (NSF OCE 87-16469). Some of our work in South San Francisco Bay was supported by the San Francisco Foundation, and the State of California Department of Water Resources supported some of our work in Suisun Bay. This work could not have been accomplished without the cooperation of the San Francisco Bay project, U.S. Geological Survey. In addition to supplying ship-time on the R/V *Polaris*, USGS scientists provided most of the environmental data presented here. We also thank Alan Jassby for comments on a draft of the manuscript.

LITERATURE CITED

Albright, L. J. (1983). Influence of river-ocean plumes upon bacterioplankton production of the Straits of Georgia, British Columbia. *Marine Ecol. Progr. Ser.* 12:107-113.

Alldredge, A. L. (1979). The chemical composition of macroscopic aggregates in two neritic seas. *Limnol. Oceanogr.* 24: 855-866.

Alldredge, A. L., and J. L. Cox. (1982). Primary productivity and chemical composition of marine snow in surface waters of the Southern California Bight. *Jour. Marine Res.* 40:517-527.

Alldredge, A. L., J. J. Cole, and D. A. Caron. (1986). Production of heterotrophic bacteria inhabiting macroscopic organic aggregates (marine snow) from surface waters. *Limnol. Oceanogr.* 31:68-78.

Alpine, A. E., and J. E. Cloern. (1992). Trophic interactions and direct physical effects control phytoplankton biomass and production in an estuary. *Limnol. Oceanogr.* 37:946-955.

Azam, F., T. Fenchel, J. G. Field, J. S. Gray, L. A. Meyer-Riel, and F. Thingstad. (1983). Ecologic role of water column microbes in the sea. *Marine Ecol. Progr. Ser.* 10: 257-263.

Azam, F., and J. W. Ammerman. (1984). The cycling of organic matter by bacterioplankton in pelagic marine ecosystems: Microenvironmental considerations. Pages 345-360 *in* M. J. R. Fasham, ed., *Flows of Energy and Materials in Marine Ecosystems: Theory and Practice.* NATO Conference Series IV: Marine Sciences, Vol. 13. Plenum Press, New York.

Biddanda, B., S. Opsahl, and R. Benner. (1994). Plankton respiration and carbon flux through bacterioplankton on the Louisiana shelf. *Limnol. Oceanogr.* 39:1259-1275.

Carlson, C. A., and H. W. Ducklow. (in press). Growth of bacterioplankton and consumption of dissolved organic carbon in the oligotrophic Sargasso Sea. *Aquatic Microbial Ecol.*.

Chin-Leo, G., and R. Benner. (1992). Enhanced bacterioplankton production and respiration at intermediate salinities in the Mississippi River plume. *Marine Ecol. Progr. Ser.* 87:87-103.

Cloern, J. E. (1982). Does the benthos control phytoplankton biomass in South San Francisco Bay? *Marine Ecol. Progr. Ser.* 9:191-202.

Cloern, J. E., B. E. Cole, R. L. J. Wong, and A. E. Alpine. (1985). Temporal dynamics of estuarine phytoplankton: A case study of San Francisco Bay. *Hydrobiologia* 129:152-176.

Cloern, J. E. (1987). Turbidity as a control on phytoplankton biomass and productivity in estuaries. *Continental Shelf Res.* 7:1367-1381.

Cole, B. E., and J. E. Cloern. (1984). Significance of biomass and light availability to phytoplankton productivity in San Francisco Bay. *Marine Ecol. Progr. Ser.* 17:15-24.

Cole, J. J., S. Findlay, and M. L. Pace. (1988). Bacterial production in fresh and saltwater ecosystems: a cross system overview. *Marine Ecol. Progr. Ser.* 43:1-10.

Cole, J. J., N. F. Caracao, D. L. Strayer and C. Ochs. (1989). Bacterial production in freshwaters: Ecosystem level constraints on conversion factors. *Limnol. Oceanogr.* 34:286-296.

Conomos, T. J. (1979). Properties and circulation of San Francisco Bay waters. Pages 47-84 *in* T. J. Conomos, ed., *San Francisco Bay: The Urbanized Estuary.* Pacific Division, American Association for the Advancement of Science, San Francisco, CA.

Conomos, T. J., R. E. Smith, and J. W. Gartner. (1985). Environmental setting of San Francisco Bay. *Hydrobiologia* 129:1-12.

Coveney, M. F., and R. G. Wetzel. (1988). Experimental evaluation of conversion factors for [^{3}H]thymidine incorporation assay of bacterial secondary production. *Appl. Environ. Microbiol.* 54:2018-2026.

DeLong, E. F., D. G. Franks, and A. L. Alldredge. (1993). Phylogenetic diversity of aggregate-attached vs free-living marine bacterial assemblages. *Limnol. Oceanogr.* 38:924-934.

Ducklow, H. W. (1983). Production and fate of bacteria in the oceans. *BioScience* 33:494-499.

Ducklow, H. W., and S. M. Hill. (1985a). The growth of heterotrophic bacteria in the surface waters of warm core rings. *Limnol. Oceanogr.* 30:239-259.

Ducklow, H. W., and S. M. Hill. (1985b). Tritiated thymidine incorporation and the growth of heterotrophic bacteria in warm core rings. *Limnol. Oceanogr.* 30:260-272.

Ducklow, H. W., D. A. Purdie, P. J. LeB. Williams, and J. M. Davies. (1986). Bacterioplankton: A sink for carbon in a coastal plankton community. *Science* 232:865-867.

Ducklow, H. W., and F.-K. Shiah. (1993). Chapter 11. Bacterial production in estuaries. Pages 261-288 *in* T. Ford, ed., *Aquatic Microbiology: An Ecological Approach.* Blackwell, Cambridge, MA.

Ferguson, R. L., E. N. Buckley, and A. V. Palumbo. (1984). Response of marine bacterioplankton to differential filtration and confinement. *Appl. Environ. Microbiol.* 47:49-55.

Fuhrman, J. A., and F. Azam. (1980). Bacterioplankton secondary secondary production estimates for coastal waters of British Columbia, Antarctica, and California. *Appl. Environ. Microbiol.* 39:1085-1095.

Fuhrman J. A., and F. Azam. (1982). Thymidine incorporation as a measure of heterotrophic bacterial production in marine surface waters: evaluation and field results. *Marine Biol.* 66:109-120.

Harvey, R. W., and L. Y. Young. (1980a). Enrichment and association of bacteria and particulates in salt marsh surface water. *Appl. Environ. Microbiol.* 39:894-899.

Harvey, R. W., and L. Y. Young. (1980b). Enumeration of particle bound and unattached respiring bacteria in the salt marsh environment. *Appl. Environ. Microbiol.* 40:156-160.

Hollibaugh, J. T. (1979). Metabolic adaptation in natural bacterial populations supplemented with selected amino acids. *Estuarine Coast. Marine Sci.* 9:215-230.

Hollibaugh, J. T. (1988). Limitations of the [^{3}H]thymidine method for estimating bacterial productivity due to thymidine metabolism. *Marine Ecol. Progr. Ser.* 43:19-30.

Hollibaugh, J. T., and P. S. Wong. (1992). Ethanol-extractable substrate pools and the incorporation of thymidine, L-leucine, and other substrates by bacterioplankton. *Canadian Jour. Microbiol.* 38:605-613.

Hollibaugh, J. T. (1994). Relationship between thymidine metabolism, bacterioplankton community metabolic capabilities and sources of organic matter. *Microbiol. Ecol.* 28:117-131.

Iriberri, J., M. Unanue, I. Barcina, and L. Egea. (1987). Seasonal variation in population density and heterotrophic activity of attached and free-living bacteria in coastal waters. *Appl. Environ. Microbiol.* 53:2308-2314.

Jassby, A. D., J. E. Cloern, and T. M. Powell. (1993). Organic carbon sources and sinks in San Francisco Bay: variability induced by river flow. *Marine Ecol. Progr. Ser.* 95:39-54.

Karl, D. M., G. A. Knauer, and J. H. Martin. (1988). Downward flux of particulate organic matter in the ocean: a particle decomposition paradox. *Nature* 332:438-441.

King, K. R., J. T. Hollibaugh, and F. Azam. (1980). Predator-prey interactions between a larvacean, *Oikopleura dioica* and bacterioplankton in enclosed water columns. *Marine Biol.* 56:49-57.

Kirchman, D. L., H. W. Ducklow, and R. Mitchell. (1982). Estimates of bacterial growth from changes in uptake rates and biomass. *Appl. Environ. Microbiol.* 44:1296-1307.

Kirchman, D. L., E. K'nees, and R. E. Hodson. (1985). Leucine incorporation and its potential as a measure of protein synthesis by bacteria in natural aquatic systems. *Appl. Environ. Microbiol.* 49:599-607.

Kirchman, D. L., S. Y. Newell, and R. E. Hodson. (1986). Incorporation versus biosynthesis of leucine: Implications for measuring rates of protein synthesis and biomass production by bacteria in marine systems. *Marine Ecol. Progr. Ser.* 32:47-59.

Kirchman, D. L., and M. P. Hoch. (1988). Bacterial production in the Delaware estuary estimated from thymidine and leucine incorporation rates. *Marine Ecol. Progr. Ser.* 45:169-178.

Kirchman, D. L., Y. Suzuki, C. Garside, and H. W. Ducklow. (1991). High turnover rates of dissolved organic carbon during a spring phytoplankton bloom. *Nature* 352:612-614.

Landry, M. R., L. W. Hass, and V. L. Fagerness. (1984). Dynamics of microbial plankton communities: Experiments in Kaneohoe Bay, Hawaii. *Marine Ecol. Progr. Ser.* 16:127-133.

Landry, M. R., and R. P. Hassett. (1982). Estimating the grazing impact of marine microzooplankton. *Marine Biol.* 67:283-288.

Lee, S., and J. A. Fuhrman. (1987). Relationships between biovolume and biomass of naturally derived marine bacterioplankton. *Appl. Environ. Microbiol.* 53:1298-1303.

Lee, S., and J. A. Fuhrman. (1991). Species composition shift of confined bacterioplankton studied at the level of community DNA. *Marine Ecol. Progr. Ser.* 79:1277-1287.

Li, W. K. W., and P. M. Dickie. (1984). Rapid enhancement of heterotrophic but not photosynthetic activities in arctic microbial plankton at mesobiotic temperatures. *Polar Biol.* 3:217-226.

Logan, B. E., and J. R. Hunt. (1987). Advantages to microbes of growth in permeable aggregates in marine systems. *Limnol. Oceanogr.* 32:1034-1048.

Nichols, F. H. (1985). Increased benthic grazing: An alternative explanation for low phytoplankton biomass in Northern San Francisco Bay during the 1976-1977 drought. *Estuarine Coastal Shelf Sci.* 21:379-388.

Nixon, S. W. (1988). Nutrient dynamics, primary production and fisheries yields of lagoons. Proceedings of the International Symposium on coastal lagoons SCOR/IABO/UNESCO, Bordeaux, 8-14 September 1981. *Oceanol. Acta* 1982:357-371.

Parsons, T. R., Y. Maita, and C. L. Lalli. (1984). *A Manual of Chemical and Biological Methods for Seawater Analysis.* Pergamon Press, Oxford. 173 pp.

Peterson, D. H., T. J. Conomos, W. W. Broenkow, and P. C. Doherty. (1975). Location of the non-tidal current null zone in Northern San Francisco Bay. *Estuarine Coastal Shelf Sci.* 3:1-11.

Peterson, B. J., and J. E. Hobbie. (1978). *Daphnia* grazing on natural bacteria. *Limnol. Oceanogr.* 23:1039-1044.
Pomeroy, L. R. (1974). The ocean's foodweb, a changing paradigm. *BioScience* 24:499-504.
Pomeroy, L. R. (1979). Secondary production mechanisms of continental shelf communities. Pages 163-186 *in* R. J. Livingston, ed., *Ecological Processes in Coastal and Marine Systems.* Plenum Publ. Corp., New York.
Pomeroy, L. R. (1980). Microbial roles in aquatic food webs. Pages 85-109 *in* R. R. Colwell and J. Foster, eds., *Aquatic Microbial Ecology.* University of Maryland Sea Grant, College Park, MD.
Porter, K. G., and Y. S. Feig. (1980). The use of DAPI for identifying and counting aquatic microflora. *Limnol. Oceanogr.* 25:943-948.
Riemann, B., P. K.Bjornsen, S. Newell, and R. Fallon. (1987). Calculation of cell production of coastal marine bacteria based on measured incorporation of [^{3}H]thymidine. *Limnol. Oceanogr.* 32:471-476.
Robarts, R. D., and T. Zohary. (1993). Fact or fiction – bacterial growth rates and production as determined by [^{3}H-methyl]thymidine? *Adv. Microb. Ecol.* 13:371-425.
Shiah, F-K., and H. W. Ducklow. (1994a). Temperature and substrate regulation of bacterial abundance, production and specific growth rate in Chesapeake Bay, USA. *Marine Ecol. Progr. Ser.* 103:297-308.
Shiah, F-K., and H. W. Ducklow. (1994b). Temperature regulation of heterotrophic bacterioplankton abundance, production, and specific growth rate in Chesapeake Bay. *Limnol. Oceanogr.* 39:1243-1258.
Simon, M. (1988). Growth characteristics of small and large free- living and attached bacteria in Lake Constance. *Microbiol. Ecol.* 15:151-163.
Simon, M., and F. Azam. (1989). Protein content and protein synthesis rates of planktonic marine bacteria. *Marine Ecol. Progr. Ser.* 51:201-213.
Simon, M., A. L. Alldredge, and F. Azam. (1990). Bacterial carbon dynamics on marine snow. *Marine Ecol. Progr. Ser.* 65:205-211
Smith, D. C., M. Simon, A. L. Alldredge, and F. Azam. (1992). Intense hydrolytic enzyme activity on marine aggregates and implications for rapid particle dissolution. *Nature* 359:139-142.
Tokuda, W., and R. Hall. (1992). *Humphrey, the Lost Whale: A True Story.* Heian International, Inc., Torrance, CA. 32 pp.
Valdes, M., and L. J. Albright. (1981). Survival and heterotrophic activities of Fraser River and Strait of Georgia bacterioplankton within the Fraser River plume. *Marine Biol.* 64:231-241.
Walters, R. A., R. T. Cheng, and T. J. Conomos. (1985). Time scales of circulation and mixing processes of San Francisco Bay waters. *Hydrobiologia* 129:13-36.
Werner, I., and J. T. Hollibaugh. (1993). *Potamocorbula amurensis*: Comparison of clearance rates and assimilation efficiencies for phytoplankton and bacterioplankton. *Limnol. Oceanogr.* 38:949-964.
White, P. A., J. Kalff, J. B. Rasmussen, and J. M. Gasol. (1991). The effect of temperature and algal biomass on bacterial production and specific growth rate in freshwater and marine habitats. *Microbiol. Ecol.* 21:99-118.
Wienke, S. M., and J. E. Cloern. (1987). The phytoplankton component of seston in San Francisco Bay. *Neth. Jour. Sea Res.* 21:25-33.
Wright, R. T. (1978). Measurement and significance of specific activity in the heterotrophic bacteria of natural waters. *Appl. Environ. Microbiol.* 36:297-305.
Wright, R. T., R. B. Coffin, C. P. Ersing, and D. Peterson. (1982). Field and laboratory measurements of bivalve filtration of natural bacterioplankton. *Limnol. Oceanogr.* 27: 91-98.
Wright, R. T., and R. B. Coffin. (1983). Planktonic bacteria in estuaries and coastal waters of northern Massachusetts: Spatial and temporal distributions. *Marine Ecol. Progr. Ser.* 11:205-216.

Wright, R. T., and R. B. Coffin. (1984). Measuring microheterotrophic grazing of marine bacterioplankton with differential filtration and incubation. *Microbiol. Ecol.* 10:137-150.

Yoon, W. B., and R. Rosson. (1990). Improved method of enumeration of attached bacteria for study of fluctuation in the abundance of attached and free-living bacteria in response to diel variation in seawater turbidity. *Appl. Environ. Microbiol.* 56:595-600.

PLANKTONIC RESPIRATION RATES IN SAN FRANCISCO BAY

JOSEPH RUDEK[1] AND JAMES E. CLOERN
U.S. Geological Survey, 345 Middlefield Road, Menlo Park, CA 94025

Dark bottle respiration rates were measured in San Francisco Bay over a one year period. Respiration rates varied over nearly two orders of magnitude from 15 to 1050 nmol 0_2/l*h. Highest rates were associated with the spring phytoplankton bloom in South Bay. Significant relationships were found between respiration rates and chlorophyll *a*, particulate carbon, and particulate nitrogen concentrations. The ratio of respiration rates to primary production rates indicated a greater proportion of planktonic respiration derived from heterotrophic activity in San Pablo and Suisun Bays than in South Bay.

In the last decade, the belief that planktonic microheterotrophs play only a minor role in the cycling of organic matter in aquatic ecosystems has changed. Evidence from both carbon assimilation (Ducklow & Shiah 1993) and oxygen consumption in respiration (Hopkinson *et al.* 1989; Jensen *et al.* 1990) indicate that significant quantities of organic matter can pass through the microheterotrophs. Large portions of primary production have been found to be processed through microbial food webs (Azam *et al.* 1983; Fenchel 1988). However, the ultimate fate of the microheterotrophically-processed organic matter remains unclear (Ducklow & Shiah 1993).

Estuaries in general are considered highly productive ecosystems supported largely by phytoplankton production. For San Francisco Bay (SFB) however, recent evidence suggests that phytoplankton production may be declining following the invasion of the bivalve, *Potamocorbula amurensis* (Alpine & Cloern 1992). A carbon budget analysis by Jassby *et al.* (1993) suggests that heterotrophy in the northern portions of SFB (Suisun and San Pablo Bays) is fueled by allochthonous sources of organic matter. With the success of introduced benthic filter feeders, such as *Potamocorbula*, in SFB, increasing proportions of allochthonous organic matter could be directly consumed by bivalves, perhaps to the detriment of pelagic heterotrophs. Under such a scenario, planktonic respiration, as an index of pelagic metabolism, could be a useful indicator of such a shift in the processing of organic matter. At present, the patterns of variability in planktonic respiration rates in SFB are unknown. The last respiration rate measurements of plankton made in SFB were almost 20 years ago (Cole & Herndon 1979). The O_2 respiration rates in SFB are expected to vary temporally and spatially due to the complex interaction of seasonal phytoplankton production and numerous physical factors (*e.g.*, tidal mixing, turbidity, water column stratification, day length, river discharge).

San Francisco Bay, the estuary of the Sacramento and San Joaquin Rivers, is comprised of two distinct estuaries. In the northern reach, freshwater input from the rivers results in a partially mixed estuary with well-developed gravitational

[1] Present Address: 324 Umstead Drive, Chapel Hill, NC 27516.

circulation and a turbidity maximum during summer (Conomos *et al.* 1985). In contrast, South Bay with no large, direct freshwater input, is a lagoonal estuary, that is typically well mixed. The differing ecology of these two estuarine systems within SFB and their impact on the planktonic respiration rates are discussed.

The planktonic respiration rates described in this report were conducted as part of a pilot regional monitoring program in SFB supported by the local dischargers group. The many other measurements conducted contemporaneously with the respiration measurements proved to be very useful in the interpretation of the results. A full account of the pilot monitoring program may be found in Caffrey *et al.* (1994).

METHODS

Respiration measurements were conducted bi-monthly in San Francisco Bay (SFB) during 1993 as part of a pilot monitoring project (Caffrey *et al.* 1994). Respiration measurements were conducted with samples from the channel that runs down the spine of the bay, through the major sub-embayments of the SFB system (Fig. 1). These sub-embayments include the southern end of the South Bay (station 32), the central South Bay (station 27), the Central Bay (station 18), San Pablo Bay (station 13), Suisun Bay (station 6), and the lower Sacramento River at Rio Vista (station 657). Respiration was measured on the following dates in 1993: 24 February, 15 April, 14-15 June, 10 August, 5-6 October, 7 December. Additional measurements were made in the South Bay only, on 16 February, 18 March, and 30 March.

All water was collected via an impeller pump on the bow of the *R.V. Polaris* (intake port approximately 1.5 meters below the water surface), passed through a 190 µm-mesh screen and collected in a 20-liter carboy. Comparison of respiration rates from grab samples collected near the pump intake with samples simultaneously collected via the bow pump did not show significant differences or consistent trends (Caffrey *et al.* 1994). Water samples were dispensed into 300 ml BOD bottles through tubing connected to a spigot at the bottom of the carboy. To maintain sample homogeneity, a stainless steel paddle was inserted to the bottom of the carboy and turned at a rate of 100 rpm while the water was being dispensed. Each BOD bottle had been acid soaked (10% HCl), rinsed three times with distilled water, and rinsed once with sample water. BOD bottles were carefully filled from the bottom and allowed to overflow to avoid the addition of oxygen. All bottles were filled within 30 minutes of collection. Twelve bottles were filled at each station.

Six of the bottles were immediately fixed with Winkler reagents (Strickland & Parsons 1972) to provide a measurement of dissolved oxygen concentration at time zero (t_0). The remaining six (dark BOD) bottles were wrapped with aluminum foil, as an additional precaution against exposure to light, and placed in a dark incubation chamber through which ambient water flowed to maintain sample temperature. Maximum ambient water temperature change in transit across the bay averaged 3.0°C when all experiment dates were considered, with the maximum difference in temperature of 5.5°C in December. Samples were incubated for 24 hours (t_{24}),

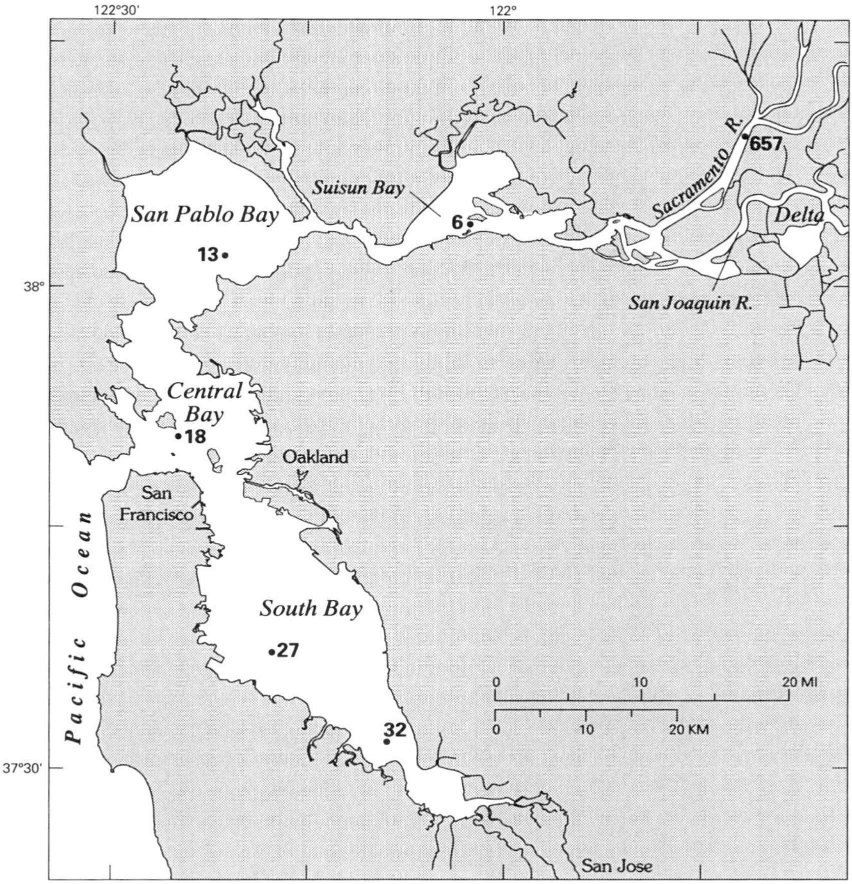

FIGURE 1. Station locations in San Francisco Bay.

then fixed with Winkler reagents. Fixed samples were stored (no longer than 2 weeks) in their BOD bottles with water in the cap-wells and topped by a snap-fit plastic cap.

A sample aliquot was collected by autopipette, and titrated with 0.01N sodium thiosulfate dispensed by a Metrohm Autotitrater using the potentiometric titration method of Graneli & Graneli (1991). Potassium iodate standardization of the sodium thiosulfate was conducted under identical procedures to eliminate problems associated with iodine volatilization (Knapp *et al.* 1991).

The highest and lowest values among 6 replicates were eliminated to calculate a trimmed mean oxygen concentration for each treatment. The respiration rate was calculated by subtracting the mean t_0 concentration from the mean t_{24} concentration. A confidence interval was calculated for each cruise date by pooling variance about the means from all DO concentration measurements. This was done to determine error associated with the technique because individual bottles are not true replicates.

Chlorophyll *a* (CHL*a*) concentrations were determined on aliquots of water collected from the same carboy used for the respiration samples. Samples were filtered at less than 12 cm Hg on to Gelman A/E glass fiber filters and immediately frozen. The air-dried filters were ground in 90 percent acetone within 1 week of collection. After extraction for 12-24 hours at -10°C, samples were centrifuged and absorbances of the extracts were read on a Hewlett Packard 8452A diode array spectrophotometer. The acetone extracts were acidified to measure phaeopigments (Riemann, 1978). Chlorophyll *a* and phaeopigment values were calculated using the equations of Lorenzen (1967).

Temperature, salinity, and suspended particulate material (SPM) were measured on a CTD (Sea-Bird Electronics) which was lowered through the water column at each station. Suspended particulate material was measured with an optical backscatter sensor (OBS) and calibrated against discrete SPM samples collected on filters. Description of the components of the CTD unit are included in Caffrey *et al.* (1994).

RESULTS

Overall, measured respiration rates varied from a low of 15.3 nmol O_2/l*h, in February at Station 13, to a maximum of 1063.6 nmol O_2/l*h in March at Station 32 (Fig. 2). The maximum respiration rate coincided with the spring phytoplankton bloom in the South Bay. Considering all measurements made in 1993, mean respiration rates were highest in South Bay (Station 32 = 345.2 nmol O_2/l*h). The

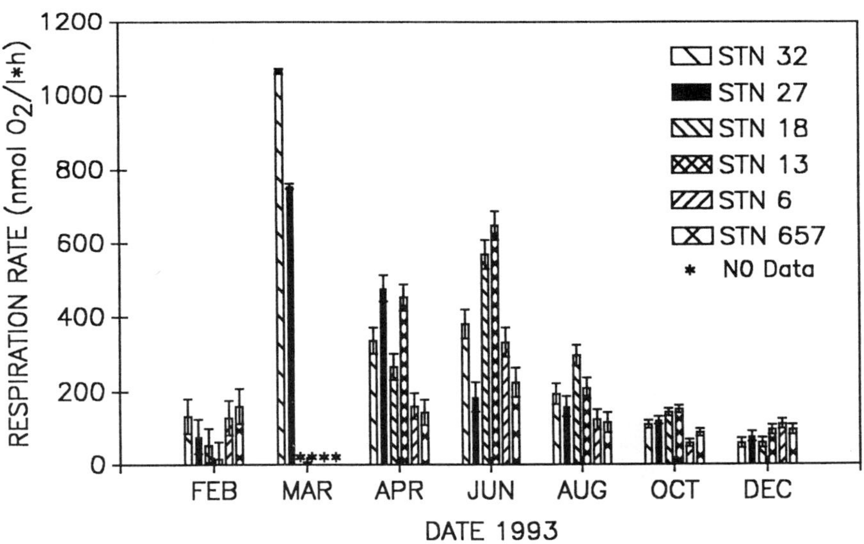

FIGURE 2. Planktonic respiration rates (nmol O_2/l*h) at six sites in the San Francisco Bay system in 1993. Error bars = 95 % confidence interval.

bi-monthly frequency of measuring respiration rates for the Regional Monitoring Program was increased in the South Bay during the spring in an effort to capture the effect of the anticipated spring phytoplankton bloom. If only the 6 bimonthly measurements common to all stations are considered, average respiration rates at Station 32 drop to 202.9 nmol O_2/l*h and Central Bay (Station 18) average respiration rates (311.6 nmol O_2/l*h) represent the highest average baywide. The highest measured respiration rates (based on the 6 bimonthly measurements) were found in all stations, except Station 27, on 15 June (Fig. 2). Occurrences of the lowest measured rates at each station were found to be geographically grouped, with the marine dominated stations in Central and San Pablo Bay (18 and 13; respectively, hereafter referred to as CB) showing lowest rates on 24 February, the North Bay river dominated stations (657 and 6; hereafter referred to as NB) on 5 October, and the South Bay stations (32 and 27; hereafter referred to as SB) on 8 December (Fig. 2).

Water column conditions varied both seasonally and spatially (Caffrey *et al.* 1994). Temperatures ranged from a low of 9.4°C at Station 657, in the Sacramento River during February to a high of 21°C in the SB during August (Caffrey *et al.* 1994). Chlorophyll *a* varied from a low of 0.7 μg/l at Station 6 in Suisun Bay during October to a high of 44.3 μg/l at Station 32 in the South Bay during March. SPM concentrations varied from a low of 1 mg/l at Station 18 in the Central Bay during June to a high of 120.5 mg/l at Station 657 in the Sacramento River in February. Salinity varied from a low of 0.08 psu at Station 657 in the Sacramento River in December to a high of 31.96 psu at Station 18 in the Central Bay in October.

In preliminary experiments (data not shown), three trial incubations produced respiration rates over 24 hours that averaged approximately 90% of 7, 8, and 20 hour incubations. As discussed in Hopkinson *et al.* (1989), reports regarding the linearity of pelagic respiration rates are inconsistent, with some researchers finding constant rates of microbial metabolism during the first 24 hours of an incubation while others did not. We chose to use 24-hour incubation periods to facilitate comparisons with other work conducted in SFB.

DISCUSSION

Planktonic Respiration Rates in San Francisco Bay and Other Coastal Waters

In SFB, planktonic respiration rate measurements are relatively scarce (Table 1). A comparison of respiration rates measured in 1976-1977 (Cole & Herndon 1979) with those reported in this study reveal respiration rates were in the same range or lower in 1993. Station locations during the two studies were similar. All respiration rates measured in the fall or at Station 657 in the Sacramento River were lower in 1993 relative to earlier rates. Only during the spring in the CB were rates higher in 1993 relative to measurements in 1976-1977. The highest respiration rates measured in both studies were in the spring, with the highest overall rate in South Bay during 1993. However, the comparison with the 1976-1977 data is limited by

TABLE 1. Planktonic respiration rates (μmol O_2/l*d)

	San Francisco Bay				
Date	South Bay	Central Bay	Suisun Bay	Rio Vista	References
Spr 1976-77	9.0	2.0-4.9	2.8-9.8	6.5	Cole & Herndon 1979
Sum 1976-77	–	6.8-11.6	–	5.0	"
Fall 1976-77	5.0	4.1	4.9-6.1	5.4-7.2	"
Win 1993[1]	1.4-16.4	0.4-2.3	2.7-3.1	2.3-3.8	This study
Spr 1993	4.5-25.5	6.4-13.7	3.9-8.0	3.5-5.4	"
Sum 1993	3.9-4.7	5.0-7.1	3.0	2.8	"
Fall 1993	2.7-2.9	3.5-3.7	1.4	2.1	"

[1] Including December 1993

lack of any winter data throughout the Bay and lack of summer data in South Bay and Suisun Bay.

The average of all 24 hour respiration rate measurements reported by Cole & Herndon (1979) was 6.1 μmol O_2/l*d. It should be cautioned that this average was calculated from Cole & Herndon (1979) for comparative purposes from measurements not balanced across the bay or throughout the year. Peterson *et al.* (1987) reported average respiration rates in northern SFB of 6.4 μmol O_2/l*d (based on studies in 1976-1977). The average respiration rate for measurements made across the bay in this study was 5.8 μmol O_2/l*d, and for northern SFB, 5.9 μmol O_2/l*d. Given the differences in the respiration rates measured in this study and historic measurements (*e.g.,* location, season, number of measurements, annual variability, methodology), no differences between these values can be assumed.

Respiration rates in SFB tended to be at the low end of the range of rates measured in other estuarine and coastal waters (Table 2). Considering the eutrophic nature of SFB (Nichols *et al.* 1986) this may be unexpected. Several factors are likely to contribute to the low respiration rates measured in SFB. While SFB does experience significant nutrient loading, the phytoplankton are light-limited resulting from high turbidity levels and mixed depths often below the compensation point (Alpine & Cloern 1988; Cloern 1987). A significant portion of SFB is very shallow providing high light conditions for phytoplankton. Shoal areas, defined as shallower than 2 m relative to MLLW, occupy 60% of the NB, 47% of the SB, and 27% of CB (Jassby *et al.* 1993). However, proliferation of introduced benthic filter feeders in these shoals, especially *Potamocorbula amurensis* in Suisun Bay, has been estimated to have caused large decreases in primary production (Alpine & Cloern 1992).

Spatial and Temporal Variability Within San Francisco Bay

Respiration rates showed a variety of responses both spatially and temporally (Fig. 2). This is not surprising considering the different types of estuarine conditions that exist in the south, central, and northern sections of SFB. The SB had the highest rates measured during this study, associated with the spring bloom. Chlo-

TABLE 2. Representative planktonic respiration rates (μmol O_2/l*d)

Roskilde Fjord, Denmark		6.9-124.4	Jensen *et al.* 1990
Western English Channel	summer; euphotic zone	2-54	Holligan *et al.* 1984
Wassaw Estuary, Mississippi	annual mean summer	14 15-60	Turner 1978[1]
Chesapeake Bay	annual mean summer	16 31-47	Peterson *et al.* 1987 Jonas 1992
Duplin River Estuary, Georgia	summer winter maximum	22.5-37.5 16.5	Hopkinson *et al.* 1989
Georgia Bight (nearshore)	summer winter	30-34 8-15	Hopkinson *et al.* 1989
Dollart Estuary, Netherlands	summer	7-30	Van Es & Ruardij 1982[1]
Potomac Estuary	outer	20-25	Peterson *et al.* 1987
San Francisco Bay	mean: all regions	0.4-25.5	This study
Southampton Estuary, England		5-23	Lima & Williams 1978[1]
North Sea		1.2-20.0	Iriarte *et al.* 1991
Newport River Estuary, Rhode Island	winter mean summer mean	7 19	Williams & Murdock 1966
Ems, Netherlands		6-15	Van Es & Ruardij 1982[1]
Georgia Bight (offshore)	summer	11.9-14.6	Hopkinson *et al.* 1991
San Francisco Bay	mean: northern bay	6.4-12	Peterson *et al.* 1987
San Francisco Bay	mean: all regions	2-11.6	Cole & Herndon 1979
Saanich Inlet, Canada	mean; mesocosms	10.1	Williams 1982
Narragansett Bay, Rhode Island	annual mean	10	Kremer & Nixon 1978[1]
NW Atlantic	off Virginia	1-6	Packard & Williams 1981[1]

[1] As reported in Peterson *et al.* 1987

rophyll *a* concentrations in the channel of the SB increased from around 2 μg/l in February to 70 μg/l in March (Caffrey *et al.* 1994). The bloom CHL*a* concentrations in the SB during the March respiration rate measurements were from 25 to 45 μg/l. Respiration rates measured during the bloom had increased from below 150 nmol O_2/l*h in February to a maximum greater than 1000 nmol O_2/l*h (Fig. 2). The CHL*a* concentrations and respiration rates in SFB decreased nearly as quickly as they rose. Chlorophyll *a* concentrations decreased to below 10 μg/l (Caffrey *et al.* 1994) and respiration rates decreased to less than 500 nmol O_2/l*h by April (Fig. 2). Respiration rates in SB generally continued to decline, although much less dramatically, for the remainder of the year, falling to below 100 nmol O_2/l*h in December.

In CB (Stations 18 and 13), respiration rates measured in February ranged from 15 to 50 nmol O_2/l*h, the lowest rates measured in this study (Fig. 2). Rates increased through spring to June when the maximum rate was measured in CB, ranging from 570 to 650 nmol O_2/l*h, possibly supported by organic matter input from coastal waters. The respiration rates in CB in June were also the maximum rates measured baywide during the summer. For the remainder of the year, respiration rates in CB declined, dropping to 60 to 100 nmol O_2/l*h in December.

The variation in CHL*a* concentrations in CB loosely paralleled the respiration rates. The maximum for both CHL*a* concentrations (Caffrey *et al.* 1994) and respiration rates in CB were measured in April and June.

In the river-dominated NB Stations (6 and 657), respiration rates were less variable on an annual basis relative to the rest of SFB (Fig. 2). For all measurements, except June, rates ranged from 60 to 160 nmol O_2/l*h. During June, rates ranged from 225 to 330 nmol O_2/l*h. The relative uniformity of respiration rates in the NB is perhaps a reflection of the CHL*a* concentrations which were less than 2 μg/l in all measurements except that measured at Station 6 in June (Caffrey *et al.* 1994). Chlorophyll *a* concentrations in June at Station 6 (11.1 μg/l) were the highest measured throughout the Bay at that time.

Relation between Plankton Respiration Rates and Water Column Variables

Chlorophyll *a* concentrations, as estimates of phytoplankton biomass, have been found to be correlated with respiration rates by some authors while not by others (see Iriarte *et al.* 1991). Peterson *et al.* (1987) estimated phytoplankton respiration in central SFB to be 75 percent of total planktonic respiration (4.8 μmol O_2/l*d) using an allomeric equation based on average CHL*a* concentrations and cellular carbon content. Similar efforts to estimate phytoplankton respiration from phytoplankton counts in this study were unsuccessful (data not shown), resulting in estimates in excess of total measured respiration a majority of the times.

Iriarte *et al.* (1991) noted that the strength of the correlation between CHL*a* and respiration rates in the North Sea was influenced by the concentration of CHL*a* present in the sample. In both Iriarte *et al.* (1991) and in this study, the near constant relation between CHL*a* concentrations and CHL*a*-specific respiration rates found at high CHL*a* concentrations did not hold for low CHL*a* concentrations (Fig. 3). Iriarte *et al.* (1991) postulate that at low algal biomass levels the correlation between CHL*a* and community respiration will be weak because microheterotrophs will make up a major portion of the plankton community respiration. At higher CHL*a* concentrations, phytoplankton respiration becomes more important, dominating the plankton respiration rates at exceptionally high CHL*a* concentrations. The SFB data were divided into groups by CHL*a* concentration as clustered in Fig. 3 and examined for linear regression. These groupings appeared to be well correlated with 92% and 59% of the variance explained for the high and low CHL*a* concentration groups, respectively (Fig. 4).

Particulate carbon (PC) and particulate nitrogen (PN) measurements of samples from the water column are a measure of the total seston organic matter, including the entire planktonic community, autotrophic and heterotrophic. For a subset of the respiration rate measurements (March, June, and October: $n = 11$), PC and PN concentrations were available (E. Canuel, pers. commun.). Particulate carbon was significantly correlated ($r = 0.955$, $P < 0.001$) with plankton respiration rates (Fig. 5). Particulate nitrogen also appeared correlated, but contained an outlier (Fig. 6). With the outlier removed, the correlation between PN and respiration was significant ($r = 0.915$, $P < 0.001$). These correlations provide validation of the planktonic

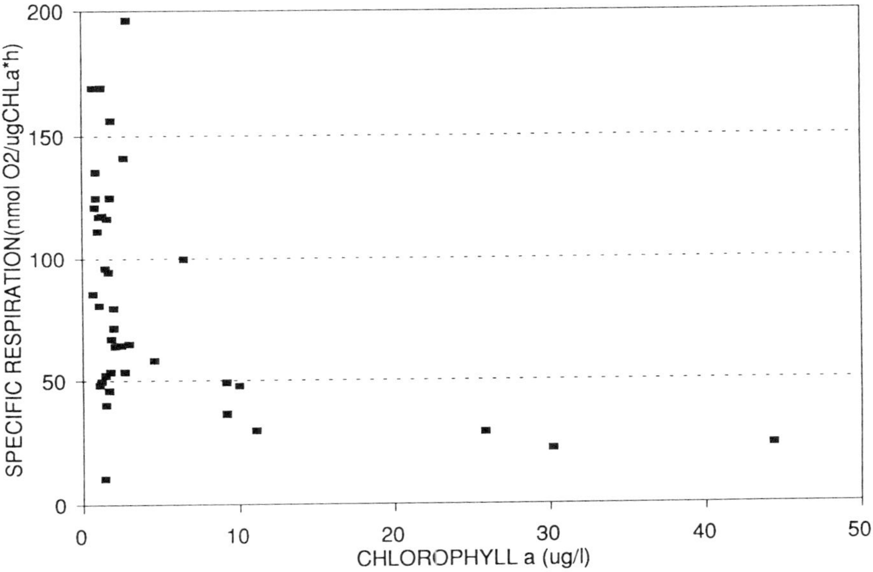

FIGURE 3. Chlorophyll *a* (μg/l) concentrations versus chlorophyll-specific planktonic respiration (nmol O_2/μg Chl*a**h) in the San Francisco Bay system in 1993.

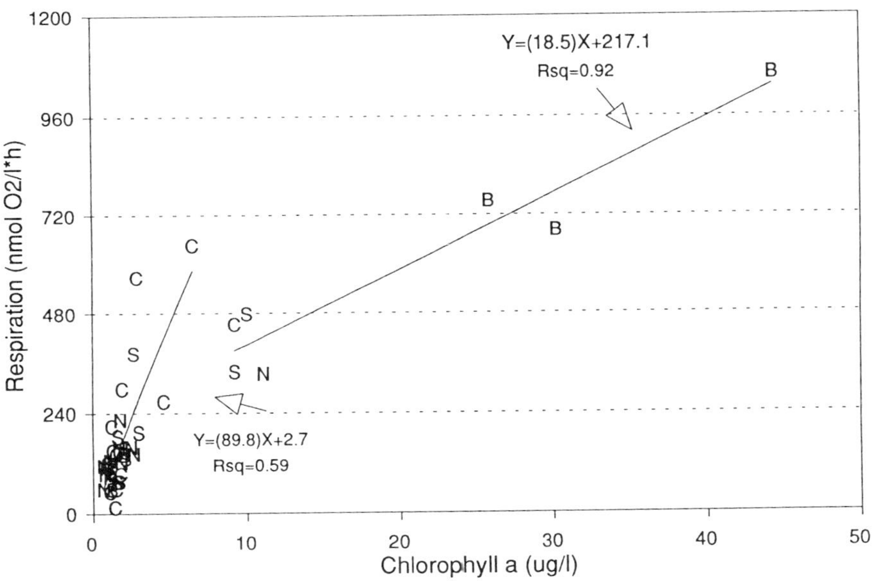

FIGURE 4. Chlorophyll *a* (μg/l) concentrations versus planktonic respiration rate measurements (nmol O_2/l*h) in the San Francisco Bay system in 1993. B = South Bay spring bloom; S = South Bay non-bloom; C = Central Bay; N = North Bay. Lines and equations indicate regressions for the data divided into groups greater than or less than 9 μg Chlorophyll

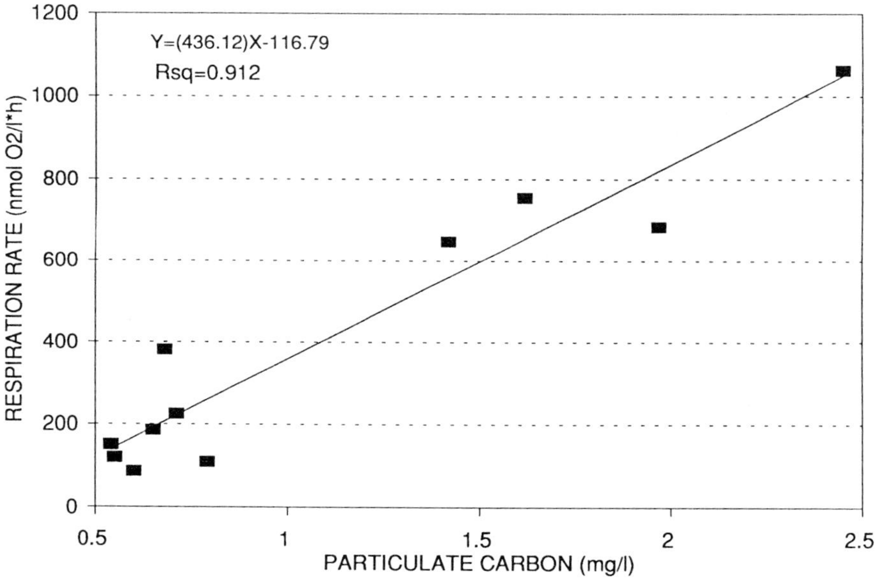

FIGURE 5. Particulate carbon (mg/l) versus planktonic respiration rates (nmol O_2/l*h) in the San Francisco Bay system in 1993. The line is the plot of the regression equation.

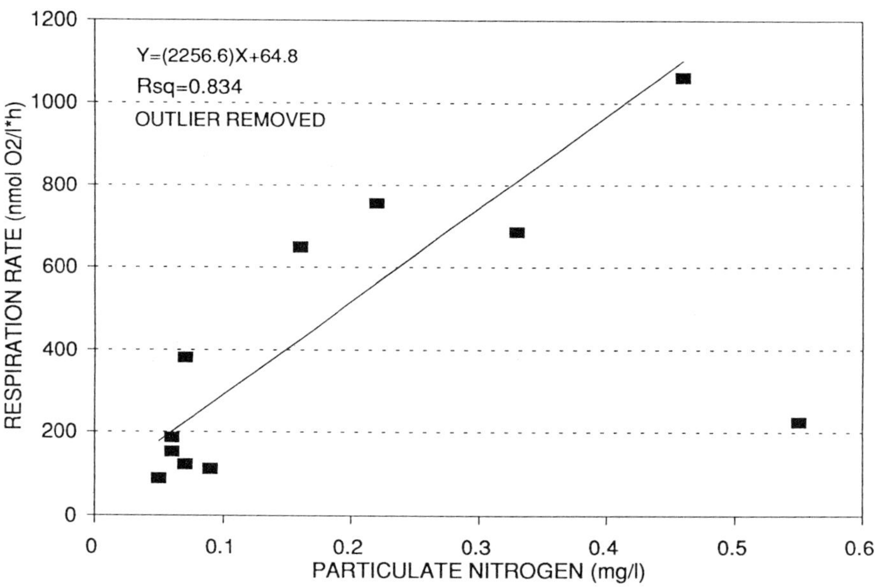

FIGURE 6. Particulate nitrogen (mg/l) versus planktonic respiration rates (nmol O_2/l*h) in the San Francisco Bay system in 1993. The line is the plot of the regression equation (with the outlier removed).

respiration results and suggest that PC and PN could serve as excellent indices of planktonic respiration rates.

Over the course of the study, temperature, salinity, and SPM were measured over wide ranges. None of these variables was significantly correlated with plankton respiration rates. Certainly temperature, salinity, and SPM all impact the environment of the plankton community and might be expected to co-vary with respiration rates of the plankton. However, the wide range of conditions found on any particular date, due to the large differences in basic ecosystem structure and function from one end of the SFB estuary to the other, probably confound the relation between these variables and respiration rate.

Net Metabolism of the San Francisco Bay Water Column:

As a result of diverse measurement programs, we have a reasonable understanding of the magnitude of organic matter inputs to the San Francisco Bay ecosystem (see the review of Jassby *et al.* 1993). However there is great uncertainty about the rates of organic matter consumption within San Francisco Bay, so fundamental questions about system metabolism remain unanswered. The respiration measurements described here provide sufficient resolution to allow first-order estimates of pelagic metabolism in different spatial domains of the estuary. We did this by comparing rates of primary production P with simultaneous measures of total pelagic respiration R.

Daily, depth-integrated primary productivity P was calculated as:

$$P\ (\text{mmol C/m}^2\text{-}d) = \int_0^H \int_0^{24} [BP_m \tan h(\alpha I_{z,t} / P_m)]\ dt\ dz$$

where the photosynthetic parameters P_m and α were determined from short-term measurements of ^{14}C assimilation using aliquots from the same samples used to measure respiration (methods are given in Caffrey *et al.* 1994). B is phytoplankton biomass as chlorophyll *a* concentration; $I_{z,t}$ is irradiance at depth z and time t, from measures of surface irradiance and the light extinction coefficient; and h is depth of the water column, taken as the basin-mean water depth at mean tide level (Jassby *et al.* 1993; Conomos 1979). The integrations with respect to depth and time (t) were done by trapezoidal quadrature using time steps of 0.1 h and depth increments of 0.05 m. Daily productivity (as carbon assimilation) was then converted to an equivalent rate of oxygen evolution using a photosynthetic quotient PQ of 1.3, typical of phytoplankton growing with ammonium as the nitrogen source (Langdon 1988). Depth-integrated respiration R was calculated as

$$R\ (\text{mmol O}_2\text{/m}^2\text{-}d) = 24\ rh,$$

where r is the hourly rate of oxygen consumption.

Although this approach for calculating P and R is based on many simplifying assumptions (*e.g.*, vertical homogeneity of r; constant PQ), it provides a useful index for comparing the relative amplitudes of the production and consumption of organic matter. Ratios R:P >1 indicate that total pelagic respiration exceeds the rate of primary production (the water column is net heterotrophic); ratios of R:P <1 indicate that the local rate of phytoplankton primary production exceeds the total rate of aerobic respiration (the water column is net autotrophic). Results of these analyses are shown in Fig. 7, which compares the mean and seasonal range of *R*:*P* at six sites in San Francisco Bay.

Several patterns are evident in Fig. 7, and we can consider these as hypotheses that remain to be tested with more rigorous investigation. First, the variability (range) of *R*:*P* was largest at the most landward stations (32 in the South Bay, and 6 and 657 in the upper estuary). This is a result of the large seasonal fluctuations in the turbidity, and hence relative depths of the photic and aphotic zones, at these landward sites. For example, the maximum *R*:*P* ratio (25.5) was estimated at Station 657 during February, a period of high river flow and high turbidity in the Sacramento River.

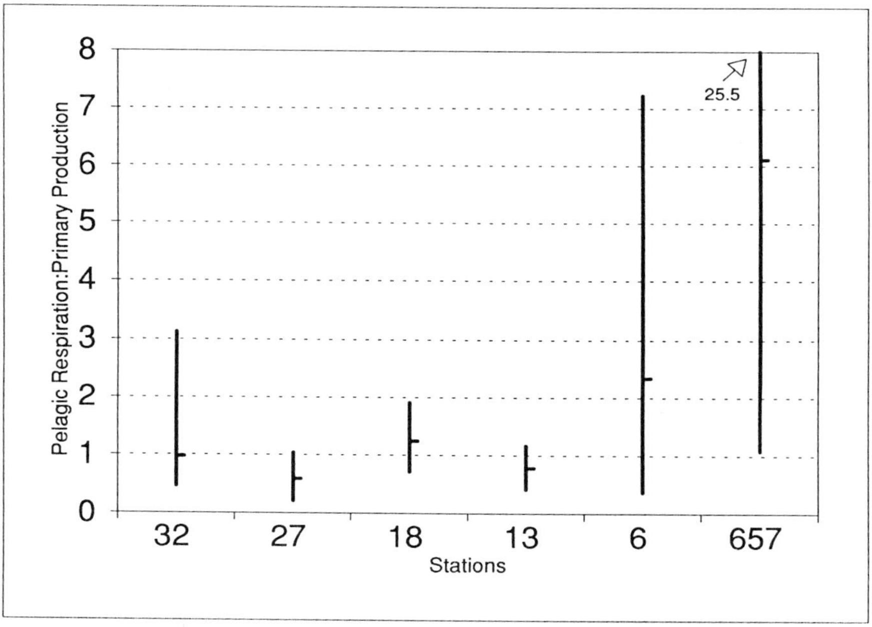

FIGURE 7. Mean (horizontal lines) and range (vertical lines) of estimated ratios of pelagic respiration (mmol O_2/m^2-d) to gross primary production (mmol O_2/m^2-d) at six sites in San Francisco Bay system during 1993. Upper range of R:P ratio is off scale for station 657 (maximum value = 25.5) as is indicated by arrow.

Second, there appears to be a strong spatial gradient in the mean balance between pelagic production and consumption of organic matter. Every measurement of *R*:*P* in the lower Sacramento River (Station 657) was greater than one, and the mean *R*:*P* there was 6.1 (Fig. 7). The mean *R*:*P* in Suisun Bay (Station 6) was 2.3, and this index was usually greater than one (exceptions occurred in April and June). These results suggest that the lower Sacramento River is always net heterotrophic and the Suisun Bay water column is typically net heterotrophic. This conclusion is consistent with the carbon budget of Jassby *et al.* (1993) and the results of Canuel *et al.* (1995) who used biomarker analyses to infer an allochthonous origin of organic matter in the river-influenced upper estuary. At the other end of the spectrum, the mean *R*:*P* at Station 27 was 0.6, and there the *R*:*P* ratio was typically less than one. This result is also consistent with Jassby *et al.* (1993) and Canuel *et al.* (1995) who used different approaches but both concluded that the sources of organic matter to the South Bay are dominated by phytoplankton primary production (the water column is net autotrophic). At the other three sites, in the lower South Bay (Station 32), Central Bay (Station 18), and San Pablo Bay (Station 13), there were seasonal shifts in *R*:*P* ratios indicating seasonal changes in the balance between production and consumption.

Plankton and Benthic Respiration

Organic matter that is not respired in the water column can be transported from the system or settle to the bottom where it can be respired or buried. Benthic respiration rates from San Francisco Bay were available from 14 measurements made between October 1991 and October 1993 from a station between Stations 27 and 32 (J. Caffrey, pers. commun.). These benthic respiration rates were averaged by month and compared to average monthly plankton respiration rates from measurements made in 1993 at Stations 27 and 32 (Fig. 8). Since station location and sample dates were different, the average planktonic and benthic respiration rates presented in Fig. 8 should not be quantitatively compared but rather should be considered a qualitative comparison of these measures.

In SFB, as was found in Chesapeake Bay (Kemp *et al.* 1992), plankton respiration rates were much more variable than those measured on the benthos. Plankton respiration rates associated with the spring bloom were high but decreased, rapidly at first, then more gradually through the remainder of the year, dropping below benthic respiration rates in October and December. Average benthic rates increased from winter through May, possibly in response to a combination of increasing temperatures and increased supply of organic matter from the annual spring bloom, then generally showed a small decline after May. Benthic respiration rates remained relatively constant through summer and fall until reaching another peak in December. The variations in plankton and benthic respiration rates likely result, in part, from the more rapid cycling of organic matter in the water column relative to the sediments (Kemp *et al.* 1992). Organic inputs to the benthos support a lower rate of respiration over a longer period of time.

The response of both planktonic and benthic respiration rates to the spring bloom highlights its importance in the SB. In contrast, in Tomales Bay which lacked a large spring bloom, benthic respiration rates did not increase substantially until summer after primary production had increased (Dollar *et al.* 1991). The low spring benthic respiration rates in Tomales Bay were hypothesized to be the result of a limited supply of labile organic matter.

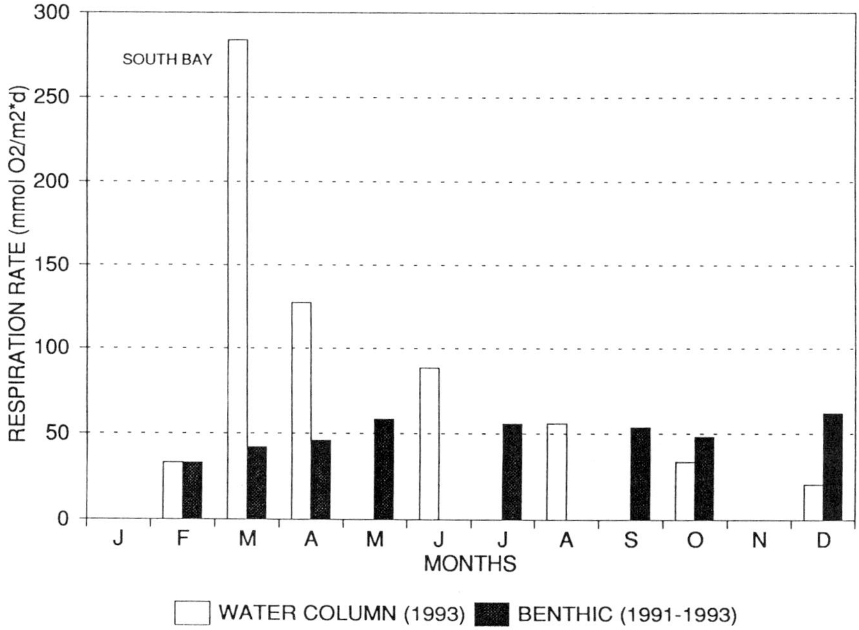

FIGURE 8. The 1993 average monthly planktonic and 1991-1993 average monthly benthic respiration rates (mmol O^2/m^2*d) in the South Bay. Benthic respiration rates (J. Caffrey, pers. commun.) are averages of available measurements and are not evenly spaced over the 1991-1993 time period. Comparisons between planktonic and benthic rates should be considered qualitative.

CONCLUSIONS

Plankton respiration rate measurements in SFB show strong relationships with a number of biomass indices and biological process rates. These relationships, particularly between plankton respiration rates, and rates of primary production and benthic respiration provide insights into the sources and utilization of organic matter in SFB. In particular, the *R:P* results reported here indicate that the South Bay is generally net autotrophic and the North Bay, generally net heterotrophic, consistent with the organic C budget of Jassby *et al.* (1993). The relationships between plankton respiration rates and instantaneously-measured variables (*e.g.*, particulate carbon, particulate nitrogen, chlorophyll) also suggest that information about the utilization of organic matter by plankton could be inferred from easily measured discrete measurements

ACKNOWLEDGMENTS

The major portion of the funding for this work came from the U.S. Geological Survey Toxic Substances Hydrology Program. This work was part of a pilot study for the Regional Monitoring Program, coordinated by the Regional Water Quality Control Board and managed by the Aquatic Habitat Institute. The Regional

Monitoring Program is supported by a number of participants, including municipal and industrial dischargers, cooling water and storm water dischargers, and dredgers. Additional funding was provided by the California Department of Water Resources. We thank them for their financial support. We also wish to thank Jane Caffrey, Liz Canuel, Brian Cole, Alan Jassby, Francis Parchaso, Christie Tyler, Sally Wienke, and the crew of the *R.V. Polaris* for all of their assistance.

REFERENCES

Alpine, A. E., and J. E. Cloern. (1988). Phytoplankton growth rates in a light-limited environment, San Francisco Bay. *Marine Ecol. Progress Ser.* 44:167-173.

Alpine, A. E., and J. E. Cloern. (1992). Trophic interactions and direct physical effects control phytoplankton biomass and production in an estuary. *Limnol. Oceanogr.* 37:946-955.

Azam, F. T. Fenchel, J. G. Field, J. S. Gray, I. A. Meyer-Reil, and F. Thingstad. (1983). The ecological role of water-column microbes in the sea. *Marine Ecol. Progr. Ser.* 10:257-263.

Caffrey, J. M., B. E. Cole, J. E. Cloern, J. Rudek, A. E. Tyler, and A. D. Jassby. (1994). *Studies of the San Francisco Bay, California, Estuarine Ecosystem: Pilot Regional Monitoring Results, 1993.* U.S. Geol. Surv., Open-File Rept. 94-82. 411 pp.

Canuel, E. A., J. E. Cloern, D. B. Ringelberg, J. B. Guckert, and G. K. Rau. (1995). Application of molecular and isotopic tracers to understanding sources of organic matter and its incorporation into the food webs of San Francisco Bay. *Limnol. Oceanogr.* 40:67-81.

Cloern, J. E. (1987). Turbidity as a control on phytoplankton biomass and productivity in estuaries. *Continental Shelf Res.* 7:1367-1381.

Cole, B. E., and R. E. Herndon. (1979). *Hydrographic Properties and Primary Productivity of San Francisco Bay Waters.* U.S. Geol. Surv., Open-File Rept. 79-983. 97 pp.

Conomos, T. J. (1979). Properties and circulation of San Francisco Bay waters. Pages 47-84 *in* T. J. Conomos, ed., *San Francisco Bay: The Urbanized Estuary.* Pacific Division, American Association for the Advancement of Sscience, San Francisco, CA.

Conomos, T. J., R. E. Smith, and J. W. Gartner. (1985). Environmental setting of San Francisco Bay. *Hydrobiologia* 129:1-12.

Dollar, S. J., S. V. Smith, S. M. Vink, S. Obrebski, and J. T. Hollibaugh. (1991). Annual cycle of benthic nutrient fluxes in Tomales Bay, California, and contribution of the benthos to total ecosystem metabolism. *Marine Ecol. Progr. Ser.* 79:115-125.

Ducklow, H. W., and F. Shiah. (1993). Bacterial production in estuaries. Pages 261-287 *in* T. E. Ford, ed., *Aquatic Microbiology: An Ecological Approach.* Blackwell Scientific Publ., Oxford, UK.

Fenchel, T. (1988). Marine plankton food chains. *Ann. Rev. Ecol. Systems* 19:19-38.

Graneli, W., and E. Graneli. (1991). Automatic potentiometric determination of dissolved oxygen. *Marine Biol.* 108:341-348.

Holligan, P. M., P. J. LeB. Williams, D. Purdie, and R. P. Harris. (1984). Photosynthesis, respiration and nitrogen supply of plankton populations in stratified, frontal, and tidally mixed shelf waters. *Marine Ecol. Progr. Ser.* 17:201-213.

Hopkinson, Jr., C. S., R. D. Fallon, B. Jansson, and J. P. Schubauer. (1991). Community metabolism and nutrient cycling at Gray's reef, a hard bottom habitat in the Georgia Bight. *Marine Ecol. Progress Ser.* 73:105-120.

Hopkinson, Jr., C. S., B. Sherr, and W. J. Wiebe. (1989). Size fractionated metabolism of coastal microbial plankton. *Marine Ecol. Progr. Ser.* 51:155-166.

Iriarte, A., G. Daneri, V. M. T. Garcia, D. A. Purdie, and D. W. Crawford. (1991). Plankton community respiration and its relationship to chlorophyll *a* concentration in marine waters. *Oceanologica Acta* 14:379-388.

Jassby, A. D., J. E. Cloern, and T. M. Powell. (1993). Organic carbon sources and sinks in San Francisco Bay: Variability induced by river flow. *Marine Ecol. Progr. Ser.* 95:39-54.

Jensen, L. M., K. Sand-Jensen, S. Marcher, and M. Hansen. (1990). Plankton community respiration along a nutrient gradient in a shallow Danish estuary. *Marine Ecol. Progr. Ser.* 61:75-85.

Jonas, R. (1992). Microbial Processes, organic matter and oxygen demand in the water column. Pages 113-148 *in* D. E. Smith, M. Leffler, and G. Mackiernan, eds., *Oxygen Dynamics in the Chesapeake Bay*. Maryland Sea Grant, College Park, MD.

Kemp, W. M., P. A. Sampou, J. Garber, J. Tuttle, and W. R. Boynton. (1992). Seasonal depletion of oxygen from bottom waters of Chesapeake Bay: Roles of benthic and plankton respiration and physical processes. *Marine Ecol. Progr. Ser.* 85:137-152.

Knapp, G. P., M. C. Stalcup, and R. J. Stanley. (1991). Iodine losses during Winkler titrations. *Deep-Sea Res.* 38:121-128.

Kremer, J. N., and S. W. Nixon. (1978). A coastal marine ecosystem: Simulation and analysis. *Ecological Studies*, vol. 24. Springer-Verlag, Heidelberg. 217 pp.

Langdon, C. (1988). On the causes of interspecific differences in the growth-irradiance relationship for phytoplankton. II. A general review. *Jour. Plankt. Res.* 10:1291-1312.

Lima, H. de S., and P. J. Williams. (1978). Oxygen consumption by the plankton population of an estuary – Southampton waters. *Estuarine Coastal Marine Sci.* 6:515-521.

Lorenzen, C. J. (1967). Determination of chlorophyll and phaeopigments: Spectrophotometric equations. *Limnol. Oceanogr.* 12:343-346

Nichols, F. H., J. E. Cloern, S. N. Luoma, and D. H. Peterson. (1986). The modification of an estuary. *Science* 231:567-573.

Packard, T. T., and P. J. Williams. (1981). Rates of respiratory oxygen consumption and electron transport in surface seawater from the northwest Atlantic. *Oceanol. Acta* 4:351-358.

Peterson, D. H., L. E. Schemel, R. E. Smith, D. D. Harmon, and S. W. Hager. (1987). The flux of particulate organic carbon in estuaries: Phytoplankton productivity and oxygen consumption. Pages 41-49 *in* S. Subitzky, ed., *Selected Papers in the Hydrologic Sciences*. U.S. Geol. Surv. Water-Supply Pap. 2330.

Riemann, B. (1978). Carotenoid interference in spectrophotometric determinations of chlorophyll degradation products from natural populations of phytoplankton. *Limnol. Oceanogr.* 23:1059-1066.

Strickland, J. D. H., and T. R. Parsons. (1972). A practical handbook of seawater analysis, 2nd ed. *Bull. Fish. Bd. Canada* 167:21-26.

Turner, R. E. (1978). Community plankton respiration in a salt marsh estuary and the importance of macrophytic leachates. *Limnol. Oceanogr.* 23:442-451.

Van Es, F. B., and V. P. Ruardij. (1982). The use of a model to assess factors affecting the oxygen balance in the water of the Dollart. *Netherlands Jour. Sea Res.* 15:313-330.

Williams, P. J. L. (1982). Microbial contribution to overall plankton community respiration – Studies in enclosures. Pages 305-321 *in* G. D. Grice & M. R. Reeve, eds., *Marine Mesocosms*. Springer-Verlag, New York.

Williams, R. B., and M. B. Murdoch. (1966). Phytoplankton production and chlorophyll concentration in the Beaufort Channel, North Carolina. *Limnol. Oceanogr.* 11:73-82.

REGIONAL DIFFERENCES IN THE ORIGINS OF ORGANIC MATTER IN THE SAN FRANCISCO BAY ECOSYSTEM

EVIDENCE FROM LIPID BIOMARKERS

ELIZABETH A. CANUEL[1] AND JAMES E. CLOERN
U.S. Geological Survey, 345 Middlefield Rd., MS-496
Menlo Park, CA 94025

Differences in the sources and reactivity of particulate matter in the northern and southern regions of San Francisco Bay (SFB) arise from spatial variations in the processes controlling the production and delivery of organic materials. This study used lipid biomarker compounds (fatty acids and sterols) associated with suspended particulate matter to evaluate temporal and compositional variations in the quality of organic matter delivered to sites representing these two domains: Stn 657 on the Sacramento River and Stn 27 in southern SFB. Particulate matter was collected from the surface waters several times during 1992-1993, a year with large seasonal fluctuations in river flow. Particulate carbon concentrations associated with events of freshwater inflow to northern SFB were higher (368 µM) than those associated with the spring phytoplankton bloom in southern SFB (135 µM). However, the composition of the organic matter associated with these two sources was very different. Fatty acid concentrations were generally higher in the South Bay (3-75 µg L^{-1}) than at the Sacramento River site (3-9 µg L^{-1}) and contained a larger percentage of polyunsaturated components (28-56% versus 20-25%), labile constituents that are indicative of phytoplankton. Although total sterol concentrations were comparable at both sites, concentrations of individual sterols derived from terrestrial vascular plants were higher at the Sacramento River site (Stn 657) while concentrations of sterols derived from phytoplankton were higher at Stn 27. These results indicate that within a single complex estuarine system such as SF Bay, large and quantifiable spatial variations exist in the chemical character and origin of the organic matter that is available to fuel ecosystem processes.

The dynamic and complex nature of estuarine environments makes it difficult to quantify spatial and temporal variations in the inputs of organic matter. Organic matter originates from a number of marine and terrestrial sources, with the relative importance of each of these sources generally varying along the freshwater-seawater continuum (Spiker & Schemel 1979; Cifuentes *et al.* 1988; Lucotte *et al.* 1991; Jassby *et al.* 1993). Together with spatial variations in organic matter inputs, temporal variations in the delivery of organic matter result from seasonal or inter-annual fluctuations in processes such as river flow and primary production. Although variations in organic matter inputs may have profound effects on system metabolism or production at higher trophic levels, only recently have there been studies that evaluated the natural variability in the delivery or form and composition of the suspended organic matter within estuaries (Zhang *et al.* 1992; Laureillard and Saliot, 1993; Fichez *et al.* 1993; Canuel *et al.* 1995).

In addition to spatial and temporal variations in the amount of organic matter delivered to estuarine environments, the nature of this organic matter also varies.

---

[1] Present Address: School of Marine Science/Virginia Institute of Marine Science, The College of William and Mary, P.O. Box 1346, Gloucester Point, VA 23062

Organic matter is derived from a number of sources originating in the surrounding watershed or from riverine, wetland, and tidal flat environments, which exchange directly with the estuary. Sources of organic matter include primary producers such as phytoplankton, benthic microalgae, macroalgae, and vascular plants, as well as materials transformed by heterotrophic organisms such as bacteria. In urban areas, sewage and municipal discharge supply additional organic matter (Kanazawa & Teshima 1978; LeBlanc *et al.* 1992; Quemeneur & Marty 1992). These inputs cover a range in chemical reactivity and hence suitability for utilization by heterotrophic organisms. Thus, quantitative information of changes in organic matter quality (*i.e.*, source and reactivity) is fundamental to understanding how estuarine ecosystems function.

Although the pool of organic matter in estuaries is typically large, a substantial fraction of the organic matter in estuaries is refractory (Day *et al.* 1989). Refractory constituents include materials derived from terrigenous environments. For example, terrestrial sources of organic matter make up over 20% of the organic inputs to estuaries on a global basis (Woodwell *et al.* 1973). Detrital inputs resulting from degradative processes that occur before delivery to the estuary or within the estuary itself also contribute to refractory carbon pools. As a result, although the delivery of carbon to estuarine environments can be large, only a small fraction of this carbon may be available to heterotrophic organisms over short time scales (*i.e.*, seasonal to annual). Unreactive constituents are either preserved in sediments accumulating in the estuary or are exported and buried along the continental margins. The ability to separate and quantify the reactive and refractory components of the organic matter delivered to and exported from estuaries is an important, albeit little understood, aspect of the carbon cycle and is essential to the development of accurate global carbon budgets.

A number of approaches have been used to identify sources of organic matter. The specificity of these different approaches ranges from simple bulk measurements, such as elemental and isotopic techniques, to the use of source-specific compounds (biomarkers) covering a wide spectrum in chemical reactivity (*e.g.*, fatty acids to lignin oxidation products). Biomarker compounds provide information about the biological origins of organic matter. Consequently, these molecules (or their degradation products when pathways are well-established) have been used to infer the origins of organic matter associated with particulate matter, sediments and petroleum source rocks (Hunt 1979; Wakeham *et al.* 1984; Brassell & Eglinton 1986). Molecular information is more specific, and the methods are more sensitive, than either elemental or isotopic techniques. As a result, these techniques can provide better resolution in identifying the biological origins of organic matter, while allowing for the simultaneous evaluation of multiple sources (Hedges & Prahl 1993).

In this study we used two classes of biomarker compounds (fatty acids and sterols) to compare temporal and compositional variations in the suspended organic matter at two sites in the San Francisco Bay-Delta System: USGS station 657 in the lower Sacramento River, the major source of freshwater to the Bay; and station 27, located centrally in southern SF Bay. Our selection of these two sites was based on the analysis of Jassby *et al.* (1993), who demonstrated that the two most important sources of organic carbon to the Bay-Delta system are riverine inputs (dominant in Suisun Bay and the upper reaches of northern SF Bay) and microalgal production (dominant in the South Bay). We collected samples on five dates during 1992-1993, a wet year (rainfall ~ 20% above normal) that followed six years of drought in California. Particulate matter associated with periods of low (baseline) vs. high freshwater inflow (Fig. 1), and seasonally varying primary production (bloom vs. non-bloom) was sampled. This approach for characterizing temporal variability was based on our previous demonstration (Canuel *et al.* 1995) that the lipid and carbon isotopic composition of organic matter in SF Bay changes in response to both hydrologic (floods) and biological (algal blooms) events. The flow events sampled during 1993 enabled us to take advantage of a natural occurrence in order to effectively characterize organic matter delivered to the estuary from the surrounding watershed.

METHODS

Samples of surface water (~ 2 m depth) were collected using a peristaltic pump. The water was filtered under low pressure through a "sandwich" consisting of 59μm polyester monofilament (PeCap) and pre-combusted (450°C; 12 hours) glass fiber filters (Gelman, A/E^2; 1.0 μm nominal pore size). The monofilament screen and glass fiber filters were then placed in separate jars, methylene chloride:methanol (2:1; v:v) was added, and samples were stored at - 80°C. Before extraction, the filters were shredded using forceps. Sample jars were placed in an ice bath and sonicated for 10 min using an ultrasonic probe (8 kHz; 5 min pulsed mode; 5 min

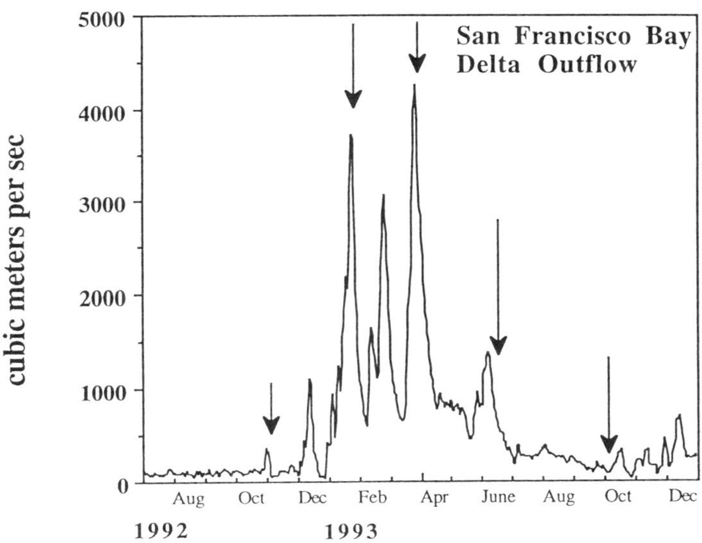

FIGURE 1. Delta outflow (cubic meters per second), an index of freshwater inflow to SF Bay, for July, 1992 through December, 1993 (from Sheila Greene, CA Department of Water Resources). Arrows indicate sampling dates.

continuous). The extracts were decanted following centrifugation. The methylene chloride:methanol was renewed and the ultrasonic extraction was repeated twice following the initial extraction. A small volume of 5% NaCl solution (~ 30 ml) was added to the combined extracts, samples were allowed to separate into two phases and the organic (lower) phase was collected. The filter was extracted a final time with methylene chloride that was then used to re-extract the aqueous phase. The combined organic phases were dried overnight in the presence of anhydrous Na_2SO_4. The extract was concentrated using turbo-evaporation (Zymark TurboVap 500 sample concentrator[2]) and the weight of the extract determined gravimetrically.

A portion (30-50%) of the lipid extract was saponified using 1N KOH (Canuel & Martens, 1993 and references therein). The saponified extract was extracted under basic and acidic conditions yielding neutral (SAP-N) and acidic (SAP-A) fractions respectively. The acids (SAP-A) were methylated with 3% BF_3-CH_3OH and purified using column chromatography. Neutral lipids (SAP-N) were separated into constituent classes on a silica gel column using solvents of increasing polarity. Sterols were eluted with 15% and 20% ethyl acetate in hexane. The combined 15 and 20% ethyl acetate in hexane fractions were concentrated to 1 ml using turbo-evaporation, dried under N_2 and converted to trimethylsilyl (TMS) ethers using BSTFA (bis(trimethylsilyl) trifluoroacetamide). Fatty acids (as methyl esters) and sterols (as trimethylsilyl ethers) were analyzed by gas chromatography according to conditions given in Canuel & Martens (1993). Individual peaks were identified based on relative retention times of known standards and peak areas were quantified relative to internal standards added prior to GC analysis (methyl heneicosanoate for fatty acids and 5α-cholestane for sterols). Peak identifications for selected samples were verified using gas chromatography-mass spectrometry (GC-MS).

In addition to the lipid analyses, total suspended particulate matter (SPM), particulate carbon and nitrogen (PC and PN) and chlorophyll *a* concentrations were determined for separate water samples collected from the pump at the same time as the lipid samples. Water samples were filtered through 47 mm Nuclepore filters and SPM was quantified gravimetrically according to the method of Hager (1993). Chlorophyll samples were filtered on 47 mm GF-AE filters and frozen immediately after collection. The filters were ground and extracted in 90% acetone. Absorbances of the extracts were measured with a Hewlett Packard 8452A diode array spectrophotometer and concentrations of chlorophyll *a* and phaeopigments calculated according to Lorenzen (1967) and Riemann (1978). Particulate carbon (PC) and nitrogen (PN) concentrations were measured with a Perkin Elmer 240 elemental analyzer (Wienke & Cloern 1987). These samples were not treated to remove inorganic carbon, so PC concentrations include contributions from carbonate. However, comparison of POC and PC measurements from previous studies conducted at these sites indicates that a minor fraction (< 5%) of the PC is inorganic (Cloern *et al.* 1993; Hager, 1994).

---

[2] Mention of brand name products does not constitute endorsement by the USGS.

RESULTS

Bulk Parameters

Particulate carbon (PC) and nitrogen (PN) concentrations were higher, on average, and exhibited a larger range in the lower Sacramento River (Stn 657) than in the South Bay (Stn 27), with the highest concentrations following the onset of seasonal rains in January (Table 1). Concentrations measured during the January flood event were 3-7 times higher than during baseline or low flow conditions (*e.g.*, November and October). In the South Bay, PC and PN concentrations were highest (3x baseline) during the spring phytoplankton bloom that occurred in March. C:N ratios were consistently higher at the Sacramento River site than in the South Bay (15-18 vs. 10-14, respectively). Chlorophyll *a* concentrations ranged from 1.3-23.3 $\mu g\ L^{-1}$ in the South Bay, reaching a maximum during March. In the lower Sacramento River, by contrast, concentrations of chlorophyll were less than 3 $\mu g\ L^{-1}$ throughout this study. Suspended particulate matter (SPM) was consistently higher at the Sacramento River (14-200 $mg\ L^{-1}$) than in the South Bay (7-17 $mg\ L^{-1}$) with the maximum occurring in January following the onset of high freshwater flows. Carbon concentrations (PC) were highly correlated with SPM at the Sacramento River site ($r^2 = 0.98$) in contrast to the South Bay where PC was more strongly correlated with chlorophyll *a* ($r^2 = 0.98$). Total extractable lipid concentrations ranged from 50-762 $\mu g\ L^{-1}$ in the Sacramento River and from 48-332 $\mu g\ L^{-1}$ in the South Bay. Lipid was positively correlated with PC; overall concentrations were higher in the Sacramento River with a maximum during the January floods while concentrations were highest during March in the South Bay (Table 1). Particulate matter in the South Bay was enriched in lipid and carbon relative to Stn 657 (Fig. 2). The lipid and particulate carbon fractions of the SPM (% by weight) ranged from 0.18-0.62 and 2.2-4.6, respectively at Stn 657, and 0.5-2.0 and 4.8-10.4, respectively at Stn 27.

TABLE 1. Bulk parameters (as defined in text)

Parameter	Station	4-5 Nov 92	26-27 Jan 93	30-31 Mar 93	14-15 Jun 93	5-6 Oct 93
PC (μM)	SACR[1]	100.0	367.5	185.0	59.2	50.0
	SB	47.5	60.8	135.0	54.2	45.8
PN (μM)	SACR	5.6	21.9	11.9	3.4	3.1
	SB	3.8	4.4	13.8	3.8	4.4
C/N (atom)	SACR	17.9	16.8	15.5	17.4	16.1
	SB	12.5	13.8	9.8	14.3	10.4
Chl *a* (μg/L)	SACR	1.5	1.1	2.1	2.6	1.2
	SB	1.3	1.6	23.3	1.8	1.5
SPM (μg/L)	SACR	26	200	74	20	14
	SB	12	7	17	7	8
Lipid (μg/L)	SACR	161	762	132	53	50
	SB	56	107	332	48	70

[1] SACR refers to Sacramento River (Stn 657) and SB to South Bay (Stn 27)

Lipid Biomarker Compounds

The highest concentrations of fatty acids were associated with particulate matter collected from the South Bay during the phytoplankton bloom of March 1993 (Table 2). Total fatty acid concentrations were approximately an order of magnitude higher at this time than at any other time throughout the study period. In addition to these fluctuations in total fatty acid concentrations, there were significant changes in the relative abundance (%) of different components within this compound class. For example, fatty acids associated with particulate matter at the Sacramento River site generally contained a higher percentage of monounsaturated acids than those at the South Bay site (Fig. 3). In contrast, polyunsaturated fatty acids (C_{18}, C_{20}, C_{22}) had higher concentrations and comprised a larger fraction of the total fatty acids at Stn 27 than at Stn 657, with the highest levels associated with the March phytoplankton bloom (Table 2, Fig. 3). The highest concentrations of methyl-branched fatty acids (iso- and anteiso-) were measured during the March bloom in southern SF Bay (Stn 27; Table 2). However, the highest relative

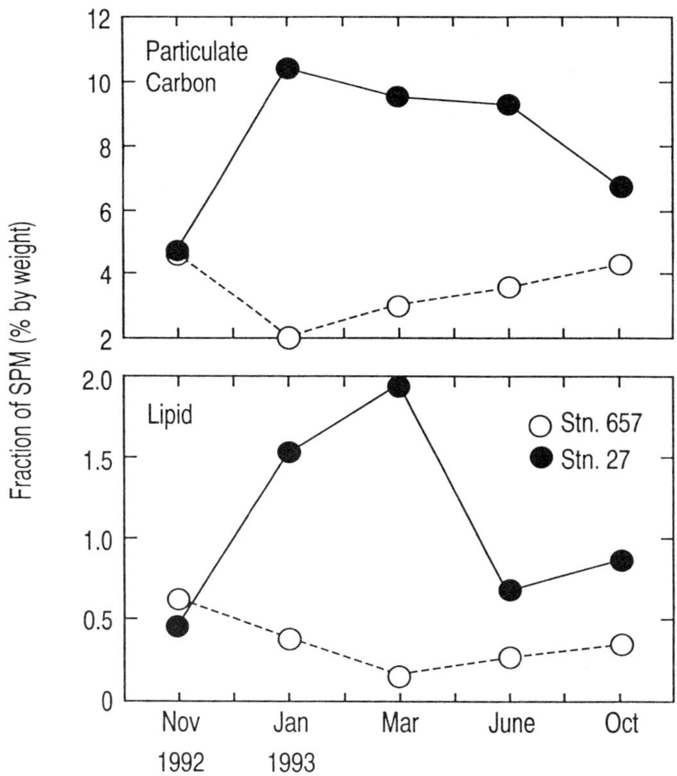

FIGURE 2. Concentration of particulate carbon (PC) and total extractable lipid normalized to suspended particulate matter (SPM) for Station 657, on the Sacramento River and Station 27 in south San Francisco Bay.

TABLE 2. Concentration of selected lipid compounds associated with suspended particulate matter.

	STN 27					*STN 657*				
Component	*5 Nov 92* *ng/L*	*26 Jan 93* *ng/L*	*30 Mar 93* ng/L	*14 Jun 93* *ng/L*	*5 Oct 93* *ng/L*	*5 Nov 92* *ng/L*	*26 Jan 93* *ng/L*	*30 Mar 93* *ng/L*	*14 Jun 93* *ng/L*	*5 Oct 93* *ng/L*
Fatty Acid[1]										
14:0	161	266	6406	337	278	177	346	300	205	311
iso-15:0	35	46	445	63	35	42	232	132	70	58
anteiso-15:0	37	41	306	48	31	32	244	106	43	40
16:1ω7	476	882	13994	647	711	490	1619	1104	812	992
16:0	642	1410	677	1524	766	485	1356	855	975	894
18:4ω3	263	1286	4899	1646	711	157	193	418	422	345
18:1ω7	287	582	2450	528	180	142	860	314	282	151
18:0	148	329	334	153	49	70	454	113	122	109
20:5ω3	214	736	19759	1138	682	194	373	691	n.d.	383
22:6ω3	140	464	4675	7	338	34	132	193	57	70
Total[2]	**3241**	**7864**	**74731**	**9176**	**4904**	**2742**	**9027**	**7592**	**6151**	**5043**
Sterol[3]										
S-1	186	220	373	188	116	448	803	403	344	208
S-2	151	216	552	373	127	203	276	139	211	103
S-3	108	177	2105	136	124		*tr*	*tr*		
S-4			*tr*			228	595	299	193	139
S-5	30	36	107	46	25	91	372	187	150	63
S-6	87	125	412	61	63	262	1388	697	226	124
Total[2]	**668**	927	**5500**	**1031**	**587**	**1648**	**4214**	**2117**	**1481**	**758**

[1] Fatty acid nomenclature A:BωC where A = number of carbon atoms, B=number of double bonds, and ωC indicates position of double bond from the aliphatic end.
[2] Total of all components (not limited to those presented in this table).
[3] S1 = cholest-5-en-3β-ol (cholesterol), S2 = 24-methylcholesta-5,22-dien-3β-01 (brassicasterol), S3 = 24-methylcholesta-5,24(28)-dien-3β-ol (24-methylenecholesterol), S4 = 24-methylcholest-5-en-3β-ol (campesterol), S5 = 24-ethylcholest-5,22-dien-3β-ol (stigmasterol), S6 = 24-ethylcholest-5-en-3β-ol (sitosterol).

abundance (%) of branched fatty acids was measured in the Sacramento River (Stn 657) during the first flow event in January (Fig. 3). This is due to the significantly higher concentrations of total fatty acids measured at Stn 27 during March. Branched fatty acids, however, made up only a small percentage (2-6 %) of the total fatty acid concentrations at all sites throughout the study.

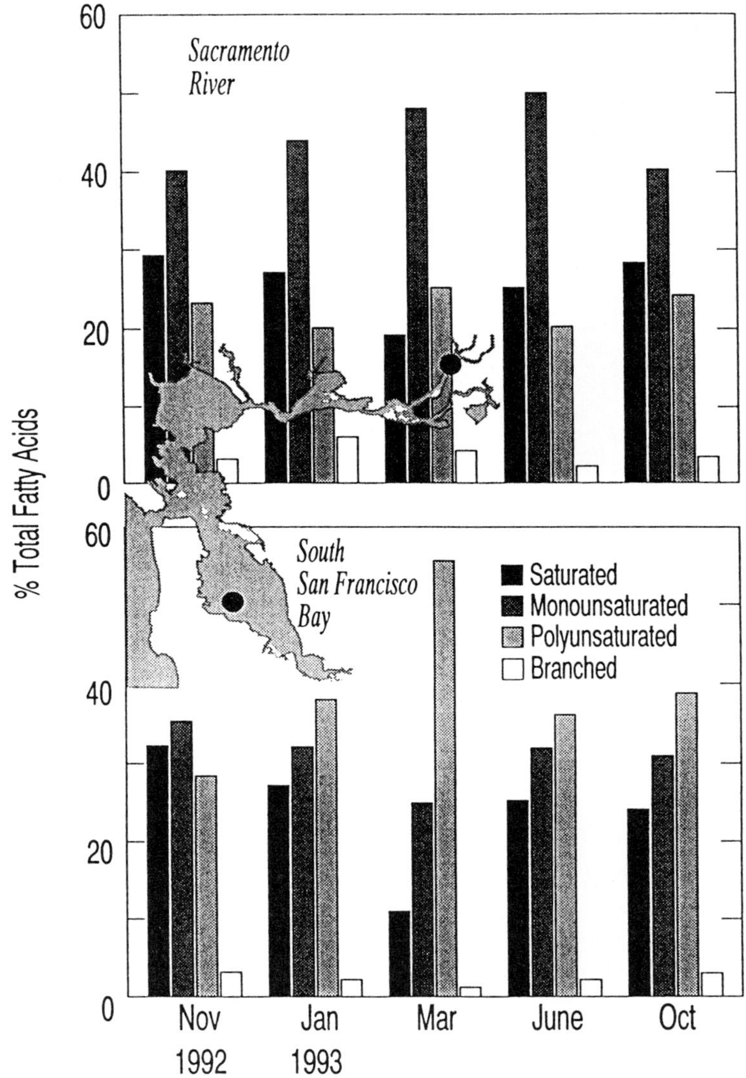

FIGURE 3. Percent abundance of saturated, monounsaturated, polyunsaturated and branched fatty acids relative to total particulate fatty acid concentrations. Generally speaking, fatty acid reactivity decreases with decreasing levels of unsaturation (*i.e.*, double bonds): polyunsaturated > monounsaturated > saturated and branched.

The concentrations of individual steroid compounds also varied spatially and temporally with the highest total sterol concentrations associated with particulate matter collected during March at Stn 27 (Table 2). With the exception of the March bloom period, sterol concentrations were higher at Stn 657 than at Stn 27, particularly during periods of high freshwater flow (January and June). Cholesterol (cholest-5-en-3β-ol) was often the most prevalent sterol component (Fig. 4; Table 2). However, concentrations of diatom-derived sterols (*e.g.*, 24-methylcholesta-5,22-dien-3β-ol and 24-methylcholesta-5,24(28)-dien-3β-ol) were 4-6x higher during bloom than during baseline conditions at Stn 27. Sterols derived from vascular plants (*e.g.*, 24-ethylcholest-5-en-3β-ol, 24-ethylcholest-5,22-dien-3β-ol) had the highest concentrations of all sterol components at Stn 657 during the high flows of January and March (Fig. 4; Table 2).

DISCUSSION

The use of multiple biomarkers (*e.g.*, source-specific molecules from a number of compound classes) improves the power of discrimination among the various biological sources of organic matter. Although the presence of a specific biomarker compound does not allow for a quantitative extrapolation to the amount of organic matter from that source, molecular tracers provide a good qualitative measure for the input of organic matter from the biomarker's established source (Hedges & Prahl 1993). In this study, source assignments were based on the distribution of sterol and fatty acid components — two of the most studied classes of biomarkers. Fatty acids are found in all organisms and fulfill numerous biochemical functions (*e.g.*, membrane components, energy storage, and protective coatings). Despite the fact that more than 500 different fatty acids have been identified, only a small number of these are abundant in most organisms (Killops & Killops 1993). Sterols function to increase rigidity in cell membranes and in lipoproteins. Compounds from these two classes have been used as indicators of organic matter derived from bacteria (fatty acids because steroids are rare in bacteria), fungi, algae, sewage and higher plants.

Sources of Organic Matter

Both classes of biomarker compounds used here, fatty acids and sterols, support previous studies that showed there are spatial and temporal variations in the sources of organic matter delivered to the northern versus southern regions of San Francisco Bay (Jassby *et al.* 1993; Canuel *et al.* 1995). Sterols associated with particulate matter delivered by the Sacramento River are dominated by components characteristic of vascular plants, including 24-methylcholest-5-en-3β-ol (campesterol), 24-ethylcholest-5,22-dien-3β-ol (stigmasterol) and 24-ethylcholest-5-en-3β-ol (sitosterol) (Goad & Goodwin 1972; Huang & Meinschein 1976; Volkman 1986). Concentrations of these three compounds were consistently higher at the Sacramento River site (Stn 657) than in the South Bay (Stn 27), suggesting that a greater fraction of the organic matter in the Sacramento River is derived from vascular plants. These three sterol components constituted approximately half of the total

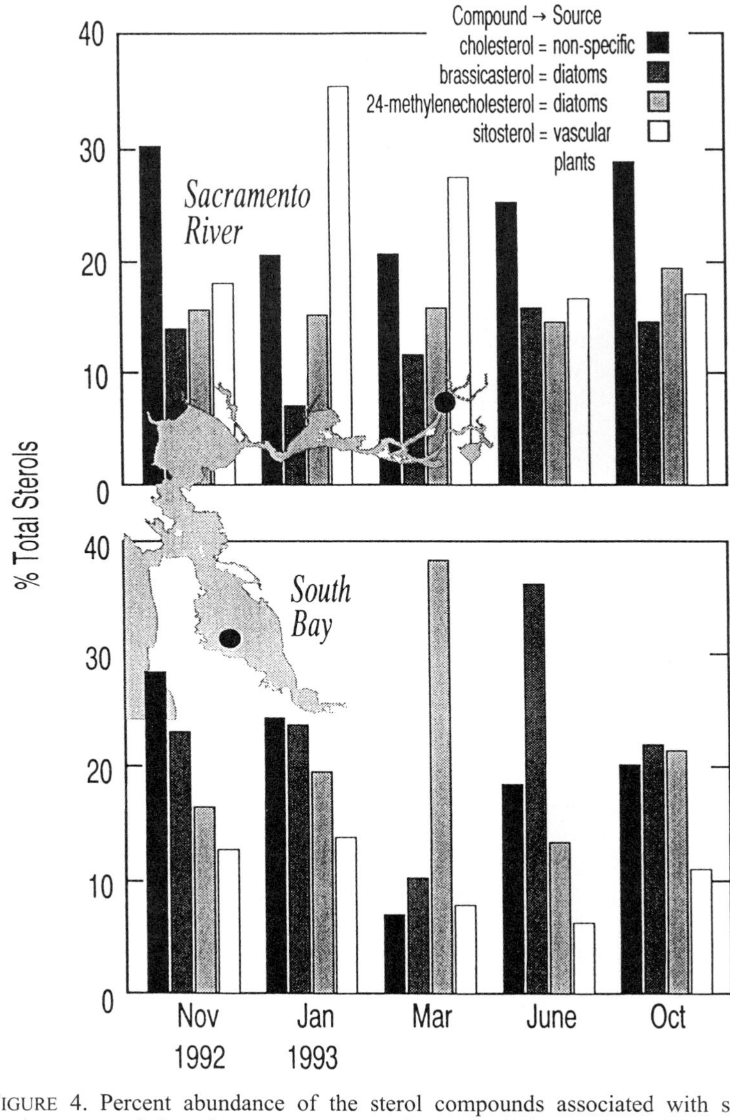

FIGURE 4. Percent abundance of the sterol compounds associated with suspended particulate matter. Selected components were chosen for their source-specificity.

sterol concentration measured at Stn 657 during January and March when freshwater flows were highest. It should be noted, however, that these components are not unambiguous biomarkers for terrigenous organic matter because they have been found in environments, such as Antarctic lake sediments, where terrigenous inputs are unlikely (Volkman, 1986). Cholesterol and 24-ethylcholest-5-en-3β-ol have also been measured at low concentrations in cyanobacteria (DeSouza and Nes, 1968). Likewise, cholesterol and sterols generally associated with phytoplankton have been measured at low concentrations in some vascular plants. Nonetheless, the large differences between the two stations in the concentrations of these components are likely the result of spatial variations in the relative contributions of organic matter derived from terrestrial plants.

In contrast, the sterol distribution in the South Bay (Stn 27) was dominated by phytoplankton-derived components, particularly those synthesized by diatoms. Generally, the major component was 24-methylcholest-5,22-dien-3β-ol, a sterol derived from diatoms (Volkman 1986). However, during the spring bloom, 24-methylenecholesterol was the most abundant sterol (Fig. 4; Table 2). This compound is also synthesized by diatoms, particularly those of the genus *Thalassiosira* (Volkman & Hallegraeff 1988; Gladu *et al.* 1991). High concentrations of this component are consistent with microscopic examination of phytoplankton samples collected at the same time that indicate that *Thalassiosira rotula* was a dominant organism during the March bloom (Caffrey *et al.* 1994). The absence of this genus in the freshwater regions of SFB accounts for the lack of this compound in particle samples collected from the Sacramento River site. As in the Sacramento River, concentrations of the terrestrial sterol 24-ethylcholest-5-en-3β-ol (sitosterol) at Stn 27 were elevated during periods of high freshwater inputs (January and March), possibly due to runoff from local streams. However, other sterols indicative of vascular plants (*e.g.*, campesterol) were either absent or only present at trace concentrations at the SB site (Table 2). Thus, sterol biomarkers can be sensitive indicators of temporal changes in the relative importance of different sources of organic matter at one location, just as they reflect spatial differences between locations within the estuary.

Although there is considerable similarity in the fatty acid composition of algae, bacteria and vascular plants, parameters such as chain-length, degree of unsaturation (*i.e.*, number of double bonds) and the occurrence of methyl-branched members can be used as indicators of organic matter source. In general, fatty acid distributions derived from vascular plants contain long-chain ($> n\text{-}C_{20}$) components with few double bonds. In contrast, shorter chain fatty acids ($< n\text{-}C_{20}$) with high levels of unsaturation are characteristic of algae. Branched fatty acids (iso- and anteiso-) in the C_{13} to C_{17} range and short-chain, saturated and monounsaturated *n*-fatty acids are usually abundant in bacteria (Gillan & Johns 1986).

Particulate matter collected from SF Bay contained fatty acids in the C_{12}-C_{24} range. During the March bloom, concentrations of polyunsaturated fatty acids at Stn 27 increased to 5-46x the prebloom and baseline concentrations (Table 2). High concentrations of polyunsaturated fatty acids indicate an enrichment in the contribution of algae to the total particulate matter. Furthermore, because polyunsaturated acids degrade readily in the environment, this algal material must be relatively fresh or even living (Shaw and Johns, 1985). High concentrations of 20:5, 16:1 and 14:0 fatty acids (Table 2) are consistent with the sterol distributions that indicated that diatoms, the dominant organisms during the bloom, were a probable source for much of the particulate matter at Stn 27 during this time (Kates & Volcani 1966; Orcutt & Patterson 1975).

Branched fatty acids, indicative of bacteria, also increased at Stn 27 during the spring bloom six- to eight-fold (Table 2). So it is likely that the biomass of heterotrophic bacteria increased as a response to the increased availability of algal-derived substrates during the March bloom; this is consistent with previous measurements by Hollibaugh (this volume) who documented elevated bacterial abundance and biomass during the South Bay spring bloom of 1991. Likewise, pelagic respiration (oxygen consumption) rates increased to about 10x the baseline rate during March (Rudek & Cloern, this volume). These are indicators of an ecosystem-level response to changes in the abundance and reactivity of organic matter as reflected by lipid biomarker compounds.

Saturated and monounsaturated compounds were the most abundant fatty acids associated with particulate matter at Stn 657 (Fig. 3, Table 2). These components are characteristic of marine and terrestrial detritus. Low concentrations (< 70 ng L^{-1}) of long-chain n-C_{20} and C_{22} acids were measured at Stn 657; however, these compounds were not detected at the South Bay site. These fatty acid data add to the evidence from the analysis of sterols that particulate organic matter at the Sacramento River site is more reworked (*i.e.*, detrital) and contains a larger contribution from vascular plants than at Stn 27. However, despite the dominance of saturated and monounsaturated fatty acids at Stn 657, polyunsaturated acids made up a substantial fraction (~ 20%) of the total fatty acids, albeit much lower than in the South Bay (Fig. 3). Previous work has indicated that polyunsaturated acids, most likely derived from freshwater algae, and the phytoplankton component of the suspended particulate matter were preferentially incorporated into the tissues of the Asian clam, *Potamocorbula amurensis*, one of the dominant macrofaunal species in the northern region of SF Bay (Canuel *et al.* 1995). Thus, even low levels of polyunsaturated fatty acids and other algal-derived components may be important to the overall ecosystem, as it is this portion of the particulate matter that supplies a large fraction of the assimilable carbon to heterotrophic organisms. However, biomarker distributions are limited in their ability to provide quantitative information of carbon (or energy) flow.

Controls on Organic Matter Delivery

Processes controlling the delivery of organic matter are significantly different in the northern vs. southern regions of SF Bay. Sources of organic matter are predominantly allochthonous in the North Bay where delivery is controlled by freshwater inflow from the Sacramento and San Joaquin Rivers (Jassby *et al.* 1993). In contrast, organic matter is derived primarily from autochthonous sources (phytoplankton and perhaps benthic microalgae) in the South Bay. In addition to these regional variations in the sources of particulate matter, the quality of the organic matter associated with suspended particulate matter is also very different. For example, concentrations of particulate carbon, total lipid, sterols and fatty acids were highest during the January sampling at Stn 657 (Tables 1 and 2). However, the concentration of total suspended particulate matter was also highest at this time (Table 1). When concentrations of these organic parameters are normalized to particulate matter concentrations, we find that suspended particulate matter collected from the Sacramento River region is depleted in organic matter (*i.e.*, carbon and lipid) relative to the South Bay (Fig. 2). Except for the initial high flow period in January, the lipid concentrations were lower during periods of high flow than during baseline or low flow conditions. The lowest particulate carbon fractions of the SPM were measured during the high-flow sampling periods of January and March. These data suggest that high concentrations of particulate organic matter associated with periods of freshwater inflow are controlled by the abundance of suspended particulate matter during periods of high flow rather than the organic content of the particles. As a result, the concentration of organic matter increases during high flow conditions simply as a result of the large influx of suspended matter. However, the quality of the organic matter as a resource for heterotrophs may be poor during these high-flow events. This is consistent with what has been found in other studies (Hager & Schemel, this volume).

In contrast, though concentrations of suspended particulate matter were substantially lower in the South Bay (Stn 27), this particulate matter was enriched in its lipid and carbon content (Fig. 2). Except for November, the lipid fraction of the SPM was higher in the South Bay than at the Sacramento River site. Although the lipid fractions were highest (2.0 %) during the March bloom, they were also high in January (1.5 %) following the onset of seasonal rains (Fig. 2). In contrast to the Sacramento River station, particulate carbon (% by weight SPM) was highest during periods of high flow (January, March and June) in the South Bay. Thus, in the South Bay, it appears that particulate matter is relatively high in organic matter content. This is particularly true during the spring bloom when phytoplankton comprise approximately half of the particulate carbon as estimated from the concentration of chlorophyll *a* (Wienke & Cloern 1987; Canuel *et al.* 1995). As a result, organic matter inputs to the South Bay are largely controlled by changes in the quality rather than the quantity of particulate matter.

Implications of Temporal Variation in River Flow

To evaluate interannual variations in organic matter inputs, we compared sterol and fatty acid distributions for particulate matter samples collected during a wet year, 1992-1993, with samples collected at the Sacramento River and South Bay sites during drought conditions in 1991 (Fig. 5). Although concentrations of total fatty acids were not available for the 1991 samples, we were able to take advantage of the availability of concentration data for ester-linked phospholipid fatty acids (PLFAs). PLFAs represent the largest pool of fatty acids in most organisms, sometimes making-up 65% of the total fatty acid abundance in plants (Killops & Killops 1993). For comparison purposes, we have assumed that PLFAs comprise about 50% of the total fatty acid abundance in SF Bay (Canuel, unpublished data).

Qualitatively, sterol and fatty acid components were similar for both periods, indicating that the biological sources of organic matter delivered to these regions of SF Bay do not change substantially between wet and dry years. Biomarker compounds suggest that particulate organic matter is derived from a mixture of "fresh" and detrital phytoplankton, bacteria and higher plants (Canuel *et al.* 1995). Phytoplankton sources of particulate organic matter are important throughout the bay system, although the size of these inputs is higher in the South Bay due to the magnitude of the spring phytoplankton bloom. In the Sacramento River, allochthonous inputs of bacterial and terrestrial vascular plants are also important (Jassby *et al.* 1993; Canuel *et al.* 1995).

Previous studies addressing the composition of organic matter in SF Bay (*e.g.*, Spiker & Schemel 1979; Jassby *et al.* 1993; Canuel *et al.* 1995) have not had the opportunity to characterize organic matter in the lower river/upper estuary during flood events such as those that occurred in January and March 1993. However, although our sampling spanned a "wet" period, we wish to make note of the fact that the "flood" events sampled here were small events in comparison to flood events observed during wet years (5000 m^3s^{-1}; see Schemel *et al.*, this volume). Qualitatively, the sources of organic matter delivered to SF Bay appear to be consistent from year to year. However, interannual variations in freshwater inflow and size of the spring bloom (*i.e.*, rates and spatial distribution of primary production) result in temporal variations in the magnitude of organic matter inputs associated with these events. For example, concentrations of sitosterol, a sterol often attributed to vascular plants, were three-fold higher at Stn 657 in March 1993 than during 1991 (Fig. 5). Furthermore, following the onset of seasonal rains in January 1993, concentrations of sitosterol rose to 1390 ng L^{-1}, the highest concentration measured at any time during 1991-1993. These high concentrations of sitosterol and other sterols derived from vascular plants indicate that wet years and

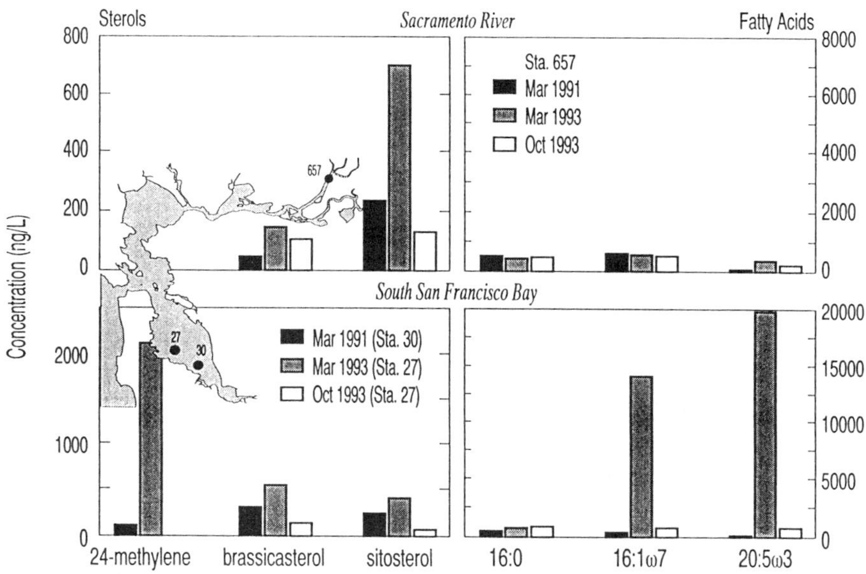

FIGURE 5. Comparison of sterols and fatty acids associated with suspended particulate matter collected during dry (March 1991 and October 1993) and wet (March 1993) conditions. Fatty acid data for 1991 are for ester-linked phospholipid fatty acids (PLFAs), a particular class of compounds comprising ~50% of the total fatty acid abundance. Note that different scales are used for the sterol and fatty acid concentrations.

seasons are characterized by enhanced inputs of terrestrial organic matter, particularly at the freshwater end-member. Concentrations of bacterial fatty acids (*e.g.*, iso- and anteiso-15:0; Table 2) were also highest in the Sacramento River during this initial period of high flow suggesting there may be some import of soil-associated microbes with this event. Seasonal variation in the nature of organic matter for some of the major river systems (*e.g.*, Amazon, Indus and Ganges and Brahmaputra) has shown that organic matter associated with high sediment discharge periods contains more terrestrial material and is less enriched in labile constituents (Zhang *et al.* 1992 and references cited therein). Similarly, the organic carbon content of suspended material has been shown to depend on the turbidity of the water. For example, particulate matter collected in turbid waters of the Chang Jiang Estuary, China was depleted in organic carbon relative to that which was collected in clear and more productive waters (Sicre *et al.* 1994).

The southern region of SF Bay is characterized by large interannual variations in the magnitude and spatial distribution of the spring phytoplankton bloom (Cloern and Jassby, 1994). This interannual variability is partly related to annual fluctuations in river flow, because large inputs of freshwater permit intense stratification of the water column — a physical phenomenon that promotes rapid growth of phytoplankton populations (Cloern 1991). For example, the spring bloom of 1991 (a dry year) was a minor event (maximum chlorophyll *a* ~ 10 $\mu g\ L^{-1}$) compared to the large spring bloom of 1993 when chlorophyll *a* concentrations exceeded 45 $\mu g\ L^{-1}$ in the lower reaches of South Bay (Caffrey *et al.* 1994). Sterol and fatty acid concentrations from this study also reflect this interannual variability. For example, diatom-derived sterols (24-methylenecholesterol and brassicasterol) were most abundant during the 1993 bloom (Fig. 5). Similarly, concentrations of specific fatty acids indicative of algae (*e.g.*, 14:0, 16:1ω7 and 20:5ω3) increased dramatically during the March 1993 bloom (Table 2, Fig. 5). Taken together, these data confirm ideas presented by Jassby *et al.* 1993 that there are substantial interannual variations in the supply of labile organic matter to the SF Bay ecosystem.

Comparison with Other Estuarine Systems

In order to evaluate how representative the variations in organic matter origins to SF Bay are to estuarine systems in general, concentrations of sterols associated with particulate matter collected from other estuaries were compared. Particulate sterol concentrations in SF Bay were within the range of values reported for other estuarine systems (summarized in Laureillard & Saliot, 1993). Although the sterol data are limited, these systems include river and marine coastal waters, covering a range in organic production. We were unable to find fatty acid data for comparison. The paucity of data reflects the fact that biomarker compounds have only recently been used to identify the sources of suspended particles in estuarine environments (*e.g.*, Bayona *et al.* 1989; Lajat & Saliot, 1990; Sicre *et al.* 1993).

The SF Bay sterol concentrations were lower than in heavily polluted systems such as the Ariake Sea (Japan) where sterol concentrations ranged from 6200-84500 ng L^{-1} (Kanazawa & Teshima, 1978), but similar to the Bedford Basin

(300-5000 ng L^{-1}; Pocklington *et al.* 1987). Of the reported data, the lowest and highest river concentrations were measured in the Rhone delta (Mediterranean Sea) (17-60 ng L^{-1} in July 1988 and 380-3100 in December 1981 and January 1989; Scribe *et al.* 1989; Scribe *et al.* 1991). Particulate sterol concentrations on the Sacramento River (Stn 657) were comparable to the winter Rhone delta values.

With the exception of the Rhone delta data (July 1988), sterol concentrations generally overlap in the river vs. coastal areas with higher concentrations in the rivers. In SF Bay, sterol concentrations were similar at the river (Stn 657) and bay (Stn 27) sites; however, the highest concentrations were found at Stn 27 during the spring bloom (5500 ng L^{-1}). The variability of sterol concentrations was also greater in the South Bay (590-5500 ng L^{-1}) than at the Sacramento River site (760-4200 ng L^{-1}).

Concentrations of sterols in both the dissolved and particulate phases have been reported for only one site, the Krka estuary (Laureillard & Saliot 1993). Dissolved sterols were significantly higher in the riverine vs. marine waters of this system. Sterol concentrations in the river waters of the Krka estuary ranged from 441-724 ng L^{-1} in the particulate phase and from 1220-1566 ng L^{-1} in the dissolved phase. In comparison, sterol concentrations were lower in the marine waters; 77-497 ng L^{-1} and 432 ng L^{-1} for the particulate and dissolved phases, respectively. Although dissolved sterol concentrations were higher than particulate, sterols made up a small fraction (~ 1%) of the total dissolved organic carbon. Dissolved organic carbon from river inflow is the largest source of carbon (62%) to the northern region of SF Bay (Jassby *et al.* 1993). Although we did not measure dissolved sterol and fatty acid concentrations, we can speculate that these compound classes comprise only a small fraction of this dissolved organic carbon.

CONCLUSIONS

In San Francisco Bay, early work by Spiker & Schemel (1979) set the framework for addressing the sources and dynamics of carbon input using stable carbon isotopes. Despite the limitations imposed by overlapping isotopic signatures of many of the potential sources of organic matter, this work established that *in situ* algal production and riverine carbon were the primary sources of carbon to SF Bay. Jassby *et al.* (1993) took this further by synthesizing information from previous studies identifying the primary sources of organic matter and spatial variations (North Bay vs. South Bay) in these sources, in addition to addressing freshwater flow induced variability. Together, these studies provided the foundation for designing our study by providing insight into the processes controlling spatial and temporal variability in the delivery of organic matter to this complex system.

The application of biomarker compounds has allowed us to begin to characterize the sources of organic matter to SF Bay with greater detail by allowing for the simultaneous identification of multiple sources. In this study we have taken the first step in evaluating regional differences in the composition of this organic matter and its availability to higher trophic levels. We have also addressed temporal variability, at time scales of less than a year, in the processes controlling these inputs. Specifically, we are able to draw the following conclusions:

(1) Analysis of lipid biomarker compounds supports the conclusions of Jassby et al. (1993) that controls on organic matter delivery vary spatially and temporally in SF Bay. Freshwater inflow associated with winter/spring rains is important in the northern region; autochthonous production during the spring phytoplankton bloom is an important source of organic matter in the lagoonal South Bay.

(2) Particulate matter delivered by each of these two processes differs in composition. During algal blooms, suspended loads are low but the particles are enriched in carbon. In contrast, during periods of freshwater inflow, suspended particulate matter concentrations are high, but the carbon content of this particulate matter is low.

(3) Biomarker compounds indicate that there are regional variations in the sources and reactivity of particulate organic matter delivered to SF Bay. High concentrations of total fatty acids and levels of unsaturation within this compound class suggest that particulate matter in the South Bay is more reactive than that delivered to the northern region from the Sacramento River. Sterol and fatty acid distributions indicate that particulate matter in the South Bay is largely derived from phytoplankton and algal detritus. In contrast, source-specific compounds indicate that particulate matter in riverine inputs to the North Bay is more reworked and is derived from a mixture of algal, vascular plant and bacterial sources.

ACKNOWLEDGMENTS

This work was done during a National Research Council post-doctoral fellowship with the U.S. Geological Survey. Funding was provided by the USGS Toxic Substances Hydrology Program. We thank Jane Caffrey, Brian Cole, Alan Jassby, Francis Parchaso, Joe Rudek, and Jan Thompson for assistance in sample collection and Byron Richards and Scott Conard of the R/V Polaris for their stellar efforts in accommodating the special needs of this project. Special thanks are extended to Scott Shapley and A. Christy Tyler for their dedicated assistance in the lab and Fran Hostettler who conducted the GC-MS analyses. We also wish to thank Jeanne Dileo-Stevens for preparing the illustrations. Reviews by J. Emmett Duffy, Wilfred E. Pereira, Fran Hostettler, and three anonymous reviewers improved the manuscript.

REFERENCES

Bayona, J. M., A. Farran, and J. Albaiges. (1989). Steroid alcohols and ketones in coastal waters of the Western Mediterranean: Sources and seasonal variability. *Marine Chem.* 27:79-104.

Brassell, S.C., and G. Eglinton. (1986). Molecular geochemical indicators in sediments. Pages 10-32 *in* M. Sohn, ed., *Organic Marine Geochemistry*. American Chemical Society, Washington, D.C.

Caffrey, J. M., B. E. Cole, J. E. Cloern, J. R. Rudek, A. C. Tyler, and A. D. Jassby. (1994). *Studies of the San Francisco Bay, California, Estuarine Ecosystem: Pilot Regional Monitoring Results, 1993*. U.S. Geol. Surv. Open-File Rept. 94-82.

Canuel, E. A., and C. S. Martens. (1993). Seasonal variations in the sources and alteration of organic matter associated with recently-deposited sediments. *Organic Geochem.* 20:563-577.

Canuel, E. A., J. E. Cloern, D. B. Ringelberg, J. B. Guckert, and G. K. Rau. (1995). Using molecular and isotopic tracers to examine sources of organic matter and its incorporation into the food webs of San Francisco Bay, *Limnol. Oceanogr.* 40:67-81.

Cifuentes, L. A., J. H. Sharp, and M. L. Fogel. (1988). Stable carbon and nitrogen isotope biogeochemistry in the Delaware estuary. *Limnol. Oceanogr.* 33:1102-1115.

Cloern, J. E. (1991). Tidal stirring and phytoplankton bloom dynamics in an estuary. *Jour. Marine Res.* 49:203-221.

Cloern, J. E., E. A. Canuel, and S. M. Wienke. (1993). *Particulate Organic Matter in the San Francisco Bay Estuary California: Chemical Indicators of its Origin and Assimilation into the Benthic Food Web.* U.S. Geol. Surv., Open-File Rept. 93-146. 46 pp.

Cloern, J. E., and A. D. Jassby. (1994). Year-to-year fluctuation of the spring phytoplankton bloom in South San Francisco Bay: An example of ecological variability at the land-sea interface. Pages 139-149 *in* J. H. Steele & T. M. Powell, eds., *Ecological Time Series.* Chapman and Hall, New York.

Day, Jr., J. W., C. A. S. Hall, W. M. Kemp, A. Yanez-Arancibia, and R. R. Christian. (1989). Microbial ecology and organic detritus in estuaries. Pages 257-308 *in* J. W. Day, Jr., C. A. S. Hall, W. M. Kemp, & A. Yanez-Arancibia, eds., *Estuarine Ecology.* John Wiley & Sons, New York.

DeSouza, N. J. and W. R. Nes. (1968). Sterols isolated from a blue-green alga. *Science* 162:363.

Fichez, R., P. Dennis, M. F. Fontaine, and T. D. Jickells. (1993). Isotopic and biochemical composition of particulate organic matter in a shallow water estuary (Great Ouse, North Sea, England). *Marine Chem.* 43:263-276.

Gillan, F. T. and R. B. Johns. (1986). Chemical markers for marine bacteria: Fatty acids and pigments. Pages 291-309 *in* R. B. Johns, ed., *Biological Markers in the Sedimentary Record.* Elsevier, Amsterdam, Netherlands.

Gladu, P. K., G. W. Patterson, G. H. Wikfors, D. J. Chitwood, and W. R. Lusby. (1991). Sterols of some diatoms. *Phytochemistry* 30:2301-2303.

Goad, L. J., and T. W. Goodwin. (1972). The biosynthesis of plant sterols. *Progr. Phytochem.* 3:113-198.

Hager, S. W. (1993). *Dissolved Nutrient and Suspended Particulate Matter Data for the San Francisco Bay Estuary, October, 1988 Through September, 1991.* U.S. Geol. Surv. Open-File Rept. 93-57. 58 pp.

Hager, S. W. (1994). *Particulate Carbon and Nitrogen and Suspended Particulate Matter in the Sacramento River at Rio Vista California, January 3-April 9, 1983 and October 31, 1983-November 29, 1984.* U.S. Geol. Surv., Open-File Rept. 94-470. 35 pp.

Hager, S. W., and L. E. Schemel. (1996). The supply of suspended sediment from the Sacramento River to the San Francisco Bay: II. Carbon and nitrogen contents and transports. Pages 189-215 *in* J. T. Hollibaugh, ed., *San Francisco Bay: The Ecosystem.* Pacific Division, American Association for the Advancement of Science, San Francisco, CA.

Hedges, J. I., and F. G. Prahl. (1993). Early diagenesis: Consequences for applications of molecular biomarkers. Pages 237-253 *in* M. Engle & S. Macko, eds., *Organic Geochemistry: Principles and Applications.* Plenum Press, New York.

Hollibaugh, J. T., and P. S. Wong. (1996). Distribution and activity of bacterioplankton in San Francisco Bay. Pages 263-288 *in* J. T. Hollibaugh, ed., *San Francisco Bay: The Ecosystem.* Pacific Division, American Association for the Advancement of Science, San Francisco, CA.

Huang, W.-Y. and W. G. Meinschein. (1976). Sterols as source indicators of organic materials in sediments. *Geochim. Cosmochim. Acta* 40:323-330.

Hunt, J. (1979). *Petroleum Geochemistry and Geology.* W. H. Freeman, San Francisco, CA. 617 pp.

Jassby, A. D., J. E. Cloern, and T. M. Powell. (1993). Organic carbon sources and sinks in San Francisco Bay: Variability induced by river flow. *Marine Ecol. Progr. Ser.* 95: 39-54.

Kanazawa, A., and S. I. Teshima. (1978). The occurrence of coprostanol, an indicator of faecal pollution, in seawater and sediments. *Oceanol. Acta* 1:39-44.

Kates, M., and B. E. Volcani. (1966). Lipid composition of diatoms. *Biochim. Biophys. Acta* 116:264-278.

Killops, S. D., and V. J. Killops. (1993). *An Introduction to Organic Geochemistry*. Longman Scientific and Technical (publisher), UK. 265 pp.

Lajat, M. and A. Saliot. (1990). Sterol composition of suspended matter from the Changjiang Estuary and adjacent East China Sea. Pages 642-666 *in* G. H. Yu, J. M. Martin, & J. Y. Zhou, eds., *Biogeochemical Study of the Changjiang Estuary*. China Ocean, Beijing.

Laureillard, J., and A. Saliot. (1993). Biomarkers in organic matter produced in estuaries: a case study of the Krka estuary (Adriatic Sea) using the sterol marker series. *Marine Chem.* 43:247-261.

LeBlanc, L. A., J. S. Latimer, J. T. Ellis, and J. G. Quinn. (1992). The geochemistry of coprostanol in waters and surface sediments from Narragansett Bay. *Estuarine Coastal Shelf Sci.* 34:439-458.

Lorenzen, C. J. (1967). Determination of chlorophyll and phaeopigments. Spectrophotometric equations. *Limnol. Oceanogr.* 12:343-346.

Lucotte, M., C. Hillaire-Marcel, and P. Louchouarn. (1991). First-order organic carbon budget in the St. Lawrence Lower Estuary from ^{13}C data. *Estuarine Coastal Shelf Sci.* 32:297-312.

Orcutt, D. M. and G. W. Patterson. (1975). Sterol, fatty acid and elemental composition of diatoms grown in chemically defined media. *Comp. Biochem. Physiol.* 50B:579-583.

Pocklington, R., J. D. Leonard, and N. F. Crewe. (1987). Le coprostanol comme indicateur de la contamination fecale dans l'eau de mer et les sediments marins. *Oceanol. Acta* 10:83-89.

Quemeneur, M., and Y. Marty. (1992). Sewage influence in a macrotidal estuary: Fatty acid and sterol distributions. *Estuarine Coastal Shelf Sci.* 34:347-363.

Riemann, B. (1978). Carotenoid interference in spectrophotometric determination of chlorophyll degradation products from natural populations of phytoplankton. *Limnol. Oceanogr.* 23:1059-1066.

Rudek, J. and J. E. Cloern. (1996). Respiration rates in San Francisco Bay. Pages 289-304 *in* J. T. Hollibaugh, ed., *San Francisco Bay: The Ecosystem*. Pacific Division, American Association for the Advancement of Science, San Francisco, CA.

Schemel, L. E., S. W. Hager, and D. Childers, Jr. (1996). Carbon content and supply of suspended sediment from the Sacramento River to the San Francisco Bay. Pages 237-260 *in* J. T. Hollibaugh, ed., *San Francisco Bay: The Ecosystem*. Pacific Divison, American Association for the Advancement of Science, San Francisco, CA.

Scribe, P., C. Pepe, C. Fuche, C. Barreau, R. Sempere, I. Bouloubassi, and A. Saliot. (1989). Application of the biogeochemical marker approach for determining the inputs and fate of biogenic and anthropogenic organic matter in the Rhone delta: Non-aromatic hydrocarbons and sterols associated to particles. *Water Pollution Res. Rept.* 13:173-178.

Scribe, P., C. Barreau, C. Pepe, C. Fuche, J. Dagaut, and A. Saliot. (1991). Fluvial inputs of organic matter from the Rhone river to the northwestern basin of the Mediterranean sea: Sterol composition of organic matter. *Water Pollution Res. Rept.* 20:259-267.

Shaw, P.M. and R. B. Johns. (1985). Organic geochemical studies of a recent Inner Great Barrier Reef sediment-I. Assessment of input sources. *Organic Geochem.* 8:147-156.

Sicre, M.-A., R. C. Tian, I. Broyelle, and A. Saliot. (1993). Aquatic distribution of 4-desmethyl sterols in the Chang Jiang Estuary, China. *Marine Chem.* 42:11-24.

Sicre, M.-A., R. C. Tian, A. and Saliot. (1994). Distribution of sterols in the suspended particles of the Chang Jiang Estuary and adjacent East China Sea. *Organic Geochem.* 21:1-10.

Spiker, E. C. and L. E. Schemel. (1979). Distribution and stable isotope composition of carbon in San Francisco Bay. Pages 195-212 *in* T. J. Conomos, ed., *San Francisco Bay: The Urbanized Estuary*. Pacific Division, American Association for the Advancement of Science, San Francisco, CA.

Volkman, J. K. (1986). A review of sterol markers for marine and terrigenous organic matter. *Organic Geochem.* 9:83-99.

Volkman, J. K. and G. M. Hallegraeff. (1988). Lipids in marine diatoms of the genus *Thalassiosira*: Predominance of 24-methylenecholesterol. *Phytochemistry* 27:1389-1394.

Wakeham, S. G., C. Lee, J. W. Farrington, and R. B. Gagosian. (1984). Biogeochemistry of particulate organic matter in the oceans: Results from sediment trap experiments. *Deep Sea Res.* 31:509-528.

White, D. C., W. M. Davis, J. S. Nickels, J. D. King, and R. J. Bobbie. (1979). Determination of the sedimentary microbial biomass by extractable lipid phosphate. *Oecologia* 40:51-62.

Wienke, S. M., and J. E. Cloern. (1987). The phytoplankton component of seston in San Francisco Bay. *Neth. Jour. Sea Res.* 21:25-33.

Woodwell, G.M., Rich, P.H. and Hall, C.A. (1973) Carbon in estuaries. Pages 221-239 *in* G. M. Woodwell & E. V. Pecner, eds., *Carbon in the Biosphere*. NTIS:CONF-720510.

Zhang, S., W.-B. Gan, and V. Ittekkot. (1992). Organic matter in large turbid rivers: The Huanghe and its estuary. *Marine Chem.* 38:53-68.

PROCESSES UNDERLYING PHYTOPLANKTON VARIABILITY IN SAN FRANCISCO BAY

ALAN D. JASSBY[1], JEFFREY R. KOSEFF[2], AND STEPHEN G. MONISMITH[2]
[1] Division of Environmental Studies, University of California, Davis, CA 95616
[2] Department of Civil Engineering, Stanford University, Stanford, CA 94305

Phytoplankton are responsible for most of the primary production within San Francisco Bay and constitute a large fraction of the total organic carbon supply from autochthonous and allochthonous sources combined. Phytoplankton variability may therefore play a role in the large year-to-year fluctuations at higher trophic levels observed over the last few decades, although strong evidence for such a link has been obtained only for the benthos.

In the past, spatial regionalization of phytoplankton processes has been deduced on the basis of surface topography but, at least in the case of South Bay, submarine features are as important. Primary productivity is less variable spatially than either chlorophyll or suspended particulate matter and a spatial mean for productivity may therefore require the fewest samples. Distortion of transect data by semidiurnal tides can be corrected, at least in principle, in the case of channel chlorophyll; tidally-driven sources and sinks, however, make such corrections impossible for suspended particulate matter and primary productivity, as well as for shoal chlorophyll.

Interannual phytoplankton variability in Suisun and South Bays has been described well elsewhere and our current understanding of the underlying mechanisms is summarized here. Variability in Suisun Bay is part of a large-scale spatial pattern that is determined by both exports from the Sacramento-San Joaquin Delta into water projects and by outflow from the Delta into the Bay. On average, more Delta phytoplankton flows into water projects than into Suisun Bay. Several mechanisms linking Delta outflow and chlorophyll variability in Suisun Bay have been proposed – shoal-channel exchange, grazing by benthic macroinvertebrates, turbidity – but their relative importance is still in question.

South Bay usually exhibits a single large spring diatom bloom, which has been investigated in several numerical and statistical modeling studies. A one-dimensional vertical model suggests that the onset of a bloom is contingent upon stratification during periods of weak tidal and wind mixing. During stratification, mixed-layer phytoplankton experience higher mean irradiance and are isolated from grazing pressures exerted by benthic macroinvertebrates. In years of extreme precipitation, freshwater inflow can also influence the South Bay bloom size, presumably by enhancing stratification. The one-dimensional model has also been extended by incorporating shoal-channel linkages. The extended model illustrates how benthic grazing and wind-driven resuspension of sediments in the shoals can control the magnitude of channel blooms and how shoal-channel exchange results in a "buffering" effect that prolongs bloom duration.

Data for North and Central Bays are comparatively sparse, but sufficient to motivate some speculation on mechanisms underlying chlorophyll variability. At least some of the Central Bay phytoplankton maxima may be connected to offshore blooms of neritic diatoms driven by periodic upwelling events.

San Francisco Bay offers a diverse environment for the development of phytoplankton communities. The northern reach is composed of two major subembayments, Suisun and San Pablo Bays, which are strongly affected by outflow from the Sacramento-San Joaquin Delta (the delta-shaped complex of channels and islands just upstream of the confluence of the Sacramento and San Joaquin rivers; Fig. 1). South Bay, on the other hand, is lagoon-like in character with much weaker inflows having more local effects. Central Bay, connecting the southern and northern reaches with the coastal ocean, is a deep mixing basin for water from all three sources. Both the northern and southern reaches have extensive shoals and more than 50% of the Bay is shallower than 2 m at low tide. River flow, winds and tides combine in different ways throughout the Bay to create a diversity of hydrodynamic conditions, superimposed on a wide array of nutrient sources, including rivers and streams, municipal, industrial and agricultural discharges, and precipitation.

Quantitative studies of the San Francisco Bay phytoplankton community originated with the work of Whedon in 1939, although his observations were based on collections from a San Francisco pier and therefore limited to Central Bay. The first description of seasonal changes that encompassed all the major subembayments was due to Storrs *et al.* (1966), who measured phytoplankton abundance as part of a multi-year study of water quality in the Bay. Since that time, numerous reports and published papers have appeared on phytoplankton seasonality and

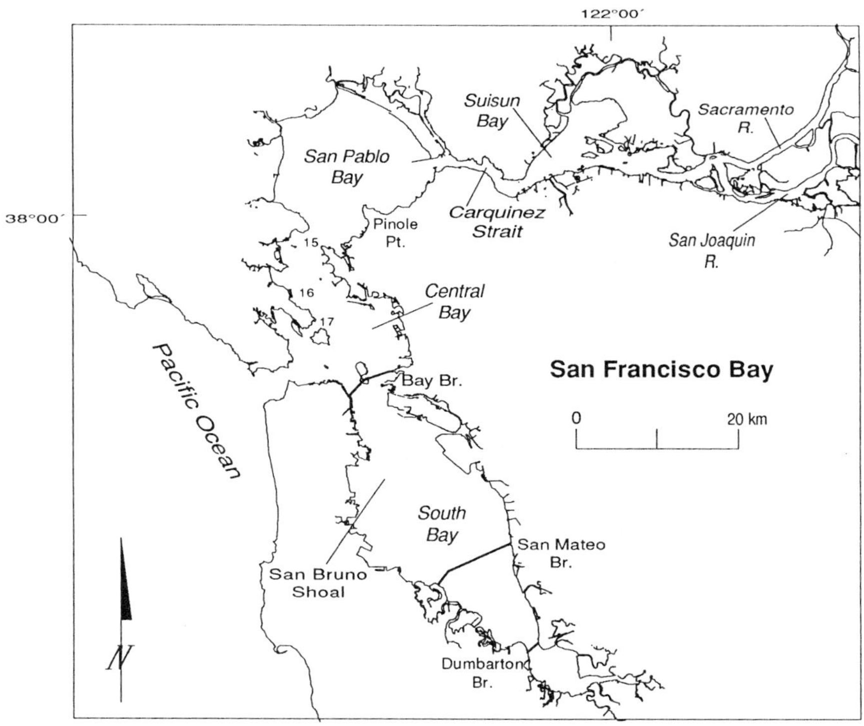

FIGURE 1. San Francisco Bay, showing the four main subembayments, the extreme western region of the Sacramento-San Joaquin Delta and other selected features referred to in the text.

interannual variability in San Francisco Bay and the Delta, many of which have been summarized by Cloern (1979, Cloern *et al.* 1985) in two previous articles. Here, we survey research that has taken place in the last decade to highlight the most important implications and the most pressing needs for further investigation.

Our main focus is variability in phytoplankton biomass and the mechanisms associated with this variability. Phytoplankton variability takes place on many time scales: the scale of weather patterns, of seasons and of years, as well as longer periods. In estuaries, the scale of semidiurnal tides is also important, and uniquely so. The mechanisms responsible for variability at different scales are often linked. If seasonality in river flow, for example, can induce seasonality in phytoplankton biomass, then it is likely that interannual variability in flow patterns will also induce year-to-year changes in biomass. For San Francisco Bay, most of the available information concerns variability on the scale of seasons to years, although there will be an opportunity to comment on other time frames as well.

Before launching into an overview of phytoplankton variability, it is well to ask why phytoplankton matter. Of course, one can always invoke the global role that phytoplankton have in the cycling of matter and transfer of energy into the food web, and little doubt exists that they have played a fundamental role in maintaining a planet suitable for animal life. What, however, are the practical implications of variability specifically in the San Francisco Bay phytoplankton? One basic assumption motivating work on phytoplankton in the Estuary has been that this variability propagates through the food web and ultimately affects organisms such as fish that are of more immediate concern to people. This assumption actually has two parts: (1) phytoplankton are a major source of energy and nutrients for the food web; and (2) variability at higher trophic levels is driven at least in part by variability at lower trophic levels.

A recent organic carbon budget for North and South Bays in 1980 enables us to address the role that phytoplankton play compared to other potential food sources (Jassby *et al.* 1993). The budget was comprehensive in the sense that all conceivable organic carbon sources were considered, although little could be deduced about net transport at the seaward boundary of the Bay and many of the other estimates were highly uncertain. The analysis suggested that seagrasses, macroalgae, bacterial autotrophs, runoff, atmospheric deposition, spills, groundwater and biotic transport were negligible sources. Phytoplankton, benthic microalgae, riverborne organic loads, tidal marsh export and point sources, on the other hand, each contributed at least 5% of the total organic carbon supply to either North or South Bays, or both. For South Bay, phytoplankton production was dominant and constituted about 60% of the total organic carbon for 1980 (Fig. 2). North Bay sources, in contrast, were dominated by the loading of organic carbon from the Sacramento and San Joaquin rivers and autochthonous phytoplankton production provided only about 20% of the total. The calculation of particulate organic carbon loading was based on the data of Schemel (1984), which did not cover the entire year, and so an estimate of January-March loading was required. A more recent study by Schemel *et al.* (this volume) suggests that this estimate was too low and that allochthonous sources are probably even more important.

The availability of this river-borne organic matter to the North Bay's food web is unknown. According to the revised estimates of Schemel *et al.* (this volume) combined with the original analysis by Jassby *et al.* (1993), 21% of it is in the form of particulate matter. Sixteen percent of this particulate organic matter (*i.e.*, 3.3% of the total organic carbon) appears to be phytoplankton and phytoplankton-derived detritus from upstream. Most of this phytoplankton-associated organic matter was produced in the Delta (Jassby & Powell 1994) and is probably not very old, in which case it might be comparable to autochthonous phytoplankton production in terms of availability. The remainder, however, is of largely unknown composition (*cf.* Canuel & Cloern, this volume). BOD_5 measurements in the Sacramento River over many years correspond on average to only about 10% of the total organic carbon concentration (Jassby *et al.* 1993), which suggests that most of the organic matter is not readily utilizable. A further complication is the timing of river-borne loads: most of the annual particulate load is transported during short-term events of days to weeks in winter (Schemel *et al.*, this volume). Compared to the load during other seasons, this material therefore has a shorter residence time in the Bay during a period of colder temperatures, with a resulting smaller probability of being incorporated into the food web.

It is a useful exercise to reconsider the North Bay carbon budget under the assumption that only 10% of the river-borne load can be assimilated into the food web (Fig. 2). In this case, the situation in North Bay is more similar to that of South Bay, in terms of both total organic carbon supply and the importance of phytoplankton production. The availability of organic matter imported from tidal marsh and other sources is probably less than that of microalgal production as well (although in these cases we have even less basis on which to make a correction). Within the range of uncertainty regarding the total amount of river-borne organic matter, its seasonality and its bioavailability, autochthonous phytoplankton production may very well play a major role in the base of the food web in North Bay, as well as in South Bay. Peterson *et al.* (1987) also suggested, on the basis of oxygen consumption rates, that autochthonous phytoplankton production was a major source of bioavailable organic matter in northern San Francisco Bay.

One of the clearest indications that variability at higher trophic levels is driven by variability at the base of the food web was provided by an experimental study of the bivalve *Macoma balthica* (Thompson & Nichols 1988). The four experimental sites were located on upper intertidal mudflats, two in South Bay, one in San Pablo Bay and one in the Carquinez Strait. Over a two-year period, growth rates were measured for individuals implanted in experimental containers at each site. Growth was not strongly affected by either salinity or temperature. Rather, differences in growth rates and tissue weights at the various sites could be attributed most strongly to differences in water column and/or benthic chlorophyll *a* levels. Water column chlorophyll *a* in particular was a significant predictor of growth rate for the South Bay and San Pablo Bay stations. Food-limited growth has also been reported for the bivalve *Corbicula fluminea* located upstream in the tidal freshwater portion of the estuary (Foe & Knight 1985).

Limitation of the pelagic food chain by phytoplankton availability is less

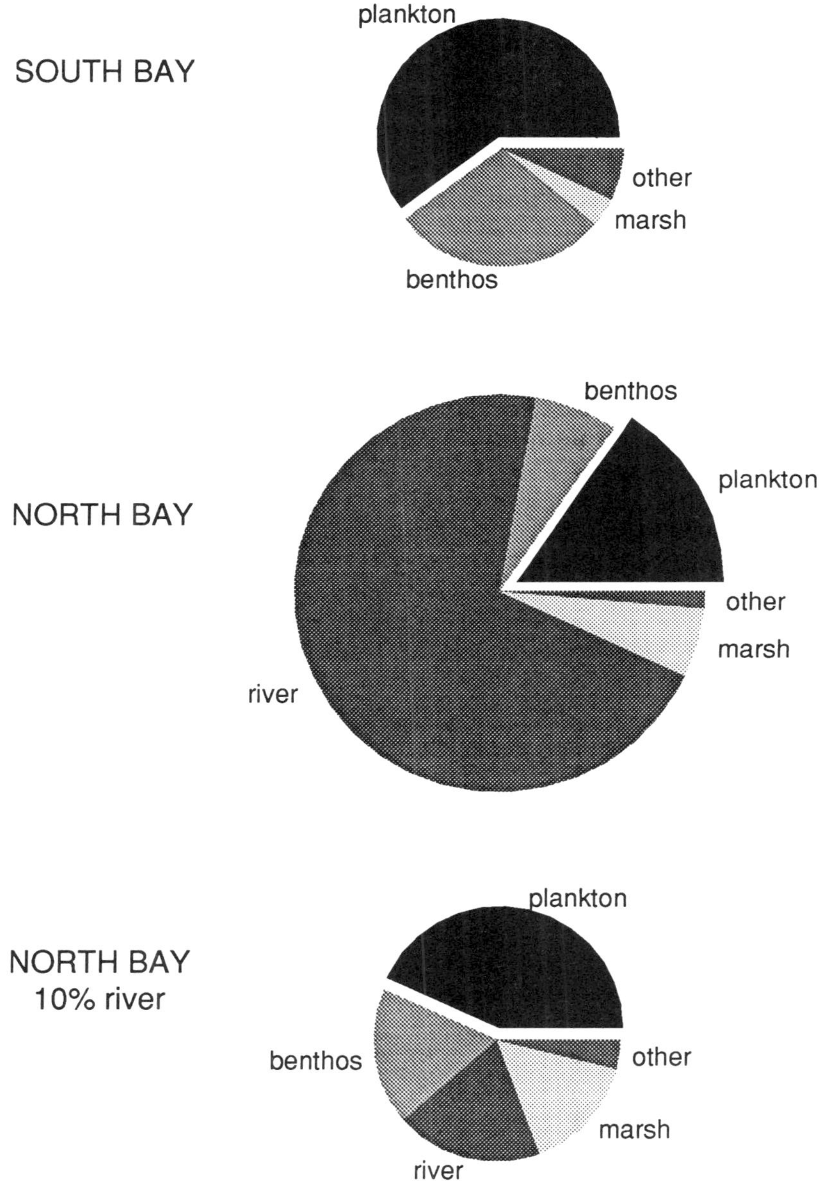

FIGURE 2. Organic carbon sources in San Francisco Bay during 1980. The area of each circle is proportional to the total supply of organic carbon (g C $m^2 yr^{-1}$). *Top panel,* South Bay; *middle panel,* North Bay; *bottom panel,* North Bay, assuming only 10% of the river-borne organic matter is available. *Plankton,* phytoplankton production; *benthos,* benthic microalgal production; *river,* river-borne organic carbon load; *marsh,* imports from tidal marsh; *other,* mostly point sources and runoff (data from Jassby *et al.* 1993, incorporating a more recent estimate of river-borne particulate loading by Schemel *et al.*, this volume).

well-established. Although positive statistical associations between chlorophyll levels and zooplankton abundance can be shown, most estuarine populations are simultaneously affected by salinity distributions (Jassby *et al.* 1995) and the correlations could be an artifact. Zooplankton gut contents certainly illustrate the dominance of phytoplankton as a food source (Kost & Knight 1975), but there is almost no experimental evidence that low ambient phytoplankton concentrations control growth rates (*cf.* Kimmerer 1991). Recent declines in the planktonic copepod *Eurytemora affinis* are due to direct consumption by clams (*Potamocorbula amurensis*) rather than by competition between zooplankton and clams for phytoplankton (Kimmerer *et al.* 1994). Declines in *Neomysis mercedis*, on the other hand, a mysid shrimp too large to be eaten by *Potamocorbula*, may be due to competition for phytoplankton (Kimmerer, pers. commun.). The significance of food limitation for pelagic primary consumers remains an important issue in this estuary and needs to be resolved further.

SPATIAL REGIONALIZATION OF PHYTOPLANKTON PROCESSES

The physiography and hydrodynamics of the various subembayments are quite different from one another, and one can infer that the mechanisms underlying phytoplankton variability must differ spatially in the Bay. Where to draw the spatial boundaries corresponding to these mechanisms is not so clear, however. Traditionally, the Bay has been divided into South, Central, San Pablo and Suisun Bays (Fig. 1). These boundaries reflect surface physiographic features, but do not necessarily coincide with natural spatial domains for phytoplankton.

One of the few studies to directly address this issue of phytoplankton spatial domains was conducted in the South Bay channel during 1984-1985 (Powell *et al.* 1986). Three distinct spatial regimes were identified from continuous near-surface transects of chlorophyll *a* and salinity between the Dumbarton and Bay bridges. These regimes differed in mean value, in variance and in small-scale spatial structure (0.1-1 km) as revealed by their respective autocorrelograms. The boundaries between the three regions coincided with major topographic features, namely a transverse shoal (the San Bruno shoal) and a constriction in the vicinity of the San Mateo Bridge. These features were assumed to induce variability in physical processes. A counterclockwise gyre, for example, has been inferred for the central regime between the shoal and the constriction (Walters *et al.* 1985). The regime north of the San Bruno shoal is characterized by highly variable salinity, probably resulting from incomplete mixing of coastal ocean water and estuarine waters derived from both North and South Bays. The South Bay therefore cannot be viewed as a single domain. Moreover, at least one of the salient topographical features controlling spatial variability, namely the San Bruno shoal, is a completely submarine feature that has no relationship to the traditional boundaries used to divide the estuary. Similar studies for the northern estuary are sorely needed.

A further study of phytoplankton variability in the South Bay in 1987 included measurements in the shallows and estimates of primary productivity as well as of biomass (Powell *et al.* 1989). Primary productivity can be derived from measurements of chlorophyll, suspended particulate matter (SPM) and the incident light field (Cole & Cloern 1984, 1987). Because of the spatial correlation between SPM and chlorophyll, primary productivity was much less variable spatially than either SPM or chlorophyll alone. Turbidity and biomass both tended to be low in the channel and high in the shallows. Light penetration was therefore low when biomass was high and vice-versa, so that the positive effect of high chlorophyll levels was partially canceled by the negative effects of a small photic depth. This spatial covariability between SPM and chlorophyll implies that fewer measurements may be needed for reliable estimates of the spatial mean of productivity, compared to SPM or chlorophyll. As a result, primary productivity may serve as a more practical indicator of interannual variability in the phytoplankton community than biomass, providing that the covariability between SPM and chlorophyll decreases sufficiently at longer time scales. This point is a significant one for regional monitoring programs concerned with status and trends, as the spatial variability in phytoplankton biomass interferes with attempts to discern long-term trends from chlorophyll levels alone.

In estuaries and shallow coastal waters, changes on the tidal time scale are a serious confounding factor in sampling programs to determine spatial variation. The 1987 study included sufficient temporal resolution to characterize the effect of tidally-induced fluctuations (Cloern *et al.* 1989). Spatially-averaged chlorophyll *a* and SPM varied inversely with tidal height, ranging about 50% from their tidal means, although the relation of SPM to tidal height was weaker than for chlorophyll. As in the previous case, derived daily productivity varied less (ca. 15% of the tidal mean) due to short-term covariability between SPM and chlorophyll. In the channel, chlorophyll variability was strongly correlated with advective flux and the intratidal variability here was primarily due to the tidal advection of spatial biomass gradients. In contrast, chlorophyll was weakly correlated with advective flux over the shoals, suggesting effects from local sources and sinks. SPM variability was only weakly related to advective flux, both in the channel and shoals. A strong correlation with current speed in both deep and shallow waters suggested that tidally-driven resuspension rather than advection was the primary source of SPM variability. These observations have important implications for transect data: As chlorophyll behaves like a conservative substance on tidal time scales in the channel, it is possible in principle to correct it for tidal variation during a spatial transect; the same is not true of derived primary productivity, however, because of its dependence on both chlorophyll and SPM.

PROCESSES UNDERLYING PHYTOPLANKTON VARIABILITY

Suisun Bay

Suisun Bay is the site of an estuarine turbidity maximum (ETM) in summer (Peterson *et al.* 1975a,b) that is often closely associated with a spatial maximum in chlorophyll and netplankton (> 22 μm) as well. The main feature of interannual chlorophyll variability in Suisun Bay *and* the Delta combined is the size of this chlorophyll peak. For example, when chlorophyll data are binned by season and salinity range, highest year-to-year change is associated with the 2-10 mS cm^{-1} range in summer, which corresponds spatially and temporally with the fullest development of the ETM (Fig. 3).

In the late 1970s, it was proposed that the accumulation of netplankton and of non-living particles in the ETM resulted from the same process, namely density-selective retention of particles within an estuarine circulation cell (Arthur & Ball 1979; Ball & Arthur 1979).

A major mechanism invoked to account for interannual variability in the chlorophyll maximum is the effect of river flow on positioning of this circulation cell (Cloern *et al.* 1983). According to this hypothesis, when river discharge falls within a certain intermediate range, the circulation cell is positioned close to broad

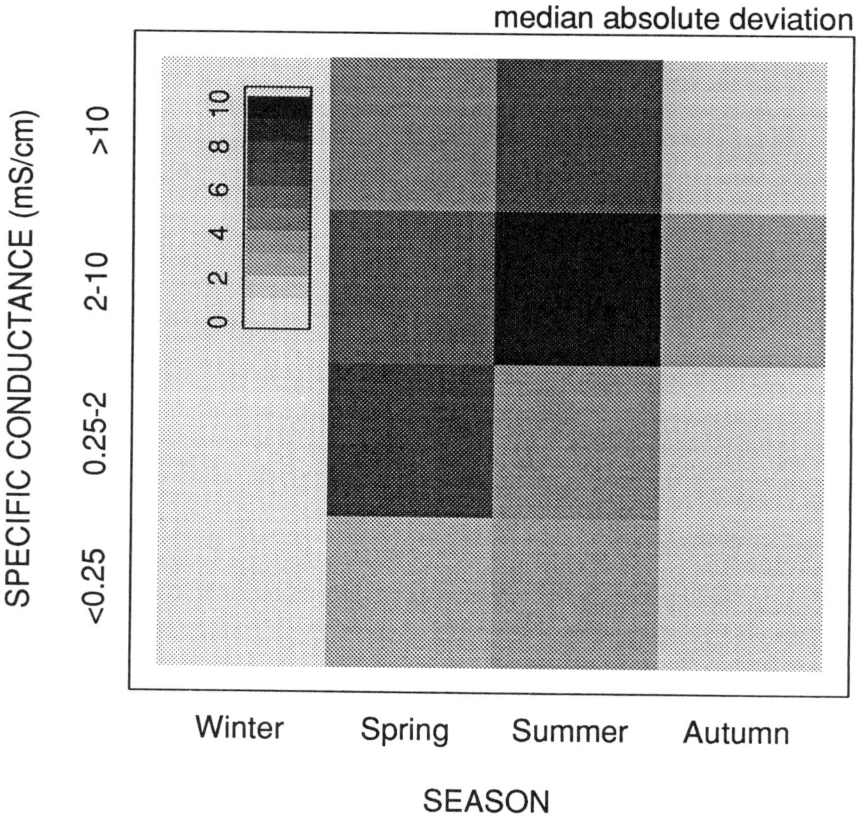

FIGURE 3. Median absolute deviations of annual median chlorophyll concentrations for the conductivity intervals and seasons indicated. Data were collected at 24 stations throughout Suisun Bay and the Delta. Data were available for all bins only for the following calendar years: 1971-1972, 1975-1976, 1979-1981, 1984-1986, and 1988-1989; other years could therefore not be included (after Jassby & Powell 1994).

expanses of lateral shoals where the mean light field is higher than in the channels. The netplankton associated with the ETM (typical taxa include *Thalassiosira decipiens, Skeletonema costatum,* and *Coscinodiscus* spp.; Cloern *et al.* 1985) can then exchange with shoal areas where the growth rates are higher and phytoplankton biomass accumulates. When river flows are too high or too low, the circulation cell is positioned in deep turbid channels upstream or downstream of these lateral shoals and growth rates are low or even negative due to the low ratio of photic to mixing depth (Cloern 1987; Alpine & Cloern 1988).

Recent measurements of currents in Suisun Bay have failed to detect gravitational circulation during much of the summer and cast doubt on the role of density mechanisms (Jon Burau, pers. commun.). The historical observations of a chlorophyll maximum in Suisun Bay are not in question and some mechanism of shoal-channel exchange dependent on river flow rates is probably responsible. Remote sensing studies with a multispectral scanner, for example, have demonstrated the tidal advection of high-chlorophyll water from the northern shoals into the main body of Suisun Bay (Catts *et al.* 1985; also see Burau *et al.*, this volume). The role traditionally ascribed to an estuarine circulation cell, however, is in doubt. The resolution of this issue awaits further field measurements and a three-dimensional simulation.

Peterson & Festa (1984; Peterson *et al.* 1989) constructed a two-dimensional (vertical and longitudinal along the channel), steady-state model of phytoplankton biomass for partially-mixed estuaries that demonstrated a great sensitivity on the part of both phytoplankton biomass and productivity to grazing pressures. Although the model precluded shoal-channel exchange, their conclusion is consistent with a process that comes into play during persistent extreme flow conditions. When flows are unusually low for periods of more than a year, the resulting equable salinity conditions in Suisun Bay permit colonization by and higher densities of estuarine bivalves. Nichols (1985) hypothesized that the failure of the Suisun Bay bloom during the drought of 1977 was due to the upstream migration of benthic invertebrates to this normally brackish area. Estimated filtration rates derived from equations in the literature for one of the species, the suspension-feeding bivalve *Mya arenaria*, suggested that all of the particles from the water column could be filtered on the order of once per day. A similar event began in 1986 with the onset of a prolonged drought and the invasion and rapid growth of *Potamocorbula amurensis* (Nichols *et al.* 1990). At the observed *Potamocorbula* population densities, suspension-feeding rates determined from laboratory studies (Cole *et al.* 1992) were greater than phytoplankton reproduction rates. Alpine & Cloern (1992) provided further evidence for these top-down impacts on the phytoplankton by detailing how chlorophyll concentrations are inversely related to benthic grazer abundance in the long-term record for Suisun Bay, including both drought periods.

Advection of phytoplankton and phytoplankton-derived detritus is an additional source of variability in the ETM-related chlorophyll peak. In their organic carbon budget for 1980, *Jassby et al.* (1993) showed that the loading of algal-derived organic carbon from the Delta into Suisun Bay actually exceeded autochthonous production within Suisun Bay; the importance of upstream phytoplankton has since been verified by independent methods (Canuel & Cloern, this volume). Jassby & Powell (1994) examined this source of variability in more detail. On the basis of statistical evidence and material budgets, they divided the influence of river flow on interannual variability into two separate variables: net Delta *outflow* into the Bay and *export* from the Delta into the state and federal water projects. As noted above, outflow probably affects chlorophyll through a variety of mechanisms. Outflow also partially determines chlorophyll loading from upstream but export plays an additional role. For a given outflow, higher export implies higher river *inflow* into the Delta, a lower residence time and lower phytoplankton densities within the Delta. Export also "short-circuits" the passage of algae-laden water from the San Joaquin River through the Delta and into the water projects before it can flow into the Bay. Median chlorophyll export to water projects exceeded chlorophyll outflow into the Bay by 60% during the period 1975-1989. An important implication of this analysis is that, by ignoring exports, the significance of Delta outflow to chlorophyll maxima in Suisun Bay has been overestimated in the past. Annual outflow and export are correlated and ignoring one leads to a biased view of the other.

A final source of interannual variability is turbidity and light penetration. The partially-mixed estuary model of Peterson & Festa (1984) demonstrated strong sensitivity of both biomass and productivity to suspended material. The statistical evidence also supports a direct role for turbidity in year-to-year chlorophyll variability. Lehman (1992) constructed a time series-regression model for monthly chlorophyll in Suisun Bay. An abstracted variable representing light penetration was formed by taking the first principal component of turbidity and Secchi depth. This variable entered into the model as a significant predictor at several lags. The proportion of the overall variability attributable to this light penetration effect was small, however. As Lehman (1992) pointed out, the performance of the models was restricted by assumptions of linearity; such assumptions do not apply to the ETM where the historical biomass peak is a distinctly nonlinear function of flow.

San Pablo Bay

Compared to Suisun and South Bays, almost no published information on year-to-year change is available for San Pablo Bay. Cloern *et al.* (1985) described the seasonal cycle for 1980 in detail. Typical numerical dominants included the neritic diatoms *Coscinodiscus* spp., although the cryptophyte *Chroomonas minuta* and other more oligohaline diatoms were also abundant. The primary features were chlorophyll peaks in spring and late summer-early fall. The spring maximum was the largest and both peaks were higher in the shoals than in the channel. This description is qualitatively similar to observations of the seasonal cycle during the 1970s (Ball & Arthur 1979), although the fall bloom exceeded the spring bloom at both shoal and channel sites in 1971 and the timing of the peaks varied by at least one month from year to year.

Cloern *et al.* (1985) suggested that phytoplankton levels in San Pablo Bay are controlled by advection, turbidity and grazing. The channel may be a net sink on an annual basis, with brief periods during spring and summer-fall when turbidity decreases sufficiently for positive net growth. At other times, growth rate falls and losses to advection and zooplankton grazing suppress chlorophyll levels. Shoal areas, in contrast to the channel, are a net source of phytoplankton. Restricted mixing depth in the shoals implies a higher growth rate than in the channel for equivalent turbidity conditions. Bloom levels also increase toward the northern shoals, where the water residence time is presumably higher than at locations nearer to the channel. As in Suisun Bay, benthic grazing may be a major sink for shoal phytoplankton. Despite this wealth of possible mechanisms, the true causal factors underlying both seasonal and interannual variability remain largely speculative.

One commonly observed feature of the spatial distribution in the San Pablo Bay channel is a chlorophyll maximum in spring near Pinole Point where the channel narrows and becomes relatively shallow (Fig. 1). Williams & Hollibaugh (1987) suggested that this maximum represents an estuarine circulation cell and that the biomass associated with it could be regulated by freshwater inflow from the Delta. Ball (1987), on the other hand, maintained that this spatial peak arises from "tidal pumping." The evidence is insufficient to throw further light on the actual mechanisms, but this chlorophyll feature may be ecologically important and invites further investigation.

Central Bay

Relative to the other subembayments, particularly South and Suisun Bays, the study of Central Bay phytoplankton has been neglected. The early work by Whedon (1939) and Storrs *et al.* (1966) suggested that phytoplankton achieved a seasonal maximum during the period March-June. The seasonal dynamics of chlorophyll, however, are not as well-behaved as suggested by the early studies, in which only cell numbers were determined. Assuming that chlorophyll is a better indicator of relative community biomass, transects of near-surface water during the years 1978-1981 indicated large interannual variability in the total phytoplankton biomass of Central Bay (Fig. 4). The correspondence between the two Central Bay stations implies that this variability had spatial coherence and is not attributable merely to measurement or sampling variability for individual stations. Year-to-year differences occurred in both the timing and the size of seasonal peaks.

During the period March-September, northerly winds along the California coast generate offshore currents and periodic upwelling events. The upwellings are often associated with blooms of neritic diatoms and these same species form some of the seasonal maxima within Central Bay, with dinoflagellates often dominating the autumn maxima (Cloern 1979; Jassby *et al.* 1994). A tempting conclusion is that at least some of the Central Bay blooms are connected to these offshore events. Is Central Bay merely a passive recipient of dispersed phytoplankton from offshore events, or do these neritic diatoms act primarily as seed for populations growing within the Bay? How does estuarine circulation interact with upwelling events to determine transport of coastal blooms into the Bay? Unfortunately, no simultaneous studies of the Bay and coastal events have yet been carried out and these interesting questions remain unexplored.

When Central Bay stations are compared with a site at the mouth of San Pablo

Bay, a further aspect of chlorophyll variability becomes apparent (Figs. 1 and 4). Some of the seasonal peaks decrease from Central Bay north while others increase in the northerly direction. For example, the large autumn bloom of 1979 drops by a factor of four between stations 17 and 15, while the smaller spring bloom increases by a factor of three. It is unlikely that these two events arise from the same causal factors and that a single mechanism such as dispersion from offshore blooms can account for Central Bay's variability. In fact, even the correspondence in species composition between Central Bay and offshore blooms is not conclusive, as some of the same dominant species bloom in San Pablo Bay (Cloern *et al.* 1985).

Toxic dinoflagellate blooms have never been recorded in the Bay. Species known elsewhere to form toxic red tides do occasionally occur in the Bay plankton. For example, during 1993, *Prorocentrum minimum* formed a small percentage of the autumnal bloom in Central Bay (Caffrey *et al.* 1994). This species has been associated with paralytic shellfish poisoning (PSP) incidents at other locations. Recently, it was found that diatoms can also produce a potent neurotoxin, domoic acid (DA), which is the causative agent of amnesic shellfish poisoning (ASP). In late summer and fall of 1991, a series of deaths of brown pelicans and Brandt's cormorants in Monterey Bay was traced to the consumption of anchovies that had, in turn, been feeding on large concentrations of *Pseudonitzschia australis*, a DA-producing diatom (Fritz *et al.* 1992). This species has not been identified as such in San Francisco Bay. It is interesting to note, however, that the very similar *Pseudonitzschia seriata* (formerly *Nitzschia seriata:* Hasle 1993) has been observed in Central Bay, forming large blooms at times (Cloern 1979). Some researchers believe that *P. seriata* is actually a cold northern species and that records of it in California may actually be misidentifications of *P. australis* (Taylor 1993). If so, then the DA-producing species could be entering Central Bay from the coastal ocean. Because northern anchovy (*Engraulis mordax*) is a major component of total fish biomass in the Bay, DA-forming diatoms have the potential to affect higher trophic levels of the estuary.

South Bay

The Spring Bloom. Spring phytoplankton blooms are a persistent feature in South Bay. In a matter of a few weeks, depth-averaged values of chlorophyll *a* concentration in the main channel between San Bruno Shoal and the Dumbarton Bridge can rise from less than 5 mg m^{-3} to more than 40 mg m^{-3} (Fig. 5). They can also return to background levels at the same rate, remaining there until the following spring. This spring bloom provides a significant fraction of annual phytoplankton production and therefore of the organic carbon input into South Bay (Fig. 2).

The spring bloom in the South Bay channel roughly coincides with a spring minimum in tidal energy. The working hypothesis for most of the past decade (Cloern 1984) has been that the spring bloom develops when the South Bay channel is stratified. Unlike the northern reach, which can stratify year-round due to Strain-Induced Periodic Stratification (Burau *et al.*, this volume), stratification in South Bay is episodic. Generally, significant stratification develops in South Bay only when local inflows and Delta outflow are large (Walters *et al.* 1985) and when

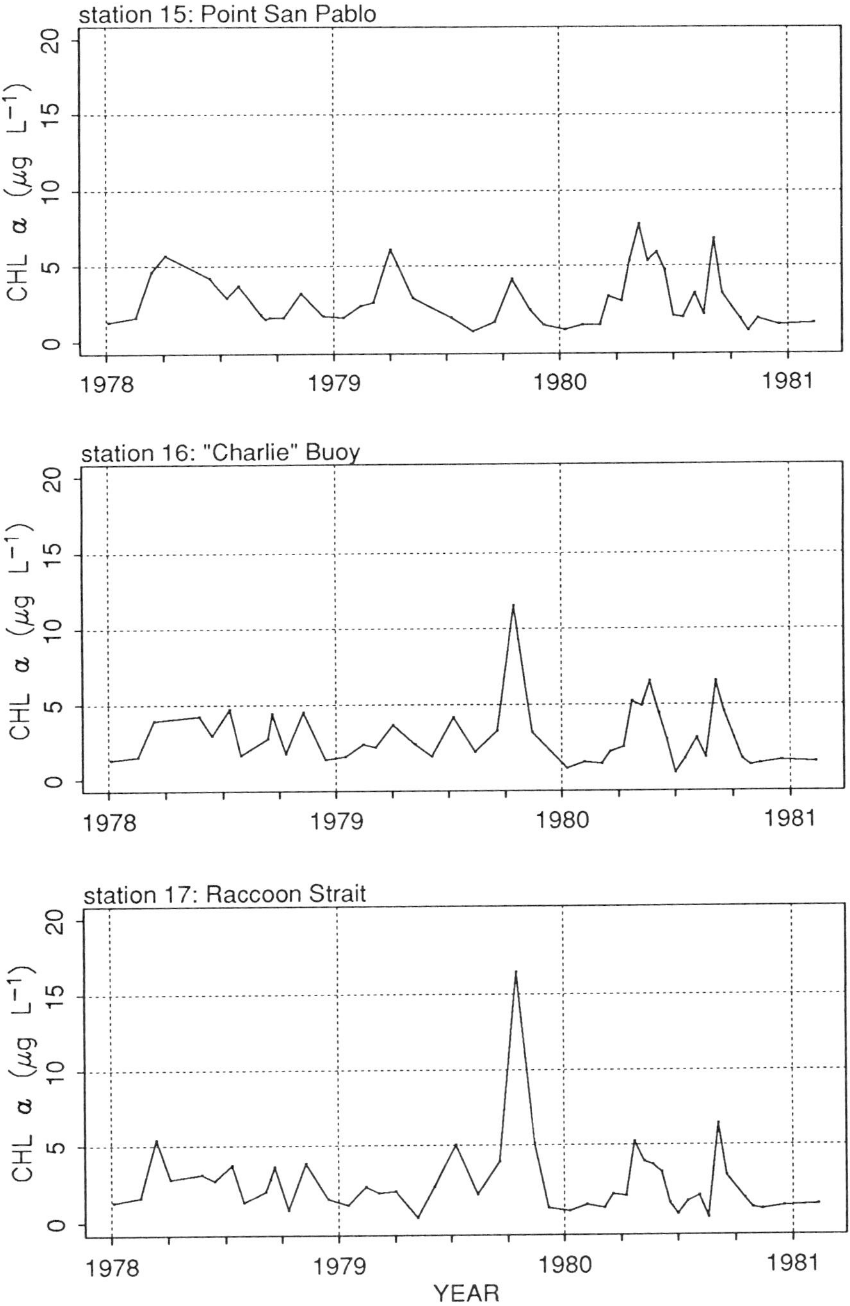

FIGURE 4. Chlorophyll *a* concentrations at 1 m depth at two Central Bay stations (16 and 17) and a station at the mouth of San Pablo Bay (15) (see Fig. 1 ; data from USGS).

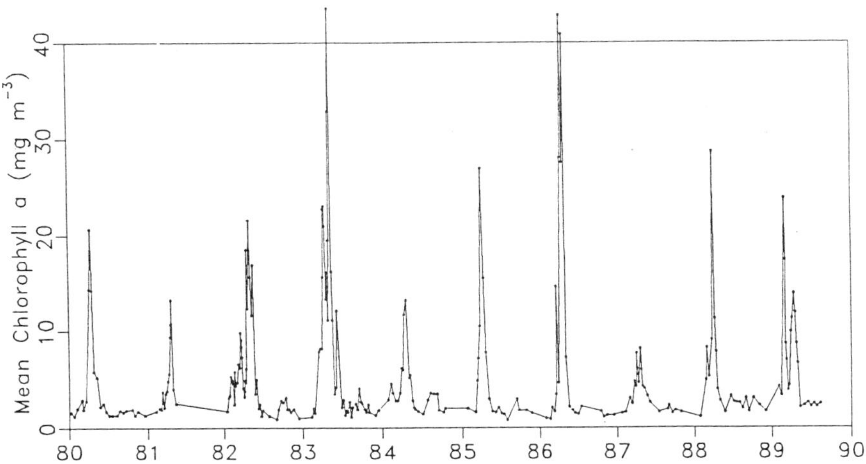

FIGURE 5. Chlorophyll *a* concentrations in South Bay illustrating the onset and evolution of the spring bloom as the dominant feature of variability (after Cloern 1991a).

tidal mixing is weak (Huzzey *et al.* 1990; Cloern 1984). As observed in 1982 (Huzzey *et al.* 1990), stratification can be associated with either Delta outflow, in which case low salinities are at the northern end of South Bay, or with local runoff (*e.g.*, from Coyote and Alameda creeks), in which case salinities tend to be lower at the southern end of South Bay. It is important to note that these two cases may create different residual flow patterns: in the first, surface waters of South Bay flow south, which probably serves to keep phytoplankton biomass within South Bay; in the second, surface waters flow out into Central Bay (Walters *et al.* 1985) and transport biomass from South Bay.

A One-Dimensional Model. The link between mixing and phytoplankton dynamics in South Bay was first explored in a numerical model by Cloern (1991a). He used a simple one-dimensional vertical model to suggest how bloom dynamics in shallow estuaries, such as South San Francisco Bay, might be controlled by spring-neap fluctuations in the intensity of tidal stirring. The model did not include many of the other complex mixing processes in estuaries, such as vertical variability in turbulent mixing intensity, effects of density stratification, wind stirring or high-frequency components of variability associated with the semidiurnal tides.

This work was extended by Koseff *et al.* (1993; KHMC hereafter), who treated mixing processes more realistically to reflect the dynamic nature of turbulence generation in estuaries. KHMC used numerical simulations, as well as scaling of the transport equation, in an attempt to determine what combinations of coupled physical-biological processes are necessary for a phytoplankton biomass increase in shallow turbid estuaries. The analysis further allowed them to describe the role of the various source and physical transport terms in the problem, as well as the role of spatial variability in the turbulent mixing and temporal variability in the tidal forcing.

KHMC found that the addition of semidiurnal tidal variation to their model greatly reduced biomass increases, indicating that variations of mixing on the time scale of hours are crucial. With physically realistic values for the turbulent mixing coefficient, their model did not simulate a bloom unless the effects of stratification were included. By greatly reducing vertical mixing (Denman & Gargett 1983), stratification serves to keep phytoplankton cells in the photic zone (typically the upper 2 m of the water column), thereby minimizing respiratory losses and enhancing net productivity. Furthermore, as suggested by Cloern (1982), reduced vertical mixing minimizes losses to benthic grazers, which appear capable of consuming all pelagic production when the water column is well-mixed. The occurrence of large blooms is, in general, a result of stratification providing a favorable light environment, while the eventual disappearance of the biomass is due to benthic grazing coming into play once stratification is eliminated (Fig. 6).

KHMC also found that stratification itself is not sufficient to ensure that a bloom will develop: mixing rates in the photic zone above the pycnocline must be small enough to allow significant biomass increases, implying that minimal wind stirring is a further prerequisite to bloom development in shallow turbid estuaries, especially those with abundant populations of benthic suspension feeders.

A Statistical Appraisal. In general, the picture of South Bay phytoplankton dynamics derived from modeling seems to support the view that blooms and stratification are linked in South Bay. An alternative, statistical approach to testing this hypothesis was taken by Cloern & Jassby (1995; *cf.* Cloern 1991b) who used principal components analysis (PCA) to analyze the spatial-temporal structure of the channel bloom. Examining 14 years of data, they found that variability in chlorophyll *a* concentration was dominated by two modes that explained 87% of the total variance. The first of these modes (PC_1, 51% of the variance) describes an early spring bloom (March-April) that is most intense at the southern end of South Bay and weakens monotonically towards the north. The second mode (PC_2, 36% of the variance) has a spatial structure nearly identical to that of the first; it describes a late spring bloom (May-June) that basically represents a prolongation of the earlier bloom during certain years.

To address the role of freshwater inflows, and hence indirectly stratification, Cloern and Jassby then looked at the PC amplitudes (scores), which measure the intensity of the response in a given year, as a function of mean salinity during the corresponding bloom periods, March-April for PC_1 and May-June for PC_2 (Fig. 7). Mean salinity serves as a proxy for freshwater flow. Except for two very wet years, 1983 and 1986, they found that freshwater flow had little systematic effect on the variability of either mode.

Although mean salinity is not an accurate indicator of vertical stratification intensity, this result raises some questions regarding the strength of the link between stratification and bloom intensity. While the one-dimensional modeling effort of KHMC revealed the importance of an additional factor — minimal wind stirring — to the occurrence of channel blooms in a shallow estuary, blooms in South Bay tend to last longer than predicted by the model. The channel bloom may last for several weeks in the presence of significant benthic grazing, whereas the KHMC

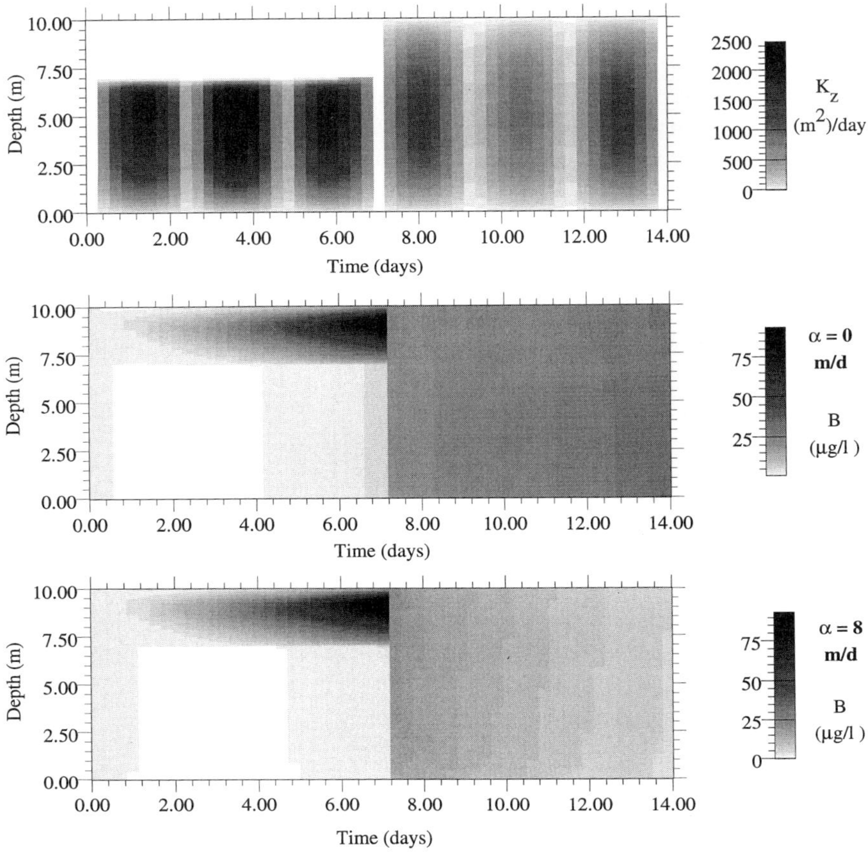

FIGURE 6. *Top panel,* assumed profiles of turbulent diffusivity K_Z (m^2d^{-1}); *middle panel,* phytoplankton chlorophyll B ($\mu g\ L^{-1}$) resulting from assumed profiles of K_Z with no benthic grazing; *bottom panel,* phytoplankton chlorophyll resulting from assumed profiles of K_Z and benthic grazing rate $\alpha = 8\ m^3m^{-2}d^{-1}$ (after Koseff *et al.* 1993).

model predicts that the bloom should disappear about one week after stratification breaks down. These discrepancies have prompted a consideration of shoal-channel exchange and its effect on bloom formation and decay.

A Two-Dimensional Model. The bulk of field observations of phytoplankton biomass have been made in the channel. A somewhat more complex view than that sketched above emerges when the more limited set of observations on shoal biomass is taken into account. First and foremost, as in other parts of the Bay, the shoals are much more productive than the channels, with net growth rates approaching 1 d^{-1} compared to 0.1 d^{-1} in the channel (Cloern *et al.* 1985). Second, the spring bloom sometimes appears to start earlier in the shoals than in the channel. For example, in 1980, a year of normal precipitation, elevated chlorophyll *a* values were seen in January even though the channel bloom did not start until April (Cloern 1982). Likewise in 1982, a very wet year, three separate blooms were

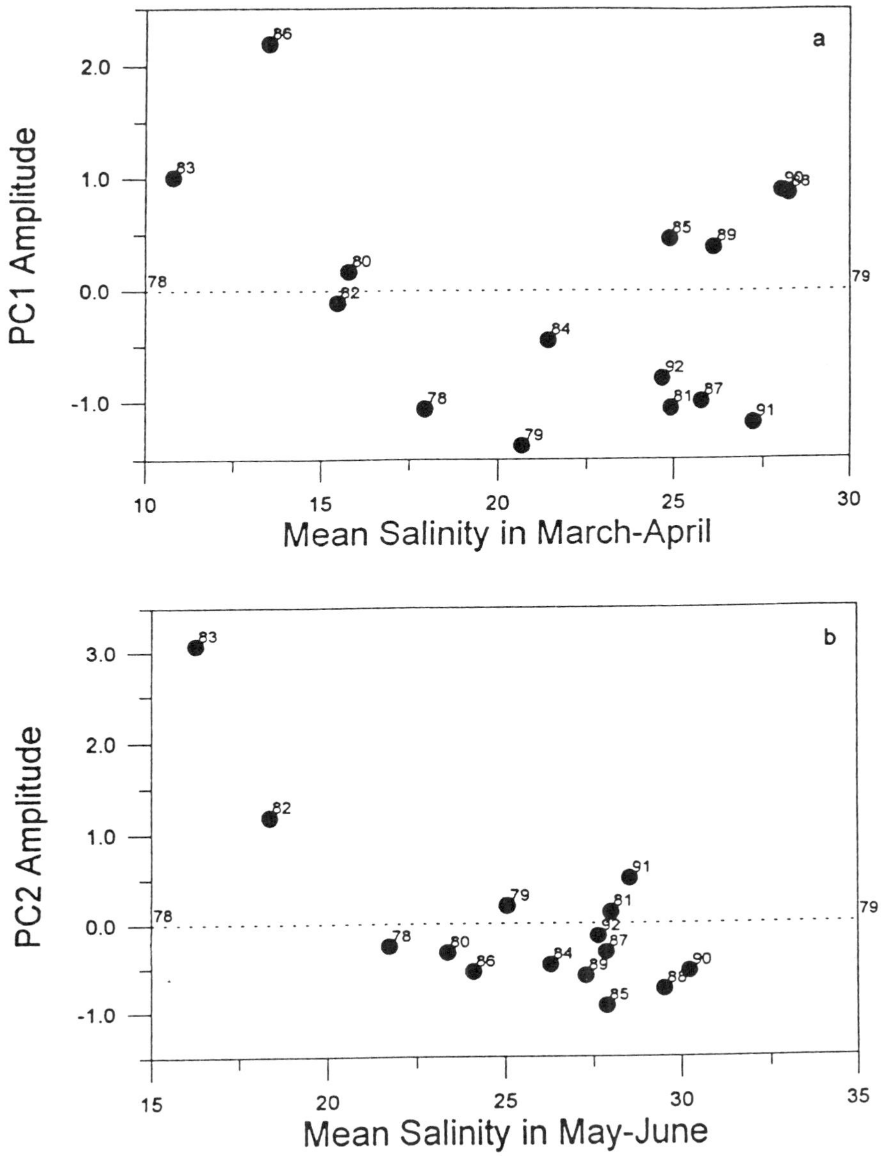

FIGURE 7. *Top panel,* amplitude of first principal component of chlorophyll spatial-temporal patterns as a function of mean March-April salinity in South Bay; *bottom panel,* amplitude of second principal component as a function of mean May-June salinity (after Cloern & Jassby 1995).

observed to start on the eastern shoals (Huzzey et al. 1990; Fig. 8). Finally, a similar pattern is evident in the data from 1987 (Powell *et al.* 1989), a dry year, albeit at considerably lower concentrations of chlorophyll *a*. The 1980 and 1987 data also suggest an intriguing pattern during certain years: The bloom starts on the eastern shoal and migrates in a clockwise fashion around the southern end of South Bay. With this flow pattern, the southern end of South Bay will receive biomass from the shoals on the east and from the largely shoal region south of the Dumbarton Bridge; hence, channel bloom dynamics may largely reflect processes occurring on the shoals.

A first step towards including shoal-channel linkages in a model was taken by Vidergar *et al.* (1993), who extended the KHMC model by incorporating horizontal biomass transport in a model of the transverse cross-section of an estuary. This model essentially superimposes horizontal mixing effects onto the processes modeled in the one-dimensional code of KHMC while retaining all other features of the one-dimensional model. Vidergar *et al.* modeled a channelized region with a neighboring shoal and allowed for differential treatment of each region. For example, the channel region may be stratified while the shoal is vertically well-mixed, and different benthic grazing rates can be specified for each region. They found that, when the benthic grazing rate in the shoal is zero, even an unstratified channel can experience a bloom. Furthermore, when the grazing rate in the shoals is nonzero but small, a stratified channel can experience a muted yet more sustained bloom relative to the one-dimensional case of KHMC. In contrast, if the shoal grazing rate is substantial, no bloom will occur in the shoal or the channel. Finally, they found that, regardless of the benthic grazing pressure in the shoals, a bloom will not occur in the channel (or shoals) if light conditions are not favorable.

This last result accords with field evidence that wind effects on sediment resuspension might be important to bloom dynamics. In 1991, for example, the bloom started at the period of minimum tidal energy but never fully developed. Apparently, this failure was the result of anomalously high winds that occurred one week after the start of the bloom, leading to extinction depths as small as 2 cm at a channel station near the Dumbarton Bridge (Wienke *et al.* 1992).

Some Hypotheses. The field observations described above, the principal component analysis of Cloern & Jassby (1995) and the numerical modeling work of Koseff *et al.* (1993) and Vidergar *et al.* (1993) can be synthesized into the following hypotheses regarding bloom dynamics in South Bay.

A phytoplankton bloom will occur initially in the shoal regions of the South Bay provided the light climate is favorable and benthic grazing is weak. If benthic grazing is large enough in the shoals, however, no bloom will occur either in the shoals or channels. While the clam community in the shoals has a very low biomass in South Bay in the spring, it is often substantial by late summer (Jan Thompson, pers. commun.). Benthic grazing may therefore prevent a bloom from occurring in the shoals at that time. Furthermore, even if a bloom begins in the channel in late summer, perhaps facilitated by some thermal stratification and weak tides, lateral transport of the biomass to the shoals would cause a rapid demise of the bloom in the channel. The observation of low benthic grazer biomass in the spring explains

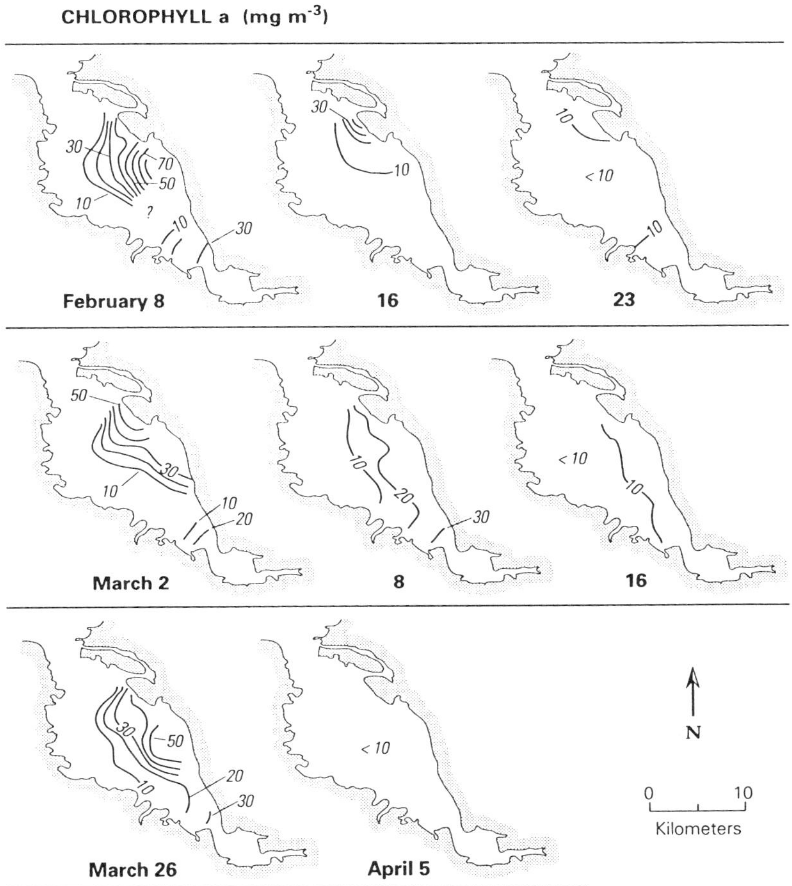

FIGURE 8. Phytoplankton blooms originating in the eastern shoals of South Bay during the spring of 1982 (after Huzzey *et al.* 1990).

why blooms are typically observed in April and usually not in September, despite the fact that favorable conditions for phytoplankton blooms (minimal tidal energy and stratification) occur at the end of September as well.

Why does the observed bloom last longer than predicted by the one-dimensional model? The numerical results of KHMC show that a bloom will last about one week after stratification breaks down (which also takes about one week), yet the blooms are observed to last two to four weeks. From the results of the two-dimensional simulations, it appears that the shoal-channel exchange process may produce a "buffering" effect, enabling the bloom to sustain itself in the channel region for a much longer period than without lateral exchange. The buffering effect works in the following way. When the channel is stratified and benthic grazing is low enough in the shoals, biomass produced in the shoals will be transported into the channel region, thereby promoting the bloom process in the channel. Because the channel is stratified and its photic zone protected from benthic grazing, biomass accumulation in the channel may exceed that in the shoals, when benthic grazing is present in the shoals. In that case net biomass will eventually be transported from the channel back into the shoals. Once stratification breaks down in the channel, net biomass will be transported from the shoals to the channel, prolonging the bloom in the channel. The channel thus acts as a buffer for the shoals while it is stratified, and the shoals then act as a buffer for the channel once stratification has broken down.

Phytoplankton blooms in South Bay may therefore depend on several conditions: channel stratification, little or no grazing in the shoals and light winds in order that turbidity and hence light extinction remain relatively low. Viewed from a perspective that recognizes the multiplicity of factors affecting bloom dynamics in South Bay, Cloern & Jassby's (1995) deduction that bloom intensity depends on flow (and its proxies) only during extreme hydrologic conditions seems, in fact, quite reasonable.

Secondary Blooms. The spring bloom in South Bay is dominated by diatoms, including *Thalassiosira* and *Coscinodiscus* spp. (Cloern *et al.* 1985; Jassby *et al.* 1994). The diatom peak is sometimes followed immediately by a "red tide" due to *Mesodinium rubrum*, a nontoxic ciliate that contains a small pigmented alga as an endosymbiont (Cloern *et al.* 1994). Circumstances preceding *Mesodinium* blooms include high freshwater runoff, neap tides and warm stable weather, *i.e.*, conditions promoting stratification of the water column. A further advantage may be conferred by the low levels of inorganic N and Si following the diatom peak. *Mesodinium* has no need for Si in large quantities and its motility permits it to migrate vertically and exploit N pools below the pycnocline; it can thereby enjoy both high light and adequate nutrients.

Not only are toxic dinoflagellate blooms absent from South Bay, but dinoflagellates play almost no role at all in the South Bay phytoplankton community. Circumstantial evidence suggests that ambient trace metal concentrations may be the reason. Kuwabara *et al.* (1989) examined spatial distributions of Cu, Zn and Cd in South Bay during 1985. Estimated free ion activities of 10^{-10}, 10^{-8}, and 10^{-10} mol L^{-1}, respectively, were similar to threshold toxicity concentrations for phytoplankton determined in other studies. Kuwabara *et al.* hypothesized that elevated free metal ion activities affect community structure in favor of metal-tolerant species. As neritic dinoflagellates were generally found to be more sensitive to Cu and Cd than neritic diatoms in these other studies, this hypothesis would account for the small contribution of dinoflagellates to the South Bay plankton.

CONCLUDING REMARKS

Simple single-mechanism explanations of phytoplankton dynamics in San Francisco Bay have gradually evolved to include more and more processes. In most cases, earlier hypotheses are incomplete rather than incorrect. Further observations eventually turn up contradictions of earlier hypotheses that can be resolved by invoking additional mechanisms. The South Bay spring bloom is an excellent example, where benthic grazing, tidal stirring, freshwater discharge, wind mixing, channel-shoal exchange and turbidity have been successively incorporated into an understanding of bloom evolution. At this level of complexity the relative importance of different processes probably can be resolved only by numerical modeling experiments; with a database of only two decades or so, a statistical approach to interannual variability can play only a secondary role and experimental manipulation of the South Bay system is largely infeasible. Similar comments apply to Suisun Bay where, as we have already discussed, a three-dimensional simulation is now required to resolve the significance of various mechanisms.

Aside from these modeling activities, what further research on phytoplankton processes is warranted? In terms of simple observation, San Pablo and Central Bays, as well as the shoals of all subembayments, need more attention. The U.S. Geological Survey and the San Francisco Estuary Institute Regional Monitoring Program for Trace Substances currently support such observations in the channels, including community composition as well as chlorophyll (Caffrey *et al.* 1994), but shoal data are comparatively sparse. With a truly Bay-wide sampling effort that includes chlorophyll and turbidity in both channel and shoals, it will be possible to estimate Bay-wide annual primary productivity as an index of autochthonous food availability for the food web, although the inclusion of remotely sensed data will be necessary. The spatial and temporal frequencies of ground measurements required to define the status of the Bay's phytoplankton community and to enable comparison among years remain to be determined, as does the extent to which transect data are distorted by sampling at different tidal stages. These problems too are unlikely to be resolved without the aid of new remote observing platforms and sensors.

A further pressing need is delineation of the role that phytoplankton play in food limitation of higher organisms. Current data support a direct link with variability of benthic macroinvertebrates but not necessarily with that of pelagic invertebrates. In general, the entire subject of food-limited abundance is in need of more intense investigation for San Francisco Bay. If phytoplankton abundance does indeed limit the food web, then the strong effects that freshwater flow regulation has on phytoplankton populations (Jassby & Powell 1994) is of practical significance and not only of intellectual interest.

Finally, it should be recalled that excessive phytoplankton populations are of increasing concern in many other estuaries and coastal environments (Atwood *et al.* 1994). Although noxious blooms do not occur at present in the Bay, phytoplankton abundance in the Delta can reach levels where the taste and odor of municipal drinking water is impaired. Many of the seasonal peaks in the Delta coincide with levels of nitrate and phosphate that could limit uptake rates (Ball & Arthur 1979; unpublished observations) and it might be inferred that maximum bloom sizes are ultimately nutrient-limited. In South Bay, the spring bloom maximum also sometimes coincides with very low levels of nitrate or silica. In view of the size of bloom peaks in the South Bay — at times reaching 80 mg m^{-3} chlorophyll *a* near the surface — the role of nutrient limitation and the consequences of future increases in nutrient discharge should be delineated.

ACKNOWLEDGMENTS

We thank Jim Cloern for stimulating, facilitating and guiding our interest in the San Francisco Bay phytoplankton. Two anonymous reviewers inspired a much-needed revision of an earlier draft. This work was supported by the U.S. Environmental Protection Agency (R819658) Center for Ecological Health Research at the University of California, Davis. The document does not necessarily reflect the views of the Agency and no official endorsement should be inferred.

REFERENCES

Alpine, A. E., and J. E. Cloern. (1988). Phytoplankton growth rates in a light limited environment, San Francisco Bay. *Marine Ecol. Progr. Ser.* 44:167-173.

Alpine, A. E., and J. E. Cloern. (1992). Trophic interactions and direct physical effects control phytoplankton biomass and production in an estuary. *Limnol. Oceanogr.* 37:946-955.

Arthur, J. F., and M. D. Ball. (1979). Factors influencing the entrapment of suspended material in the San Francisco Bay-Delta estuary. Pages 143-174 *in* T. J. Conomos, ed., *San Francisco Bay: The Urbanized Estuary*. Pacific Division, American Association for the Advancement of Science, San Francisco, CA.

Atwood, D. K., A. Bratkovich, M. Gallagher, and G. L. Hitchcock (eds.). (1994). Papers from NOAA's Nutrient Enhanced Coastal Ocean Productivity. *Estuaries* 17:729-903.

Ball, M. D. (1987). *Phytoplankton Dynamics and Planktonic Chlorophyll Trends in the San Francisco Bay-Delta Estuary*. U.S. Dept. Interior, Bur. Reclamation, Sacramento, CA. 10 pp.

Ball, M. D., and J. F. Arthur. (1979). Planktonic chlorophyll dynamics in the northern San Francisco Bay and Delta. Pages 265-285 *in* T. J. Conomos, ed., *San Francisco Bay: The Urbanized Estuary*. Pacific Division, American Association for the Advancement of Science, San Francisco, CA.

Caffrey, J. M., B. E. Cole, J. E. Cloern, J. R. Rudek, A. C. Tyler, and A. D. Jassby. (1994). *Studies of the San Francisco Bay, California, Estuarine Ecosystem: Pilot Regional Monitoring Results, 1993*. U.S. Geol. Surv., Menlo Park, CA. Open-file Rept. 94-82. 412 pp.

Catts, G. P., S. Khorram, J. E. Cloern, A. W. Knight, and S. D. DeGloria. (1985). Remote sensing of tidal chlorophyll-a variations in estuaries. *Internat. Jour. Remote Sensing* 6:1685-1706.

Cloern, J. E. (1979). Phytoplankton ecology of the San Francisco Bay system: The status of our current understanding. Pages 247-164 *in* T. J. Conomos, ed., *San Francisco Bay: The Urbanized Estuary*. Pacific Division, American Association for the Advancement of Science, San Francisco, CA.

Cloern, J. E. (1982). Does the benthos control phytoplankton biomass in South San Francisco Bay? *Marine Ecol. Progr. Ser.* 9:191-202.

Cloern, J. E. (1984). Temporal dynamics and ecological significance of salinity stratification in an estuary (South San Francisco Bay, USA). *Oceanol. Acta* 7:137-141.

Cloern, J. E. (1987). Turbidity as a control on phytoplankton biomass and productivity in estuaries. *Continental Shelf Res.* 7:1367-1381.

Cloern, J. E. (1991a). Tidal stirring and phytoplankton bloom dynamics in an estuary. *Jour. Marine Res.* 49:203-221.

Cloern, J. E. (1991b). Annual variations in river flow and primary production in the South

San Francisco Bay Estuary (USA). Pages 91-96 *in* M. Elliott & J. P. Ducrotoy, eds., *Estuaries and Coasts: Spatial and Temporal Intercomparisons*. Olsen & Olsen, Fredensborg, Denmark.

Cloern, J. E., and A. D. Jassby. (1995). Year-to-year fluctuation of the spring phytoplankton bloom in South San Francisco Bay: an example of ecological variability at the land- sea interface. Pages 139-149 *in* T. M. Powell & J. H. Steele, eds., *Ecological Time Series*. Chapman and Hall, New York, NY.

Cloern, J. E., A. E. Alpine, B. E. Cole, R. L. J. Wong, J. F. Arthur, and M. D. Ball. (1983). River discharge controls phytoplankton dynamics in the northern San Francisco Bay Estuary. *Estuarine Coastal Shelf Sci.* 16:415-429.

Cloern, J. E., B. E. Cole, R. L. J. Wong, and A. E. Alpine. (1985). Temporal dynamics of estuarine phytoplankton: A case study of San Francisco Bay. *Hydrobiologia* 129:153-176.

Cloern, J. E., T. M. Powell, and L. M. Huzzey. (1989). Spatial and temporal variability in south San Francisco Bay (USA). II. Temporal changes in salinity, suspended sediments, and phytoplankton biomass and productivity over tidal time scales. *Estuarine Coastal Shelf Sci.* 28:599-613.

Cloern, J. E., B. E. Cole, and S. W. Hager. (1994). Notes on a *Mesodinium rubrum* red tide in south San Francisco Bay. *Jour. Plankt. Res.* 16:1269-1276.

Cole, B. E., and J. E. Cloern. (1984). Significance of biomass and light availability to phytoplankton productivity in San Francisco Bay. *Marine Ecol. Progr. Ser.* 17:15-24.

Cole, B. E., and J. E. Cloern. (1987). An empirical model for estimating phytoplankton productivity in estuaries. *Marine Ecol. Progr. Ser.* 36:299-305.

Cole, B. E., J. K. Thompson, and J. E. Cloern. (1992). Measurement of filtration rates by infaunal bivalves in a recirculating flume. *Jour. Marine Res.* 113:219-225.

Denman, K. L., and A. E. Gargett. 1983. Time and space scales of vertical mixing and advection of phytoplankton in the upper ocean. *Limnol. Oceanogr.* 28:801-815.

Foe, C., and A. Knight. (1985). The effect of phytoplankton and suspended sediment on the growth of *Corbicula fluminea* (Bivalvia). *Hydrobiologia* 127:105-115.

Fritz, L., M. A. Quilliam, J. L. C. Wright, A. M. Beale, and T. Work. (1992). An outbreak of domoic acid poisoning attributed to the pennate diatom *Pseudonitzschia australis*. *Jour. Phycol.* 28:439-442.

Hasle, G. R. (1993). Nomenclatural notes on marine planktonic diatoms. The family Bacillariaceae. *Beih. Nova Hedwigia* 106:315-321.

Huzzey, L. M., J. E. Cloern, and T. M. Powell. (1990). Episodic changes in lateral transport and phytoplankton distribution in south San Francisco Bay. *Limnol. Oceanogr.* 35:472-478.

Jassby, A. D., and T. M. Powell. (1994). Hydrodynamic influences on interannual chlorophyll variability in an estuary: upper San Francisco Bay-Delta (California, U.S.A.). *Estuarine Coastal Shelf Sci.* 39:595-618.

Jassby, A. D., J. E. Cloern, and T. M. Powell. (1993). Organic carbon sources and sinks in San Francisco Bay: variability induced by river flow. *Marine Ecol. Progr. Ser.* 95:39-54.

Jassby, A., J. Cloern, J. Caffrey, B. Cole, and J. Rudek. (1994). Plankton and water quality studies. Pages 117-128 *in San Francisco Estuary Regional Monitoring Program for Trace Substances: 1993 Annual Report*. San Francisco Estuary Institute, Richmond, CA.

Jassby, A. D., W. J. Kimmerer, S. G. Monismith, C. Armor, J. E. Cloern, T. M. Powell, J. R. Schubel, and T. J. Vendlindski. (1995). Isohaline position as a habitat indicator for estuarine populations. *Ecol. Applications* 5:272-289.

Kimmerer, W. (1991). Tests on effects of food limitation on reproduction in two copepod species important in the diet of larval striped bass. Pages 7-17 *in 1990 Working Papers of the Food Chain Group*. Interagency Ecol. Studies Progr., California Dept. Water Resources, Sacramento, CA.

Kimmerer, W. J., E. Gartside, and J. J. Orsi. (1994). Predation by an introduced clam as the probable cause of substantial declines in zooplankton in San Francisco Bay. *Marine Ecol. Progr. Ser.* 113:81-93.

Koseff, J. R., J. K. Holen, S. G. Monismith, and J. E. Cloern. (1993). Coupled effects of vertical mixing and benthic grazing on phytoplankton populations in shallow, turbid estuaries. *Jour. Marine Res.* 51:843-868.

Kost, A. L. B., and A. W. Knight. 1975. The food of *Neomysis mercedis* Holmes in the Sacramento-San Joaquin Delta ecosystem. *California Fish & Game* 61:34-41.

Kuwabara, J. S., C. C. Y. Chang, J. E. Cloern, T. L. Fries, J. A. Davis, and S. N. Luoma. (1989). Trace metal associations in the water column of south San Francisco Bay, California. *Estuarine Coastal Shelf Sci.* 28:307-325.

Lehman, P. W. (1992). Environmental factors associated with long- term changes in chlorophyll concentration in the Sacramento-San Joaquin Delta and Suisun Bay, California. *Estuaries* 15:335-348.

Nichols, F. H. (1985). Increased benthic grazing: An alternative explanation for low phytoplankton biomass in northern San Francisco Bay during the 1976-1977 drought. *Estuarine Coastal Shelf Sci.* 21:379-388.

Nichols, F. H., J. K. Thompson, and L. E. Schemel. (1990). Remarkable invasion of San Francisco Bay (California, USA) by the Asian clam *Potamocorbula amurensis*. 2. Displacement of a former community. *Marine Ecol. Progr. Ser.* 66:95-101.

Peterson, D. H., and J. H. Festa. (1984). Numerical simulation of phytoplankton productivity in partially mixed estuaries. *Estuarine Coastal Shelf Sci.* 19:563-589.

Peterson, D. H., T. J. Conomos, W. W. Broenkow, and P. C. Doherty. (1975a). Location of the non-tidal current null zone in northern San Francisco Bay. *Estuarine Coastal Marine Sci.* 3:1-11.

Peterson, D. H., T. J. Conomos, W. W. Broenkow, and E. P. Scrivani. (1975b). Processes controlling the dissolved silica distribution in San Francisco Bay. Pages 153-187 *in* L. E. Cronin, ed., *Estuarine Research*, vol. 1. Academic Press, New York, NY.

Peterson, D. H., L. E. Schemel, R. E. Smith, D. D. Harmon, and S. W. Hager, (1987). The flux of particulate organic carbon in estuaries: phytoplankton productivity and oxygen consumption. Pages 41-49 *in* S. Subitzky, ed., *Selected Papers in the Hydrologic Sciences, 1987.* U.S. Geol. Surv. (Reston, VA.) Water-Supply Pap. 2330.

Peterson, D. H., D. R. Cayan, J. F. Festa, F. H. Nichols, R. A. Walters, J. V. Slack, S. E. Hager, and L. E. Schemel. (1989). Climate variability in an estuary: effects of river flow on San Francisco Bay. Pages 419-442 *in* D. H. Peterson, ed., *Aspects of Climate Variability in the Pacific and the Western Americas*. Geophys. Monogr. 55. American Geophysical Union, Washington, DC.

Powell, T. M., J. E. Cloern, and R. A. Walters. (1986). Phytoplankton spatial distributions in south San Francisco Bay: mesoscale and small-scale variability. Pages 369-383 *in* D. A. Wolfe, ed., *Estuarine Variability*. Academic Press, Orlando, FL.

Powell, T. M., J. E. Cloern, and L. M. Huzzey. (1989). Spatial and temporal variability in South San Francisco Bay (USA). I. Horizontal distributions of salinity, suspended sediments, and phytoplankton biomass and productivity. *Estuarine Coastal Shelf Sci.* 28:583-597.

Schemel, L. E. (1984). *Salinity, Alkalinity, and Dissolved and Particulate Organic Carbon in the Sacramento River Water at Rio Vista, California, and at Other Locations in the Sacramento-San Joaquin Delta, 1980*. U.S. Geol. Surv. (Sacramento, CA.) Water Resour. Inv. Rept. 83-4059. 45 pp.

Storrs, P. N., E. A. Pearson, and R. E. Selleck. (1966). *A Comprehensive Study of San Francisco Bay, A Final Report. Vol. V. Summary of Physical, Chemical and Biological Water and Sediment Data*. Rept. No. 67-2, Sanitary Engineering Res. Lab., University of California, Richmond, CA. 140 pp.

Taylor, F. J. R. (1993). Implications of the North American west coast experience. *Harmful Algae News* (UNESCO IOC Newsletter) 6:2-3.

Thompson, J. K., and F. H. Nichols. (1988). Food availability controls seasonal cycle of growth in *Macoma balthica* (L.) in San Francisco Bay, California. *Jour. Experi. Marine Biol. Ecol.* 116:43-61.

Vidergar, L. L., J. R. Koseff, and S. G. Monismith. (1993). Numerical models of phytoplankton dynamics for shallow estuaries. Pages 1628-1634 *in* H. W. Shen, S. T. Su, and F. Wen, eds., *Hydraulic Engineering '93*. ASCE, New York, NY.

Walters, R. A., R. T. Cheng, and T. J. Conomos. (1985). Time scales of circulation and mixing processes of San Francisco Bay waters. *Hydrobiologia* 129:13-36.

Whedon, W. F. (1939). A three-year survey of the phytoplankton in the region of San Francisco. *Internat. Rev. gesamten Hydrobiol.* 38:459-476.

Wienke, S. M., B. E. Cole, J. E. Cloern, and A. E. Alpine. (1992). *Plankton Studies in San Francisco Bay. XIII. Chlorophyll Distributions and Hydrographic Properties of San Francisco Bay, 1991*. U.S. Geol. Surv., Menlo Park, CA. Open-File Rept. 92-158. 116 pp.

Williams, P. B., and J. T. Hollibaugh. (1987). *A Flow Standard to Maximize Phytoplankton Abundance by Positioning an Entrapment Zone in San Pablo Bay*. Philip Williams & Associates, San Francisco, CA. 35 pp.

CHANGES IN CHLOROPHYLL *a* CONCENTRATION AND PHYTOPLANKTON COMMUNITY COMPOSITION WITH WATER-YEAR TYPE IN THE UPPER SAN FRANCISCO BAY ESTUARY

PEGGY W. LEHMAN
California Department of Water Resources
3251 S Street, Sacramento, CA 95816

Spatial and temporal changes in chlorophyll *a* concentration and phytoplankton community composition were a function of interannual changes in water-year type for the upper San Francisco Bay estuary. Changes in chlorophyll *a* concentration and phytoplankton community composition were determined from monthly or semi-monthly samples taken at 16 to 25 stations within the upper San Francisco Bay estuary between 1970 and 1993. Water year types were categorized as wet, normal, dry or critical based on inflows to the upper estuary. Among water-year types, average chlorophyll *a* concentration in the upper estuary increased by 11% in normal years, 8% in wet years and 4% in dry years but decreased by 17% in critical years. Within water-year types, chlorophyll *a* concentration often differed between upstream and downstream regions, with high chlorophyll *a* concentration upstream during critical years and downstream during wet years. Phytoplankton community composition also varied with water-year type between upstream and downstream regions. Upstream, diatoms, greens and cryptophytes were abundant during normal and critical years while flagellates were abundant during critical and wet years. Downstream, diatoms and greens were abundant during normal and wet years while bluegreens and flagellates were abundant during critical years. The coincident change in phytoplankton size structure and cellular carbon content with community composition suggests the influence of water-year type on community composition may be an important factor influencing the estuarine food web because of its influence on zooplankton food quality and quantity.

Research has demonstrated the influence of streamflow or variables related to streamflow on chlorophyll *a* concentration and phytoplankton community composition. Streamflow affects the accumulation of chlorophyll *a* and size distribution of species due to processes associated with tides and gravitational circulation in the turbidity maximum (Arthur & Ball 1979; Cloern *et al.* 1985), accumulation of biomass through effects on residence time within rivers (Lehman 1992) and transport of phytoplankton carbon downstream (Jassby & Powell 1994). High streamflows carry nutrients needed for phytoplankton growth (Conomos 1979; Peterson *et al.* 1985) and suspended sediments, which affect the light available within the water column for photosynthesis (Peterson & Festa 1984; Peterson *et al.* 1989). Recent studies suggest streamflow may control the removal of phytoplankton biomass by influencing the density of benthic grazers (Nichols 1985). Long-term studies also suggest streamflow is associated with interannual changes in environmental factors that affect phytoplankton biomass and community succession (Lehman & Smith 1991; Lehman 1992).

Streamflows in the upper estuary and their influence on estuarine ecology are primarily controlled by climate, despite water management practices. Cayan & Peterson (1989) demonstrated the association between climate and streamflow along the west coast of North America, including streamflow in the Sacramento River. Peterson *et al.* (1989) developed this relationship into a conceptual model of how climate, through effects on streamflow, can influence phytoplankton production and productivity in San Francisco Bay. Climate was subsequently correlated with long-term changes in streamflow and other environmental factors that control chlorophyll *a* concentration (Lehman 1992). The long- term influence of climate on phytoplankton ecology in the upper estuary was further suggested by an association between the 1976 climate shift in the eastern Pacific (Quinn & Neal 1985) and phytoplankton community composition and environmental variables (Lehman & Smith 1991). Water management practices can influence streamflows in the upper estuary (Jassby & Powell 1994), but management is largely based on water-year type. Despite an awareness of the influence of streamflow on phytoplankton ecology, the influence of water-year type on chlorophyll *a* concentration and phytoplankton community composition and its potential influence on the estuarine food web has not been studied.

The purpose of this study is to determine if chlorophyll *a* concentration and phytoplankton community composition vary spatially and temporally with water-year type and how this might affect food availability at the base of the food web. These findings are discussed in relation to long-term climate change and provide insight into the role of natural processes on shaping estuarine ecology.

METHODS

Description of Study Area. The upper San Francisco Bay estuary is bounded by the lower reach of the Sacramento River on the north, the lower reach of the San Joaquin River on the south, and San Pablo Bay on the west (Fig. 1). Together, the Sacramento and San Joaquin Rivers and their tributaries drain approximately 153,000 km^2, or 40% of California. The region upstream of the confluence of the Sacramento and San Joaquin Rivers is a convex-shaped delta containing a network of interconnected channels and islands. The region downstream of the confluence consists of deep channels and two shallow bays. Yearly average outflows from the Sacramento and San Joaquin Rivers range from about 120 to 1740 m^3s^{-1}.

Chlorophyll a Concentration and Phytoplankton Community Composition. Chlorophyll *a* concentration and phytoplankton community composition were determined from water samples collected monthly or semi-monthly at 1m depth between October 1970 and September 1993 by the California Department of Water Resources and the U.S. Bureau of Reclamation. Chlorophyll *a* concentrations were measured at 25 stations (Fig. 1) between 1970 and 1993 using the spectrophotometric methods of Strickland and Parsons (1968). Water samples for phytoplankton analysis were collected using a Van Dorn bottle at 15 stations between 1975 and 1993 and at San Pablo Bay between 1980 and 1993. Samples were preserved with Lugol's solution and prepared for analysis using the Utermohl method (1958). Phytoplankton cells and colonies (*e.g.*, filaments) were enumerated and identified to genus, or to species for common forms. Phytoplankton genera and species were also grouped into eight groups based on taxonomy and morphology (Table 1).

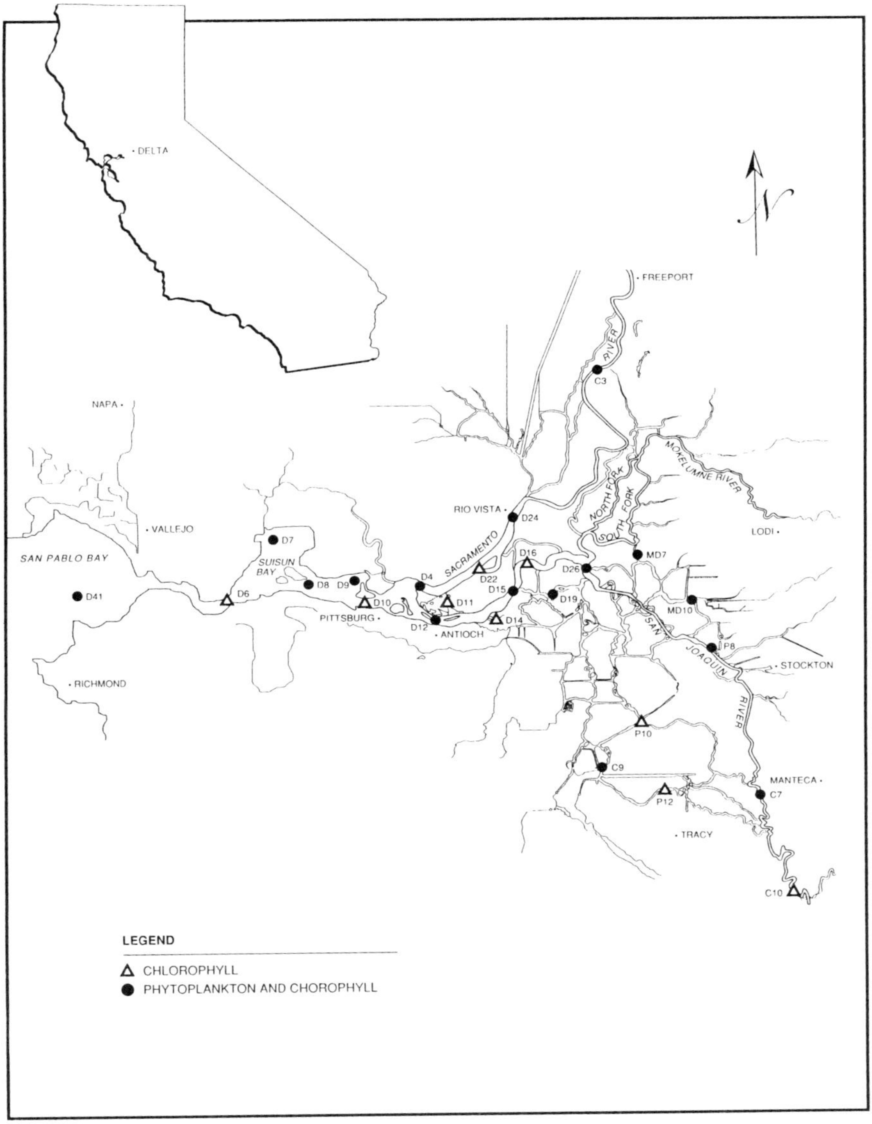

FIGURE 1. Sampling stations for chlorophyll *a* concentration and phytoplankton community composition in the upper San Francisco Bay estuary.

Changes in methodology and personnel occurred during the 19-year phytoplankton study period. Magnification was 280X between 1975 and 1981, 350X in 1982 and 1983 and 750X between 1984 and 1993. Personnel changed in 1977, 1982, 1988, 1989 and 1990 (G. Weber, pers. commun.). These methodology differences did not affect the long term trends in the results, since similar percent

TABLE 1. Eight groups used to summarize phytoplankton species composition. Representative genera and species within each group are listed.

Group	*Representative Species*
1 Diatom	*Synedra actinastroides, Melosira granulata, Actinella* spp., *Neidium* spp., *Skeletonema potamos, Thalassiosira eccentrica, Ceratoneis arcus, Cyclotella glomerata, Coscinodiscus* spp.
2 Green	*Micractinium* spp., *Actinastrum hantzschii, Westella botryoides, Sphaerocystis schroeteri, Chlorella* spp., *Carteria Klebsii, Pediastrum simplex, Chlamydomonas* spp.
3 Chrysophyte	*Chromulina* spp., *Chrysochromulina parva, Chrysococcus* spp., *Synura uvella, Mallomonas* spp.
4 Cryptophyte	*Rhodomonas* spp., *Rhodomonas lacustris, Chroomonas* spp., *Cryptomonas erosa, Cryptomonas* spp.
5 Bluegreen	*Anabaena circinalis, Agemenellum elegans, Anacystis limnecticus, Agmenellum* spp., *Anacystis* spp., *Aphanizomenon* spp., *Ocillatoria* spp.
6 Dinoflagellate	*Glenodinium* spp., *Gymnodinium* spp., *Glenodinium quadridens, Hemidium* spp., *Gyrodinium* spp.
7 Green Flagellate	*Euglena deses, Lepocinclis* spp., *Phacus* spp., *Euglena* spp., *Trachelmonas* spp.
8 Misc Flagellate	unidentified flagellates

diatom composition was measured for 1979 to 1981 by independent investigators (Wong & Cloern 1981; 1982).

Phytoplankton densities were converted to relative densities because individual cells within colonies were not counted separately. Relative density was calculated as the phytoplankton cell count for each species divided by the maximum count for that species at each station. Analyzing the data in this fashion created a value that was more quantitative than presence or absence. Phytoplankton densities were used in conjunction with cell dimensions to produce estimates of phytoplankton cell carbon concentrations using equations developed by Strathmann (1967). Approximately 90% of the phytoplankton had cell dimensions that could be used to calculate cell carbon.

Data Analysis. Water year type was used as an index of streamflow. A water-year describes precipitation for a period beginning October 1 and ending September 30 of the following year and is an appropriate classification for the upper estuary where precipitation normally falls between October and June. Water years are classified in the Sacramento River Basin as wet, above normal, below normal, dry and critical water-year types based on the Sacramento River Basin index. The index is developed from stream inflow data and is available for years since 1906 (State of California 1993). For this study, water years were divided into wet, normal, dry and critical water-year types. Above and below normal water-year types were combined into one category, since there were only 3 years in these two water-year types during the 23 water years of this study. The normal water-year type was maintained because as a commonly occurring water-year type before 1970, it might provide insight into conditions in the past.

Chlorophyll *a* concentrations and phytoplankton densities among water-year types were compared using monthly percent deviations from the long-term mean. Percent deviations from the mean normalized the data and provided a common base from which to compare the relative magnitude of change among water-year types. Positive deviations describe values higher than the long-term mean, while negative deviations describe values lower than the long-term mean. Monthly percent deviations from the mean for chlorophyll *a* concentration and phytoplankton density were calculated for each station as the monthly mean minus the long term mean divided by the long term mean times 100.

Standard deviation units were used to compare time series of phytoplankton groups. Standard deviation units were calculated for each species as the monthly mean relative density minus the long-term mean relative density divided by the standard deviation of the long-term mean relative density.

Percent deviations among water years were compared using the non-parametric, multiple comparison Kruskal-Wallis test. A non-parametric statistical method was chosen because percentages were not normally distributed and sample sizes were unequal (Zar 1984). It is recognized that because of the large number of comparisons, some comparisons may be significant by chance. This problem is reduced by the conservative nature of non-parametric methods and the large sample size.

Sampling stations within the upper estuary were grouped into regions for analysis based on combined and individual hierarchical cluster analysis of monthly data for 14 physical and chemical variables and chlorophyll *a* concentration (Table 2). An independent analysis for phytoplankton community composition produced a similar grouping (Lehman & Smith 1991).

RESULTS

Water-Year Type

The 24-year period 1970 to 1993 was characterized by decreasing inflow in the Sacramento River Basin. Only wet and normal years occurred between 1970 and

TABLE 2. Regions of the upper Estuary and their associated sampling stations.

	Region	*Sampling Stations*
Upstream		
1	Northern	C3
2	Western	D11, D12, D 14, D15
3	Lower Sacramento River	D4, D22, D24
4	Southern	C7, C9, C10, P8, P10, P12
5	Eastern	MD10, MD7
6	San Joaquin River	D16, D19, D26
Downstream		
7	Suisun Bay	D6, D7, D8, D9, D10
8	San Pablo Bay	D41

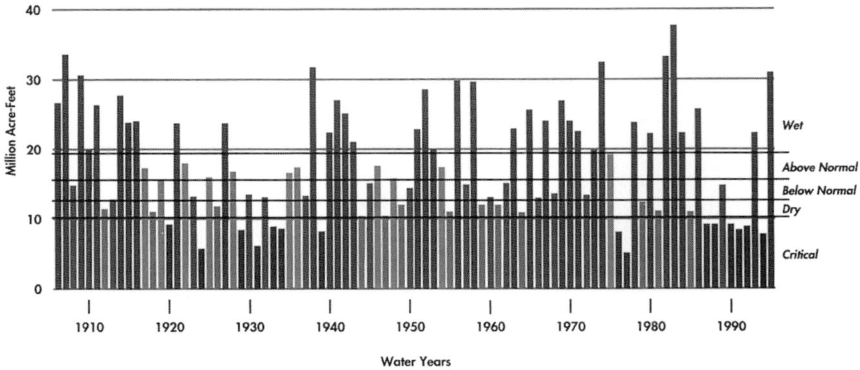

FIGURE 2. Sacramento River Basin index for water years between 1906 and 1993.

1975 (Fig. 2). These years were followed by two critical years and then a 9-year period 1978 to 1986 that alternated between wet and dry years. The driest period was between 1987 and 1993, when 5 out of 7 years were critical.

Water-year types between 1970 and 1993 differed from previous 24-year periods. Nearly half of the years between 1970 and 1993 were critical and dry. In contrast, a quarter of the years between 1946 and 1969 were dry, with no critical years. The dry period 1922 to 1945 was probably the period most similar to 1970 through 1993 because it also had an extended drought. Between 1922 and 1945, however, there were twice as many wet and normal years than dry and critical years. Prior to 1922, the pattern of water years was similar to other periods before 1970, with the total number of dry and critical years comprising less than half of the total number of wet and normal years.

Chlorophyll *a* Concentration

Long-term Trend. Chlorophyll *a* concentrations in the upper estuary were highly variable among water years but, on the average, decreased after 1976. Over time, average monthly chlorophyll *a* concentrations in the upper estuary ranged from 1 to 47 *ug*/l (Fig. 3) (CV = 84%) and encompassed a wide range of monthly values among stations (0.0 - 336 *ug*/l). Among regions, chlorophyll *a* concentrations were significantly higher in the southern region where average values reached 160 *ug*/l. In contrast, concentrations in the northern region rarely exceeded 5 *ug*/l. Despite high regional variability, average chlorophyll *a* concentrations decreased after 1977 by at least a factor of 2. Chlorophyll *a* concentrations decreased further in 1987 through 1990, before returning to 1977-1986 values.

Water year Type Variation. Among water-year types, higher chlorophyll *a* concentrations occurred in normal years. Average chlorophyll *a* concentrations were 11% higher than the mean in normal years and 17% lower than the mean in critical years (Fig. 4). Normal years also had higher chlorophyll *a* concentrations than dry and wet years which only increased by 4% and 8% above the mean,

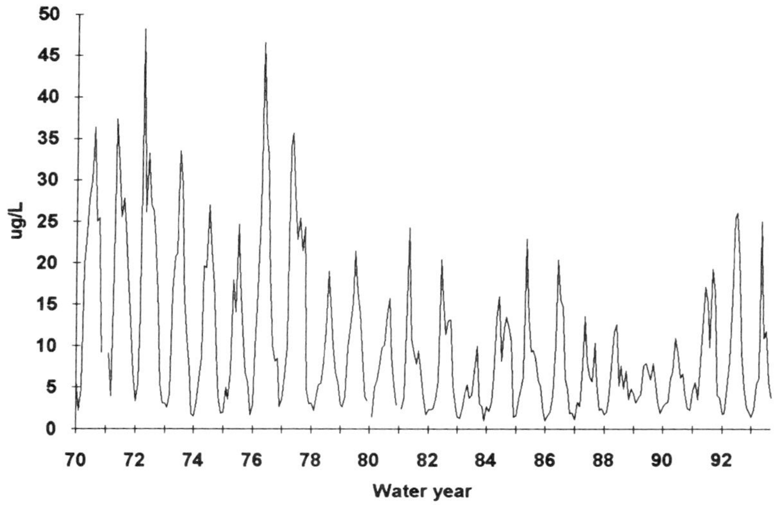

FIGURE 3. Average monthly chlorophyll *a* concentrations in the upper San Francisco Bay estuary for water years 1971 through 1993.

respectively. Among years, only concentrations during wet and dry years were not significantly different.

Chlorophyll *a* concentrations within water-year types were variable among stations. During normal years, chlorophyll *a* concentrations were 16 to 175% higher than the mean for most stations (Fig. 4). During wet years, however, chlorophyll *a* concentrations were lower than the mean upstream but higher than the mean downstream. Chlorophyll *a* concentrations during wet years were opposite to those during critical years, which were higher than the mean at the eastern margin of the upper estuary and decreased to below the mean downstream. During dry years, chlorophyll *a* concentrations were lower than the mean for most stations, but values were highly variable. Positive and negative deviations alternated among stations with distance downstream.

Seasonal Variation. Spring and summer had high chlorophyll *a* concentrations during wet and normal years. Chlorophyll *a* concentrations were higher than the mean during the spring in all water years, but concentrations were significantly lower in critical years (Fig. 5; Table 3). Concentrations were also higher than the mean in the summer during wet and normal years. Summer differed from spring, however, by having only low positive deviations for dry years and negative deviations for critical years.

Unlike the spring and summer, fall was characterized by higher chlorophyll *a* concentrations than the mean during dry years. Otherwise, the pattern of deviations among water years was similar in the fall to those in the spring and summer. Chlorophyll *a* concentrations were not significantly different between normal and wet years and the lowest concentrations occurred during critical years. During the winter, chlorophyll *a* concentrations were lower than the mean for all water years. However, chlorophyll *a* concentrations in critical years, which usually had the

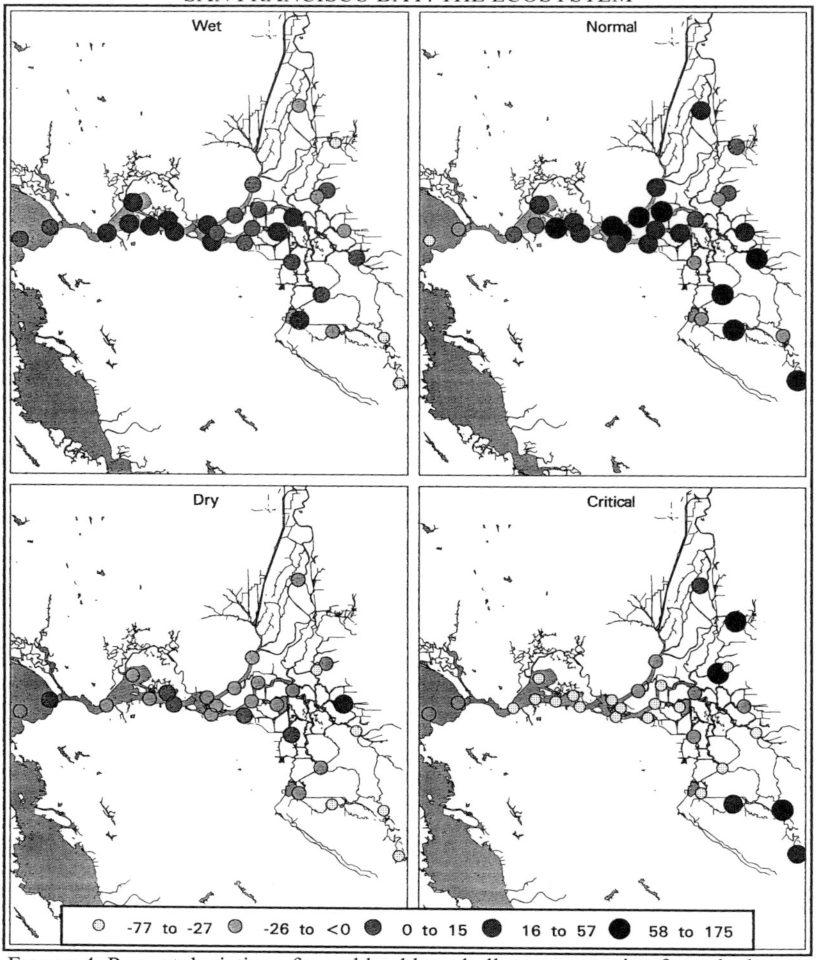

FIGURE 4. Percent deviation of monthly chlorophyll *a* concentration from the long-term mean for water-year types at sampling stations throughout the upper San Francisco Bay estuary.

lowest chlorophyll *a* concentrations among water years, were equal to those in normal and wet years.

Regional Variation. Upstream and downstream chlorophyll *a* concentrations differed among water-year types during the spring and summer. Normal and wet years had higher chlorophyll *a* concentrations than the mean in the spring or summer for all regions except San Pablo Bay Fig. 6, Table 4). These concentrations were significantly higher upstream in normal years and downstream in wet years. Critical years were characterized by chlorophyll *a* concentrations that were higher than the mean upstream but lower than the mean downstream during the spring. During the summer of critical years, however, chlorophyll *a* concentrations were lower than the mean at most upstream and downstream regions. Dry years had

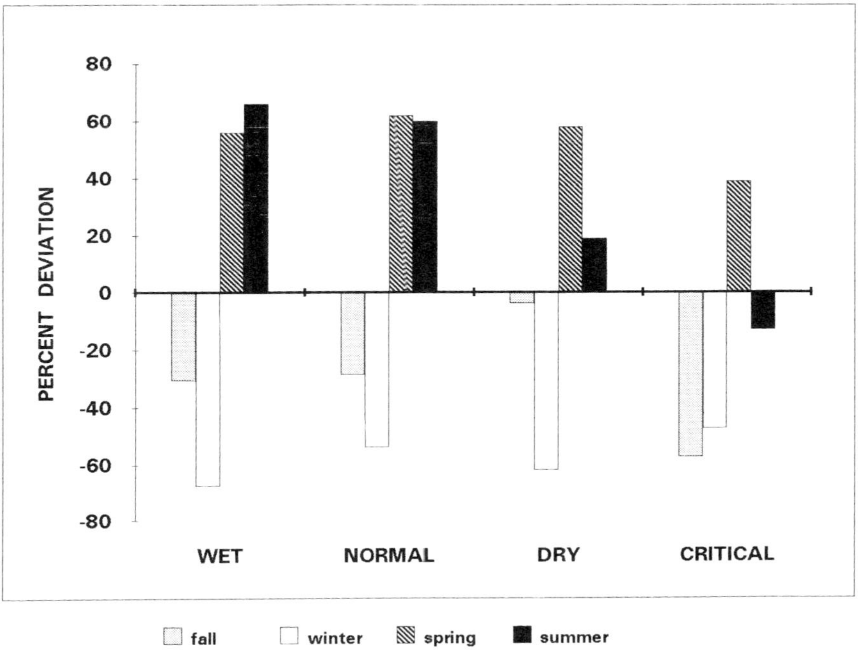

FIGURE 5. Seasonal average percent deviation of monthly chlorophyll *a* concentration from the long-term mean among water-year types.

higher chlorophyll *a* concentrations than the mean for some upstream and downstream regions during the spring and summer. However, these concentrations were only occasionally higher than those for critical years.

For both upstream and downstream regions, differences in chlorophyll *a* concentrations among water-year types were few during the fall and winter. During the fall, negative deviations for normal and wet years were not significantly different for most upstream and downstream regions (Fig. 6; Table 4). Normal years, however, had relatively higher chlorophyll *a* concentrations upstream and downstream in Suisun Bay than in critical years. Only in dry years were chlorophyll *a* concentrations higher than the mean. Positive deviations, combined with low

TABLE 3. Seasonal differences in chlorophyll *a* concentrations among water-year types. Statistical significance at the .05 level or higher was determined using the Kruskal-Wallis Test and is indicated by an 'X'.

	Fall	*Winter*	*Spring*	*Summer*
Wet/Normal		x		
Wet/Dry	x			x
Wet/Critical	x		x	x
Normal/Dry	x			x
Normal/Critical	x		x	x
Dry/Critical	x		x	x

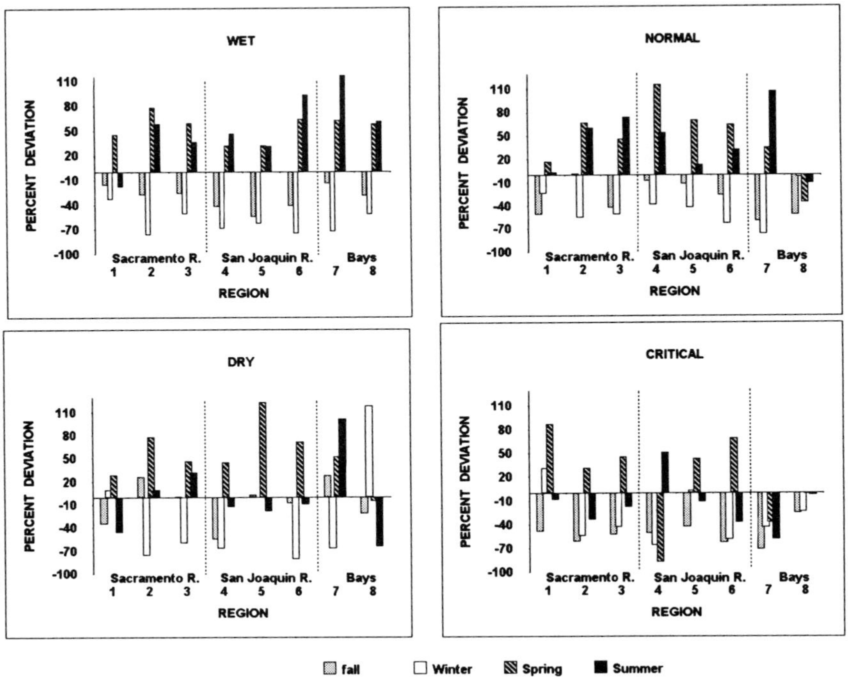

FIGURE 6. Regional average percent deviation of monthly chlorophyll *a* concentrations from the long-term mean among water-year types by season.

negative deviations, often gave dry years the highest chlorophyll *a* concentrations at both upstream and downstream regions. During the winter, chlorophyll *a* concentrations were consistently higher in normal years at both upstream and downstream regions.

Phytoplankton Community Composition

Long-term Trend. Phytoplankton community composition between 1975 and 1993 was characterized by a decrease in the percentage of diatoms. Diatoms comprised approximately 80-100% of the phytoplankton community between 1975 and 1979 (Fig. 7). Greens, which were the second most abundant group of phytoplankton, comprised up to 30% of the community. The phytoplankton community shifted to a more mixed assemblage, with fewer diatoms and more greens and flagellates, between 1980 to 1983. Diatoms reached a minimum during the 1987 to 1990 drought before returning to 1981-1986 levels in the early 1990s. The large decrease in the percentage of diatoms over time was a function of both a decrease in diatoms and an increase in other types of phytoplankton (Fig. 8). Changes in phytoplankton community composition for the entire upper estuary were similar to those for each region (not shown).

Water-Year-Type Variation. Most phytoplankton increased during normal and critical years, but diatoms increased only in normal years (Fig. 9, Table 5). Diatoms

TABLE 4. Regional differences in chlorophyll *a* concentrations among water-year-types by season. Significant differences at the .05 level were determined using the Kruskal-Wallis Test and are indicated by region number.

	Fall	*Winter*	*Spring*	*Summer*
Wet/Normal	2, 6	4	4	2, 3, 4, 7, 8
Wet/Dry	2, 5, 7	5, 8		2, 6, 8
Wet/Critical	7		4, 7	2, 3, 4, 5, 7, 8
Normal/Dry	4, 7	4, 6		1, 2, 3, 4, 6
Normal/Critical	2, 4, 6, 7	2, 4	2, 4, 7	2, 3, 4, 6, 7
Dry/Critical	2, 3, 6, 7		7	2, 3, 7

were also relatively more abundant during critical years than wet years, even though deviations were negative. Greens and bluegreens increased in both normal and critical years. However, more greens occurred in normal years, while more bluegreens occurred in critical years. Like greens and bluegreens, chrysophytes and cryptophytes also increased in normal and critical years. Other flagellates decreased in normal years. Miscellaneous flagellates, green flagellates and dinoflagellates were more abundant in wet or critical years.

Seasonal Variation. For the summer, large and positive deviations characterized most phytoplankton groups during normal and critical years and were usually higher in normal years (Fig. 10; Table 6). In contrast, bluegreens and miscellaneous

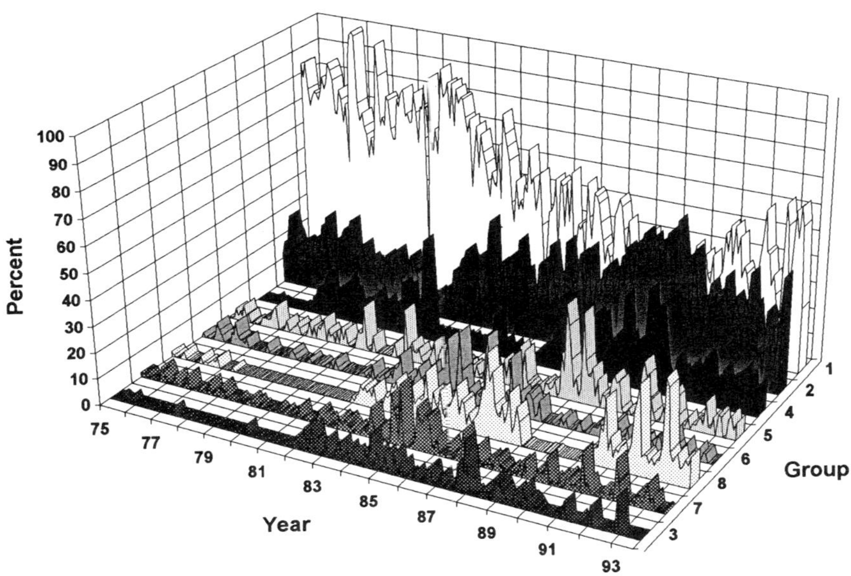

FIGURE 7. Average percent composition of phytoplankton groups among water years 1975 and 1993. Groups are: diatoms (1), greens (2), chrysophytes (3), cryptophytes (4), bluegreens (5), dinoflagellates (6), green flagellates (7) and miscellaneous flagellates (8).

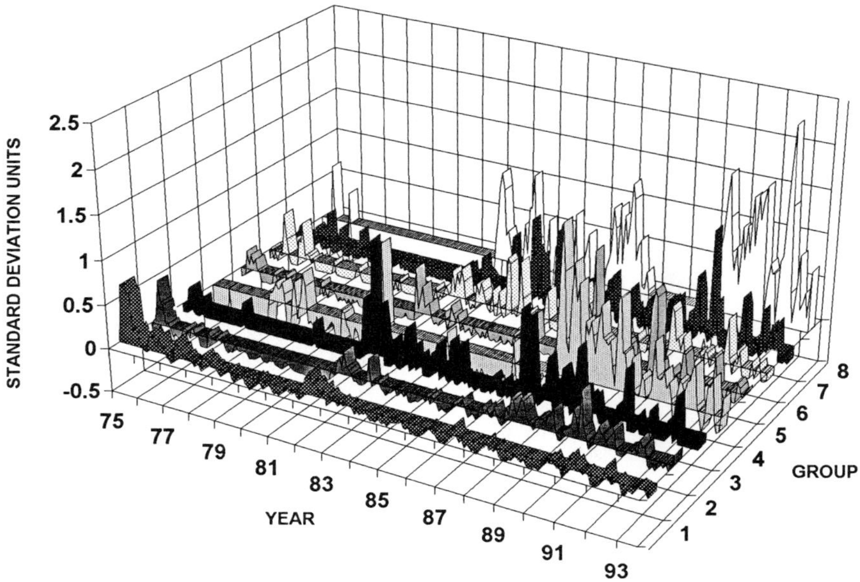

FIGURE 8. Monthly average standard deviation of phytoplankton group density from the long-term mean. Phytoplankton groups are described in Fig. 7.

flagellates increased more in the summer during critical and wet years. For the spring, diatoms and greens increased in normal years. Diatoms and greens, along with cryptophytes, also increased more during critical years than wet years. For the fall and winter, normal years had more greens, bluegreens and cryptophytes in the fall, and diatoms in the winter.

Regional variation. Phytoplankton community composition in upstream and downstream regions varied with water-year type during the spring and summer. During normal years, diatoms and greens increased upstream and downstream, as far as Suisun Bay (Fig. 11, Table 7). Diatoms and greens in the upstream regions were also accompanied by bluegreens and cryptophytes. The second largest positive deviations occurred during critical years, when diatoms, greens and bluegreens increased upstream and cryptophytes increased downstream. For wet years, miscellaneous flagellates had positive deviations upstream, but these were rarely different among water years. In contrast, miscellaneous flagellates, green flagellates and dinoflagellates increased more downstream during wet years than during normal or critical years. None of the positive deviations calculated for phytoplankton groups in dry years exceeded those of other years.

Phytoplankton community composition in upstream and downstream regions also varied with water-year type in the fall and winter. During normal years, diatoms, greens and cryptophytes increased upstream, while diatoms increased downstream in Suisun Bay (Fig. 12; Table 7). During critical years, bluegreens,

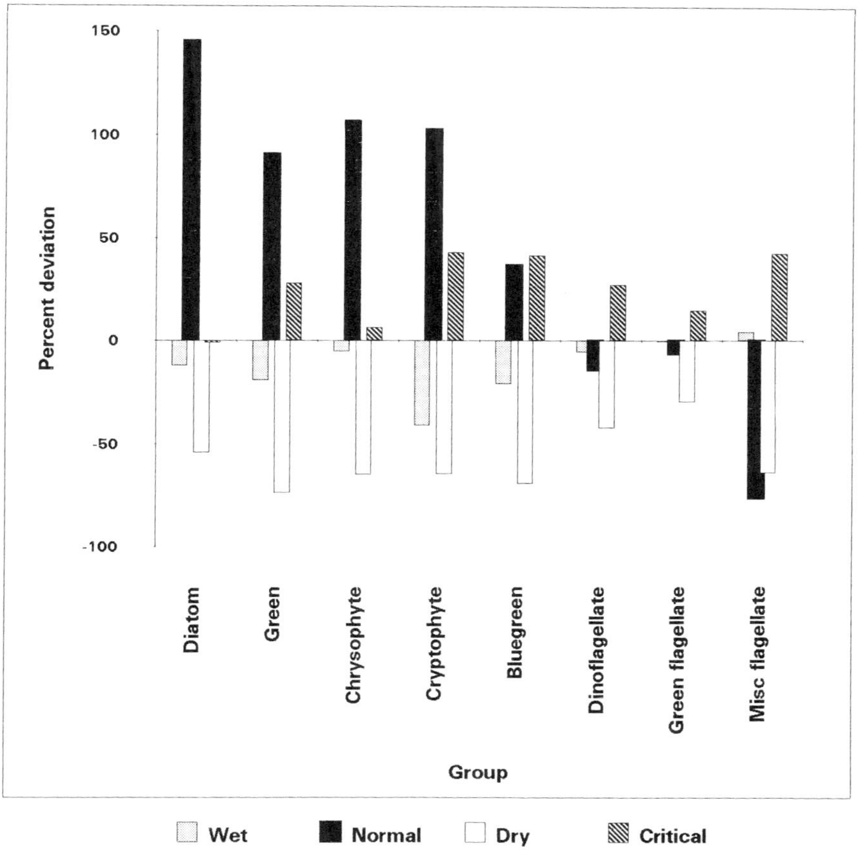

FIGURE 9. Average percent deviation of monthly phytoplankton density from the long-term mean among water-year types.

miscellaneous flagellates and green flagellates increased in both upstream and downstream regions. In wet and dry years, negative deviations characterized most phytoplankton upstream, with the exception of diatoms and cryptophytes. Downstream phytoplankton communities had more diatoms during both wet and dry years and more miscellaneous flagellates during wet years.

Phytoplankton Community Size Structure and Carbon Content. Changes in phytoplankton community composition with water-year type affected the size structure and carbon content of food particles. The average cell diameter of single-celled diatoms was significantly larger than other single-celled phytoplankton (Table 8). Green flagellates had the second largest cell diameter. Both diatoms and green flagellates contrasted with miscellaneous flagellates, which had the smallest diameter cells. Differences in cell diameter were also significant among most of the other single-celled phytoplankton, with the exception of the pairs bluegreens/chrysophytes and cryptophytes/dinoflagellates. Differences in size

TABLE 5. Differences in phytoplankton group densities from the long-term mean among water-year types. Significant differences at the .05 level or higher were determined using the Kruskal-Wallis Test and are marked with an 'X'. Water-year types are indexed as Wet (W), Normal (N), Dry (D) and Critical (C).

	W/N	*W/D*	*W/C*	*N/D*	*N/C*	*D/C*
Diatom	x	x	x	x	x	
Green	x	x	x	x	x	x
Chrysophyte		x		x		
Cryptophyte	x		x	x		x
Bluegreen	x	x	x	x	x	x
Dinoflagellate		x				
Green flagellate	x		x			
Misc flagellate	x	x			x	x

structure among phytoplankton groups were accompanied by changes in cellular carbon content, despite adjustments for plasma volume. Dinoflagellates, cryptophytes, green flagellates and diatoms had the highest cellular carbon content among groups.

Diatoms probably had the largest influence on the spatial and temporal variation of food particle size and nutrition. About 94% of the diatoms had cell diameters within the 10 *u*m or larger size range consumed by adult zooplankton. Although dinoflagellate, bluegreen and green flagellate cells were also larger than 10 *u*m in diameter and had high cellular carbon content, they occurred less frequently and differed less spatially and temporally among water-year types.

TABLE 6. Seasonal differences in phytoplankton group densities from the long-term mean among water-year types. Significant differences at the .05 level or higher were determined using the Kruskal-Wallis Test. Only Seasons with significant differences are listed. Seasons are Fall (f), Winter (w), Spring (sp) and Summer (sm). Water-year types are Wet (W), Normal (N), Dry (D) and Critical (C).

	W/N	*W/D*	*W/C*	*N/D*	*N/C*	*D/C*
Diatom	f,w	f	f,w	f	f,w	f,w
	sp,sm		sp	sp,sm	sp,sm	
Green	f,w	f	f	f,w	f,w	f,w
	sp,sm	sp,sm	sp,sm	sp,sm	sp,sm	sp,sm
Chrysophyte						
				sp,sm		
Cryptophyte	f		w	f	f	w
			sp,sm	sp,sm		sp,sm
Bluegreen	f		f	f,w	f	f
	sp,sm	sp		sp,sm	sp,sm	sm
Dinoflagellate						
		sm				
Green Flagellate		f	f		f,w	
	sm		sm		sp,sm	sm
Misc Flagellate	f,w	f,w	f			f,w
	sp,sm	sp,sm				sp,sm

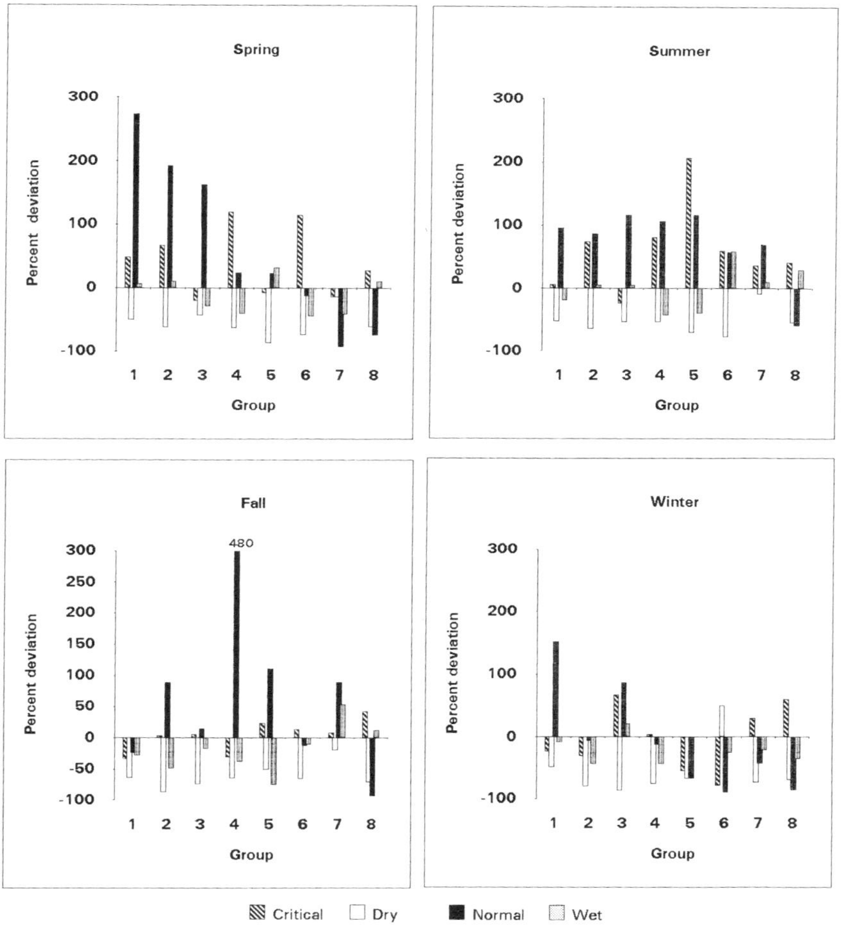

FIGURE 10. Seasonal average percent deviation of monthly phytoplankton group densities from the long term mean among water-year types. Phytoplankton groups are described in Fig. 7.

DISCUSSION

The spatial and temporal variation of chlorophyll *a* concentration with water-year type differed for upstream and downstream regions. During normal and critical years, high chlorophyll *a* concentrations upstream were a function of low or moderate streamflows, which allowed accumulation of phytoplankton biomass upstream (Conomos 1979; Peterson *et al.* 1989; Lehman 1992). Lower chlorophyll *a* concentrations during critical years were probably caused by reduced transport of phytoplankton biomass throughout the upstream region with low streamflows.

Upstream regions contrasted with downstream regions where higher chlorophyll *a* concentrations occurred during wet and normal years. During these years, phytoplankton biomass is transported downstream with high to moderate stream-

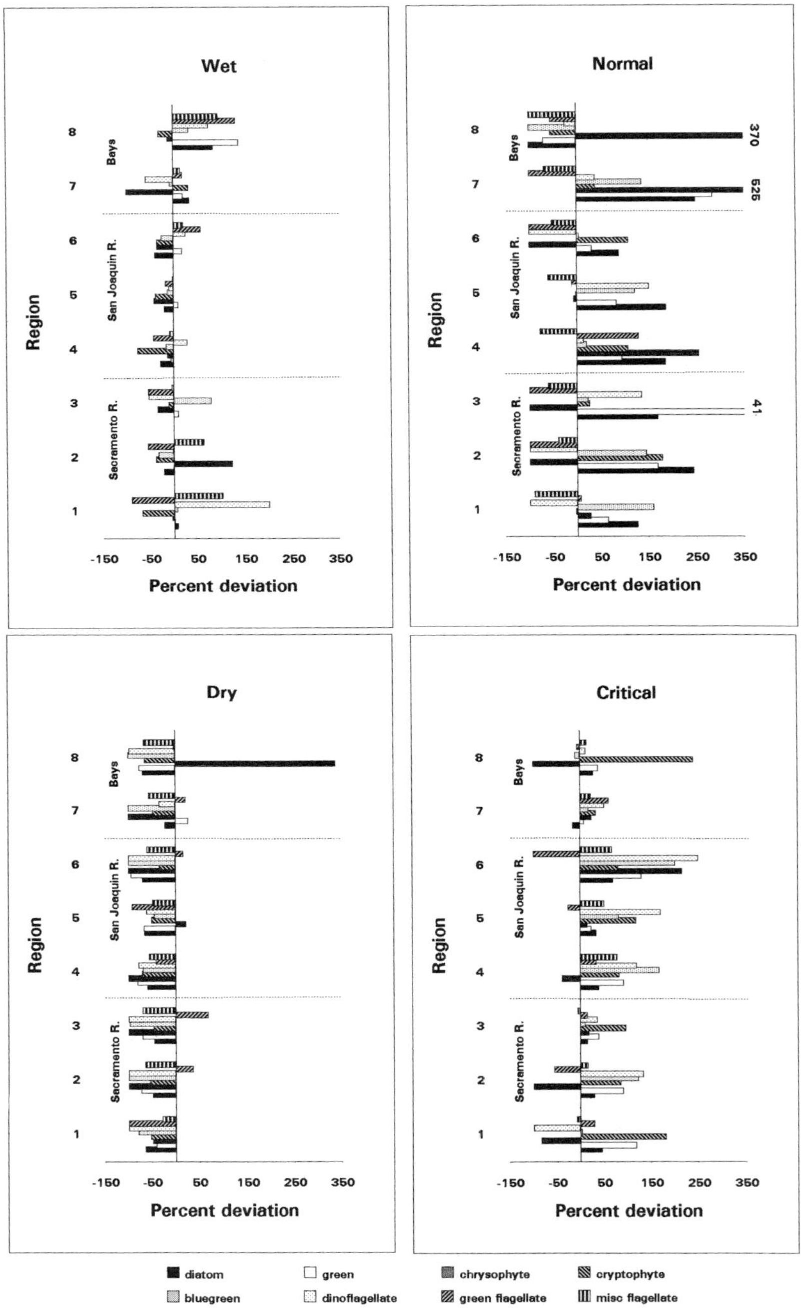

FIGURE 11. Regional average percent deviations of monthly phytoplankton group densities from the long-term mean among water-year types during the spring and summer.

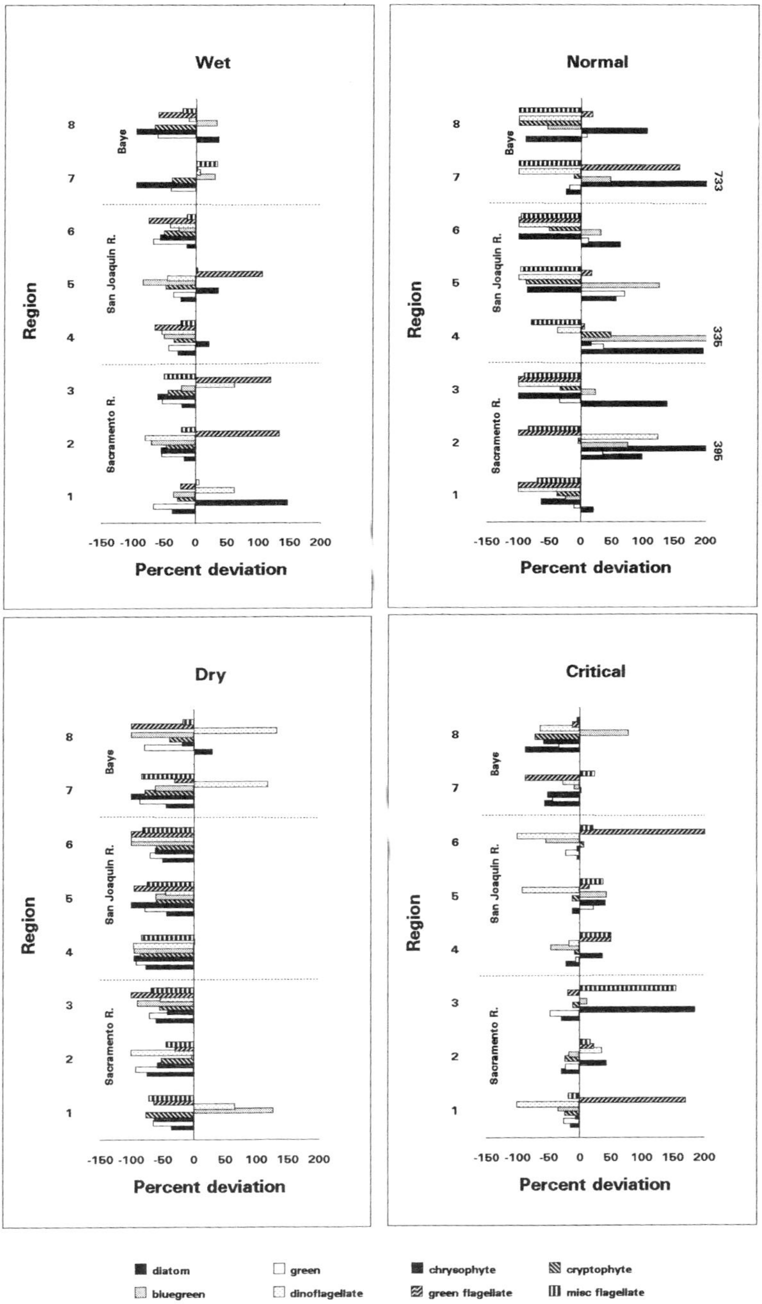

FIGURE 12. Regional average percent deviations of monthly phytoplankton group densities from the long-term mean among water-year types during the fall and winter.

TABLE 7. Regional differences in phytoplankton group densities from the long-term mean among water-year types by season. Significant differences at the .05 level were determined using the Kruskal-Wallis Test. Only regions with significant differences are listed. Water-year types are Wet (W), Normal (N), Dry (D) and Critical (C).

	W/N	*W/D*	*W/C*	*N/D*	*N/C*	*D/C*
			Fall			
Diatom	1,3,5,7,8		1,3,7,8	7	7	7
Green	3,4,5	4	4	1,3,4,5	4,5	4
Chrysophyte						
Cryptophyte	4,5			4,5,7	4,5	
Bluegreen	4		2,5	4	4	4,6
Dinoflagellate						
Green flagellate			3,7			
Misc flagellate		4,7	4,5		4,5	4,7,5
			Winter			
Diatom	4,8		3,7,8	4,8	2,3,4,7	3,7,8
Green	4,5			4		
Chrysophyte						
Cryptophyte			4			
Bluegreen	4		7	4	4	
Dinoflagellate						
Green flagellate			7			
Misc flagellate	7				7	4
			Spring			
Diatom	2,4,5,6,7,8		7,8	2,3,4,5,6,7	2,4,5,7	4
Green	2,3,4,5,7	6,4	4	2,3,4,5,6,7	2,3,4,5,7	4
Chrysophyte				4	1	
Cryptophyte			3,4		8	3,4
Bluegreen	2,3,4,5	3,7	3,7	2,3,4,5	2,3,4,5	
Dinoflagellate						
Green flagellate						
Misc flagellate		2,4				4
			Summer			
Diatom	4,8	8	4,7,8	4,5	2,3,4,7	4,7
Green	2,3,4,7	8	4,8	2,3,4,5	2,3,7	4,5
Chrysophyte						
Cryptophyte	2,4		2,3,4,5	2,4		2,3,4
Bluegreen	2,4,5		2,4,6	2,4		4,6
Dinoflagellate		8	8			
Green flagellate	7		7,8			
Misc flagellate	1,2,8	1,2,3,8	1		4	3,4,8

flow (Peterson *et al.* 1989). Once downstream, high to moderate streamflows plus tides may enhance the accumulation of phytoplankton biomass near the turbidity maximum (Peterson *et al.* 1975; Arthur & Ball 1979). During dry and critical years, low chlorophyll *a* concentrations in the downstream region were probably also affected by the turbidity maximum. The turbidity maximum is located farther upstream during dry and critical years where biomass accumulation and shoal to channel transport are reduced and net productivity is low (Cloern *et al.* 1983). The reason for lower phytoplankton biomass during dry years compared with critical years is unclear, but may be related to water management practices. The larger

TABLE 8. Average cell dimensions and cellular carbon for phytoplankton groups. (The letters s and f indicate single and filamentous forms.)

Group	n	Length µm	SD	Width µm	SD	Depth µm	SD	Log cell carbon pg	SD	% > or = 10 µm
Diatoms - s	10995	29.19	22.11	8.73	5.37	4.18	4.60	1.85	0.28	94
Diatoms - f	7932	5.94	3.60	6.62	2.50	21.20	34.94	1.50	0.50	7
Green - s	6602	14.66	13.50	5.97	3.33	1.27	1.83	1.63	0.28	42
Green - f	6	29.33	39.26	29.66	39.02	98.19	108.38	2.25	0.83	33
Chrysophytes	185	9.88	8.16	5.25	3.19	0.52	1.97	0.17	0.30	27
Cryptophytes	3838	14.01	4.24	7.22	1.04	3.89	2.62	1.83	1.27	69
Bluegreen - s	640	8.50	7.10	5.59	5.66	1.82	1.57	1.30	0.57	15
Bluegreen - f	455	4.76	0.62	4.76	0.62	67.71	14.15	2.46	0.13	0
Dinoflagellates	402	15.60	11.68	11.23	4.89	0.04	0.71	1.11	0.28	100
Green Flagellates	554	21.25	11.52	9.83	3.48	0.53	1.63	0.93	0.31	85
Misc. Flagellates	2182	6.43	0.55	4.24	0.10	0.06	0.25	0.07	0.05	0.05

percentage of inflow diverted for agriculture during dry years than critical years removes phytoplankton biomass and decreases residence time.

The spatial and temporal variation of phytoplankton community composition at upstream and downstream regions was also a function of changes in water-year type. During normal and critical years, diatoms and greens increased upstream during the spring and summer. Diatoms commonly occur early in the year when water temperature is low, turbidity is high and wind mixing is sufficient to prevent cells from settling to the bottom. Greens increase later in the spring when water temperatures are warmer. Their increase was facilitated in both normal and critical years by low to moderate streamflows which enabled the accumulation of cells. Moderate streamflows during normal years also dispersed these phytoplankton throughout the upstream regions. The coincident increase of cryptophytes upstream during critical years was probably a direct response to increased phytoplankton biomass, since they often function as mixotrophs. In addition, their buoyancy and mobility may protect them from sinking in the slow moving waters of critical years when they would otherwise settle on the bottom or be removed by benthic herbivores. Flagellates usually occur in warm and slow moving waters where their ability to migrate vertically allows them to avoid high light near the surface and access nutrients near the bottom. During wet and dry years, negative and small deviations from the mean for most phytoplankton were probably a function of the flushing of cells downstream with high streamflow or export of cells due to water management practices.

Downstream, diatoms and greens increased during normal and wet years. High streamflows transport cells downstream to Suisun Bay in normal years and to San Pablo Bay in wet years. Once these cells are downstream, they may be retained by tidal or gravitational circulation patterns (Peterson *et al.* 1975; Arthur & Ball 1979). In contrast, diatoms and greens were few downstream during critical years when low streamflows reduce the transport of cells downstream. Factors other than streamflow may contribute to the decrease of diatoms and greens downstream during critical years. Cells are lost from sedimentation by reduced turbulence (Thompson *et al.* 1981), increased benthic grazing by marine species (Nichols 1985) and increased retention of cells in deep channels upstream where net primary productivity is low (Arthur & Ball 1979; Cloern *et al.* 1983). In contrast, these factors probably contribute to the increase of flagellates downstream during critical years, since flagellates are mobile. The decrease of most phytoplankton groups during dry years is probably related to water management practices that result in export of phytoplankton-rich San Joaquin River water.

Spatial and temporal differences in the upstream and downstream distribution of chlorophyll *a* concentration with water-year type are important factors controlling the estuarine food web. Phytoplankton biomass can be an important source of carbon to the upper estuary (Jassby *et al.* 1993), where it is correlated with zooplankton biomass (Lehman 1992). Accumulation of phytoplankton biomass upstream during critical years provides food for the survival of freshwater and freshwater-tolerant species. However, accumulation of phytoplankton biomass upstream reduces the amount of phytoplankton biomass transported to the food web downstream. During wet years, the transport of phytoplankton biomass downstream provides food at the base of the food web in downstream nursery areas for fish, but leaves little for upstream organisms. Mass balance calculations suggest that Delta phytoplankton provide about 38% of the carbon load to Suisun Bay (Jassby & Powell 1994). Normal years probably provide the best combination of conditions for both upstream and downstream components of the food web. Moderate streamflows enable accumulation of phytoplankton biomass for use upstream and yet are high enough to transport some of that biomass downstream for use in Suisun Bay. In contrast, dry years appear to provide the poorest conditions for the food web, since both upstream and downstream chlorophyll *a* concentrations are low.

Changes in phytoplankton community composition with water-year type are important for understanding the ecology of the upper estuary because they affect the quality and size structure of food particles available at the base of the food web. Phytoplankton are not used equally by zooplankton in the aquatic food web. Although adult zooplankton can consume phytoplankton of varying cell size, they often feed selectively on phytoplankton larger than 8 to 11 *u*m in diameter (Paffenhoffer & Knowles 1978; Peterson *et al.* 1991; Kiorboe *et al.* 1990).

Availability of phytoplankton of the appropriate cell diameter and composition influences zooplankton population survival through intra- and inter-species competition for food quantity (Paffenhoffer & Knowles 1978) and quality (Peterson *et al.* 1991).

The linear size ratio between predators and their optimal prey is 18:1 for copepods (Hansen *et al.* 1994). This ratio is similar to the 22:1 ratio calculated for the copepod *Eurytemora affinis*, an important zooplankton food for juvenile fish in this estuary, and an average sized diatom in the upper estuary. Because of this size relationship, diatoms, which have large cell diameters and high cellular carbon content, may influence the food quality of large zooplankton among water-year types. In fact, the decrease of diatoms in the early 1980s coincided with the decline in abundance of many large zooplankton during the same period (Obrebski *et al.* 1992). A decrease in the average size of food particles may also have contributed to the success of introduced zooplankton species during the 1987 to 1992 drought (Orsi & Walter 1991). The influence of phytoplankton size structure on food quality is probably not through differences in phytoplankton growth rates, since growth rates do not differ among phytoplankton size classes (Cole & Cloern 1986).

Because of the strong influence of water-year type on phytoplankton, the long-term sequence of changes in water-year types was an important factor controlling the long-term trends in chlorophyll *a* concentration and phytoplankton community composition. Normal and wet years were associated with high chlorophyll *a* concentrations and diatom densities in the early 1970s. In contrast, wet and dry years coincided with low chlorophyll *a* concentrations and diatom densities between 1978 and 1986. Chlorophyll *a* concentrations and diatom densities continued to decrease downstream after 1986 when critical years were common. The magnitude of the decrease in critical years was enhanced downstream by the loss of phytoplankton to grazing by the clam *Potamocorbula amurensis*, which became established in the estuary in 1987 (Nichols 1985). Average chlorophyll *a* concentrations and diatom densities increased in the early 1990s in association with critical and wet years.

The long-term trend in chlorophyll *a* concentration and phytoplankton community composition with water-year type was partly a function of long-term climate change. World-wide climate is a function of the frequency and intensity of El Niño and southern oscillation (ENSO) events (Quinn & Neal 1985; Horel & Wallace 1981). Climatic conditions affect precipitation patterns in western North America (Cayan & Peterson 1989), which influence biological and environmental factors in San Francisco Bay (Peterson *et al.* 1989).

Determining the effect of individual ENSO events is often difficult because ENSO events in the upper estuary are associated with both wet and critical water-year types. The difficulty of determining their effects is compounded by the fact that the frequency and intensity of ENSO events have increased since 1977 (Quinn & Neal 1985). In the upper estuary, this 1977 climate shift resulted in an increase in the number of wet and critical water years. It also coincided with the decline in chlorophyll *a* concentration and diatom density in the early 1980s. Previous research has also demonstrated a correlation between the 1977 climate shift and phytoplankton community composition, chlorophyll *a* concentration and environmental variables in the upper estuary (Lehman & Smith 1991; Lehman 1992) and environmental and biological variables in the nearby Pacific Northwest (Ebbesmeyer *et al.* 1991).

Understanding the influence of water-year type on phytoplankton biomass and community composition may provide a way to gain insight into past and future changes in the estuarine food web. The large number of normal years and small number of critical years before 1970 suggest that average chlorophyll *a* concentrations and diatom densities were high early in the 20th century. It is also possible that phytoplankton communities were more stable early in the century since the interval between normal years was smaller than after 1970. That past changes in phytoplankton biomass and community composition can be accurately inferred from water-year types is suggested by phytoplankton data collected by Allen in 1920. These data indicate diatoms comprised at least 80 to 90% of the phytoplankton community during the normal year of 1913. This is similar to the percent composition of diatoms measured for normal years in the early 1970s. Similarly, it should be possible to use projected changes in water-year types to determine how the estuarine food web will be affected by climate and water management practices in the future.

ACKNOWLEDGMENTS

I wish to thank Tim Hollibaugh, Harlan Proctor, Randy Brown, Vera Tharp and two anonymous reviewers for their thoughtful reviews of the manuscript. Dave Sullivan assisted with the GIS mapping.

The views of this paper do not necessarily reflect the views of the California Department of Water Resources.

LITERATURE CITED

Allen, W. E. (1920). *A Quantitative and Statistical Study of the Plankton of the San Joaquin River and its Tributaries in and Near Stockton, California, in 1913*. Univ. Calif. Publ. Zool., vol. 22, pp. 1-292. University of California Press, Berkeley, CA.

Arthur, J. F., and M. D. Ball. (1979). Factors influencing the entrapment of suspended material in the San Francisco Bay-Delta estuary. Pages 143-174 *in* T. J. Conomos, ed., *San Francisco Bay: The Urbanized Estuary*. Pacific Division, American Association for the Advancement of Science, San Francisco, CA.

Cayan, D. R., and D. H. Peterson. (1989). The influence of north Pacific atmospheric circulation on stream flow in the west. Pages 375-397 *in* D. H. Peterson, ed., *Aspects of Climate Variability in the Pacific and Western Americas*. Geophys. Monogr. 55. American Geophysical Union, Washington, DC.

Cloern, J. E., A. E. Alpine, B. E. Cole, R. L. J. Wong, J. F. Arthur and M. D. Ball. (1983). River discharge controls phytoplankton dynamics in the northern San Francisco Bay Estuary. *Estuarine Coastal Shelf Sci.* 16:415-429.

Cloern, J. E., B. E. Cole, R. L. J. Wong, and A. E. Alpine. (1985). Temporal dynamics of estuarine phytoplankton: A case study of San Francisco Bay. *Hydrobiologia* 129:153-176.

Cole, B. E. and J. E. Cloern. (1986). Biomass and productivity of three phytoplankton size classes in San Francisco Bay. *Estuaries* 9:117-126.

Conomos, T. J. (1979). Properties and circulation of San Francisco Bay waters, Pages. 47-84 *in* T. J. Conomos, ed., *San Francisco Bay: The Urbanized Estuary*. Pacific Division, American Association for the Advancement of Science, San Francisco, CA.

Ebbesmeyer, C. C., D. R. Cayan, D. R. McLain, F. H. Nichols, D. H. Peterson, and K. T. Redmond. (1991). 1976 step in the Pacific Climate: Forty environmental changes between 1968-1975 and 1977-1984. Pages. 115-126 *in* J. L. Betancourt & V. L. Tharp, eds., *Proceedings of the Pacific Climate (PACLIM) Workshop*. Tech. Rept. 26. Interagency Ecological Studies Program for the Sacramento-San Joaquin Estuary, Sacramento, CA.

Hansen, B. H., P. K. Bjornsen, and P. J. Hansen. (1994). The size ratio between planktonic predators and their prey. *Limnol. Oceanogr*. 39:395-403.

Horel, J. D., and J. M. Wallace. (1981). Planetary-scale atmospheric phenomena associated with the southern oscillation. *Monthly Weather Rev*. 109:813-829.

Jassby, A. D., J. E. Cloern, and T. M. Powell. (1993). Organic carbon sources and sinks in San Francisco Bay: Variability induced by river flow. *Marine Ecol. Progr. Ser*. 95:39-54.

Jassby, A. D., and T. M. Powell. (1994). Hydrodynamic influences of interannual chlorophyll variability in an estuary: Upper San Francisco Bay-Delta (California, U.S.A.). *Estuarine Coastal Shelf Sci*. 39:595-618.

Kiorboe, T., H. Kaas, B. Kruse, F. Mohlenberg, P. Tiselius, and G. AErtebjerg. (1990). The structure of the pelagic food web in relation to water column structure in the Skagerrak. *Marine Ecol. Progr. Ser*. 59:19-32.

Lehman, P. W. (1992). Environmental factors associated with long term-changes in chlorophyll concentration in the Sacramento-San Joaquin Delta and Suisun Bay, California. *Estuaries* 15:335-348.

Lehman, P. W., and R. W. Smith. (1991). Environmental factors associated with phytoplankton succession for the Sacramento-San Joaquin Delta and Suisun Bay Estuary, California. *Estuarine Coastal Shelf Sci*. 32:105-128.

Nichols, F. J.(1985). Increased benthic grazing: An alternative explanation for low phytoplankton biomass in northern San Francisco Bay during the 1976-1977 drought. *Estuarine Coastal Shelf Sci*. 21:379-388.

Obrebski, S., J. J. Orsi, and W. Kimmerer. (1992). *Long-term Trends in Zooplankton Distribution and Abundance in the Sacramento-San Joaquin Estuary*. Tech. Rept. 32. Interagency Ecological Studies Program for the Sacramento-San Joaquin Estuary. Sacramento, CA. 42 pp.

Orsi, J. J., and T. C. Walter. (1991). *Pseudodiaptomus forbesi* and *P. marinus* (Copepoda:Calanoida), the latest copepod imigrants to California's Sacramento-San Joaquin Estuary. *Bull. Plankt. Soc. Japan*, Spec. Vol. 1991:553-562.

Paffenhofer, G., and S. C. Knowles. (1978). Feeding of marine planktonic copepods on mixed phytoplankton. *Marine Biol*. 48:143-152.

Peterson, D. H., D. R. Cayan, J. F. Festa, F. H. Nichols, R. A. Walters, J. V. Slack, S. E. Hager, and L. E. Schemel. (1989). Climate variability in an estuary: effects of river flow on San Francisco Bay. Pages 419-442 *in* D. H. Peterson, ed., *Aspects of Climate Variability in the Pacific and Western Americas*. Geophys. Monogr. 55. American Geophysical Union, Washington DC.

Peterson, D. H., T. J. Conomos, W. W. Broenkow, and P.C. Doherty. (1975). Location of the non-tidal current null zone in northern San Francisco Bay. *Estuarine Coastal Marine Sci*. 3:1-11.

Peterson, D. H., and J. F. Festa. (1984). Numerical simulation of phytoplankton productivity in partially mixed estuaries. *Estuarine Coastal Shelf Sci*. 19:563-589.

Peterson, D. H., R. E. Smith, S. W. Hager, D. D. Harmon, R. E. Herndon, and L. E. Schemel. (1985). Interannual variability in dissolved inorganic nutrients in northern San Francisco Bay Estuary Pages 37-58 *in* J. E. Cloern & F. N. Nichols, eds., *Temporal Dynamics of an Estuary: San Francisco Bay*. Dr. W. Junk Publishers, Dordrecht, Netherlands.

Peterson, W. T., P. Tiselius, and T. Kiorboe.(1991). Copepod egg production, molting and growth rates, and secondary production, in the Skagerrak in August 1988. *Jour. Plankt. Res.* 13:131-154.

Quinn, W. H., and V. T. Neal. (1985). Long-term variations in the southern oscillation, El Niño and Indonesian drought. *Fish. Bull.* 76:663-678.

Strathmann, R. R. (1967). Estimating the organic carbon content of phytoplankton from cell volume or plasma volume. *Limnol. Oceanogr.* 12:411-418.

Strickland, J. D. H. and T. R. Parsons. (1968). *A Practical Handbook of Seawater Analysis*. Fish. Res. Bd. Canada, Bull. 167. 311 pp.

Thompson, J. K., F. H. Nichols, and S. M. Wienke. (1981). *Distribution of Benthic Chlorophyll in San Francisco Bay, California. February 1980-February 1981*. U. S. Geol. Surv., Open File Rept. 81-1134. 36 pp.

Utermohl, H. (1958). Zur Vervollkommung der quantitätiven Phytoplankton-methodik. *Mitt. Int. Ver. Theor. Angew. Limnol.* 9:1-38.

Wong, R. L. J., and J. E. Cloern. (1981). *Plankton studies in San Francisco Bay. II. Phytoplankton Abundance and Species Composition, July 1977-December 1979*. U. S. Geol. Surv. (Menlo Park, CA), Open-File Rept. 81-214. 103 pp.

Wong, R. L. J., and J. E. Cloern. (1982). *Plankton studies in San Francisco Bay. IV. Phytoplankton Abundance and Species Composition, January 1980-February 1981*. U.S. Geol. Surv. (Menlo Park, CA), Open File Rep. 82-443. 152 pp.

Zar, J. H. (1984). *Biostatistical Analysis*. Prentice Hall, Inc. Englewood Cliffs, New Jersey. 718 pp.

FOOD LIMITATION AS THE PROBABLE CAUSE OF A LONG-TERM DECLINE IN THE ABUNDANCE OF *NEOMYSIS MERCEDIS* THE OPOSSUM SHRIMP IN THE SACRAMENTO-SAN JOAQUIN ESTUARY

JAMES J. ORSI AND WALTER L. MECUM
California Department of Fish and Game, Bay-Delta and Special Water Projects Division, 4001 N. Wilson Way, Stockton, CA 95205.

Neomysis mercedis **was sampled from 1968 to 1993 throughout Suisun Bay and the Delta. Adults alternated between large ones in winter and spring and small ones in summer and fall. Annual abundance tended to reach a maximum from May to July. A long-term downtrend in abundance occurred in spring, summer and fall after 1975. Abundance was lowest during the drought years 1987-1992 but did not rise when the drought ended in 1993. Females tended to be smaller in the years of lowest abundance and the percent of females carrying eggs was low in those years. Five hypotheses for the decline were investigated: (1) reduced concentration by estuarine gravitational flow and increased seaward loss to tidal pumping, (2) food limitation, (3) rice herbicides, (4) high temperature, (5) export pumping losses. Food limitation of juveniles in the form of reduced phytoplankton concentrations best explained the decline. Food appears to be permanently limited by the grazing of the introduced Asian clam,** ***Potamocorbula amurensis.*** **Competition for food by two introduced Asian mysid shrimps may also hamper the recovery of the native mysid population.**

Neomysis mercedis is a member of the malacostracan order Mysidacea, a group of small, shrimp-like animals that are commonly known as opossum shrimp because of the ventral brood pouch in which they carry their developing young. Mysidaceans are found in a variety of habitats from deep oceans and coastal seas to estuaries, rivers and lakes. They are often found in shoals or schools numbering thousands of individuals (Clutter 1969).

In the Sacramento-San Joaquin Estuary there are eight species, of which one, *Acanthomysis aspera*, was introduced from Japan in 1992, presumably in ballast water, and another but undescribed *Acanthomysis* species was introduced from an unknown part of the Far East in the same year. *Neomysis mercedis* is located primarily in Suisun Bay and the Delta (Heubach 1969; Siegfried *et al.* 1979; Knutson & Orsi 1983) where it is by far the most abundant mysidacean. It is an important food for juvenile striped bass (Heubach *et al.* 1963; Stevens 1966a and b), and is also consumed by many other fish species (Radtke 1966; Stevens 1966b), and by the caridean shrimps, *Crangon franciscorum* and *Palaemon macrodactylus* (Sitts & Knight 1979).

Sampling for *N. mercedis* began in 1964 as part of the original Department of Fish and Game (CDFG) ecological studies (Turner & Heubach 1966). Sampling was suspended in 1967 and resumed in June 1968.

N. mercedis was found to be most abundant from freshwater to 2.0 ‰ chlorinity (3.6 ‰ surface salinity) (Heubach 1969) or from 1.2 to 5.6 ‰ salinity (Knutson & Orsi 1983). These salinities place it in what the literature on this estuary has traditionally termed the entrapment zone. This is the section of the salinity gradient where particle settling and estuarine flow patterns concentrate particulate matter. It has been given an operational definition of 1.2-6.0 ‰ salinity (specific conductance 2-10 mS/cm (Arthur & Ball 1980; Kimmerer 1992). Turbidity tends to maximize in this area, making it roughly equivalent to the maximum turbidity zone identified in other estuaries.

Temperature > 22°C in combination with low dissolved oxygen concentration was associated with low mysid abundance (Heubach 1969) but temperature was not a factor affecting interannual abundance variations (Knutson & Orsi 1983). The upper lethal temperature was established as 24.2 to 25.5°C. in 48 hour bioassays (Hair 1971).

Regression analysis of 1968-1981 data indicated that annual July-October population size was significantly and negatively related to salinity intrusion and positively related to the abundance of the copepod *Eurytemora affinis*, a food item. Salinity intrusion is an indirect measure of the location of the entrapment zone and the location of this zone is associated with variations in mysid abundance. Highest abundance from 1972 to 1988 was associated with entrapment zone positions in Suisun Bay < 72 km from the Golden Gate and lowest abundance when the entrapment zone was at > 92 km and in the western Delta (Kimmerer 1992). The reason for the relationship between abundance and entrapment zone position has not been established.

Mysids migrate vertically. Adults tend to have a midwater location and the neonates are closer to surface than bottom in the turbid entrapment zone (Orsi 1986). In clearer water seaward from the entrapment zone, the adults were closer to bottom than to surface. The vertical migration pattern may interact with two-layered estuarine flow to concentrate the mysids in the entrapment zone and prevent their downstream loss from the estuary (Orsi 1986). Losses upstream to the Federal and State water export pumping plants in the south Delta might also occur (Knutson & Orsi 1983).

Food habits investigations showed that detritus and diatoms were the most abundant items in mysid gut contents (Baldo Kost & Knight 1971). But Siegfried & Kopache (1980) found that although diatoms were the most abundant prey identified, predation on rotifers and copepods accounted for > 80% of the energy consumed by adults. Mysids fed as herbivores when small, as carnivores when large (Siegfried & Kopache 1980).

Toxic chemicals may have adverse effects on *N. mercedis*. Sacramento Valley rice fields are treated with several herbicides and pesticides and the fields are drained from April through June into the Sutter Bypass, which enters the Sacramento River upstream from the city of Sacramento. The toxicity of some of these chemicals to *N. mercedis* has been determined in bioassays (Bailey 1985; Faggella *et al.* 1990).

The purpose of this paper is to present relevant life history information, to describe mysid abundance trends from 1968 to 1993, and to evaluate hypotheses to explain the trends. Several hypotheses can be constructed based on the above literature results. These are:

(1) Greater seaward loss of mysids when estuarine gravitational circulation is lessened during periods of low freshwater outflow. This hypothesis envisions an interaction between the predominant vertical positions of juvenile and adult mysids in the water column and the ebb-flood asymmetry in vertical velocity profiles that causes slightly greater seaward displacement of surface water on the ebb tide and slightly greater upstream displacement of bottom water on the flood over a tidal cycle (Pritchard 1956). Juvenile mysids are positioned above mid-depth (Orsi 1986) and should experience a net downstream movement over a tidal cycle. As they mature they drop into the bottom layer and experience a net upstream movement. The downstream movement of one life stage is thus balanced by the upstream movement of the other to minimize population loss from the entrapment zone. During periods of low freshwater outflow tidal circulation dominates over gravitational flow and tidal pumping may remove mysids from the estuary.

(2) Food limitation. Food limitation has been extensively investigated in copepods (Checkley 1980; Ambler 1986; Beckman & Peterson 1986) but no work has been done on mysidaceans. However, the biological effects of food limitation should be similar in mysids and copepods. At low food levels copepod egg production declines sharply and growth is reduced. These conditions will affect birth rate and hence, population size.

(3) Mortality from high temperature. In the laboratory, mysid mortality increased sharply at temperatures > 22°C (Hair 1971). Field abundance in 1965 and 1966 showed a decline at such temperatures (Heubach 1969). Obviously, temperature mortality will occur only in summer and possibly early fall.

(4) Mortality from rice herbicides. This would also be a seasonal phenomenon restricted to the months when such chemicals are in the river – April, May and June.

(5) Loss to the south Delta export pumping plants. These plants draw water out of Old River and send it to the San Joaquin Valley and southern California. In summer and fall the volume they extract is so large – a maximum of > 300 $m^3 sec^{-1}$ – that water to supply them is pulled out of the Sacramento River, down the Mokelumne River, across the San Joaquin River, and up Old and Middle rivers. Much of the water in the San Joaquin River up and downstream from the mouth of Old River flows to the pumps in summer and fall. When mysids are located in the Delta they can conceivably become entrained in these pumping flows and be exported from the estuary. Clearly, mysids would be most vulnerable to export when the entrapment zone is in the Delta.

METHODS

Field

Mysids were collected in a conical plankton net attached to a rectangular towing frame of steel pipe designed to protect the net from snags. Mesh size and net dimensions have varied over the years (Knutson & Orsi 1983). The most important change was from 1 mm mesh to 0.505 mm mesh in 1974 to reduce net escapement by juvenile mysids < 4 mm in total length. Because of this change, abundance calculations in this paper are for ≥ 4 mm mysids. Net mouth diameter has been 30 cm and net length 1.48 m since 1974. Ten minute oblique, stepwise, bottom to surface tows were made at a varying number of stations from San Pablo Bay throughout the Delta. Of these, 35 in Suisun Bay and the Delta have been sampled consistently since 1968 and are the ones used in this analysis (Fig. 1). Sampling was generally done once monthly from November to March and twice monthly the rest of the year. But no December to February sampling was done after 1983. Surface specific conductivity and surface temperature were measured and Secchi disk readings were taken at each station. Starting in 1976, chlorophyll *a* was measured by the acidification method from water collected at 1 m depth. For years prior to 1976 chlorophyll *a* data from the Department of Water Resources monitoring program was used. Starting in 1972, a Clarke-Bumpus net and a small pump were used to collect copepods, cladocerans and rotifers at the same stations. Field sampling and laboratory processing methods for *N. mercedis* are detailed in Knutson & Orsi (1983) and for zooplankton in Orsi & Mecum (1986).

Analysis

Abundance of an organism in the salt field of an estuary is partly a function of salinity. High freshwater outflow in spring can push most of the mysid population downstream from the sampling sites so the tail of the distribution is sampled and a biased estimate of abundance is obtained. Temperature also has a large influence on population size. To determine if changes in abundance over time have occurred due to other factors, the effects of salinity and temperature must be removed from

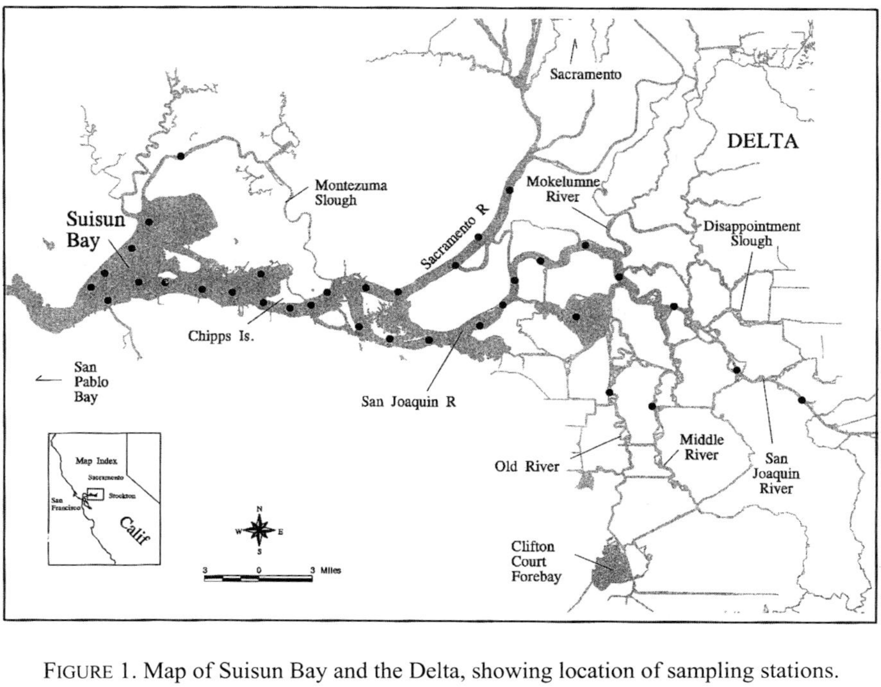

FIGURE 1. Map of Suisun Bay and the Delta, showing location of sampling stations.

the measurements of abundance. To do this, abundance calculations in numbers per cubic meter for all years were combined for all stations in Suisun Bay and the Delta and divided into 20 specific conductivity classes of nearly equal sample size (Table 1). Mean abundance was then calculated for each combination of specific conductivity class and month and each abundance value at each date and station was subtracted from the class and month mean to yield a corrected or anomaly value. The anomalies are expressed as unitless numbers positive or negative to the all-over mean. The anomalies are the abundance values actually subjected to statistical analysis.

RESULTS

Life History

Neonates are released from the brood pouch at a total length of 2 or 3 mm and grow to a maximum length of 16-18 mm for females and 14-16 mm for males. Adult length varies seasonally (Fig. 2). Females mature at ~10 mm and reach their maximum length in March. The largest females then begin to disappear and young females begin to mature at smaller sizes so that female length becomes progressively shorter until it ranges between 7-8 mm and 11-12 mm in July and August. Length begins to increase again in late October as water temperatures drop. The

TABLE 1. Mean salinity and specific conductance in each salinity class used in the abundance-salinity relationship.

Salinity Class	*Mean Salinity (%)*	*Mean Specific conductance (mS/cm)*
1	0.04	0.07
2	0.08	0.15
3	0.10	0.17
4	0.11	0.20
5	0.12	0.22
6	0.15	0.26
7	0.16	0.28
8	0.25	0.42
9	0.30	0.57
10	0.45	0.80
11	0.70	1.21
12	1.01	1.89
13	2.02	2.98
14	3.05	4.53
15	4.08	6.42
16	5.13	8.66
17	6.19	11.27
18	8.34	14.23
19	10.53	18.11
20	15.03	26.85

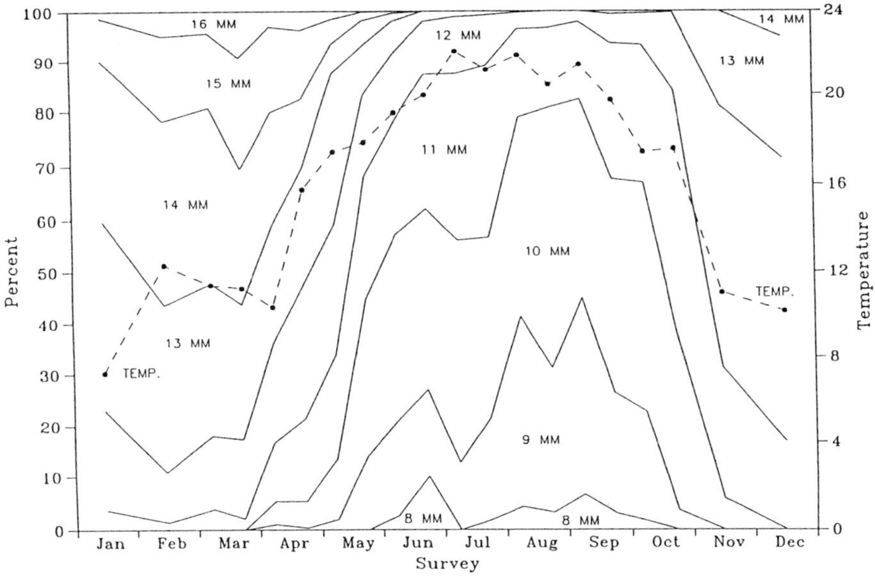

FIGURE 2. Female length frequency composition and mean temperature (C) on each 1982 sampling survey.

largest changes in length frequencies occur in April when temperature rises rapidly. The population thus cycles between large females in winter and small ones in summer. This also occurs in *N. mercedis* in the Fraser River, British Columbia (Johnston & Northcote 1988).

Because the number of young carried per female is a function of female size, fecundity varies with season (Figs. 3 and 4). The number of young per brood ranges from a mean of ~11 for 8 mm females to a mean of ~ 80 for 17 mm females. Population fecundity is highest in spring, declines rapidly to a summer low and increases slightly in fall. The number of developing young in the population depends not only on female size and abundance but on the percent of females gravid. This varies over time but has no apparent seasonal pattern (Fig. 5). Adult density may be a factor regulating percent gravid as samples with > 500 mysids usually have a higher percent gravid than those with < 500 mysids. The percent gravid was very low in 1976 and 1977 and from 1989 to 1993, years characterized by low mysid abundance (Fig. 7).

Prior to 1981 three cohorts of young appeared in most years, large ones in May to July and a small one in October (Fig. 6). The May cohort was presumably produced by overwintering females and young born during winter and early spring and maturing in May. Young from the May cohort should mature in about two months (H. C. Bailey, Univ. California, Davis, pers. commun.) and should produce the July cohort.

The length frequencies for adult females show that 13 to 15 mm females dominated from January to March but by the time the first cohort appeared, 10 to

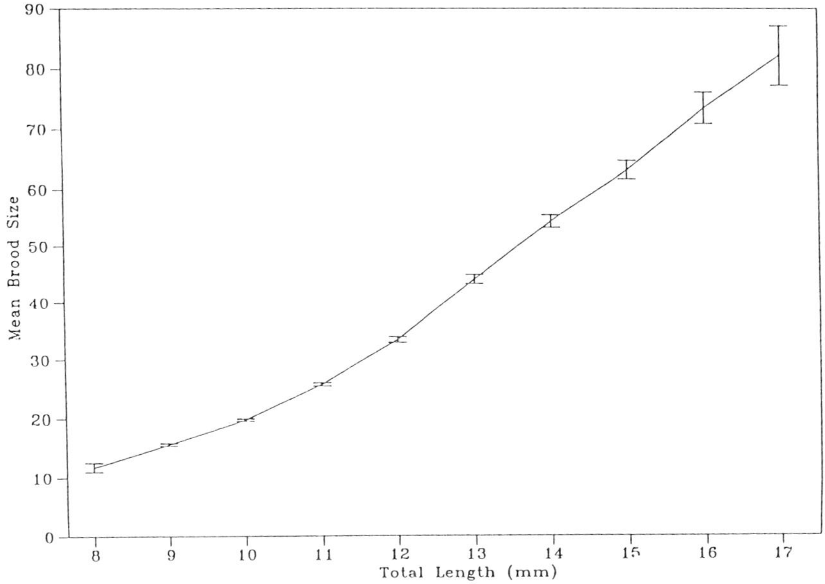

FIGURE 3. Female length-brood size relationship. Bars mark the 95% confidence intervals of the 1976-93 means for each length group.

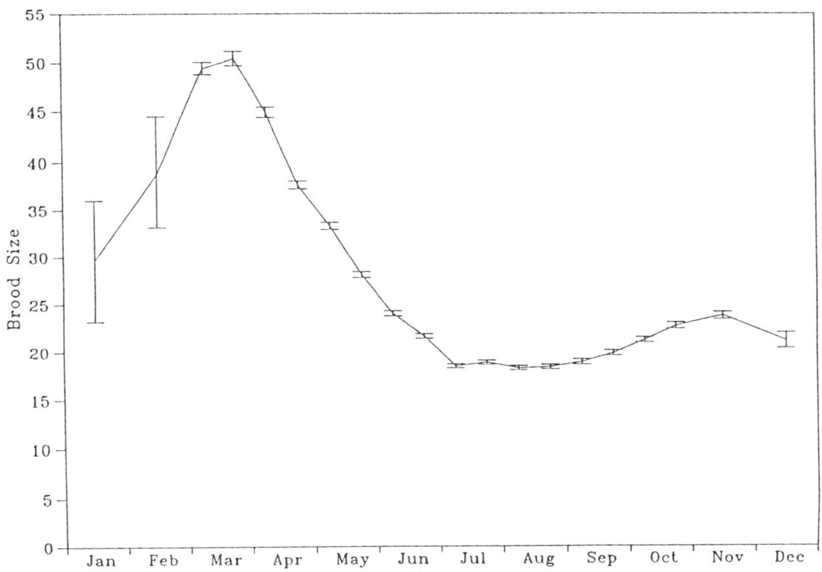

FIGURE 4. Mean brood size of all gravid females on each sampling survey from 1976 to 1993. Bars mark the 95% confidence intervals of the means.

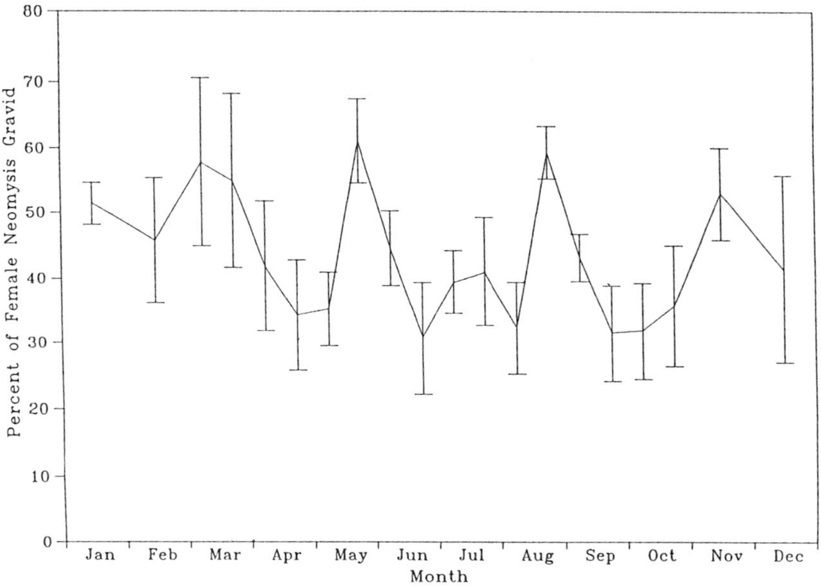

FIGURE 5. Percent of females gravid on each sampling survey of 1982.

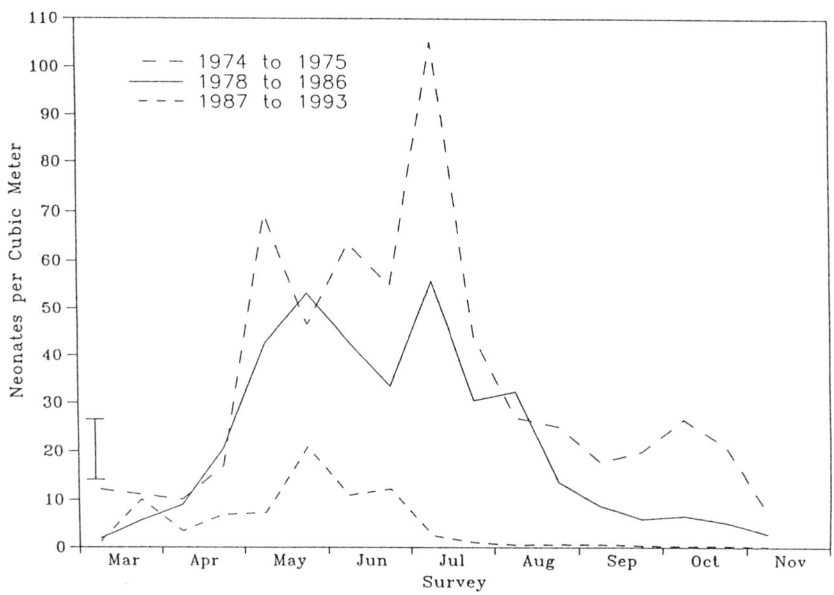

FIGURE 6. Mean abundance of neonates (2-3 mm mysids) on each March to November sampling survey in 3 groups of years, 1974-1975, 1978-1986 and 1987-1993. Bar at the lower left is the mean of the 95% confidence intervals for all points in the figure.

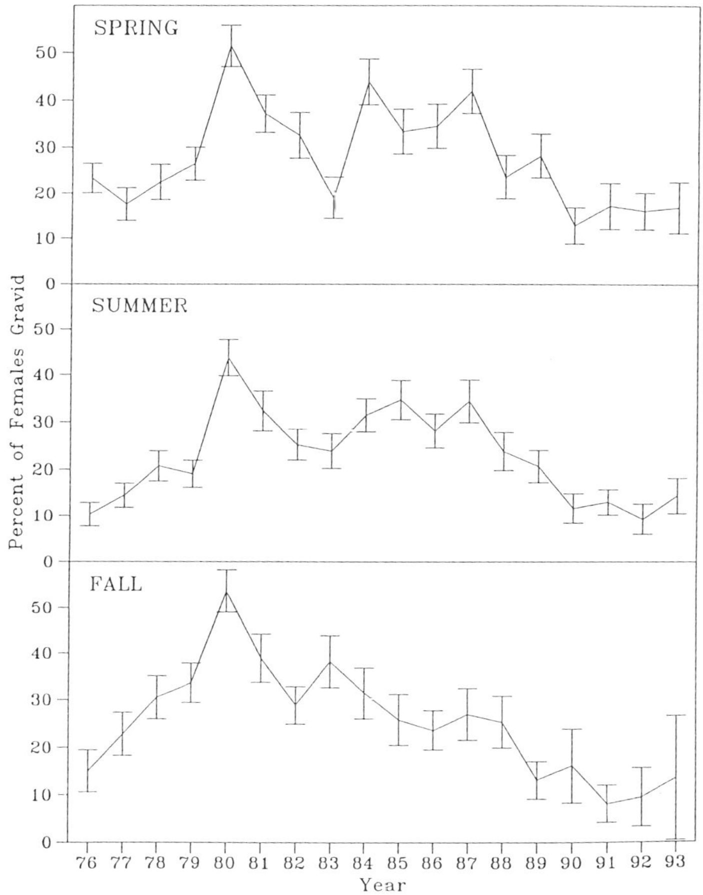

FIGURE 7. Mean percent of females gravid in spring, summer and fall of each year, 1976-1993. Bars mark the confidence intervals.

11 mm females were dominant (Fig. 2). The large females died off sequentially in spring and early summer, the largest and hence, presumably, the oldest, first. The October cohort should be produced by the females maturing from the July cohort. The small size of the October cohort and its absence after 1980 suggests that the young born in July experience high mortality during the summer.

Maximum adult length rarely exceeds 12 mm during summer (Fig. 2), presumably because life span is shorter at high temperatures and mysids die before they can reach greater lengths (Asttorsson & Ralph 1984; Toda *et al.* 1983), or because their growth slows at temperatures > 20°C (Toda *et al.* 1987).

Distribution and Abundance

As previously reported, mysids are concentrated in the entrapment zone, although they extend seaward of the entrapment zone and well downstream from it

(Fig. 8). Because the entrapment zone oscillates up and downstream with tide and freshwater outflow, the location of the mysid population also shifts between Suisun Bay and the Delta and during very high outflow events is located in Carquinez Strait and San Pablo Bay.

Mysid abundance is usually highest during May to July and lowest during winter (Fig. 9). Fall abundance was high from 1968 to 1975 but low in most years thereafter.

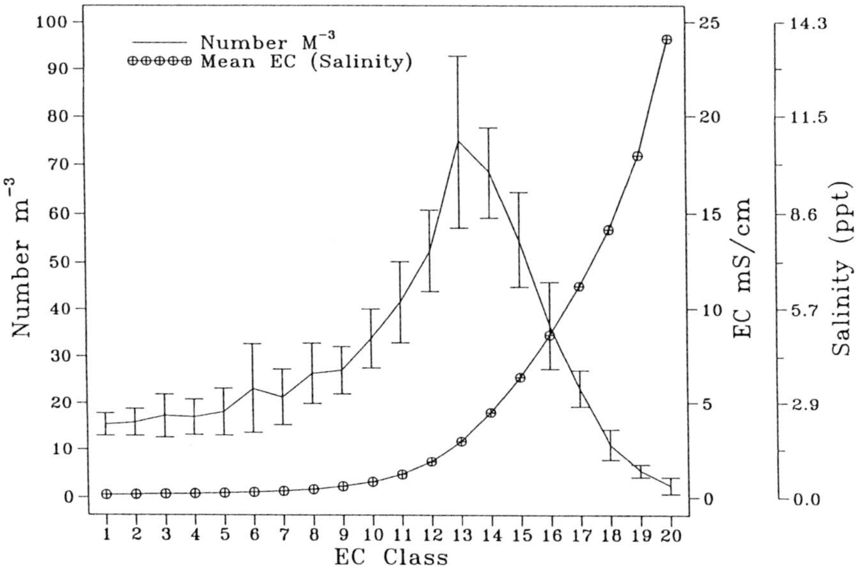

FIGURE 8. Mean abundance of mysids ≥ 4 mm in each specific conductivity class, 1968-1993. Bars mark the 95% confidence intervals of the means. Also shown is the mean specific conductivity and salinity in each class.

Long-term Trends in Abundance

Numbers per cubic meter as well as abundance anomalies were calculated for each season and for the March-November period for comparison (Fig. 10). The two measures of abundance were in good agreement, especially for March-November but long-term downtrends in spring and fall were apparent only in the anomalies. Mean densities ranged from > 100 m^{-1} to < 1 m^{-1}.

The March to November abundance anomalies show that abundance was highest and rather stable from 1968 to 1975 (Fig. 10). Abundance fell very low in the drought year, 1977, rose in 1978 but not to previous levels and oscillated without a trend from 1978 to 1986. The drought years, 1987-1992, were characterized by very low and generally decreasing anomalies. Abundance did not increase in 1993, a wet year. The seasonal patterns were similar to the annual results with some variations.

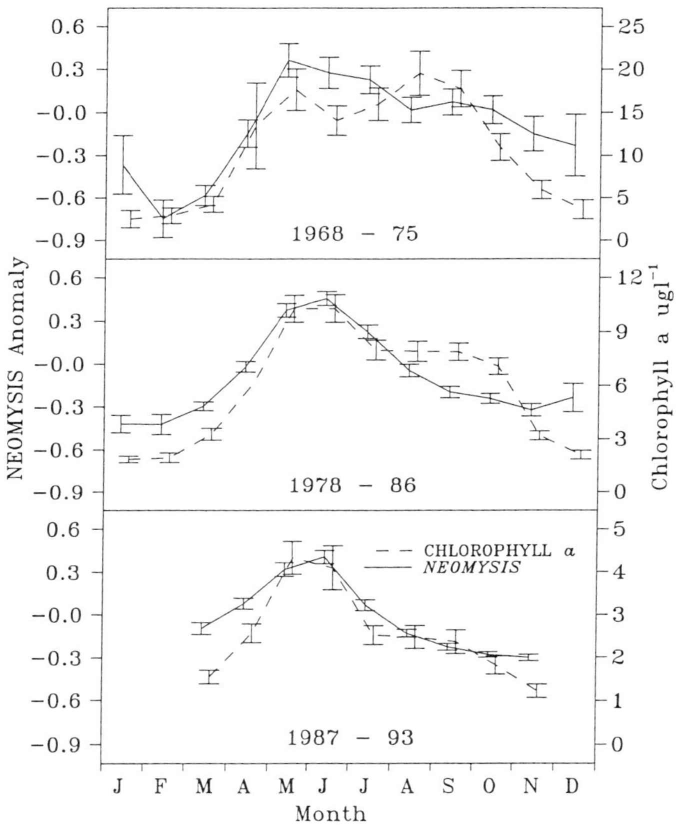

FIGURE 9. Mean abundance anomalies of mysids ≥ 4 mm and mean chlorophyll *a* concentration for each month in three groups of years: 1968-1975, 1976-1986, 1987-1993. Bars mark the 95% confidence intervals of the means.

Abundance Anomalies and Delta Outflow

The anomalies appear to be flow related because they are lowest in drought years. The outflow-abundance relationship can be explored by correlating the anomalies with the log of Delta outflow at Chipps Island in each season (Fig. 11). The variance in the anomalies accounted for by the correlations (R^2) ranges from 53.2% in summer to 56.1% in fall. The relationship is curvilinear due to low anomalies in the very high outflow years 1982 and 1983. By summer of 1983, outflow had dropped enough for the entrapment zone to enter Suisun Bay but the anomaly value was still low. Outflow explains a substantial portion of the variance but not enough to be the only factor affecting abundance.

Multiple Regression

A stepwise multiple regression analysis was done to apportion the variance in anomalies to factors related to the hypotheses. These factors are: Delta outflow,

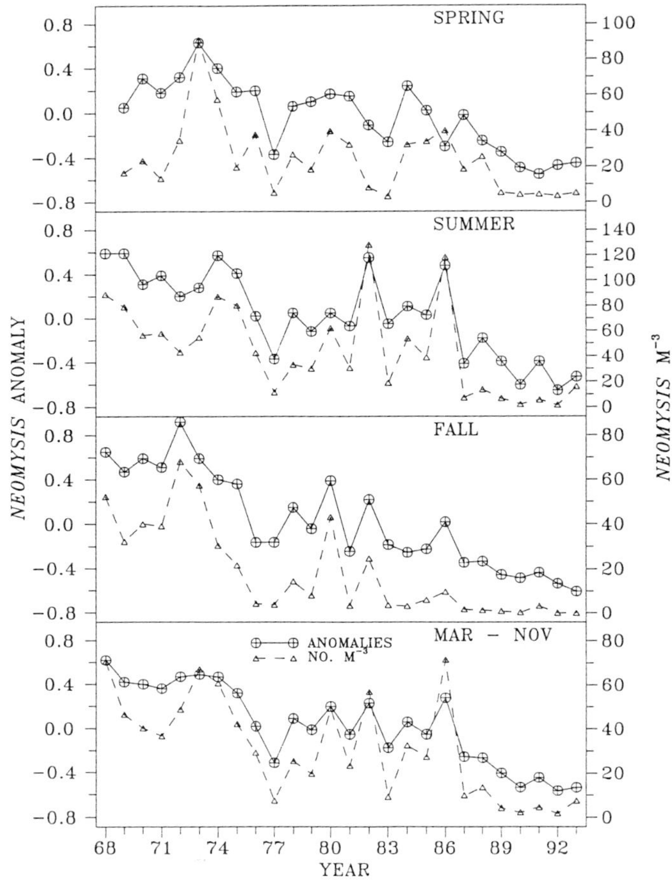

FIGURE 10. Abundance anomalies and numbers m^{-3} for mysids ≥ 4 mm for March to November and for spring, summer and fall of each year, 1968-1993. There was no sampling in spring 1968.

temperature, the rice herbicide molinate, and indices of food abundance – chlorophyll *a* and logs of copepod and rotifer densities. Cladocerans were excluded from the analysis because they are usually abundant upstream from *N. mercedis*. The analysis was restricted to April, May and June when the rice fields were drained and molinate was in the river. The years are those for which a complete data sequence was available, 1973-1992.

Chlorophyll *a* was the single most important variable and explained 77.6% of the variance (Table 2). Molinate was not significant. The best two variable model contained chlorophyll *a* and copepods and accounted for only a small additional amount of the variance, 1.2%. The best 3-variable model contained chlorophyll *a*, outflow and outflow squared and increased the variance explained to 87.7%. Addition of more variables to the model added little to its explanatory power.

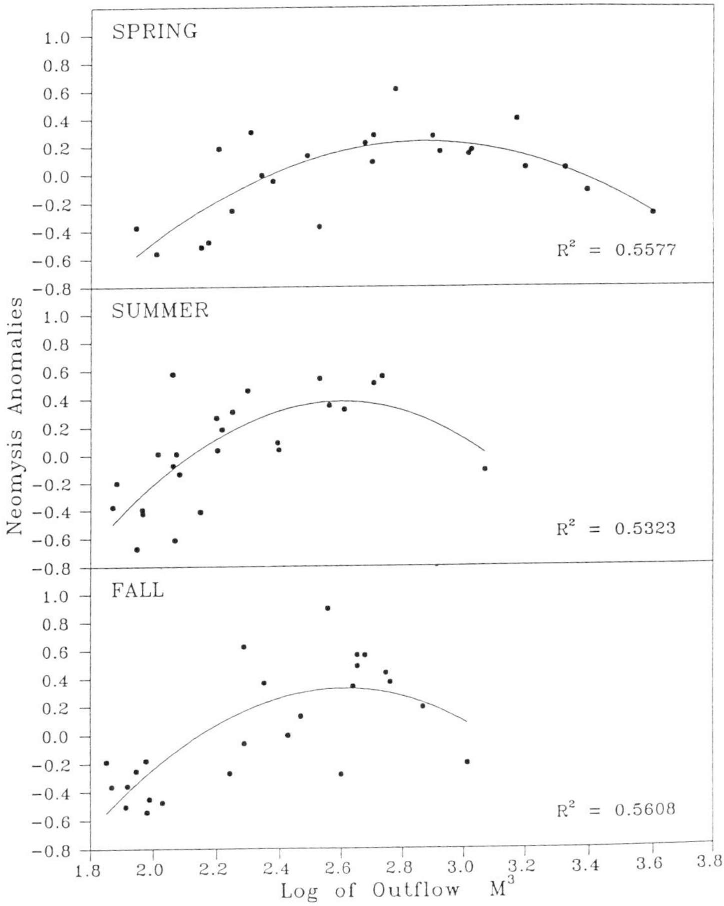

FIGURE 11. Abundance anomalies for mysids ≥ 4 mm vs. Delta outflow at Chipps Island in spring, summer and fall, 1968-1993. Each point represents a year.

Additional regression analyses were run for spring, summer and fall to determine if seasonal changes would occur in the significant variables. These regressions were run for small (4-7 mm) and large (> 8 mm) mysids because literature results indicate that phytoplankton is an important food source for small mysids and zooplankton for large mysids (Baldo-Kost & Knight 1971; Siegfried & Kopache 1980). Molinate was excluded from these analyses because it was not significant during the months when its in-river concentrations were highest.

The regression results were similar for both size groups and all seasons (Table 3). For both sizes in spring and summer, chlorophyll *a* was the most important variable, followed by outflow. In fall, chlorophyll *a* and rotifers were the most important variables for both sizes. Addition of more than two variables only slightly increased the variance explained except in spring when outflow squared increased

TABLE 2. *R*-square analysis results for mysids ≥ 4 mm vs. chlorophyll *a*, Molinate, outflow squared, temperature, log rotifers m^{-3}, log copepods m^{-3} in April, May and June 1973-1992

Variables in Model[1]	*R-squared Values*
chl	0.776
chl, copes	0.788
OF, OF2, chl	0.877
OF, OF2, chl, copes	0.888
Molinate, OF, OF2, chl, copes	0.891
Molinate, OF, OF2, temp, chl, copes	0.893
All	0.894
Single Variables	
chlorophyll a	0.776
rotifers	0.363
molinate	0.100
copepods	0.041
outflow	0.018
temperature	0.002
outflow squared	0.002

[1] chl = chlorophyll *a*; OF = outflow; OF2 = outflow squared; temp = temperature; copes = copepods

it significantly for both small and large mysids. The percent of the variance explained rose from spring to fall.

Mysid abundance actually had a curvilinear relationship with chlorophyll *a* in spring and summer and a linear relationship in fall (Fig. 12). Since both small and large mysids had similar *R*-square values in the regressions, they were combined in Fig. 12. Data points for 1968-1971 were added to this figure because chlorophyll *a* data was available for those years but copepod and rotifer sampling did not begin until 1972. The percent of the variance explained by these correlations was higher than in the *R*-square analysis.

Rotifer abundance experienced a long-term decline but copepod abundance did not (Figs. 13 and 14). Copepod abundance actually increased during the falls of 1987-93 relative to earlier years due to the presence of introduced exotic species.

Temperature was not a significant variable even in summer. To investigate temperature further, mysid abundance in the San Joaquin River at Stockton, where the highest temperatures in the sampling area are recorded, was compared with temperature for the years of record (Fig. 15). Abundance peaked at 21°C but was generally higher at temperatures > 21°C than at temperatures < 21°C. Temperature did not show a long-term rise in summer and fall but did so in spring (Fig. 16).

Export Pumping

The significant correlation with outflow may indicate a concentration effect as hypothesized by Orsi (1986) or it may mean that when the population is located in

TABLE 3. *R*-squared analysis results for small (4-7 mm) mysids and large (≥ 8 mm) mysids vs. chlorophyll *a*, outflow, outflow squared, temperature, log rotifers m-3, copepods m-3 during Spring, Summer and Fall, 1972-1992

	SPRING	
	R-squared Values	
Variables	*Small Mysids*	Large Mysids
chl	0.444	0.536
OF, chl	0.530	0.638
OF, OF2, chl	0.795	0.842
OF, OF2, chl, copes	0.797	0.849
OF, OF2, chl, temp, copes	0.798	0.852
All	0.798	0.852
	SUMMER	
	Small Mysids	*Large Mysids*
chl	0.740	0.813
OF, chl	0.798	0.846
OF, OF2, chl	0.820	0.864
OF, OF2, chl, rots	0.836	0.889
OF, OF2, chl, rots, temp	0.852	0.897
All	0.860	0.897
	FALL	
	Small Mysids	*Large Mysids*
chl	0.802	0.811
chl, rots	0.832	0.841
chl, copes, rots	0.874	0.878
OF, chl, copes, rots	0.905	0.902
OF, chl, copes, rots, temp	0.906	0.904
All	0.907	0.905

[1] chl = chlorophyll a; OF = outflow; OF2 = outflow squared; temp = temperature; rots = rotifers; copes = copepods

Suisun Bay it is too far downstream to be vulnerable to export pumping. To test the effects of export pumping from the south Delta pumping plants, an analysis was done using data from CDFG monitoring at two stations in the south Delta in Old and Middle rivers from 1972 to 1988 and from Department of Water Resources samples taken at the same locations during the spring and summer of 1989 with a Clarke-Bumpus net. Population size was calculated by multiplying the monthly abundance at each station by the water volume each station was considered to represent. Export loss rates were then calculated for each month by multiplying the mean of the abundance values at the two stations and the export flow rate for the

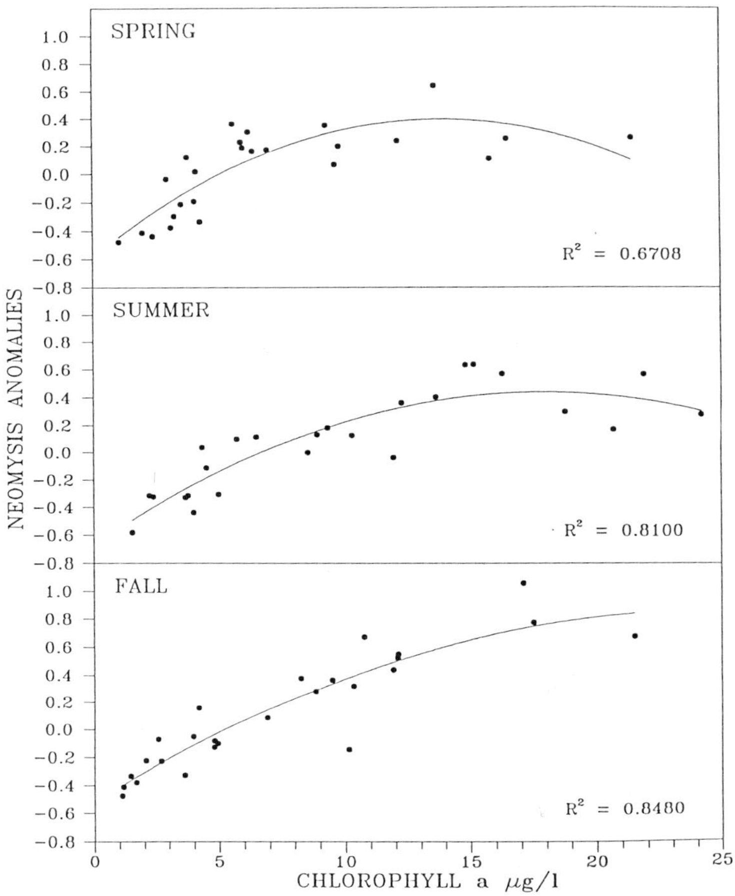

FIGURE 12. Abundance anomalies for mysids ≥ 4 mm vs. mean chlorophyll *a* concentrations in spring, summer and fall, 1968-1993. Each point represents a year.

same month. The proportion of the population exported was then determined by dividing the export loss rate by the population size. The resulting median daily loss rates were 0.01% day^{-1}. The means were somewhat larger but only a few values exceeded 1% day^{-1} (Table 4) (W. Kimmerer, pers. commun.).

Food Limitation

If phytoplankton is limiting *N. mercedis*, abundance patterns within years should be related to chlorophyll *a* concentrations. To test for this the years were divided into 3 groups based on abundance: 1968-1975, 1976-1986, and 1987-1993. Abundance anomalies were then calculated for each month of these periods, correcting the anomalies only for salinity. The anomalies were then plotted against mean monthly chlorophyll *a* concentrations. The plots showed that in all 3 year groups in spring, both abundance and chlorophyll *a* rose rapidly from winter lows to a peak

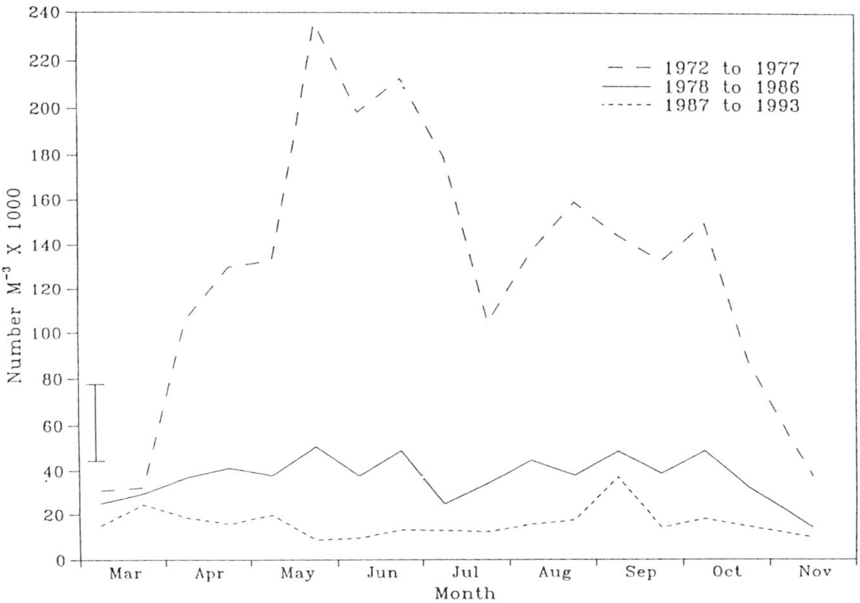

FIGURE 13. Mean rotifer abundance (in thousands per cubic meter) on each sampling survey for three periods: 1972-1977, 1978-1986 and 1987-1993. Bar at lower left marks the mean 95% confidence interval of all points on the figure.

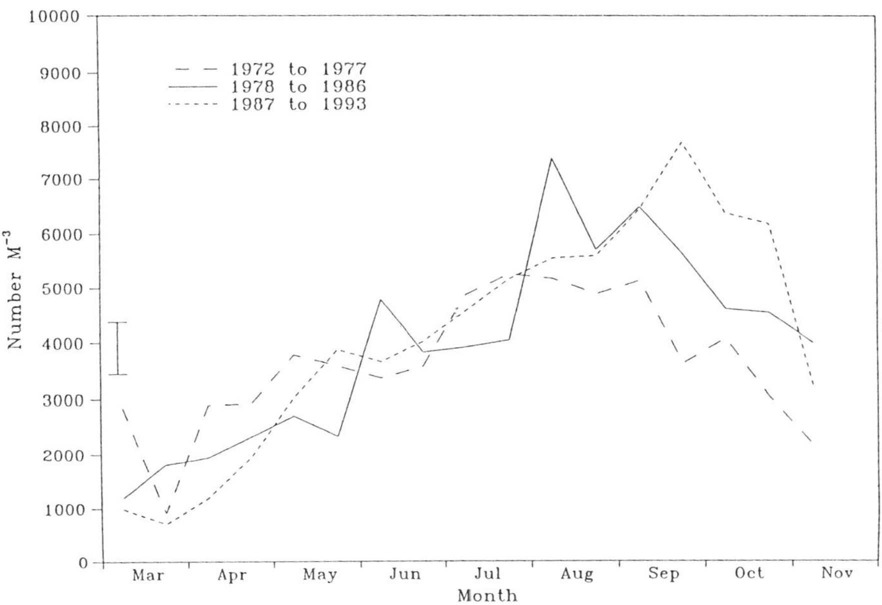

FIGURE 14. Mean copepod abundance (number per cubic meter) on each sampling survey for three periods: 1972-1977, 1978-1986 and 1987-1993. Bar at left marks the mean 95% confidence interval of all points on the figure.

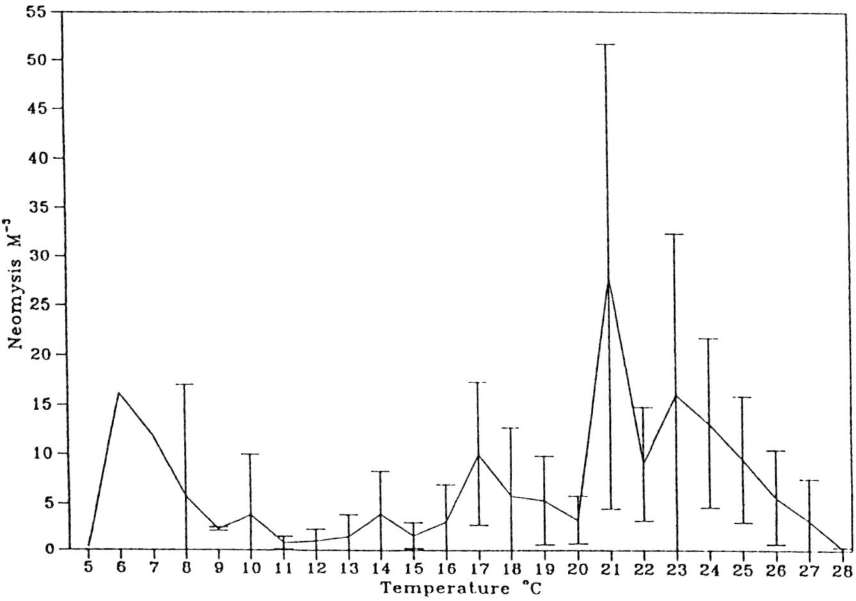

FIGURE 15. Mean abundance of mysids ≥ 4 mm in each degree Celsius temperature class. Data are for 1968-1993. Bars mark the 95% confidence intervals of the means.

in May or June and then declined (Fig. 9). But in the first group of years a second and higher chlorophyll *a* mode occurred in August and a second but small abundance mode appeared in September. In the second group of years, chlorophyll *a* leveled off in August and September but abundance declined steadily during summer and fall. In the last group of years, abundance and chlorophyll *a* were in good agreement from March to November.

If mysids are food limited, their abundance should be related to chlorophyll *a* concentrations under low flow conditions when gravitational circulation is weak

TABLE 4. Frequency table of the percent daily mortality rate due to export pumping on *Neomysis mercedis*.

Daily Percent Loss	*Percent Frequency*
0.00	41.4
0.005	9.2
0.01	10.7
0.02	17.9
0.05	6.4
0.10	7.1
0.20	5.3
0.50	1.0
1.00	1.0

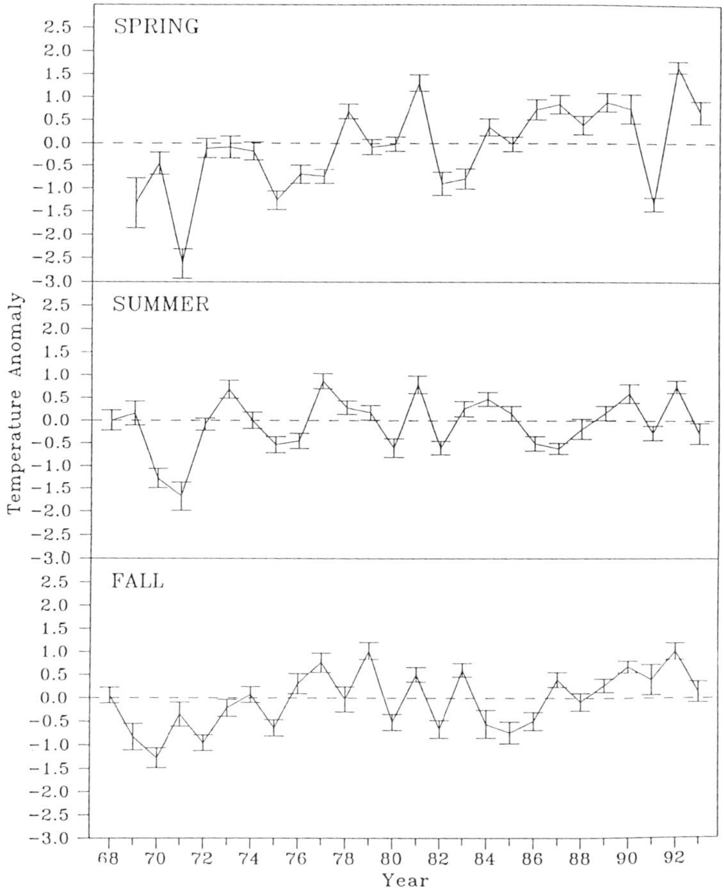

FIGURE 16. Temperature anomalies (corrected for salinity) in spring, summer and fall from 1968 to 1993. There was no sampling in spring 1968. Bars mark the 95% confidence intervals of the annual means.

or non-existent. To test for this, log mysid abundance in the entrapment zone (1-5.6‰ surface salinity) was plotted against chlorophyll *a* concentrations for sampling surveys when Delta outflow was < 170 m^3sec^{-1}. The years were divided into three groups: 1968-1975, 1976-1986, and 1987-1993, corresponding to the abundance trends in Fig. 10. Abundance in the entrapment zone showed a significant correlation with chlorophyll *a* under low flow conditions (Fig. 17). Even at higher flows (≥ 170 m^3sec^{-1}) the relationship between abundance and chlorophyll *a* was significant (Fig. 18).

When food is limiting, egg production in copepods and cladocerans is reduced (Checkley 1980; Ambler 1986; Beckman & Peterson 1986; Urabe 1991). To see if *N. mercedis* egg production declined when chlorophyll *a* did, fecundity anomalies, corrected for female length, were calculated for each season, spring, summer

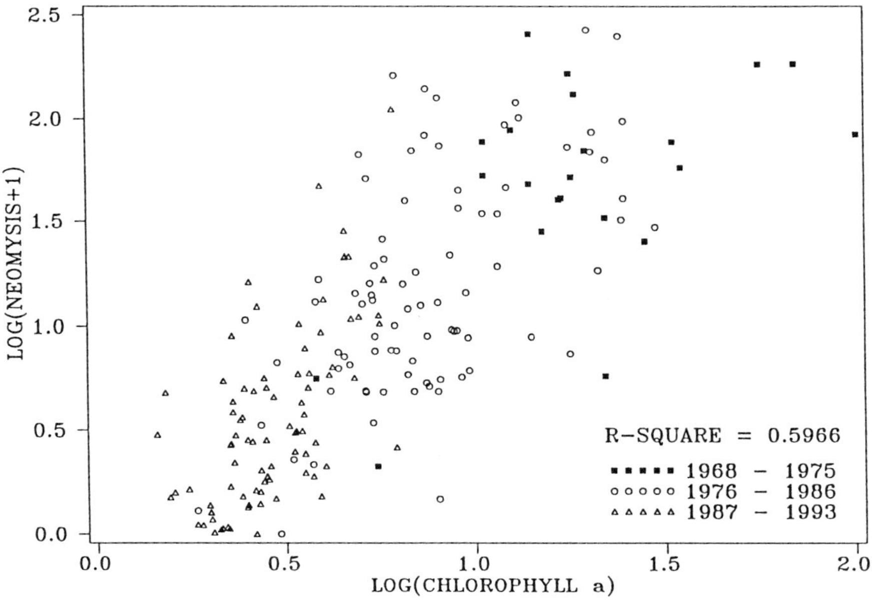

FIGURE 17. Log abundance of ≥ 4 mm mysids vs chlorophyll *a* concentrations in the entrapment zone on surveys when Delta outflow was < 70 $m^3 sec^{-1}$. Points represent three groups of years.

and fall for each year for which data was available, 1976-1993 (Fig. 19). The anomalies were then correlated with mean chlorophyll *a* in each season. None of the seasonal correlation coefficients were significant at the 5% level.

Another test for food limitation is reduced adult size. In copepods and cladocerans growth rates were lower or adults were smaller when food was limiting (Checkley 1980; Urabe 1991; Durbin *et al.* 1992). To see if this happened in *N. mercedis* the mean percent of females > 11 mm in July and August when adults reach their minimum size (Fig. 2), was calculated for each year, 1968-1993. The percent > 11 mm was very variable from 1968 to 1984 with a low point in 1979 (Figs. 20 and 21). After 1984, the percent of large mysids declined and was very low from 1988 to 1993. This could not have been a temperature effect because temperature showed no upward trend over this time period. Correlations with chlorophyll *a* were not run because they would obviously not be significant.

Birth rates can be used to detect declines in reproduction between groups of years. Because accurate birth rates could not be calculated from the data collected by the study, neonate abundance was used as a proxy. Neonates grow out of the 2-3 mm length class within a day or two (H. C. Bailey, Univ. California, Davis, pers. commun.) and hence are unlikely to suffer high mortality between birth and sampling times. Mean neonate abundance was calculated for each survey during the March to November period for the following groups of years: 1974-1975, 1978-1986, and 1987-1993 (Fig. 6). (Mysids < 4 mm were not caught efficiently

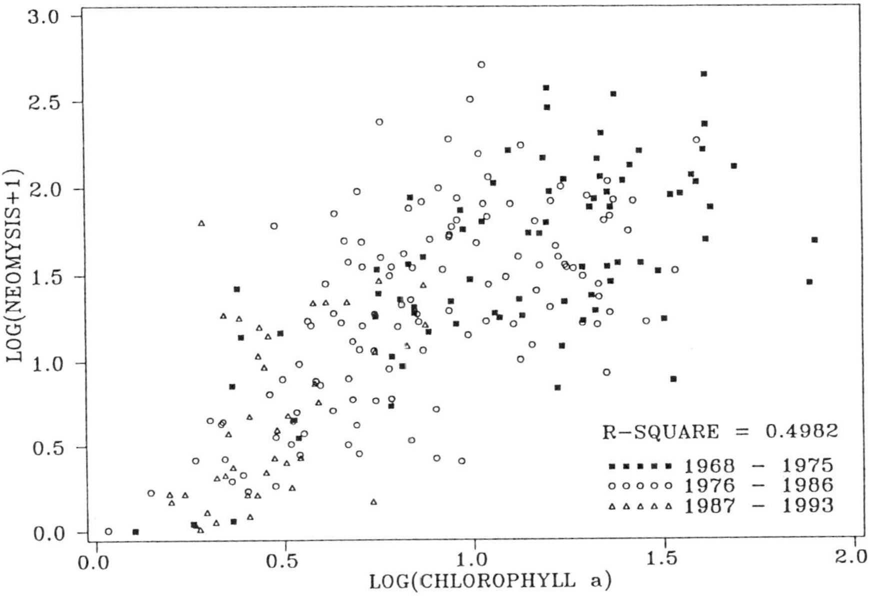

FIGURE 18. Log abundance of ≥ 4 mm mysids vs chlorophyll *a* concentrations in the entrapment zone on surveys when Delta outflow was ≥170 m^3sec^{-1}. Points represent three groups of years.

prior to 1974). Neonate abundance was highest for 1974-1975 and showed several large modes from May to July and a small one in October that mark the cohorts. Neonate abundance dropped sharply after July. In the 1978-1986 period cohorts were present in May and July but were smaller than in 1974-1975. Abundance decreased in August and the fall cohort was scarcely detectable. In the 1987-1993 period, abundance was greatly reduced. A May cohort was present, abundance decreased in July and no fall cohort appeared. In all three periods neonate abundance declined in summer. In the last group of years no significant reproduction occurred after June.

DISCUSSION

Neomysis mercedis abundance underwent a long-term downtrend with the greatest decline occurring in fall and in the years 1987-1993 (Fig. 10). These years encompass a long drought but also a high outflow year, 1993. Downtrends also occurred in some food organisms important to *N. mercedis*: phytoplantkon and rotifers. But another type of food, copepods, did not decline and actually showed an increase in the fall of the years 1987-1993.

Five hypotheses to explain the *N. mercedis* decline were proposed and examined. The evidence to support three of them appears weak. These are the toxics, export pumping and temperature hypotheses. The stepwise regression analysis showed that the herbicide molinate did not explain a significant amount of the

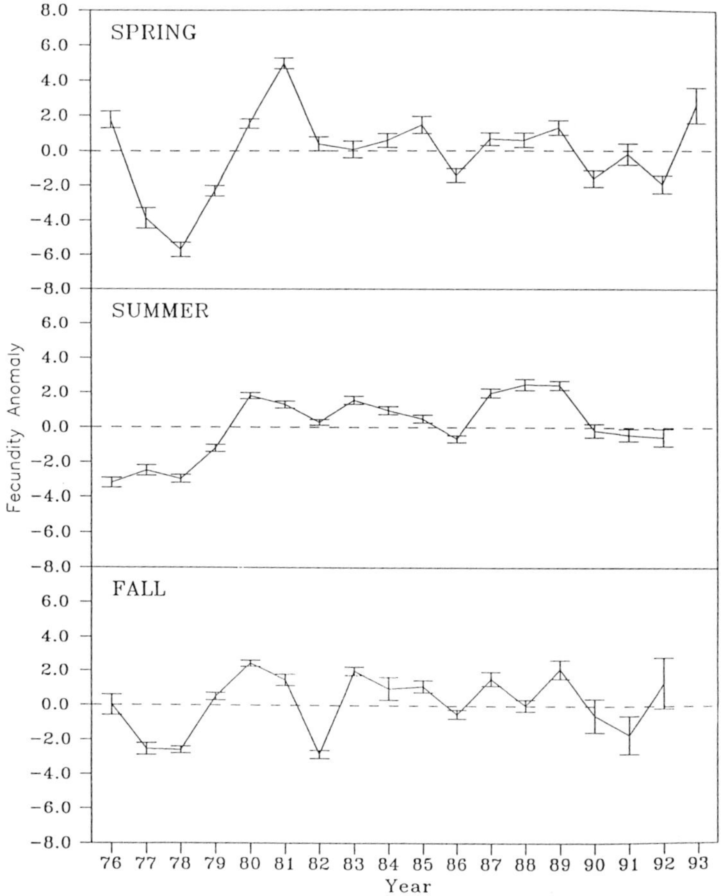

FIGURE 19. Fecundity anomalies corrected for female length in spring, summer and fall, 1976-1993. Not enough females were taken in summer and fall 1993 for calculation. Bars mark the 95% confidence intervals of the annual means.

variance in the abundance anomalies during the months when it was in the system. In addition, the CDFG tolerance criteria for molinate do not ever appear to have been exceeded in the area where *N. mercedis* is abundant. Criteria for molinate have been set at 26 $\mu g\ l^{-1}$ when present alone and 13 $\mu g\ l^{-1}$ in the presence of thiobencarb, another rice herbicide that is usually present with molinate (Harrington 1990). The highest recorded in-river concentrations for molinate were 27 $\mu g\ l^{-1}$ in the Sacramento River upstream from the city of Sacramento in 1983 (Harrington & Lew 1989). Since 1986, molinate concentrations have not exceeded 11 $\mu g\ l^{-1}$ in this section of the Sacramento River, which is well upstream from where *N. mercedis* is found. Dilution of molinate would occur downriver in the

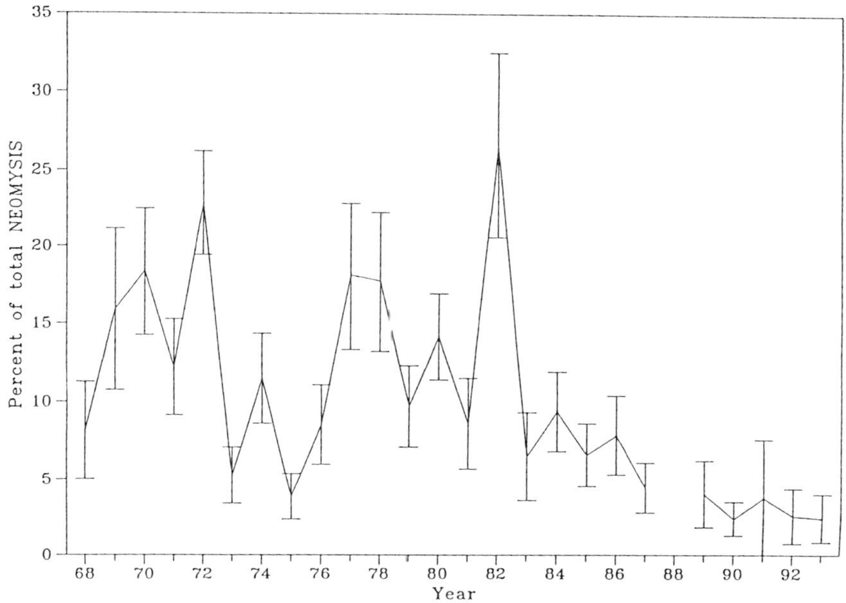

FIGURE 20. Mean percent of adult mysids >11 mm in length in July of each year, 1968-1993. Bars mark the 95% confidence intervals of the means. There was no sampling in July 1988.

Delta and Suisun Bay. This should have reduced molinate concentrations well below the criteria levels even in 1983.

Temperature also failed to explain a significant amount of the variance in the abundance anomalies. Temperature did rise somewhat from 1968 to 1993 but not in summer when high temperatures should impact survival. Additionally, the data from the San Joaquin River at Stockton did not show a strong decrease in abundance at temperatures ≥ 22°C.

Export pumping losses from the south Delta are simply not high enough to remove much of the population. Even at very low rates of population increase the calculated median daily loss rates of 0.01% day^{-1} are too small to explain the population decline. Losses to another type of export, agricultural pumping, which removes water from Delta channels to irrigate the farming islands cannot be evaluated because of a lack of data. However, it is unlikely that irrigation pumping increased during the study years. Hence, while some mysids must undoubtedly be lost to these pumps, it is unlikely that this source of mortality can explain the downtrend in mysid abundance.

Of the two hypotheses remaining, the food limitation one seems to be the strongest. The evidence for food limitation is (a) smaller adult size in July and August of recent drought years when chlorophyll *a* was low, (b) a significant correlation between abundance in the entrapment zone and chlorophyll *a* at both low and high outflows, (c) the good agreement between seasonal trends in abun-

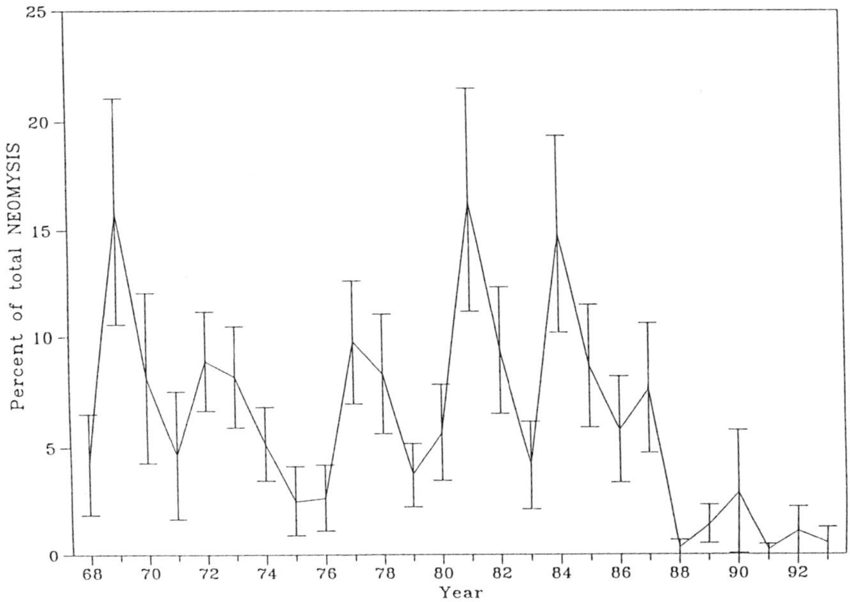

FIGURE 21. Mean percent of adult mysids >11 mm in length in August of each year, 1968-1993. Bars mark the 95% confidence intervals of the mean.

dance anomalies and chlorophyll *a*, (d) the correlations between abundance anomalies and chlorophyll *a* in all seasons.

Brood size was not reduced but this can be explained if adults are not food limited because they feed primarily on zooplankton (Siegfried & Kopache 1980) and copepod abundance did not decline Fig. 14). Only two other investigators have tried to find a relationship between mysid fecundity and productivity at lower trophic levels. The results were not conclusive. In Lake Washington, Seattle, the fecundity of *N. mercedis* was found to be unrelated to differences in copepod abundance between 1979 and 1980 (Murtaugh 1989). In Lake Tahoe, California, the number of eggs per brood of *Mysis relicta* was significantly higher in the more productive water of Emerald Bay compared to the main lake, but the number of stage 5 (the last stage) embryos was similar in both areas (Morgan 1980).

If food limitation of juveniles was the primary factor behind the long-term downtrend in mysid abundance, mysid population size is likely to be controlled by survival of young rather than by reproduction of adults.

The primary factor now affecting phytoplankton abundance where *N. mercedis* is most abundant is grazing by the introduced Asian clam, *Potamocorbula amurensis* (Kimmerer *et al.* 1994). So long as *P. amurensis* remains abundant in Suisun Bay, the abundance of *N. mercedis* should also remain low. Additionally, the introduction of two Asian mysids in 1992 should provide competition for *N. mercedis*. Neither of these species has become abundant and perhaps both are also food limited but if the food resource was already limiting *N. mercedis* then the added competition for that resource can only have further negative effects on its population size.

Estuarine circulation could have important but secondary effects on mysid abundance. For a better understanding of how the complex hydrodynamics of the estuary interact with mysids, specific studies involving *in situ* measurements of vertical velocity profiles and mysid vertical distribution need to be done.

ACKNOWLEDGEMENTS

Many people contributed to this study in many ways over the many years of its existence. We express gratitude to all of them. G. Schmidt deserves special recognition for braving the waters in fair weather and foul, year after year to bring back many hundreds of crates containing thousands of jars and millions of mysids for S. Skelton to count and measure. Of such routine tasks is science made.

REFERENCES

Ambler, J. W. 1986. Effect of food quantity and quality on egg production of *Acartia tonsa* Dana from East Lagoon, Galveston, Texas. *Estuarine Coastal Shelf Sci.* 23:183-196.

Arthur, J. F. and M. D. Ball. 1979. Factors influencing the entrapment of suspended material in the San Francisco Bay-Delta Estuary. Pages 143-174 *in* T. J. Conomos, ed., *San Francisco Bay: The Urbanized Estuary*. Pacific Division, American Association for the Advancement of Science, San Francisco, CA.

Astthorsson, O. S. and R. Ralph. 1984. Growth and moulting of *Neomysis integer* (Crustacea, Mysidacea). *Marine Biol.* 79:55-61.

Bailey, H. C. 1985. Chronic toxicity of rice herbicides to *Neomysis mercedis*. SRI Project LSU-7578. SRI International, Menlo Park, CA. (CDFA Library Document No.228-055). 28 pp.

Baldo Kost, A. and A. W. Knight. 1975. The food of *Neomysis mercedis* in the Sacramento-San Joaquin Estuary. *Calif. Fish & Game* 61(1):35-46.

Beckman, B. R. and W. T. Peterson. 1986. Egg production by *Acartia tonsa* in Long Island Sound. *Jour. Plankton Res.* 8(5):917-925.

Checkley, Jr., D. M. 1980. Food limitation of egg production by a marine, planktonic copepod in the sea off southern California. *Limnol. Oceanogr.* 25(6):991-998.

Clutter, R. L. 1969. The microdistribution and social behavior of some pelagic mysid shrimps. *Jour. Exp. Marine Biol. Ecol.* 3:125-155.

Durbin, E. G., A. G. Durbin, T. J. Smayda, P. G. Verity. 1983. Food limitation of production by adult *Acartia tonsa* in Narragansett Bay, R.I. *Limnol. Oceanogr.* 28(6):1199-1213.

Faggella, G. A., B. J. Finlayson, and O. M. Brandt. 1990. *Standardized Testing Program 1988 Progress Report*. California Dept. Fish & Game, Environ. Serv. Div., Elk Grove, CA (January 1990). 56 pp.

Hair, J. R. 1971. Upper lethal temperature and thermal shock tolerances of the opossum shrimp, *Neomysis awatschensis* from the Sacramento-San Joaquin Estuary, California. *California Fish & Game* 57(1):17-27.

Harrington, J. M. 1990. *Hazard Assessment of the Rice Herbicides Molinate and Thiobencarb to Aquatic Organisms in the Sacramento River System*. California Dept. Fish & Game, Environ. Serv. Div., Admin. Rept. 90-1. 97 pp.

Harrington, J. M. and T. L. Lew. 1989. *Rice Herbicide Concentrations in the Sacramento River and Associated Agricultural Drains, 1987-88*. California Dept. Fish & Game, Environ. Serv. Br., Admin. Rept. 89-1. 88 pp.

Heubach, W. 1969. *Neomysis awatschensis* in the Sacramento-San Joaquin River estuary. *Limnol. Oceanogr.* 14(4):533-546.

Heubach, W., R. J. Toth and A. M. McCready. 1963. Food of young-of-the-year striped bass (*Roccus saxatilis*) in the Sacramento-San Joaquin River system. *California Fish & Game* 49(4):224-239.

Jassby, A. D., W. J. Kimmerer, S. G. Monismith, C. Armor, J. E. Cloern, T. M. Powell, J. R. Schubel, and T. J. Vendlinski. 1995. Isohaline position as a habitat indicator for estuarine populations. *Ecol. Applications* 5(1):272-289.

Johnston, N. T. and T. G. Northcote. 1988. Life-history variation in *Neomysis mercedis* Holmes (Crustacea, Mysidacea) in the Fraser River estuary, British Columbia. *Canadian Jour. Zool.* 67(2):363-372.

Kimmerer, W. 1992. *An Evaluation of Existing Data in the Entrapment Zone of the San Francisco Bay Estuary*. Interagency Ecological Studies Program for the Sacramento-San Joaquin Estuary. Tech. Rept. 33, 49 pp. Sacramento, California.

Kimmerer, W., E. Gartside and J. J. Orsi. 1994. Predation by an introduced clam as the likely cause of substantial declines in zooplankton of San Francisco Bay. *Marine Ecol. Progr. Ser.* 113:81-93.

Knutson, Jr., C. A. and J. J. Orsi. 1983. Factors regulating abundance and distribution of the shrimp *Neomysis mercedis* in the Sacramento-San Joaquin Estuary. *Trans. American Fish. Soc.* 112:476485.

Morgan, M. D. 1980. Life history characteristics of two introduced populations of *Mysis relicta*. *Ecology* 61(3):551-561.

Murtaugh, P. A. 1989. Fecundity of *Neomysis mercedis* Holmes in Lake Washington (Mysidacea). *Crustaceana* 57(2):194-200.

Orsi, J. J. 1986. Interaction between diel vertical migration of a mysidacean shrimp and two-layered estuarine flow. *Hydrobiologia* 137:79-87.

Orsi, J. J. and W. L. Mecum. 1986. Zooplankton distribution and abundance in the Sacramento-San Joaquin Delta in relation to certain environmental factors. *Estuaries* 9(4b):326-339.

Pritchard, D. W. 1956. The dynamic structure of a coastal plain estuary. *Jour. Marine Res.* 15:33-42.

Radtke, L. D. 1966. Distribution of smelt, juvenile sturgeon, and starry flounder in the Sacramento-San Joaquin Delta with observations on the food of sturgeon. Pages 115-126 *in* Ecological Studies of the Sacramento-San Joaquin Delta. Part II. *California Fish & Game, Fish Bull.* 136.

Siegfried, C. A. and M. S. Kopache and A. W. Knight. 1979. The distribution and abundance of *Neomysis mercedis* in relation to the entrapment zone in the western Sacramento-San Joaquin Delta. *Trans. American Fish. Soc.* 108:262-268.

Siegfried, C. A. and M. E. Kopache. 1980. Feeding of *Neomysis mercedis* (Holmes). *Biol. Bull.* 159:193-205.

Sitts, R. M. and A. W. Knight. 1979. Predation by the estuarine shrimps, *Crangon franciscorum* Stimpson and *Palaemon macrodactylus* Rathbun. *Biol. Bull.* 156:356-368.

Stevens, D. E. 1966a. Food habits of striped bass, *Roccus saxatilis*, in the Sacramento-San Joaquin Delta. Pages 68-96 *in* J. L. Turner & D. W. Kelley, eds., *Ecological Studies of the Sacramento-San Joaquin Delta. Part II.* California Fish & Game, Fish Bull. 136.

Stevens, D. E. 1966b. Distribution and food habits of the American shad, *Alosa sapidissima*, in the Sacramento-San Joaquin Delta. Pages 97-103 *in* J. L. Turner & D. W. Kelley, eds., *Ecological Studies of the Sacramento-San Joaquin Delta. Part II.* California Fish & Game, Fish Bull. 136.

Toda, H., S. Nishizawa, M. Takahashi and S. Ichimura. 1983. Temperature control on the post-embryonic growth of *Neomysis intermedia* Czerniawsky in a hypereutrophic temperate lake. *Jour. Plankt. Res.* 5(3):377-392.

Toda, H., T. Arima, M. Takahashi and S. Ichimura. 1987. Physiological evaluation of temperature effect on the growth processes of the mysid, *Neomysis intermedia* Czerniawsky. *Jour. Plankt. Res.* 9(1):51-63.

Turner, J. L. and W. Heubach. 1966. Distribution and concentration of *Neomysis awatschen-*

sis in the Sacramento-San Joaquin Delta. Pages 105-112 *in* D. W. Kelley, ed., *Ecological Studies of the Sacramento-San Joaquin Delta. Part I.* California Fish & Game, Fish Bull. 133.

Urabe, J. 1991. Effects of food concentration on growth, reproduction and survivorship of *Bosmina longirostris* (Cladocera): An experimental study. *Freshwater Biol.* 25:1-8.

CHANGES IN THE ZOOPLANKTON OF THE SAN FRANCISCO BAY ESTUARY SINCE THE INTRODUCTION OF THE CLAM *POTAMOCORBULA AMURENSIS*

W. J. KIMMERER
Romberg Tiburon Center for Environmental Studies
San Francisco State University, 3150 Paradise Drive
Tiburon CA 94920

and

J. J. ORSI
California Department of Fish and Game
4001 N. Wilson Way, Stockton CA 95205

The zooplankton of the northern San Francisco Bay estuary has been characterized over the last two decades by long-term declines in abundance of some species, and introductions of others. The most striking change occurred in 1987-88, concurrent with the spread of the introduced clam *Potamocorbula amurensis*. Previous work presented evidence suggesting that declines in some copepod species were caused by the clam, more by predation on nauplii than by competition for food. Here we show that the clam probably caused a decline in other species of zooplankton, and that the large decline in *Neomysis mercedis* was probably due to competition with *P. amurensis* for food. At the time of these declines in zooplankton the copepod *Pseudodiaptomus forbesi* became abundant over much of the region formerly occupied by the previously common copepod *Eurytemora affinis*. Based on present data, we cannot explain this apparent coincidence in timing, but assume it was due to a competitive interaction.

Although much of the post-clam period was characterized by drought, the low freshwater flows of that period do not explain the reduction in chlorophyll or zooplankton abundance. The high flows of 1993 caused a temporary reduction in abundance of *P. amurensis*, but the larger size of the clams made up for the reduction in numbers, and by summer the zooplankton populations were again low. The change in zooplankton distribution and abundance may persist into the future; however, it is not clear that higher trophic levels are being adversely affected.

Long-term monitoring of environmental conditions and populations can permit inferences to be made that would be unavailable from a shorter-term study. For example, Colebrook (1978) used data from repeated transects across the North Sea with continuous plankton recorders to show variability in abundance of common plankton species at all temporal scales. Some long-term trends have been associated with trends in weather, implying a causative relationship between climate and plankton abundance (Venrick *et al.* 1987; Aebischer *et al.* 1990).

In the San Francisco Bay Sacramento-San Joaquin River Delta estuary ("Bay-Delta"), monitoring has been conducted by the California Interagency Ecological Program over the past 2-3 decades for abundance of fish and plankton and several water quality variables. The zooplankton data set comprises nearly 400 surveys of approximately 40 stations, mainly in the freshwater reaches of the delta (Fig. 1). This rich data set can provide useful insight into long-term trends in abundance of important species.

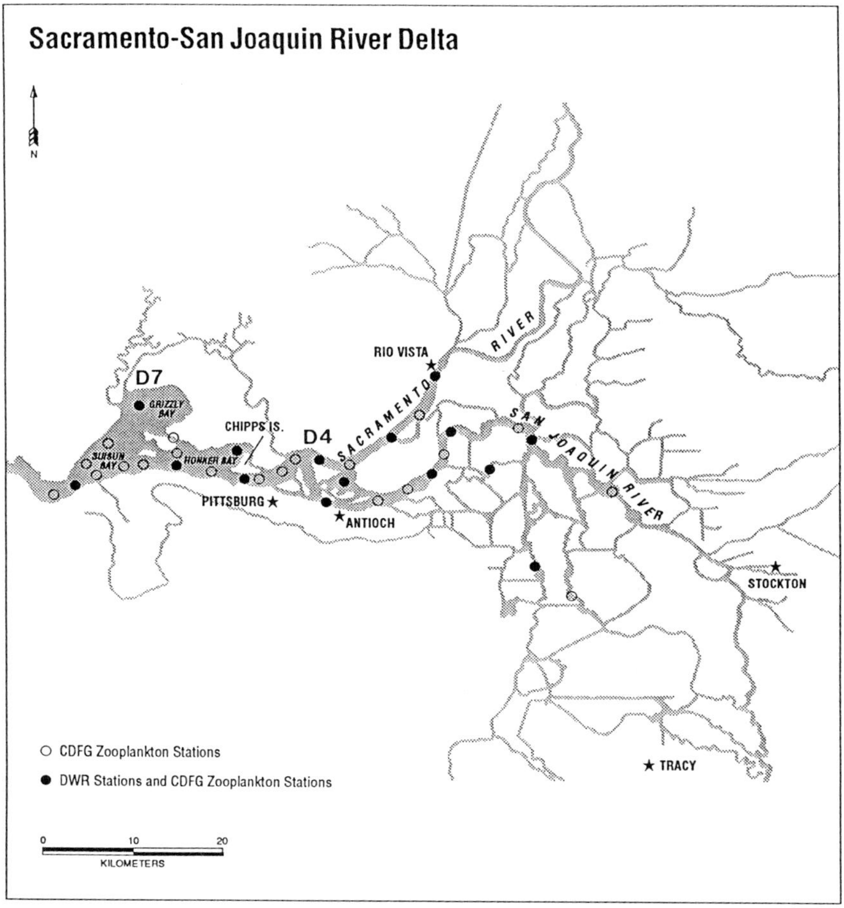

FIGURE 1. Map of the northern San Francisco Bay-Delta estuary showing sampling stations including benthic sampling stations D7 (Grizzly Bay) and D4 (river confluence).

The data in this set may be divided roughly into three time periods separated by two events in which abundances declined. The first event, which occurred in 1976-1978, consisted of a decline in chlorophyll and in about 2/3 of the planktonic taxa of the upper estuary (Obrebski *et al.* 1992). The second event occurred in 1987-88, concurrent with the spread of the introduced clam *Potamocorbula amurensis* (Kimmerer *et al.* 1994). Between 1978 and 1987 there was relatively little change in abundance in the zooplankton compared to the earlier or later periods.

P. amurensis appeared in San Francisco Bay in 1986, probably as a result of accidental introduction in ship ballast water (Carlton *et al.* 1990; Nichols *et al.* 1990). It spread very rapidly, becoming abundant throughout the northern bay by mid-summer 1987 (Carlton *et al.* 1990). Several independent studies have described substantial effects of filtering by *P. amurensis* on all lower trophic levels in the bay. Alpine & Cloern (1992) showed that phytoplankton biomass declined sharply in late 1987, and attributed the decline to grazing by *P. amurensis*. Werner and Hollibaugh (1993) found high experimental clearance rates of *P. amurensis* on phytoplankton and bacteria and, considering the high densities of this clam in northern San Francisco Bay, concluded that the potential existed for high grazing pressure on both phytoplankton and bacteria. Kimmerer *et al.* (1994) showed also that *P. amurensis* was capable of consuming nauplii of several species of copepod. They also indicated that several species of copepod had declined and that the decline of *Eurytemora affinis* was probably due more to predation on nauplii than competition for food. They presented evidence that: (1) the location and timing of the decline coincided with the appearance of the clam; (2) clams are capable of consuming nauplii at a rate sufficient to cause the observed effect; (3) nauplii are exposed to predation by virtue of their vertical distribution; and (4) ratios of eggs to females did not decline, but ratios of nauplii to females did, implying that food was not limiting egg production and that mortality of nauplii had increased.

The benthic fauna of the northern Bay-Delta has followed a characteristic pattern in the past. During droughts, marine species expand into the northern bay (Suisun Bay, Fig. 1) as freshwater habitat shrinks (Nichols 1985; Alpine & Cloern 1992). High freshwater flow eliminates marine species from this region. Our expectation had been that *P. amurensis* would be similarly eliminated or reduced from this region by high outflows, and that increased stratification due to higher outflows might insulate some of the zooplankton from predation.

From the time of the introduction of *P. amurensis* until the end of 1992, California experienced the most severe drought in recorded history. The drought broke in 1993, a year of above-normal precipitation. The period 1978-1993 also encompassed introductions and population increases of several other zooplankton species. Most significant for this discussion are the copepods *Sinocalanus doerrii* (became abundant in 1979, Orsi *et al.* 1983), *Oithona davisae* (1982, Ferrari & Orsi 1984), and *Pseudodiaptomus forbesi* (1988, Orsi & Walter 1991). *Eurytemora affinis*, widely regarded as native, may also have been introduced in the last century (J. Chapman, OSU, pers. commun.).

In this paper we examine further the long-term data record, focusing mainly on the 1987-88 decline. The period between 1978 (the end of the previous period of decline for most species) and 1987 is used as a reference period since populations were fairly stable during that period. We answer the following questions: How did the abundance of *P. amurensis* and its zooplankton prey/competitors change in 1993? What other species declined besides those already identified? What other changes have occurred in the zooplankton fauna? We also attempt to explain these patterns.

METHODS

Zooplankton abundance data were obtained from the California Department of Fish and Game zooplankton monitoring program, for which methods are described fully in Orsi & Mecum (1986). Samples were taken by day near high slack water at ca. 35-40 stations throughout the upper estuary (Fig. 1), monthly in March and November and twice monthly in April-October, 1972-1993. Samples were taken with a 154 μm mesh, 10 cm diameter Clarke-Bumpus net towed obliquely from the bottom to the surface, and with a pump whose intake was slowly raised through the water column. The net sample and a 1.7 liter composite subsample of the pump discharge were preserved in 5% formaldehyde. At each station surface specific conductance and temperature were measured with a conductivity meter, secchi disk depth was recorded, and surface samples were taken for chlorophyll *a* analysis.

In the laboratory, the net sample was concentrated on a 154 μm mesh screen and subsampled to obtain at least 200 animals for counting. The pump sample was passed through a 154 μm mesh screen and collected on a 43 μm mesh screen, and the entire sample was counted. Adults of all copepods and nauplii of *Eurytemora affinis* were identified to species or genus. Other taxa (mostly cladocerans and rotifers) were identified mainly to the generic level, and many of these probably contain more than one species. Data were generally log-transformed for analysis, with a number added that was close to the minimum non-zero value: 0.1 for the mysid *Neomysis mercedis*, 1 for cladocerans, 10 for copepods, and 100 for rotifers. We used data only from a set of core stations sampled throughout the period, plus several stations in San Pablo Bay, west of the normally sampled area, that have been sampled only during high-flow periods.

Zooplankton species resident in brackish water are most abundant in a limited salinity range (Miller 1984), so their distributions are best referred to salinity rather than to geographic position. However, spatial distributions of species residing upstream of the limit of penetration of oceanic salt cannot be referred to salinity. Agricultural return flow adds salt upstream, especially in the San Joaquin River, blurring the relationship between salinity and position. We therefore devised a measure of distance along the axis of the estuary, or "distance parameter," based on salinity for estuarine species and geographic position for upstream species. The geometric mean regression of distance vs. mean log specific conductance at stations up to 100 river kilometers was

$$\text{Distance (km)} = 168.7 - 27.34\ [\text{Log}_{10}\ (\text{Specific conductance, mS/cm})], \qquad (1)$$

($r^2 = 0.98$, 17 df). The distance parameter was set equal to actual station position for specific conductance below 2 mS cm^{-1}, where the source of salt can become ambiguous. Above 2 mS cm^{-1}, the distance parameter was calculated from the above regression. Data were aggregated into means by distance parameter in 5- or 10-km bins and by season (3 months each) for statistical and graphing purposes.

For graphical depictions and statistical tests of the magnitude of changes, data for each taxon were averaged within years and seasons for a range of specific conductance or distance values depending on their distribution during 1978-86, a period of relatively stable abundances. In general, the range of distances or conductances corresponded to the abundance maximum of each taxon. Although using annual means for each season reduced the power of these tests, it also eliminated the problem of non-independence of data, because the generation time of these taxa is far shorter than a year. The analysis gave sets of annual means for each season. Differences between data in 1978-86 and after 1987 were tested using *t*-tests on the seasonal data. These results were divided into variables that changed in 1987-1990 vs. those that changed earlier; timing of the change was determined as the first branch in a regression tree with year as the independent variable (Venables & Ripley 1994). Time-series methods were not suitable for analyzing this data set.

Abundance data for *P. amurensis* were obtained from the results of monitoring by the California Department of Water Resources (DWR) for 1988-1993. Ponar grab samples (0.053 m^2) were taken monthly at several stations through the delta and in Grizzly Bay (Fig. 1); we used data from the Grizzly Bay station and from the confluence of the rivers (river kilometer 81). Additional data on spatial distributions of *P. amurensis* were obtained from a series of several hundred samples taken over several days in August-September 1990 (Hymanson 1992) and again in July 1993 (P. Lehman, DWR, pers. commun.). The spatial data sets included size distributions as well as abundance. These data were used to estimate a relative clearance rate for the population ($m\ d^{-1}$, *i.e.*, $m^3\ m^{-2} d^{-1}$) using the relationship between shell length and clearance rate for phytoplankton of Werner and Hollibaugh (1993).

Estimates of freshwater flow rate into the estuary were obtained from the DWR Dayflow index of net delta outflow. We also used an index of estuarine habitat, the longitudinal distance from the mouth of the estuary to where near-bottom salinity is 2 psu ("X_2"), determined by interpolation from continuous salinity monitors (Jassby *et al.* 1995), and averaged for each month. X_2 is linearly related to the log of freshwater flow, and is useful as an index of the position of other isohalines, and of a persistent estuarine turbidity maximum (ETM) and its associated fauna. X_2 is used here as an index of flow conditions because it integrates over a period of weeks, and because it is a measured quantity (Jassby *et al.* 1995). X_2 is close to a distance parameter value (eq. 1) of 74 km. The ETM is assumed to persist between 1 and 6-10 psu salinity (Arthur & Ball 1979).

RESULTS

Monthly mean X_2 varied between 44 and 92 km as freshwater outflow fluctuated, but for most months X_2 was between 60 and 85 km (Fig. 2). Generally in each year, X_2 decreased in winter-early spring, the period of highest natural runoff, then gradually increased through summer and fall as delta outflow declined. During the drought years 1985 and 1987-92, X_2 was rarely below 70 km in any month.

The mean abundance of *P. amurensis* at DWR Station D7 in Grizzly Bay was 2900 m^{-2} and that at Station D4 near the confluence of the Sacramento and San Joaquin Rivers was 900 m^{-2} from July 1987 through 1992 (Fig. 3). Abundance at Station D7 was fairly steady during that time, while at Station D4 abundance initially increased to about 9000 m^{-2}, then decreased to fluctuate around 1000 m^{-2}, except for a period of low abundance in 1989-90 (Fig. 3). Abundance decreased during the high flows of 1993, but the mean size of clams was greater, so biomass did not decrease as much as abundance (Thompson & Nichols 1994).

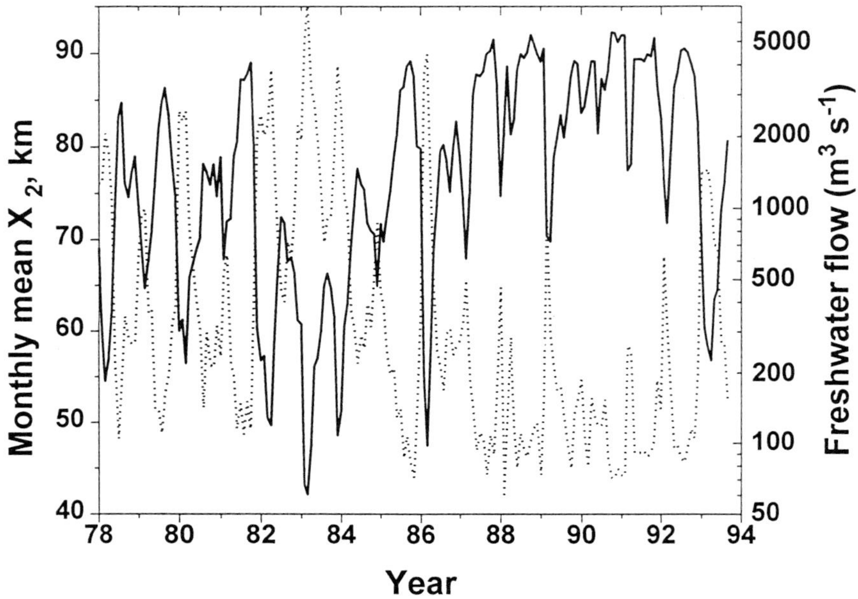

FIGURE 2. Monthly mean X_2 and net freshwater outflow from the delta for 1978-1993.

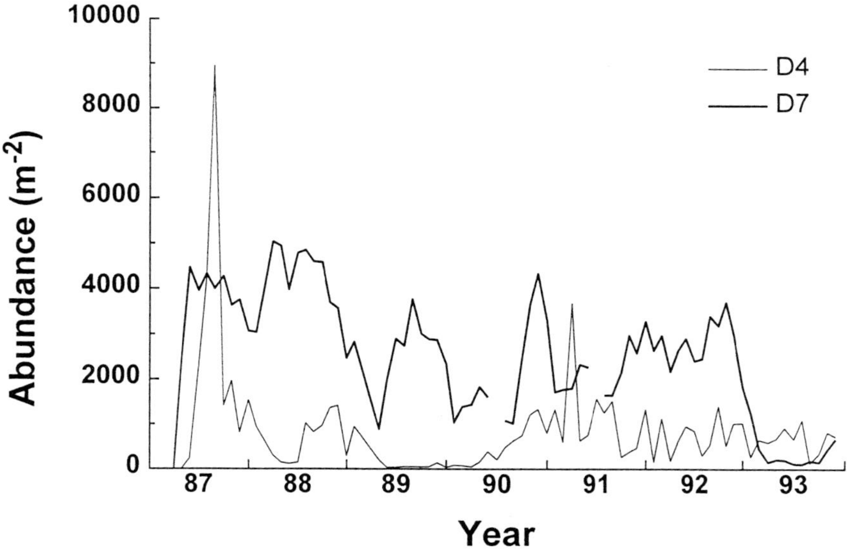

FIGURE 3. *P. amurensis*. Monthly abundance at monitoring stations in upper Grizzly Bay (D7) and at the confluence of the Sacramento and San Joaquin Rivers (D4).

The broad-scale surveys showed considerable difference in abundance of *P. amurensis* between 1990 and 1993, with lower abundance and larger clams in 1993 (Fig. 4). The median population clearance rate was reduced from 5 md^{-1} in 1990 to 0.9 md^{-1} in 1993 in the Suisun-Grizzly-Honker Bay complex, while rates increased in San Pablo Bay and in the lower reaches of the rivers.

Each of the taxonomic groups of plankton examined has a characteristic distribution with respect to distance along the longitudinal axis of the upper estuary for freshwater taxa, or salinity for estuarine taxa. The distributions of common taxa with respect to the distance parameter in 1978-86 (Fig. 5) illustrate the degree to which these taxa differ in spatial distribution and opportunity for contact with the *P. amurensis* population. Contact would be greatest for taxa with abundances substantially or entirely downstream of 90 km (*i.e.*, the first 9 taxa in Fig. 5). Contact would be somewhat less for taxa with population centers at or upstream of 90 km (*e.g.*, *Pseudodiaptomus forbesi* and *Sinocalanus doerrii*). The remaining taxa would have infrequent opportunity for direct contact with *P. amurensis*. Total calanoids, harpacticoids, and the rotifers *Synchaeta* spp. and *Trichocerca* spp. had multiple abundance peaks, each presumably representing a different species or group of species. Time trends of these taxa were examined separately for each of the peaks.

Long-term trends in chlorophyll concentration, turbidity as measured by the inverse of Secchi disk depth, and abundances of several taxa had two periods of changing values: one in 1976-78, and the other in 1987-88 (Fig. 6). The first period included apparent declines in turbidity, chlorophyll, and abundance of *Eurytemora affinis*, *Neomysis mercedis*, and two genera of cladocerans; *Acartia* spp. appeared to increase at that time, and *Sinocalanus doerrii* was introduced. The later period encompassed a more radical change in chlorophyll and all of the planktonic taxa with downstream distributions (Fig. 6).

The changes occurring in 1987-88 are revealed more clearly in contour plots of abundance *vs.* season and the distance parameter (Fig. 7), and by the results of statistical tests (Table 1). All of the variables that changed in 1987-90 declined, while the direction of change was about equally split between increases and decreases for taxa that changed in 1982-85. The log of chlorophyll concentration declined significantly in all seasons in the estuarine turbidity maximum (ETM; approximately 1-15 mS cm^{-1}), in summer and fall upstream of river kilometer 90, and in spring and summer at river kilometer 136 (Stockton on the San Joaquin River). Turbidity declined significantly in the ETM in summer and fall and upstream of river km 115 in all seasons, but these declines occurred after 1983, and may be related to the frequency of drought conditions in 1984-1992.

The spatial distribution of chlorophyll averaged over 1978-1986 and 1988-1993 for each season (Fig. 8) shows that the decline in chlorophyll was greatest below 90 km. The chlorophyll maximum formerly found close to the turbidity maximum has been eliminated. Chlorophyll in the ETM (1-15 mS cm^{-1}) declined significantly within three categories of X_2 (< 75 km, 75-85 km, and 85-95 km; $p < 0.001$, *t*-tests). Chlorophyll upstream of 90 km declined significantly when X_2 was at 85-95 km ($p = 0.031$) or above 95 km ($p < 0.001$), but not when X_2 was below 75

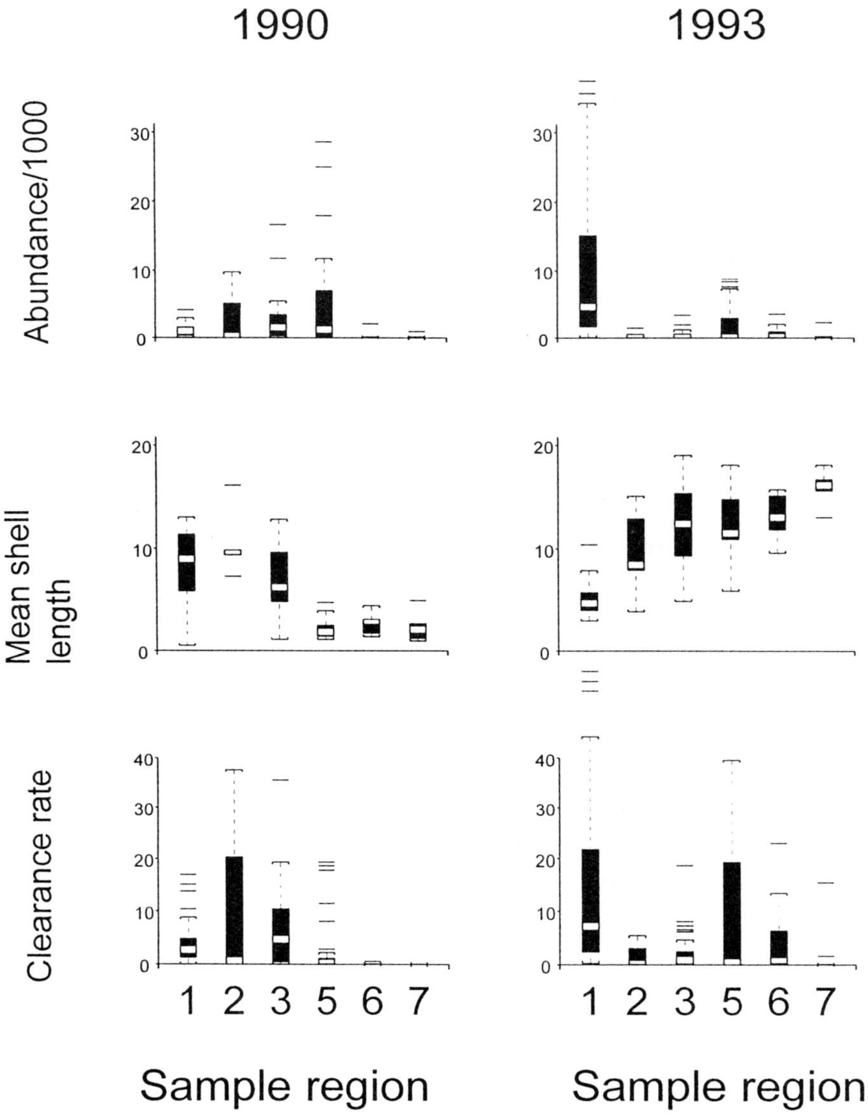

FIGURE 4. *P. amurensis*. Comparison of results of broad-scale surveys in 1990 and 1993: Abundance (clams $m^{-2}/1000$), mean shell length (mm), and calculated clearance rate (md^{-1}) for sample regions: 1, San Pablo Bay (west of the area in Fig. 1); 2, Carquinez Strait; 3, Suisun, Grizzly, and Honker Bays; 5, channels of Suisun Marsh; 6, Sacramento River; and 7, San Joaquin River. Boxplots show the median as a horizontal bar, the quartiles at the ends of the large black bar, and extreme values as individual horizontal lines.

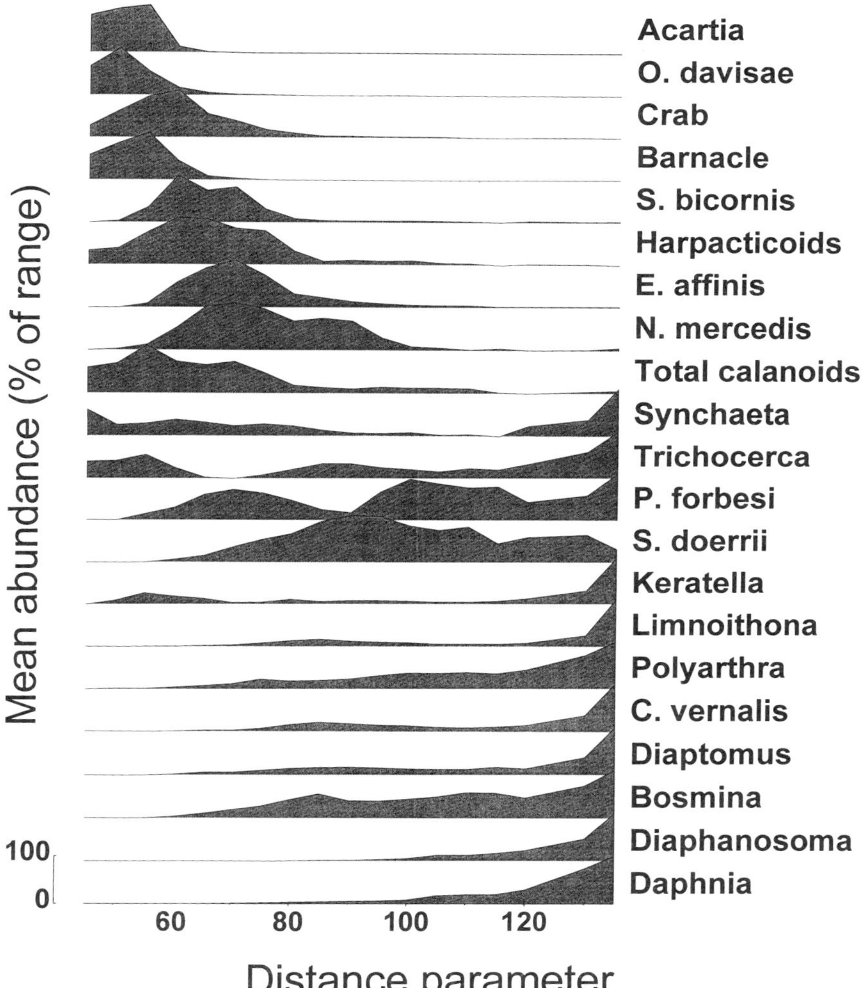

FIGURE 5. Abundance patterns of common zooplankton taxa relative to the distance parameter. Abundances are geometric means for each 5-km band of the distance parameters during 1978-1986, scaled to the maximum value. The distance parameter is equal to river kilometer upstream, and calculated from specific conductance downstream.

km ($p = 0.07$); that is, chlorophyll declined in the upstream region only when X_2 was in the delta.

Taxa that Declined

Taxa that declined significantly ($p < 0.05$) throughout their nominal salinity range (Table 1) during at least two seasons included the three most common calanoid copepods, *Acartia* spp., *Eurytemora affinis*, and *Sinocalanus doerrii*, harpacticoid copepods, the mysid *Neomysis mercedis*, the rotifers *Synchaeta bicornis*, *Synchaeta* spp., and *Trichocerca* spp., and the cladoceran *Bosmina* spp. and *Daphnia* spp.

Peaks in abundance that occurred at predictable locations and seasons before 1987 were absent, greatly reduced in space or time, or relocated after 1987 (Fig. 7). *Acartia* spp. (Fig. 7B) has historically been centered downstream of the sampled area, but has been very abundant in samples from salinities over 10 psu. Since 1987, abundance was high only in spring of 1992 and 1993, and abundance in summer and autumn has been extremely low. The mean change in log abundance (Table 1) translates to a reduction in abundance by 88%, 90%, and 96% in the three seasons. Although abundance of *Acartia* spp. was lower before 1978 than in 1978-1986, the decline in summer and fall is still significant if 1972-77 is included in the pre-clam period ($p < 0.001$, Mann-Whitney *U*-test).

Eurytemora affinis (Fig. 7C) had an abundance peak close to the ETM, but since 1988 this peak has virtually disappeared. Declines in abundance were significant (Table 1) at 89%, 98%, and 97% during the three seasons. Since 1988, *E. affinis* has been abundant in spring upstream at river kilometer 120-136 where it had previously been uncommon; however, this peak has disappeared each year by summer. *Sinocalanus doerrii*, which became abundant in 1979, had a broad maximum in abundance slightly upstream of the ETM before 1988, and extended well into freshwater (Fig. 7E). The decline in *S. doerrii* was confined to autumn (by 70%) and to a lesser extent summer (by 60%; Table 1).

The distribution of *Neomysis mercedis* (Fig. 7D) before 1987 was similar to that of *E. affinis* except for slightly higher abundance upstream of the ETM. The reduction of this species was 87%, 89% and 95% in spring, summer, and autumn, similar to that of *E. affinis*. Abundance of *N. mercedis* declined upriver as well as in the population maximum.

In addition to *Synchaeta bicornis*, other rotifers lumped into the genus *Synchaeta* declined significantly at both ends of their bimodal distribution in all seasons, but only the downstream decline occurred shortly after 1987 (Table 1). *Trichocerca* spp., with a similar distribution, declined significantly at both ends of its distribution except for upstream in fall; however, its abundance was very low upstream in the drought year 1985 and high in 1986, a wet year; thus, the upstream response may be to flow conditions.

Of all the taxa with population maxima at the upstream end of the sampled area (Fig. 5), only the cladocerans *Bosmina* spp. and *Daphnia* spp. showed significant change (Fig. 9). For both taxa the principal change seemed to be a shrinkage of abundance patterns into the upper estuary; proportional reductions were less at the far upstream stations than at stations between 100 and 130 km. Declines were confined to summer and autumn and were substantial for *Bosmina* spp., which declined by 94% and 90% in those seasons.

Increases in chlorophyll and some of the affected taxa (particularly *Acartia*) were observed during high flows in spring of 1993, but these increases did not last far into summer.

To some extent the pre- and post-clam periods are confounded by flow condi-

TABLE 1. Changes in mean values of variables including log abundances of zooplankton taxa by season before (1978-86) *vs.* after (1988-93) the spread of *Potamocorbula amurensis* (*t*-test: **bold**, $p < 0.05$; ***bold italic***, $p < 0.01$; 12 or 13 degrees of freedom except for *O. davisae*, 7 df for spring and summer, 9 for autumn). Data in the first block are for taxa that changed in abundance in 1987-1990; those in the second block changed in earlier years, based on regression tree analysis. The regions are high-salinity (> 5 mS/cm or 3 psu), ETM (1-15 mS/cm or 0.5-8 psu), and regions upstream of various river kilometer indices.

Taxon	*Region*	*Spring*	*Summer*	*Autumn*
Variables that changed in 1987-1990				
Acartia spp.	High salinity	***-0.93***	***-0.99***	***-1.38***
Total calanoids	High salinity	***-0.95***	***-0.75***	***-0.82***
Harpacticoids	High salinity	***-0.44***	**-0.30**	***-0.73***
Synchaeta bicornis	High salinity	-0.26	***-0.94***	***-1.05***
Sybchaeta spp.	High salinity	***-0.67***	***-0.60***	**-0.48**
Trichocerca spp.	ETM	***-0.59***	***-0.76***	**-0.78**
Chlorophyll	ETM	***-0.58***	***-0.64***	***-0.65***
Eurytemora affinis	ETM	***-0.96***	***-1.69***	***-1.50***
Total calanoids	ETM	***-1.05***	-0.33	-0.29
Sinocalanus doerrii	ETM	0.05	**-0.39**	***-0.50***
Neomysis mercedis	ETM	***-0.88***	***-0.96***	***-1.29***
Chlorophyll	> 90 km	-0.14	**-0.30**	**-0.31**
Sinocalanus doerrii	> 90 km	0.14	-0.48	**-0.66**
Neomysis mercedis	> 90 km	-0.18	***-0.76***	***-0.50***
Bosmina spp.	> 90 km	-0.24	***-1.23***	***-0.98***
Daphnia spp.	> 90 km	0.03	***-0.76***	***-0.46***
Trichocerca spp.	> 115 km	***-0.56***	***-0.39***	-0.31
Chlorophyll	> 135 km	***-0.44***	***-0.36***	-0.18
Diaptomus spp.	> 135 km	0.11	**-0.53**	***-0.58***
Variables that changed in 1982-1985				
Oithona davisae	High salinity	0.36	***1.32***	**0.97**
Crab zoeae	High salinity	0.12	***0.59***	**0.17**
Barnacle nauplii	High salinity	0.48	-0.38	-0.56
1/Secchi depth	ETM	-0.56	**-0.68**	***-0.91***
Total calanoids	> 90 km	**0.51**	0.36	0.29
Harpacticoids	> 90 km	0.38	0.12	0.13
Diaptomus sp.	> 90 km	-0.05	-0.23	-0.15
1/Secchi depth	> 115 km	***-0.55***	***-0.59***	***-0.46***
Synchaeta spp.	> 115 km	***-0.73***	***-0.76***	**-0.52**
Limnoithona sinensis	> 115 km	0.06	0.50	0.74
Cyclops vernalis	> 125 km	**0.59**	-0.44	-0.17

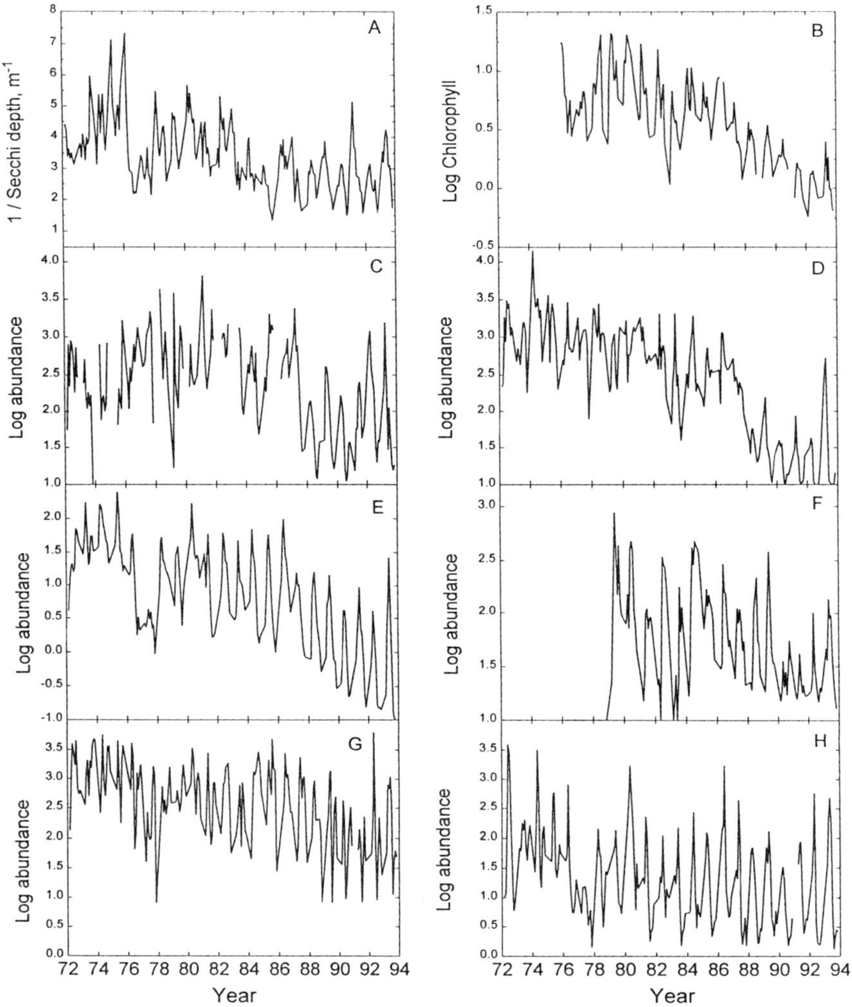

FIGURE 6. Monthly mean values of: A, inverse Secchi disk depth; B, $\log_{10}$ of chlorophyll concentration; C-H, $\log_{10}$ abundance of: C, *Acartia* spp.; D, *Eurytemora affinis*; E, *Neomysis mercedis*; F, *Sinocalanus doerrii*; G, *Bosmina* spp.; H, *Daphnia* spp. See text for details of calculations.

tions, in that drought persisted during most of the latter period (Fig. 2). We calculated means by month of the variables in Fig. 5 for selected ranges of specific conductance or distance (see Table 1), and plotted them against monthly mean X_2 (Fig. 10). The pre-clam data have moderate responses to X_2, but most of the post-clam values fall below the pre-clam values for the same X_2 values, particularly for high X_2 (*i.e.*, low flow). The major exceptions were turbidity, which responded mainly to X_2, and *Daphnia* spp.

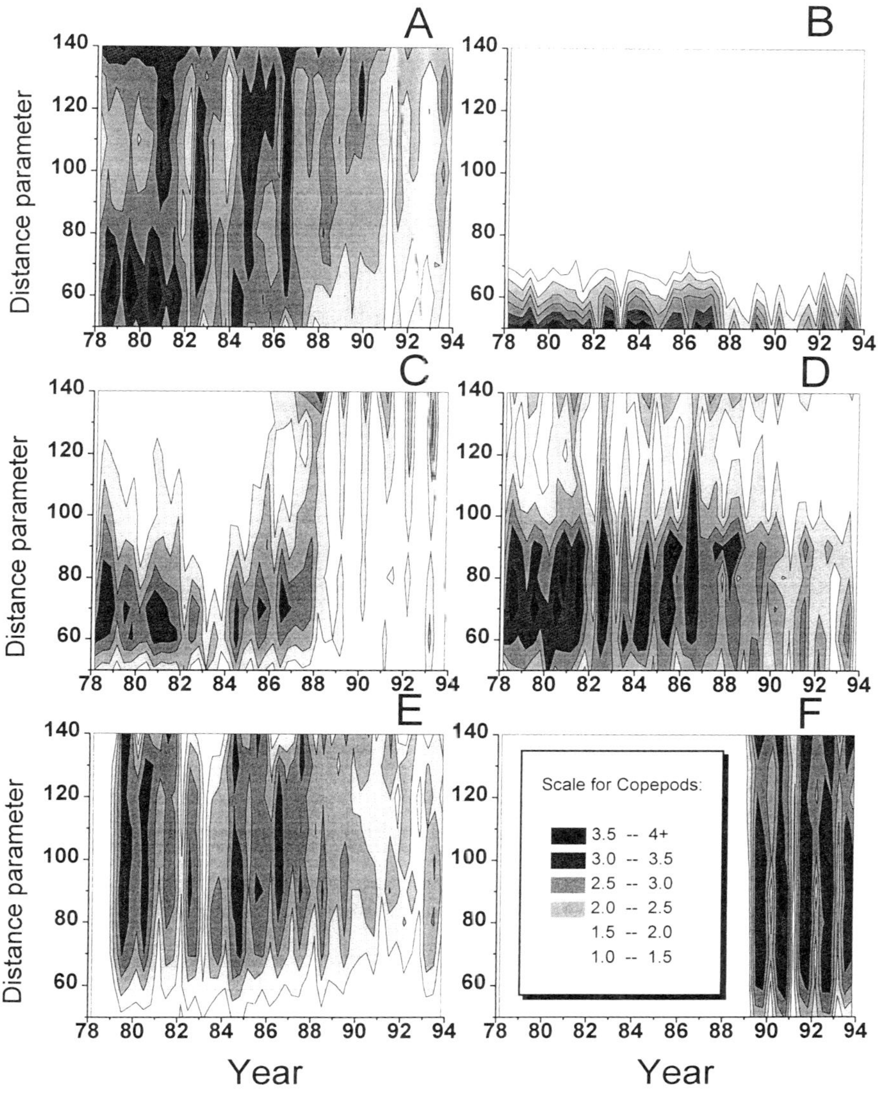

FIGURE 7. Contour plots of log chlorophyll concentration and log zooplankton abundance as a function of year and distance parameter. Scaling is shown for copepods only; for *Neomysis mercedis* the increments are the same over the range -1 to 3, while log chlorophyll is contoured from -0.3 to 1.2 in increments of 0.5. A, Chlorophyll concentration; B, *Acartia* spp.; C, *Eurytemora affinis*; D, *Neomsysis mercedis*; E, *Sinocalanus doerrii*; F *Pseudodiaptomus forbesi*.

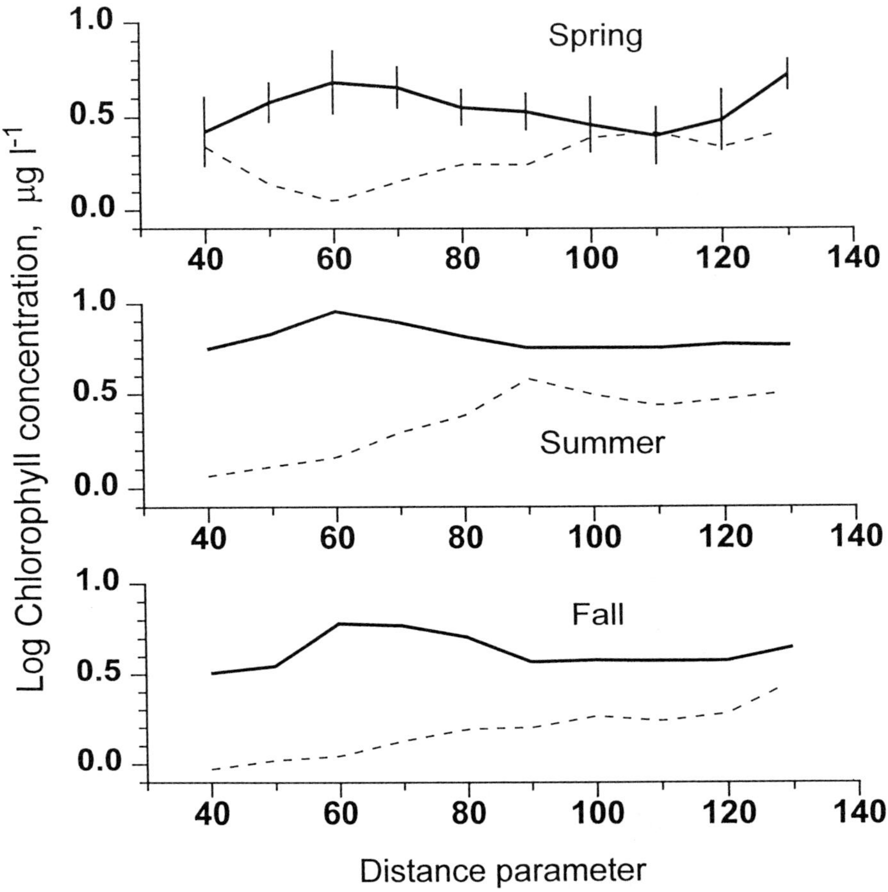

FIGURE 8. Spatial distribution of chlorophyll relative to the distance parameter for each season before (solid) and after (dashed) 1987. Vertical lines are 95% confidence intervals of the mean.

Taxa that did not Decline

The taxa that did not decline, or that declined in different periods (Table 1) included most of the taxa found almost exclusively in freshwater: the cladoceran *Diaphanosoma* sp., rotifers except as noted above, and freshwater copepods.

There was little change in abundance of several euryhaline marine taxa. These included the copepod *Oithona davisae*, introduced in 1981, crab zoeae, and barnacle nauplii. The latter two meroplanktonic groups had highly variable abundance patterns throughout the record.

The copepod *Pseudodiaptomus forbesi* became abundant in 1988-89 and has remained abundant during summers since then (Fig. 7F). Its abundance has a broad peak extending from the ETM to about 100 km. The summer peak in abundance

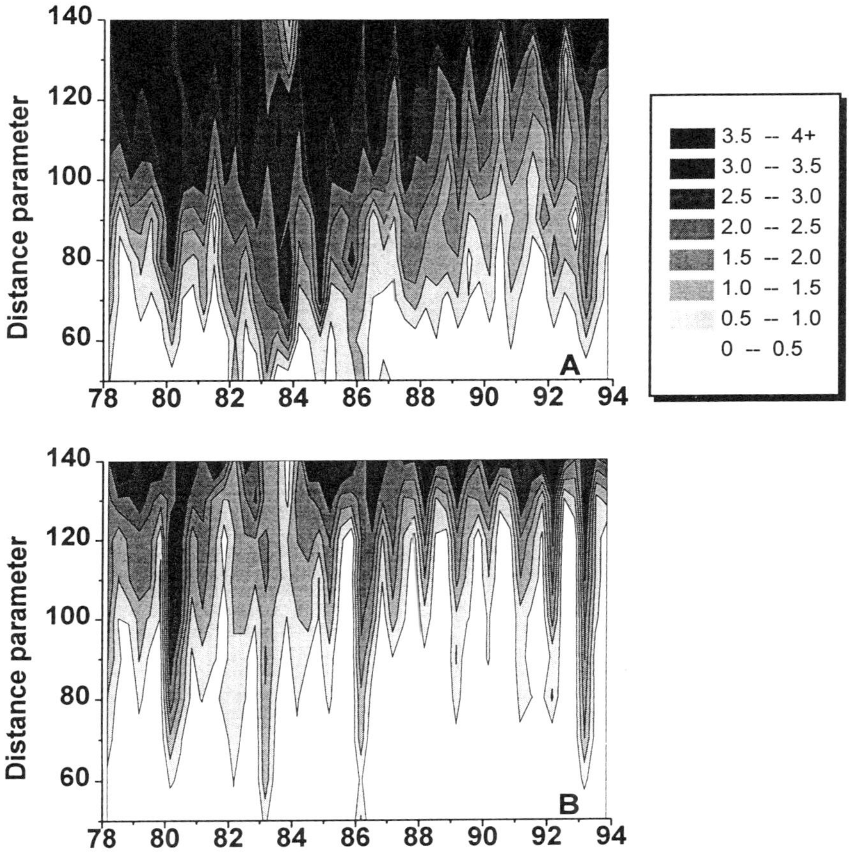

FIGURE 9. Contour plots as in Figure 7 for two cladocerans: A *Bosmina* spp.; B. *Daphnia* spp.

of this species in the ETM is similar in magnitude to that of *E. affinis* before 1987, and its upstream peak is somewhat higher than that of *Sinocalanus doerrii.*

E. affinis adults were usually uncommon or even absent in summer when *P. forbesi* adults were abundant (Fig. 11). Nauplii (Fig. 11) and copepodites (not shown) of *E. affinis* were also rare in summer, but increased again in October-November; these stages occur at the same salinity range as the adults. The steepest decrease in *E. affinis* nauplii and adults coincided with the steepest increase in *P. forbesi*; however, the increase of *E. affinis* nauplii in autumn was not accompanied by a large decrease in *P. forbesi.*

The total abundance of all calanoid copepods decreased significantly in the high-salinity area, but the decrease in the ETM was smaller and significant only in spring because of the increase of *P. forbesi* (Table 1). Total calanoids increased significantly upstream in spring.

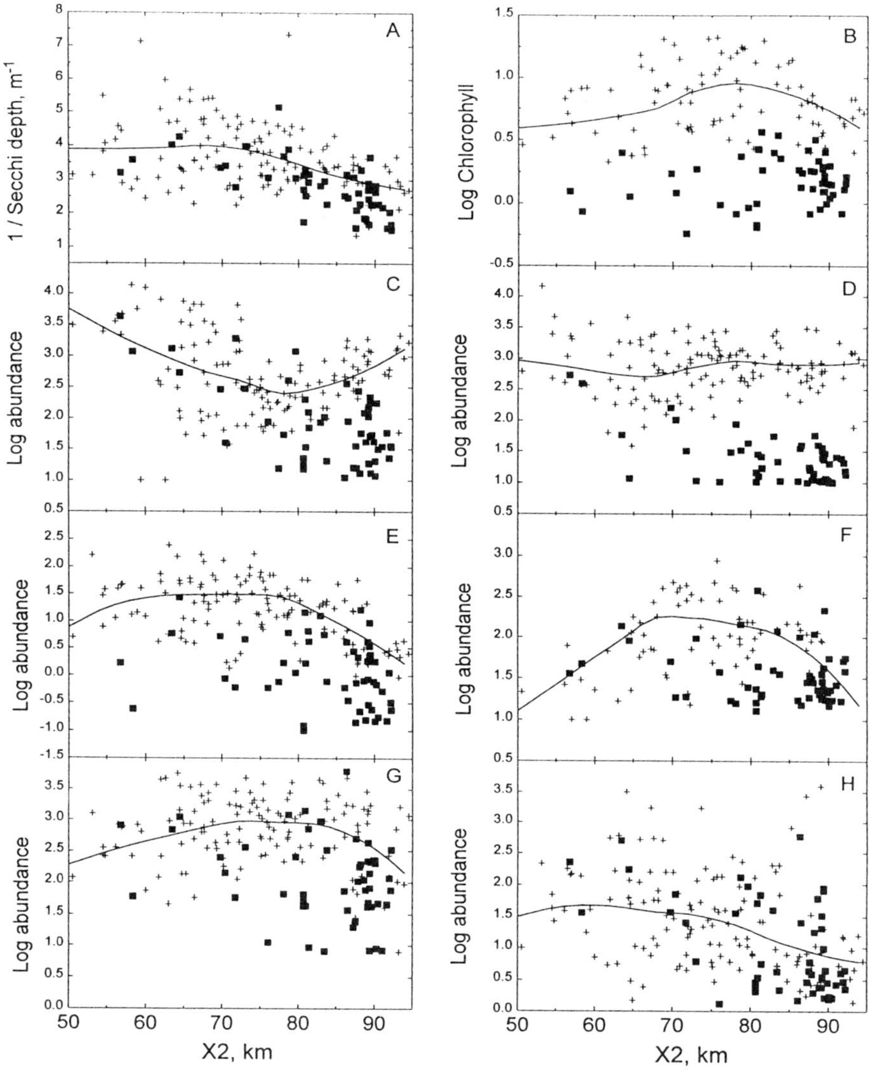

FIGURE 10. Variables in Figure 6 plotted against monthly mean X_2. Crosses, data from 1978-1986; solid squares, data from 1988-1993. The lines are locally weighted least-squares regression lines fitted to the medians for each 2-km interval of X_2 from the 1978-1986 period.

DISCUSSION

The general pattern of decline in phytoplankton and zooplankton is sufficiently coincident in time with the spread of *P. amurensis* to suggest a causal link. Kimmerer *et al.* (1994) presented evidence consistent with a causal relationship between the spread of *P. amurensis* and the decline in chlorophyll concentration and abundance of three copepod species.

In the present study, we show declines in most taxa found only in brackish water, including those whose distributions extend upstream, and 2 of 8 taxa with maximum abundance in freshwater. One species of brackish-water copepod (*Pseudodiaptomus forbesi*) became abundant after 1987, and another (*Oithona davisae*) did not change in abundance. Of the taxonomic groups (*e.g*., calanoids, harpacticoids, etc.) having broad or polymodal distributions, all declined at least at the brackish end of their distribution.

Because 1988-92 were drought years, the declines could be due to low-flow conditions in those years. Nutrient concentrations generally do not appear to limit chlorophyll concentrations (Cloern *et al.* 1983), but chlorophyll responds to freshwater flow (Alpine & Cloern 1992), and is generally highest at intermediate flows (Cloern *et al.* 1983). However, chlorophyll was lower in post-clam years even if flow (represented by X_2) is taken into account (Fig. 10). Furthermore, the rebound in chlorophyll during high flows of 1993 was short-lived. Therefore, the decline in chlorophyll can be attributed at least partly to the clam. The mechanism for the declines in zooplankton is unknown, although predation was suggested as the predominant cause of the decline of *Eurytemora affinis* (Kimmerer *et al.* 1994).

Predation can be ruled out in the case of *Neomysis mercedis*, of which all life stages are too large to be ingested in clam siphons (see also Orsi & Mecum, this volume). *N. mercedis* feeds on phytoplankton, rotifers, and small planktonic crustaceans (Siegfried & Kopache 1980). This species at least must have been affected by competition for food, especially in view of the ca. 80% decline in chlorophyll concentration. Other zooplankton species were probably also affected by the reduced chlorophyll concentration.

For most of the taxa that declined in the downstream or ETM regions, the greatest decline was seen in summer or fall and the least in spring. Although differences in thermal responses between clams and zooplankton could be important, it is more likely that changes in clam grazing rates within the seasons are responsible for the difference. Growth rates in terms of increase in biomass are generally highest in the summer months, tapering off in fall (Thompson & Nichols 1994). This should translate to higher ingestion rate and higher consumption rate of both phytoplankton and zooplankton.

Given the above, our results lend some support to the predation theory in the case of *E. affinis*. In Fig. 11 an increase in abundance of nauplii is evident in fall of each year that greatly exceeds the increase in adult abundance; that is, the ratio of nauplii to adults is low in summer and increases again in fall. However, chlorophyll concentrations in fall are still quite low where *E. affinis* is abundant (Figs. 7 and 8). Therefore, food limitation of egg production by adults of *E. affinis* would be as severe as in summer, and the increase in abundance of nauplii can be explained as a response to a reduction in predation rate below the rate of production of nauplii.

The influence of *Pseudodiaptomus forbesi* is still unknown, however. Its abundance increased sharply in late spring in each year, when that of *E. affinis* was dropping (Fig. 11). A similar pattern was observed at the Stockton station (river kilometer 136). *P. forbesi* has an abundance peak in summer and fall in the ETM, and a larger peak upstream (Fig. 7).

Pseudodiaptomus forbesi could affect *Eurytemora affinis* through competition for food or predation. However, its abundance appears too low to have had either effect. Mean summer abundances are on the order of 1-2 liter^{-1} for adults. Although *Pseudodiaptomus* species are not known to be predators, most copepods will feed on animals that they can capture (Paffenhöfer & Knowles 1980). The clearance rate of *Acartia* spp. feeding on nauplii is on the order of 20-40 ml female^{-1}d^{-1} (Landry 1978). *Pseudodiaptomus* species probably grow and therefore feed more slowly than *Acartia* spp. (Fancett & Kimmerer 1985) and are less predaceous based on morphology of mouthparts. If the clearance rate of *P. forbesi* is on the order of 10 ml copepod^{-1}d^{-1}, the mortality rate imposed on prey nauplii would be 1-2% d^{-1}. This clearance rate is about 10-20% of that estimated for grazing by *P. amurensis* (Kimmerer *et al.* 1994), and seems too low to remove all the nauplii in a month as implied by the data presented in Fig. 11.

Competition between *Pseudodiaptomus forbesi* and *Eurytemora affinis* or other resident species implies reduction in the available food supply by one or more of these species resulting in elimination or reduction in abundance of the other. The clearance rate of *P. forbesi* on chlorophyll would be about the same as that on nauplii, and little measurable effect on phytoplankton above that imposed by *P. amurensis* would be expected. However, the striking coincidence of appearance of *P. forbesi* soon after *E. affinis* was suppressed by *P. amurensis* implies that *P.*

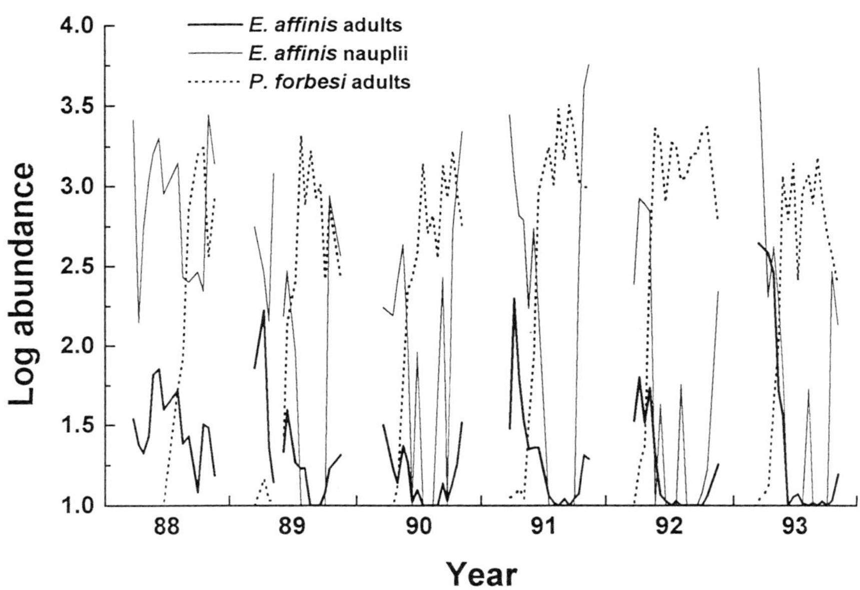

FIGURE 11. Abundance of *Eurytemora affinis* adults and nauplii and *Pseudodiaptomus forbesi* adults by date in 1988-1993 for the salinity range of the estuarine turbidity maximum, 1.2-15.3 mS cm^{-1}.

forbesi may previously have been suppressed by *E. affinis*, either by predation or, more likely, competition.

The spring floods of 1993 resulted in an extended period of downstream position of X_2 compared to the other post-clam years. Although a partial recovery was apparent in chlorophyll and abundance of *Acartia* spp., *Eurytemora affinis*, and *Sinocalanus doerrii*, these recoveries did not last into summer. The partial recoveries may have resulted from the decrease in abundance and clearance rate of *P. amurensis*, which may in turn have resulted from the decreased salinity in Suisun Bay. There was no evidence that zooplankton were insulated from the effects of *P. amurensis* by stratification. Furthermore, the calculated clearance rate for phytoplankton in 1993 was about 10-fold higher than that calculated by Kimmerer *et al.* (1994) for predation on *E. affinis* nauplii; correcting for the ca. 8-fold lower clearance rate on nauplii relative to phytoplankton, the clearance rate in 1993 should still have been sufficient to suppress the copepod population. Finally, with X_2 more downstream than in previous years, plankton populations were probably more exposed to the higher populations of *P. amurensis* in San Pablo Bay to the west than in other years.

The exceptions to the general pattern of decline may prove more interesting than the rule. They fall into two categories (other than *P. forbesi*): upstream species that declined, and downstream species that failed to decline. Among the upstream species, the cladocerans *Bosmina* spp. and *Daphnia* spp. were the most striking. Their abundances declined substantially in summer and fall, particularly toward the downstream limits of abundance. Since their distributions are well upstream of that of *P. amurensis*, any effects of *P. amurensis* on these cladocerans must have been indirect, probably through the reduction in chlorophyll.

The decrease in chlorophyll upstream of 90 km may have been due to mixing. Previous analyses have shown that phytoplankton concentrations in the delta are not mixed strongly downstream because of the lack of a substantial concentration gradient (Jassby & Powell 1994). However, since 1987 a strong downstream gradient has existed in all seasons, particularly summer and fall (Fig. 8). This gradient may suffice to mix chlorophyll downstream from the delta, reducing concentrations in the delta. This reduction could, in turn, be responsible for the decrease in abundance of *Sinocalanus doerrii*, *Neomysis mercedis*, and the two cladoceran genera upstream of 90 km. Better information is needed on the time scale of this mixing.

Among the euryhaline marine taxa, the larvae of barnacles and crabs did not decline in 1987-88. These larvae may be more affected by the vagaries of recruitment from the coastal ocean and the bay (Roughgarden *et al.* 1988) than by food conditions in the estuary. This is suggested by the higher degree of variability of these taxa than most others.

The reason for the lack of decrease in abundance of the copepod *Oithona davisae* remains unclear, however. This copepod is another introduced species with an abundance maximum in summer. It carries its eggs rather than releasing them into the water column as *Acartia* species do; however, *Eurytemora affinis* also carries its eggs, yet appears vulnerable to predation by *P. amurensis* (Kimmerer *et al.*

1994). The failure of *O. davisae* to respond could be due to a combination of relatively strong escape response of the nauplii (Kimmerer *et al.* 1994) and lower growth rate reducing vulnerability to poor food conditions (Paffenhöfer 1993).

We conclude that there have been substantial shifts in abundance patterns of zooplankton resulting directly or indirectly from *P. amurensis*, and that these shifts may be permanent. However, it is not yet clear that any of these changes have affected higher trophic levels. Many of the fish species native to the estuary have declined (Herbold *et al.* 1992), and abundance of many of these species is related to flow patterns (Jassby *et al.* 1995). In addition, the larvae and juveniles of most fish species are planktivorous. Yet, no evidence has yet been presented showing a continuing depression of abundance of any fish species that can be attributed to the reduction in food supply. Two reasons for this might be that most species of zooplankton did not decline in spring, when most fish reproduce; and that declines in some zooplankton species were made up by increases in others such as *Pseudodiaptomus forbesi*. Thus, the profound effect of *P. amurensis* on phytoplankton and zooplankton may not be felt at higher trophic levels.

From the results to date it appears that *Potamocorbula amurensis* is now a permanent and abundant resident of this estuary. Furthermore, its effect on lower trophic levels appears to persist over years of different hydrology. Several questions remain about its effects: (1) How do *Pseudodiaptomus forbesi* and *Oithona davisae* avoid suppression by clam predation; (2) Does the predatory impact of *P. amurensis* apply to other species than *Eurytemora affinis*; and (3) What effect does the reduction in lower trophic levels have on fish and other populations depending on these lower trophic levels for their food?

LITERATURE CITED

Aebischer, N. J., J. C. Coulson, and J. M. Colebrook. (1990). Parallel long-term trends across 4 marine trophic levels and weather. *Nature* 347:753-755

Alpine, A. E., and J. E. Cloern. (1992). Trophic interactions and direct physical effects control phytoplankton biomass and production in an estuary. *Limnol. Oceanogr.* 37:946-955.

Arthur J. A., and M. D. Ball. (1979). Factors influencing the entrapment of suspended material in the San Francisco bay-delta estuary. Pages 143-174 *in* T. J. Conomos, ed., *San Francisco Bay: The Urbanized Estuary.* Pacific Division, American Association for the Advancement of Science, San Francisco, CA.

Carlton, J. T., J. K. Thompson, L. E. Schemel, and F. H. Nichols. (1990). Remarkable invasion of San Francisco Bay (California, USA) by the Asian clam *Potamocorbula amurensis*. 1. Introduction and dispersal. *Marine Ecol. Progress Ser.* 66:81-94.

Cloern, J. E. (1982). Does the benthos control phytoplankton biomass in south San Francisco Bay (USA)? *Marine Ecol. Progr. Ser.* 9:191-202.

Cloern, J. E., A. Alpine, B. Cole, R. Wong, J. A. Arthur, and M. D. Ball. (1983). River discharge controls phytoplankton dynamics in the northern San Francisco Bay estuary. *Estuarine Coastal Shelf Sci.* 16:415-429.

Colebrook, J. M. (1978). Continuous plankton records: zooplankton and environment, North-East Atlantic and North Sea, 1948-1975. *Oceanol. Acta.* 1:9-23

Fancett, M. S., and W. J. Kimmerer. (1985). Vertical migration of the demersal copepod

Pseudodiaptomus as a means of predator avoidance. *Jour. Exper. Marine Biol. Ecol.* 88:31-43

Ferrari, F. D. and J. J. Orsi. (1984). *Oithona davisae*, new species, and *Limnoithona sinensis* (Burkhardt, 1912 (Copepoda: Oithonidae) from the Sacramento-San Joaquin estuary, California. *Jour. Crust. Biol.* 4:106-126

Herbold, B., A. D. Jassby, and P. M. Moyle. (1992). *Status and trends report on aquatic resources in the San Francisco Estuary*. San Francisco Estuary Project, U.S. Environmental Protection Agency, San Francisco, CA.

Jassby, A. D. and T. M. Powell. (1994). Hydrodynamic influences on interannual chlorophyll variability in an estuary: upper San Francisco Bay-Delta (California, U.S.A.). *Estuarine Coastal Shelf Sci.* 39:595-618.

Jassby, A. D., W. J. Kimmerer, S. G. Monismith, C. Armor, J. E. Cloern, T. M. Powell, J. R. Schubel, and T. J. Vendlinski. 1995. Isohaline position as a habitat indicator for estuarine populations. *Ecol. Applications* 5:272-289

Hymanson, Z. P. (1991). *Results of a Spatially Intensive Survey of* Potamocorbula amurensis *in the Upper San Francisco Bay Estuary*. California Interagency Ecol. Studies Progr. Tech. Rept. 30. Sacramento, California. 21pp.

Kimmerer, W. J., E. Gartside, and J. J. Orsi. (1994). Predation by an introduced clam as the probable cause of substantial declines in zooplankton in San Francisco Bay. *Marine Ecol. Progr. Ser.* 113:81-93

Landry, M. R. (1978). Population dynamics and production of a planktonic marine copepod, *Acartia clausii*, in a small temperate lagoon on San Juan Island, Washington. *Internat. Rev. ges. Hydrobiol.* 63:77-119.

Miller C. B. (1983). The zooplankton of estuaries. Pages 103-149 *in* B. H. Ketchum, ed., *Estuaries and Enclosed Seas*. Elsevier, New York, NY.

Nichols, F. H. (1985). Increased benthic grazing: an alternative explanation for low phytoplankton biomass in northern San Francisco Bay during the 1976-1977 drought. *Estuarine Coastal Shelf Sci.* 21:379-388.

Nichols, F. H., J. K. Thompson, and L. E. Schemel. (1990). Remarkable invasion of San Francisco Bay (California, USA) by the Asian clam *Potamocorbula amurensis* .2. Displacement of a former community. *Marine Ecol. Progr. Ser.* 66:95-101.

Obrebski, S., J. J. Orsi, and W. J. Kimmerer. 1992. *Long-term Trends in Zooplankton Distribution and Abundance in the Sacramento-San Joaquin Estuary of California*. California Interagency Ecol. Studies Progr. Tech. Rept. 32. Sacramento, CA. 42 pp.

Orsi, J. J., and W. L. Mecum. (1986). Zooplankton distribution and abundance in the Sacramento-San Joaquin Delta in relation to certain environmental factors. *Estuaries* 9:326-339.

Orsi, J. J., T. E. Bowman, D. C. Marreli, and A. Hutchinson. (1983). Recent introduction of the planktonic calanoid copepod *Sinocalanus doerrii* (Centropagidae) from mainland China to the Sacramento-San Joaquin Estuary of California. *Jour. Plankt. Res.* 5:357-375

Orsi, J. J., and T. E. Walter. (1991). *Pseudodiaptomus forbesi* and *P. marinus* (Copepoda: Calanoida), the latest copepod immigrants to California's Saccramento-San Joaquin Estuary. Pages 553-562 *in* S. I. Uye, S. Nishida, & J.-S. Ho, eds., *Proceedings of the Fourth International Conference on Copepoda*. Bull. Plankt. Soc. Japan, Spec. Vol. Hiroshima.

Paffenhöfer, G.-A. (1993). On the ecology of marine cyclopoid copepods (Crustacea, Copepoda). *Jour. Plankt. Res.* 15:37-55

Paffenhöfer, G.-A., and S. C. Knowles. (1980). Omnivorousness in marine planktonic copepods. *Jour. Plankt. Res.* 2:355-365

Roughgarden, J., S. Gaines, and H. Possingham. (1988). Recruitment dynamics in complex life cycles. *Science* 241:1460-1466

Siegfried, C., and M. Kopache. (1980). Feeding of *Neomysis mercedis* (Holmes). *Biol. Bull.* 159:193-205.

Thompson, J. K., and F. H. Nichols. (1994). Control of a phytoplankton bloom in San Francisco Bay, California by the filter-feeding bivalve *Potamocorbula amurensis. Proc. Pacific Div., American Assoc. Advance. Sci.* 13(1[June]):98. [abstract]

Venables, W. N., and B. D. Ripley. (1994). *Modern applied statistics with S-Plus.* Springer, New York. 462 pp.

Venrick, E. L., J. A. McGowan, D. R. Cayan, and T. L. Hayward. (1987). Climate and chlorophyll a: Long-term trends in the central North Pacific Ocean. *Science* 238: 70-72

Werner, I., T, and J. Hollibaugh. (1993). *Potamocorbula amurensis*: Comparison of clearance rates and assimilation efficiencies for phytoplankton and bacterioplankton. *Limnol. Oceanogr.* 38:949-964.

BENTHIC PROCESSES IN SOUTH SAN FRANCISCO BAY: THE ROLE OF ORGANIC INPUTS AND BIOTURBATION

J. M. CAFFREY[1], D. E. HAMMOND[2], J. S. KUWABARA[3], L. G. MILLER[3], AND R. R. TWILLEY[4]

[1] **Institute of Marine Science, University of California, Santa Cruz, CA 95064;**
[2]**Department of Geology, University of Southern California, Los Angeles, CA 90089;**
[3]**U.S. Geological Survey, 345 Middlefield Rd., Menlo Park, CA 94025;**
[4]**Department of Biology, University of Southwestern Louisiana, Lafayette, LA**

Measurements of benthic processes over the last 13 years in South San Francisco Bay have spanned a range of hydrologic and water quality conditions. However, benthic fluxes of O_2, NH_4^+ and NO_3^- have shown little variation. NH_4^+ and NO_3^- flux during this period were relatively constant in magnitude and direction. Sediment O_2 consumption was lower in the early 1980s than in the early 1990s. The reason for this difference is unclear, but may be a result of methodology or a lower density of the tube dwelling *Asychis elongata*. Although dissolved inorganic phosphate fluxes generally have been close to zero, in fall, there is a net flux into the sediments. Dissolved Si fluxes were consistently out of the sediments, while dissolved organic carbon fluxes were large and exhibited shifts in direction. Compared with other estuarine and coastal systems, temperature was not a major control on benthic fluxes because of the small annual temperature range. Irrigation by macrofauna enhances O_2 consumption and may control NH_4^+ fluxes in the shoals, where NH_4^+ flux was correlated with the biomass of mollusks. Porewater profiles of NH_4^+, TCO_2 and radon suggest that irrigation is important in controlling dissolved constituent concentrations. Organic inputs such as deposition of phytoplankton may also affect benthic fluxes. In the channel, NH_4^+ flux was positively correlated to phaeopigment concentrations in sediments. Ratios of sediment O_2 consumption to DIN (NH_4^+ + NO_3^-) or DIP flux were very different than ratios of remineralization of organic matter with a Redfield ratio, and suggest that both denitrification and phosphorus burial were occurring at both locations.

Fluxes across the sediment-water interface represent an important linkage between benthic and pelagic environments. Nutrient fluxes out of sediments are often a significant nutrient source for water column phytoplankton production (Nixon 1981). Benthic fluxes may also respond to seasonal variations in productivity. This flux of materials is a result of a complex set of interdependent physical (diffusion, resuspension), chemical (sorption/desorption, complexation/dissociation, redox reactions, precipitation/dissolution), and biological (microbial transformations, bioturbation by benthic invertebrates and inputs of organic matter) components. Deposition of phytoplankton production and enhancement of benthic fluxes has been observed in many estuarine and marine ecosystems (Kemp & Boynton 1984; Jensen *et al.* 1990; Graf 1992; Grebmeier 1993). For example, O_2 consumption by sediments is affected by decomposition of organic matter and respiration by benthic invertebrates (Grebmeier 1993; Tahey *et al.* 1994). Similarly, NH_4^+ fluxes also reflect sediment processes such as NH_4^+ remineralization, nitrification and NH_4^+ excretion by macrofauna (Blackburn & Henriksen 1983; Henriksen *et al.* 1993), while NO_3^- fluxes are affected by nitrification, denitrification and NO_3^- concentrations. Phosphorus fluxes have been related to bottom water oxygen concentrations (Kemp & Boynton 1992). Dissolved silica fluxes are controlled by bottom water silica concentrations and the chemical dissolution of biogenic silica (Conley *et al.* 1993).

San Francisco Bay is a highly urbanized estuary, surrounded by a population of approximately 5.8 million people. South San Francisco Bay (South Bay) is a lagoonal system in which the major input of freshwater and nutrients to the system comes from sewage treatment plants (Nichols *et al.* 1986; Hager & Schemel, this volume). South Bay is characterized by a broad expanse of shallow subtidal and intertidal areas. Approximately 80% of South Bay is ≤ 2 m deep (Fig. 1).

One of the most notable features of the South Bay ecosystem is the spring phytoplankton bloom. Phytoplankton growth is normally light-limited except in spring when tidal energy is reduced and persistent density stratification leads to phytoplankton blooms when chlorophyll a concentration exceeds 10 $\mu g\ l^{-1}$ (Cloern 1991a). Phytoplankton, along with benthic algae, are the source of about 90% of organic carbon inputs to South Bay (Jassby *et al.* 1993). The benthic community seems to be major sink of phytoplankton production (Cloern 1982; Koseff *et al.* 1993). Although recent research in San Francisco Bay has emphasized this grazing link, less is known about the link between pelagic production and microbial processes in the sediments. This paper reviews benthic flux measurements made during three studies conducted in South San Francisco Bay between 1980 and 1993 and uses this record of benthic flux measurements to investigate the links between phytoplankton production and benthic processes.

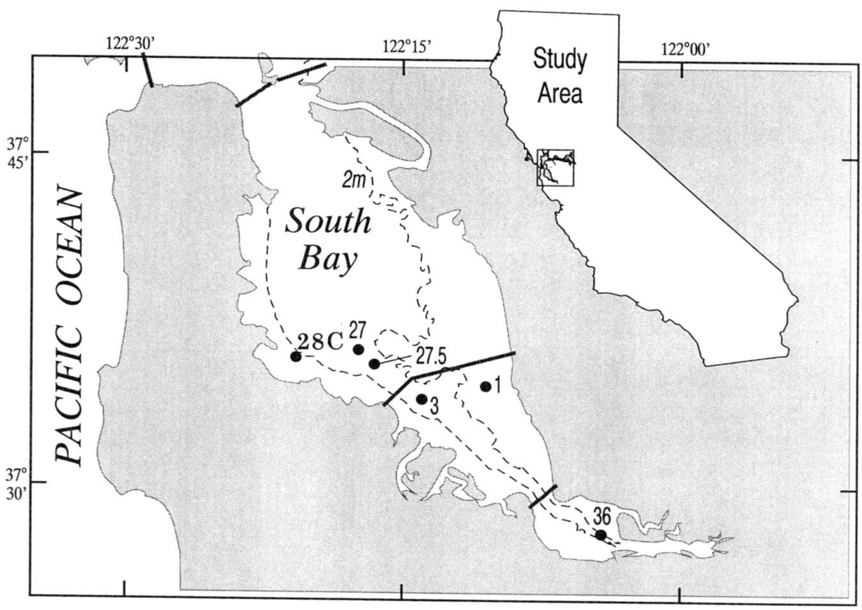

FIGURE 1. South San Francisco Bay location of stations and 2 m depth contour. Benthic flux measurements were made a stations 27.5 and 28C between 1980-1981, at stations 27, 28C and 36 in 1984, and at stations 1 and 3 between 1991-1993.

MATERIALS AND METHODS

Benthic fluxes in 1980-1981 were measured at two stations: 27.5 in the channel and 28C in the shoals of South Bay (Fig. 1). *In-situ* plexiglas chambers were used (Hartman & Hammond 1984) that trapped a 20 cm-high column of water overlying 1600 cm^2 of sediments. Samples were drawn three times during a deployment (0.2, 4 and about 20 hrs after emplacement). Measurements were also made in 1992 at stations 1 and 3 using a slightly different chamber design (Berelson *et al.* 1987). This device was PVC, trapped a 10 cm-high water column, covered 700 cm2 of sediment, drew 6 samples during a 20 h deployment and measured oxygen with an electrode. In both chamber types, water was stirred continuously with a paddle. During a study in 1984, benthic fluxes were measured in cores at three stations (27, 28C, and 36) using a flow-through system as described in Miller-Way (1994). Benthic fluxes during 1991-1993 were measured as described below.

Benthic cores (12.1 cm diameter) were collected on the shoals (station 1) and in the channel (station 3) in South Bay (Fig. 1) with a modified dart corer. The corer and attached acrylic core tube were slowly lowered into the sediments to reduce the loss of the surface flocculent layer. Bottom water was collected at each station with a peristaltic pump from 0.5-1 m above the bottom. The cores were brought back to the lab in dark containers filled with water from the sampling site to minimize temperature changes. In the lab, overlying water was replaced with the bottom water, and cores were sealed with tops equipped with magnetic stirrers and gas-tight sampling ports. The height of water overlying the sediment ranged from 25-35 cm and sediment thickness ranged from 15-20 cm. Cores were incubated in the dark at ambient temperatures. Four water samples (30 ml) were removed at one hour intervals following an hour pre-incubation period. Water removed from the cores was replaced with bottom water. This resulted in a less than 1% dilution of the overlying water.

Water samples for O_2 analysis were stored unfiltered in glass bottles (7 ml) with Winkler reagents. O_2 concentrations were measured within 2 hrs of collection with an automated titration system (Metrohm 686™) using the potentiometric titration method of Graneli & Graneli (1991). Water samples for nutrient analysis were filtered (GF/F). NO_3^- and NH_4^+ samples were initially frozen and analyzed on a Technicon II autoanalyzer within 2 months of collection. NO_3^- and NO_2^- (referred to as NO_3^-) were analyzed by the automated Cd reduction method and NH_4^+ by a modification of the salicylate-hypochlorite method (Hager 1993). Dissolved organic carbon concentrations were determined by ultraviolet/oxygen/persulfate oxidation (Hunter & Kuwabara 1994). Subsamples of wet sediment, approximately 8 g, from the top 1 cm were weighed, dried at 60°C for a minimum of 48 h and reweighed for calculation of percent water content. Porosity was determined using percent water content and assuming a sediment density of 2.7 g cm^{-3}. Particulate C and N content were determined as described in Caffrey (1995). Following incubation, samples were taken from the cores for benthic chlorophyll *a* and phaeopigments. Cores were sieved through 0.55 mm screens to remove the infauna and the animals were preserved in 10% formalin for identification and weight determination. Wet weights were converted to organic content using factors determined by Lie (1968).

RESULTS AND DISCUSSION

Benthic fluxes in South Bay have been measured periodically at shallow and deep stations between 1980 and 1993 (Table 1). Sediments at these stations tend to be fine grained with porosities ranging from 0.70-0.82 (Table 1). The particulate carbon (PC) content of the sediments at these stations ranged between 1.1 and 1.6%, while the range of particulate nitrogen (PN) content was between 0.12 and 0.27%. Station 1 is characterized by muddy sands with a lower PC and PN content, while stations 27, 28C, 27.5, and 3 have fine-grained sediments and higher PC and PN contents. Station 36 is closest to the major sewage treatment outfall in South Bay and has the highest PC and PN content.

TABLE 1. Years that benthic flux were measured in South San Francisco Bay, stations and methods, sediment characteristics: porosity, % carbon, % nitrogen

Year	*Station*	*Water Depth (m)*	*Incubation Method*	*Average Porosity*	*% Carbon*	*% Nitrogen*
1980-1981[1]	27.5	14	*in-situ* chambers	0.75	1.38	0.16
	28C	2	"	0.70	1.41	0.15
1984	27	14	cores – flow through	0.81	1.43	0.18
	28C	2	"	0.77	1.41	0.16
	36	10	"	0.82	1.59	0.27
1991-1993	1	2	cores-batch[2]	0.74	1.09	0.12
	3	13	"	0.80	1.44	0.14

[1] 1980-1981 data from Hammond *et al.* (1985).
[2] On one occasion in 1992, *in-situ* chambers were used in addition to core incubations.

During the 13-year period spanned by these studies, freshwater input from the Sacramento-San Joaquin Rivers to San Francisco Bay ranged from inputs less than 1000 m^3s^{-1} (between January and April) during the 1987-1992 drought and in 1981, to very high flow during 1982, 1983, 1986, and 1993 (Sheila Greene, pers. commun.). Benthic flux measurements have been made during low, intermediate and high flow years (as defined by Peterson *et al.* 1985, Table 2). River flow and increased stratification have been positively correlated with the spring phytoplankton biomass and productivity in South Bay (Cloern 1991b). Phytoplankton and benthic algae comprise about 90% of the organic carbon inputs to this area (Jassby *et al.* 1993). The highest levels of phytoplankton occur in spring when chlorophyll a levels increase from background levels of about 2 $\mu g\ l^{-1}$ to 25-50 $\mu g\ l^{-1}$ (Cloern 1987, 1991a and b, Table 2). Because the spring bloom is a well defined event which generally lasts between 2-4 weeks (Cloern 1991a), sedimentation of the spring bloom is most likely a relatively short, pulse input of fresh material to the sediments. The highest phaeopigment concentrations in the surface sediments as well as the greatest NH_4^+ regeneration in sediments occur about a month after peak

TABLE 2. Year of benthic flux measurement, mean Sacramento and San Joaquin River Flow (m^3s^{-1}) between January and April, and chlorophyll *a* concentrations ($\mu g\ l^{-1}$) averaged between March and April for the South Bay channel (USGS Hydrostations 24-36).

Year	*River Flow* $m^3 s^{-1}$	*Mean Chlorophyll a concentration*[1], $\mu g\ l^{-1}$
1980	2800	11.9
1981	500	5.4
1984	1300	7.6
1991	280	6.2
1992	380	8.0
1993	1570	16.5

[1] Chlorophyll *a* values are from Cloern (1991), Wienke *et al.* (1992, 1993), and Caffrey *et al.* (1994).

chlorophyll *a* concentrations in spring (Caffrey 1995). These results support the idea that sedimentation of the spring bloom is a fairly well defined event and that sediment properties may respond to it.

Sediment O_2 Consumption

O_2 consumption in sediments at the shoal stations were similar during the 1980-1981 and 1991-1992 sampling pericds, while oxygen consumption was lower in the channel during the 1980-1981 period than during the 1991-1993 period (Fig. 2a). The average ± s.e. oxygen consumption in 1980-81 at channel and shoal stations was 16 ± 7 and 30 ± 7 mmol $O_2\ m^{-2}d^{-1}$, respectively, and in 1991-93, it was 49 ± 3 and 47 ± 6 mmol $O_2\ m^{-2}d^{-1}$, respectively. The lack of sediment oxygen flux measurements for the period between 1981 and 1991 make it impossible to determine a reason for the higher oxygen consumption at the channel station in the 1990s. However, there are a number of factors which could contribute to these observed differences: different station locations, sediment type, organic inputs, types of incubations (*in-situ* chambers versus cores), and macrofaunal communities.

The channel stations 27.5 and 3 are approximately 8 km from one another. Water depths at these are similar, as are sediment porosity, PC, PN (Table 1) and percent mud (Nichols & Pamatmat 1988), suggesting that the physical characteristics of the sediments at these two station are similar. In addition, average spring chlorophyll a values for both time periods were similar, so it seems reasonable to assume that organic inputs from phytoplankton were comparable.

In April 1992, the comparison between *in-situ* chambers and core incubations at stations 1 and 3 demonstrated mixed results. Sediment O_2 consumption as measured with *in-situ* chambers was 28 ± 1 and 46 ± 7 mmol $m^{-2}d^{-1}$ at channel and shoal stations respectively, while rates measured with core incubations were 43 ± 5 and 12 ± 2 mmol $m^{-2}d^{-1}$ at the same channel and shoal stations respectively. No single factor can explain these differences. Fluxes measured with *in-situ* chambers may be higher than core fluxes because the cores can cut off or exclude deep

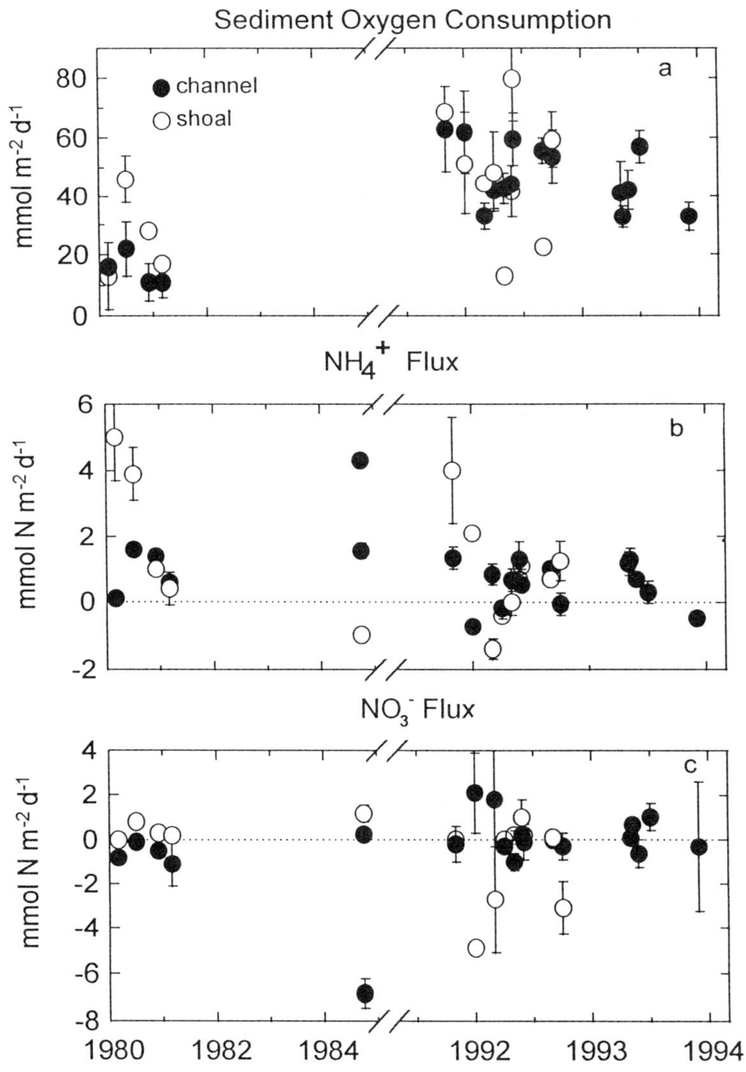

FIGURE 2. Sediment oxygen consumption in mmol O_2 $m^{-2}d^{-1}$ (a), NH_4^+ flux in mmol N m^{-2} d^{-1} (b), and NO_3^- in mmol N $m^{-2}d^{-1}$ (c) at shoal (open circles) and channel (closed circles) stations in South San Francisco Bay between 1980 and 1993. 1980-1981 data is replotted from Hammond *et al.* (1985). + value is flux out of sediment for NH_4^- or NO_3^- flux.

burrowing macrofauna. Irrigation by macrofauna such as *Asychis elongata* could be responsible for the higher fluxes from the chambers. If *in-situ* chambers should give higher rates of sediment O_2 consumption than core incubations because the in-situ chambers include macrofauna, this does not explain the difference between

1980s and 1990s values since the higher 1990s measurements were made with core incubations, not in-situ chambers.

Another difference between in-situ chambers and core incubations is the duration of the incubation. Samples are collected over a 20 hr period from the in-situ chambers, but usually over a 5-hr period from the cores. Sediment O_2 consumption is fairly linear over 5 hrs, but not for 20 hrs, so using a linear model to calculate SOC for a 20-hr period could underestimate the actual rate. For the longer incubation periods, a quadratic model may be more appropriate and was used to calculate rates for the *in-situ* chambers in April 1992. However, in 1980-81, SOC was calculated using a linear model, so this may explain the lower values during this period.

In addition, macrofauna densities were different between the 1980s and 1990s data, with greater densities of *Asychis elongata* at station 3 than station 27.5. Hammond *et al.* (1985) report that *Heteromastus filiformis* was the dominant tube dwelling polychaete at station 27.5. In San Francisco Bay, *Heteromastus* is considerably smaller than *Asychis* (Caffrey, unpublished data), so a community dominated by *Heteromastus* may not have the ability to enhance sediment O_2 consumption as much as *Asychis*.

There was no consistent seasonal pattern of O_2 consumption at either channel or shoal stations. Over the 13-year period, temperature was not significantly correlated with sediment O_2 consumption at either channel ($r = 0.39$, $p = 0.1$) or shoal stations ($r = 0.27, p = 0.28$). In contrast, O_2 consumption was correlated with sediment chlorophyll a concentrations ($r = 0.38, p = 0.009$, Table 3) and sediment phaeopigment concentrations ($r = 0.32, p = 0.03$) during 1991-1993. O_2 consumption was also related to the number, but not biomass, of annelids at the channel

TABLE 3. Correlation Coefficients between Benthic Fluxes: Sediment O_2 Consumption (SOC), NH_4^+, NO_3^- and Selected Parameters: Temperature, Sediment Chlorophyll *a*, Sediment Phaeopigments, Macrofauna Biomass, Overlying water NH_4^+ and NO_3^- Concentrations at Channel (3) and Shoal (1) Stations during 1991-1993. NS – not significant; first line – correlation coefficient; second line – *P* value.

	Channel			*Shoals*		
	SOC	*NH_4^+ Flux*	*NO_3^- Flux*	*SOC*	*NH_4^+ Flux*	*NO_3^- Flux*
SOC		0.27 0.02	NS		NS	NS
Temperature	NS	NS	NS	NS	0.81 0.001	0.45 0.005
Chlorophyll a	0.38 0.009	0.37 0.01	NS	NS	NS	NS
Phaeopigment	0.32 0.03	0.30 0.04	NS	NS	-0.51 0.02	NS
Macrofauna Biomass	NS	0.27 0.02	NS	NS	0.73 0.001	NS
NH_4^+ Concentration		NS			-0.66 0.001	
NO_3^- Concentration			0.44 0.002			-0.49 0.003

station ($r = 0.36, p = 0.005$, data not shown). Annelids were the dominant invertebrate at this location, particularly *Asychis elongata* which can build tubes as long as 50 cm. Macrofauna enhance O_2 consumption through respiration and by increasing the sediment-water interface area through building tubes (Kristensen *et al.* 1991).

NH_4^+ Flux

NH_4^+ measurements usually indicated a flux out of the sediments (Fig. 2b) and were higher, on average, at the shoal stations (average 1.61 ± 0.47 mmol N $m^{-2}d^{-1}$) than at channel stations (average 0.57 ± 0.13 mmol N $m^{-2}d^{-1}$). In contrast to sediment O_2 consumption, NH_4^+ fluxes were similar over the 13-year period. NH_4^+ fluxes did not have consistent seasonal trends at either station (Fig. 2b). During 1991-1992 at the shoal station, the highest rates occurred in October, declined throughout the winter, became negative in February, and increased during spring, becoming positive in late April.

Although there was no apparent seasonal pattern at the channel station, NH_4^+ flux between 1991-1993 was correlated with concentration of benthic phaeopigments ($r = 0.30, p = 0.04$, Table 3) and benthic chlorophyll a concentrations ($r = 0.37, p = 0.01$, Table 3). Phaeopigment concentrations were particularly high in 1993 following the spring bloom (Table 4). Accumulation of phaeopigments in the channel may have been the result of trapping material from the shoals in the channel (Huzzy *et al.* 1990). Only a fraction of the organic matter that is remineralized in the sediments diffuses out as NH_4^+ flux. NH_4^+ remineralization in South Bay sediment was significantly correlated with water column chlorophyll a concentrations, suggesting a relationship between organic inputs (from phytoplankton production) and NH_4^+ remineralization (Caffrey 1995). However, on average, NH_4^+ flux out of the sediment was only 22% of NH_4^+ remineralization in the channel and 26% at the shoal station, suggesting that nitrification is a significant process in these sediments.

NO_3^- Flux

Mean NO_3^- fluxes varied from +2 to -6 mmol N $m^{-2}d^{-1}$ and were negligible for 18 of the 34 measurements (Fig. 2c). High NO_3^- fluxes into the sediment occurred at station 36 during the 1984 sampling period. This station is close to the discharge of the largest sewage treatment plant in South Bay. Nutrient concentrations in South Bay are usually highest at this station (Hager & Schemel, this volume) and were 115 μM during this sampling period. High NO_3^- fluxes into the sediment also occurred during December 1991 and February 1992 at shoal station 1 when bottom water NO_3^- concentrations were high, about 40 μM (Fig. 2c). NO_3^- fluxes were related to temperature and bottom water NO_3^- concentration at the shoal station (Table 3). This is consistent with the pattern of NO_3^- flux into the sediment during the winter when NO_3^- concentrations are high.

DIP and DSi Fluxes

Dissolved inorganic phosphate (DIP) fluxes were close to zero except during September 1984 (Fig. 3a). In September, DIP fluxes were negative with the highest

TABLE 4. Temperature, benthic chlorophyll *a*, phaeopigment, and macrofauna biomass from benthic flux stations in 1980-81 and 1991-93. Chlorophyll *a* and phaeopigment data for 1980-81 are from Thompson *et al.* (1981). n.d. – no data

Station	*Season*	*Temperature* °C	*Chlorophyll a* $\mu g\ cm^{-2}$	*Phaeopigment* $mg\ cm^{-2}$	*Macrofauna Biomass* $gC\ m^{-2}$
27.5	Winter '80	12	n.d.	n.d.	n.d.
	Summer '80	17	0.0	25.4	n.d.
	Fall '80	17	0.0	9.5	n.d.
	Winter '81	12	0.0	13.0	n.d.
28	Winter '80	12	n.d.	n.d.	n.d.
	Summer '80	17	4.7	37.9	n.d.
	Fall '80	16	7.4	42.5	n.d.
	Winter '81	12	1.1	19.4	n.d.
3	Fall '91	20	n.d.	n.d.	7.8
	Winter '91	13	2.5	33.4	6.3
	Spring '92	18	4.5	40.6	4.1
	Summer '92	21	7.2	36.4	14.3
	Fall '92	20	1.3	27.8	15
	Spring '93	16	4.4	131.2	12.6
	Fall '93	18	3.2	33.4	7.7
1	Fall '91	20	n.d.	n.d.	31.9
	Winter '91	13	3.7	63.5	7.3
	Spring '92	18	11.1	49.7	2.6
	Summer '92	21	7	51.4	10.3
	Fall '92	20	18.6	70.6	14.7

rates at station 36. This is consistent with the negative NO_3^- fluxes at this station and probably the result of high nutrient concentrations in the water column. DIP concentrations at station 36 were 35 μM, while concentrations at the other benthic flux stations in September 1984 were between 4-7 μM. The consistently low or negative DIP fluxes contrasts with the patterns observed in other estuaries, where high DIP fluxes out of sediment occur when sediment redox conditions drop (Kemp & Boynton 1984; Hopkinson 1987; Koop *et al.* 1990; van Duyl *et al.* 1993).

Dissolved silica (DSi) fluxes were positive with values ranging between 2 and 12 mmol Si $m^{-2}d^{-1}$, except in February 1980 when fluxes were into the sediments (Fig. 3b). No seasonal pattern was apparent in the channel. In the shoals, DSi fluxes were highest in June. This may be because biogenic silica regeneration in sediments is a predominantly chemical process which is determined by a combination of factors including porewater silica concentrations, temperature, pH, and salinity (Conley *et al.* 1993). DSi fluxes also reflect irrigation by macrofauna (Hammond *et al.* 1985).

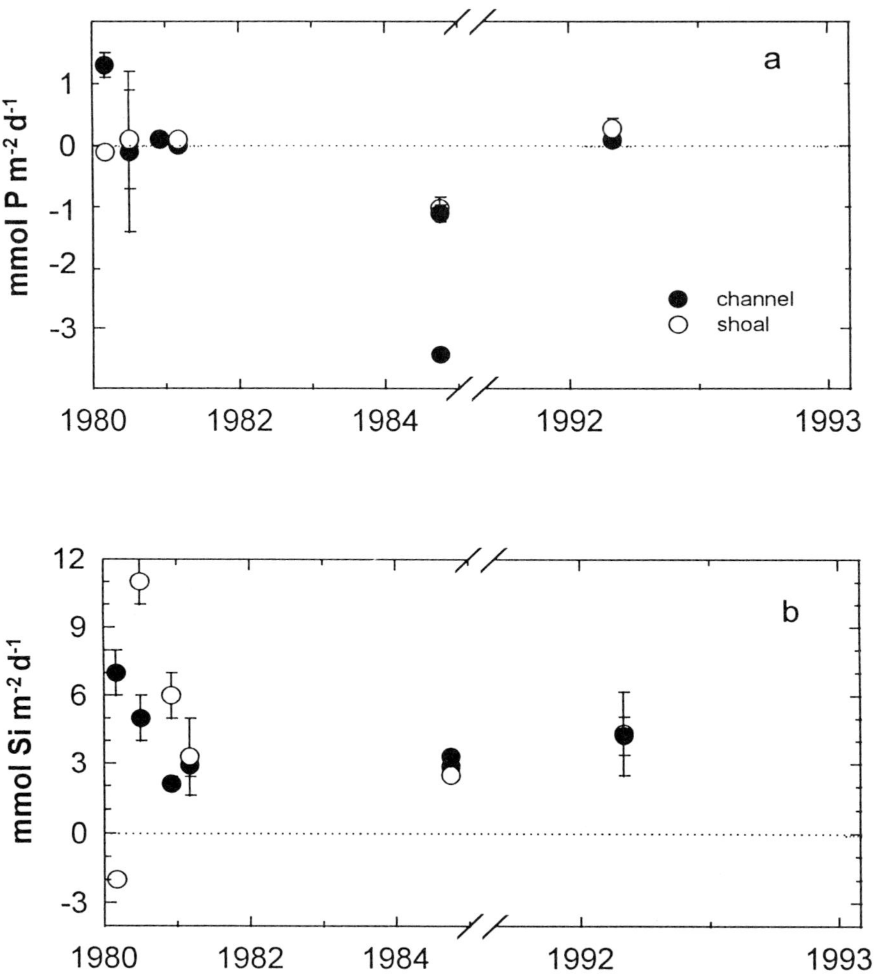

FIGURE 3. Seasonal variation in DIP flux in mmol P $m^{-2}d^{-1}$ (a), DSi flux in mmol Si $m^{-2}d^{-1}$ (b) at shoal (open circles) and channel stations (closed circles) in South San Francisco Bay. 1980-1981 data is replotted from Hammond *et al.* (1985). + value is flux out of sediment.

DOC Flux

The seasonal pattern in dissolved organic carbon (DOC) fluxes was quite variable between 1992 and 1993. During spring of 1992, DOC fluxes were negative in February and changed to positive fluxes by April with highest positive DOC flux at the channel station in the fall (Fig. 4). The seasonal pattern of DOC fluxes at the shoal station was similar to the channel station except that fluxes were near zero in September. The pattern of DOC fluxes in the channel was quite different during 1993. During this year, fluxes decreased during the spring and were negative in the

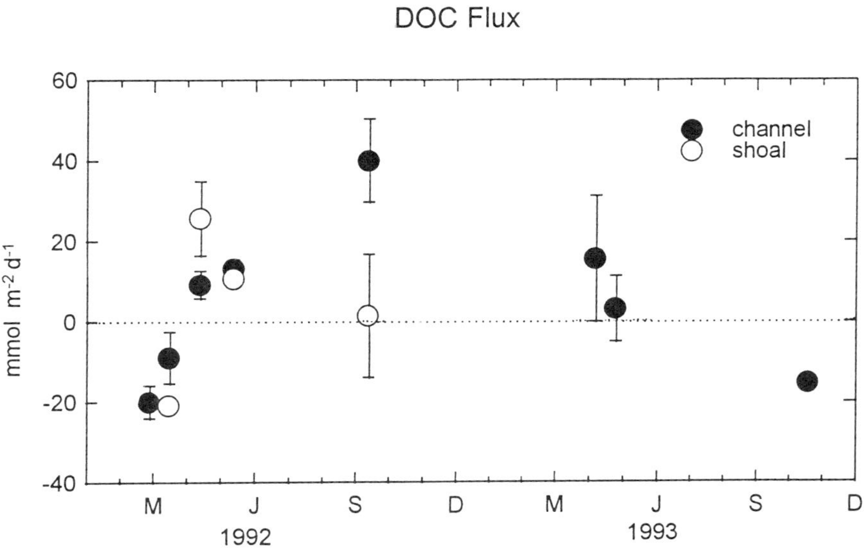

FIGURE 4. Sediment DOC fluxes in mmol $m^{-2}d^{-1}$ at shoal and channel stations in South San Francisco Bay between 1992 and 1993. + value is flux out of sediment

fall. Despite the higher water column chlorophyll *a* concentrations in 1993 than 1992, DOC fluxes at the channel station in April were similar. Average DOC fluxes in the shoals and channel are about 5% and 10% of the average sediment O_2 consumption at these stations, respectively.

Uptake of DOC by sediments has not been reported for other marine sediments (Burdige *et al.* 1992, and references therein). DOC fluxes in other coastal environments range from 0.55 to 6.0 mmol $m^{-2}d^{-1}$ (Burdige *et al.* 1992) and have usually been determined based on porewater gradient calculations. In contrast, the 1992 and 1993 flux core experiments: (1) include bioturbation effects, (2) do not impose diffusion coefficients for flux determinations, but (3) assume that physical stirring of the core overlying water reasonably represents natural hydrodynamic conditions. Uptake of DOC by heterotrophic bacteria has been observed in the open ocean and lakes (Kirchman *et al.* 1991; Sundh 1992). Although DOC uptake by heterotrophic bacteria has not been reported in sediments, this process may be responsible for the observed uptake measurements from San Francisco Bay. In addition, macrofauna, including cnidarians, polychaetes and bivalves, may utilize DOC (Wotton 1991; Neira & Hopner 1994), so uptake by benthic macrofauna represents another potential mechanism to explain our measurements.

Benthic-Pelagic Coupling

Deviations of flux ratios, such as O:N or N:P, from Redfield ratios have often been attributed to processes such as denitrification, or enhanced burial of phosphorus (Nixon 1981; Boynton & Kemp 1985; Hammond *et al.* 1985; Hopkinson 1987; Teague *et al.* 1988). Benthic remineralization of organic matter with a Redfield ratio would lead to benthic fluxes with a ratio of 106 C: 16 N: 1 P. N:Si ratio of benthic fluxes is another useful estimate of the relative regeneration of these nutrients. Uptake N:Si ratios for diatoms are quite variable (Brezinski 1985). In San Francisco Bay, these uptake ratios have been estimated to be about 1:1 (Peterson *et al.* 1985; Hager & Schemel, this volume). In San Francisco Bay, high average O:N ratios (where N is DIN) in the channel (O:N = 66) and in the shoals (O:N = 58) and low N:Si ratios of 0.2 in both channel and shoals suggest the enhanced removal of nitrogen by denitrification. O:P ratios were 932 in the channel and negative in the shoals, while N:P ratios were 14.1 in the channel and negative in the shoals because of net DIP uptake. These ratios are strong evidence for denitrification and phosphorus burial in both channel and shoal sediments. Another explanation for the extreme O:P ratios in sediment fluxes is P remineralization in the water column and deposition of organic material to the sediments that does not have a Redfield ratio.

In other estuaries, benthic NH_4^+ fluxes can support a significant fraction of phytoplankton nitrogen requirements (Kemp & Boynton 1984; Nixon & Pilson 1983; Boynton & Kemp 1985). Benthic NH_4^+ fluxes in South Bay contributed between 0 and 100% of the nitrogen requirements for phytoplankton production in the channel depending on the season (Table 5). In the spring (March-May) when phytoplankton N demands are greatest, NH_4^+ fluxes were able to satisfy between 6 and 12% of phytoplankton N demand. In contrast, NH_4^+ fluxes during the fall period could often meet > 80% of potential phytoplankton nitrogen demand. This contrasts with the pattern observed in the Patuxent River where regeneration could supply over 60% of phytoplankton nitrogen requirements, except in the winter (Kemp & Boynton 1984). Similar estimates can be made for P and Si demand by phytoplankton. DSi fluxes can supply about 22% of the Si requirement of phytoplankton in the spring (assuming a C:Si ratio of 5.3), and about 100% of phytoplankton requirements in most other months. DIP fluxes were quite variable, and able to supply between 0 and > 00% of potential phytoplankton P requirements.

Influence of Macrofauna

A major change in the macrofauna community occurred between the early 1980s and 1990s with the introduction of *Potamocorbula amurensis* (Carlton *et al.* 1990). *Potamocorbula* was able to displace much of the resident bivalve community (Nichols *et al.* 1990), although it seems to have had less of an effect in channel sediments dominated by *Asychis elongata* (J. Thompson, pers. commun.). Stations 27.5 and 28C, occupied during the early 1980s, had very high densities of *Heteromastus filiformis* and *Asychis elongata*, respectively. Burrow densities of 1,000 to 7,000 m^{-2} were reported at these stations (Hammond *et al.* 1985).

At shoal station 1, NH_4^+ flux was correlated with the biomass of mollusks (r = 0.79, $p < 0.001$, Fig 5). This suggests that NH_4^+ excretion or irrigation of sediments by mollusks is a significant component of NH_4^+ flux out of sediments. In the channel during the 1991-1993 sampling period, NH_4^+ fluxes and macrofauna biomass were also correlated ($r = 0.27, p = 0.02$). Similar relations between macrofauna biomass and NH_4^+ flux have been observed in other systems (Blackburn &

TABLE 5. Estimated contribution of benthic flux to nutrient demands of phytoplankton in South San Francisco Bay

Date	*Phytoplankton Production*[1] (mg C m-2d-1)	*Phytoplankton Requirements*[2] ($mmol\ m^{-2}d^{-1}$)			*Benthic Flux* ($mmol\ m^{-2}d^{-1}$)			*Nutrient Supplied* %		
		N	*P*	*Si*	NH_4^+	*DIP*	*DSi*	*N*	*P*	*Si*
Winter (Dec-Feb)										
Feb 80	124	1.57	0.1	2.0	0.1	1.3	7.9	6	> 100	> 100
Feb 81	99	1.25	0.1	1.6	0.6	0.0	2.9	48	0	> 100
Dec 91	78	0.98			-0.74	n.d.	n.d.	< 0		
Feb 92	120	1.51			0.84	n.d.	n.d.	56		
Spring (Mar-May)										
Mar 92	332	4.19			-0.18	n.d.	n.d.	< 0		
Apr 92	1213	13.32	1.0	19.1	0.99	0.1	4.2	6	10	22
May 92	439	5.54			0.53	n.d.	n.d.	10		
Apr 93	678	8.56			1.05	n.d.	n.d.	12		
May 93	303	3.83			0.3	n.d.	n.d.	8		
Summer (Jun-Aug)										
Jun 80	341	4.31	0.3	5.4	1.6	-0.1	5.0	37	< 0	93
Jul 92	497	6.27			1.01	n.d.	n.d.	16		
Fall (Sep-Nov)										
Nov 80	113	1.42	0.1	1.8	1.4	0.1	2.1	98	100	> 100
Sep 84	334	4.22	0.3	5.2	-0.98	-1.1	3.3	< 0	< 0	64
Oct 91	187	2.36			1.84	n.d.	n.d.	78		
Sep 92	205	2.59			-0.06	n.d.	n.d.	< 0		

[1] Net phytoplankton production in photic zone calculated using equation 5 from Cloern (1991a) with average monthly chlorophyll *a* concentrations, a mean attenuation coefficient of 1.8 and average monthly irradiance from Redwood Creek (B. Cole, pers. commun.).
[2] Phytoplankton nutrient requirements were estimated using Redfield rations (C:N = 6.6, C:P = 106, C:Si = 5.3). [3] n.d. = no data.

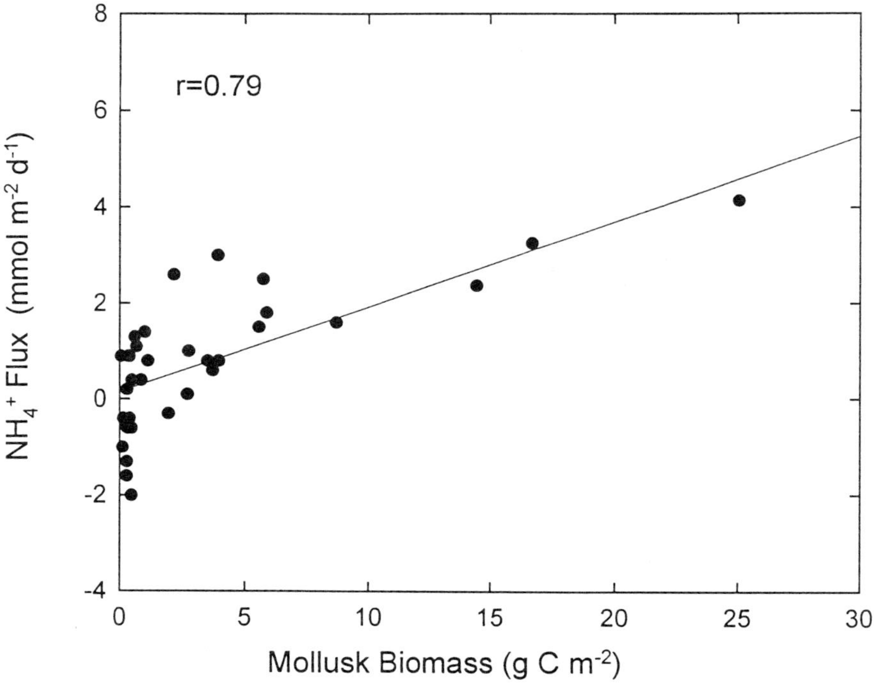

FIGURE 5 Ammonium flux versus mollusk biomass at shoal station (1) between 1991-1992.

Henriksen 1983; Matisoff *et al.* 1984; Lomstein *et al.* 1989; Kristensen *et al.* 1991; Henriksen *et al.* 1993). Excretion of NH_4^+ by mollusks can also have a significant impact on benthic fluxes and porewater profiles (Yamada & Kayama 1987). Porewater profiles of NH_4^+ and radon throughout South Bay change little with increasing depth in sediment, particularly in areas with higher macrofauna densities (Hammond *et al.* 1985; Caffrey 1995). Mollusks dominate benthic biomass in South Bay, although densities of tube dwelling polychaetes like *Asychis elongata* can be high (Nichols & Pamatmat 1988; Hammond *et al.* 1985). During the 1991-1993 study period, mollusks dominated at station 1 in the shoals, while polychaetes dominated station 3 in the channel (Caffrey, unpublished data). Irrigation of tubes and burrows can have a great influence on porewater concentrations (Aller 1982; Hammond *et al.* 1985; Aller 1988).

SUMMARY

In South Bay, the response of benthic fluxes to the spring phytoplankton bloom and phytoplankton deposition seems muted, compared to other estuarine and marine systems (Jensen *et al.* 1990; Grebmeier 1993). South Bay also seems to function quite differently than Tomales Bay where benthic fluxes had a strong seasonal pattern, with highest rates in later summer during peak temperatures

(Dollar *et al.* 1991). Tomales Bay differs from South Bay by having a summer peak in primary productivity, runoff from an agricultural watershed, smaller percentage of shallow water, and no large riverine inputs. These differences may contribute to the differences in benthic flux pattern.

It is difficult to determine whether benthic fluxes have changed significantly over this 13 year period because of the few numbers of measurements. NH_4^+, NO_3^-, DIP and DSi fluxes are all quite similar despite changes in station location and method of sampling. In contrast, SOC is much higher in recent years than in the 80s. Carbon inputs from phytoplankton have not changed during this time period (Table 2). Differences in methodology, specifically the length of incubation, may be responsible for these differences, as well as a difference in biomass of tube-dwelling macrofauna such as *Asychis elongata*.

In 1992, there was no enhancement of sediment oxygen consumption following peak chlorophyll *a* levels, although sediment oxygen consumption did increase following the spring bloom of 1993 (Fig. 2a). DOC fluxes during the 1992 and 1993 spring blooms were similar despite much higher productivity in 1993. In addition, the significant correlations between sediment oxygen consumption, NH_4+ flux and benthic chlorophyll *a* at channel station 3 suggests a link between phytoplankton production and benthic processes (Table 3). However, it seems clear that better spatial and temporal coverage of benthic fluxes is necessary in order to understand the coupling between benthic and pelagic systems in South Bay.

REFERENCES

Aller, R. C. (1982). The effects of macrobenthos on chemical properties of marine sediment and overlying water. Pages 53-102 *in* P. L. McCall & M. J. S. Tevesz, eds., *Animal-Sediment Relations*. Plenum, New York, NY.

Aller, R. C. (1988). Benthic fauna and biogeochemical processes in marine sediments: the role of burrow structures. Pages 301-338 T.H. Blackburn & J. Sorensen, eds., *Nitrogen Cycling in Coastal Marine Environments*. John Wiley & Sons,Chichester, NY.

Berelson, W. M., D. E. Hammond, K. L. Smith, Jr., R. A. Jahnke, A. H. Devol, K. R. Hinga, G. T. Rowe, and F. Sayles. (1987). *In situ* benthic flux measurement devices:bottom lander technology. *Marine Technol. Soc. Jour.* 21:26-32.

Blackburn, T. H., and K. Henriksen. (1983). Nitrogen cycling in different types of sediments from Danish waters. *Limnol. Oceanogr.* 28:477-493.

Boynton, W. R., and W. M. Kemp (1985). Nutrient regeneration and oxygen consumption by sediments along an estuarine salinity gradient. *Marine Ecol. Progr. Ser.* 23:45-55.

Brezinski, M. (1985). The Si:C:N ratio of marine diatoms: interspecific variability and effect of some environmental variables. *Jour. Phycol.* 21:347-357.

Burdige, D. J., M. J. Alperin, J. Homstead, and C. S. Martens. (1992). The role of benthic fluxes of dissolved organic carbon in oceanic and sedimentary carbon cycling. *Geophysical Res. Lett.* 19:1851-1854.

Caffrey, J. M., B. E. Cole, J. E. Cloern, J. R. Rudek, A. C.Tyler, and A. D. Jassby. (1994). *Studies of the San Francisco Bay, California Estuarine Ecosystem. Pilot Regional Monitoring Program Results 1993*. U.S. Geol. Surv., Open-File Rept. 94-82.

Caffrey, J. M. (1995). Spatial and seasonal patterns in sediment ammonium concentrations and regeneration rates in San Francisco Bay, CA. *Estuaries* 18:219-233.

Carlton, J. T., J. K. Thompson, L. E. Schemel, and F. H. Nichols. (1990). Remarkable invasion of San Francisco Bay (California, USA) by the Asian clam *Potamocorbula amurensis*. I. Introduction and dispersal. *Marine Ecol. Progr. Ser.* 66:81-94.

Cloern, J. E. (1982). Does the benthos control phytoplankton biomass in South San Francisco Bay. *Marine Ecol. Progr. Ser.* 9:191-202.

Cloern, J. E. 1987. Turbidity as a control on phytoplankton biomass and productivity in estuaries. *Continental. Shelf. Res.* 7:1367-1381.

Cloern, J. E. (1991a). Tidal stirring and phytoplankton bloom dynamics in an estuary. *Jour. Marine Res.* 49:203-221.

Cloern, J. E. (1991b). Annual variations in river flow and primary production in the South San Francisco Bay Estuary (USA). Pages 91-96 in M. Elliott & J-P Ducrotoy, eds., *Estuaries and Coasts: Spatial and Temporal Intercomparisons*. Olsen and Olsen, Denmark.

Conley, D. J., C. L. Schelske, and E. F. Stoermer. (1993). Modification of the biogeochemical cycle of silica with eutrophication. *Marine Ecol. Progr. Ser.* 101:179-192.

Dollar, S. J., S. V. Smith, S. M. Vink, S. Obrebski, and J. T. Hollibaugh. (1991). Annual cycle of benthic nutrient fluxes in Tomales Bay, California, and contribution of the benthos to total ecosystem metabolism. *Marine Ecol. Progr. Ser.* 79:115-125.

Graf, G. (1992). Benthic-pelagic coupling: A benthic view. *Ocean. Marine Biol. Ann. Rev.* 30:149-190.

Graneli, W., and E. Graneli. (1981). Automatic potentiometric determination of dissolved oxygen. *Marine Biol.* 108:341-348.

Grebmeier, J. M. (1993). Studies of pelagic-benthic coupling extended onto the Soviet continental shelf in the northern Bering and Chukchi Seas. *Continental Shelf. Res.* 13:653-668.

Hager, S. W. (1993). *Dissolved Nutrient and Suspended Particulate Matter Data for the San Francisco Bay Estuary. October 1988 through September 1991*. U.S. Geol. Surv., Open-File Rept. 93-57.

Hager, S. W., and L.E . Schemel. (this volume). Dissolved inorganic nitrogen, phosphorus, silicon in South San Francisco Bay. I. Major factors affecting distributions.

Hammond, D. E., C.Fuller, D.Harmon, B. Hartman, M. Korosec, L. G. Miller, R. Rea, S.Warren, W. Berelson, and S. W. Hager. (1985). Benthic fluxes in San Francisco Bay. *Hydrobiologia* 129:69-90.

Hartman, B., and D. E. Hammond. (1984). Gas exchange rates across the sediment-water and air-water interfaces in South San Francisco Bay. *Jour. Geophys. Res.* 89:3593-3603.

Henriksen, K., T. H. Blackburn, B. Aa. Lomstein, and C. P. McRoy. (1993). Rates of nitrification, distribution of nitrifying bacteria and inorganic N fluxes in northern Bering-Chukchi shelf sediments. *Continental Shelf Res.* 13:629-651.

Hopkinson, C. S., Jr. (1987). Nutrient regeneration in shallow-water sediments of the estuarine plume region of the nearshore Georgia Bight, USA. *Marine Biol.* 94:127-142.

Hunter, Y. R., and J. S. Kuwabara. (1994). Ionic strength and DOC determinations from various freshwater sources to the San Francisco Bay. *Bull. Environ. Contam. Toxicol.* 52:311-318.

Huzzey, L. M., J. E. Cloern, and T. M. Powell. (1990). Episodic changes in lateral transport and phytoplankton distribution in South San Francisco Bay. *Limnol. Oceanogr.* 35:472-478.

Jassby, A. D., J. E. Cloern, and T. M. Powell. (1993). Organic carbon sources and sinks in San Francisco Bay: variability induced by river flow. *Marine Ecol. Progr. Ser.* 95:39-54.

Jensen, M. H., E. Lomstein, and J. Sorensen. (1990). Benthic NH_4^+ and NO_3^- flux following sedimentation of a spring phytoplankton bloom in Aarhus Bight, Denmark. *Marine Ecol. Progr. Ser.* 61:87-96.

Kemp, W. M., and W. R. Boynton. (1984). Spatial and temporal coupling of nutrient inputs to estuarine primary production: The role of particulate transport and decomposition. *Bull. Marine Sci.* 35:522-535.

Kemp, W.M., and W.R. Boynton (1992). Benthic-Pelagic Interactions: Nutrient and Oxygen Dynamics. Pages. 149-222 *in* D. E. Smith, M. Leffler, & G. Mackiernan, eds., *Oxygen*

Dynamics in the Chesapeake Bay: A Synthesis of Recent Research. Maryland Sea Grant College, College Park, MD.

Kirchman, D. L., Y. Suzuki, C. Garside, and H. W. Ducklow. (1991). High turnover rates of dissolved organic carbon during a spring phytoplankton bloom. *Nature* 352:612-614.

Koop, K., W. R. Boynton, F. Wulff, and R. Carman. (1990). Sediment-water oxygen and nutrient exchanges along a depth gradient in the Baltic Sea. *Marine Ecol. Progr. Ser.* 63:65-77.

Koseff, J., J. K. Holen, S. G. Monismith, and J. E. Cloern. (1993). Coupled effects of vertical mixing and benthic grazing on phytoplankton populations in shallow, turbid estuaries. *Jour. Marine Res.* 51:843-868.

Kristensen, E., M. H. Jensen, and R. C. Aller. (1991). Direct measurement of dissolved inorganic nitrogen exchange and denitrification in individual polychaete (*Nereis virens*) burrows. *Jour. Marine Res.* 49:355-377.

Lie, U. (1968). A quantitative study of benthic infauna in Puget Sound, Washington, USA, in 1963-1964. *FiskDir. Skr. Ser. HavUnders.* 14:229-556.

Lomstein, B. Aa., T. H. Blackburn, and K. Henriksen. (1989). Aspects of nitrogen and carbon cycling in the northern Bering Shelf sediment I. The significance of urea turnover in the mineralization of NH_4^+. *Marine Ecol. Progr. Ser.* 57:237-247.

Matisoff, G., J. B. Fisher, and S. Matis. (1985). Effects of benthic macroinvertebrates on the exchange of solutes between sediments and freshwater. *Hydrobiologia* 122:19-33.

Miller-Way, T. (1994). *The Role of Suspension Feeding Macrofauna in Benthic-Pelagic Coupling in a Southeastern Estuary*. Ph.D. Dissertation. Louisiana State University, Baton Rouge, LA.

Neira, C., and T. Hopner. (1994). The role of *Heteromastus filiformis* (Capitellidae, Polychaeta) in organic carbon cycling. *Ophelia* 39:55-73.

Nichols, F. G., and M. M. Pamatmat. (1988). The ecology of the soft-bottom benthos of San Francisco Bay: a community profile. *U.S. Fish Wildl. Serv. Biol. Rept.* 85(7.19), 73 pp.

Nichols, F. H., J. E. Cloern, S. N. Luoma, and D. H. Peterson. (1986). The modification of an estuary. *Science* 231:567-573.

Nichols, F. H., J. K. Thompson, and L. E. Schemel. (1990). Remarkable invasion of San Francisco Bay (California, USA) by the Asian clam *Potamocorbula amurensis*. II. Displacement of a former community. *Marine Ecol. Progr. Ser.* 66:95-101.

Nixon, S. W. (1981). Remineralization and nutrient cycling in coastal marine ecosystems. Pages 111-138 in B. J. Neilson & L. E. Cronin, eds., *Nutrients and Estuaries*. Humana Press, New Jersey.

Nixon, S. W., and M. E. Q. Pilson. (1983). Nitrogen in estuarine and coastal marine ecosystems. Pages 565-648 *in* E. J. Carpenter & D. G. Capone, eds., *Nitrogen in the Marine Environment*. Academic Press, New York, NY.

Peterson, D. H., R. E. Smith, S. W. Hager, D. D. Harmon, R. E. Herndon, and L. E. Schemel. (1985). Interannual variability in dissolved inorganic nutrients in Northern San Francisco Bay estuary. *Hydrobiologia* 129:37-58.

Sundh, I. 1992. Biochemical composition of dissolved organic carbon derived from phytoplankton and used by heterotrophic bacteria. *Applied Environ. Microbiol.* 58:2938-2947.

Tahey, T. M., G. C. A. Duineveld, E. M. Berghuis, and W. Helder. (1994). Relations between sediment-water fluxes of oxygen and silicate and faunal abundance at continental shelf, slope and deep-water stations in the northwest Mediterranean. *Marine Ecol. Progr. Ser.* 104:119-130.

Teague, K. G., C. J. Madden, and J. W. Day, Jr. (1988). Sediment-water oxygen and nutrient fluxes in a river-dominated estuary. *Estuaries* 11:1-9.

Thompson, J. K., F. H. Nichols, and S. M. Wienke. (1981). *Distribution of Benthic Chlorophyll in San Francisco Bay, California. February 1980-1981*. U.S. Geol. Surv., Open-File Rept. 81-1134. 53 pp.

van Duyl, F. C., W. van Raaphorst, and A. J. Kop. (1993). Benthic bacterial production and nutrient sediment-water exchange in sandy North Sea sediments. *Marine Ecol. Progr. Ser.* 100:85-95.

Wienke, S. M., B. E. Cole, J. E. Cloern, and A. E. Alpine. (1992). *Plankton Studies in San Francisco Bay, XIII. Chlorophyll Distributions and Hydrographic Properties of San Francisco Bay, 1991.* U.S. Geol. Surv., Open-File Rept. 92-158. 116 pp.

Wienke, S. M., B. E. Cole, and J. E. Cloern. (1993). *Plankton Studies in San Francisco Bay, XIV. Chlorophyll Distributions and Hydrographic Properties of San Francisco Bay, 1992.* U.S. Geol. Surv., Open-File Rept. 93-423. 175pp.

Wooton, R. S. (1991). Particulate and dissolved organic material as food. Pages 213-261 *in* R. S. Wooton, ed., *The Biology of Particles in Aquatic Systems.* CRC Press, Boca Raton, LA.

Yamada, H., and M. Kayama. (1987). Liberation of nitrogenous compounds from bottom sediments and effect of bioturbation by small bivalve, *Theora lata* (Hinds). *Estuarine Coastal Shelf Sci.* 24:539-555.

ECOSYSTEM MANAGEMENT

SUMMARY OF FEDERAL AND STATE WATER PROJECT ENVIRONMENTAL IMPACTS IN THE SAN FRANCISCO BAY-DELTA ESTUARY, CALIFORNIA

JAMES F. ARTHUR, MELVIN D. BALL, AND SHERYL Y. BAUGHMAN
U.S. Bureau of Reclamation, Mid-Pacific Region
2800 Cottage Way, Sacramento, CA 95825-1898

This paper provides a brief summary of early and post Federal and State water project development and summarizes major changes in physical, chemical, and biological constituents that have occurred as the result of direct water transfer through the Delta (fresh-brackish water portion of the San Francisco Bay-Delta Estuary). Transfer of increasing amounts of Sacramento River water across the Delta channels to the Federal and State water project export pumps in the South Delta over the last 45 years has resulted in several major environmental impacts. A positive impact of water project operations has been the increase in fresh water in many Delta channels during the summer and fall. This increase in fresh water comes from water released from reservoirs to reduce salinity intrusion into the Delta and protect water quality. Most of the negative project-related impacts result from transferring large quantities of water across the Delta in existing channels. The ever-increasing demand for project export water has resulted in net flow reversals during most months of the year in the central and southern Delta. Flow reversal has resulted in: the recycling of large quantities of salt from the San Joaquin Valley back into the valley; scouring of Delta channels; increases in trihalomethane (THM) precursors from Delta sources in export water designated for municipal use; flushing of Delta aquatic habitat resulting from decreased residence times; and entrainment of plankton and various life stages of fish in the project intakes. Other nonproject impacts are mentioned. Potential short- and long-term structural and operational management solutions are also discussed.

This paper provides a brief historical perspective of water development as it relates to the Delta and summarizes major physical, chemical, and biological trends associated with Federal and State water project development in the Delta. The discussion includes our views on specific beneficial and detrimental effects associated with the direct transport of project water through Delta channels. Potential short- and long-term solutions are suggested that may better balance water allocation between environmental, agricultural, and urban needs.

Managing California's water supply has become increasingly difficult as the demand for water escalates. Two-thirds of the State's fresh water originates north of the Delta, while about 70 percent is used south of the Delta. The problem is compounded by the fact that most precipitation occurs during only half of the year – late fall through early spring. Over the last 50 years that the U.S. Bureau of Reclamation's (Reclamation) Central Valley Project (CVP) has operated and the 30 years that the California Department of Water Resources (Water Resources)

[1] DISCLAIMER. The opinions and conclusions expressed in this paper are those of the authors. They do not represent the official position of the Bureau of Reclamation, a Department of the Interior agency, or the agencies participating in the Interagency Ecological Program for the Sacramento-San Joaquin estuary.

State Water Project (SWP) has operated, there has been a wide range of water year types, including several very wet years in a row (*e.g.*, 1969-71 and 1982-84) to as many as six continuous drought years (*e.g.*, 1987-92).

The two major sources of freshwater inflow to the Delta are the Sacramento River, which enters the Delta from the north, and the San Joaquin River, which enters the Delta from the south (Fig. 1). The area referred to as the Delta includes about 700,000 acres and nearly 700 miles of waterways. About 150 years ago, the Delta consisted of large areas of marshlands and meandering river channels. Since then, the Delta has undergone extensive structural and ecological change.

Levees in the Delta, originally constructed to allow agricultural production on fertile peat soils, are critical to maintaining a system of channels conveying river water to repel salinity and protect water quality for in-Delta use and project exports. The interior of many Delta islands has subsided from 10 to 20 feet below sea level due to peat oxidation resulting from agricultural practices. Levee failures and island flooding occur frequently because many of the Delta levees were not designed or constructed to handle the present differences in elevation between the channels and interior. A major earthquake, high tides combined with strong winds, or a flood could result in catastrophic levee failure and a corresponding loss of a major portion of the State's water supplies.

To provide a dependable water supply for California, most of the major rivers and streams that flow to the estuary have been dammed for flood control, power generation, recreation, and local and out-of-basin water supply. Presently 14 major water entities, in addition to the Federal and State water projects, have reservoirs with capacities exceeding 100,000 acre-feet. The total cumulative storage with date of construction is illustrated in Fig. 2. Currently Reclamation's CVP, the largest water project in California, has a storage capacity of about 8.8 million acre-feet. The SWP has a storage capacity of 3.5 million acre-feet. Thus, a total of about 58 percent of water storage capacity is under Federal and State control. Other major upstream water diversion facilities have a total storage capacity in excess of 9 million acre-feet, representing about 42 percent of the combined capacity in the estuary's drainage basin. Numerous other small water users, with storage capacities under 100,000 acre-feet, also divert water that historically would have flowed to the estuary.

Water diverted to the high-use areas south of the Delta and to the Bay area presently flow directly through existing Delta channels to the Federal and State export facilities. Water began to be moved across the Delta in 1937, with construction of Reclamation's Contra Costa Canal. In the early 1950s, Reclamation constructed the Delta Cross Channel to facilitate water movement from the Sacramento River to its pumping plant near Tracy, in the southern Delta. The Tracy Pumping Plant and the Delta Mendota Canal were completed in 1951, with a total export capacity of 4,600 cfs. The Tracy Fish Collection Facility was completed in 1956.

Ability to divert water out of the Delta was about doubled in 1967 with completion of the Water Resources' Harvey O. Banks Delta Pumping Plant and John E. Skinner Fish Protective Facility. The State's export facility is about 1 mile north of Reclamation's facilities and transports water south via the California

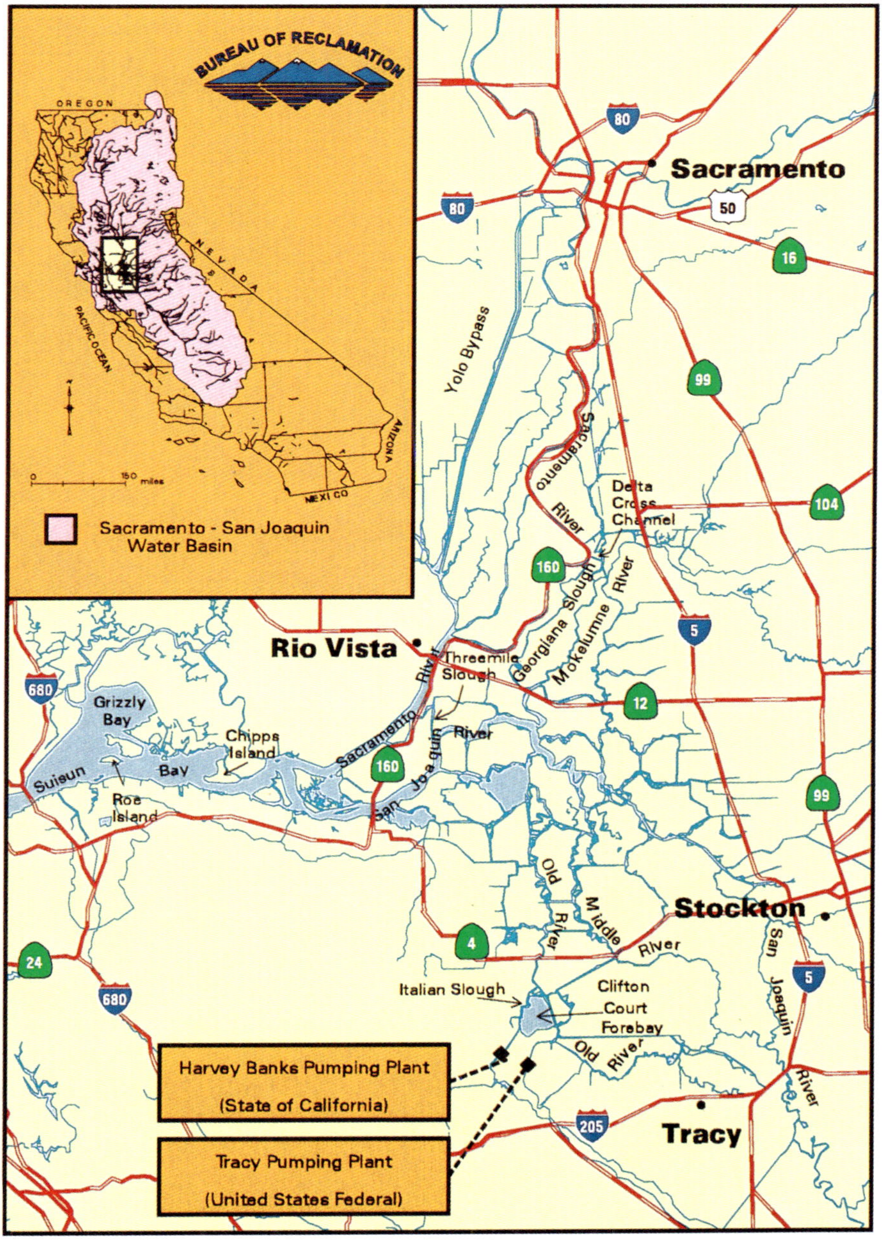

FIGURE 1. Delta portion of the Sacramento-San Joaquin Estuary.

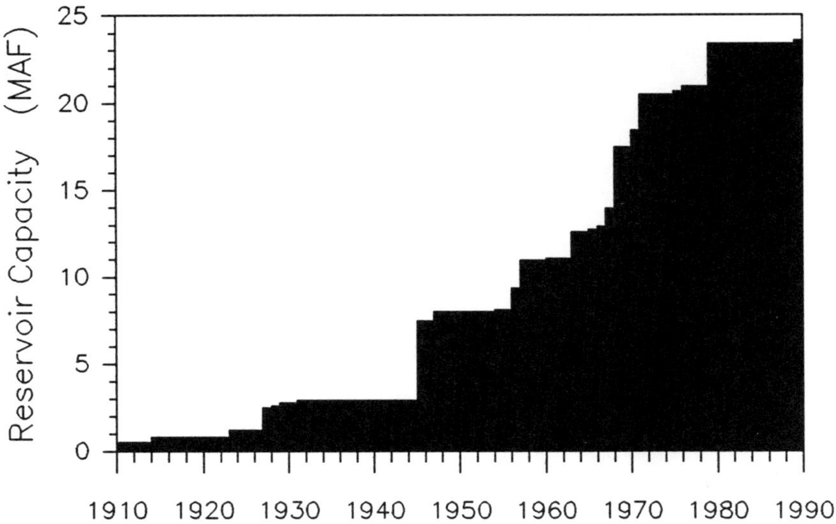

FIGURE 2. Cumulative capacity of selected Central Valley reservoirs. (Obtained from Water Resources.)

Aqueduct and to the Bay area via the South Bay Aqueduct. Banks Pumping Plant has an export capacity of about 10,300 cubic feet per second, but is limited to about 6,400 cubic feet per second under present regulations.

The CVP and SWP have been instrumental in the economic growth of California. About 20 million Californians receive at least part of their water from the Delta. The project reservoirs are multipurpose — providing water for municipal, agricultural, and industrial uses; flood protection during high runoff periods; hydroelectric power generation; various environmental benefits — and are one of the largest sources of recreation in California.

The water projects have also caused major environmental problems. Upstream of the Delta, many of the salmon and steelhead spawning grounds have been blocked and submerged by construction of federal, state, and local dams. Water temperatures, critical for salmon migration and survival, are elevated by reservoir storage. Much of the natural sediment, which has numerous positive environmental impacts, is trapped in reservoirs. Although not the direct result of water projects, much of the natural vegetation, used as habitat along water channels and in the estuary, has been eliminated and replaced by levees to reclaim farmland and for flood control.

The issues in the estuary are complex. Analysis and discussion in this paper are related to major environmental impacts in the Delta that are related to water projects. We believe the greatest negative impacts are largely associated with altered flow patterns and hydrodynamic changes in the Delta resulting from direct transport of CVP/SWP export water through existing Delta channels. Impacts include: reductions in sediment and associated increases in water transparency; recycling of Delta and San Joaquin River salts into the San Joaquin Valley;

additions of Delta-derived organic materials that form THM precursors in water diverted from the upper estuary; and significant alterations of Delta habitat (via decreased residence time and alteration of flow patterns) with corresponding reductions in some fish and their food web.

HISTORICAL PERSPECTIVE

In 1941, Reclamation investigated several alternatives for the present Delta Cross Channel (Carr 1941). At that time the preferred plan included an isolated channel with fish screens at the head. The proposed Delta Cross Channel extended from Freeport along the eastern Delta to the proposed head of the DMC, at that time south of Tracy. This proposed canal was even longer than the Peripheral Canal, recommended later in the 1960s. Following World War II, the Federal water project operators decided to transport water south through natural Delta channels, a decision based strictly on economics.

The actual Delta Cross Channel eventually constructed is only 2 miles long and is used to connect the Sacramento River and Mokelumne River systems. This facility increases the flow exchange out of the Sacramento River to the central Delta-San Joaquin River system and facilitates water delivery to the Federal and State export pumps.

As the Federal export facilities were put on line, it became evident that using the Delta's natural channels for water transfer would create several major environmental problems. Consequently, during the early 1960s, Federal and State agencies formed the Interagency Delta Committee (IDC) to study plans for a joint water transfer facility to the southern Delta (IDC, 1965). Four plans were considered:

- The then-existing and present-day operation using hydraulic barriers;
- A physical barrier (a dam with flood gates and ship locks at Chipps Island);
- Delta waterway control (isolated transfer facility through the central Delta); and
- The Peripheral Canal (isolated channel around the eastern and southern Delta).

The Peripheral Canal plan was selected as the most environmentally sound method of transporting water south. IDC water project biologists, hydrologists, and engineers who had studied the four options for water transfers predicted many of the negative impacts observed today by continued use of a hydraulic barrier to protect project export water transfers, as currently done. Major engineering and environmental studies were initiated in the 1970s to determine how to best construct and operate the Peripheral Canal. However, the plan was not approved by California voters. Work on the Peripheral Canal was canceled after its 1982 defeat at the polls. The lack of a Peripheral Canal, which was designed to solve many of the present environmental problems, has resulted in a stalemate on how to best protect the environment while still exporting water for urban and agricultural uses.

Several alternative Delta water transfer plans have been proposed. One plan called for enlargement of natural Delta channels in the northern and eastern portion of the Delta, without fish protective measures. Installation of channel barriers and a siphon under the San Joaquin River were also evaluated by Water Resources and Reclamation. These plans were part of Senate Bill 1369, which was shelved in 1984 due to a lack of overall support. In cooperation with other agencies and local interests, Reclamation and Water Resources have continued to study structural and water management plans to improve water transport within the Delta.

Numerous legal actions and major events in the Delta that have taken place are summarized in Table 1 (Argent 1989). The intent of all the recent actions and standards has been to protect and/or enhance estuarine resources, including endangered and threatened species. However, we believe that none of these actions directly considered the major cause of the problem — the use of Delta channels to transport water to the export facilities.

Finding an environmentally sound water transfer system is complicated by the complex and dynamic nature of the estuarine biotic community. In addition to the Federal and State water projects, several nonproject factors may be negatively affecting the estuarine food web and fisheries. Although the nonproject impacts have been recognized and discussed for many years, they have been impossible to quantify and difficult to control. Nonproject impacts include: toxicity, introduction of exotic species, entrainment of fish and their food into Delta diversions, illegal take of fish, and direct and indirect dredging effects. Cumulatively, nonproject impacts are clearly having detrimental effects on the estuary's biota, and these impacts should be understood and controlled to the extent possible.

Many alternatives being considered that would significantly enhance environmental resources under the present Delta configuration require reductions in water project exports and corresponding increases in Delta outflow. The consequences of reducing water supply could be devastating to California's economy. We believe some type of isolated water project transfer facility needs to be evaluated as a long-term solution for protecting both water supply and environmental resources.

METHODS AND MATERIALS

This paper uses existing data to summarize major physical, chemical, and biological trends in the San Francisco Bay-Delta estuary since the Central Valley Project (1953) and State Water Project (1968) began exporting water out of the Delta. Where data are available (some flow data extend back to 1930), comparisons are made to pre-project conditions. Most of the information comes from data records collected under the auspices of the Interagency Ecological Program (IEP), formerly the Interagency Ecological Study Program.

The Interagency Ecological Study Program was formed in 1970 via a cooperative agreement between the U.S. Bureau of Reclamation, U.S. Fish and Wildlife Service (FWS), California Department of Fish and Game (Fish & Game), and California Department of Water Resources. Since then, five additional agencies have become members: U.S. Geological Survey, U.S. Army Corps of Engineers, U.S. Environmental Protection Agency, State Water Resources Control Board (State Board), and National Marine Fisheries Service. A primary objective of the present Interagency Ecological Program is to collect and evaluate environmental data for the estuary, particularly in regard to the potential impact of water project operations. An overview of all the data collected by each program element can be found in the Interagency Program's *Data Summary Report* (1993).

TABLE 1. Chronology of San Francisco Bay-Delta Estuary

1772	First recorded sighting of the Delta by Fray Juan Crespi and Captain Pedro Farges.
1849	Settlers begin arriving in the Delta region to farm its rich land, one year after the discovery of gold in California.
1861	California Legislature authorizes the Reclamation District Act allowing drainage of Delta lands and construction of sturdier levees to protect the area from flooding.
1914	Passage of the Water Commission Act.
1930	Completion of a comprehensive State Water Plan calling for major transfers of northern California water to the Central Valley.
1940	Contra Costa Canal, the first unit of the federal Central Valley Project (CVP), is completed and use of Delta channels to convey water for export begins.
1944	Completion of Shasta Dam and reservoir as the key feature of the CVP, adding water to Delta channels during naturally low-flow periods.
1951	The State Feather River Project (now State Water Project) authorized by the California Legislature.
	Diversions from the Delta begin for the CVP's Delta-Mendota Canal.
1959	State Delta Protection Act approved by the state Legislature.
1960	California voters approve the Burns-Porter Act to assist in financing the State Water Project (SWP), including Delta facilities for water conservation, water supply in the Delta, transfer of water across the Delta, flood and salinity control and related functions.
1961	State Department of Water Resources (DWR) initiates the Interagency Delta Committee, consisting of DWR, U.S. Bureau of Reclamation (Bureau) and the U.S. Army Corps of Engineers, to find solutions to Delta problems.
1965	Interagency Delta Committee releases its report which contains a variety of proposals designed to offset adverse effects of increasing use of water from the Delta. The proposal included a plan for a Peripheral Canal.
	DWR officially selects the Peripheral Canal as the Delta water facility of the SWP.
1967	Construction begins of SWP's Clifton Court Forebay.
	Initial diversions begin from the Delta to the California and South Bay aqueducts of the SWP.
1969	U.S. Department of Interior adopts the Bureau's Peripheral Canal Feasibility Report and recommends the project be a joint-use facility of the CVP and SWP with shared costs.
1970	State endorses a joint-use facility, urges Congressional authorization on the condition that Delta water requirements have priority over exports.
1971	State Water Resources Control Board (SWRCB) adopts its Delta Water Rights Decision 1379 establishing water quality standards to be met by both CVP and SWP.
1972	DWR announces that the state is proceeding with planning of the Peripheral Canal as a state-only project.
	First SWP deliveries to souther California.
1973	Delta Environmental Advisory Committee forms. As part of a three-point solution, the committee concludes in January 1977 that a properly designed, built, and operated facility such as the Peripheral Canal is necessary.
	State Department of Fish and Game (DFG) concludes a 10-year study which probes the Delta's problems. Report concludes a Peripheral Canal is the most desirable plan for a Delta water facility.
1974	DWR, DFG, the Bureau, and U.S. Fish and Wildlife Service sign a statement of intent that the agencies will provide protection of fish and wildlife resources in the Delta.
1975	DWR calls for complete reappraisal of alternative possibilities for Delta water management problems.
	U.S. Department of the Interior releases its opinion that the Federal Water Pollution Control Act does not require the Bureau to release water for salinity repulsion in the Delta.

TABLE 1. Chronology of San Francisco Bay-Delta Estuary (*continued*)

DWR releases a legal opinion stating the Federal Water Pollution Control Act *does* apply to the operation of the CVP.

1977 After reviewing nearly 40 alternative courses of action in the Delta, DWR reaffirms that building the Peripheral Canal is the best answer to Delta problems.

1978 SWRCB issues SWP-CVP Water Rights Decision 1485 regarding CVP and SWP operation to provide water quality control in the Delta.

1979 Secretary of the Interior Cecil D. Andrus announces the CVP will be operated to voluntarily meet state Delta water quality standards (Decision 1485) until legal questions of mandatory federal compliance are resolved. Negotiations proceed between DWR and the Bureau.

Senator Ruben Ayala introduces Senate Bill (SB) 200 specifying the Peripheral Canal as the Delta transfer facility, not requiring federal participation.

1980 Voters pass Proposition 8 insuring more Delta protection unless SB 200 is defeated.

1982 Voters defeat Proposition 9, which includes the Peripheral Canal SB 200 package, by a 3-2 margin. Northern Californians vote 9-1 against SB 200 and Southern Californians vote 3-2 for the bill.

1983 DWR releases a report analyzing four through-Delta alternatives to the Peripheral Canal.

1984 Attention focuses on Governor Deukmelian's through-Delta plan utilizing natural Delta channels and reconstructed levees. By June "Duke's Ditch" (SB 1369) is shelved.

1986 Congress passes DWR and Bureau historic accord, the Coordinated Operation Agreement (COA).

California Supreme Court affirms state Court of Appeal ruling (Racanelli decision) strengthening powers of the SWRCB to protect all uses of Delta water and potentially San Francisco Bay.

DWR and DFG sign Delta Pumping Plant fishery mitigation agreement for direct fish losses.

1987 The SWRCB begins the San Francisco Bay/Sacramento-San Joaquin Delta Estuary Hearing (Bay-Delta Proceedings)

1988 Senate Bill 34 provides S120 million over 10 years for DWR to rebuild levees, enlarge channels and to help reclamation districts make levee improvements.

Suisun Marsh facilities (tide gates) begin operation to provide water quality for waterfowl protection.

Construction begins on four additional pumping units at the Delta Pumping Plant.

An engineering study by the California Urban Water Agencies examines options for improving drinking water quality for users of Delta water.

1991[1] In May the SWRCB adopts and submits a Bay-Delta water quality plan to the Environmental Protection Agency (EPA).

In September, EPA disapproves the fish and wildlife objectives.

1994[1] In January, EPA publishes draft standards in the Federal Register.

In March, SWRCB commences reviewing requirements of the 1978 and 1991 Bay-Delta plans.

On December 15, an historic agreement is signed by Federal, State, and special interest groups endorsing a revised Bay-Delta water quality plan. SWRCB releases the draft water quality control plan.

[1] Items added by Reclamation staff. (From Argent 1989).

Most flow evaluations used data from Water Resources' DAYFLOW model. Data prior to 1978 were hand calculated, and data prior to 1955 used a different calculation for some of the flows due to unavailability of input flows. The five DAYFLOW parameters used in this report are:

- Delta Inflow — Sum of Delta inflows from the Sacramento, San Joaquin, Cosumnes, and Mokelumne Rivers, plus miscellaneous stream flows;
- Exports — Sum of State Water Project, Central Valley Project at Tracy, Contra Costa Canal, and miscellaneous diversions;
- Delta Outflow — Delta inflow minus exports and Delta consumptive depletions;
- Reverse Flow (Qwest) — Delta inflows plus Georgiana Slough and Delta Cross Channel minus exports and 65% of Delta depletions; and
- Percent Diverted — Exports plus depletions/inflow times 100.

It should be noted that Delta depletion is an estimated total for about 1,800 miscellaneous diversions, evaporation, and rainfall, varying weekly but not between years.

As part of the Interagency Ecological Program, Fish and Game conducts several long-term monitoring programs in the estuary. Three were originally established to detect striped bass at different life stages:

- Egg and larval sampling at 30 sites in the Delta and Suisun Bay;
- Young-of-the-year (Y-O-Y) tow-net survey at 30 sites in the Delta and Suisun Bay (not necessarily the same as egg and larval sites); and
- Midwater trawl at 180 sites in the Delta, Suisun Bay, and San Pablo Bay to index striped bass abundance in the fall of their first year.

Other species are also captured in each of these studies. Fish and Game also studies food organism abundance at 50 sites in the Delta, Suisun Bay, and San Pablo Bay (*Neomysis*/zooplankton survey). These data bases were used to examine the biological impacts of project exports.

Analytical tools used for evaluations in this report were SAS, a statistical and graphical analysis system, and ArcInfo, a Geographical Information System software.

RESULTS AND DISCUSSION

We first characterized water project development by examining long-term trends in outflow, inflow, and exports. The impacts on estuarine ecology of through-Delta water transfers are divided into three categories — physical, chemical, and biological. Where data are available and appropriate, we based our analysis of project impacts on major human alterations in the Delta during five periods:

- Prior to 1944 — Before completion of the first large Federal and State dams;
- 1944 to 1951 — After Shasta and Friant dams were in place but prior to significant Federal exports from the Delta;
- 1951 to 1967 — A period of Federal exports only;
- 1968 to 1977 — A period of both Federal and State exports and when State Board Decision 1379 standards were in place; and
- 1978 to 1992 — When State Water Resources Control Board Decision 1485 standards were in place.

Water Project Development

The Federal, State, and local water projects have altered the Delta environment in two major ways. First, Federal and State dams, as well as numerous nonproject diverters upstream of the Delta, have changed the timing and quantity of inflow reaching the Delta. Second, pumping by the two water project export facilities in the southern Delta has changed the magnitude and direction of flow within and through the Delta. The following discussions look at changes in these flows over time.

Delta Outflow

The term "Delta outflow" is used to characterize river discharge from the Sacramento-San Joaquin River basins into Suisun Bay. Historically, Delta outflow has been calculated for net river discharge at the downstream end of the Delta at Chipps Island. Direct measurement of Delta outflow has been impossible because of large tidal flows (peaking near 300,000 cfs at Chipps Island) relative to net river discharge.

Data using different calculations for two periods (1930-1954 and 1955-present) from the DAYFLOW model were combined to provide a continuous record of total annual Delta outflow from 1930 through 1993, as illustrated in Fig. 3. Also indicated in this figure are major changes to the system.

Total annual Delta outflow has been highly variable over the relatively short period of record and is directly related to water year type: wet, above normal, below normal, dry, and critical based on unimpaired flow (as defined by the State Board). On an annual basis, there is no apparent major relationship to human events. Since 1930, both the low and high annual outflows occurred in the last decade. The lowest annual outflow, less than 4 million acre-feet, occurred during the 1977 water year. By far the highest annual outflow, nearly 65 million acre-feet, occurred during the

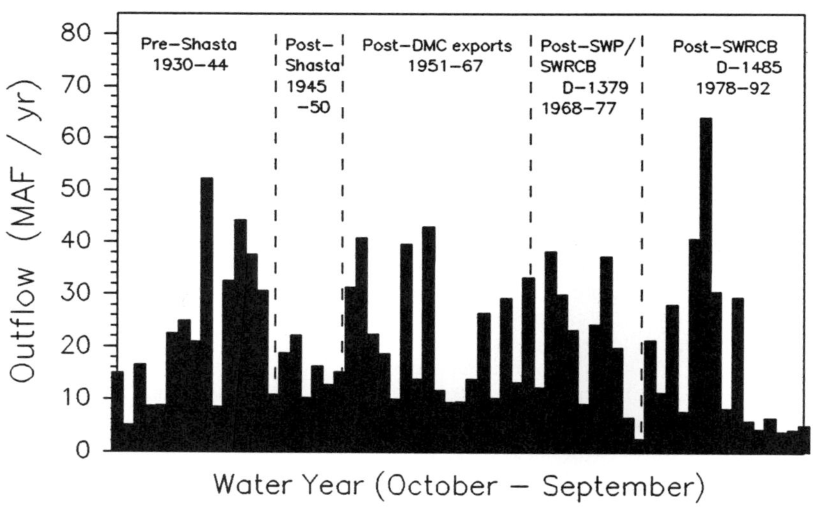

FIGURE 3. Total Delta outflow by water year and periods of major manmade impacts.

1983 water year. Although there were some low outflow years from 1928-1934, the longest duration of drought (below-normal water years) occurred from 1987 to the present. Although major changes in Delta outflow resulting from human impacts are not apparent on an annual basis, evaluating seasonal trends in Delta inflow is enlightening.

Delta Inflow

During the 1955-1993 period when water projects were exporting water, the Sacramento River averaged 84 percent of the total fresh water inflow, the San Joaquin River averaged 13 percent, and the eastern peripheral streams contributed 3 percent. Greater water project impacts occurred in the Delta in the below-normal water years. For the same period and considering only below-normal water years, the Sacramento River averaged 88 percent of the total inflow, the San Joaquin River 10 percent, and the eastside streams only 2 percent.

As the demand for water increased, more large storage reservoirs (Fig. 2) and major delivery systems were built. This was followed by an increase in the quantity of water exported from the Delta and a change in the seasonal pattern and quantity of inflow to the Delta. Since the greatest impacts occur in drier years, only below-normal, dry, and critical water year were used to calculate the changing inflow patterns, 1930-1992 (Fig. 4). Presented are long-term averages for the major manmade changes in the system.

The general trend indicates that year-round Delta inflows were actually lower prior to construction of Shasta Dam until major exports began in the 1950s. Summer and fall inflows have increased especially during the 1951-1992 period because the water projects began exporting water from the Delta. Conversely, winter and spring have been reduced about 40 percent since the State began exporting in 1968 and since Decisions 1379 and 1485 were implemented. However, Delta inflows have become more stable during the spring through fall since exports began.

Water Project Exports

The Federal export pumps have historically been operated on a continuous, 24-hour basis, drawing water directly out of Old River. In about 9 out of 10 years, water in the San Joaquin River is typically pulled into upper Old River and exported by the Federal export pumps. Little of this water reaches the State export pumps located farther downstream. Thus, most of the San Joaquin River flow ends up being pumped back into the San Joaquin Valley.

The balance of water exported by the Federal pumps is Sacramento River water, flowing upstream through the central Delta out of the lower San Joaquin River via Old River and Middle River. During below-normal water years, most of the water throughout the Delta is Sacramento River water. As an illustration, the San Joaquin River may be flowing at 2,000 cubic feet per second, Tracy pumping at 4,000 cubic feet per second; thus, about half of the exported water would be from the Sacramento River.

The State's export pump intake is just north of the Federal intake. Water is drawn from Old River and Middle River into a small regulating reservoir, Clifton Court

Forebay. Unlike Federal export operations, which continuously pump water, the forebay intake gates are generally open only on higher tides (about half the time). The forebay is operated in this manner to allow much of the pumping to be done at night, using off-peak power rates. Opening the gates on higher tides results in nearly all water exported (when San Joaquin River inflow is low) to be of Sacramento River origin via lower Old River, Middle River, and the central Delta. This type of operational scenario also results in large quantities of water being drawn into Clifton Court Forebay over relatively short periods. Arthur (1987)

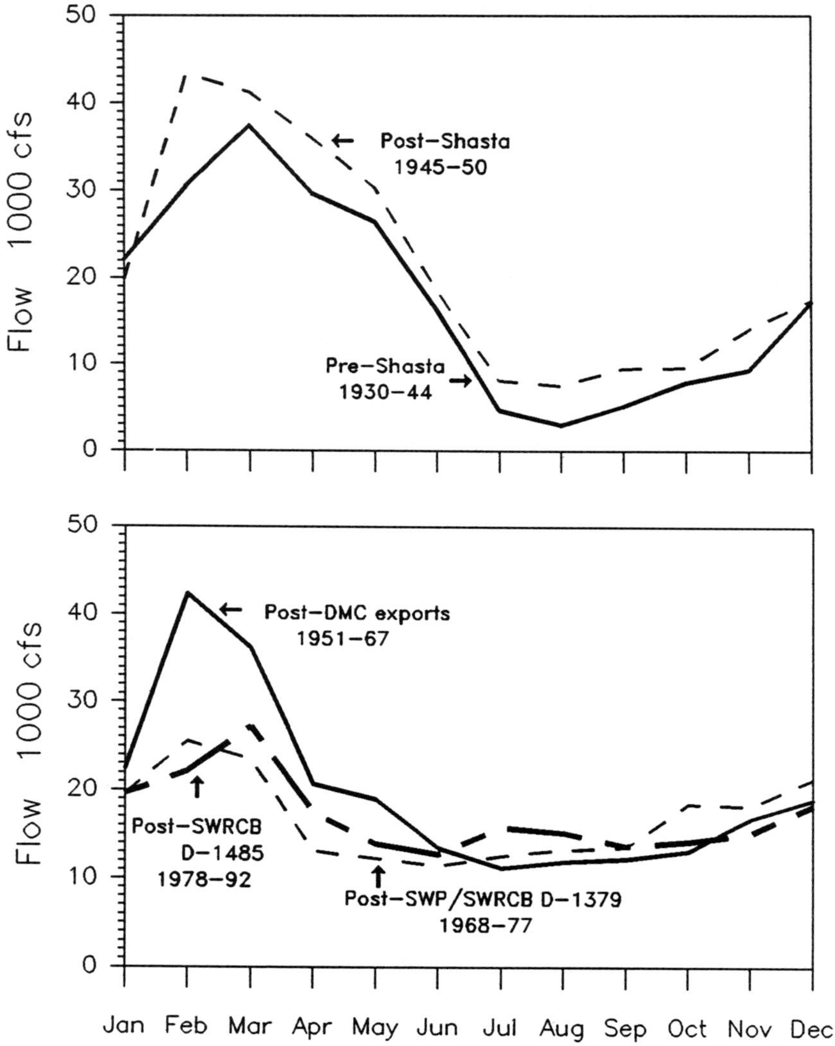

FIGURE 4. Change in average monthly Delta inflows in below average water years for major manmade impacts.

calculated historical export flows based on operations of the radial gates into Clifton Court Forebay, referring to the calculated flows as "instantaneous flow." Typically, actual "instantaneous" export flows are double those reported as daily export flows in DAYFLOW. At times, instantaneous export flows are in excess of 25,000 cubic feet per second, with corresponding large increases in flows and decreased residence times in Old and Middle Rivers.

To evaluate changes in exports, the 1951-1992 period (following the start of Delta-Mendota Canal exports) was subdivided into the applicable periods (Fig. 5). As expected, total exports have increased as demands have increased over the years. In the earlier years, Delta exports peaked in the summer, with minimal amounts pumped in the winter. After the State export pumps came on-line in 1968, in addition to the increase in total pumping, the monthly variation in pumping has been reduced. Part of the year-round increase in exports was made possible by construction of reservoirs south of the Delta such as San Luis and other offstream storage reservoirs, which allow storage of high winter flows.

With a greater quantity of water being transported across the southern and central Delta to the export pumps, channel residence times have decreased. As discussed later, we believe this has resulted in increased flushing of pelagic (open water), food chain and fish habitat in the Delta.

Physical Impacts

Relating physical impacts to biological impacts is not simple. The timing and location of the biological life stage being evaluated, relative to the area of influence, is critical. Evaluating the impacts of water project export pumping on the fate of a particular life stage of a fish is further complicated by constantly changing flows — flows that determine where an organism is transported. At a given export rate,

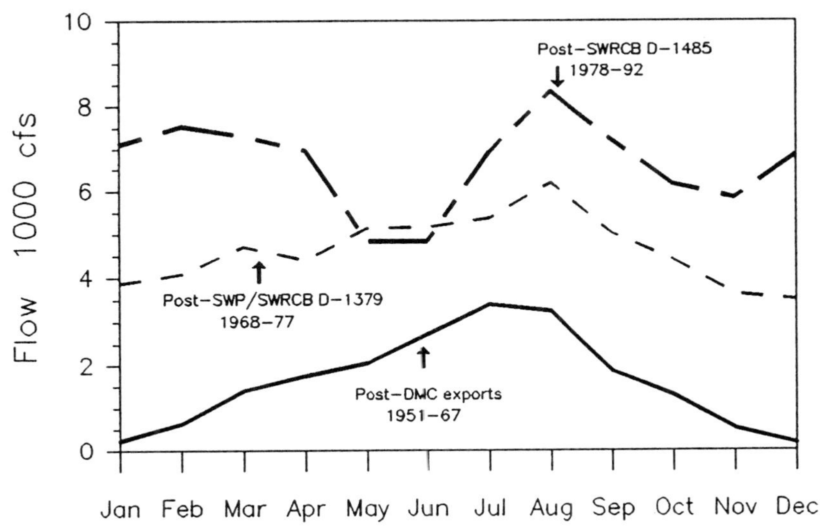

FIGURE 5. Change in average monthly Delta exports in all water year types for major manmade impacts.

impacts can be quite variable, depending on flows in the Sacramento and San Joaquin Rivers relative to the export rate being evaluated. Thus, the net transport rates of suspended materials or organisms are dependent on the interaction of varying streamflows in 700 miles of waterways; ocean water intrusion; agricultural, municipal, and industrial Delta entrainments; and water project export rates plus other factors such as toxins. Understanding the basic hydrodynamic flow regime in the Delta is essential to evaluating biological and chemical interactions related to flow.

Hydrodynamics

Over the past 50 years, several numerical and physical models have been developed that are driven primarily by differences in stage between areas. These models have provided a general understanding of flow patterns in the Delta, but information on specific net flow rates in the various channels and sloughs requires further field verification. Actual measurements of flow in key channels and diversion points have been difficult and limited over the years; however, recent technological advances now allow us to obtain continuous measurement of net channel flow and direction over extended periods. Some of this new technology is used in our analysis. In attempting to understand the Federal and State water project impacts on the fisheries, it is important to understand the changes they have caused to net Delta flow patterns.

Sacramento River System. A conceptual schematic of typical net flow patterns before and after the water projects began exporting water out of the Delta is illustrated in Figs. 6A and 6B. Prior to water project exports (or during very high San Joaquin River inflows under present conditions), all net flows in the major open channels were toward the Pacific Ocean (Fig. 6A). However, a typical Delta flow pattern under lower flow condition since the water projects began exporting shows flow reversals in some of these channels as Sacramento River water is pulled toward the pumps (Fig. 6B). Sacramento River water is diverted into the central and southern Delta at three locations: Delta Cross Channel; Georgiana Slough, a natural channel immediately downstream from the Delta Cross Channel; and Threemile Slough, several miles downstream of Rio Vista.

As mentioned earlier, Sacramento River water can also move south through Sherman Lake or around Sherman Island under conditions of high export pumping and low San Joaquin River inflow to the Delta.

A Reclamation study (Collins *et al.* 1992) suggests that flow rates from the Sacramento River to the Delta Cross Channel and Georgiana Slough are primarily driven by differences in stage and tidal phase between the Sacramento and San Joaquin Rivers.

San Joaquin River System. In most years, San Joaquin River flows are relatively low compared to flows in the Sacramento River. Part of the San Joaquin River splits downstream of where it enters the Delta to form upper Old River (Fig. 6B). Under conditions of high exports, a relatively large portion of this flow is drawn toward the Federal and State export intakes north of Tracy. Much of the San Joaquin River flow not drawn off at upper Old River continues downstream in the main channel, is intercepted, and is drawn south to the export facilities via lower Old River, Middle River, and smaller channels (how the water enters the export facilities is discussed in greater detail in the section on water project exports).

San Joaquin River flow not diverted to the export pumps continues downstream past the head of Old River to Stockton. In low-flow years, the San Joaquin River flow often reverses in this area, with net flow being upstream. Water Resources places a rock barrier at the head of Old River in many years during the early fall to return positive downstream flow in that reach, which assists upstream migration of adult salmon. The Old River barrier also helps to reduce the annual dissolved oxygen sag in the Stockton area. The problem is caused by settling of organic sediments and discharges from the municipal waste treatment plant in the Stockton Ship Channel and turning basin.

The treatment plant was upgraded in 1979, to provide tertiary treatment during the late summer-early fall canning season. At the same time, New Melones Reservoir (Reclamation) was completed and Stanislaus River flow releases to the San Joaquin River flow were reallocated. This combination has greatly improved dissolved oxygen levels and allowed upstream fall migration of salmon. However, in recent dry years, the rock barrier has also been installed in the spring to reduce diversion of downstream migrating salmon smolts to the export facilities, enhancing their survival.

The authors believe the cause of the most serious environmental impacts results from direct diversion of water out of southern Delta channels. This method of diversion both lowers the quality of export water and negatively affects estuarine fish and their food web. These impacts are mainly the result of reversal of net nontidal flows that occur to varying degrees throughout the San Joaquin River system of the western, central and southern Delta. For this evaluation, the term "reverse flow" is used to describe flow conditions when net nontidal streamflow direction is opposite of pre-project flows.

To estimate the extent of flow reversal, Water Resources staff adds together all Delta inflows, then subtracts 65% the Delta depletion and project export pumping. A negative number signifies net flows calculated to be reversed; a positive number signifies net downstream flows. We must emphasize that Qwest (estimate of net flow in the lower San Joaquin River) is a calculated value and does not represent actual flows, although it may provide reasonable estimates of net flows for periods of a month or more.

Another consideration is the effect of spring and neap tides on net flow in the lower San Joaquin River. Several years ago, Water Resources (Chung, pers. commun.) estimated that an approximate 4,000 cubic foot per second net flow, upstream or downstream, is attributable to spring and neap tides. The net flow (non-tidal) direction is upstream during spring tides and downstream during neap tides. In other words, the Delta is filling and emptying on about a 2-week cycle. This neap/spring cycle becomes increasingly important relative to salinity intrusion and standards during periods of low net outflow and high water demand. Neap/spring tidal conditions are not factored into the Qwest calculation and under

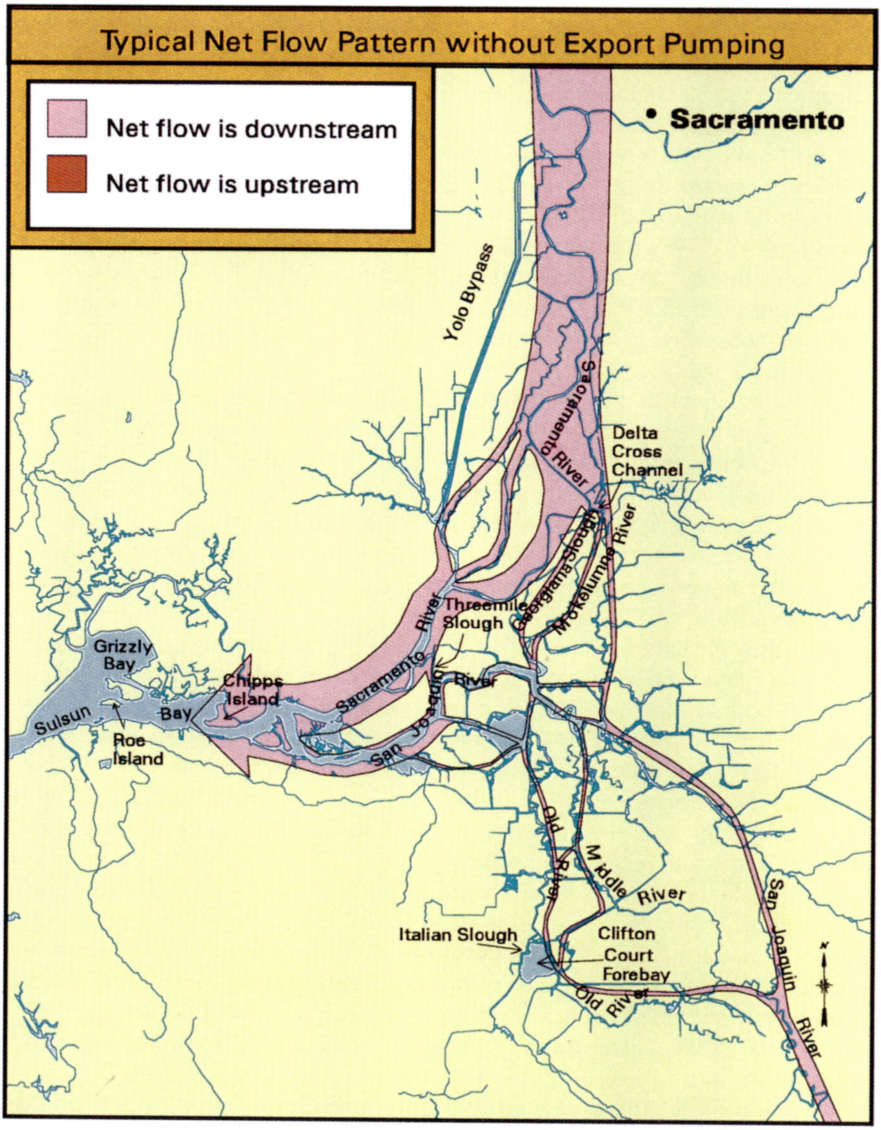

FIGURE 6A. Net flows in the Sacramento-San Joaquin Delta Estuary under low inflow conditions without export pumping.

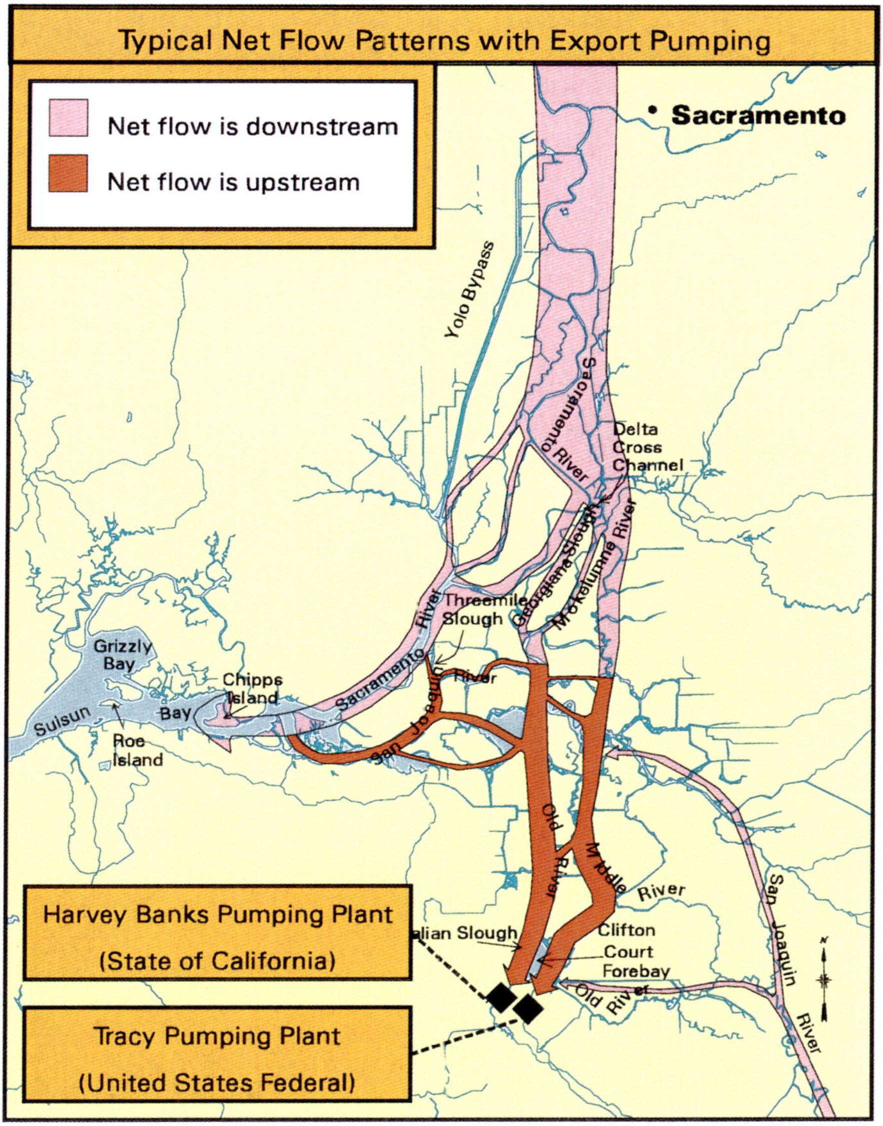

FIGURE 6B. Net flows in the Sacramento-San Joaquin Delta Estuary under low inflow conditions with export pumping.

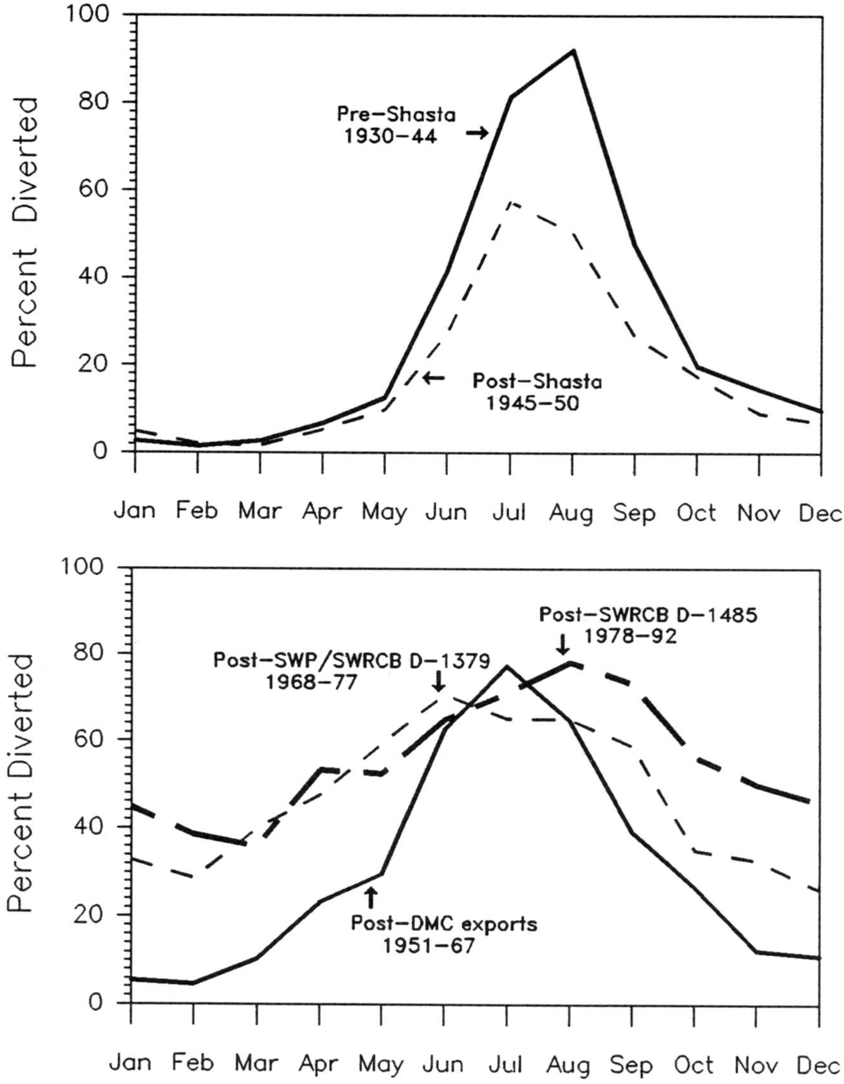

FIGURE 7. Change in average monthly percent of inflow diverted in below average water years for major manmade impacts.

low streamflow conditions (discussed later) often dominates hydrodynamic transport. Periods of sustained high westerly winds can also affect filling of the Delta.

For evaluation of long-term trends of water project reservoir releases and export pumping impacts on net flow patterns in the Delta, we examined the phenomenon of reverse flows using diverse data sets:

- Percent of Delta inflow diverted by project exports;
- Acoustical velocity measurements in Old River and Middle River;

- Reverse flows (Qwest) in the lower San Joaquin River; and
- Upward looking Doppler profile data obtained near Antioch in 1994.

Percent of Delta Inflow Diverted. The percent of Delta inflow diverted has changed greatly as the result of the Federal and State water project operations. The percent diverted is computed by summing Delta inflows and subtracting channel depletions plus Federal and State export pumping. For this analysis, we assumed Delta depletions to be constant between years for the period of record.

The data indicate about a 30 percent decrease in percent diverted during the summer comparing pre-Shasta to post-Shasta conditions (prior to Delta water project exports) (Fig. 7). Since Delta depletions were assumed constant, this decrease was largely the result of increased summer inflows via reservoir releases.

The data also demonstrate that the percent diverted in the fall/winter/spring period has increased from less than 10 percent in the 1930-1950 period to an excess of 30 to 60 percent diverted under more recent conditions, 1968-1992. There has also been a significant increase in the percent diverted as water demands have increased.

Comparison of the summer period indicates that the percent diverted has increased about 20 percent since the late 1940s, when the projects began upstream reservoir storage and Delta diversion. However, the post-project percent diverted is still lower than pre-Shasta conditions. Overall, there has been a shift from a high percent of inflow being primarily diverted in the summer to a year-round high rate of diversion.

Comparison of Delta inflows (Fig. 4) with percent diverted (Fig. 7) shows there has actually been a decrease of inflows in winter and spring, even though water exports have increased with time. Winter inflows prior to the State exporting water (1930-1967) are generally in the same range. After 1968, winter and spring inflows decreased as much as 50 percent, while demands increased (Fig. 5). December and January inflows generally remained constant, even though exports increased over time.

We draw two conclusions from the data presented in Figs. 4, 5, and 7. First, since the water projects began diverting water from the Delta, much greater quantities of water are being transported across the central Delta to the export pumps. At the same time, inflows during the winter and spring have decreased and summer/fall inflows have increased. Less inflow and higher pumping have caused much shorter residence time and consequent flushing of habitat in areas critical to certain fish and their food web. Second, implementation of State Water Resources Control Board standards has not improved conditions in this part of the estuary, as intended.

Acoustical Velocity Measurements in Old and Middle Rivers. In 1987, as part of the Interagency Ecological Program, the Geological Survey installed AVMs (acoustical velocity meters) in lower Old and Middle Rivers (about halfway between the lower San Joaquin River and the export facilities). Combining Old River and Middle River AVM measurements (USGS, open file) into monthly averages, the data were plotted against total export pumping and San Joaquin River inflow at Vernalis to determine if there was a relationship. The results demonstrate

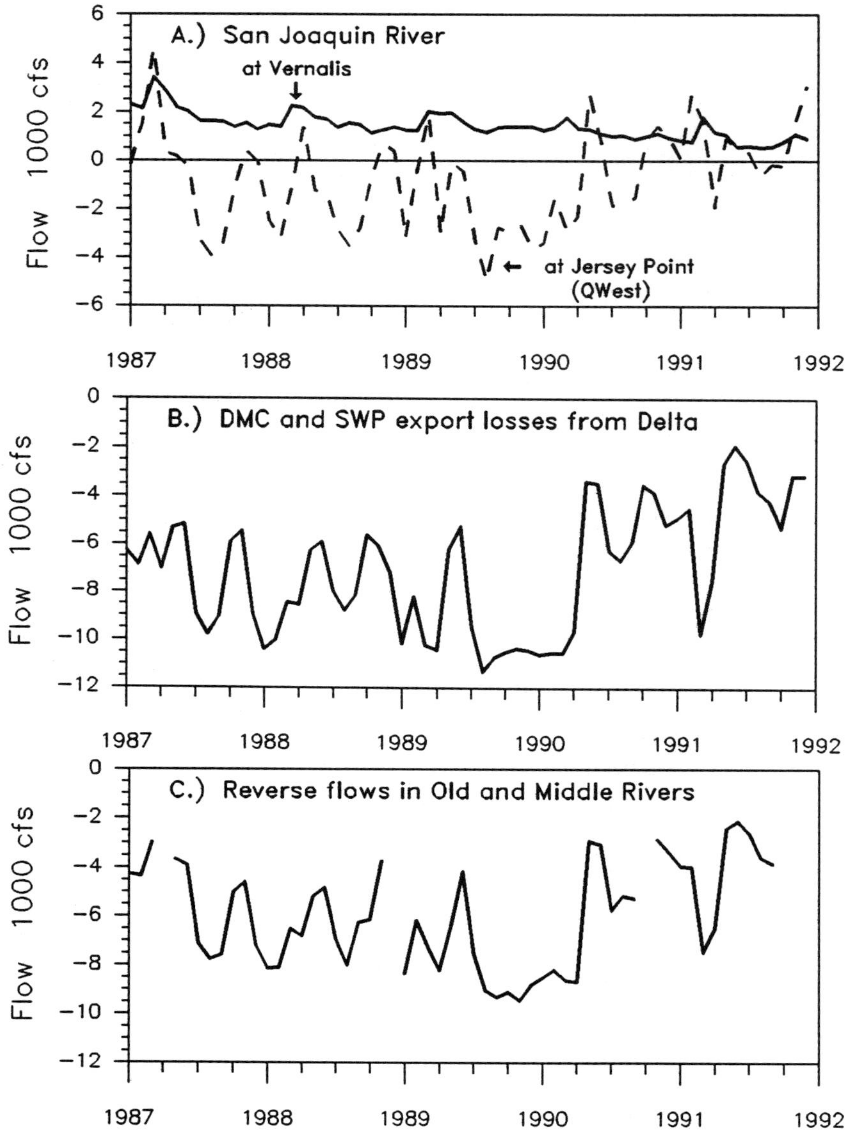

FIGURE 8. Monthly average flows: (A) San Joaquin River at Vernalis; (B) DMC and SWP exports; and (C) sum of Old and Middle River UVM measurements. (USGS data).

that over a 5-year period, 1987-1991, flows were always toward the export pumps (Fig. 8). Average net flows for both rivers were generally greater than -3,000 cfs and, at times, exceeded -9,000 cfs. Under the low San Joaquin inflow occurring in those drought years, the export pumping greatly influenced net nontidal flows in Old and Middle Rivers.

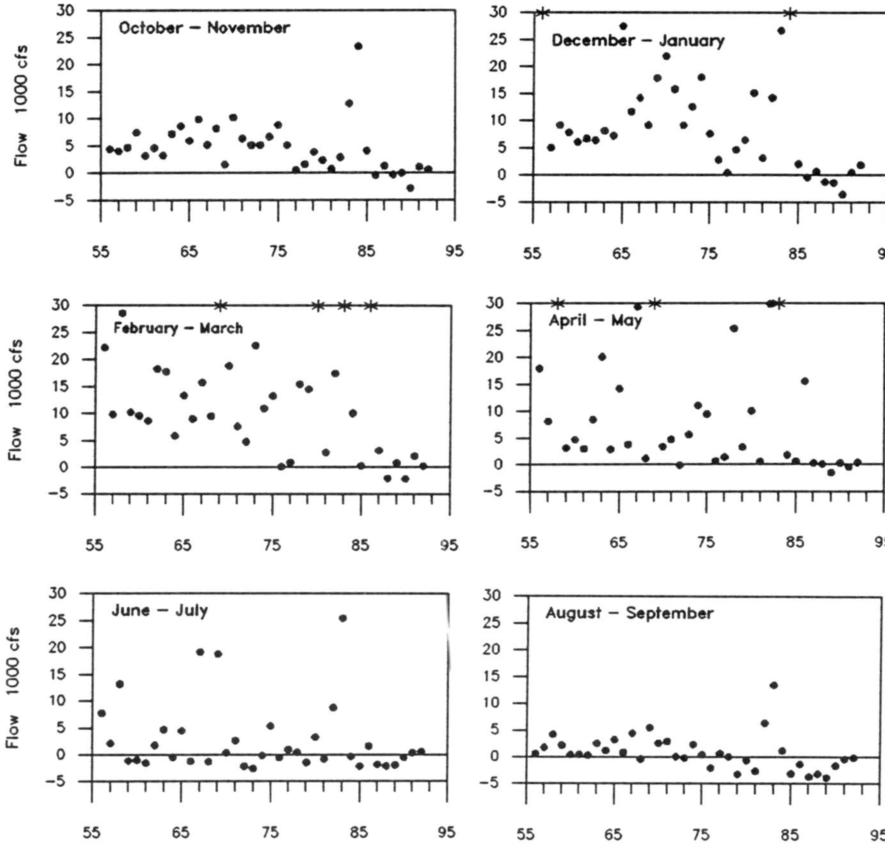

FIGURE 9. Average bimonthly net nontidal flow (QWest) computed for Jersey Point.

Lower San Joaquin River. Calculated flows at Jersey Point (Qwest), averaged for 2-month intervals, were graphed to show long-term trends (Fig. 9). The reference line, zero, means there is neither upstream nor downstream net flow. Negative or reverse flows are upstream in the lower San Joaquin River, positive flows are downstream.

The data indicate that net flows generally have been positive over the 35-year period of record from October through May, with a few exceptions in very dry years. Highest positive flows generally occurred during the high runoff months, December through May.

Net flows are most often negative during June through September, with the exception of a few wet years. Average monthly reverse flows first occurred in the 1950s, during June and July, a few years after the Delta-Mendota Canal started exporting water. In the mid-1970s, after the State Water Project had been operating a few years, mean monthly reverse flows in August and September became the rule except in very wet years. Since the last drought, starting in 1987, reverse flows have become common in all months of the year.

Overall, in the San Joaquin River at Jersey Point there has been an increased tendency in net flow reversal since the early 1970s corresponding to increases in water project export demands. However, high variability in water year types makes the trend difficult to ascertain.

Direct Measurements of Net Flows in the Lower San Joaquin River. The Interagency Ecological Program conducted a special entrapment zone study in the spring of 1994. The primary objective was to collect hydrodynamic (physical) and biological information under low Delta outflow conditions. The study area was from Suisun Bay to slightly upstream of the confluence of the Sacramento and San Joaquin Rivers.

The hydrodynamic study element included deployment by the Geological Survey of five upward-looking acoustic Doppler current profiler (ADCP) sensors, as well as other current and water quality instruments throughout the study area. The ADCP sensors were located about 1 meter off the bottom at each site and continuously measured horizontal current velocity and direction at 1-meter depth intervals. Measurements were taken from March through the first week of June 1994. The Geological Survey low pass filtered the continuous data to remove the tidal component of flow, which allows these data to be plotted as long-term net velocity and direction in the horizontal at 1-meter depth intervals at each site.

Data were obtained from Geological Survey (Jon Burau, pers. commun.) for the ADCP deployed near Antioch, on the lower San Joaquin River. Figure 10a shows net current at five depths in the water column over the 3-month deployment period. Bin 1 is the measured residual velocity near the channel bottom 2 meters above the bed, and bin 11 is near-surface residual current measured 13 meters above the bed. The uppermost plot is the measured (nonfiltered) current velocity at bin 11 and is provided to show clearly the spring/neap tidal cycle. All the velocities shown have been rotated into the direction of the prevailing channel, known as the principal direction. Therefore, net flows (shown as sticks) that are drawn above the zero line (positive) represent up-estuary residual currents. Conversely, sticks that are drawn below the zero line (negative) represent net currents that are directed down-estuary toward Suisun Bay.

The net residual flow data clearly demonstrate continuous upstream net flows in the lower third of the water column over the entire 3-month study period. Net residual flows in the upper two-thirds of the water column were downstream. This pattern was consistent during a number of spring and neap tidal cycles. Also noteworthy is that Corps of Engineers current-meter data collected between Collinsville and Rio Vista (Fig. 10b) demonstrated that during this same period the net bottom-to-surface flow was downstream in the lower Sacramento River (Jon Burau, personal communications).

The circulation pattern observed at Antioch occurred during a period of relatively low Delta outflow and low water project exports. We are not certain of the cause of the circulation pattern. However, suspended materials including organisms that spend time in the lower portion of the water column are affected by

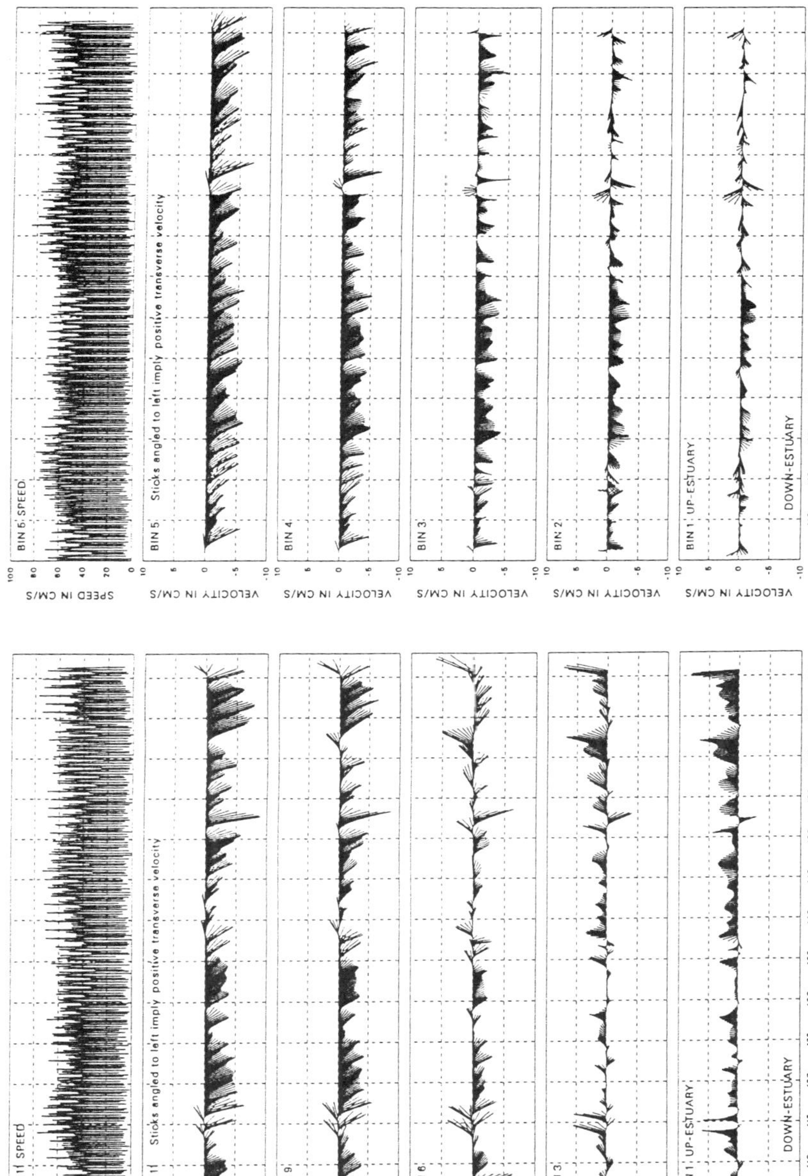

FIGURE 10. (A) Acoustic Doppler current profiler data at San Joaquin River near Antioch. USGS data (Jon Burau). (B)Acoustic Doppler current profiler data at Sacramento River near Emmaton. USGS data (Jon Burau).

residual currents at this site and could have been transported upstream into the Delta. Further studies are warranted.

In summary, the available information indicates that during periods of low inflow and high Delta demands and exports, suspended particles, including pelagic fish and food-web organisms influenced by flow, in the vicinity of the lower San Joaquin River may be drawn to the export pumps. The data strongly suggest a trend toward increased net flow reversals as water demands have gone up, particularly during the drought years of the mid-1970s and the latter part of the 1980s through early 1990s. Residence times in the central Delta have increasingly been reduced as more water is required to meet demands. We believe reduced residence time has resulted in the flushing of a significant part of the estuary's habitat for organisms requiring a long residence time to maintain their abundance.

Although there are valid technical concerns about reverse flows as denoted by Qwest, *i.e.*, peak tidal flows are large (150,000 cfs) compared to net reverse flows (0 to -5,000 cfs), we believe QWest is still a very important factor. For example, if streamflows are low (particularly in the San Joaquin River) and Delta water demands are high, salinity intrusion will increase and organisms being actively or passively transported downstream through the Delta will have greater exposure to in-Delta diversions in lower Old and Middle Rivers. Furthermore, if this occurs when Qwest is negative and the Delta is filling due to a spring tide, particles or organisms may be moved upstream or remain in this part of the system. Because there is about a 6-mile tidal excursion in this area, they will be constantly exposed to the diversions from the San Joaquin River towards the export pumps.

Delta Outflow and Entrapment Zone Location. The hydrodynamic trends described above for the Delta may also affect conditions in the downstream embayments. We limited our analysis to long-term trends in Delta outflow and location of the entrapment zone, an area of significant biological importance.

Water development in California, including the Federal and State water projects has greatly altered mean monthly trends of Delta outflow. To evaluate the trends, outflow from 1930-1992 (period of record) was subdivided into the five periods of major events using only the below-normal, dry, and critical water years (Fig. 11).

Prior to operation of Shasta Dam, Delta outflow during midsummer, on the average, was slightly negative in August during below-normal water years. Generally, outflows were highest for all months (below normal years) after completion of Shasta Dam but before exports started. Following the onset of combined project exports, early spring into late fall became the low outflow period. Overall, average outflows in the winter and spring months have been cut about 50 percent in below-normal water years since the period of combined project exports (1968-present).

Outflows affect the location of the entrapment zone, which occurs in the area where surface salinities are about 1-6 ppt. The entrapment zone is an area where suspended particles and certain food-web organisms and fish concentrate as the result of chemical/physical and biological interactions with estuarine hydrodynamics.

In the 1960s, the U.S. Geological Survey conducted a study in San Francisco

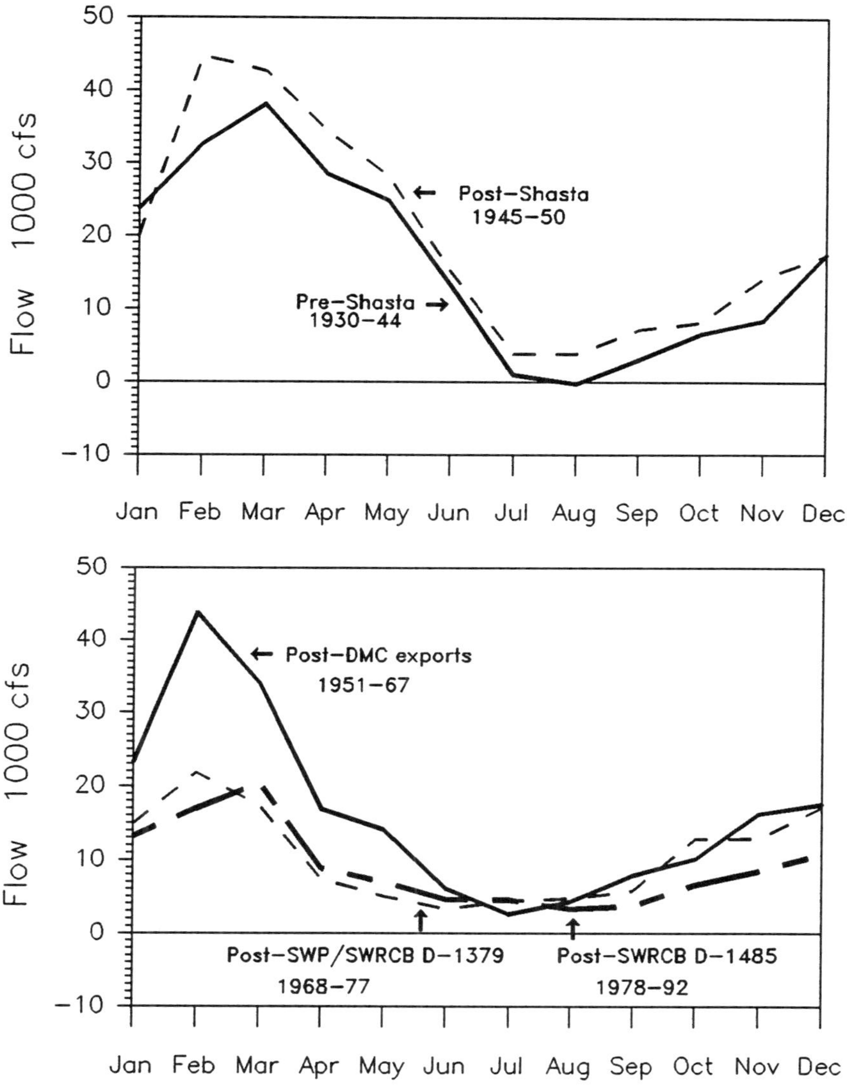

FIGURE 11. Change in average monthly Delta outflows in below average water years for major manmade impacts.

Bay in which surface and bottom drifters were released at different points in the estuary (Conomos *et al.* 1970). Many of the bottom drifters released in San Francisco Bay moved a considerable distance upstream. These data provided strong evidence that net upstream bottom flow caused by gravitational circulation (induced by differences in density between fresh water and salt water) was a prominent component of circulation in this estuary.

In the early 1970s, Reclamation theorized that observed phytoplankton, certain zooplankton, and larval striped bass peaks in Suisun Bay were the result of biological interactions with estuarine circulation (Arthur 1975). Over the next 5 years, Reclamation conducted a series of field studies, which resulted in development of the present conceptual model of the entrapment zone. Study results were published in a series of detailed reports and in the literature (Arthur & Ball 1978, 1979a, 1979b; 1980; Ball & Arthur 1979).

Our analysis was limited to evaluating historical changes in the entrapment zone. Maximum annual entrapment zone intrusion was obtained from maximum salinity intrusion records for 1920-1978 presented in the *Delta Atlas* (DWR 1993). Daily mean tidally averaged location for 1968-1992 was obtained from Russ Brown (Jones and Stokes), which used a model to calculate entrapment zone location from mean daily water project salinity records and the X2 relationship developed by Kimmerer (1992). The computed mean monthly values were adjusted 3 miles (½ mean tidal excursion) upstream to represent the maximum intrusion each month.

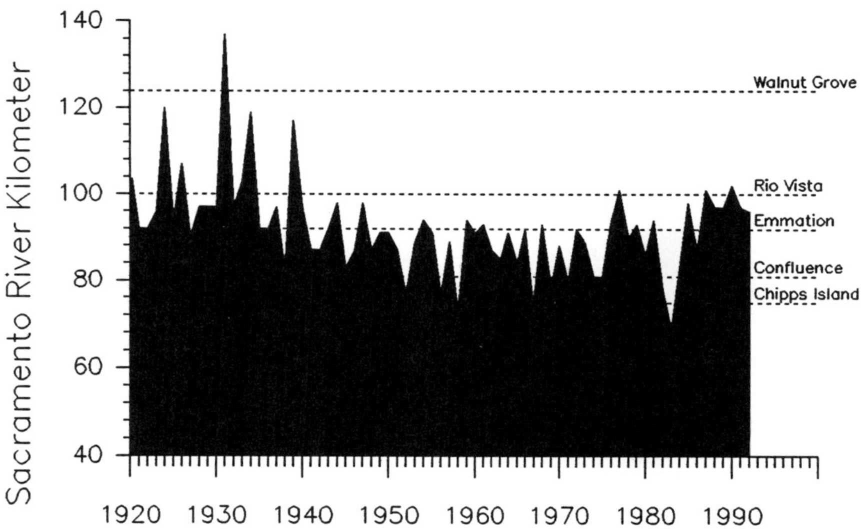

FIGURE 12. Yearly maximum intrusion of the entrapment zone estimated from salinity data (Delta Atlas, 1993).

The period of maximum easterly intrusion of the entrapment zone during 1920-1992 occurred during drought years prior to completion and operation of Shasta Reservoir (Fig. 12). From the mid-1940s (after completion of Shasta Dam) to the early 1970s, intrusion was held to a minimum. During drought years after the mid-1970s, intrusion increased, but not to the levels of the pre-Shasta years.

Mean monthly entrapment zone intrusion was also evaluated. Using the relationship between Jones and Stokes data and mean monthly Delta outflow, approximate locations for earlier years were developed (Fig. 13). The breaks in the data are for years when mean monthly Delta outflow was negative for up to three

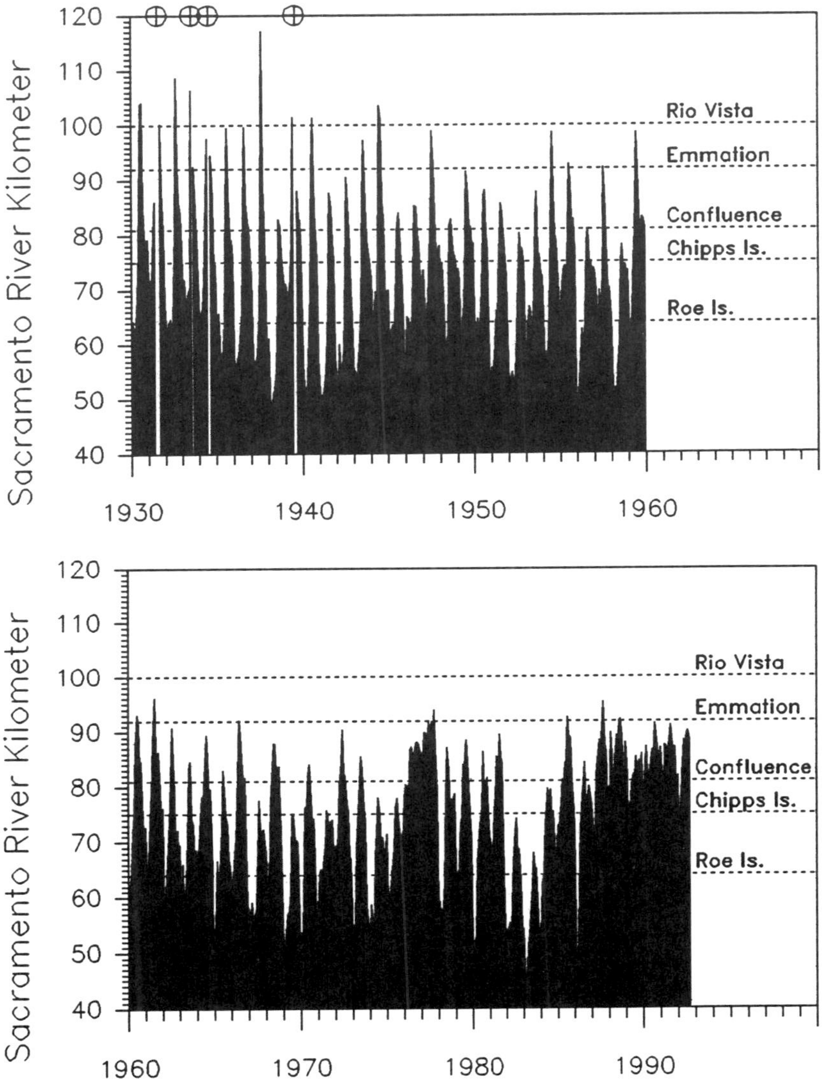

FIGURE 13. Average monthly location of the entrapment zone estimated from Delta outflow.

consecutive months. In those cases, the regression would not predict a location, and a circle with a cross was placed at the top of the graph. The main conclusions are that maximum intrusions have not been as great as prior to Shasta, however, since the mid-1970s, high-intrusion periods have been of longer duration. This same period is when reverse flows in the lower San Joaquin River (QWset) have become increasingly negative for August and September.

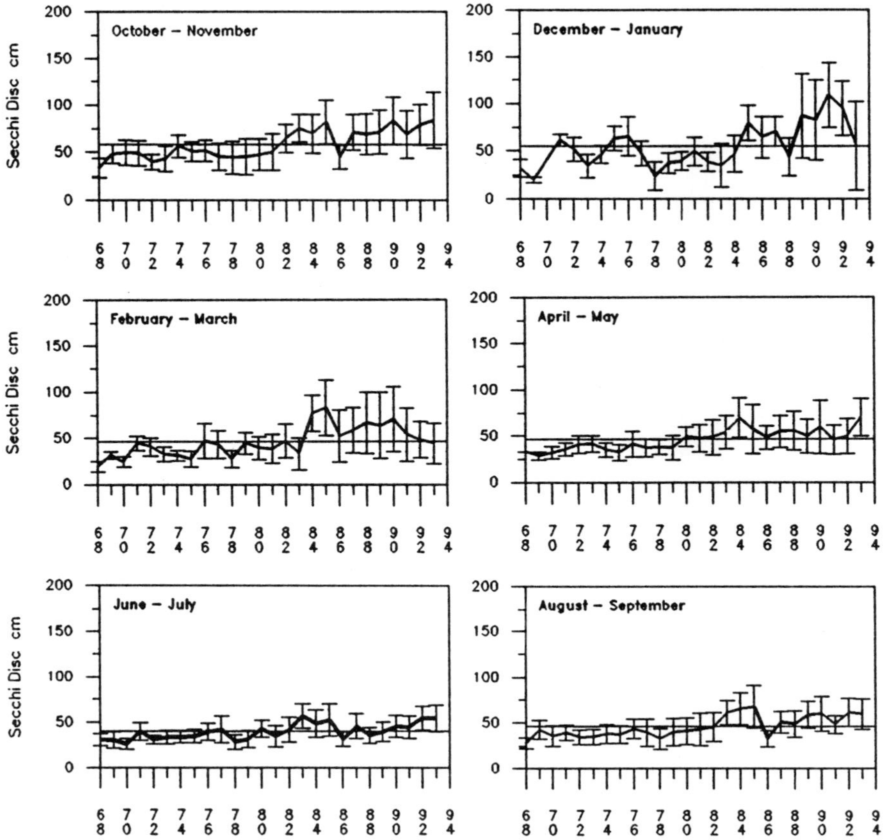

FIGURE 14. Average bimonthly water transparency (std dev.) showing mean of all years in the western and central Delta.

Suspended Sediments

The water projects have decreased the load of suspended sediment to the estuary (Arthur 1987). The decreased load has been partially responsible for increased water transparency within the estuary (Ball 1987). Numerous reservoirs constructed by Federal, State, and local agencies have trapped sediments. Some of the fine silts and clay particles, however, manage to pass through the reservoirs, depending on the water residence time.

Water transparency measurements in the western/central Delta have indicated a significant clearing trend since the mid-1970s (Fig. 14). During intense blooms in the mid-1980s, water transparency in the area would often drop by 50 percent in less than a month as phytoplankton levels peaked.

The clearing trend for water transparency was greater for the western/central Delta than for the inflow water of the Sacramento River in the northern Delta.

Water transparency in Suisun Bay area has also experienced a major increase since the mid-1980s. There were two main reasons for this. First, there was an extended period of very low flow during the drought, and second, the newly introduced Asian clam has greatly reduced phytoplankton levels. The clam also filters suspended sediments.

There are several possible causes or combination of causes for this clearing trend in the estuary, five of which were discussed by Ball (1987):

- Sediments are being trapped in Central Valley reservoirs;
- Scouring below the reservoirs removed the suspended sediment deposited in the beds of valley river systems as a result of hydraulic mining (banned in the late 1880s), and now the suspended materials have been washed out of the system, allowing the stream bed to return to a pre-hydraulic mining condition;
- Resuspended sediments in the Delta shoals and channels are being lost to export pumping;
- Sediments that cycle between the Delta and downstream bays with the entrapment zone and two-layered flow circulation are being lost to export pumping, and
- Agricultural practices on the valley floor have changed, resulting in a reduction of suspended sediments to the rivers.

One consequence of this clearing appears to be the increased intensity and dominance of phytoplankton blooms by *Melosira granulata* in the western/central Delta region (Ball 1987). This diatom produces taste and odor problems in water supplies destined for municipal and industrial use. It also is not desirable food for zooplankton.

There are many unanswered questions involving sediment resuspension, relocation, and loss from the estuary by entrainment to the export pumps through the reverse flow patterns in the lower San Joaquin River system. Sediments are constantly being settled, resuspended, and translocated by tides, wind, and net streamflows. During periods of high flows, the upstream estuarine areas are scoured, and the sediments move downstream. In the summer, under low streamflows, sediments are resuspended in the bays by wind and tidal action, move upstream with two-layered flow circulation patterns, and are redeposited in the adjacent shoals. This was documented in the upper estuary in the mid-1970s when the Corps of Engineers added iridium (a tracer) to dredge spoil dumped in San Pablo Bay. Later that year, iridium was found upstream in Suisun Bay sediments.

Chemical Impacts

Human actions that can alter estuarine water quality include: actions that change the quantity or quality of water within the Sacramento-San Joaquin drainage basin; unregulated agricultural returns; municipal, industrial, and unregulated discharges within the Delta and San Francisco Bay; airborne pollutants and water project exports. Water quality in this estuary is generally good (DWR annual reports to SWRCB, 1975-present) relative to many estuaries throughout the world. Many of the historical problems, such as the low dissolved oxygen and smell that were characteristic of South Bay into the 1970s, have largely been eliminated through regulatory actions and improvements in waste water treatment. Recent actions by the U.S. Environmental Protection Agency, San Francisco Bay Regional Water

Quality Control Board, and others to form the San Francisco Estuary Institute continue to address concerns involving various toxicants within the estuary.

Delta Water Quality

Water quality concerns in the estuary have changed over time. For example, a major potential water project impact identified in State Water Resources Control Board Decision 1379 hearings in 1967-68 was massive algal blooms. University experts testified that the projects would remove sediment via export pumping and increase nutrients via discharge through the proposed San Luis Drain, resulting in massive algal blooms throughout the system.

The Decision 1379 hearings resulted in formation of the Interagency Ecological Study Program and development over the next 10 years of phytoplankton and water quality models, with accompanying field studies. Although some of proposed impacts proved to be correct (*e.g.*, water transparencies have increased as sediments have been removed from the system), others did not (*e.g.*, additional nutrients may not cause higher phytoplankton production). Today's major concern is too little phytoplankton as the result of the introduced clam. Toxins are also of some concern to the food web in the Delta, but a direct cause-and-effect relationship has yet to be established.

Reservoir releases have prevented extreme salinity intrusion during drought years, such as occurred prior to Shasta Dam. The increased flow of fresh water through the Delta to the export pumps has provided a flushing action for water quality in the Delta. However, for this paper we have concentrated on what we believe are the two most important negative water quality impacts of exporting project water out of the natural Delta channels — namely, recycling of salts and additions of trihalomethane (THM) precursors.

Recycling of Salts

Salt buildup on San Joaquin Valley agricultural lands as a result of evapotranspiration and poor drainage has been a major problem since farming began. Historical civilizations have been lost as a consequence of salt buildup and today it is still a major problem worldwide. The west side of California's San Joaquin Valley is an area of high salt buildup resulting from poor drainage. In portions of the valley, one solution has been to drain the salts to the San Joaquin River (FWPCA 1967 and 1968; DWR 1965 and 1969; and Interagency San Joaquin Valley Drainage Program 1990). San Joaquin River salt loads combined with substantial agricultural return flows from Delta islands are recycled back into the valley by the water project export pumps.

As part of Reclamation's testimony in the SWRCB Decision 1630 hearings, Arthur (1987) evaluated salt loads entering and leaving the Delta via the water project export pumps. Among other things, these data indicated that:

- Central Valley Project exports generally had higher salt concentrations than the State Water Project, indicating interception of San Joaquin River water via upper Old River;

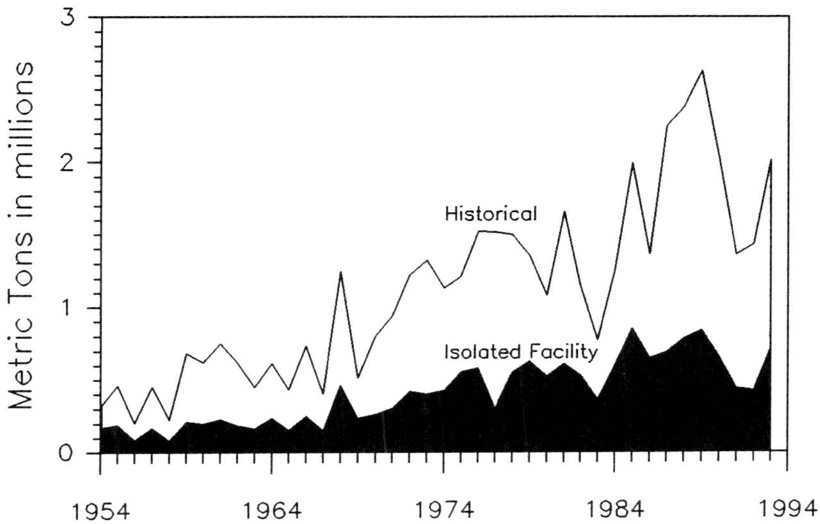

FIGURE 15. Annual combined SWP/CVP salt load exported vs. the possible salt load with an isolated facility diverting at Greens Landing on the Sacramento River. (Calculated from daily salinity and flow data.)

- Salt loads were generally higher at both export facilities in the summer/fall period, particularly in below-normal water years; and
- The salt load being recycled back into the San Joaquin Valley by the export pumps often exceeded the salt load entering the Delta via the rivers and streams (up to 300 percent).

Arthur concluded from these data that much of the salt load that was in excess of river inflow loads probably originated from Delta island return flows. Ocean-derived salt also contributes to the salt load leaving the Delta via the project export pumps. However, the projects are required to meet State Water Resources Control Board salinity standards at varying locations downstream of the pumps in the San Joaquin River, which reduces the amount of ocean salt diverted at the pumps.

A comparison was made between actual CVP/SWP export salt loads observed (1954-1993) with what would have occurred if an isolated transfer facility had been in place (Fig. 15). Annual historical export salt loads were calculated by converting mean daily records of specific conductance to total dissolved solids and multiplying the concentration by the daily export rate to determine annual loads. Projected salt loads for an isolated transfer facility were derived using mean daily data for the Sacramento River at Greens Landing near Hood (proposed diversion site for the Peripheral Canal) and multiplying by the daily export rates to obtain an annual load.

Total annual salt loads exported via the Delta-Mendota Canal ranged from about 0.2 million to 0.7 million metric tons per year in 1954-1967, increasing since 1968 to a peak of about 2.7 million metric tons per year in 1989. The increase is the result of increased water demand and export capacity when the SWP export pumps came on line in 1968. Periods of lower annual salt loads are generally associated with very wet years, such as 1982 and 1983, when water demand and salt concentrations are relatively low.

Based on the calculations, if an isolated facility had been in place during the 1954-1994 period, corresponding salt loads exported would have ranged from a low of 0.2 million to a high of almost 0.8 million metric tons per year. Total salt loads for the entire period of project exports was estimated to be about 45 million metric tons. Total salt load for the same period, if from an upstream Sacramento River diversion, would have been about 16 million metric tons, or 35 percent of the actual salt load exported.

Trihalomethane Precursors

Trihalomethanes are disinfection byproducts (DBP) formed during water processing at municipal treatment plants. Chlorine and chloramines have been the preferred disinfectant because of lower cost and high effectiveness in controlling bacterial growth in water distribution systems. New Environmental Protection Agency regulations have resulted in utilities initiating research on more expensive water treatment technology to limit DBPs (DWR 1994).

Delta waters, southern Delta in particular, have elevated concentrations of organic matter and bromides, which contribute to the formation of DBPs. Water diverted by the State Water Project, Central Valley Project, and Contra Costa Water District are generally higher in organic matter, bromide, and other mineral salts than water from the northern Delta. Sea water has been identified as the major source of bromides in the southern Delta.

Studies conducted in the 1980s (DWR 1994) documented sources of organic materials being exported by the Central Valley Project and State Water Project. These included streamside vegetation, decaying crop residues, algae, and sewage. The largest source appears to be the region's soils. Delta islands, consisting of high peat deposits, are constantly being eroded. Seasonal farming activities affect the amount of organic matter leached and drained from island soils, which eventually discharges into the Delta channels. These return flows are largely unregulated, and no treatment is required to remove organic materials (or any other substance) at the source of the discharge.

THMs would largely be eliminated (as would treatment problems and expenses incurred by municipal treatment plants south of the Delta) if Sacramento River water were not directly transported through the Delta.

Biological Impacts

When biological sampling was first initiated in the 1960s, abundance levels of important food-web organisms — phytoplankton, zooplankton, and many fish species — were relatively high in the Delta and Suisun Bay. These included periods of below-normal water years when the entrapment zone was located in the western Delta. In the early 1970s, the abundance of striped bass and certain food chain organisms began to decline in the Delta, followed by a further decline during the 1976-77 drought. After the initial Delta decline, abundance of striped bass and certain entrapment zone organisms began to decline in Suisun Bay. The

Delta/Suisun Bay decline continued after implementation of Decision 1485 and became greatly accentuated during the extended drought of the 1980s (Herbold *et al.* 1992; Monroe & Kelly, 1992).

Phytoplankton

We believe the Federal and State water projects have had major impacts on estuarine phytoplankton (the primary producers at the base of the aquatic food web). Phytoplankton dynamics in the Delta and Suisun Bay have been extensively evaluated in the past (Arthur & Ball 1978, 1979a, 1979b, 1980; Ball 1977, 1987; Ball & Arthur 1979; Cloern *et al.* 1983; Stewart *et al.* 1984). Conclusions have been mainly that phytoplankton blooms in the Delta tend to occur when net flows through the central Delta drop, creating long residence times. Since the early 1970s, with increased water export through the Delta, the duration of bloom conditions has greatly declined.

Under the same conditions of Delta inflow and diversions from the Delta, recent evaluations by Jassby & Powell (1994) and Jassby (1995) of chlorophyll and flow in the upper estuary suggest that primary production in the Delta and supply of phytoplankton to the entrapment zone would be greater if the water were diverted from the upper Delta and not through the Delta. Such isolated diversion conditions would enhance the upper estuarine food web.

Phytoplankton growth dynamics in Suisun Bay and the entrapment zone have been studied and reviewed more extensively (previously referenced). Results of these studies and evaluations have all concluded that phytoplankton abundance is enhanced when the entrapment zone location is centered at the upstream end of Suisun Bay. This entrapment zone location retains entrapped phytoplankton adjacent to the shallow productive shoals, where adequate levels of light stimulate highest production levels. Longer duration of low outflows have moved the entrapment zone upstream into the lower San Joaquin River. Phytoplankton in this area are subject to less light causing lower production rates and also loss to entrainment caused by the export pumps.

The accidental introduction of the Asiatic clam, *Potamocorbula amurensis*, first observed in 1986, has had a devastating effect on the food web in areas of the estuary where the clam has become abundant (Nichols *et al.* 1990). Since 1987, in the Suisun Bay area and upstream to the freshwater boundary, the clam has reduced phytoplankton levels by more than an order of magnitude. Evaluations of the clam's filtering rate, assimilation efficiencies of phytoplankton and bacterioplankton, and their impact on the food web were studied by Hollibaugh & Werner (1991).

The major decline in phytoplankton abundance in Suisun Bay and the entrapment zone in recent years appear to have been caused by introduced species. However, the State and Federal water project operations are also impacting phytoplankton levels in Suisun Bay and the entrapment zone by flushing much, if not most, of the phytoplankton standing crop in the San Joaquin River portion of the Delta out the export canals. Based on our understanding of the estuary, we believe this is caused by:

- Increased flushing of Delta habitat (via reduced residence times);
- Delta entrainments; and
- Net reverse flow in the lower San Joaquin River, which increases the exposure of organisms to diversions leading to the export pumps and reduces the amount of phytoplankton being transported downstream and out of the Delta.

Zooplankton

Zooplankton are near the base of the food web and are consumed by larger zooplankton, immature fish, and some mature fish. Many species of zooplankton have also declined in recent years. For our analysis, we evaluated water project impacts on their abundance by reviewing zooplankton trends recently compiled by Obrebski *et al.* (1992) and Kimmerer (1992).

Obrebski examined the abundance trends (1972-1988) of 20 species of zooplankton in several families during spring, summer, and fall for distribution relative to salinity and for changes in geographical distribution (Obrebski *et al.* (1992). The more marine species (occurring downstream of the entrapment zone) with the exception of *Acartia* sp. had no downward trend in abundance. In the fall of 1987 *Acartia* sp. numbers declined by nearly one order of magnitude following the introduction of the Asian clam, *Potamocorbula amurensis*. This suggests that the water projects had little or no impact on their abundance.

Kimmerer (1992) evaluated the *Neomysis* sp. and *Eurytemora* sp. data relative to entrapment zone location as related to Delta outflow. He compared abundance differences in the entrapment zone between the Sacramento River side with similar differences up the San Joaquin River side (at locations relative to the confluence, Emmaton, and Jersey Point) for similar levels of outflow. In general, the declines were much greater in the lower San Joaquin River than in the lower Sacramento River, particularly for *Neomysis* sp., which have a much longer generation time.

Kimmerer also calculated losses of these species to export entrainment using a simple dilution model and concluded no significant loss. We believe the fact that modeling the loss as he did without incorporating the effects of two-layer flow and pumping mainly on higher tidal phases would not yield valid results.

We also believe the data relate well to what one would expect if reverse flows in the San Joaquin River had a negative impact on abundance. With the farthest upstream location for the entrapment zone, there was a slightly lower peak abundance for *Eurytemora* sp. but a drop in *Neomysis* sp. abundance of nearly an order of magnitude. Also, for both species the peak was in slightly higher salinity water (1.5 ppt higher), suggesting entrainment losses from the upstream end of their habitat range. The other significant observation was that when the entrapment zone was below Chipps Island, concentrations of both species were much higher in the San Joaquin River above the confluence than at equal distances up the Sacramento River. The opposite occurred (concentrations were much lower in the San Joaquin River) when the entrapment zone location was above the confluence. This effect was even more pronounced when the entrapment zone was above Emmaton.

When the entrapment zone is below Chipps Island, the net advective downstream flow in the Sacramento River is greater than in the San Joaquin River, and zooplankters tend to be flushed farther downstream. When the entrapment zone is upstream of the confluence and farther on, up to Emmaton, in the Sacramento River, the entrapment zone concentrates the two species, with opposing net bottom flows acting on their net vertical swimming behavior.

The opposite occurs in the lower San Joaquin River. Net bottom flows in the San Joaquin River (with similar upstream entrapment zone locations as the Sacramento River) are different under present conditions of export pumping. With high export pumping, the net flows are toward the central Delta and the export pumps. This results in net upstream movement and a dilution effect on organisms that concentrate in the entrapment zone.

Orsi (1986) observed that *Neomysis* sp. move up vertically in the water column during the day only on flood tides. The surface waters travel faster than the bottom waters. This characteristic of their swimming behavior also would accelerate their upstream movement in the San Joaquin River under reverse flow conditions. This also further explains the major decline in *Neomysis* sp. abundance when the entrapment zone is located above the confluence, and there is higher export pumping.

Fish

Information amassed over the years by the Interagency Ecological Program and others documenting water project impacts on estuarine fish is voluminous and extensive. For example, for many years Fish and Game has related striped bass abundance and survival to a combination of Delta outflow and export pumping. Results have been presented in a number of hearings, scientific publications, and Interagency Ecological Program reports (Turner & Chadwick 1972; Chadwick *et al.* 1977; Stevens 1979, 1980; Stevens *et al.* 1985; IESP 1987; Turner 1992). Most recently, Fish and Game developed a striped bass model for the Decision 1630 hearings (Kholherst *et al.* 1992) based on 20 years of data used to predict adult striped bass levels that would be obtained at various Delta outflows and export levels.

The U.S. Fish and Wildlife Service has also evaluated water project operations on salmon migration and survival. Studies based on the release of marked salmon smolts at various locations in the Delta, with subsequent recaptures (of a small percentage) at Chipps Island to determine survival rates, have documented many of the problems caused by directly diverting water from the Delta. These studies have indicated, among other things, that salmon diverted off the mainstem Sacramento or San Joaquin Rivers within the Delta have higher mortality rates than salmon smolts not diverted by water project operations (Fish & Wildlife Service, 1992).

Thousands of pages have been written over the years relating to the impacts of water project operations on the estuarine biota. This includes testimony presented in the various State Water Resources Control Board Bay-Delta hearings (D1379; D1485; D1630), numerous Interagency Ecological Program reports, scientific publications, and project planning evaluations. Therefore, we have limited our evaluation of biological impacts to a few examples of direct water project entrainment, changes in distribution and abundance of several fish species and a zooplankter that were historically abundant in the Delta, and modeling results evaluating water project operations on biological transport in the Delta.

Fish Entrainment. Both the Central Valley Project and State Water Project screen out most fish longer than about an inch prior to exporting water from the Delta. Louvers are used to direct fish to holding tanks, where they are counted and trucked to release sites in the Delta. Food web organisms, fish eggs, and larvae are essentially pumped out of the Delta via the Delta-Mendota Canal and the California Aqueduct.

For this evaluation, fish salvage records from the project fish facilities were obtained for 1978-1991. Also included in our analysis are total entrainment losses based on estimated predation losses.

Water Resources and Fish and Game have been conducting predation tests in Clifton Court Forebay over the last several years. Marked hatchery-reared salmon smolts were released near the entrance to Clifton Court, and the numbers captured at the fish screens were recorded. Losses were presumably the result of predation by striped bass and other predators. Predation losses were estimated at about 75 percent for Clifton Court. Although no field verification has been conducted, biologists have assigned a prescreen loss rate of 15 percent for the CVP intake.

Based on the these tests, Fish and Game has developed a correction factor for predation and louver efficiency at both the CVP and SWP that is applied to actual salmon salvage measurements to obtain entrainment numbers. The correction factor (for salmon) for the CVP is 0.568 and for the SWP is 4.33. However, predation and screen efficiency tests have not been done on other species because of insufficient larval fish for testing. Understanding the limitations, we assumed for this analysis that the predation loss estimates are applicable to the other species evaluated.

Average monthly entrainment records at the fish protective facilities for all fish species, 1978-1991, are summarized in Fig. 16. Most entrainment of fish occurs during May through August. Overall, entrainment at the SWP is many times higher than at the CVP fish facilities.

Entrainment peaks of all species combined (Fig. 16a) correspond to periods with high percent of inflow diverted (Fig. 7). Obviously, many fish species are highly susceptible to Delta export entrainment and export during this period.

Entrainment records for some key fish species indicate peaks occur in different periods. Chinook salmon entrainment, for example, is high from April through June, with a pronounced peak in May (Fig. 16c). Delta smelt (Fig. 16d) were abundant from May through July, with the peak in June. Striped bass, on the other hand, peak in June (Fig. 16b) and represent nearly all of the total entrainment estimates illustrated in Fig. 16a.

Export salvage records provide a good picture of when fish present in the Delta are susceptible to water project entrainment. Species of concern need to be evaluated relative to their time of maximum exposure to the export pumps and corrective actions should be taken to minimize their losses.

Effects on Distribution and Abundance. To evaluate the possible impacts of

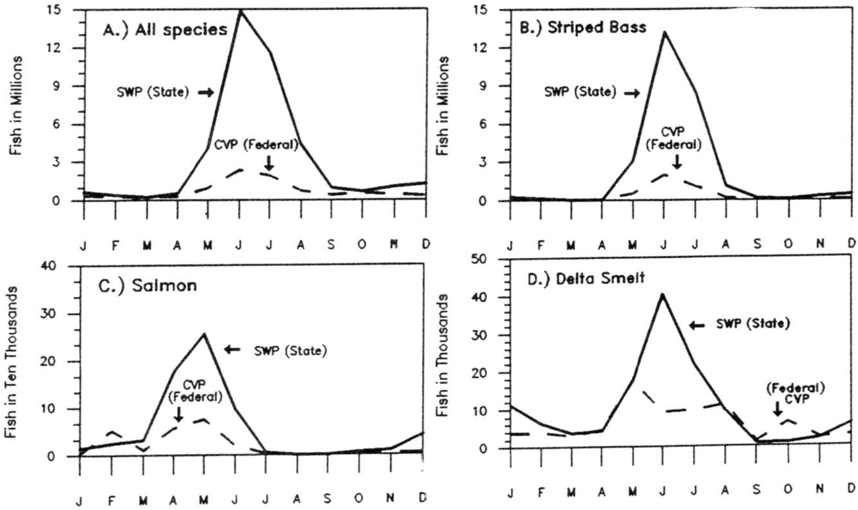

FIGURE 16. Average yearly entrainment of juvenile and adult fish at the DMC and SWP export facilities: (A) all species; (B) striped bass; (C) salmon and (D) Delta smelt.

through-Delta water transport on the biota, we examined long-term trends in the distribution and abundance of several zooplankters and key fish species that have experienced major declines in recent years. Three example species first began to decline in the Delta prior to their overall systemwide decline. *Trichocerca* sp. is a rotifer whose peak abundance historically occurred in the central and southern Delta. Striped bass, the focus of Fish and Game studies over the last 30 years, has the most complete data record. Delta smelt was listed as threatened in 1993. Salmon were not evaluated because their life history indicates they only pass through the Delta in a relatively short time on their outmigration to the ocean.

Reclamation's geographical information system (GIS) was used to graphically display changes in long-term average abundance and distribution of the three example species throughout the estuary prior to and following their decline. For example, GIS was used to plot the average concentration of *Trichocera* sp. (plotted as circles in Figures 17a-b) prior to its decline, 1967-1976, and following its decline, 1977-1992. These data illustrate a major reduction of this rotifer in the central and southern Delta and in the upstream end of Suisun Bay. We believe this major decline is at least partly attributable to reduced residence times and Delta entrainment resulting from direct transport of project water through Delta channels.

Although the timing of the decline for the two fish species differed (1981-1992 for delta smelt and 1977-1992 for striped bass), both species experienced little pre- to post-decline differences in average abundance on the Sacramento River side of the western Delta (Figs. 18a-c, 19a-c). The major declines in average abundance for both species occurred on the San Joaquin River side of the western, central, and southern Delta and the southern edge of Suisun Bay (although all of Suisun Bay showed a significant decrease). Again, we feel these observations are the results of

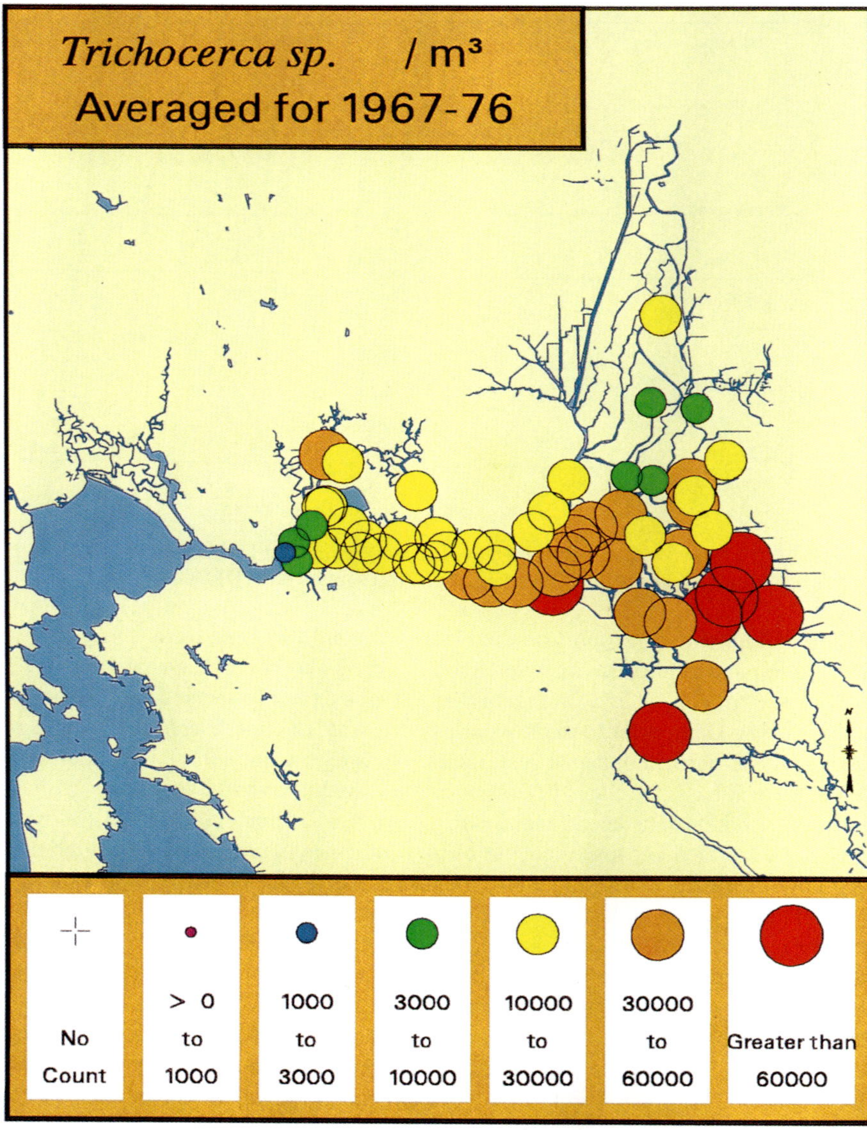

FIGURE 17A. Average density of the rotifer, *Trichocerca* sp., for each station of zooplankton survey prior to decline. Data from California Department of Fish and Game.

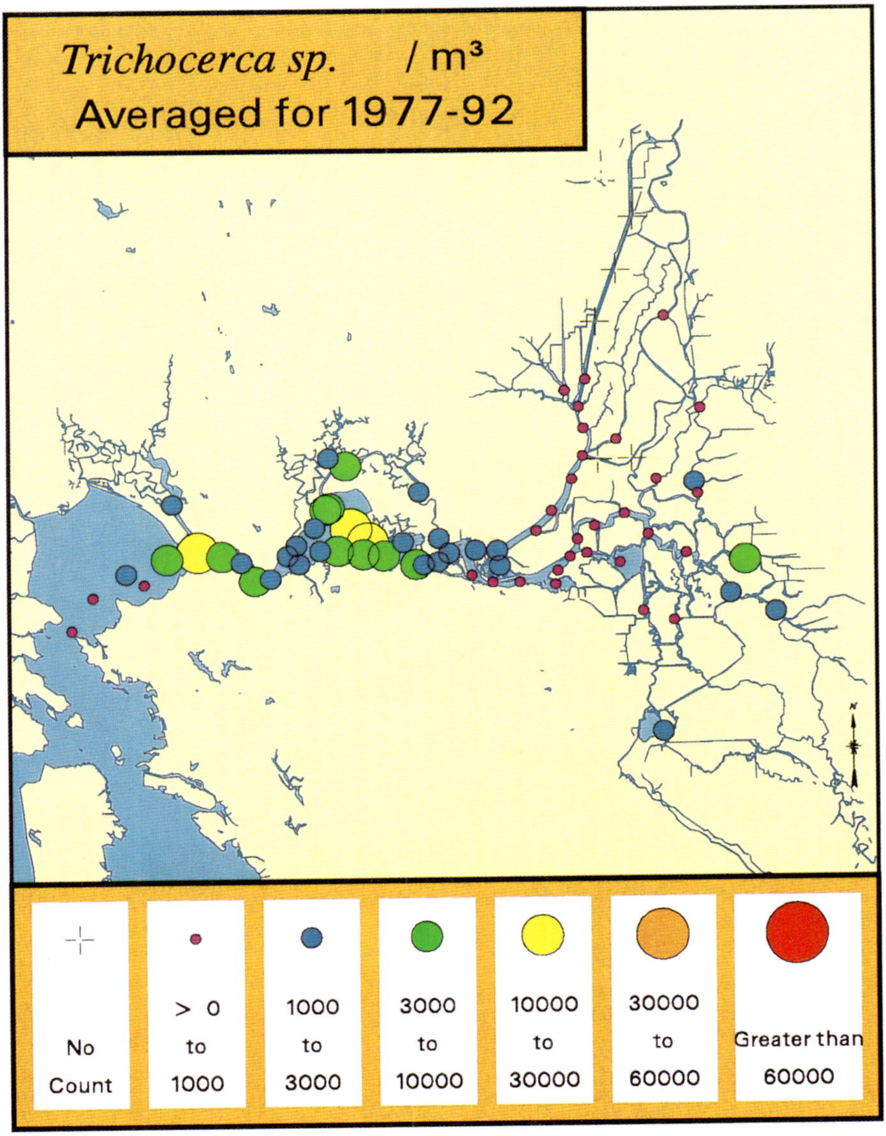

FIGURE 17B. Average density of the rotifer, *Trichocerca* sp., for each station of zooplankton survey following decline. Data from California Department of Fish and Game.

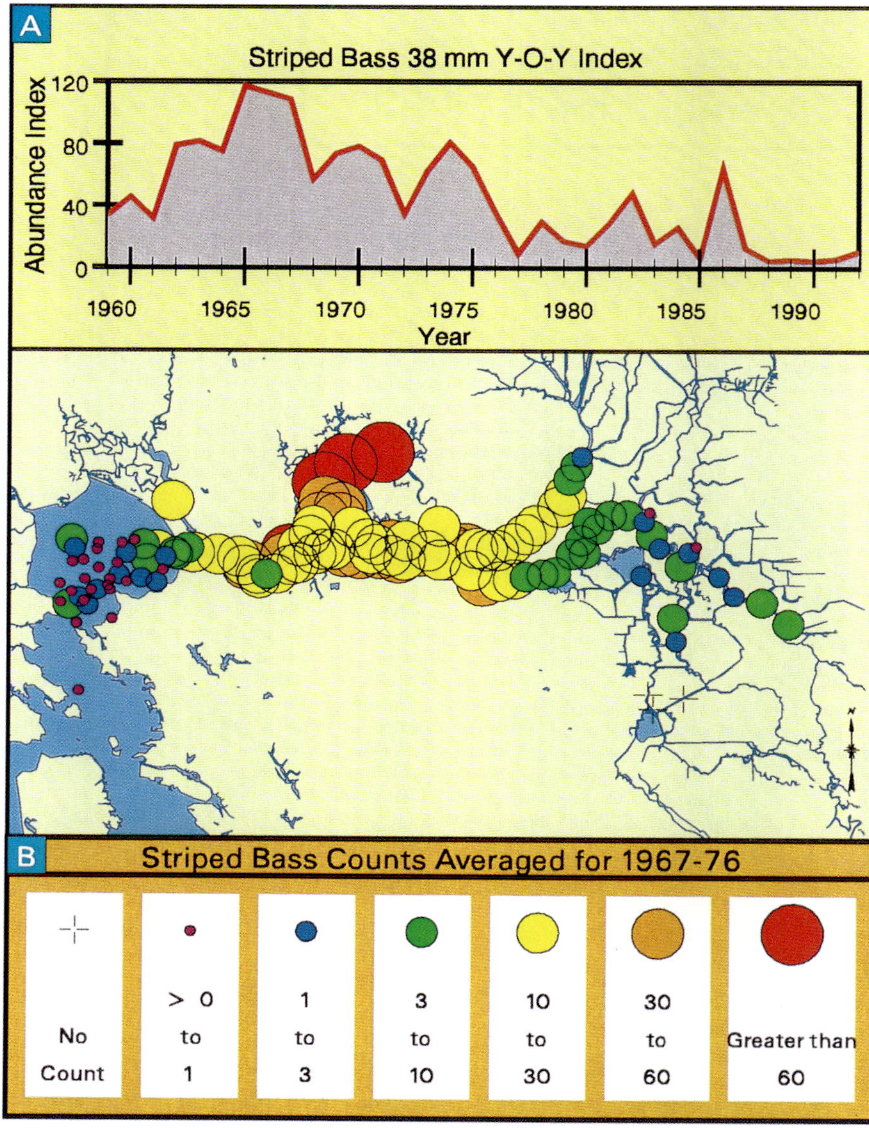

FIGURE 18(A/B). (A) Yearly Y-O-Y striped bass index from summer townet survey depicting declining abundance. (B) Average counts per station from midwater trawl survey prior to decline. Data from California Department of Fish and Game.

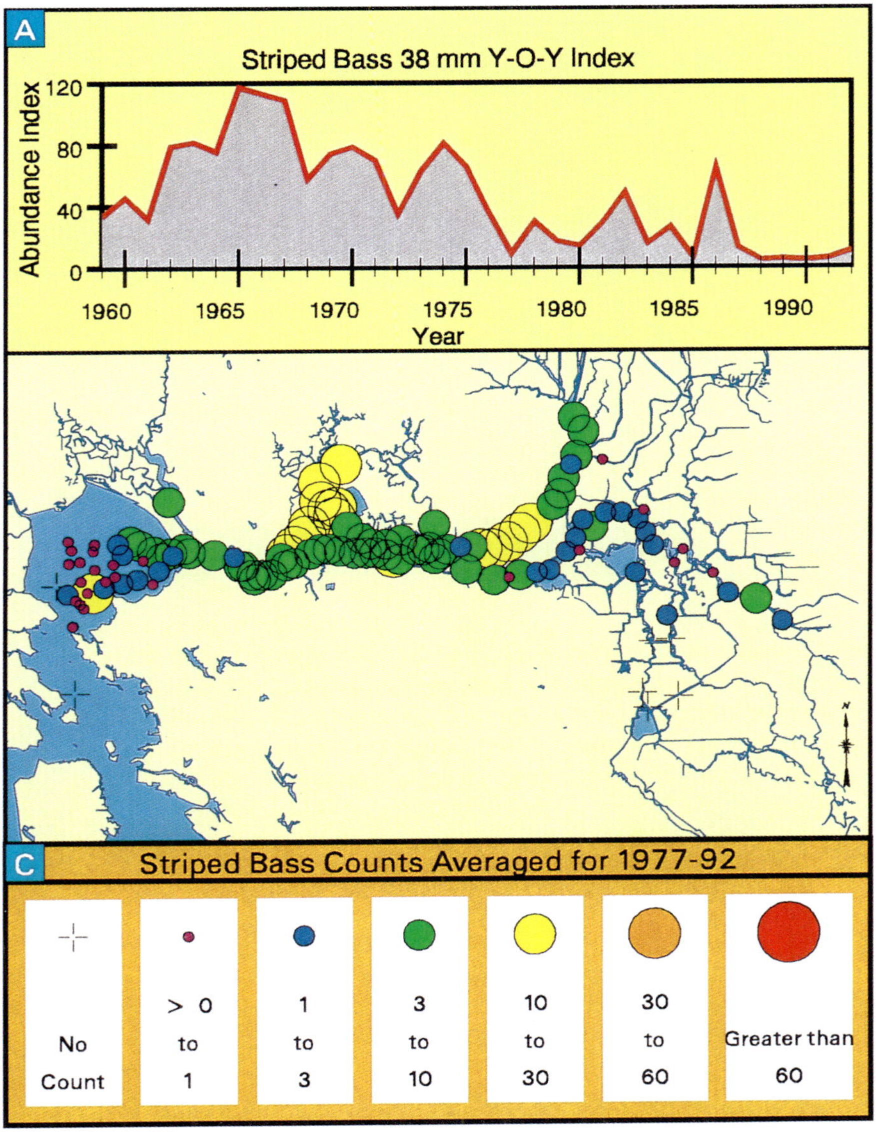

FIGURE 18(A/C). (A) Yearly Y-O-Y striped bass index from summer townet survey depicting declining abundance. (C) Average counts per station from midwater trawl survey following decline. Data from California Department of Fish and Game.

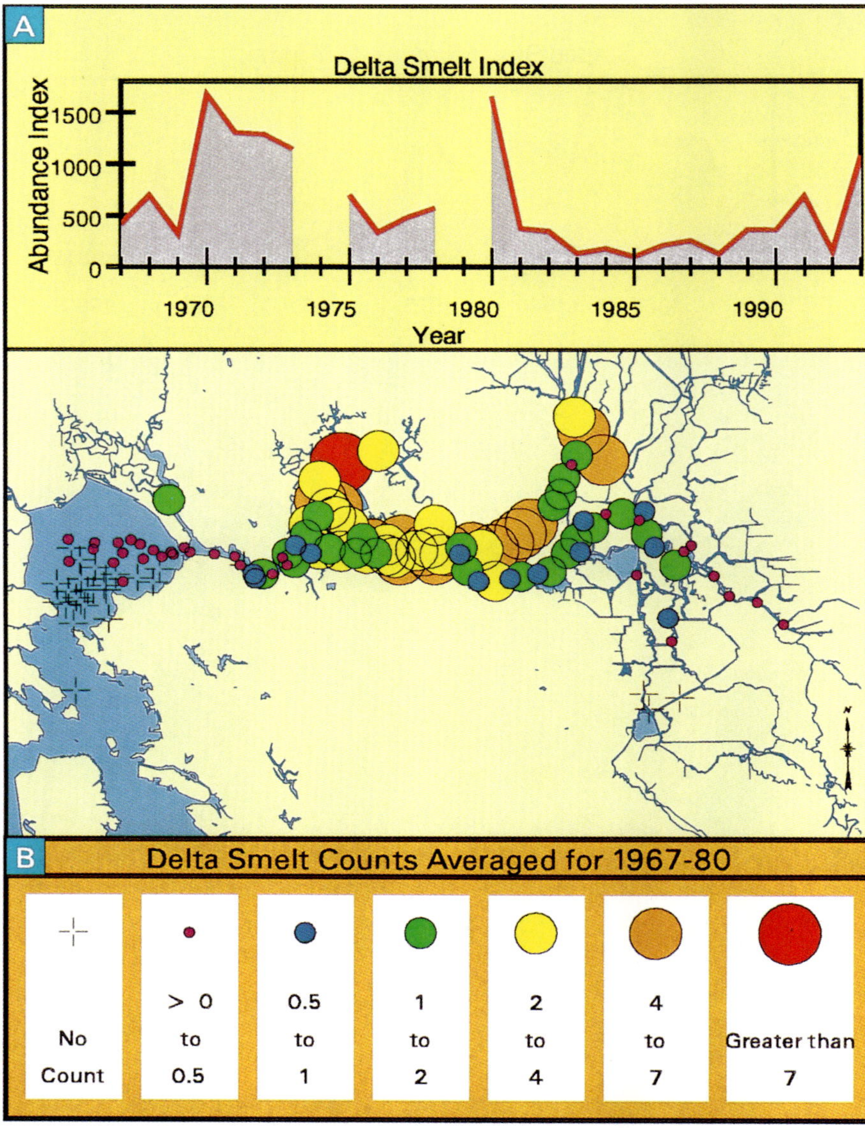

FIGURE 19(A/B). (A) Yearly delta smelt index from fall midwater trawl survey depicting declining abundance. (B) Average counts per station from midwater trawl survey prior to decline. Data from California Department of Fish and Game.

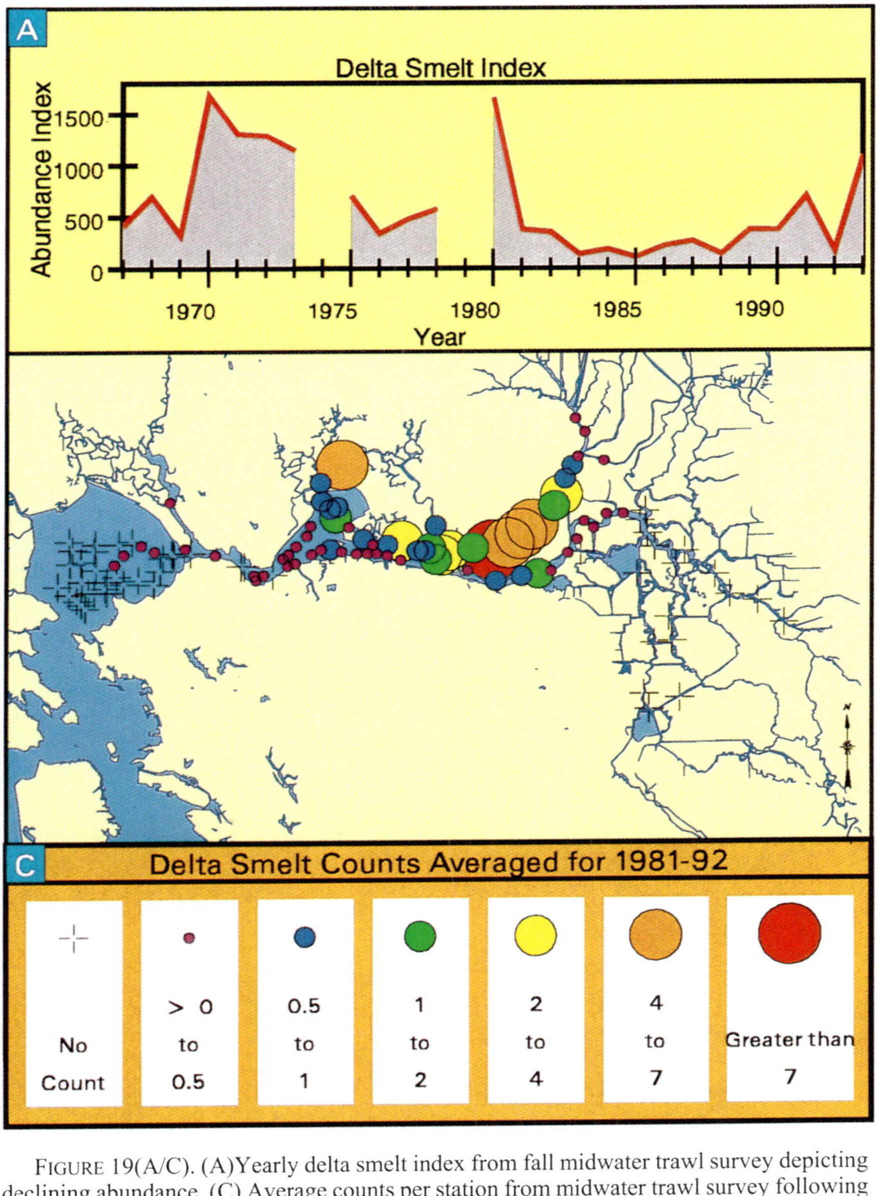

FIGURE 19(A/C). (A)Yearly delta smelt index from fall midwater trawl survey depicting declining abundance. (C) Average counts per station from midwater trawl survey following decline. Data from California Department of Fish and Game.

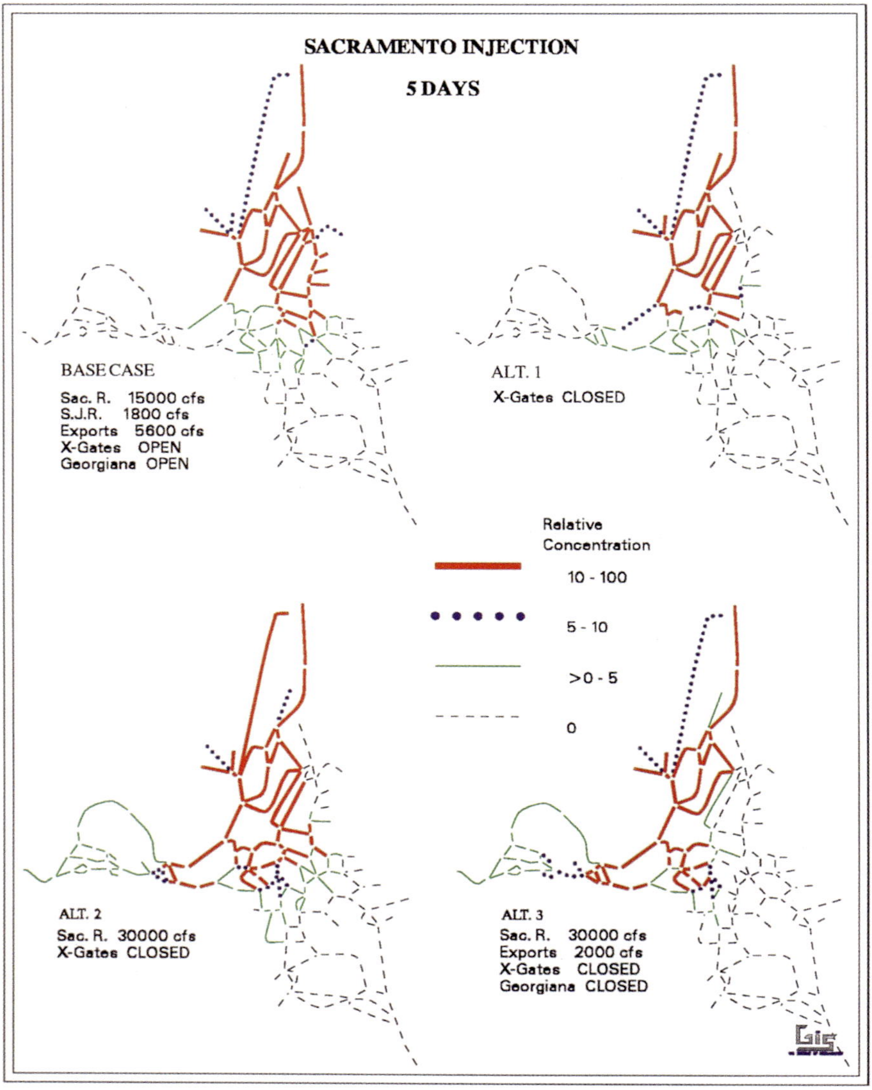

FIGURE 20A. Numeric model estimates of fish transport in the Delta under different management alternatives at 5 days.

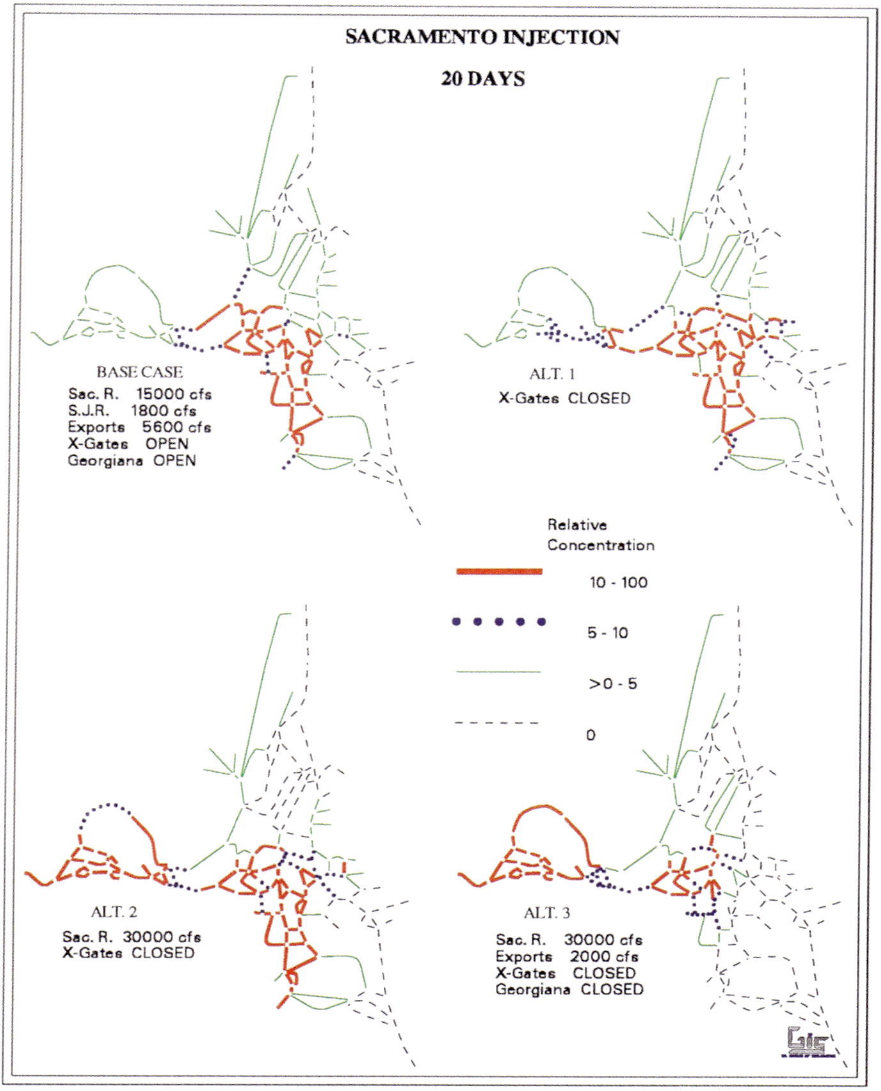

FIGURE 20B. Numeric model estimates of fish transport in the Delta under different management alternatives at 20 days.

the process discussed throughout this paper. Although these dynamic physical/biological interactions are difficult to quantify, further evaluations are needed on fish and food chain organisms that are subject to Delta hydrodynamic conditions as influenced by water project operations.

Numeric Transport Modeling. In 1989, Reclamation initiated a program to evaluate management options that could be taken to enhance the transport of larval striped bass through the Delta (where they are exposed to unscreened diversions and the project export pumps) to their downstream nursery area in Suisun Bay (Arthur *et al.* 1990, 1991a). A component of these studies used a salt transport model developed by Wong (Reclamation) to evaluate several physical and operational alternatives that would enhance downstream transport of striped bass eggs and larvae when they were most exposed to Delta entrainment (Arthur *et al.* 1991b). As a base case for the model evaluations, we chose average flow and export levels observed under Decision 1485, which were then compared over time to several management alternatives.

The results (examples shown in Figs. 20 and 21 for two periods) indicate that under average Decision 1485 flow and export conditions, salt and presumably planktonic organisms subject to transport are primarily transported toward the export pumps via Old and Middle Rivers. These modeling results appear to substantiate trends we observed in our evaluation of physical changes in the Delta. Comparisons with other available Delta hydrodynamic and salt transport models yielded similar results and indicated net transport toward the export facilities under average (and below-average) water years.

As a result of the model studies conducted thus far, Water Resources is developing a "particle tracking model" that will better represent the transport of biota in the Delta. The authors believe this will be the best tool available for evaluating short- and long-term management alternatives in the Delta.

SUMMARY AND CONCLUSIONS

Our analysis included long-term trends of project impacts based on major human alterations in the Delta. Where data are available and appropriate, we based our analysis of project impacts on major changes: prior to 1944, before the large Federal and State dams were first constructed; 1944 to 1951, following construction of Shasta and Friant Dams but before Federal exports in the Delta; 1951 to 1967, period of Federal exports only; 1968 to 1977, period of both the Federal and State exports and when Decision 1379 standards were in place; and 1978 to 1992, when Decision 1485 standards were in place.

Trends indicate that water project operation has impacted Delta hydrodynamics in below-average water years in the following ways:

- Since Delta exports began, outflows have been reduced but more stable from the spring through fall;
- Delta inflows have increased in the summer and fall;
- Winter and spring inflows have decreased;
- Percent of Delta inflow diverted has been much higher in the winter, spring, and fall since the combined SWP and CVP exports;

- Reverse flows in the lower San Joaquin, Old, and Middle River systems show increased duration during the extended drought and have occurred at increased frequency in recent years;
- Maximum entrapment zone intrusion into the Delta occurred most frequently prior to Shasta Dam;
- Least entrapment zone intrusion occurred following Shasta Dam construction until the mid-1970s; and
- Summer entrapment zone intrusions have been of greater duration since mid-1970s.

The effect of water project operations on water chemistry has resulted in:

- A major recycling and addition of salts into the San Joaquin Valley; and
- The addition of THM precursors, generated in Delta water, to urban water supplies.

Water project impacts on the estuarine biology have included:

- Increased suspended sediments being diverted;
- A corresponding increase in water transparency in the Delta;
- Increased intensity of *Melosira granulata* phytoplankton blooms (causing taste and odor problems and being a poor food source) in the western Delta apparently resulting from increased water transparency and hydrodynamic conditions created by operations;
- Reduction of food-web organisms from the Delta to Suisun Bay due in part to increased reverse flows and losses to project and Delta entrainment;
- Peak entrainment of fish by the water projects occurs during periods of high percent inflow diverted; and
- Declines of several key species (examples present: striped bass, Delta smelt, and *Trichocerca* sp. in the San Joaquin River side of the Delta, resulting in further declines in Suisun Bay.

We conclude that the observed declines in some fish and food chain organisms are most likely the result of some combination of: changes in flow patterns and transport in the lower San Joaquin River and central Delta (toward the water project export facilities); flushing of available central and southern Delta habitat (due to decreased residence time); and Delta islands and water project entrainment. These factors are exacerbated during below-normal water years. Directly exporting water through the Delta is much more of a problem than the amount of water exported. We believe a properly designed and operated isolated diversion facility with appropriate upstream monitoring and fish screening facilities would largely restore the Delta as good habitat.

In addition, exporting water directly out of the southern Delta results in the recycling of salts from the San Joaquin Valley, plus adding additional salts from Delta islands and ocean sources to the San Joaquin Valley. Forty years of exporting water with a high salt load has greatly added to major environmental and agricultural problems in the valley. An upstream isolated water diversion facility would have reduced the salt load by about 65 percent.

Organic materials, largely originating from Delta island discharge sources, form THM precursors. THM precursors are a major problem to urban water users that soon will require expensive treatment to remedy. Transporting water from upstream of the Delta would largely reduce this problem.

POTENTIAL SOLUTIONS

The estuary is a complex, highly dynamic system, that is under constant change. Numerous factors, some of which have been summarized in this paper, are also negatively affecting the estuarine water quality and biota. The focus of this paper, however, has been on the Federal and State water projects and how they impact the estuary.

The evidence for negative impacts of water project diversions in the Delta is indisputable and has been recognized and documented over the last 30 years. The early project engineers and biologists recognized the potential problems with direct water transfer through the Delta and predicted the general environmental impacts observed today. The question is more than when and how much do exports have to be reduced under the present system to prevent problems related to fish and water quality. We believe the preponderance of available information indicates that by continuing the direct transfer of water through natural Delta channels as presently done, there is probably no level that does not create some problems. Thus, *how* water is diverted is more of a problem than the quantity diverted.

Numerous advances have been made in technology and techniques to continuously monitor different fish life stages and flows in the Delta since the proposal to construct the Peripheral Canal was defeated on the 1982 referendum. Use of a new particle tracking model being developed by DWR as well as Delta and Suisun Bay 1-dimensional and 3-dimensional hydrodynamic models being developed by the IEP, combined with specific fish and water quality studies, will give project planners much better tools to assess alternatives before implementing any solution. For example, continuous monitoring of early fish life stages upstream of the Delta Cross Channel could provide sufficient warning to close the gates until peaks of fish are past the diversion point. There have also been advances in "fish friendly" methods of diverting water and fish screen technology.

Studies to assess all available options that will optimize resources levels while balancing all the beneficial uses in California are essential before any major changes are made in the system. The official position of the Department of the Interior and, thus, the Bureau of Reclamation is that *all* potential solutions to the Delta problems will be fully evaluated. On December 15, 1994, Reclamation and other Federal and State agencies involved in the Bay/Delta signed an agreement to provide a Bay/Delta Protection Plan (CAL/FED). CAL/FED, working with other interested parties, has provided the standards that are to be adopted by the State Water Resources Control Board in the next few months. These standards are expected to provide short-term measures to reduce the decline in fish abundance. Among other things, the Bay/Delta water quality plan requires evaluation of alternative facilities to transport water through the Delta.

LITERATURE CITED

Argent, G. (1989). *Layperson's Guide to California Water*, R. S. Sudman, ed. Water Education Foundation, Sacramento, CA. 20 pp.

Arthur, J. F., *et al.* (1990). *Continuous Monitoring of Striped Bass Eggs and Larvae in the San Francisco Bay-Delta Estuary: A Potential Management Tool.* U.S. Bur. Reclamation Rept., Sacramento, CA. 55 pp.

Arthur, J. F. (1987). *River Flows, Water Project Exports, and Water Quality Trends in the San Francisco Bay-Delta Estuary*, Exhibit 111, U.S. Bur. Reclamation, Mid-Pacific Region, Sacramento, CA. 57 pp.

Arthur, J. F. (1975). Preliminary studies on the entrapment of suspended materials in Suisun Bay, San Francisco Bay-Delta Estuary. Pages 17-36 *in* R. L. Brown, ed., *Proceedings of a Workshop on Algal Nutrient Relationships in the San Francisco Bay and Delta*, November 1973. California State Dept. Water Resources, Sacramento, CA.

Arthur, J. F., and M. D. Ball. (1980). *The Significance of the Entrapment Zone Location to Phytoplankton Productivity in the San Francisco Bay-Delta Estuary*, U.S. Dept. Interior, Water & Power Resources Serv. 109 pp.

Arthur, J. F., and M. D. Ball. (1979a). Factors influencing the entrapment of suspended material in the San Francisco Bay-Delta Estuary. Pages 143-174 *in* J. T. Conomos, ed., *San Francisco Bay: The Urbanized Estuary*. Pacific Division, American Association for the Advancement of Science, San Francisco, CA.

Arthur, J. F., and M. D. Ball. (1979b). *Factors Influencing the Low Phytoplankton Standing Crop in Suisun Bay during the 1976-77 Drought*. U.S. Bur. Reclamation, Mid-Pacific Region, Sacramento, CA. 84 pp.

Arthur, J. F., and M. D. Ball. (1978). *Entrapment of Suspended Materials in the San Francisco Bay-Delta Estuary*. U.S. Bur. Reclamation, Sacramento, CA. 106 pp.

Arthur, J. F., M. D. Ball, L. Hess, C. Liston, S. Hiebert, and G. Collins. (1991a). *Egg and Larvae Management Studies San Francisco Bay-Delta Estuary*. Executive Summary, U.S. Dept. Interior, Bur. Reclamation, Mid-Pacific Region, Sacramento, CA. 18 pp.

Arthur, J. F., H. F. N. Wong, M. D. Ball, and L. Hess. (1991b). *Egg and Larvae Management Studies San Francisco Bay-Delta Estuary: Evaluation of Potential Striped Bass Management Scenarios by Use of a Numerical Salt Transport Model*. U.S. Dept. Interior, Bur. Reclamation, Sacramento, CA. 11 pp.

Ball, M. D. (1987). *Phytoplankton Dynamics and Planktonic Chlorophyll Trends in the San Francisco Bay-Delta Estuary*, Exhibit 103. U.S. Bur. Reclamation, Sacramento, CA. 95 pp.

Ball, M. D. (1977). *Phytoplankton Growth and Chlorophyll Levels in the Sacramento-San Joaquin Delta through San Pablo Bay*. U.S. Bur. Reclamation, Sacramento, CA. 96 pp.

Ball, M. D., and J. F. Arthur. (1979). Planktonic chlorophyll dynamics in the Northern San Francisco Bay and Delta. Pages 265-286 *in* T. J. Conomos, ed., *San Francisco Bay: The Urbanized Estuary*. Pacific Division, American Association for the Advancement of Science, San Francisco, CA.

Carr, J. K. (1941). *A Memorandum Report of Investigations in the Sacramento-San Joaquin Delta Bearing on the Location of the Proposed Delta Cross Channel*, Dept. Interior, Bur. Reclamation, Mid-Pacific Region, Sacramento, CA. 10 pp plus maps and appendix.

Chadwick, H. K., D. E. Stevens, and L. W. Miller. (1977). Some factors regulating the striped bass population in the Sacramento-San Joaquin Estuary, California. Pages 18-35 *in* W. van Winkle, ed., *Proceedings of the Conference on Assessing the Effects of Powerplant Induced Morality on Fish Populations*. Pergamon Press, New York.

Cloern, J. E., A. E. Alpine, B. E. Cole, R. L. J. Wong, J. F. Arthur, and M. D. Ball. (1983). River discharge controls phytoplankton dynamics in the northern San Francisco Bay Estuary. *Estuarine Coastal Shelf Sci.* 16:415-429.

Collins, G. M., J. F. Arthur, M. D. Ball, S. Hiebert, and L. Hess. (1992). *Egg and Larvae Management Studies San Francisco Bay-Delta Estuary: Reconnaissance Surveys of*

Striped Bass Egg and Larval Distribution at the Delta Cross-Channel. U.S. Dept. Interior, Bur. Reclamation, Mid-Pacific Region, Sacramento, CA. 14 pp.

Conomos, T. J., D. H. Peterson, P. R. Carlson, and D. S. McCulloch. (1970). *Movement of Seabed Drifters in the San Francisco Bay Estuary Adjacent Pacific Ocean*. U.S. Geol. Survey Circ. 637B. 7 pp.

Department of Water Resources. (1994). *Five-Year Report of the Municipal Water Quality Investigation Program*, Executive Summary. California State Dept. Water Resources, Sacramento, CA. 20 pp.

Department of Water Resources. (1993). *Sacramento Delta San Joaquin Atlas*. California State Dept. Water Resources, Sacramento, CA. 115 pp.

Department of Water Resources. (1969 [April]). *Waste Water Quality, Treatment, and Disposal*, Bull. No. 127, Appendix D. California State Dept. Water Resources, Sacramento, CA. pp. 47-54.

Department of Water Resources. (1965 [January]). *San Joaquin Master Drain*. Bull. No. 127. California State Dept. Water Resources, Sacramento, CA. pp. 33-41.

Federal Water Pollution Control Administration. (1968 [August]). *Effects of the San Joaquin Master Drain on Water Quality of the San Francisco Bay and Delta*, Central Pacific Basins Comprehensive Water Pollution Control Project, Appendix Part C, Nutrients and Biological Response. U.S. Dept. Interior, Federal Water Pollution Control Adm., Southwest Region.

Federal Water Pollution Control Administration. (1967 [January]). *Effects of the San Joaquin Master Drain on Water Quality of the San Francisco Bay and Delta*, Central Pacific Basins Comprehensive Water Pollution Control Project, U.S. Dept. Interior, Federal Water Pollution Control Adm., Southwest Region, San Francisco, CA..

Fish and Wildlife Service. (1992 [July]). *Measures to improve the protection of Chinook salmon in the Sacramento/San Joaquin River Delta: Expert Testimony of United States Fish and Wildlife Service on Chinook Salmon Technical Information for Water Rights Phase of the Bay/Delta Estuary Proceedings*. FWS Testimony to State Water Resources Control Board, WRINT-USFWS-7.

Herbold, B., A. D. Jassby, and P. B. Moyle. (1992). *Status and Trends Report on Aquatic Resources in the San Francisco Estuary*. Report to the San Francisco Estuary Project. 257 pp plus appendices.

Hollibaugh, J. T., and I. Werner. (1991). *Potamocorbula amurensis: Comparison of Clearance Rates and Assimilation Efficiencies for Phytoplankton and Bacterioplankton*. Interagency Ecological Studies Program for the Sacramento-San Joaquin Estuary. Tech. Rept. 29. Sacramento, CA. 42 pp.

Interagency Delta Committee. (1965 [January]). *Plan of Development-Sacramento-San Joaquin Delta: Recommendation Prepared as a Recommendation for a Plan of Action to the California Department of Water Resources*. U.S. Bur. Reclamation and U.S. Corps of Engineers. 32 pp.

Interagency Ecological Studies Program. (1993). *IESP 1993 Data Summary Report*, prepared by Olof Hansen, IESP Technical Information Specialist. Sacramento, CA. 91 pp.

Interagency Ecological Studies Program. (1987 [August]). *Factors Affecting Striped Bass Abundance in the Sacramento-San Joaquin River System*, Tech. Rept. 20. California State Dept. Water Resources, Sacramento, CA.

Interagency San Joaquin Valley Drainage Program. (1990). *A Management Plan for Agricultural Subsurface Drainage and Related Problem on the Westside San Joaquin Valley*. Final report of the San Joaquin Valley Drainage Program. U.S. Dept. Interior and California Resources Agency. 183 pp.

Jassby, A. D. (1995). Hydrodynamic influences on interannual chlorophyll variability. Pages 1-2 *in* R. L. Brown, ed., *Interagency Ecological Program for the Sacramento-San Joaquin Estuary*. California Dept. Water Resources, Sacramento, CA.

Jassby, A. D., and T. M. Powell. (1994). Hydrodynamic influences on interannual chlorophyll variability in an estuary: Upper San Francisco Bay-Delta (California, U.S.A.). *Estuarine Coastal Shelf Sci. 39*(6):595-618.

Kholhorst, D. W, D. E. Stevens, and L. Miller. (1992). *A Model for Evaluating the Impacts of Freshwater Outflow and Export on Striped Bass in the Sacramento-San Joaquin Estuary*. Bay-Delta and Special Water Projects Div., California Dept. Fish & Game, Stockton, CA. 17 pp.

Kimmerer, W. (1992). *An Evaluation of Existing Data in the Entrapment Zone of the San Francisco Bay Estuary*. IESP Tech. Rept. 33. 49 pp.

Monroe, M. W., and J. Kelly. (1992). *State of the Estuary*. San Francisco Estuary Project, San Francisco, CA. 235 pp. plus appendices.

Nichols, F. H., J. K. Thompson, and L. E. Schemel. (1990). Remarkable invasion of San Francisco Bay (California, USA) by the Asian clam *Potamocorbula amurensis* .2. Displacement of a former community. *Marine Ecol. Progr. Ser*. 66:95-101.

Obrebski, S., J. J. Orsi, and W. J. Kimmerer. (1992). *Long-Term Trends in Zooplankton Distribution and Abundance in the Sacramento-San Joaquin Estuary of California*. IESP Tech. Rept. 32. Sacramento, CA. 42 pp.

Orsi, J. J. (1986). Interactions between diel vertical migration of a mysidacean shrimp and two-layered flow. *Hydrobiologia* 137:79-87.

Stevens, D. E. (1980). Factors affecting the striped bass fisheries of the West Coast. Pages 15-28 *in* H. Clepper, ed., *Proceedings of the Fifth Annual Recreational Fisheries Symposium*. Sport Fishing Institute, Washington, D.C.

Stevens, D. E. (1979). Environmental factors affecting striped bass *Morone saxatilis* in the Sacramento-San Joaquin Estuary. Pages 469-478 *in* T. J. Conomos, ed., *San Francisco Bay: The Urbanized Estuary*. Pacific Division, American Association for the Advancement of Science, San Francisco, CA.

Stevens, D. E., D. W. Kohlhorst, L. W. Miller, and D. W. Kelley. (1985). The decline of striped bass in the Sacramento-San Joaquin Estuary, California. *Trans. American Fish. Soc.* 114:15-28.

Stewart, D., J. F. Arthur, and M. D. Ball. (1984). *Analysis of the Effects of Hydrodynamics on Phytoplankton Standing Crops in the Sacramento-San Joaquin Western Delta*. Final Rept., Hydraulic Subcommittee of the Phytoplankton Task Force, U.S. Bur. Reclamation, Sacramento, CA. 70 pp.

Turner, J. L. (1992 [July]). *Effects of Geographic Distribution of Larval Striped Bass in Determining Year Class Size of Striped Bass in the Sacramento-San Joaquin Estuary: Environmental and Energy Consultants*. U.S. Bur. Reclamation, Exhibit No. 100 for Water Rights Phase of the Bay-Delta Estuary Proceedings.

Turner, J. L., and H. K. Chadwick. (1972). Distribution and abundance of young striped bass *Morone saxatilis* in relation to riverflow in the Sacramento-San Joaquin Estuary. *Trans. American Fish. Soc.* 101(3);442-452.

AN EVALUATION OF THE EFFECTIVENESS OF FISH SALVAGE OPERATIONS AT THE INTAKE TO THE CALIFORNIA AQUEDUCT, 1979-1993

RANDALL BROWN AND SHEILA GREENE
California Department of Water Resources
and
PAT COULSTON AND SCOTT BARROW
California Department of Fish and Game

We describe the results of the last 14 years of experience gained in operating a large fish protection facility at the intake to the California Aqueduct in the Sacramento-San Joaquin Delta. The Department of Water Resources constructed the behavioral barrier (louver) system in the late 1960s for a maximum flow of about 180 m^3/s. The facilities have since been modified extensively, and the capacity has been increased to about 290 m^3/s. Although more than 40 species of fish have been entrained in the diversion, we focus on the facilities' effectiveness at screening Chinook salmon. Screening efficiency varies with fish length, ranging from 70 to 80 percent for juvenile salmon about 100 mm in length. Significant prescreen losses appear to be caused by subadult striped bass predation in a forebay in front of the fish facility. Small additional losses occur due to handling in the salvage process and when trucking the salvaged fish to Delta release sites about 40 kilometers from the intake. We also briefly describe mitigation measures to offset direct losses of Chinook salmon at the intake as well as actions that might be taken to reduce losses of these and other fish.

California voters authorized the State Water Project in 1960 to provide water from the Feather River, a tributary to the Sacramento River, to urban and agricultural users in the San Francisco Bay area, the San Joaquin Valley, and Southern California. Key project features include Oroville Dam and Reservoir on the Feather River, Harvey O. Banks Delta Pumping Plant in the Sacramento-San Joaquin Delta, the 714-kilometer California Aqueduct, and several storage reservoirs south of the Delta (Fig. 1). The general premise of the State Water Project was to store flows in Oroville Reservoir during the winter and spring months and move water to south-of-Delta storage when flows were naturally high or when water was released from storage in the summer and fall. A more complete description of the State Water Project can be found in DWR 1994.

State Water Project planners included the Feather River Hatchery and a fish protective facility at the intake to the California Aqueduct as the project's two principal fish mitigation and protection features. Our objective is to provide an overview of how the John E. Skinner Fish Protective Facility (Skinner Fish Facility in this report) operates, the species and numbers of fish captured at the facility during 1979 to 1993, major factors contributing to fish losses, and ideas that have been advanced to minimize losses associated with State Water Project diversions from the Delta. (Information on the Feather River Hatchery can be found in Brown & Greene 1994).

Although more than 40 species of fish are captured at the Skinner Fish Facility, in this article we focus on Chinook salmon, *Oncorhynchus tshawytscha*. We

FIGURE 1. Major features of the California State Water Project.

selected this species because more is known about how it fares in the salvage process and because of its widespread abundance and economic, environmental, and cultural importance in California and the Pacific Northwest.

The article briefly describes the Bay/Delta and State Water Project diversions, life history of the Chinook salmon, the salvage process, estimated losses of salmon during each step of the process, and measures taken to minimize or offset these losses. A final section discusses some additional actions being taken or considered to resolve problems associated with State Water Project diversions from the Delta.

During the past 10 to 15 years, several fish species using the estuary have declined in abundance, including striped bass juveniles, delta smelt, and winter-run Chinook salmon. There are several possible reasons for the observed declines, such as: direct and indirect losses at the State, Federal, and agricultural diversions in the Delta; decreases in Delta inflow and outflow resulting from upstream and in-Delta diversions and droughts; effects of toxic materials such as agricultural pesticides; and changes in the estuarine ecosystem caused by accidental and purposeful introductions of non-indigenous fish and invertebrate species.

Although there is no consensus about the relative importance of these and other factors in controlling the distribution and abundance of a particular fish species, almost everyone agrees that losses due to diversions may be contributing or limiting factors for some important species. Attempts to "fix the Delta" almost always include measures designed to minimize or offset fish losses at State Water Project and Central Valley Project diversions. A December 15, 1994, Bay/Delta protection agreement between representatives of agricultural, urban, and environmental water interests and State and Federal agencies provided for seasonally-adjusted pumping limitations (export/inflow ratios) to minimize direct losses of fish at the SWP and CVP Delta intakes. The State Water Resources Control Board adopted these de facto pumping limitations as part of its Water Quality Control Plan for the estuary (SWRCB 1995).

THE DELTA AND STATE WATER PROJECT PUMPING

SWP pumps are located in the southern Delta near the town of Tracy and only about a mile from the intake to the Central Valley Project's Delta-Mendota Canal (Fig. 2). To better understand fish entrainment into the SWP, it is necessary to have a basic appreciation of a few physical, hydrologic, and operational features of the Delta and SWP Delta operations. More complete descriptions are found in Brown and Greene (1992) and DWR (1993).

Most of the water diverted from the Delta by State, Federal, and local interests originates in the Sacramento Valley watershed. Water coming down the Sacramento River to Delta diversions enters the central Delta by way of the CVP's gated Delta Cross Channel and two natural channels, Georgiana Slough and Threemile Slough. Water movement through these channels is dependent on temporal variation in the differences in elevation between the Sacramento and San Joaquin river systems caused by streamflow and tides and is essentially independent of project pumping in the southern Delta (Brown & Greene 1992). When CVP and SWP pumping exceeds flow entering the Delta through these channels plus flow from the San Joaquin River, Sacramento River water can also move into the Delta by going around the tip of Sherman Island and up the lower San Joaquin River. Net upstream flow in the lower San Joaquin River is called "reverse flow."

Fish arrive at the pumps from interior Delta channels, most of them probably from the lower San Joaquin River and central Delta or, in the case of juvenile Chinook salmon, fish that are migrating down the San Joaquin River toward the ocean. A much lower percentage of fish originating in the Sacramento River system is also salvaged at the State and Federal facilities.

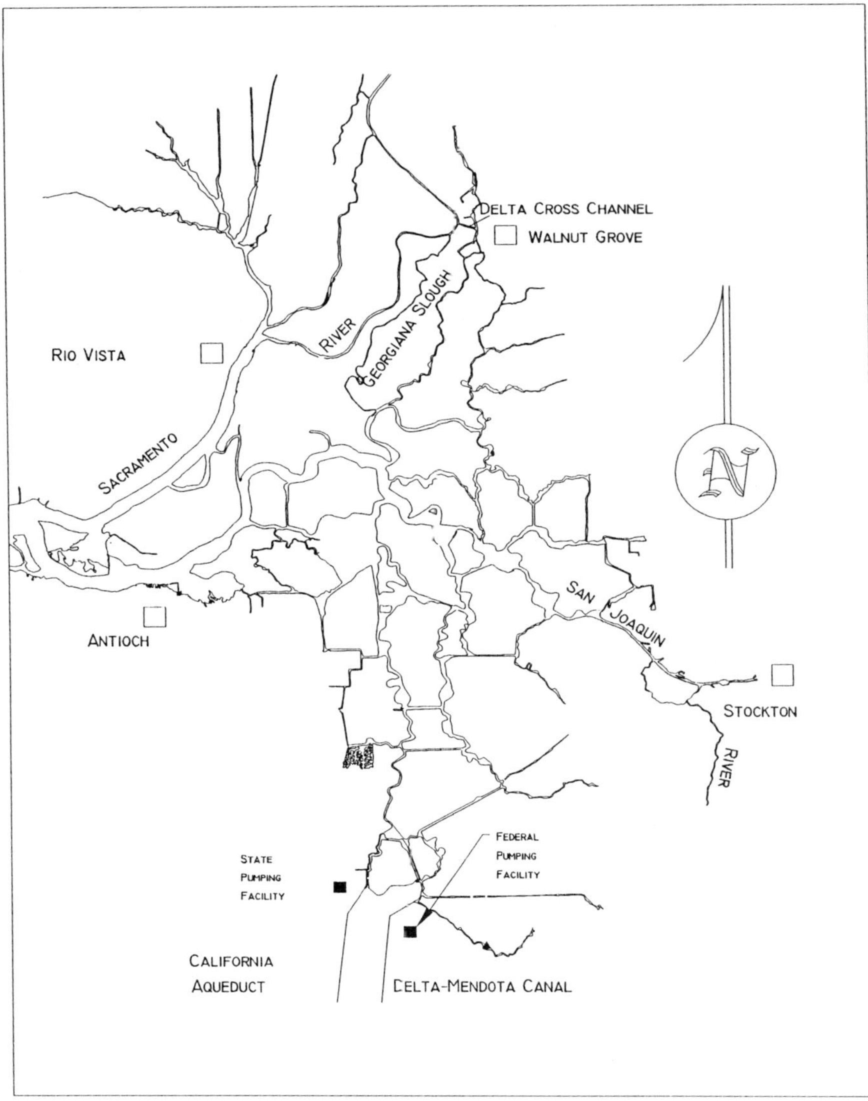

FIGURE 2. The Sacramento/San Joaquin Delta and the intake to the California Aqueduct.

One confounding feature of Delta water movement is the dominance of tidal flows. Reverse flow in the lower San Joaquin River above its confluence with the Sacramento River is often postulated as a fish transport mechanism; however, preliminary results of particle transport modeling indicate that reverse flow (also called "negative Qwest") does not appear to move eggs and larvae from the Sacramento River toward the pumps (Chung & Smith 1993). On the other hand, Wendt (1987) found that salvage of small striped bass at Skinner Fish Facility was correlated with calculated net flow in the lower San Joaquin River. A U.S. Fish and Wildlife Service (1992) regression model had Qwest as a statistically significant factor along with temperature, percent water diverted off the Sacramento River, and combined SWP/CVP pumping to explain survival of Sacramento River juvenile salmon through the Delta. Subsequent analyses indicated that the importance of Qwest relative to other factors in the model may have been overstated (Brown & Greene 1992). Whatever the mechanism, and it may even be a stochastic process, some small percentage of juvenile fish in the Sacramento River does move through the Delta and are salvaged by the SWP and CVP pumps.

The pumps have a combined capacity of about 420 cubic meters per second, with the SWP at 290 and the CVP at 130 cubic meters per second. At present, however, U.S. Army Corps of Engineers permits generally allow a maximum of only about 200 cubic meters per second at the SWP pumps.

ASPECTS OF THE LIFE HISTORY OF CENTRAL VALLEY CHINOOK SALMON IMPORTANT TO SCREENING

The following is a summary of some of the important features of Chinook salmon life history that bear on successfully screening juvenile salmon from Delta diversions. Population estimates are included to provide an idea as to why screening and other measures are needed to increase salmon numbers.

Chinook salmon are found in many of the larger streams along the Pacific Rim from California to Japan (Moyle 1976). All Chinook salmon are anadromous in that they spawn in freshwater streams, and the young move to the ocean at an early stage. During their 1-5 years of ocean residence, the salmon grow to adult size, approach sexual maturity and, using the sense of smell and other yet unknown cues, head for streams in which they hatched and reared to repeat the cycle. All Chinook salmon die after spawning.

Life history characteristics within the species and throughout its range vary considerably, with the Central Valley races having some of the greatest variability. Central Valley streams support four Chinook races, or runs — the fall, late-fall, winter, and spring — named after the time the adults move from the ocean into San Francisco Bay en route to their natal spawning grounds.

Following is a short summary of some of the important life history and other features of each race as they relate to Delta diversions. Figure 3 has recent historical escapement (run sizes) for each race. A more complete description of Central Valley salmon life cycles is found in Vogel & Marine (1991).

The fall run, presently the most numerous race in the Central Valley, occurs in the mainstem and tributaries of the Sacramento River and in tributaries to the San Joaquin River. Fall-run individuals generally reach the spawning grounds sexually mature and ready to spawn, with most of the spawning occurring at low-elevation stretches of rivers just above the valley floor. Adults mostly spawn in October and November with most of the young salmon, called "smolts," moving downstream and through the estuary in April through mid-June. Fall-run smolts are generally 70-80 mm long and have physiologically adapted to making the osmoregulatory

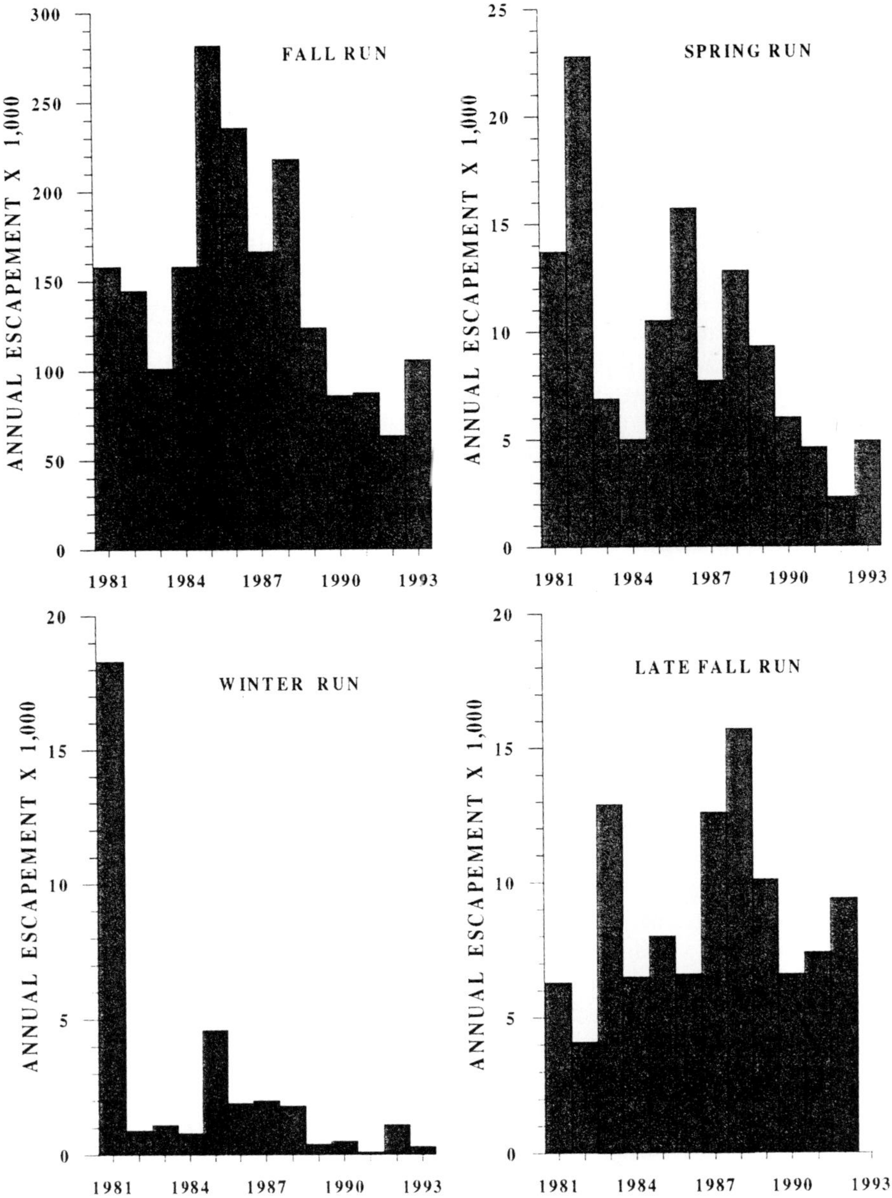

FIGURE 3. Estimated spawning escapement of four races of Central Valley Chinook salmon.

transition from fresh water to the ocean. Some young fall Chinook salmon do enter the estuary in late winter/early spring as fry; others may not leave their home stream until about a year after hatching. The early and late outmigrants can range in size from 40 to nearly 200 millimeters. Fall Chinook spawn mostly at 3 years of age, after 2 years in the ocean. Fall-run salmon are artificially propagated at several Central Valley hatcheries: Coleman National Fish Hatchery, Feather River Hatchery, Nimbus Hatchery, and Mokelumne and Merced River Fish Facilities. Smolts from the Coleman, Merced, and Mokelumne facilities are released upriver and migrate through the Delta. The Department of Fish and Game transports smolts from the Feather River and Nimbus hatcheries to San Pablo Bay in 7500 L tank trucks, so these juveniles avoid mortality factors associated with migration through the lower rivers and the Delta.

The late-fall run is relatively small numerically, but the adults achieve the largest average size of the four Central Valley races, mainly because a higher percentage of the adults spawn at 4 and 5 years than in the other three races. Most of the spawning occurs in the mainstem Sacramento River during February and March. Although not much is known about late-fall outmigration, it appears that most juvenile late-fall salmon pass through the estuary in November through January, about a year after hatching. They enter the Delta ranging in size from about 120 to 200 millimeters. There is a small late-fall run artificial propagation program at Coleman National Fish Hatchery, and all of the smolts are released upriver, generally in January.

The winter run is now the smallest of the four Central Valley races in size and abundance and is listed as endangered by both the State of California and the Federal Government. Winter Chinook mostly spawn during May through July in the mainstem Sacramento River, usually in the reach between Red Bluff and Redding. Most juveniles leave the upstream areas by the end of December; peak smolt emigration through the Delta is in late February through mid-April. Smolts are generally in the 80-150 mm length range when moving through the Delta. There is a small winter Chinook artificial propagation program at Coleman National Hatchery with the fish released up river in December and January.

Although perhaps once the most abundant of the four races of Central Valley Chinook, the spring run is now among the least abundant and a petition has been prepared to list the race under the Federal Endangered Species Act. Historically, adult spring run migrated to higher elevation streams, where they held over in cold pools during summer and spawned in late August through September. Dams have blocked access to many of these streams and in some streams, such as the Feather River, eliminated the geographic isolation between fall and spring runs. Due to the present lack of physical isolation, some interbreeding now occurs between the two races. There are still small, genetically distinct spring-run populations on Mill Creek, Deer Creek, and perhaps Butte Creek — all tributaries to the Sacramento River. There is almost no information on fish length and timing of spring run movement through the Delta. Some data indicate that spring run may move into and through the Delta as fry, smolts, and yearlings at about any time of the year except July through September. Spring run are propagated at the Feather River Hatchery, with the production released in San Pablo Bay.

Fish protection at the State Water Project Delta intakes depends on the timing of juvenile outmigration through the upper Delta as well as the size of the fish. The bulk of the outmigration, in terms of total numbers, is in April, May, and June when fall-run smolts move to the ocean. Although accurate estimates are not available, it appears that during this spring period the numbers of annual juvenile outmigrants from the Sacramento Valley are in the low tens of millions and from the San Joaquin Basin are in the low hundreds of thousands. However, other fall-run life stages and life stages of the other three races may be present at almost any time and size. This variation in size and timing makes it difficult to eliminate, reduce, or offset salmon losses at the State Water Project intake. The overall screening operation is further complicated by timing, size, and physiological makeup of the other 40 or so fish species that may encounter the screens.

Juvenile Chinook salmon all look alike and in the Delta cannot be accurately sorted by race. The 1992 and 1993 biological opinions issued by the National Marine Fisheries Service contained incidental take statements that required an estimate of the numbers of winter-run juveniles taken (killed) at the SWP/CVP Delta intakes. The opinions called for adoption of a length criteria system developed by Department of Fish and Game staff; a system that uses spawning time and growth rates to predict the range of size that fish of each race might have on a given date.

Department of Water Resources staff converted this size criteria system to a graphical representation of the length intervals on which individual fish can be plotted to determine their race. As shown in a plot of untagged salmon collected at a rotary screw trap near the intake to the Glenn-Colusa Irrigation District (Fig. 4), this size by date classification system is not completely reliable in that there are not discrete groups of fish within the size band for each race. The large overlap between the races, especially late-fall and winter, is due to the natural variation in spawning time, emergence and growth among the races and even within each race among spawning and rearing locations. It does, however, provide a qualitative estimate of run timing and has proven to be a useful tool in sorting winter run from the other races. The Department of Water Resources has contracted with a geneticist at the University of California at Davis' Bodega Marine Laboratory for a 3-year study to determine if micro-satellites in the nuclear DNA can be used to provide a more quantitative race identification system.

FISH SCREENS AND THE FISH SALVAGE PROCESS

The State Water Project Delta screening system, illustrated in Fig. 5, was patterned after a system developed for the Central Valley Project Delta intake in the early 1950s. The screens are based on the ability of fish to avoid being entrained (a behavioral barrier) as compared to a more typical screen designed to physically exclude fish (a positive barrier). The system works as follows.

Clifton Court Forebay (not shown on Fig. 5), a 37 million-cubic-meter regulating reservoir, is located between the canal intake (and the screens) and the Delta. Inflow of water and entrained fish from the Delta is controlled by three radial gates, which are opened at higher tidal elevations to fill the reservoir. The presence of the forebay helps regulate water level and allows project operators to control water

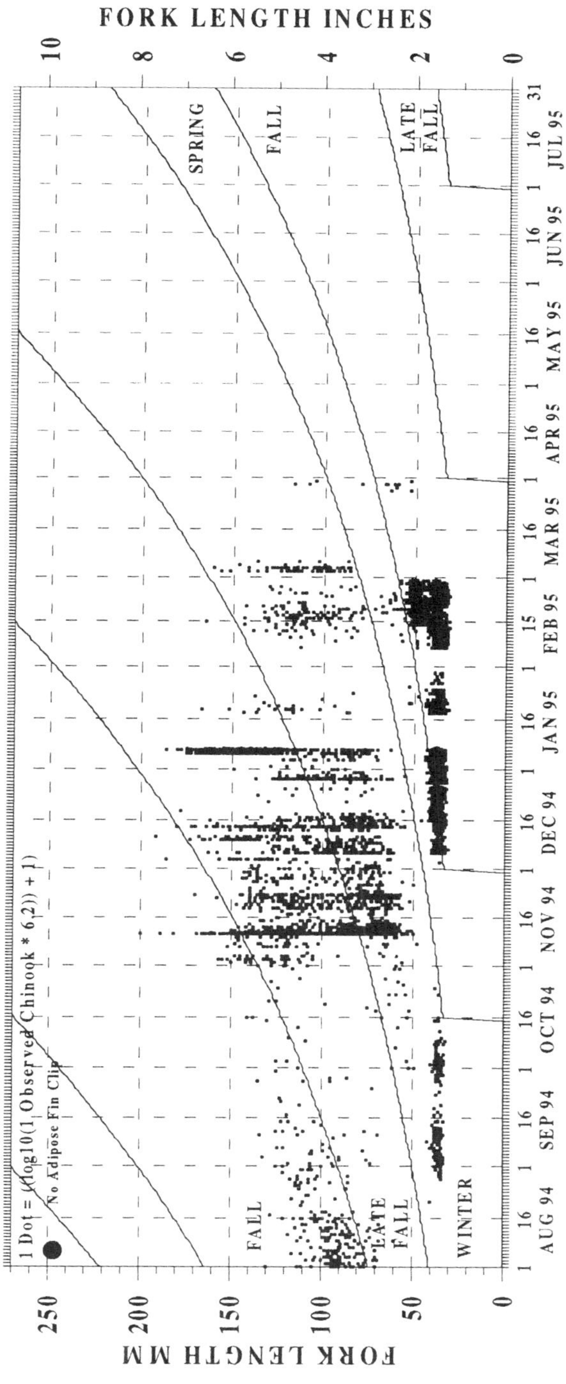

FIGURE 4. Classification by race of juvenile Chinook salmon collected at the Glenn-Colusa Irrigation District by rotary screw trap, August 1994-March 1995.

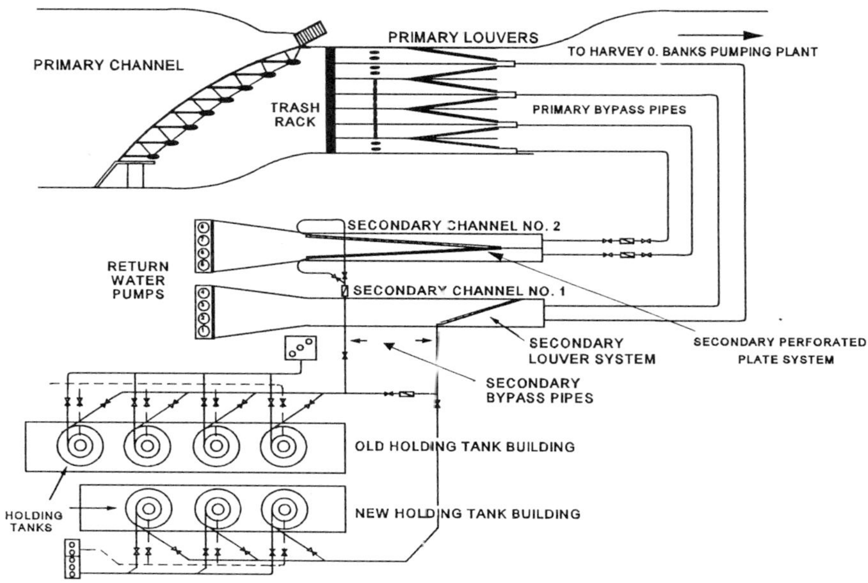

FIGURE 5. Schematic diagram of the John E. Skinner Fish Protective Facility.

velocity at the screens. Water velocity is an important determinant of screen efficiency.

A trashrack in front of the screens collects much of the floating debris before it encounters the screens themselves.

The primary bays contain louver screens (about 6 m high by 25 m long) on each side of the "V" and a solid wall in the middle, which ends just before the apex of the "V." The louver screens themselves are similar to large vertical venetian blinds, with 2.5 cm openings along the face, which is at a 15 degree angle with the flow. The operating principle is that fish will sense the turbulence along the screen face and avoid entering the spaces between each slat in the louver. The ability to avoid entering the slot depends on fish size and water velocity. In the mid-1980s, at the request of the Department of Fish and Game, the Department of Water Resources installed center walls in all bays to improve fish salvage efficiency, especially for striped bass.

Fish going through the primary louvers wind up in the California Aqueduct and may be killed in one of the pumping plants along the aqueduct. Many of the non-salmonids apparently survive the process, since there are extensive fisheries for catfish and striped bass in the aqueduct itself and in reservoirs from the Delta to Silverwood, near Riverside.

Fish making it to the apex of the "V" enter a bypass pipe leading to the secondary screening system. The purpose of the secondary screens is to further concentrate the fish before they enter the holding tank buildings. The original secondary was a louver system similar to the primary. In the mid-1980s, the Department of Water

Resources constructed another parallel, secondary system but with a positive barrier, flat-plate screen with about 6-mm openings. At the end of the secondaries, bypasses lead to holding tanks.

The five holding tanks, in two separate buildings, are about 6m diameter by 6 m deep and contain about 80 cubic meters of water at nominal operating depth of 2.75 meters. Fish enter the holding tanks from the secondary bypasses. Originally there was only one series of holding tanks but in the early 1990s, at the request of the Department of Fish and Game, the Department of Water Resources added a second series so in-tank velocities could be maintained at acceptable levels when pumping at full capacity.

Every 2 hours when the facility is operating, the bypass flow (and fish) are diverted to a counting tank, typically for 10 to 30 minutes. All fish in the counting tanks during the sampling period are collected, enumerated, and identified. Department of Fish and Game staff, who conduct the collection, counting, and transport process, convert the sampling data to estimates of total daily salvage by species. If the counting interval is at least 10 minutes of 120 minutes and the fish are reasonably abundant, the estimates are probably accurate to ± 20%. The sample counts are extrapolated to daily totals, by species, and the data (including fish length) are found on the Interagency Ecological Program's Home Page (http://wwiep.water.ca.gov). The database is updated weekly.

Periodically during the day, Department of Fish and Game staff loads all fish from the holding tanks into a 7500-liter tank truck and hauls them about 40 kilometers to release sites near the confluence of the Sacramento and San Joaquin rivers. Salt (sodium chloride) is added to reduce stress, and pure oxygen is injected to maintain acceptable dissolved oxygen levels. Hauling frequency is based on fish density and any controlling reasonable and prudent measures in the delta smelt or winter Chinook biological opinions.

HISTORICAL SALVAGE OF ALL FISH SPECIES

During the 1979-1993 period, 47 species of fish were salvaged at Skinner Fish Facility (Table 1). (We selected 1979-1993 because this represents the most reliable historical salvage database. Data actually go back to 1968, but quality control on data collection and storage was not rigorous until 1979.) Six additional species, including white bass and yellow perch, although listed in the database in small numbers, are not included because their identifications have not been confirmed. More than half (26 out of 47) of the species salvaged have been purposefully or accidentally introduced into California. Most are fresh or brackish water species, although salvage of Pacific herring and staghorn sculpin indicates that typical Bay species sometimes move into the Delta.

Not all of the 47 species are abundant in the salvage. Five introduced species accounted for more than 90 percent of the salvage (Table 2). One species alone, striped bass, dominated the salvage for the reporting period, with about 70 percent of the total. (These numbers represent fish longer than about 20 mm; larval striped bass and other fish less than about 20 mm total length are not salvaged or counted very effectively with the present system.)

TABLE 1. Fish species collected at Skinner Fish Facility, 1979-1993.

Common Name	*Scientific Name*	*Introduced* (I) *or Native* (N)
Chinook salmon	*Oncorhynchus tshawytscha*	N
Steelhead rainbow trout	*Oncorhynchus mykiss*	I
Striped bass	*Morone saxatilis*	I
White catfish	*Ameiurus catus*	I
Brown bullhead	*Ameiurus nebulosus*	I
Yellow bullhead	*Ameiurus natalis*	I
Black bullhead	*Ameiurus melas*	I
Channel catfish	*Ictalurus punctatus*	I
Blue catfish	*Ictalurus furcatus*	I
Black crappie	*Pomoxis nigromaculatus*	I
White crappie	*Pomoxis annularis*	I
Green sunfish	*Lepomis cyanellus*	I
Bluegill	*Lepomis macrochirus*	I
Largemouth bass	*Micropterus salmoides*	I
Smallmouth bass	*Micropterus dolomieu*	I
Warmouth	*Lepomis gulosus*	I
Redear sunfish	*Lepomis microlophus*	N
Tule perch	*Hysterocarpus traski*	N
Sacramento perch	*Archoplites interruptus*	N
American shad	*Alosa sapidissima*	I
Threadfin shad	*Dorosoma petenense*	I
Splittail	*Pogonichthys macrolepidotus*	N
Sacramento squawfish	*Ptychocheilus grandis*	N
Hardhead	*Mylopharodon conocephalus*	N
Golden shiner	*Notemigonus crysoleucas*	I
Carp	*Cyprinus carpio*	I
Hitch	*Lavinia exilicauda*	N
Sacramento blackfish	*Orthodon microlepidotus*	N
Goldfish	*Carassius auratus*	I
Sacramento sucker	*Catostomus occidentalis*	N
Threespine stickleback	*Gasterosteus aculaetus*	N
Longfin smelt	*Spirinchus thaleichthys*	N
Delta smelt	*Hypomesus transpacificus*	N
Wakasagi[1]	*Hypomesus nipponensis*	I
White sturgeon	*Acipenser transmontanus*	N
Green sturgeon	*Acipenser medirostris*	N
Inland silverside[2]	*Menidia beryllina*	I
Yellowfin goby	*Acanthogobius flavimanus*	I
Chameleon goby[3]	*Tridentiger trigonocephalus*	I
Prickly sculpin	*Cottus asper*	N
Staghorn sculpin	*Leptocottus armatus*	N
Riffle sculpin	*Cottus gulosus*	N
Bigscale logperch	*Percina macrolepida*	I
Starry flounder	*Platichthys stellatus*	N
Lamprey	Various Species	N
Mosquitofish	*Gambusia affinis*	I
Pacific herring	*Clupea pallasii*	N

[1] Identified by Johnson Wang. Electrophoretic confirmation pending.
[2] Also called Mississippi silverside.
[3] According to Scott Matern, University of California at Davis, two species are actually present: *T. trigonocephalus* and *T. bifasciatus*.

HISTORICAL SALVAGE OF CHINOOK SALMON

As shown in Table 2, about 1.9 million juvenile Chinook salmon were salvaged at Skinner Fish Facility during 1979 through 1993. This total is broken down by year in Table 3, which also includes total annual pumping for the period. Although there is considerable variability in the numbers, the totals after 1989 are generally lower than in previous years, reflecting the general lower abundance of Central Valley Chinook. The high salvage in 1986 probably reflects fry washed out during high February/March flows, as well as good fry-to-smolt survival as a result of high streamflows during the rearing period.

Chinook salmon juveniles can be present in the Delta during all months of the year, but they are most abundant during April, May, and June (Table 4). The April-June salvage consists mainly of wild fall-run smolts, with a few hatchery fish included. The fall fish are yearlings, which may be wild fall- or spring-run and probably some Merced River Hatchery fish.

TABLE 2. Estimated salvage of 20 most common fish species at State Water Project Intakes, 1979-1993.
(*All estimates are rounded*)

Common Name	*Total Salvage*	*Maximum Annual Salvage*	*Year Maximum Occurred*
Striped bass	71,887,000	13,038,000	1986
Threadfin shad	15,408,311	7,184,000	1993
American shad	7,852,000	900,000	1981
White catfish	3,630,000	811,000	1984
Yellowfin goby	2,963,000	1,610,000	1984
Splittail	2,186,000	1,160,000	1986
Chinook salmon	1,913,840	435,000	1986
Channel catfish	1,184,709	359,000	1986
Prickly sculpin	685,000	376,000	1986
Inland silverside	684,000	74,102	1987
Longfin smelt	328,000	140,000	1988
Delta smelt	319,000	56,110	1988
Bigscale logperch	155,000	41,000	1988
Chameleon goby	127,000	65,000	1993
Steelhead	92,000	17,000	1982
Bluegill	84,000	31,000	1988
Largemouth bass	67,000	31,000	1993
Riffle sculpin	60,598	45,962	1985
Black crappie	56,000	12,000	1991
Staghorn sculpin	37,000	28,000	1990

SOURCES OF SALMON MORTALITY IN THE SALVAGE PROCESS

The Chinook salmon salvage numbers are estimates of the numbers of juvenile salmon that wound up in the holding tanks. Some salmon (and other salvaged fish) die or are lost between the time they enter Clifton Court Forebay to the time they are discharged from the hauling trucks at the release sites. For Chinook salmon, sufficient experimental work has been done to provide a good idea of where the mortalities occur. In most cases, we have empirically-derived loss estimates at each step, but not everyone agrees as to the reliability of these loss estimates. Following

TABLE 3. Annual salvage of Chinook salmon at Skinner Fish Facility and SWP Delta Pumping, 1979-1993.

Year	*Salvage (All Races)*	*Pumping (km^3)*
1979	122,439	2.669
1980	80,133	3.104
1981	101,605	2.626
1982	278,419	3.250
1983	68,942	2.322
1984	145,041	2.032
1985	140,713	3.327
1986	435,233	3.301
1987	177,880	2.819
1988	151,908	3.338
1989	106,259	3.822
1990	35,296	3.825
1991	39,170	2.193
1992	22,190	1.940
1993	8,641	3.146

TABLE 4. Typical annual pattern of Chinook salmon salvage at State Water Project Intake (*data are for 1984*).

Month	*Total Number Salvaged*	*Number per 1000 Cubic Meters*	*Average Length (mm)*
January	—	—	—
February	79	0.001	101
March	1,659	0.01	161
April	27,259	0.10	103
May	40,077	0.20	100
June	46,131	0.21	94
July	2	—	115
August	575	0.002	136
September	—	—	—
October	10,515	0.08	195
November	8,856	0.03	179
December	9,884	0.03	192

is a summary of the loss estimates, how they were derived, and some suggestions on how losses might be reduced.

Pre-screen Losses

Artificial environments such as occur near structures in water (*e.g.*, bridge piers and fish screens) often act to concentrate predators. The predatory fish use the structures for concealment and can remain in backwater areas, which reduces their energy expenditure while waiting for prey to appear. The intake to the California Aqueduct appears to provide predator habitat, and the density of striped bass and other piscivores is generally higher than in adjacent Delta channels.

At the SWP intake, prescreen losses are synonymous with Clifton Court Forebay predation losses. The forebay contains a wide variety of fish (Table 5). More than 80 percent of the fish collected in a 1-month beach seining effort were striped bass,

TABLE 5. Numbers and sizes of 12 most abundant fish species captured by beach seining in Clifton Court Forebay, November 16 to December 21, 1992.

Common Name	*Total Catch*	*Mean Length (mm)*	*Length Range (mm)*
Striped bass*	10,852	321	53 - 1105
White catfish*	655	359	215 - 520
Channel catfish*	517	489	236 - 628
American shad	475	280	178 - 405
Carp	374	587	420 - 758
White sturgeon	72	1,091	811 - 1643
Goldfish	66	405	372 - 425
Green sturgeon	32	657	416 - 1632
Largemouth bass[1]	28	361	282 - 497
Splittail	24	327	282 - 376
Sacramento blackfish	24	433	295 - 471
Chinook salmon	10	686	510 - 887

[1] Potential predator on Chinook salmon.

most of which were large enough to prey on juvenile Chinook salmon. As part of the Interagency Ecological Program studies, the Department of Fish and Game has estimated that in March 1993 the total striped bass population in the forebay was about 200,000, even after almost 29,000 stripers had been removed in a pilot predator removal program. Other estimates have indicated similar levels of abundance, although there may be seasonal variation.

Juvenile salmon enter Clifton Court Forebay through the radial gates and, even by the most direct route, must traverse a few kilometers of open water before entering the salvage facilities. Interagency Ecological Program staff have made several attempts to determine losses of juvenile Chinook between the radial gates and the trashracks. The experimental design for each test to estimate predation losses consists of releasing two groups of marked (spray-dyed) hatchery salmon smolts — one group near the radial gates and the other in front of the trashracks immediately in front of the screens. Marked fish are recovered in the holding tanks, and the difference in recovery between the two release groups provides an estimate of predation in Clifton Court Forebay. Losses of marked hatchery salmon across the forebay are significant (Table 6).

Models to determine predation losses of juvenile salmon due to Clifton Court Forebay predation use a 75 percent prescreen loss factor, which is an average of the 1978, 1984, and 1985 estimates (*i.e.*, of 100 juveniles entering the forebay, only 25 are calculated to reach the salvage facility). The National Marine Fisheries Service specifies this factor in its winter Chinook biological opinion. The 75 percent factor was also adopted in 1986 as part of a Department of Fish and Game/Department of Water Resources mitigation agreement to offset direct losses of fish at the salvage facility.

The high estimates of losses to predators and the finding that striped bass are likely the major predators have caused considerable interest in a possible program

TABLE 6. Losses of juvenile hatchery-reared fall Chinook salmon crossing Clifton Court Forebay. (*Estimated from mark/recapture studies.*)

Month and Year of Study	*Average Size of Test Fish (mm)*	*Estimated Predation Rate (%)*
October 1976	115	97
October 1978	88	88
April 1984	79	63
April 1985	45	75
June 1992	77	98
December 1992	121	77
April 1993	66	94
November 1993	117	99

to reduce losses by catching striped bass with nets and hauling them for release in San Pablo Bay or other locations far from Clifton Court Forebay. A major predator removal program was planned for fall 1994, but opposition by angler organizations caused it to be postponed. In the meantime, results of a limited acoustic tagging program indicated that sublegal striped bass in the forebay move freely through the radial gates to the open Delta, so removal may have only limited effectiveness.

Future studies by Department of Water Resources and Department of Fish and Game personnel will refine the loss estimates and confirm preliminary indications that striped bass move between Clifton Court Forebay and the Delta more or less at will.

Screen Efficiency

Because the bars in the louver system are 2.5 cm apart, the State Water Project primary and secondary louver screening systems are not 100 percent efficient for Chinook salmon and other fish whose bodies can fit through the gaps. In 1970 and 1971, the Department of Fish and Game and Department of Water Resources evaluated the efficiency of the screening facility for several species of fish, including Chinook salmon. Several variables were tested, such as fish length, approach velocity, bypass ratio, and the presence of a solid center wall in the V-shaped screens.

The average screening efficiency derived from all salmon tests for the primary and secondary louver screens is summarized in Fig. 6 (DWR/DFG 1973). For the primary screens, efficiency ranged between about 70 and 85 percent with no apparent relationship with fish length. Efficiency of the secondary screens ranged from about 70 to 95 percent and, as would be expected, increased with fish length. The lack of a similar relationship in the primary system is probably an artifact of the testing process.

The Department of Fish and Game has combined the data to obtain an overall (combined primary and secondary) screen efficiency, which is calculated from the following equations based on fish length:

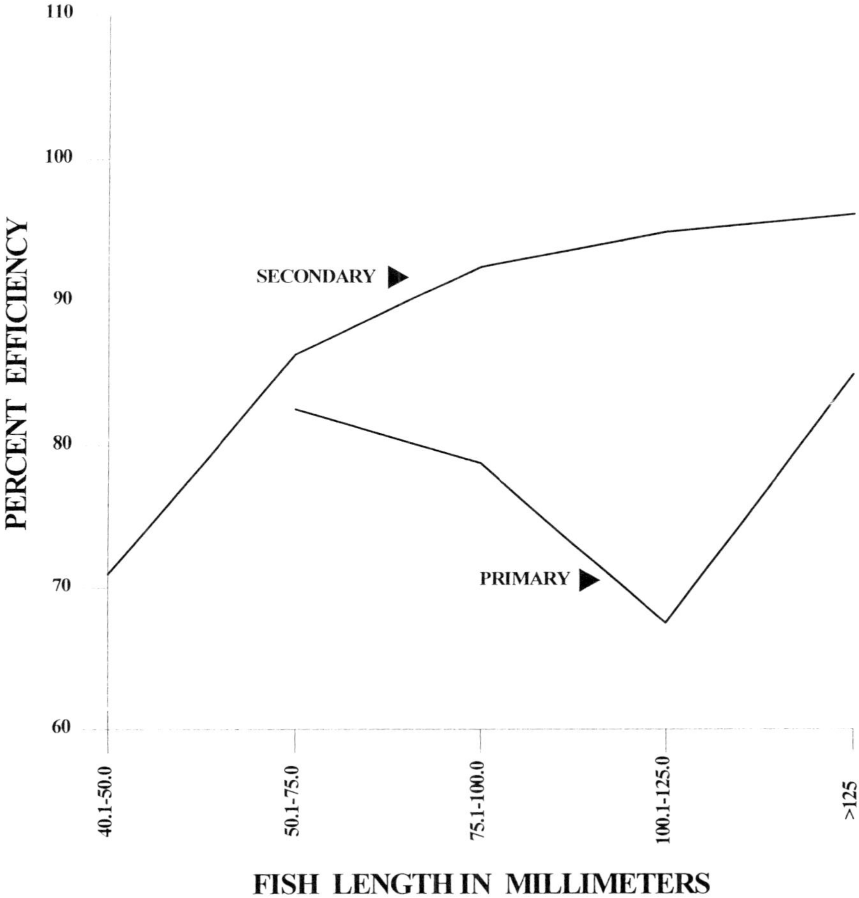

FIGURE 6. Relation between Chinook salmon length and screening efficiency, average of all test data. (Modified DWR/DFG 1973)

Length	*Efficiency* (E)
1-100 mm	E = 0.630 ÷ (0.0494) (Approach velocity)
>100 mm	E = 0.568 ÷ (0.0579) (Approach velocity)

where velocity is expressed in feet per second. During the designated salmon outmigration season, October through May 15, water project operators attempt to maintain the approach velocity near 1 m sec^{-1}, which optimizes screening efficiency for Chinook salmon. Overall, calculated screen efficiencies are typically in the range of 70-80 percent (*i.e.,* 70 to 80 of 100 salmon encountering the screens end up being salvaged).

A variable and unknown portion of the losses at the screens is probably due to predation by striped bass in the primary and secondary screen channels. For example, on March 10, 1992, Department of Fish and Game staff removed 74 striped bass (average size of 289 mm) upstream of the secondary channel louvers. Periodic removal of piscivores appeared to reduce the population, and this is included in routine maintenance activities at the screens.

Handling and Trucking

The salvage process requires that fish be removed from the holding tank, loaded into a tank truck, and hauled to a release site about 40 km from the facility. In 1984 and 1985, Department of Fish and Game staff evaluated losses of Chinook salmon from handling and trucking. The procedures involved handling or trucking the salvaged fish similarly to normal operations, then placing the fish in a small pool, and recording latent mortality for the next 24 hours. In 12 of 15 tests with Chinook salmon, survival after handling and trucking averaged 99 percent. Two tests, conducted in June, had low survival due to high water temperatures. The other test, in November, had 87.5 percent survival. Based on results of the handling and trucking studies, Fish and Game has established criteria for holding tank flows and fish density in the hauling trucks.

Post-release Mortality

There are no data available to determine if the salvage process causes Chinook salmon and other fish to be particularly vulnerable to predation at the release sites.

Aquatic Weeds

In the past 2 years, there has been a buildup of aquatic weeds in Clifton Court Forebay. The main problems have been due to an introduced pondweed, the Brazilian elodea (*Egeria densa*) and, to a lesser extent, hornwort (*Ceratophyllum demersum*).

Weeds are dislodged by wind, and some move toward the California Aqueduct, where they clog the screens, and fragments enter the holding tanks and fish transport trucks. Salvaged fish become entangled in the weed mass, both in the holding tanks and in the truck. The stress associated with the entanglement increases the direct and latent mortality of the salvaged fish.

The Department of Water Resources attempted to control the weed problem by a spring 1995 application of a selective herbicide (Komeen) over a 150-hectare section of Clifton Court Forebay where they are most dense. The first application appeared successful, and a second treatment will follow later in the year. Department of Fish and Game and Department of Water Resources staff did not observe any direct toxicological effects of the herbicide on fish in the forebay.

DISCUSSION

As is apparent from the information presented above, the fish protective facilities at the intake to the California Aqueduct salvage large numbers of fish but do not provide complete protection for juvenile Chinook salmon and other fish entrained into the California Aqueduct by way of Clifton Court Forebay. For example, calculations using procedures specified in the 4-Pumps Agreement and the winter-run biological opinion show that for every salmon salvaged more than three have been lost to predators in the forebay or through the fish screens. Even though the calculations may overestimate the losses, they do demonstrate a serious problem.

Suggested technical solutions have taken several forms, some of which have been implemented. The most commonly advanced suggestions are described below. Solutions that decrease water demand are beyond the scope of this article and are not included.

Improve Existing Facilities

During the past 10 years, the Department of Water Resources has spent several million dollars to improve Skinner Fish Facility. Improvements have ranged from major changes in the screens themselves (including a new, positive-barrier secondary system), a new holding tank building, rescreening the old holding tanks, a new tank truck and improved hauling procedures, and contracting with the Department of Fish and Game to conduct the counting and salvage operations. Also, Water Resources has invested additional millions in support studies to better understand the sources of mortality at the intake/salvage system and find ways to minimize these losses. It is doubtful that significant improvements to the existing system can be found that will substantially reduce losses, except perhaps reduction in losses to predators.

Mitigate for the Losses

In 1986, the Department of Fish and Game and Department of Water Resources signed an agreement, known as the 4-Pumps Agreement, in which Water Resources was to mitigate or offset direct fish losses at the intake. For Chinook salmon, the agreement specified that Fish and Game would calculate the annual losses and both departments, with the help of a fish advisory committee, and would propose, approve, and fund projects to increase salmon production. The general principals were that efforts should be geared toward natural production, as compared to hatcheries, and the emphasis should be on San Joaquin Valley streams. Typical projects have included gravel restoration, conjunctive use (installing pumps that allow farmers to use well water and leave streamflows in place), predator isolation (isolate large backwater ponds created by mining gravel in valley salmon streams), and improvements to the Merced Fish Hatchery. The goal is to have enough projects to offset the annual losses; however, because of the recent low run sizes in the San Joaquin system, there has not been sufficient production to offset salmon losses accumulated since 1986.

The 4-Pumps Agreement also included a $15 million lump-sum account for projects that appear to have significant benefits but that are difficult to quantify. For example, this fund was used to place about 76,500 cubic meters of spawning gravel in the upper Sacramento River near Redding. Additional approved allocations from this fund to benefit salmon include a conservation hatchery on the Tuolumne River, fish screens in Suisun Marsh and predator isolation projects on the lower tributaries.

Replace Existing Screens

Most modern fish screens are positive-barrier flat plates with openings as small as 3-4 mm to exclude fish while allowing water to pass through. To prevent fish from being impinged on the screens, the velocity of water approaching the screen face is low; for example, existing Department of Fish and Game criteria call for an approach velocity of about 0.1 m/sec. Because of space limitations, it would be physically very difficult to install the amount of screen area necessary to achieve the required approach velocity at the existing intake and maintain present export capacity.

New passive-barrier screens could be moved to the forebay intake, perhaps in a series of V-shaped screens at the present location of the radial gates. Bypasses at the narrow end of the Vs would lead the fish to a collection facility, possibly a barge to transport the fish to release sites away from the screens. A major problem with this alternative is that the fish would continue to be salvaged at the southwest corner of the Delta — not a location on their usual migratory path. In addition the screens would be designed for the maximum instantaneous flow (the flow when the gates are just opened), thus dramatically increasing the required screen area and cost. The small screen openings and the amount of aquatic plant and peat fibers in the water would also make cleaning difficult.

Reduce the Number of Salmon Entering the Forebay and Encountering the Screens

Since Chinook salmon move from the spawning grounds to the ocean in fairly well-known migration corridors and times, measures can be taken to limit the number of fish arriving at the intake. Measures underway or being considered include:

(1) Close the Delta Cross Channel from February 1 through May 20 each year. This measure is part of the December 15, 1994, Delta Agreement and is now in effect. Closing the cross channel reduces the number of salmon that leave the Sacramento River.

(2) Install an acoustical barrier at the head of Georgiana Slough during the outmigration period. Recent studies have indicated that Georgiana Slough is a natural migration path for some Sacramento Valley Chinook, that their survival when going this route is lower than if they had stayed in the main river, and that an acoustical barrier can keep at least a portion of the fish in the Sacramento River (U.S. Fish & Wildlife Service 1992; Hanson 1995). This barrier is now in the third year of testing at Georgiana Slough.

(3) Install a physical barrier at the head of Old River in the spring to help keep San Joaquin Basin smolts from leaving the river and winding up at the State Water Project and Central Valley Project pumps. The barrier has been tested with positive results, but it is not clear when or if it will be installed on a permanent basis.

(4) Pump in accordance with export/inflow ratios in the Delta Agreement and the 1995 water quality control plan for the estuary.

(5) Continue to truck much of hatchery production to San Pablo Bay.

Move the Intake to the California Aqueduct

One solution is to move the intake from the southern Delta to the Sacramento River near Hood. Two variations of this solution are to isolate the conveyance channel from Hood just downstream of Sacramento to the California Aqueduct (a peripheral canal) or to have a canal from Hood to some location in the Delta (a through-Delta system). Several technical issues remain to be resolved about screening a large intake on the Sacramento River (Odenweller & Brown 1982), as well as issues related to the size of the diversion and the amount of flow to be bypassed. The Interagency Ecological Program is addressing the fish screening technical issues associated with a possible diversion at Hood.

The next few years will see a serious examination of Delta issues from both fish and water supply viewpoints. A group of stakeholders representing agricultural, environmental, and urban water interests will be working with the recently appointed Bay/Delta Advisory Council and State and Federal agencies to review the issues and recommend actions that have the general goals of improving conditions for fish and their associated estuarine habitat as well as improving water supply reliability for the 20 million Californians who receive at least part of their water supply from the Delta.

LITERATURE CITED

Brown, R. L., and S. Greene. (1992). *Effects of Central Valley Project and State Water Project Delta Operations on Winter-Run Chinook Salmon*. California Dept. Water Resources. Sacramento, CA. 137 pp. plus appendices.

Brown, R. L., and S. Greene. (1994). An evaluation of the Feather River Hatchery as mitigation for the construction of the California State Water Project's Oroville Dam. *Proceedings of the 1993 USCID/USBR Seminar – Environmental Enhancement of Water Projects*. April 14-16, 1993. Sacramento, CA. 13 pp.

California State Department of Water Resources. (1994). *Management of the California State Water Project*. Bull. 132-93. 370 pp.

California State Department of Water Resources and Department of Fish and Game. (1973). *Evaluation Testing of Program Report for Delta Fish Protective Facility, State Water Project, California Aqueduct, North San Joaquin Division*. Memorandum Report. Sacramento, CA. 321 pp. plus appendices.

California Water Resources Control Board. (1995). *Water Quality Control Plan for the San Francisco Bay/Sacramento-San Joaquin Delta Estuary*. May 1995. Sacramento, CA. 45 pp.

Chung, F., And T. Smith. (1993). *Qwest and Transport*. California Dept. Water Resources [Unpublished Report.] 13 pp. and figures.

Hanson, C. H. 1995. *Georgiana Slough Acoustic (Underwater Sound) Barrier Applied Research Project: Results of the 1994 Phase II Field Tests*. February 1995 Draft.

Moyle, P. B. 1976. *Inland Fishes of California*. University of California Press, Los Angeles.

Odenweller, D., and R. Brown. (1982). *Delta Fish Facilities Report through 1982*. Interagency Ecol. Study Progr. Tech. Rept. 6. California Dept. Water Resources, Sacramento, CA. 89 pp.

U.S. Fish and Wildlife Service. (1992). *Measures to Improve the Protection of Chinook Salmon in the Sacramento/San Joaquin River Delta*. WRINT-USFWS-7. [Submitted to the State Water Resources Control Board, July 6, 1992.]. 84 pp.

Vogel, D., and K. Marine. 1991. *Guide to Upper Sacramento River Chinook Salmon Life*

History. CH_2M-Hill, Redding, CA. 55 pp. plus appendices. [Prepared for the U.S. Bur. Reclamation, Sacramento, CA.]

Wendt, P. G. 1987. *Preliminary Evaluation of Factors Controlling Striped Bass Salvage Loss at Skinner Fish Facility: Quantity and Direction of Flow in the Lower San Joaquin River, Striped Bass Abundance and Size, and Total Delta Exports*. California Dept. Water Resources Tech. Infor. Rec. 32 pp.

WHERE HAVE ALL THE FISHES GONE? INTERACTIVE FACTORS PRODUCING FISH DECLINES IN THE SACRAMENTO-SAN JOAQUIN ESTUARY

WILLIAM A. BENNETT
Bodega Marine Laboratory, University of California
P.O. Box 247, Bodega Bay, California 94923

and

PETER B. MOYLE
Department of Wildlife, Fish, and Conservation Biology
University of California, Davis, California 95616

Fish populations have declined markedly in the Sacramento- San Joaquin estuary. We review evidence for factors contributing to the declining abundance of fishes in the context of a conceptual model that emphasizes the interactive nature of six main pathways by which alteration of freshwater outflow to the estuary affects the survival of larval and juvenile fishes (recruitment success). Specific pathways include the influence of freshwater outflow on (1) transport and entrainment, (2) retention in and/or advection from preferred habitats, (3) the success and effects of invading species, (4) primary production and food web dynamics, (5) dilution and/or flushing of toxic compounds, and (6) the quantity and quality of shallow-water spawning/rearing habitat. The collective evidence suggests that within and among years recruitment success may be regulated by several pathways acting in concert or synergistically, indicating the futility of promoting single-factor, or single-pathway, explanations for fish declines. Clearly, ameliorating the effects of various factors (*e.g.*, reducing entrainment, toxic runoff, and improving shallow-water habitat) will improve conditions for fish. However, recovery of many depressed populations may be precluded by the continuously changing composition of the estuarine food web by exotic species. The current lack of life history information on several of the more severely affected species and the need to prevent extinctions suggests that the most pragmatic and promising solution is to ensure adequate outflow to the estuary.

The fish populations of the Sacramento-San Joaquin estuary have declined dramatically in recent years (Herbold *et al.* 1992; Jassby *et al.* 1995; Meng *et al.* 1994). The majority of species about which we have adequate information reflect this overall decline. Most dramatically, the winter-run chinook salmon is now listed as an endangered species, the delta smelt is listed as a threatened species, and Sacramento splittail has been proposed for listing. In addition, major fisheries for striped bass, chinook salmon, and others have been reduced to low levels. The simultaneous declines of so many species with different habitat requirements and life history strategies is an indication of broad problems with the estuarine environment, especially in the Delta, the uppermost part of the estuary (Fig. 1). Drought and diversion of inflowing freshwater have been implicated as the overriding factors in the fish declines (Stevens *et al.* 1985; Herbold *et al.* 1992; Moyle *et al.* 1992). However, it is difficult to elucidate how lack of freshwater outflow directly affects fish populations, as well as how other factors (*e.g.*, toxic runoff, species invasions, see Nichols *et al.* 1986; Herbold & Moyle 1989; Jassby *et al.* 1995;

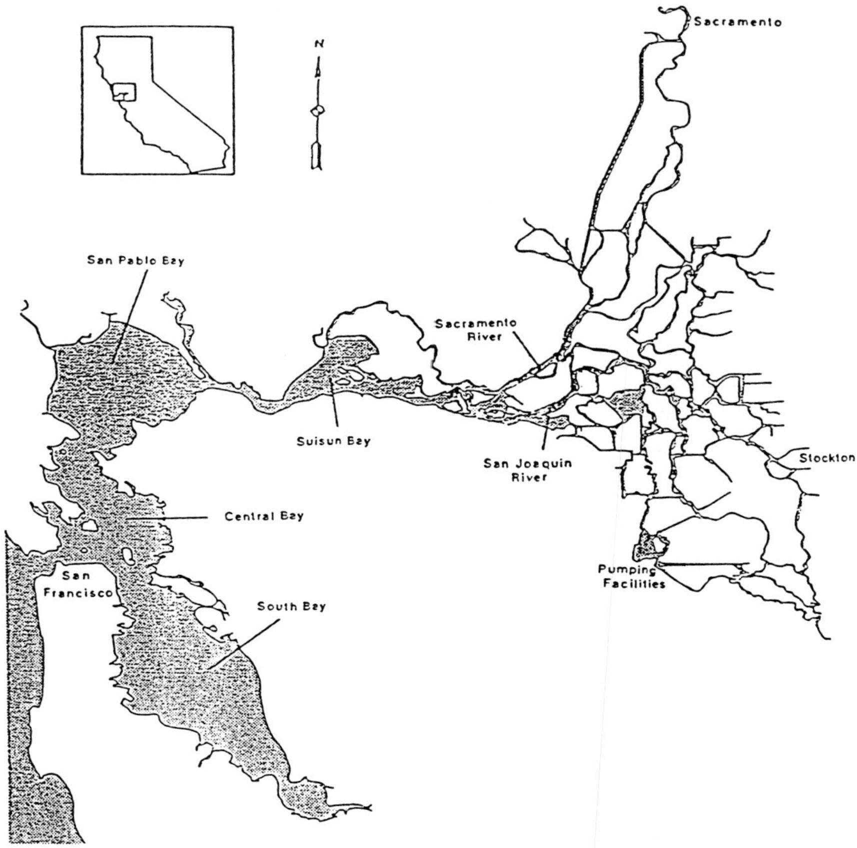

FIGURE 1. The Sacramento-San Joaquin estuary showing four sub- embayments, rivers and Delta region extending from Sacramento to Stockton and the State and Federal water projects.

Bennett *et al.* 1995) are affected by outflow to act synergistically with drought and diversions to maintain populations at low abundance.

In recent years, the majority of species appear to be recruitment limited (Herbold *et al.* 1992; Moyle *et al.* 1992), which we define operationally as poor survival through the first year of life. Because many species are short-lived (1-3 years) such recruitment limitation may regulate population dynamics. Recruitment is naturally a highly variable process, and generally considered to be influenced by the interaction of environmental factors with food web and demographic factors (Sissenwine 1984; Rothschild 1986). However, the relative importance of factors producing variability are rarely well understood (Shepherd & Cushing 1990). This is especially true for the Sacramento-San Joaquin estuary, because we know very little about the early life stages of many of the most severely affected species.

In this paper we review the factors contributing to fish declines in the Sacramento-San Joaquin estuary in the context of a conceptual model (Fig. 2) that outlines various pathways by which freshwater outflow affects recruitment. We then show how several pathways may act together to produce the poor recruitment of selected species (striped bass, Fig. 3; delta smelt, Fig. 4). Finally, by examining life history characteristics, we attempt to extrapolate to other native (Fig. 5) and introduced (Fig. 6) species about which there is considerably less biological information. Model pathways are intended to organize discussion, illuminating interactions among factors and should not be interpreted as statistically derived pathways (*e.g.*, path analysis, Wootton 1994), or as pathways of energy flow which typify many estuarine models (Kremer & Nixon 1978; Baird & Ulanowicz 1989). Clearly, various pathways could be expanded into formal models, however, suitable information is currently incomplete. Therefore, we seek to provide a framework for evaluating the relative importance of contributing factors and their interrelated pathways as well as to identify important gaps in our understanding of the causes of the fish declines.

FRESHWATER OUTFLOW AND RECRUITMENT SUCCESS
THE PROBLEM

Reduction of freshwater inflow to estuaries has often been associated with declines in productivity and fish recruitment (Stevens *et al.* 1985; Nixon *et al.* 1986; Skreslet 1986; Houde & Rutherford 1993). However, the relative importance of mechanisms underlying apparent outflow-recruitment relationships remain unclear in most cases (Sutcliff 1973; Stevens *et al.* 1985; Sinclair *et al.* 1986; Jassby *et al.* 1995). In the Sacramento-San Joaquin estuary (Fig. 1), freshwater diversions and reduced outflow are highly associated with declining recruitment and abundance of the majority of fish species (Stevens *et al.* 1985; Herbold *et al.* 1992; Jassby *et al.* 1995). While entrainment in the major state and federal water projects is the most identifiable explanation for fish declines (Turner & Chadwick 1972; Stevens & Miller 1983; Stevens *et al.* 1985; CDFG 1987), current focus has shifted to distinguishing the effects of entrainment from those of food web alteration, loss of low salinity nursery habitat, pollution, and other factors that are considerably more difficult to quantify (Herbold *et al.* 1992; Moyle *et al.* 1992; Jassby *et al.* 1995; Bennett *et al.* 1995). Progress in understanding recruitment failure, however, remains burdened by the promotion of competing single-factor hypotheses (*e.g.*, entrainment vs. toxic exposure).

In this light, we provide a flow diagram (Fig. 2) to organize discussion of several direct and indirect pathways by which freshwater outflow interacts with various factors to regulate recruitment. Our thesis is supported by the growing awareness that recruitment in marine and estuarine systems may often be regulated by the effects of 2 or more factors acting simultaneously (Lasker 1981; Rothschild 1991; Lawrence 1990; Cowan *et al.* 1993). This review is not meant to be a comprehensive history of effects, but summarizes and updates the reviews in Smith & Kato (1979), Nichols *et al.* (1986), Herbold *et al.* (1992), and Moyle *et al.* (1992) to specifically address how freshwater outflow affects the survival of young fishes.

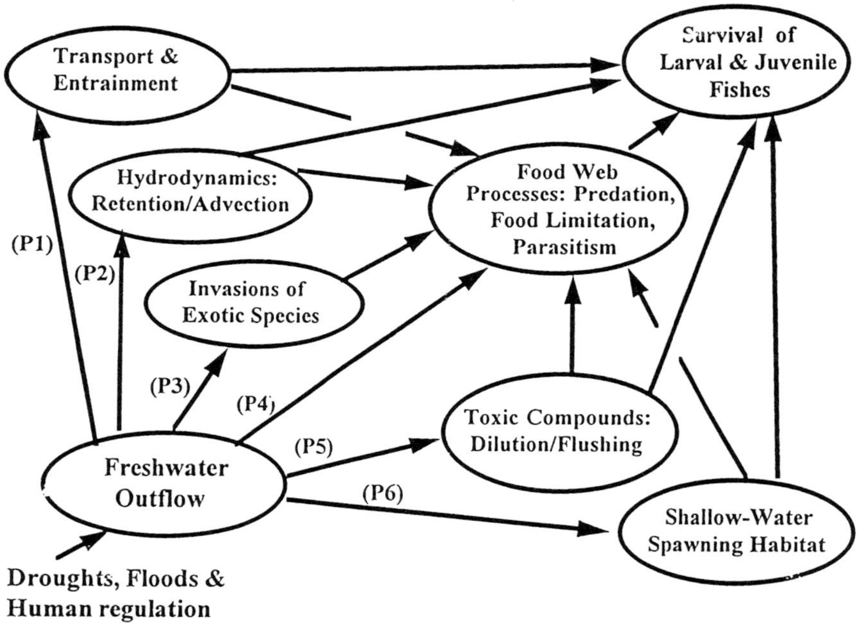

FIGURE 2. Conceptual model illustrating pathways (P1-P6) by which the amount of freshwater outflow to the estuary affects recruitment of young fishes. Various factors may affect recruitment simultaneously by direct and/or indirect pathways with various food web processes.

A CONCEPTUAL MODEL

Climatic and Human Regulation of Freshwater Outflow. — The amount and timing of freshwater entering the estuary is controlled by human response to and possible influence on the effects of natural climatic variability. The Sacramento-San Joaquin estuary receives runoff from 40% of the state's surface area and is located in a region where annually winters are cool and wet, while summers are hot and dry. High interannual variability in wintertime precipitation affects the amount of freshwater available to the estuary (Peterson *et al.* 1995). Moreover, because 70% of California's rainfall occurs north of the estuary, while 80% of it's water consumption occurs south of the estuary, considerable freshwater outflow is diverted south by an elaborate system of water development (Nichols *et al.* 1986; Moyle *et al.* 1992). Complicating this picture further is a twenty-year period (1974-1994) with much more variable precipitation than the previous 50-100 years (Peterson *et al.* 1995). 1976 and 1977 were severe drought years, while 1983 was an extreme wet year. The drought years (1985-1994) were broken by the wettest month on record (February 1986) and the wet year 1993 (Peterson *et al.* 1995). The 1986 rain event had a particularly strong effect, flushing many organisms out of the estuary that already had depleted populations (*e.g.*, delta smelt, Moyle *et al.* 1992). Subsequently, a number of new invading species became established (Path

3 below, Nichols *et al.* 1990; Kimmerer *et al.* 1995; Meng *et al.* 1994; Matern & Fleming, in press).

Diversion of freshwater is the single biggest factor altering flows through the estuary today (Moyle *et al.* 1992; Arthur *et al.*, this volume). Water is diverted in three main areas: upstream of the Delta, within the Delta, and at the south Delta pumps of the State Water Project (SWP) and the Central Valley Project (CVP). Upstream diversions range from the myriad of small riparian diversions to large diversion dams such as Red Bluff Diversion Dam, to major water storage facilities such as Shasta, Friant, and Oroville dams. The effects of these diversions include reduced stream flows, reduced variability in flows, altered channel morphology, and changes in water quality, especially water temperatures. About 30% of the mean annual unimpaired flow to the estuary is diverted upstream, although some of it eventually reaches the Delta as polluted irrigation return water. Within Delta diversions for agriculture amount to about 1 million acre feet per year in about 2,000 small diversions. How these diversions affect within-Delta hydraulics is not well understood, although the requirement that such diversions must be allowed in order to satisfy riparian water rights maintains the Delta as a largely freshwater system. The south Delta pumps have been taking increasing amounts of water from the estuary in recent decades (Moyle *et al.* 1992; Arthur *et al.*, this symposium) although actions to protect endangered fishes have reduced this take in the past several years. These pumps have taken as much as 60% of the inflowing water in some years and in most years drastically alter the way water flows through the estuary, often producing reverse flows in the San Joaquin river (Moyle *et al.* 1992; Arthur *et al.*, this volume).

A major problem for understanding the effects of water development on fish recruitment is the cumulative effects of all the diversions, big and small, such that it is often hard to separate the effects of one diversion or set of diversions from others. The effect of water development on inflow to the estuary, in many respects, reduces variability often creating prolonged drought conditions.

Oceanic conditions can also affect estuarine fish, especially anadromous forms, in ways other than by influencing freshwater outflow. El Niño events, such as those that happened in 1976-77 and 1983, can significantly reduce ocean productivity, which in turn can reduce growth and survival of fish such as chinook salmon and Pacific herring (Herbold *et al.* 1992). The 1976-77 El Niño initiated a dramatic step-change, persisting to the late 1980s, in at least 40 climatic, oceanic, and biotic variables in the Pacific (Ebbesmeyer *et al.* 1991; Kerr 1922) that may have also affected resident estuarine fishes. For example, both adult and young-of-the-year (38mm) striped bass experienced a similar drop in abundance in 1976 and 1977 which has persisted and worsened (Fig. 3). Because poor recruitment would not produce a change in the adult population for at least 3 years, the concurrent drop in adult abundance may reflect this climatic shift, perhaps by inducing migration from the estuary. However, while this well-documented decline of striped bass in the mid-1970s (Fig. 3; Stevens *et al.* 1985) does not appear to be due to recruitment failure, poor year-class success as well as large-scale stocking of juveniles during

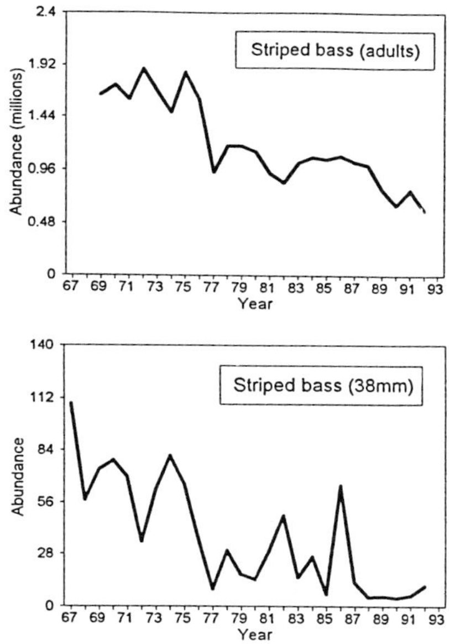

FIGURE 3. Changes in abundance of adult (406mm, age 2 yrs.) and young-of-the-year (38mm, age 3-4 months) striped bass. Data for adult abundances are Peterson population estimates, and young-of-the-year abundances are indices derived from summer tow-net surveys weighted by the volume of water at each sampling site (see Stevens *et al.* 1985).

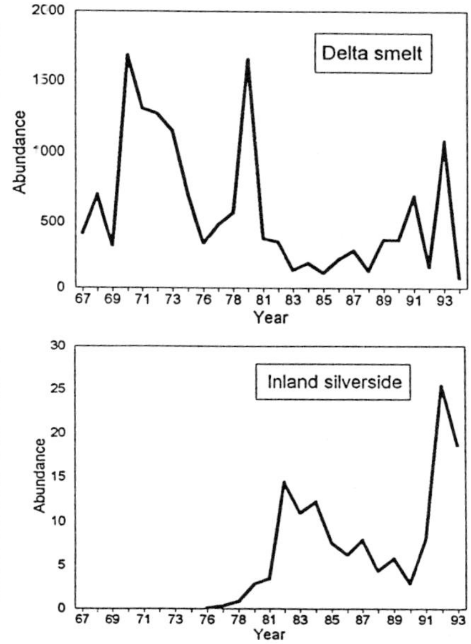

FIGURE 4. Change in abundance of delta smelt (native) and inland silversides (introduced in 1975, Moyle 1976). Data for delta smelt are abundance indicies derived from CDFG's fall mid-water trawl survey (see Moyle *et al.* 1992). Data for inland silversides are catch-per-haul estimates from USFWS's salmon smolt beach seine survey.

most of the past 15 years have not replenished the population, indicating conditions in the estuary remain poor for young fish.

PATHWAYS AFFECTING RECRUITMENT

Below we outline how alteration of the amount, variability, and timing of freshwater outflow can affect recruitment success via six main pathways (Fig. 2). All pathways can produce direct and interactive effects (with food web processes) on larval and juvenile fish survival.

Path 1. Transport and Entrainment.—The effects of altered transport and entrainment in water diversions are covered in detail by Arthur *et al.* and Brown *et al.* (this volume), therefore our discussion will be brief. Water diversions alter the typical pattern of outflow seaward, producing cross-Delta flow towards the state and federal pumping facilities located in the south Delta (Fig. 1). Consequently, millions of juvenile and larval fishes are entrained in the diversions. In addition, the migration routes of young fishes (in particular, salmon smolts) are altered such that they accumulate in the proximity of the pumping facilities (*e.g.*, Clifton Court Forebay, Fig. 1) where they experience substantial predation by aggregations of larger fishes (typically adult striped bass) in addition to entrainment (Arthur *et al.*, this volume).

Path 2. Physical processes of retention and advection.— Diversion of freshwater outflow alters estuarine hydrodynamics, which can influence the retention or advection of young fishes and their food in the upper estuary. Under conditions of reduced outflow, estuarine circulation becomes dominated by tidal mixing rather than gravitational mixing that results from stratification of freshwater outflowing over denser, inflowing saltwater (Kimmerer 1992). In the St. Lawrence estuaries, the interaction of vertical movements of fishes and zooplankton with tidal currents and gravitational circulation acts to retain populations in the upper estuary (Fortier & Leggett 1983; Laprise & Dodson 1989, 1990). This also allows organisms to co-occur with food patches (Dauvin & Dodson 1990), and it may provide an isolation mechanism during their evolution (Sinclair 1988). Domination of tidal mixing during times of low outflow may reduce the effectiveness of this process such that organisms are lost from optimal habitat by diffusive or advective processes. Recent investigations during a dry year (1994) indicate larval striped bass and juvenile longfin smelt migrate vertically, occurring near the surface on flood tides and near the bottom on ebb tides apparently facilitating maintenance of longitudinal position (Bennett, Kimmerer, Burau, Arthur, unpublished data). In addition, flood events flush estuarine organisms out of the system (*e.g.*, delta smelt, Moyle *et al.* 1992).

Path 3. Riverine Input / The Agricultural Model.— Estuaries are among the most productive of aquatic systems, in part because riverine transport of nutrients and carbon provides fertilizer, enhancing the estuarine food web. This phenomenon, called the agricultural model (Nixon *et al.* 1986; Houde & Rutherford 1993), provides an intuitively appealing mechanism for the association between river outflow and recruitment success of invertebrates and fishes, because it invokes the traditional notion of a linear food chain (Carpenter & Kitchell 1993). The basic assumption of this pathway is that larval and/or juvenile fishes are food limited during times of low freshwater outflow, such that recruitment is regulated by starvation and/or predation on malnourished individuals. However, evidence for starvation in estuarine and marine systems is rare, especially considering the number of cases in which it has been investigated (see reviews by May 1974; Leggett 1986; Houde 1987).

During the recent drought, 1987-1992, the importance of food limitation was investigated for larval striped bass (Bennett *et al.* 1995), because growing evidence suggested food limitation may be an important source of larval mortality. Drought, diversions, and grazing by introduced benthic mollusks reduced primary productivity in the larval habitat (Jassby & Powell 1994), and the preferred food for first-feeding larvae, notably the copepod *Eurytemora affinis*, also declined in abundance by about two orders of magnitude (Kimmerer *et al.* 1995; Kimmerer & Orsi, this symposium). In addition, an increasingly abundant copepod, *Sinocalanus doerii*, an exotic from Asia, appeared difficult for first-feeding larval fish to capture (Meng & Orsi 1991). However, morphological and histological evaluations of field-caught larvae from 1988-1991, indicated that all larvae examined (N = 980) had been feeding, and lacked alterations attributable to starvation (Bennett *et al.* 1995). To examine the interaction of predation with food limitation, field experiments were conducted presenting an inland silverside (*Menidia beryllina*) predator with a choice between starved and fed striped bass larvae in large enclosures (Bennett & Rogers-Bennett, submitted manuscript). Our results indicated that starved and fed larvae were consumed at similar rates, but that predation rate declined exponentially over a 5-7 mm larval size range. This suggests that slower growth will dramatically enhance the effect of predation during the larval period.

Additional evidence suggests that while larval food supplies may be recovering, larval growth appears slow. After 1988, another exotic copepod, *Pseudodiaptomus forbesi*, became abundant (see Kimmerer & Orsi, this volume) and appears to be suitable food, replacing native species in larval stomachs (Meng & Orsi 1991; L. Miller, CDFG, pers. commun.). Switching to a new and abundant food item would sufficiently augment the larval food base, decreasing the probability of slow growth. However, otolith analyses of field-caught larvae by CDFG (L. Miller pers. commun.) indicate mean larval growth ranged from 0.15 to 0.22 mm d^{-1} during the drought years with record low survival (1988-1991), whereas during a wet year (1986) larvae grew 0.25 mm d^{-1} and exhibited high survival (L. Miller, pers. commun.; Jassby *et al.* 1995; Fig. 3). In contrast, growth rates of Chesapeake Bay striped bass larvae are typically much higher (about 0.30 mm d^{-1}, Chesney 1989), suggesting larvae in the Sacramento-San Joaquin estuary may grow slowly, increasing the period when they are most vulnerable to predators (Bennett & Rogers-Bennett, submitted manuscript; Rose *et al.*, unpublished manuscript). Comparisons of results from individual-based-modeling (IBM) of Chesapeake (Cowan *et al.* 1993) and California (Rose *et al.*, unpublished manuscript) young striped bass populations also support the slow-growth hypothesis. Therefore, factors which influence growth (*e.g.*, food, temperature, pollutants) may be important, if subtle, sources of mortality.

Path 4. Success of Species Introductions. — Invasions of exotic species accompanied the massive alterations of habitats and flows during the late 1800s. With the completion of the transcontinental railroad in 1871, many carloads of marine and estuarine fishes and invertebrates from the Atlantic coast were released into the highly altered estuarine system. Three species of estuarine fish became established in large numbers almost immediately: American shad (1871), white catfish (1874), and striped bass (1879). These introductions, and many others that have taken place since the 19th century suggest that invasions of the estuary have a higher potential for success when the established assemblages are stressed or depleted from habitat alteration and/or reduced outflow (Moyle 1986; Baltz & Moyle 1993).

Further evidence for this 'rule' is the number of major invasions that occurred between 1985 and 1993, a period of severe drought whose effects were exacerbated by increasing water diversions (Nichols *et al.* 1990; Moyle *et al.* 1992; Kimmerer *et al.* 1995; Matern & Fleming, in press). Following the extreme flood event of February, 1986, populations of many estuarine organisms (particularly benthic organisms) were greatly diminished and a number of new species invaded. The most spectacular invasion was that of the Asiatic clam, *Potamocorbula amurensis*,

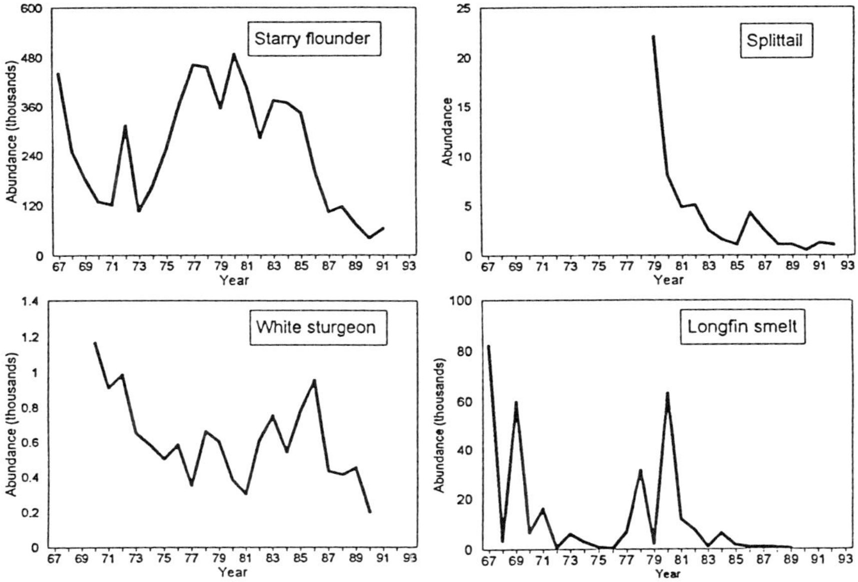

FIGURE 5. Change in abundance of selected native fishes from the Sacramento-San Joaquin estuary. Data for starry flounder (thousands of pounds) and white sturgeon (thousands of fish) are landings from commercial passenger fishing vessels. Data for Sacramento splittail are catch-per-otter trawl estimates from Moyle's Suisun Marsh survey, begining in 1979. Data for longfin smelt are abundance indices dervived from fall mid-water trawl surveys weighted by the volume of water at each site (by CDFG).

which quickly became the dominant filter-feeding organism in the estuary, with drastic effects on the plankton (Kimmerer *et al.* 1995; Kimmerer & Orsi, this volume). Other post-1986 invaders include a number of crustacean species (Carlton & Geller 1993; Hedgpeth 1993; Kimmerer & Orsi, this volume). The most recent fish invader is the shimofuri goby (*Tridentiger bifasciatus*) which arrived around 1986 (Matern & Fleming, in press), and is now one of the most abundant fish in the system (Meng *et al.* 1994; see Fig. 6). Therefore, the rate at which exotic species have become established in the estuary has increased dramatically in recent years, concurrent with and perhaps facilitated by, the maintenance of low outflow conditions by drought and diversions.

In addition to flow regulation, the high invasibility of the estuary appears to be the result of a number of interacting factors: (1) the young geologic age of the estuary and the consequent comparative lack of species specifically adapted for estuarine conditions; (2) the depauperate nature of the estuarine biota; (3) the high degree of alteration of estuarine habitats by humans; (4) the enormous extent to which introductions, both deliberate and accidental, have been made; and (5) the changing nature of estuarine conditions. As a consequence of these factors, many of the organisms that appear to dominate ecosystem processes are non-native species, new invaders appear on a regular basis, and assemblages of organisms

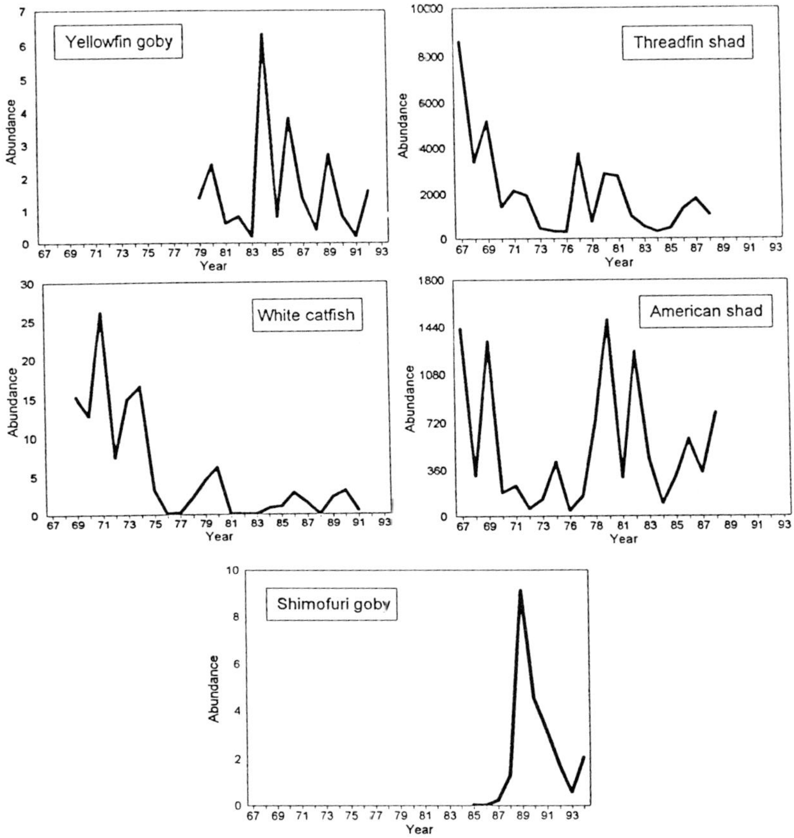

FIGURE 6. Change in abundance of selected introduced fishes from the Sacramento-San Joaquin estuary. Data for yellowfin goby (introduced about 1960, Moyle 1976) and shimofuri goby (introduced about 1986, Matern & Fleming, in press) are catch- per-otter trawl estimates from Moyle's Suisun Marsh survey, begining in 1979. Data for American shad (introduced in 1871, Moyle 1976) and threadfin shad (introduced in 1953, Moyle 1976) are annual catches from fall mid-water trawl surveys (by CDFG). Data for white catfish (introduced in 1874, Moyle 1976) are catch-per-tow estimates from CDFG s summer tow-net survey.

have developed that are made up of mixtures of native and non-native species (Herbold & Moyle 1989). Surprisingly, extinctions of native organisms have been few (*e.g.*, thicktail chub, *Gila crassicauda*, last seen in 1958, Moyle 1976), although many are greatly diminished in numbers and may face extinction in the future.

One consequence of an invasion prone food web is that the abundance and composition of prey items, competitors, and predators can change contemporaneously (Drake 1991). For the most part, fishes in this system seem to be able to quickly switch to new prey. Delta smelt and larval striped bass, for example, both began preying on non-native zooplankton when the native species declined (Meng & Orsi 1991; Moyle *et al.* 1992; see Path 3 above). However, when an invading species both competes with and preys on established species, *i.e.*, is an intraguild predator (Polis *et al.* 1989), it can produce major negative effects (Race 1982).

Such an intraguild relationship may be occurring between the exotic inland silverside and delta smelt. Inland silversides were introduced to Clear Lake, California, in 1966 to control the gnat population, and eventually made their way to the Delta (Moyle 1974, 1976). Like many successful invaders, they became established in the estuary during drought conditions (1976-1977) (Meinz & Mecum 1977), and increased sharply after 1981-1982, concurrent with the decline of delta smelt (Fig. 4). Gut content analyses indicate considerable diet overlap between silversides and delta smelt (Moyle, unpublished data), and field experiments indicated silversides are very efficient predators on fish larvae (see Path 3 above; Bennett & Rogers-Bennett, submitted manuscript). Moreover, silversides appear to be highly abundant in the shoal habitats of the western Delta (lower Sacramento River) where delta smelt spawn in low outflow years. In high outflow years, delta smelt larvae appear to be more evenly distributed in the estuary and more abundant in Suisun Bay, and thus away from shoreline areas, possibly reducing contact with silversides (Bennett, unpublished manuscript). The abundances of inland silversides and delta smelt are negatively associated in low outflow (dry) years, using a reduced major axis (Model II Sokal & Rohlf 1981) regression ($P < 0.05$, Fig. 7), supporting this hypothesis. Therefore, while the silverside-delta smelt relationship requires more study, it illustrates how freshwater outflow may regulate invasion success as well as the impact of an exotic species.

Path 5. Dilution and Flushing of Pollutants. — Pollution has been a problem in the estuary starting in the 1860s when sediments and mercury poured in from upstream mining. When mining sources of pollution diminished, urban, industrial, and agricultural sources increased. The most obvious pollutants were the domestic sewage and industrial waste which was dumped directly into the estuary with minimal treatment. In 1972, however, the federal Clean Water Act was passed, followed by the passage of the state Porter-Cologne Act. The gradual but dramatic increase in water quality that followed the decrease in sewage input in the next 20 years revealed that many problems exist, including pesticides, heavy metals, and petroleum compounds in the estuary. In 1990, the State Water Resources Control Board identified human and ecosystem health problems as cumulative pesticide loads were found in organisms as well as concentrations of arsenic, cadmium, chromium, copper, chlorinated dibenzodioxins and dibenzofurans, hydrocarbons, lead, mercury, nickel, organochlorines, selenium, silver, tribulylin, and zinc (Nichols *et al.* 1986; Setzler-Hamilton *et al.* 1988). Human health warnings have been posted against the regular human consumption of striped bass and other Delta fishes, especially those from the San Joaquin River (Setzler-Hamilton *et al.* 1988).

The effects of some pollutants are associated with the amount of freshwater outflow (Cornaccia *et al.* 1984; Foe & Connor 1989; Bailey *et al.* 1994). For example, most of the area surrounding the striped bass spawning habitat in the Sacramento River is used for rice cultivation, and consequently discharges of irrigation water containing pesticides can contribute up to 30% of the river volume, depending on the amount of outflow (Cornaccia *et al.* 1984). Bioassays indicate

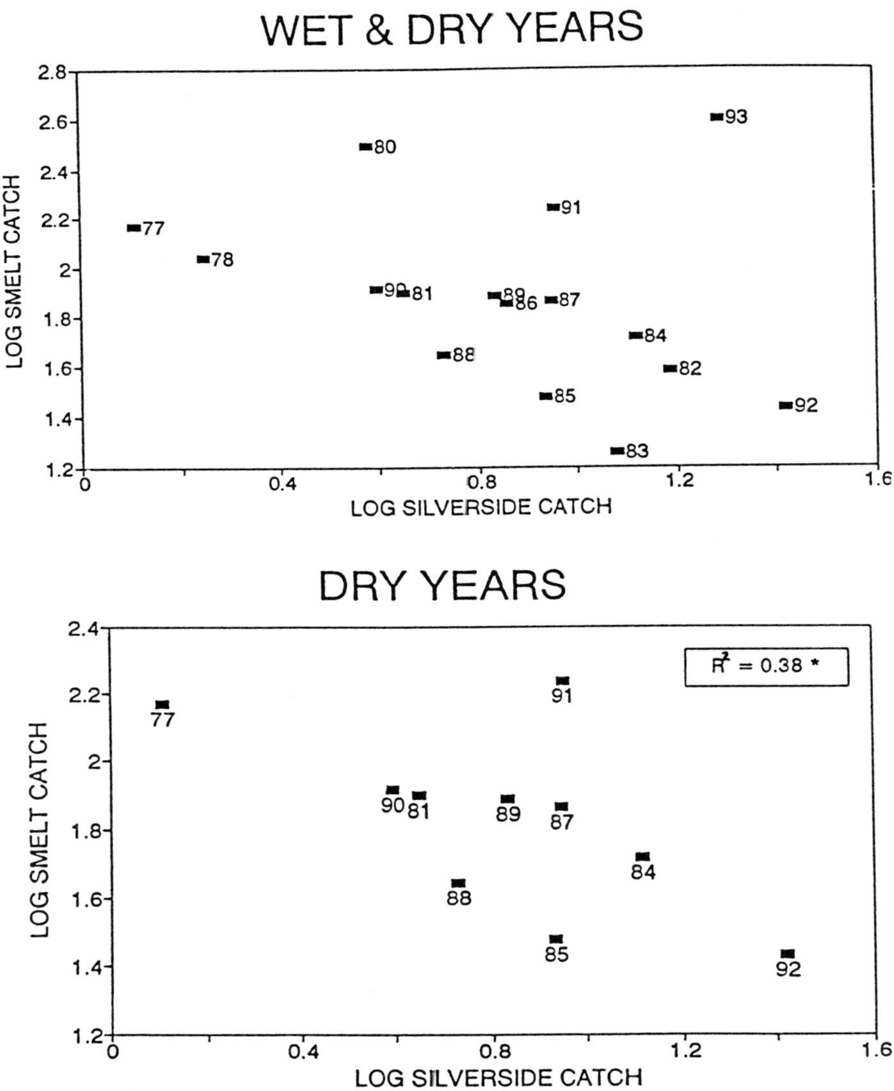

FIGURE 7. Association of delta smelt catch in CDFG's fall mid- water trawl survey with catch of inland silversides per haul in USFWS's salmon smolt beach seine survey in "wet and dry years" and in "dry years" (using DWR criteria for water year type) since 1977. Reduced major axis (Model II, Sokal & Rohlf 1981) regression is significant in "dry years" ($P < 0.05$, $R^2 = 0.38$).

that the runoff produced sublethal and lethal effects on invertebrates and striped bass larvae (Foe & Connor 1989; Heath *et al.* 1993; Bailey *et al.* 1994). In addition, histopathological evaluation of field-caught striped bass larvae during the recent drought (1988-1990) indicated that 26-30% had liver alterations consistent with exposure to rice pesticides that would slow larval growth, enhancing mortality (Bennett *et al.* 1995). In 1991, the toxicity of rice runoff in bioassays and the proportion of afflicted larvae decreased by about 50%, as a result of changes in the application of rice pesticides (Bennett *et al.* 1995). However, while the incidence of affected larvae declined by 50%, striped bass year-class success remained poor. In addition, pesticides from other agricultural activities are regularly detected in Delta water particularly during critical spawning seasons in low outflow years (C. Foe, pers. commun.), suggesting they will remain an important concern.

Path 6. Habitat Alterations. — The estuarine habitats have been experiencing considerable alteration since 1850 with the advent of hydraulic mining, which pushed millions of tons of sediment downstream. The diking of islands in the Delta for farmland began about the same time (1852) and the vast wetland system that once existed became 90% dry land, intersected with deep, narrow channels. Only 3% of the Delta still resembles its original shallow-water habitat (Herbold & Moyle 1989; Herbold *et al.* 1992). Similar kinds of diking operations occurred in lower parts of the estuary as well (*e.g.*, Suisun Marsh) although some of that area has been restored to wetland. Dredging of channels and filling in of shallow areas has also caused significant alteration of estuarine habitats. While habitat alteration in the estuary has been drastic and is still ongoing, most of the activity took place before the present precipitous declines of fish abundances. Presumably habitat alteration had equally drastic effects when it occurred in the 19th century, but there was no monitoring of fish populations at the time. Many resident native species may have started to decline in this early period because the loss of wetlands and shallow water habitat may have reduced estuarine productivity, as well as decreased residence time of water in the Delta, reducing the suitable habitat available to larval and juvenile fish. In recent times substantial shallow water habitat is created only in extremely wet years during spring, when large areas become flooded. The recruitment success of native species (*e.g.*, Sacramento splittail) is highly associated with the creation of such habitat (T. Sommer, DWR, pers. commun.), possibly due to adequate spawning substrate for their adhesive eggs and nursery habitat for young fish (Daniels & Moyle 1983).

Harvesting. — While not directly influenced by outflow (and thus not included in our model, Fig. 2), overharvesting is often cited as the primary cause of collapse in many of the world's fish populations (Rothschild 1986). However, the current role of fishing on populations in the estuary is unclear. In the 19th century, overharvesting of chinook salmon and white sturgeon was a major factor contributing to their declines and reduction of the fisheries led to a recovery (Smith & Kato 1979). Because of critically low abundance, it is likely that commercial and sport fishing for naturally spawned chinook salmon is now contributing to their decline, or at least making recovery more difficult. Depleted populations of wild fish, with relatively low levels of juvenile survival in freshwater, presumably cannot withstand even moderate levels of exploitation. Likewise, it is possible that the removal of large female striped bass by legal or illegal fishing has contributed to the downward spiral of the bass population. The largest females have the highest fecundities and can contribute a high proportion of the new recruits to the population each year (Cowan *et al.* 1993). When survival rates of larvae and juveniles are low, high production rates of young may partially compensate for the reduced survival rates. In addition, sampling of severely depressed/endangered populations has been shown to increase substantially the risk of extinction (Harrison *et al.* 1991).

DISCUSSION

Multiple Pathway Effects and Recruitment Success. — We have reviewed work indicating that several pathways can act simultaneously to affect survival of young fish. However, the relative importance of various pathways can differ among years and species. In this light, we highlight differences among pathway functions in order to provide a better understanding of their relative importance. Our discussion will focus on the problems of young striped bass (see Fig. 2 pathways 1,3,5), because we know the most about the species.

Without human intervention, fish eggs and larvae naturally experience very high mortality. A large portion of this mortality may be produced by inherent variability among larvae, leading to differential survival. Among the many factors that can affect survival of eggs and larvae, individual growth and size variation have recently been emphasized as important factors determining recruitment success (Houde 1987; Crowder *et al.* 1992; Rice *et al.* 1993). This is partly due to the diversity of factors that can influence growth and size variability (*e.g.*, parental/egg size, inherent developmental processes, food abundance, temperature, competition, parasitism), and partly to the frustration of trying to assign single factor explanations for recruitment variability (Sissenwine 1984; Leggett 1986; Lawrence 1991).

Slower growth influences recruitment success by increasing the cumulative impact of predation on a cohort (Houde 1987; Lawrence 1991). Traditionally, this has been considered to be a density-dependent process with compensatory effects, because low spawner abundance translates into reduced competition for food or habitat. This in turn produces higher larval growth rates than would occur during times of very high spawner/larval abundance (Shepherd & Cushing 1980; Sissenwine 1984; Rothschild 1986). In the Sacramento-San Joaquin estuary, however, abundance of spawning striped bass is low enough such that mortality of larvae from slower growth is density-independent, or density vague (Strong 1986), acting in a stochastic fashion (Lawrence 1991). This is because larvae are typically at low densities, thus they should not impact their food resources and they are presumably an incidental rather than sought-after prey item. This implies that finding food and being eaten can be stochastic or density-independent events, when populations are at intermediate to low abundance.

Entrainment and lethal exposure to contaminants also act in a density independent manner. Various environmental or human interventions affecting larval growth (*e.g.*, temperature, sublethal toxic exposure) become superimposed on the natural growth pattern in a stochastic fashion. This suggests that food limitation, toxic exposure, and entrainment all produce mortality in a stochastic fashion, and thus they can work simultaneously to limit recruitment in low outflow years. This is consistent with the idea that recruitment failure or success is often the product of at least 2 factors acting synergistically (Lasker 1981; Cohen *et al.* 1988; Cowan *et al.* 1993). Low outflow conditions promote these interactions.

The altered estuarine hydrodynamics in low outflow years may produce a gauntlet of stochastic effects for young striped bass. In low outflow years, water flow is typically altered and drawn south across the Delta towards the state and federal pumping facilities (Fig. 1, Moyle *et al.* 1992; Arthur *et al.*, this volume). Toxic runoff enters the main striped bass spawning areas in the upper Sacramento river, and throughout the Delta. Thus, the probability of exposure begins at the time of spawning (or before, by maternal gift of toxins accumulated in eggs). Food limitation/stage duration problems arise after about 1 week of egg and larval transport, and thus may occur most often in the central and western Delta. While the magnitude of entrainment losses in the 1,800 or so small water diversions throughout the Delta is uncertain, clearly large numbers of young fish are lost from the combined effects of predation in Clifton Court Forebay and in the pumping facilities (Fig. 1). Therefore, young striped bass must run a gauntlet of stochastic effects beginning with toxic exposure and food limitation, even before entrainment affects them. Therefore, the probabilities of mortality from toxic exposure and food limitation operate before entrainment. Essentially, if one pathway doesn't kill you, odds are the other ones will.

This simple scenario is supported by the larval condition study (Bennett *et al.* 1995). Histopathological evaluations indicated the proportion of larvae with damaged livers dropped by 50% in 1991, concurrent with an approximately equal reduction in the toxicity of rice-field runoff. However, the 1991 young-of- the-year (38mm) recruitment index remained very low (Fig. 3). This indicates that larvae spared by the cleaner effluent were killed by other factors (*e.g.*, slower growth, entrainment), such that various pathways act in concert to regulate recruitment. Therefore, we suggest that the high degree with which pathways interact, indicates the futility of determining the relative importance of single factors. Moreover, factors may alternate in importance within and among years, as well as among species.

Indeed, the transport/entrainment pathway clearly has an important impact. Simulation models indicate that high entrainment rates can severely limit young-of-the-year success of Hudson River striped bass (Christensen *et al.* 1977). Other modeling studies indicate constant harvesting (entrainment) in a variable environment that undergoes sequences of adverse conditions (*e.g.*, prolonged drought) will exacerbate greatly the harvest effects (Beddington & May 1977; Fogarty *et al.* 1991). We are, however, emphasizing the importance of interacting pathways that occur in the Delta region during times of low outflow, and in a transport sense, have the potential to affect fish even before they are entrained. This suggests that if entrainment was curtailed entirely, yet outflow remained low, recruitment may also remain low. However, because species exhibit differences in their susceptibility to entrainment and in their relationships with outflow (Jassby *et al.* 1995), elimination of entrainment could improve the recruitment of highly susceptible species (*e.g.*, striped bass, salmon).

The implication of so many pathways acting simultaneously is that the estuarine habitat has been degraded to such an extent that the potential carrying capacity (K) has been reduced significantly for many fishes. The degree of reduction may have been severe enough to have pushed many populations to alternative states of low abundance (May 1977; Peterman *et al.* 1979). An unfortunate consequence of this, however, is that a population's abundance may remain in the lower state for an indefinite period even after perturbations are removed (*e.g.*, outflow is not diverted), because ecologically similar species may acquire numerical dominance during the perturbation period and maintain dominance subsequent to its removal (Peterman *et al.* 1979; Botsford 1981). This could occur in the Sacramento-San Joaquin estuary if abundant exotic species (*e.g.*, shimofuri goby, inland silverside) can effectively reduce the recruitment success of formally dominant species (*e.g.*, delta smelt, white catfish) by competition and/or predation, precluding their return to pre- perturbation levels of abundance.

Life History and Recruitment of Less Studied Species. — While striped bass have been well studied, they differ considerably in several key life history characteristics from native fishes. This suggests they may be a poor model organism for understanding declines in other species. For example, striped bass are long-lived (20-40 years), but reproduce at about age 3, whereas most native species live less than 5 years (Table 1). This high degree of generation overlap complicates evaluations of striped bass population dynamics. In addition, larval striped bass are highly resistant to starvation (Houde 1987), and several studies indicate they can survive without food for 30 days (Eldridge *et al.* 1981; Chesney 1989; Bennett *et al.* 1995). This period for striped bass, is about twice as long as for most similarly sized fish larvae (Miller *et al.* 1988).

By comparison, delta smelt larvae are more susceptible to starvation and have recently been shown to starve in 10-14 days post-hatch (R. Mager, UCD, pers. commun.). The importance of this for delta smelt is underscored, because rotifers, a preferred food item of first-feeding larvae, have declined dramatically throughout the estuary in recent years (Kimmerer & Orsi, this volume). While estimates of larval susceptibility to starvation (point-of-no-return) have not been conducted on other larval fishes residing in the upper estuary, several studies (Table 1.) using various methods indicate many are sensitive to low food and/or susceptible to slow growth.

Overall, species in decline (see Fig. 3-6) with the exception of striped bass, tend to be pelagic, reside in the Delta, exhibit some degree of food sensitivity, and require suitable spawning substrate. This implies that they are susceptible to all pathways in Figure 2. For example, pelagic and Delta resident fishes are susceptible to transport/entrainment and toxic runoff events. Most native species (*e.g.*, delta smelt, longfin smelt, splittail, see Table 1) require suitable spawning substrate (submerged vegetation, fallen trees), because they have benthic or adhesive eggs. For example, lack of suitable substrate may preclude spawning by splittail in low outflow years (Caywood 1974; Daniels & Moyle 1983).

An alternative way to identify factors that produce fish declines is to examine the characteristics of the few species that are either not in dramatic decline (*e.g.*,

TABLE 1. Life history characteristics of selected Bay/Delta fishes.

Species	Life Span (years)	Spawning Season	Egg Type	Size at Hatch (mm)	Starvation Sensitive	Nursery Habitat: Pelagic/ Benthic	Nursery Habitat: Delta Dependent?
Striped bass *Morone saxarilis*	20-40	April-June	semi-boyant	3-5	No (25-30 dph[1])	pelagic	Yes
American shad *Alosa sapidissima*	3-4	April-July	semi-boyant	7-10	Yes (Creco & Savoy 1987)	pelagic	No
Threadfin shad *Dorosoma petenese*	2	April-Sept.	benthic	3-4	Yes (Kashuba & Matthews 1984)	pelagic	No
Delta smelt *Hypomesus transpacificus*	1	April-May	benthic	4-5	Yes (10-14 dph, R. Mager, UCD)	pelagic	Yes
Longfin smelt *Spirinchus thaleichthys*	2	Feb.-April	benthic	5-7	?	pelagic	Yes
Splittail *Pogonichthys macrolepidotus*	5-7	March-July	benthic	4-6	?	pelagic	Yes
Starry flounder *Platichthys stellatus*	4+	Nov.-Feb.	buoyant	2-3	Yes (Policanski & Sieswerda 1979)	pelagic/ benthic (7mm)	No
Yellowfin goby *Acanthogobius flavimanus*	3-4	Nov.-March	benthic/ protected	4-5	?	pelagic/ benthic (13mm)	No
Shimofuri goby *Tridentiger bifasciatus*	2-3	April-Sept.	benthic/ protected	2-3	?	pelagic/ benthic (14mm)	No
White catfish *Ictalurus catus*	10?	June-Sept.	benthic/ protected	9-10	?	benthic	Yes
White sturgeon *Acipenser transmontanus*	50-100	Feb.-May	benthic	10-11	Yes (12-17 dph, K. Kroll, UCD)	benthic	Yes
Inland silverside *Menidia beryllina*	1	April-Sept.	benthic	4-5	?	pelagic	No

[1] Days post-hatch

yellowfin goby, white sturgeon) or are actually increasing in abundance (shimofuri goby). It is significant that most of these species are benthic predators with life histories that avoid many of the problems faced by fish with a prolonged pelagic existence (pathways 1-6, Fig. 2). One advantage of benthic foraging is that benthic food organisms have generally remained abundant, despite major changes in species composition in recent years (Hymanson *et al.* 1994). The reasons for this are not clear, but reduced predation by benthic foragers that have declined (*e.g.*, splittail, white catfish) may play a role. In particular, the white sturgeon may have benefited from the abundance of benthic prey, including exotic clams, although its long life span and reliance on episodic reproduction events during extremely high outflow years (Kolhorst *et al.* 1991) also offer it a high degree of immunity from estuarine problems. The two goby species (Fig. 3) appear to be able to take advantage of the immense amount of spawning habitat created by the extensive rip-rap (concrete rubble) used to reinforce the sides of Delta channels. In addition, they have long or multi-clutch spawning seasons and begin their benthic life style early, spending a comparatively short time in the plankton as larvae (Table 1). The only pelagic fish that does not appear to have declined is the inland silverside. Its successful invasion of the estuary can probably be attributed to three factors: (1) it is primarily an inshore species, rather than a true pelagic species, (2) it is a fractional spawner, producing eggs all summer long (Table 1), and (3) it is present in large numbers in areas upstream of the estuary (Moyle 1976; Bennett, unpublished manuscript).

CONCLUSIONS

The presence of a critical pelagic stage in the early life histories of most of the declining fishes in the estuary suggests that the declines are related to repeated recruitment failure. Clearly, the factors producing failure can be multiple and/or synergistic pathways for each species (Fig. 2). The relative importance of pathways may also differ among species; pesticides, for example, may be more likely to affect striped bass than white catfish, because of their relative spawning times (Table 1). However, the best evidence available to date indicates that the single most important factor associated with recruitment failure in fishes of the upper Sacramento-San Joaquin estuary is the reduction of outflowing freshwater through the estuary (Fig. 2). This is produced by water diversions both within and upstream of the Delta and exacerbated by drought. Unfortunately, even in the unlikely case that the diversion problems were solved soon, fish populations may not exhibit an immediate resurgence as might be predicted. Solving the outflow problem may quite likely reveal additional limiting factors as the estuarine food web reshapes itself under a different hydrodynamic regime. This is suggested by the apparent lack of response in striped bass recruitment after the reduction in the toxicity of pesticide runoff into the spawning habitat in the Sacramento River. All of this indicates that we need a much better understanding of the early life stages of the estuary's fishes if their declines are to be permanently reversed. Even with the volumes of information on striped bass, we are only now beginning to understand its recruitment problems. This understanding is being facilitated by an integrated approach in which field studies (especially long-term monitoring) are used to design experiments to test process-oriented hypotheses. Particularly important in this approach have been the development of hydrodynamic and individual-based simulation models (Cowan *et al.* 1993; Rose *et al.*, unpublished manuscript). However, while such models are shedding considerable light on the highly interactive nature of mortality pathways affecting striped bass, there is too little information available to apply them to native fishes. Currently, we have to hope that the depressed populations of native fishes can persist until a better understanding of their life histories is developed.

This pessimistic statement does not mean that we are without options until more research is completed. Clearly, improving estuarine hydrodynamics, reducing entrainment, decreasing pesticide runoff, and increasing the amount of shallow-water spawning habitat, as well as other actions to ameliorate the pathways identified here (Fig. 2) will improve conditions for fishes in the estuary. Better knowledge of early life histories, however, should tell us why various actions work. If we understand why actions work, then management policies can be refined to target problem species. Ideally, more sophisticated management of fish populations could increase the amount of water for other beneficial uses as well as improve conditions in the entire estuarine ecosystem.

ACKNOWLEDGMENTS

We thank the Interagency Ecological Program, especially the Bay-Delta Division of the California Department of Fish and Game for much of the data presented here, and the California Department of Water Resources for continued support. W.A. Bennett thanks fellow members of the IEP Estuarine Ecology Team for numerous discussions. We also thank Laura Rogers-Bennett and 2 anonymous reviewers for constructive comments on an earlier draft and Tim Hollibaugh for arranging and editing this symposium.

LITERATURE CITED

Arthur, J. F., M. D. Ball, and S. Y. Baughman. (1996). Summary of federal and state water project environmental impacts in the San Francisco Bay-Delta Estuary, California. Pages 445-495 *in* James T. Hollibaugh, ed., *San Francisco Bay: The Ecosystem*. Pacific Division, American Association for the Advancement of Science, San Francisco, CA.

Bailey, H. C., C. Alexander, C. DiGiorgio, M. Miller, S. I. Doroshov, and D. E. Hinton. (1994). The effect of agricultural discharge on striped bass (*Morone saxatilis*). *Ecotoxicology* 3:123-142.

Baird, D., and R. E. Ulanowicz. (1989). The seasonal dynamics of the Chesapeake Bay ecosystem. *Ecol. Monogr.* 59:329-364.

Baltz, D. M., and P. B. Moyle. (1993). Invasion resistance to introduced species by a native assemblage of California stream fishes. *Ecol. Applications* 3:246-255.

Beddington, J. R., and R. M. May. (1977). Harvesting natural populations in a randomly fluctuating environment. *Science* 197:463-465.

Bennett, W. A., and L. Rogers-Bennett. [Submitted manuscript.] Interactive influence of starvation and size-selective predation on larval striped bass (*Morone saxatilis*).

Bennett, W. A., D. J. Ostrach, and D. E. Hinton. (1995). Larval striped bass condition in a drought-stricken estuary: Evaluating pelagic food-web limitation. *Ecol. Applications* 5:680-692.

Botsford, L. W. (1981). The effects of increased individual growth rates on depressed population size. *American Nat.* 117:38-63.

Brown R., S. Greene, P. Coulston, and S. Barrow. (1996). An evaluation of the effectiveness of fish salvage operations at the intake of the California Aqueduct, 1979-1993. Pages 497-518 *in* James T. Hollibaugh, ed., *San Francisco Bay: The Ecosystem*. Pacific Division, American Association for the Advancement of Science, San Francisco, CA.

California Department of Fish and Game. (1987). *Factors Affecting Striped Bass Abundance in the Sacramento-San Joaquin River System*. Exhibit 25, for the State Water Resources Control Board Water Quality Hearings. Tech. Rept. 20. FWQ/BIO-4ATR/89-20.

Carlton, J. T., and J. B. Geller. (1993). Ecological roulette: The global transport of nonindigenous marine organisms. *Science* 261:78-82.

Carpenter, S. R., and J. F. Kitchell. (1993). *The Trophic Cascade in Lakes.* Cambridge University Press, London, England. 385 pp.

Caywood, M. L. (1974). Contributions to the life history of the splittail, *Pogonichthys macrolepidotus* (Ayers). M.S. Thesis. California State Univ., Sacramento. 91 pages.

Chesney, E. J. (1989). Estimating the food requirements of striped bass larvae, *Morone saxatilis*: The effects of light, turbidity, and turbulence. *Marine Ecol. Progr. Ser.* 53:191-200.

Christensen, S. W., D. L. DeAngeles, and A. G. Clark. (1977). Development of a stock-progeny model for assessing power plant effects on fish populations. Pages 196-226 *in* W. Van Winkle, ed., *Assessing the Effects of Power-Plant-Induced Mortality on Fish Populations.* Pergamon, New York, NY.

Cohen, E. B., M. P. Sissenwine, and G. C. Lawrence. (1988). The "recruitment problem" for marine fish populations with emphasis on Georges Bank. Pages 373-392 *in* B. J. Rothschild, ed., *Toward a Theory on Biological-physical Interactions in the World Ocean.* NATO ASI series C. Kluwer, Dordrecht, Netherlands.

Cornacchia, J. W., D. B. Cohen, G. W. Bowes, R. J. Schnagl, and B. L. Montoya. (1984). *Rice Hericides: Molinate (Ordram) and Thiobencarb (Bolero).* Spec. Proj. Rept. No. 84-4SP. California State Water Resources Control Board, Sacramento, CA.

Cowan, Jr., J. H., K. A. Rose, E. S. Rutherford, and E. D. Houde. (1993). Individual-based model of young-of-the-year striped bass population dynamics. II. Factors affecting recruitment in the Potomac river, Maryland. *Trans. American Fish. Soc.* 122:439-458.

Crecco, V., and T. Savoy. (1987). Effects of climatic and density-dependent factors on intra-annual mortality of larval American shad. *American Fish. Soc. Symposium* 2:69-81.

Crowder, L. B., J. A. Rice, T. J. Miller, and E. A. Marschall. (1992). Empirical and theoretical approaches to size-based interactions and recruitment variability in fishes. Pages 237-255 *in* D. L. DeAngelis & L. J. Gross, eds., *Individual-based Models and Approaches in Ecology.* Chapman & Hall, New York, NY.

Daniels, R. A., and P. B. Moyle. (1983). Life history of splittail (Cyprinidae: *Pogonichthys macrolepidotus*) in the Sacramento-San Joaquin estuary. *U.S. Fish. Bull.* 81:647-654.

Davin, R. G., and J. J. Dodson. 1990. Relationship between feeding incidence and vertical distribution of rainbow smelt larvae (*Osmerus mordax*) in a turbid well-mixed estuary. *Marine Ecol. Progr. Ser.* 60:1-12.

Drake, J. A. (1991). Community-assembly mechanics and the structure of an experimental species ensemble. *American Nat.* 137:1-26.

Ebbesmeyer, C. C., D. R. Cayan, D. R. McLain, F. H. Nichols, D. H. Peterson, and K. T. Redmond. (1991). 1976 step in the Pacific climate: Forty environmental changes between 1968-1975 and 1977 -1984. Pages 115-125 *in* J. L. Betancourt & V. L. Tharp, eds, *Proceedings of the Seventh Annual Pacific climate (PACLIM) Workshop, April 1990.* Interagency Ecol. Study Progr. Tech. Rept. 26.

Eldridge, M. B., J. A. Whipple, D. Eng, M. J. Bowers, and B. M. Jarvis. (1981). Effects of food and feeding factors on laboratory-reared striped bass larvae. *Trans. American Fish. Soc.* 110:111-120.

Foe, C., and V. Connor. (1989). *1989 Rice Season Toxicity Monitoring Results.* California Regional Water Quality Control Board, Sacramento, CA. 162 pp.

Fogarty, M. J., M. P. Sissenwine, and E. B. Cohen. (1991). Recruitment variability and the dynamics of exploited populations. *Trends Ecol. Evol.* 6:241-246.

Fortier, L., and W. C. Leggett. (1983). Vertical migrations and transport of larval fish in a partially mixed estuary. *Canadian Jour. Fish. Aquatic Sci.* 40:1543-1555.

Harrison, S., J. F. Quinn, J. F. Baughmen, D. D. Murphy, and P. R. Ehrlich. (1991). Estimating the effects of scientific study on two butterfly populations. *American Nat.* 137:227-243.

Heath, A. G., J. J. Cech, J. G. Zinkl, B. Finlayson, and R. Fujimura. (1993). Sublethal effects of methyl parathion, carbofuran and molinate on larval striped bass. *American Fish. Soc. Symposium* 14:17-28.

Hedgpeth, J. W. (1993). Foreign invaders. *Science* 261:34-35.

Herbold, B., and P. B. Moyle. (1989). *The Ecology of the Sacramento-San Joaquin Delta: A Community profile.* U.S. Fish & Wildlife Serv. Biol. Rept. 85. 106pp.

Herbold, B., A. D. Jassby, and P. B. Moyle. (1992). *Status and Trends Report on Aquatic Resources in the San Francisco Estuary.* San Francisco Estuary Project, U.S. Environmental Protection Agency, Oakland, CA. 257 pp.

Houde, E. D. (1987). Fish early life dynamics and recruitment variability. *American Fish. Soc. Symposium* 2:17-29.

Houde, E. D., and E. S. Rutherford. (1993). Recent trends in estuarine fisheries: Predictions of fish production and yield. *Estuaries* 16:161-176.

Hymanson, Z., D. Mayer, and J. Steinbeck. (1994). *Long-term Trends in Benthos Abundance and Persistence in the Upper Sacramento-San Joaquin Estuary. Summary Report 1980-1990.* California Dept. Water Resources, Interagency Ecol. Progr. Tech. Rept. 38. 66 pp.

Jassby, A., and T. M. Powell. (1994). Hydrodynamic influences on interannual chlorophyll variability in an estuary: Upper San Francisco Bay-Delta (California, U.S.A.). *Estuarine Coastal Shelf Sci.* 39:595-618.

Jassby, A., W. J. Kimmerer, S. G. Monismith, C. Armor, J. E. Cloern, T. M. Powell, J. R. Schubel, and T. J. Vendlinski. (1995). Isohaline position as a habitat indicator for estuarine populations. *Ecol. Applications* 5:272-289.

Kashuba S. A., and W. J. Matthews. (19840. Physical condition of larval shad during spring-summer in a southwestern reservoir. *Trans. American Fish. Soc.* 113:199-204.

Kerr, R. A. 1992. Unmasking a shifty climate system. *Science* 255:1508-1510.

Kimmerer, W. J. (1992). *An Evaluation of Existing Data in the Entrapment Zone of the San Francisco Bay Estuary.* Tech. Rept. 33, Interagency Ecol. Study Program, Sacramento-San Joaquin Estuary. 49 pp.

Kimmerer, W. J., and J. J. Orsi. (1996). Changes in the zooplankton of the San Francisco Bay estuary since the introduction of the clam *Potamocorbula amurensis*. Pages 403-424 *in* James T. Hollibaugh, ed., *San Francisco Bay: The Ecosystem.* Pacific Division, American Association for the Advancement of Science, San Francisco, CA.

Kimmerer, W. J., E. Gartside, and J. J. Orsi. (1995). Predation by an introduced clam as the likely cause of substantial declines in zooplankton of San Francisco Bay. *Marine Ecol. Progr. Ser.* 113:81-93.

Kohlhorst, D. W., L. W. Botsford, J. S. Brennan, and G. M. Cailliet. (1991). Aspects of the structure and dynamics of an exploited central California population of white sturgeon (*Acipenser transmontanus*). Pages 277-292 *in* P. Williot, ed, *Acipenser.* Cemagref Publisher, Bordeaux, France.

Kremer, J. N., and S. W. Nixon. (1978). *A Coastal Marine Ecosystem. Simulation and Analysis.* Springer-Verlag, New York. 217 pp.

Laprise, R., and J. J. Dodson. (1989). Ontogeny and importance of tidal vertical migrations in the retention of larval smelt *Osmerus mordax* in a well-mixed estuary. *Marine Ecol. Progr. Ser.* 55:101-111.

Laprise, R., and J. J. Dodson. (1990). The mechanism of retention of pelagic tomcod, *Microgadus tomcod*, larvae and juveniles in a well-mixed part of the St. Lawrence Estuary. *Environ. Biol. Fishes* 29:293-302.

Lasker, R. (1981). The role of a stable ocean in larval fish survival and subsequent recruitment. Pages 80-87 *in* R. Lasker, ed., *Marine Fish Larvae.* University of Washington Press, Seattle, WA.

Lawrence, G. C. (1990). Growth, survival, and recruitment in large marine ecosystems. Pages 132-150 *in* K. Sherman, L. M. Alexander, & B. D. Gold, eds., *Large Marine Ecosystems: Patterns, Processes, and Yields.* American Association for the Advancement of Science, Washington, DC.

Leggett, W. C. (1986). The dependence of fish larval survival on food and predator densities. Pages 117-137 *in* S. Skreslet, ed., *The Role of Freshwater Outflow in Coastal Marine Ecosystems. Springer-Verlag,* Berlin, Germany.

Matern S. A., and K. J. Fleming. (1995). Invasion of a third Asian goby, *Tridentiger bifaciatus*, into California. *California Fish & Game* [in press].

May, R. C. (1974). Larval mortality in marine fishes and the critical period concept. Pages 3-19 *in* J. H. S. Blaxter, ed., *The Early Life History of Fish.* Springer-Verlag, New York.

May, R. M. (1977). Thresholds and breakpoints in systems with a multiplicity of stable states. *Nature* 269:471-477.

Meinz, M., and W. L. Mecum (1977). A range extension for Mississippi silversides in California. *California Fish & Game* 63:277-278.

Meng, L., and J. J. Orsi. (1991). Selective predation by larval striped bass on native and introduced copepods. *Trans. American Fish. Soc.* 120:187-192.

Meng, L. M., P. B. Moyle, and B. Herbold. (1994). Changes in abundance and distribution of native and introduced fishes of Suisun Marsh. *Trans. American Fish. Soc.* 123: 498-507

Miller, T. J., L. B. Crowder, J. A. Rice, and E. A. Marschall. 1988. Larval size and recruitment mechanisms in fishes: Toward a conceptual framework. *Canadian Jour. Fish. Aquatic Sci.* 45:1657-1670.

Moyle, P. B. (1974). Mississippi silversides and logperch in the Sacramento-San Joaquin River system. *California Fish & Game* 60:144-149.

Moyle, P. B. (1976). *Inland Fishes of California.* University of California Press, Berkeley, CA. 405 pp.

Moyle, P. B. (1986). Fish introductions into North America: Patterns and ecological impact. Pages 27-43 *in* H. A. Mooney & J. A. Drake, eds., *Ecology of Biological Invasions of North America and Hawaii.* Springer-Verlag, New York, NY.

Moyle, P. B., B. Herbold, D. E. Stevens, and L. W. Miller. (1992). Life history and status of the delta smelt in the Sacramento-San Joaquin Estuary, California. *Trans. American Fish. Soc.* 121:67-77.

Nichols, F. H., J. E. Cloern, S. N. Luoma, and D. H. Peterson. (1986). The modification of an estuary. *Science* 231:567-573.

Nichols, F. H., J. K. Thompson, and L. E. Schemel. (1990). Remarkable invasion of San Francisco Bay (California, USA) by the Asian clam *Potamacorbula amurensis.* 2. Displacement of a former community. *Marine Ecol. Progr. Ser.* 66:95-101.

Nixon, S. W., C. A. Oviatt, J. Frithsen, and B. Sullivan. (1986). Nutrients and the productivity of estuarine and coastal marine ecosystems. *Jour. Limnol. Soc. South Africa* 12:43-71.

Peterman, R. M., W. C. Clark, and C. S. Holling. (1979). The dynamics of resilence: Shifting stability domains in fish and insect systems. Pages 321-341 *in* R. M. Anderson, B. D. Turner, & L. R. Taylor, eds., *Population Dynamics.* Oxford Blackwell Scientific Publ., Oxford, UK.

Policanski, D., and P. Sieswerda. (1979). Early life history of the starry flounder, *Platichthys stellatus*, reared through metamorphosis in the laboratory. *Trans. American Fish. Soc.* 108:326-327.

Polis, G. A., C. A. Myers, and R. D. Holt (1989). The ecology and evolution of intraguild predation: Potential competitors that eat each other. *Ann. Rev. Ecol. Systemat.* 20:297-330.

Race, M. S. (1982). Competitive displacement and predation between introduced and native mud snails. *Oecologia* 54:337-347.

Rice, J. A., T. J. Miller, K. A. Rose, L. B. Crowder, E. A. Marschall, A. S. Trebitz, and D. L. DeAngelis. (1993). Growth rate variation and larval survival: Inferences from an individual-based size-dependent predation model. *Canadian Jour. Fish. Aquatic Sci.* 50:133-142.

Rose, K. A., J. H. Cowan, Jr., L. W. Miller, and D. E. Stevens. [Unpublished manuscript.] Individual-based model of young-of-the-year striped bass: Application to the Sacramento-San Joaquin River system.

Rothschild, B. J. (1986). *Dynamics of Marine Fish Populations.* Harvard University Press, Cambridge, MA. 277 pp.

Rothschild, B. J. (1991). On the causes for variability of fish populations ô the linkage between large and small scales. Pages 367-376 *in* K. Sherman, L. M. Alexander, & B. D. Gold, eds., *Food Chains, Yields, Models, and Management of Large Marine Ecosystems.* American Association for the Advancement of Science, Washington, DC.

Runge, J. A. (1988). Should we expect a relationship between primary production and fisheries? The role of copopod dynamics as a filter of trophic variability. *Hydrobiologia* 167/168:61-71.

Setzler-Hamilton, E. M., J. A. Whipple, and B. Macfarlane. (1988). Striped bass populations in Chesapeake and San Francisco Bays: Two environmentally impacted estuaries. *Marine Pollution Bull.* 19:466-477.

Shepherd, J. G., and D. H. Cushing. (1980). A mechanism for density-dependent survival of larval fish as the basis of a stock-recruitment relationship. *Jour. Conseil internat. Explor. Mer* 39:160-167.

Shepherd, J. G., and D. H. Cushing. (1990). Regulation in fish populations: Myth or mirage? *Philos. Trans. Roy. Soc. London,* Ser. B, 330:151-164.

Sinclair, M. (1988). *Marine Populations: An Essay on Population Regulation and Speciation.* Washington Sea Grant Program. Univ. Washington Press, Seattle, WA. 252 pp.

Sinclair, M., G. L. Bugden, C. L. Tang, J. C. Therriault, and P. A. Yeats. (1986). Assessment of effects of freshwater runoff variability on fisheries production in coastal waters. Pages 139-160 *in* S. Skreslet, ed., *The Role of Freshwater in Coastal and Marine Ecosystems.* Springer-Verlag, New York, NY.

Sissenwine, M. P. (1984). Why do fish populations vary? Pages 59-94 *in* R. M. May, ed., *Exploitation of Marine Communities.* Springer-Verlag, Berlin.

Skreslet, S. [ed.] (1986). *The Role of Freshwater Outflow in Coastal and Marine Ecosystems.* Springer-Verlag, Berlin. 440 pp.

Smith, S. E., and S. Kato. (1979). The fisheries of San Francisco Bay: Past, present, and future. Pages 445-468 *in* T. J. Conomos, ed., *San Francisco Bay: The Urbanized Estuary.* Pacific Division, American Association for the Advancemnt of Science, San Francisco, CA.

Sokal, R. R., and F. J. Rohlf. (1981). *Biometry,* 2nd ed. W.H. Freeman & Co., San Francisco, CA. 859 pp.

Stevens, D. E., and L. W. Miller. (1983). Effects of river flow on abundance of young chinook salmon, American shad, longfin smelt, and delta smelt in the Sacramento-San Joaquin River system. *North American Jour. Fish. Management* 3:425-437.

Stevens, D. E., D. W. Kohlhorst, L. W. Miller, and D. W. Kelley. (1985). The decline of striped bass in the Sacramento-San Joaquin Estuary, California. *Trans. American Fish. Soc.* 114:12-30.

Strong, D. R. (1986). Density vagueness: Abiding the varience in the demography of real populations. Pages 257-268 *in* J. Diamond & T. J. Case, eds., *Community Ecology*. Harper & Row, New York, NY.

Sutcliffe, Jr., W. H. (1973). Correlations between seasonal river discharge and local landings of American lobster (*Homarus americanus*) and Atlantic halibut (*Hippoglossus hippoglossus*) in the Gulf of the St. Lawrence. *Jour. Fish. Res. Bd. Canada* 30:856-859.

Turner, J. L., and H. K. Chadwick. (1972). Distribution and abundance of young-of-the-year striped bass, *Morone saxatilis*, in relation to river flow in the Sacramento-San Joaquin Estuary. *Trans. American Fish. Soc.* 101:442-452.

Wang, J. C. S. (1991). *Early life stages and early life history of the delta smelt,* Hypomesus transpacificus, *in the Sacramento-San Joaquin Estuary, with comparison of early life stages of the longfin smelt,* Spirinchus thaleichthys. Interagency Ecol. Study Program, Sacramento-San Joaquin Estuary. Tech. Rept. 28. 52 pp.

Wootton, J. T. (1994). Predicting direct and indirect effects: An integrated approach using experiments and path analysis. *Ecology* 75:151-165.